Sensor Systems

FUNDAMENTALS AND APPLICATIONS

Sensor Systems

FUNDAMENTALS AND APPLICATIONS

Clarence W. de Silva

CRC Press
Taylor & Francis Group
Boca Raton London New York

CRC Press is an imprint of the
Taylor & Francis Group, an **informa** business

CRC Press
Taylor & Francis Group
6000 Broken Sound Parkway NW, Suite 300
Boca Raton, FL 33487-2742

Printed at CPI on sustainably sourced paper
Version Date: 20160919

International Standard Book Number-13: 978-1-4987-1624-6 (Hardback)

Library of Congress Cataloging-in-Publication Data

Names: De Silva, Clarence W.
Title: Sensor systems : fundamentals and applications / Clarence W. de Silva.
Description: Boca Raton : Taylor & Francis, CRC Press, 2017. | Includes
bibliographical references and index.
Identifiers: LCCN 2016032430 | ISBN 9781498716246 (hardcover : alk. paper)
Subjects: LCSH: Detectors. | Systems engineering.
Classification: LCC TK7871.674 .D425 2017 | DDC 681/.2--dc23
LC record available at https://lccn.loc.gov/2016032430

Visit the Taylor & Francis Web site at
http://www.taylorandfrancis.com

and the CRC Press Web site at
http://www.crcpress.com

To the memory of my parents.

The true and only practicable object of a polytechnic school is, as I conceive, the teaching, not of the minute details and manipulations of the arts, which can be done only in the workshop, but the inculcation of those scientific principles which form the basis and explanation of them, their leading processes and operations in connection with physical laws.

William Barton Rogers,
Founder of the Massachusetts Institute of Technology, 1859

Contents

Preface

This book concerns not just sensors but "systems" of multiple sensors and other necessary components, be it in sensor networks or multisensor data fusion, and so on. It presents the physics and the principles of operation, ratings and performance specification, selection, necessary hardware and software for integration into an engineering system, signal processing and data analysis, parameter estimation and decision-making, and practical applications related to sensors and sensor systems. Even though this book has all the features of a course textbook, it also contains a wealth of practical information on the subject. Hence, the book is a valuable reference tool as well, particularly for the practicing professional. In the context of the book, the term *sensor system* can mean either a system of multiple sensors or a sensor and other accessories that are necessary in the implementation in a practical application.

In addition to presenting standard material on sensors in a student-friendly and interest-arousing manner, the book somewhat deviates from other books on sensors in the following ways:

- Physical principles and analytical methods are presented using simple mathematics.
- Different types of sensors in various domains (electronic, magnetic, inductive, capacitive, resistive, piezoelectric, piezoresistive, magnetostrictive, semiconductor, optical, ultrasonic, mechanical, fluid, thermal, etc.) are presented in an organized and unified manner so that users with a variety of engineering backgrounds (mechanical, electrical, civil, materials, mining, biomechanical, electrical, manufacturing, aerospace, etc.) would be able to follow and equally benefit from the information.
- Innovative and advanced topics of sensors, including wireless sensor networks, microelectromechanical system (MEMS) sensors, and multisensor data fusion, are covered.
- Practical procedures are given for the design, selection, and integration of sensors and other accessories in applications.
- A large number of worked examples, analytical examples, numerical examples, simulations, design examples, instrumentation examples, case studies, and end-of-chapter problems (with complete solutions) are provided throughout the book and are related to real-life situations and practical engineering applications.
- Key issues presented in the book are summarized in point form at various places in each chapter for easy reference, recollection, and presentation in PowerPoint form.
- Topics covered in each chapter are indicated in the beginning of the chapter. The key material and formulas in each chapter are provided in a Summary Sheet at the end of the chapter.
- Useful material that cannot be conveniently integrated into the chapters is given in a concise form as separate appendices at the end of the book.
- The book uses and illustrates available software tools, such as MATLAB® toolboxes, throughout.

- The book is concise, avoiding unnecessarily lengthy and uninteresting discussions, for easy reference and comprehension.
- There is adequate material in the book for two 14-week courses, one at the undergraduate level and the other at the graduate level.
- In view of the practical considerations, design issues, industrial techniques, and commercial information presented throughout the book and in view of the simplified and snapshot style presentation of more advanced theory, concepts, and practical information, the book serves as a useful reference tool for engineers, technicians, project managers, and other practicing professionals in industry and in research laboratories.
- References and reading suggestions are given for further information and study.
- A set of laboratory exercises and hands-on projects are presented in sufficient detail.
- Answers are provided for the numerical end-of-chapter problems. A solutions manual with comprehensive solutions to each end-of-chapter problem is available for the convenience of the instructors.

Background

Several years after my book *Control Sensors and Actuators* (Prentice Hall, 1989) was published, I received many requests for a revised and updated version. The revision was undertaken in 2000 during a sabbatical. As a result of my simultaneous involvement in the development of undergraduate and graduate curricula in mechatronics and in view of substantial new and enhanced material, which I was able to gather, the project quickly grew into one in mechatronics and led to the publication of the monumental 1300-page textbook *Mechatronics: An Integrated Approach* (Taylor & Francis Group, CRC Press, 2005). In meeting the original goal, however, the book *Sensors and Actuators: Control System Instrumentation* (1st edition) was subsequently developed as a condensed version of *Mechatronics*, while focusing on control sensors and actuators. I developed the manuscript for the original book from the notes for an undergraduate course titled *Instrumentation and Design of Control Systems* and for a graduate course titled *Control System Instrumentation* at Carnegie Mellon University. The undergraduate course was a popular senior elective taken by approximately half of the senior mechanical engineering class. The graduate course was offered for electrical and computer engineering, mechanical engineering, and chemical engineering students. The prerequisite for both courses was a conventional introductory course in feedback controls as well as the consent of the instructor. During the development of the material for that book, a deliberate attempt was made to cover a major part of the syllabuses for two courses—Analog and Digital Control System Synthesis and Computer Controlled Experimentation—offered in the Department of Mechanical Engineering at the Massachusetts Institute of Technology. At the University of British Columbia, I further developed, revised, and enhanced the original material for teaching courses in mechatronics and control sensors and actuators. The material in the book has acquired an application orientation through my industrial experience in the subject at places such as the International Business Machines Corporation, Westinghouse Electric Corporation, Bruel and Kjaer, and NASA's Lewis and Langley Research Centers.

While completing another edition of *Sensors and Actuators* and spending a sabbatical at the University of Melbourne, Australia, I was given the opportunity to teach an integrated undergraduate-graduate course in sensor systems. This is an important course in the mechatronics program at the university. The instructors who previously taught the course had used the material from my book *Sensors and Actuators*, which they indicated as the "best available" book for the course. However, they convinced me that a book on sensor systems should be different from the one on sensors. This book addresses that need as well.

Scope of This Book

The material presented in this book serves as a firm foundation for the subsequent building up of expertise in the subject on sensor systems—perhaps in an industrial setting or an academic research laboratory—with further knowledge of hardware, hands-on development and experimentation, and analytical skills gained during the process. Undoubtedly, for best results, a course in sensor systems should be accompanied by a laboratory component and/or class projects. The aspects of practical implementation, laboratory experimentation, and hands-on projects are presented and emphasized throughout the book.

This book consists of 12 chapters and 4 appendices. To maintain clarity and focus and to maximize the usefulness of the book, I present the material in a manner that will be useful to anyone with a basic engineering background, be it civil, electrical, mechanical, manufacturing, material, mining, aerospace, or biomechanical. Case studies, worked examples (analytical, numerical, and hardware/software based), and exercises are provided throughout the book. Although solutions are available for all problems included in the book, answers to only the numerical problems are given at the end of the book in order to encourage independent thinking. Complete solutions to the end-of-chapter problems are provided in a solutions manual, which is available to instructors who adopt the book. Additional material is available from the CRC website: http://www.crcpress.com/product/isbn/9781498716246.

Rationale

Sensor systems is evolving as an important subject in many curricula of engineering, including mechatronics, mechanical engineering, electrical–electronic–computer engineering, manufacturing engineering, and aeronautical–aerospace engineering. In my experience in teaching a number of undergraduate and graduate courses in sensors, actuators, instrumentation, and mechatronics, the existing books on sensors were found to have several notable weaknesses when considering them as textbooks on sensor systems. The following are some common drawbacks:

1. They present a selected group of sensors in a very few physical/application domains.
2. They do not address the multidomain nature of many engineering applications.

3. They do not treat other accessories that are needed when integrating a sensor into a practical application.
4. The topic of sensor selection is not addressed in many cases. Where addressed, it does not consider both engineering performance specifications and practical/commercial product ratings.
5. They do not adequately treat instrumentation in sensor applications.
6. They do not present laboratory experiments, engineering projects, and case studies.
7. They do not treat systems of multiple sensors, particularly networked sensors (wireless sensor networks) and multisensor data fusion.
8. Except in specialized publications, they do not treat such modern topics of sensing as MEMS sensors.
9. They do not adequately treat signal processing and parameter estimation using sensory data.
10. They do not adequately contain textbook features (detailed worked examples, comprehensive end-of-chapter problems, case studies, design problems, laboratory exercises, solutions manual, etc.).

This book is expected to be a highly readable and useful textbook that overcomes these shortcomings.

Clarence W. de Silva
Vancouver, Canada

MATLAB® is a registered trademark of The MathWorks, Inc. For product information, please contact:

The MathWorks, Inc.
3 Apple Hill Drive
Natick, MA 01760-2098 USA
Tel: 508-647-7000
Fax: 508-647-7001
E-mail: info@mathworks.com
Website: www.mathworks.com

Acknowledgments

Many individuals have assisted in the preparation of this book, but it is not practical to acknowledge all such assistance here. First, I wish to recognize the contributions, both direct and indirect, of my graduate students, research associates, and technical staff. I particularly thank Tony Teng Li, my research associate and lab manager. I am also grateful to Jonathan W. Plant, executive editor, CRC Press/Taylor & Francis Group, for his great enthusiasm and support throughout the project. This project would not have been possible if not for his constant encouragement, advice, and drive. Others at CRC Press, particularly Ed Curtis and Vijay Bose. Finally, I acknowledge the unwavering love and support of my wife and children.

Author

Clarence W. de Silva, PE, fellow ASME and fellow IEEE, is a professor of mechanical engineering at the University of British Columbia, Vancouver, Canada, and occupies the Senior Canada Research Chair Professorship in Mechatronics and Industrial Automation. Prior to that, he occupied the NSERC-BC Packers Research Chair in Industrial Automation since 1988. He served as a faculty member at Carnegie Mellon University (1978–1987) and as a Fulbright visiting professor at the University of Cambridge (1987/1988).

He earned PhD degrees at the Massachusetts Institute of Technology (1978) and the University of Cambridge, England (1998), and received an honorary DEng from the University of Waterloo, Ontario, Canada (2008). Dr. De Silva has also occupied the Mobil Endowed Chair Professorship in the Department of Electrical and Computer Engineering at the National University of Singapore; honorary professorship at the Xiamen University, China; and honorary chair professorship at the National Taiwan University of Science and Technology.

Other fellowships: Fellow at the Royal Society of Canada, fellow at the Canadian Academy of Engineering, Lilly fellow at Carnegie Mellon University, NASA-ASEE fellow, senior Fulbright fellow at the Cambridge University, fellow of the Advanced Systems Institute of British Columbia, Killam fellow, Erskine fellow at the University of Canterbury, professorial fellow at the University of Melbourne, and Peter Wall scholar at the University of British Columbia.

Awards: Paynter Outstanding Investigator Award and Takahashi Education Award, ASME Dynamic Systems and Control Division; Killam Research Prize; Outstanding Engineering Educator Award, IEEE Canada; Lifetime Achievement Award, World Automation Congress; IEEE Third Millennium Medal; Meritorious Achievement Award, Association of Professional Engineers of British Columbia; and Outstanding Contribution Award, IEEE Systems, Man, and Cybernetics Society. He also made 32 keynote addresses at international conferences.

Editorial duties: The author served on 14 journals, including *IEEE Transactions on Control Systems Technology* and *Journal of Dynamic Systems, Measurement, and Control, Transaction of the ASME*; he worked as an editor in chief for *International Journal of Control and Intelligent Systems* and *International Journal of Knowledge-Based Intelligent Engineering Systems*, a senior technical editor for *Measurements and Control*, and a regional editor in North America for *Engineering Applications of Artificial Intelligence—IFAC International Journal*.

Publications: 22 technical books, 19 edited books, 50 book chapters, 240 journal articles, and more than 260 conference papers.

Recent books: *Sensors and Actuators: Engineering System Instrumentation* (2nd edition, Taylor & Francis Group/CRC, 2016); *Mechanics of Materials* (Taylor & Francis Group/CRC, 2014); *Mechatronics: A Foundation Course* (Taylor & Francis Group/CRC, 2010); *Modeling and Control of Engineering Systems* (Taylor & Francis Group/CRC, 2009); *Vibration: Fundamentals and Practice*, 2nd edition (Taylor & Francis Group/CRC, 2007); *Mechatronics: An Integrated Approach* (Taylor & Francis Group/CRC, 2005); and *Soft Computing and Intelligent Systems Design: Theory, Tools, and Applications* (Addison Wesley, 2004).

1

Sensor Systems in Engineering

Chapter Objectives

- Role of Sensors and Sensor Systems
- Importance of Estimation in Sensing
- Innovative Sensor Technologies
- Scenarios of Engineering Application
- Human Sensory System
- Sensors in Mechatronic Engineering
- Sensing in Control Systems
- Instrumentation Process
- Application Examples
- Organization of the Book

1.1 Role of Sensors and Sensor Systems

Sensors

Sensors (e.g., semiconductor strain gauges, tachometers, RTD temperature sensors, cameras, piezoelectric accelerometers) are needed to measure (i.e., sense) unknown signals and parameters of an engineering system and its environment. Essentially, sensors are needed to monitor and "learn" about the system and possible interactions with its surroundings. This knowledge will be useful not only in operating or controlling the system but also for many other purposes such as

1. Process monitoring
2. Experimental modeling (i.e., model identification)
3. Product testing and qualification
4. Product quality assessment
5. Fault prediction, detection, and diagnosis
6. Advisory/warning generation
7. Surveillance

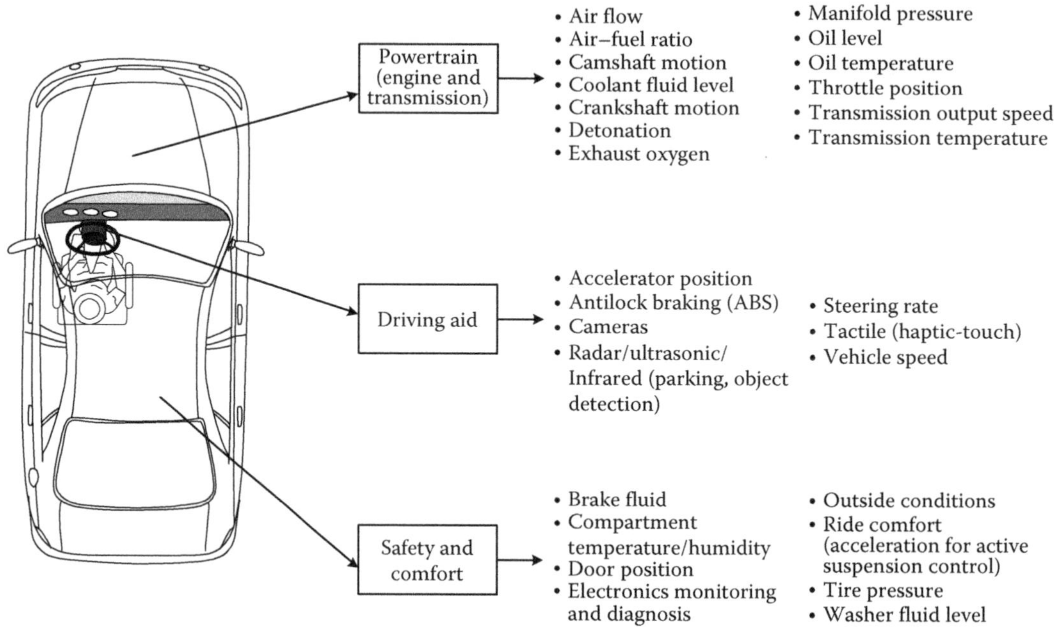

FIGURE 1.1
Sensors in an automobile.

A common application of sensors is in automobiles where a vast variety of sensors are used in the powertrain, driving assistance, safety and comfort, and so on, as presented in Figure 1.1. Some examples of commercially available sensors are given here.

Motion sensors: Potentiometer, differential transformer (LVDT), magnetostrictive (temposonic) displacement sensor, magnetic induction proximity sensor, tachometer, resolver, synchro, gyro, piezoelectric accelerometer, laser ranger, and ultrasound ranger

Force/torque sensors: Semiconductor strain-gauge and motor current sensor

Fluid flow sensors: Coriolis flow meter, pitot tube, rotameter, and orifice flow meter

Pressure sensors: Manometer, Bourdon tube, and diaphragm type

Temperature sensors: Thermocouple, thermistor, and resistance temperature detector (RTD)

Sensor Systems

A "sensor system" may mean

1. A system of multiple sensors, including sensor network and sensor/data fusion (when one sensor may not be adequate for the particular application)
2. A sensor and accessories that will be necessary in implementing it in a practical application (e.g., signal processing, data acquisition, data transmission/communication)

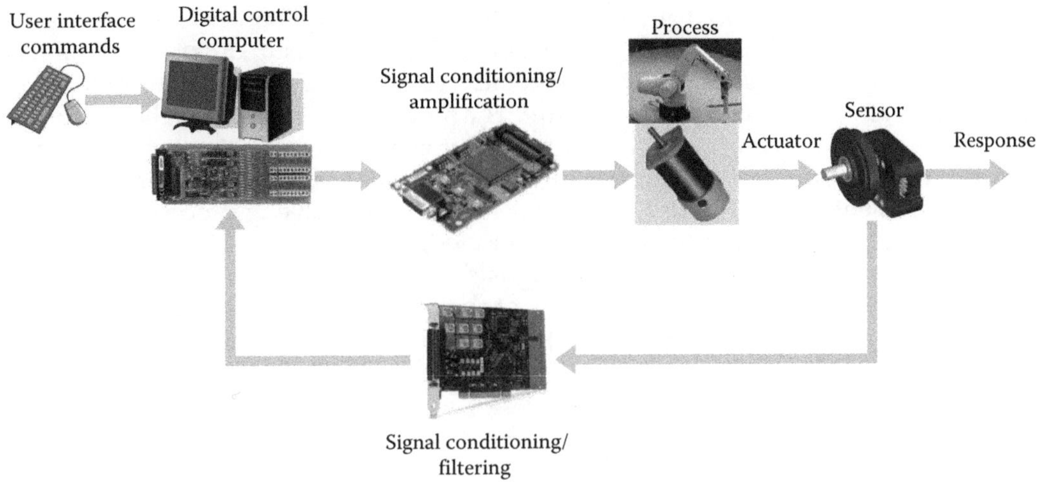

FIGURE 1.2
Sensors and other components in a feedback control system.

Several examples of sensor systems in both of these categories will be introduced in this chapter and further studied in subsequent chapters.

This book concerns both individual sensors and sensor systems. It will present the physics and the principles of operation, ratings and performance specification, selection, necessary hardware and software for integration into an engineering system, signal processing and data analysis, parameter estimation and decision making, and practical applications related to sensors and sensor systems.

Use in a Control System

Sensors and sensor systems are indispensable in a control system. A control system is a dynamic system that contains a controller as an integral part. The purpose of the controller is to generate control signals, which will drive the *process* (or *plant*) that is being controlled in a desired manner (i.e., according to some *performance specifications*; see Chapter 5), using various control devices. Specifically in a feedback control system, the control signals are generated based on the sensed response signals of the plant. Sensors and other main components in a feedback control system are schematically shown in Figure 1.2.

1.1.1 Importance of Estimation in Sensing

The measurement from a sensor may not provide the true value of the required parameter or variable for two main reasons:

1. The required quantity is not directly measured and has to be computed from the measured value (or values) using a suitable "model."
2. The sensor (or even the sensing process) is not perfect and will introduce "measurement error."

Hence, sensing may be viewed as a problem of estimation (see Chapter 7), where the "true value" of the measured quantity is "estimated" using the measured data. Two main categories of error, "model error" and "measurement error," enter into the process of estimation and will affect the accuracy of the result. The model error arises from the relationship between quantity of interest and the quantity that is measured (or the model of the system). Unknown (and random) input disturbances may also be treated under model error. The measurement error will arise from the sensor and the sensing process (e.g., how the sensor is mounted and how the data are collected, communicated, and recorded). It is clear that estimation (of parameters and signals) is an important step of sensing. Many methods are available for estimation. Some of them are presented in the book (e.g., least squares, maximum likelihood, Kalman filter, extended Kalman filter, unscented Kalman filter; see Chapter 7).

1.1.2 Innovative Sensor Technologies

Apart from conventional sensors, many types of innovative and advanced sensors are being developed (see Chapters 10 through 12). Several types are listed as follows:

1. Microminiature and embedded sensors (those based on integrated circuit [IC] technologies and microelectromechanical system [MEMS] technologies, with integrated signal processing, control, and other hardware; sensors may be *integral* with the system components)
2. Intelligent sensors (built-in information preprocessing, reasoning, and inferencing to provide high-level knowledge-based decision making; multisensor fusion to provide more reliable and accurate results)
3. Networked sensors (multiple sensor nodes [SNs] communicate with each other in a distributed sensing setup; there can be significant *geographic separation* between SNs; the node connection may be wired or wireless; typically an SN contains one or more sensors, a microcontroller, and signal conditioning hardware)
4. Hierarchical sensory architectures (low-level sensory information is preprocessed to meet higher level requirements), for example, in hierarchical control, each control layer is serviced by a corresponding sensor layer

1.2 Application Scenarios

Sensors and transducers are necessary to acquire output signals (process responses) for system monitoring; fault prediction, detection, and diagnosis; generation of warnings and advisories; feedback control; and supervisory control and to measure input signals for experimental modeling (system identification) and feedforward control and for a variety of other purposes. Since many different types and levels of signals are present in a dynamic system, *signal modification* (including signal conditioning and signal conversion; see Chapters 3 and 4) is indeed a crucial function associated with sensing. In particular, signal modification is an important consideration in component interfacing. It is clear that the subject of system instrumentation should deal with sensors, transducers, signal modification, and component interconnection. In particular, the subject should address the identification of the necessary system components with respect to type, functions, operation and interaction, and proper selection and interfacing of these components for various applications. Parameter selection (including component sizing and system tuning) is an important step as well. Design is a necessary part of system instrumentation, for it is design that enables us to build a system that meets the performance requirements—starting, perhaps, with a few basic components such as sensors, actuators, controllers, compensators, and signal modification devices.

Engineers, particularly mechatronic engineers, should be able to identify or select components, particularly sensors, actuators, controllers, and interface hardware for a system, to model and analyze individual components and the overall integrated system, and to choose proper parameter values for the components (i.e., component sizing and system tuning) for the system to perform the intended functions in accordance with some *specifications*.

Instrumentation (sensors, actuators, signal acquisition and modification, controllers, and accessories and their integration into a process) is applicable in branches of engineering. Typically, instrumentation is applicable in process monitoring; fault prediction, detection, and diagnosis; testing; and control, in practically every engineering system. Some branches of engineering and typical application situations are listed as follows:

Aeronautical and aerospace engineering: Aircraft and spacecraft

Civil engineering: Monitoring of civil engineering structures (bridges, buildings, etc.)

Chemical engineering: Monitoring and control of chemical processes and plants

Electrical and computer engineering: Development of electronic and computer-integrated devices, hard drives, etc., sensor-embedded systems, and control and monitoring of electrical and computer systems

Materials engineering: Material synthesis processes and material testing

Mechanical engineering: Vehicles and transit systems, robots, manufacturing plants, industrial plants, power generation systems, jet engines, oil and gas extraction, transportation, and refining

Mining and mineral engineering: Mining machinery and processes and raw material processing

Nuclear engineering: Monitoring and control of nuclear reactors and testing and qualification of components

We have highlighted the automotive example as a valuable application scenario for sensors and sensor systems. Several applications and their use of sensors are noted in Table 1.1. Some important areas of application are indicated here.

As noted before, transportation is a broad area where sensors have numerous applications. In ground transportation in particular, automobiles, trains, and automated transit systems use airbag deployment systems, antilock braking systems (ABS), cruise control systems, active suspension systems, and various devices for monitoring, toll collection, navigation, warning, and control in intelligent vehicular highway systems. All these devices use sensors and sensor systems. In air transportation, modern aircraft designs with advanced materials, structures, electronics, and control benefit from sophisticated sensors. Flight simulators, flight control systems, navigation systems, landing gear mechanisms, traveler/driver comfort aids, and the like use sensors and sensor systems primarily for monitoring and control.

TABLE 1.1

Sensors Used in Some Common Engineering Applications

Process	Typical Sensors
Aircraft	Displacement, speed, acceleration, elevation, heading, force pressure, temperature, fluid flow, voltage, current, global positioning system (GPS)
Automobile	Displacement, speed, force, pressure, temperature, fluid flow, fluid level, vision, voltage, current, GPS, radar, sonar
Home heating system	Temperature, pressure, fluid flow
Milling machine	Displacement, speed, force, acoustics, temperature, voltage, current
Robot	Optical image, displacement, speed, force, torque, tactile, laser, ultrasound, voltage, current
Water quality monitoring	Temperature, flow rate, pH value, dissolved oxygen, electrical conductivity, oxidation–reduction potential
Wood drying kiln	Temperature, relative humidity, moisture content, air flow

Manufacturing and production engineering is another broad field that uses various technologies of sensors. Factory robots (for welding, spray painting, assembly, inspection, etc.), automated guided vehicles (AGVs), modern computer numerical control machine tools, machining centers, rapid (and virtual) prototyping systems, and micromachining systems are examples. Product quality monitoring, machine/machine tool monitoring, and high-precision motion control are particularly important in these applications, which require advanced sensors and sensor systems.

In medical and healthcare applications, sensor technologies for patient examination, surgery, rehabilitation, drug dispensing, and general patient care are being developed and used. In that context, novel sensors and sensor systems are applied for patient transit devices, various diagnostic probes and scanners, beds, exercise machines, prosthetic and orthotic devices, physiotherapy, and telemedicine.

In a modern office environment, automated filing systems, multifunctional copying machines (copying, scanning, printing, electronic transmission, etc.), food dispensers, multimedia presentation and meeting rooms, and climate control systems incorporate advanced technologies of sensors and sensor systems.

In household applications, home security systems, robotic caregivers and helpers, robotic vacuum cleaners, washers, dryers, dishwashers, garage door openers, and entertainment centers all use a variety of sensors and associated technologies.

The digital computers and related digital devices use MEMSs, embedded and integrated sensors. The impact goes further because digital devices are integrated into a vast variety of other devices and applications.

In civil engineering applications, cranes, excavators, and other construction machinery, buildings, reservoirs, and bridges will improve their performance by adopting proper sensors and sensor systems.

In space applications, mobile robots such as NASA's Mars Exploration Rover, space station robots, and space vehicles depend on sensing for their proper operation and control.

Identification, analysis, selection matching and interfacing of components, component sizing, and tuning of the integrated system (i.e., adjusting parameters to obtain the required response from the system) are essential tasks in the instrumentation and design of an engineering system. The book addresses these issues in the context of sensors and sensor systems, starting from the basics and systematically leading to advanced concepts and applications.

1.3 Human Sensory System

A robust area in the development of "intelligent" robots that can mimic characteristics of natural intelligence concerns sensing. The main goal is to develop robotic sensors that can play the role of human sensory activities (five senses):

1. Sight (visual)
2. Hearing (auditory)
3. Touch (tactile)
4. Smell (olfactory)
5. Taste (flavor)

Sensors in the first three categories are in a more advanced stage of development, starting with basic sensors (cameras, microphones, and tactile sensors). The last two categories of sensors primarily use chemical processes and are less common.

In addition to these five senses, humans also have other types of sensory features, in particular, the sense of balance, pressure, temperature, pain, and motion. In fact, some of these sensory capabilities will involve the use of one or more of the basic five senses, simultaneously through the central nervous system (CNS).

In their development, robotic and other engineering systems have long relied on and are inspired by the sensing process of humans and other animals. The basic biological sensing process is shown in Figure 1.3. A stimulus (e.g., light for vision, sound waves for hearing) is received at the receptor where the dendrites of the neurons convert the energy of the stimulus into electromechanical impulses in the dendrites of the neurons. The axons of the neurons then conduct the corresponding action potentials into the CNS of the brain. These potentials are then interpreted by the brain to create the corresponding sensory perception. An engineering sensory process, such as that used in a robot, basically uses similar processes. As we will see in the succeeding chapters, they involve sensor and transducer, transmission, and signal conversion and processing.

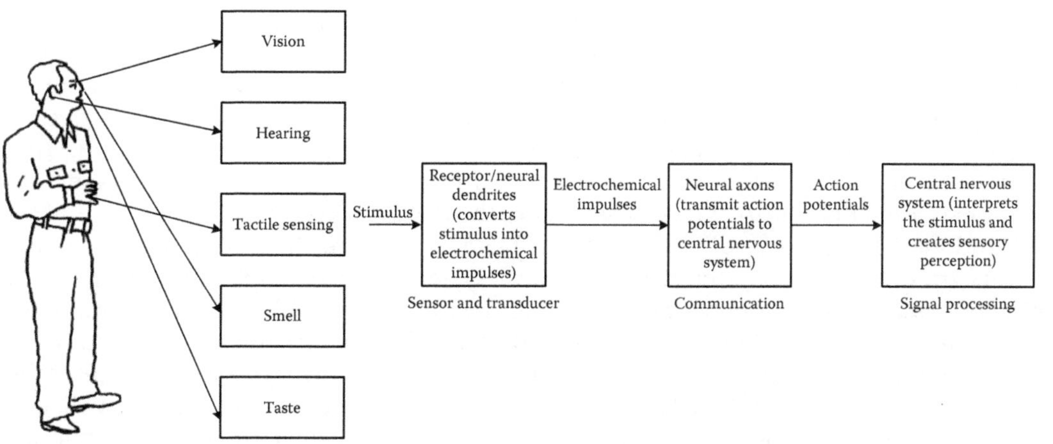

FIGURE 1.3
Biological sensing process and its analogy to engineering sensing process.

1.4 Mechatronic Engineering

The subject of mechatronics concerns the synergistic application of mechanics, electronics, controls, and computer engineering in the development of electromechanical products and systems, through an integrated design approach. Mechatronics is particularly applicable in mixed-domain (or multidomain) systems, which incorporate several physical domains such as electrical, mechanical, fluid, and thermal, in an integrated manner. For example, an ABS of an automobile may involve mechanics, electronics, hydraulics, and heat transfer and may be designed in an "optimal" manner as a mechatronic product. Mechatronic products and systems include modern automobiles and aircraft, smart household appliances, medical robots, space vehicles, and office automation devices.

A typical mechatronic system consists of a mechanical skeleton, actuators, sensors, controllers, signal-conditioning/modification devices, computer/digital hardware and software, interface devices, and power sources. Different types of sensing and information acquisition and transfer are involved among all these various types of components. For example, a servomotor, which is a motor with the capability of sensory feedback for accurate generation of complex motions, consists of mechanical, electrical, and electronic components. Its main mechanical components are the rotor, stator, bearings, mechanics of the speed sensor such as an optical encoder, and motor housing. The electrical components include the circuitry for the field windings and rotor windings (not in the case of permanent-magnet rotors) and circuitry for power transmission and commutation (if needed). Electronic components include those needed for sensing (e.g., optical encoder for displacement and speed sensing and/or tachometer for speed sensing). The overall design of a servomotor can be improved by taking a mechatronic approach, where all components and functions are treated concurrently in an integrated manner in its design.

1.4.1 Mechatronic Approach to Instrumentation

Study of mechatronic engineering should include all stages of modeling, design, development, integration, instrumentation, control, testing, operation, and maintenance of a mechatronic product or system. From the viewpoint of instrumentation, which includes sensing, a somewhat "optimal" and unified approach, not a sequential approach, has to be taken with regard to component selection, modeling, integration, evaluation, and operation. Specifically, "instrumentation" has to be treated as an integral aspect of "design." This is simply because by design, we develop systems that can carry out required functions while meeting certain performance specifications. Sensors, sensor systems, and instrumentation play a direct role in achieving the design objectives.

Traditionally, a "sequential" approach has been adopted in the design of multidomain (or mixed) systems such as electromechanical systems. For example, first the mechanical and structural components are designed, next electrical and electronic components are selected or developed and interconnected, and subsequently, a computer or a related digital device is selected and interfaced with the system, along with a digital controller. The dynamic coupling (interaction) between various components of a system dictates, however, that an accurate design of the system should consider the entire system as a whole rather than designing different domains (e.g., the electrical/electronic aspects and the mechanical aspects) separately and sequentially. When independently designed components are interconnected, several problems can arise (see Chapter 2):

1. When two independently designed components are interconnected, the original characteristics and operating conditions of the two components will change due to loading or dynamic interactions.
2. Perfect matching of two independently designed and developed components will be practically impossible. As a result, a component can be considerably underutilized or overloaded, in the interconnected system, both conditions being inefficient, possibly hazardous, and undesirable.
3. Some of the external variables in the components will become internal and "hidden" due to interconnection, which can result in potential problems that cannot be explicitly monitored through sensing and cannot be directly controlled.

The need for an integrated and concurrent design for multidomain (e.g., electromechanical) systems can be identified as a primary justification for the use of the "mechatronic" approach. In particular, when incorporating instrumentation in the design process, such a unified and integrated approach is desirable with regard to sensors, other components, and interface hardware. For example, consider the design and development of a sensor jacket (as presented as an example project at the end of this chapter) for a telemedicine system. Recent advances in sensor technologies that are applicable in human health monitoring such as biomedical nanosensors, piezoelectric sensors, force and motion sensors, and optical/vision sensors for abnormal motion detection of humans may be incorporated into the jacket. However, for optimal performance, the selection/development, location, mounting, and integration of the sensors should not be treated independently of the development of other aspects of the jacket. For example, a mechatronic design quotient (MDQ) may be employed to represent the "goodness" of the overall design of the jacket where a design index is defined with respect to each design requirement (e.g., size, structure, components, cost, accuracy, speed). Then, parameters such as sensor size, interface hardware, power requirements, and component location and configuration may be incorporated into the MDQ, which will improve/optimize the process of signal acquisition and processing, body conformability, weight, robustness, and cost and also will improve the speed, accuracy, and reliability of the information provided by the sensor jacket.

1.4.2 Bottlenecks for Mechatronic Instrumentation

Even though, in theory, the mechatronic approach is the "best" ("optimal") particularly with regard to instrumentation, it may not be practical to realize the optimal results of instrumentation as dictated by the approach. The mechatronic approach requires the entire system including the process and instrumentation to be designed concurrently. This assumes that all aspects and components of the entire system can be modified (continuously) according to the mechatronic result. However, unless the entire system (including the process) is a new design, such flexibility is often not realistic. For example, typically, the process is already available, and it is not practical, convenient, or cost effective to modify some or all of its components. Then, even if the instrumentation is chosen according to a mechatronic procedure, the overall system will not function as optimally as if we had the freedom to modify the entire plant as well. Furthermore, in practice, the available set of components are discrete and limited. Then, even if the optimality dictates a particular set of specifications for a component, we may have to select a component having a different (yet close) set of specifications.

As an example, consider an automated vehicle–guideway system of public transit. Suppose that the system already exists and that it is required to replace some of its cars. Then, it is not practical to significantly modify the guideway to accommodate a new design of cars. In fact the design freedom with regard to the cars will be limited even if it is constrained by a specific guideway design. Even if a car is designed and instrumented optimally, according to mechatronics, the operation of the overall vehicle–guideway system will not be optimal.

It is clear that true mechatronic instrumentation may not be perfectly possible for existing processes. Furthermore, since the components for instrumentation (sensors, actuators, controllers, accessory hardware) may come from different manufacturers and their availability would be limited, as noted before, it is not practical to realize a true "mechatronic" product (since the available set of components is limited and may not also be truly compatible).

1.5 Control System Architectures

Sensor systems are important components in the instrumentation of a *control system*. The *controller*, which is an essential part of any control system, makes the *plant* (i.e., the *process* that is being controlled) behave in a desired manner, according to some *specifications*. The overall system that includes at least the plant and the controller is called the control system. The system can be quite complex and may be subjected to known or unknown excitations (i.e., inputs), as in the case of an aircraft.

Some useful terminology related to a control system is listed as follows:

- *Plant or process*: System to be controlled
- *Inputs*: Commands, driving signals, or excitations (known or unknown)
- *Outputs*: Responses of the system
- *Sensors*: Devices that measure system variables (excitations, responses, etc.)
- *Actuators*: Devices that drive various parts of the system
- *Controller*: Device that generates the control signal
- *Control law*: Relation or scheme according to which the control signal is generated
- *Control system*: At least the plant and the controller (may include sensors, signal conditioning, and other components as well)
- *Feedback control*: Control signal determined according to the plant response
- *Open-loop control*: Plant response not used to determine the control action
- *Feedforward control*: Control signal determined according to the plant excitation or a model of the plant

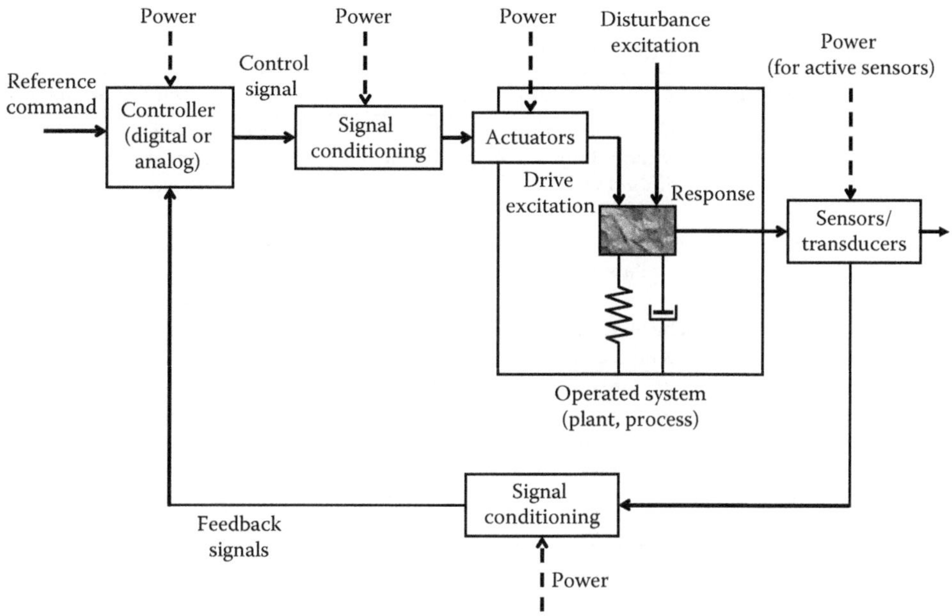

FIGURE 1.4
Key components of a feedback control system.

In Figure 1.4, we have identified key components of a feedback control system. Several discrete blocks are shown, depending on the various functions that take place in a typical control system. In a practical control system, this type of clear demarcation of components might be difficult; one piece of hardware might perform several functions or more than one distinct unit of equipment might be associated with one function. Embedded systems in particular may have distributed multifunctional components where demarcation of the functional blocks will be difficult. Nevertheless, Figure 1.4 is useful in understanding the architecture of a general feedback control system. In an *analog control system,* control signals are continuous-time variables generated by analog hardware; in particular, the controller is an analog device. On the other hand, in a *digital control system,* the controller is a digital processor (e.g., microcontroller), and hence, sampling or encoding (digital representation) of the control signal is not required.

The control problem can become challenging due to such reasons as

- Complex plant (many inputs and many outputs, dynamic coupling, nonlinear, time-varying parameters, etc.)
- Rigorous performance specifications
- Unknown or unmeasurable excitations (unknown inputs/disturbances/noise)
- Unknown or unmeasurable responses (unmeasurable state variables and outputs, measurement errors and noise)
- Unknown dynamics (incompletely known plant)

Since the operation of a control system is based on a set of performance specifications, it is important to identify key performance characteristics that a good control system

TABLE 1.2

Performance Specifications for a Control System

Attribute	Desired Value	Objective	Specifications
Stability level	High	The response does not grow without limit and decays to the desired value.	Percentage overshoot, settling time, pole (eigenvalue) locations, time constants, phase and gain margins, damping ratios
Speed of response	Fast	The plant responds quickly to inputs/excitations.	Rise time, peak time, delay time, natural frequencies, resonant frequencies, bandwidth
Steady-state error	Low	The offset from the desired response is negligible.	Error tolerance for a step input
Robustness	High	Accurate response under uncertain conditions (input disturbances, noise, model error, etc.) and under parameter variation.	Input disturbance/noise tolerance, measurement error tolerance, model error tolerance
Dynamic interaction	Low	One input affects only one output.	Cross-sensitivity, cross-transfer functions

should possess. In particular, the following performance requirements are important (see Chapter 5):

1. Sufficiently stable response (*stability*). Specifically, the response of the system to an initial-condition excitation should decay back to the initial steady state (*asymptotic stability*). The response to a bounded input should be bounded (bounded-input, bounded-output [*BIBO*] *stability*).
2. Sufficiently fast response (*speed of response* or *bandwidth*). The system should react quickly to a control input or excitation.
3. Low sensitivity to noise, external disturbances, modeling errors, and parameter variations (*sensitivity* and *robustness*).
4. High sensitivity to control inputs (*input sensitivity*).
5. Low error, for example, tracking error and steady-state error (*accuracy*).
6. Reduced coupling among system variables (*cross-sensitivity* or *dynamic coupling*).

As listed here, some of these specifications are rather general. Table 1.2 summarizes typical performance requirements for a control system. Some requirements might be conflicting. For example, fast response is often achieved by increasing the system gain, and increased gain increases the actuation signal, which has a tendency to destabilize a control system. Note further that what are given here are primarily qualitative descriptions for "good" performance. In designing a control system, however, these descriptions have to be specified in a quantitative manner. The nature of the used quantitative design specifications depends considerably on the particular design technique that is employed. Some of the design specifications are time-domain parameters, and the others are frequency-domain parameters.

1.5.1 Feedback and Feedforward Control

As seen before, in a *feedback control system*, the control loop has to be closed, where the system response is sensed and employed to generate the control signals. Hence, feedback control is also known as *closed-loop control*.

If the plant is stable and is completely and accurately known and if the inputs to the plant can be precisely generated (by the controller) and applied, then accurate control might be possible even without feedback control. Under these circumstances, a measurement system is not needed (or at least not needed for feedback into the controller) and thus, we have an *open-loop control system*. In open-loop control, we do not use current information on system response to determine the control signals. In other words, there is no feedback. Even though a sensor is not explicitly needed in an open-loop architecture, sensors may be employed within an open-loop system to monitor the applied input, the resulting response, and possible disturbance (unknown) inputs.

The significance and importance of sensors and sensor systems remain regardless of the specific control system architecture that is implemented in a given application. We now outline several architectures of control system implementation while indicating the presence of sensors in them.

Even in a feedback control system, there may be inputs that are not sensed and used in feedback control. Some of these inputs might be important variables for the plant, but more commonly, they are undesirable inputs, such as external disturbances, which are unwanted yet unavoidable. Generally, the performance of a control system can be improved by measuring these (unknown) inputs and somehow using the information to generate control signals. In feedforward control, unknown inputs are measured, and that information, along with desired inputs, is used to generate control signals that can reduce errors due to these unknown inputs or variations in them. The reason for calling this method feedforward control stems from the fact that the associated measurement and control (and compensation) both take place in the forward path of the control system. *Note*: In some types of feedforward control, the input signal is generated by using a model of the plant (and may not involve sensing).

As a practical example, consider the natural gas home heating system shown in Figure 1.5a. A simplified block diagram of the system is shown in Figure 1.5b. In conventional feedback control, the room temperature is measured and its deviation from the desired temperature (set point) is used to adjust the natural gas flow into the furnace. On/off control through a thermostat is typically used in such applications. Even if proportional or three-mode (proportional–integral–derivative [PID]) control is employed, it is not easy to steadily maintain the room temperature at the desired value if there are large changes in other (unknown) inputs to the system, such as water flow rate through the furnace, temperature of water entering the furnace, and outdoor temperature.

Better results can be obtained by measuring these disturbance inputs and using that information in generating the control action. This is feedforward control. Note that in the absence of feedforward control, any changes in the inputs w_1, w_2, and w_3 in Figure 1.5 would be detected only through their effect on the feedback signal (i.e., room temperature). Hence, the subsequent corrective action can considerably lag behind the cause (i.e., changes in w_i). This delay will lead to large errors and possible instability problems. With feedforward control, information on the disturbance input w_i will be available to the controller immediately, and its effect on the system response can be anticipated, thereby speeding up the control action and also improving the response accuracy. Faster action and improved accuracy are two very desirable effects of feedforward control.

1.5.2 Digital Control

In digital control, a digital computer (e.g., a microcontroller) serves as the controller. Virtually, any control law may be programmed into the control computer. Control

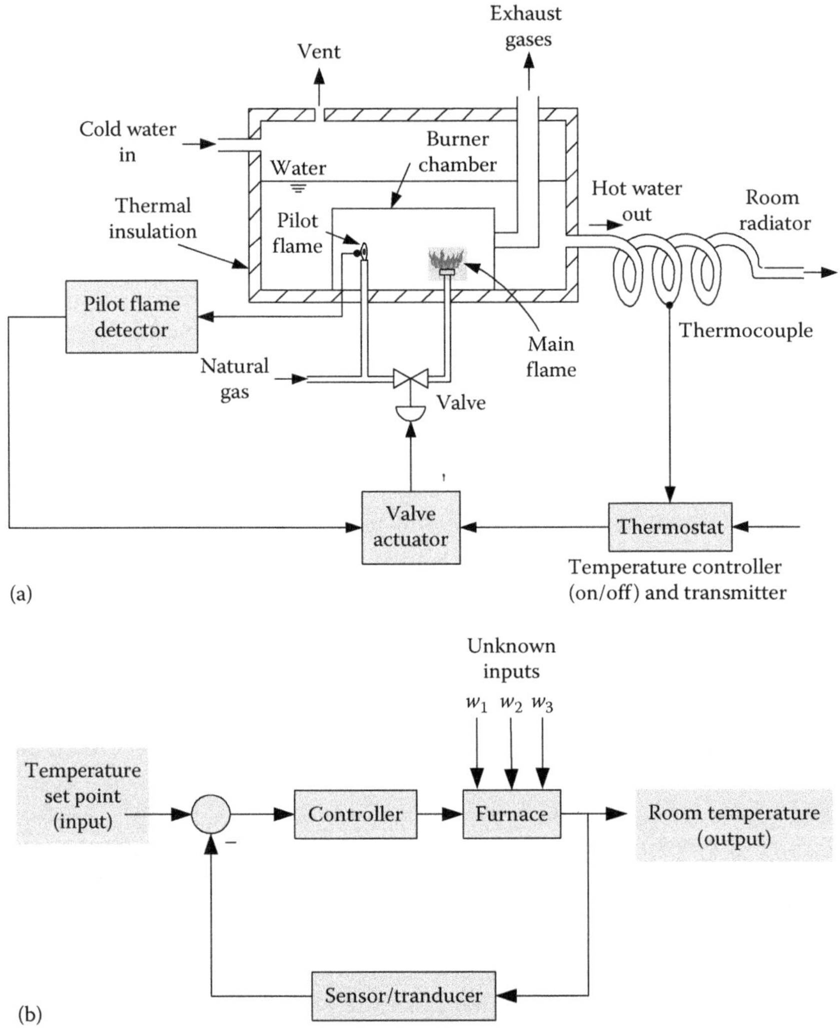

FIGURE 1.5
(a) A natural gas home heating system. (b) A block diagram representation of the system. w_1, water flow rate; w_2, temperature of cold water into furnace; w_3, temperature outside the room.

computers have to be fast and dedicated machines for real-time operations where processing has to be synchronized with plant operation and actuation requirements. This requires a real-time operating system. Apart from these requirements, control computers are basically not different from general-purpose digital computers. They consist of a processor to perform computations and to oversee data transfer, memory for the storage of programs and data during processing, mass-storage devices to store information that is not immediately needed, and input or output devices to read in and send out information (i.e., for interfacing with other components of the control system).

Digital control systems might use digital instruments and additional processors for actuating, signal-conditioning, or measuring functions. For example, a stepper motor, which responds with incremental motion steps when driven by pulse signals, can be considered as a digital actuator. It usually contains digital logic circuitry in the motor drive system. Similarly, a two-position solenoid is a digital (binary) actuator. Digital flow control may be accomplished using a digital control valve. A typical digital valve consists of a bank of orifices, each sized in proportion to a place value (2^i, $i = 0, 1, 2, \ldots, n$) of a binary word (n-bit). Each orifice is actuated by a separate rapid-acting on/off solenoid. In this manner, many combinations of discrete flow values can be obtained.

Direct digital measurement of displacements and velocities can be made using shaft encoders. These are digital transducers that generate coded outputs (e.g., in binary or gray-scale representation) or pulse signals that can be coded using counting circuitry. Such outputs can be read by a digital controller with relative ease. Frequency counters also generate digital signals that can be fed directly into a digital controller. When measured signals are in the analog form, an analog front end is necessary to interface the transducer and the digital controller. Input/output (I/O) interface cards that can take both analog and digital signals are available with digital controllers.

Analog measurements and reference signals have to be sampled and encoded before digital processing within the controller. Digital processing can be effectively used for signal conditioning as well. Alternatively, a digital signal processing (DSP) chip can function as a digital controller. However, analog signals have to be preconditioned by using analog circuitry before digitizing in order to eliminate or minimize problems due to aliasing distortion (high-frequency components above half the sampling frequency appearing as low-frequency components; see Chapter 6) and leakage (error due to signal truncation) as well as to improve the signal level and filter out extraneous noise. The drive system of a plant typically takes in analog signals. Often, the digital output from the controller has to be converted into analog form for this reason. Both analog-to-digital conversion (ADC) and digital-to-analog conversion (DAC) can be interpreted as signal-conditioning (modification) procedures (see Chapter 4). If more than one output signal is measured, each signal will have to be conditioned and processed separately. Ideally, this will require separate conditioning and processing hardware for each signal channel. A less expensive (but slower) alternative would be to time-share this expensive equipment by using a multiplexer. This device will pick one channel of data from a bank of data channels in a sequential manner and connect it to a common input device.

The current practice of using dedicated, microcontroller-based, and often decentralized (i.e., distributed) digital control systems in industrial applications can be rationalized in terms of the major advantages of digital control. The following are some of the important considerations:

1. Digital control is less susceptible to noise or parameter variation in instrumentation because data can be represented, generated, transmitted, and processed as binary words, with bits possessing two identifiable states.
2. Very high accuracy and speed are possible through digital processing. Hardware implementation is usually faster than software implementation.
3. Digital control systems can handle repetitive tasks extremely well, through programming.
4. Complex control laws and signal-conditioning algorithms that might be impractical to implement using analog devices can be programmed.
5. High reliability in operation can be achieved by minimizing analog hardware components and through decentralization using dedicated microcontrollers for various control tasks.

6. Large amounts of data can be stored using compact, high-density data storage methods.
7. Data can be stored or maintained for very long periods of time without drift and without getting affected by adverse environmental conditions.
8. Fast data transmission is possible over long distances without introducing excessive dynamic delays and attenuation, as in analog systems.
9. Digital control has easy and fast data retrieval capabilities.
10. Digital processing uses low power and low operational voltages (e.g., 0–12 V DC).
11. Digital control is cost effective.

1.5.3 Programmable Logic Controllers

There are many control systems and industrial tasks that involve the execution of a sequence of steps, depending on the state of some elements in the system and on some external input states. A programmable logic controller (PLC) is essentially a digital computer-like system that can properly sequence a complex task, consisting of many discrete operations and involving several devices that need to be carried out in a particular order. The process operation might consist of a set of two-state (on/off) actions, which the PLC can sequence in the proper order and at correct times. PLCs are typically used in factories and process plants, to connect input devices such as switches to output devices such as valves, at high speed at appropriate times in a task, as governed by a program (ladder logic). Examples of such tasks include sequencing the production line operations, starting a complex process plant, and activating the local controllers in a distributed control environment.

In the early days of industrial control solenoid-operated electromechanical relays, mechanical timers and drum controllers were used to sequence such operations. Today's PLCs are rugged computers. An advantage of using a PLC is that the devices in a plant can be permanently wired, and the plant operation can be modified or restructured by software means (by properly programming the PLC) without requiring hardware modifications and reconnection.

Internally, a PLC performs basic computer functions such as logic, sequencing, timing, and counting. It can carry out simpler computations and control tasks such as PID control. Such control operations are called *continuous-state control*, where process variables are continuously monitored and made to stay very close to desired values. There is another important class of controls, known as *discrete-state control* (*discrete-event control*), where the control objective is for the process to follow a required sequence of states (or steps). In each state, however, some form of continuous-state control might be operated, but it is not quite relevant to the task of discrete-state control. PLCs are particularly intended for accomplishing discrete-state control tasks.

As an example for PLC application, consider an operation of turbine blade manufacture. The discrete steps in this operation might be the following:

1. Move the cylindrical steel billets into furnace.
2. Heat the billets.
3. When a billet is properly heated, move it to the forging machine and fixture it.
4. Forge the billet into shape.
5. Perform surface finishing operations to get the required aerofoil shape.
6. When the surface finish is satisfactory, machine the blade root.

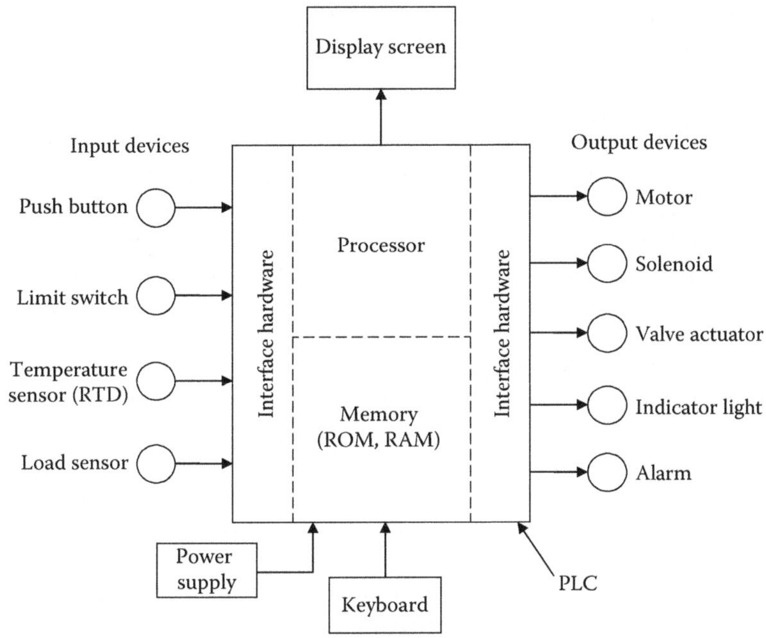

FIGURE 1.6
Schematic representation of a programmable logic controller.

Note that the entire task involves a sequence of events where each event depends on the completion of the previous event. In addition, it may be necessary for each event to start and end at specified time instants. Such *time sequencing* would be important for coordinating the current operation with other activities and perhaps for proper execution of each operational step. For example, activities of the parts handling robot have to be coordinated with the schedules of the forging machine and milling machine. Furthermore, the billets will have to be heated for a specified time, and machining operation cannot be rushed without compromising the product quality, tool failure rate, safety, and so on. The task of each step in the discrete sequence might be carried out under continuous-state control. For example, the milling machine would operate using several direct digital control (DDC) loops (say, PID control loops), but discrete-state control is not concerned with this except for the starting point and the end point of each task.

A schematic representation of a PLC is shown in Figure 1.6. A PLC operates according to some "logic" sequence programmed into it. Connected to a PLC are a set of input devices (e.g., pushbuttons, limit switches, and analog sensors such as RTD temperature sensors, diaphragm-type pressure sensors, piezoelectric accelerometers, and strain-gauge load sensors) and a set of output devices (e.g., actuators such as DC motors, solenoids, and hydraulic rams; warning signal indicators such as lights, alphanumeric LED displays, and bells; valves; and continuous control elements such as PID controllers). Each device is assumed to be a two-state device (taking the logical value 0 or 1). Now, depending on the condition of each input device and according to the programmed logic, the PLC will activate the proper state (e.g., on/off) of each output device. Hence, the PLC performs a switching function. Unlike the older generation of sequencing controllers, in the case of a PLC, the logic that determines the state of each output device is processed using software and not hardware elements such as hardware relays. Hardware switching takes place at the output port, however, for turning on/off the output devices controlled by the PLC.

1.5.3.1 PLC Hardware

As noted before, a PLC is a digital computer that is dedicated to perform discrete-state control tasks. A typical PLC consists of a microprocessor, RAM and ROM memory units, and interface hardware, all interconnected through a suitable bus structure. In addition, there will be a keyboard, a display screen, and other common peripherals. A basic PLC system can be expanded by adding expansion modules (memory, I/O modules, etc.) into the system rack.

A PLC can be programmed using a keyboard or touch screen. An already developed program could be transferred into the PLC memory from another computer or a peripheral mass-storage medium such as hard disk. The primary function of a PLC is to switch (energize or deenergize) the output devices connected to it, in a proper sequence, depending on the states of the input devices and according to the logic dictated by the program. Consider the schematic representation of a PLC that is shown in Figure 1.6. Note the sensors and actuators in the PLC.

In addition to turning on and off the discrete output components in a correct sequence at proper times, a PLC can perform other useful operations. In particular, it can perform simple arithmetic operations such as addition, subtraction, multiplication, and division on input data. It is also capable of performing counting and timing operations, usually as part of its normal functional requirements. Conversion between binary and binary-coded decimal might be required for displaying digits on an LED panel and for interfacing the PLC with other digital hardware (e.g., digital input devices and digital output devices). For example, a PLC can be programmed to make a temperature measurement and a load measurement, display them on an LED panel, make some computations on these (input) values, and provide a warning signal (output) depending on the result.

The capabilities of a PLC can be determined by such parameters as the number of input devices (e.g., 16) and the number of output devices (e.g., 12) that it can handle, the number of program steps (e.g., 2000), and the speed at which a program can be executed (e.g., 1 M steps/s). Other factors such as the size and the nature of memory and the nature of timers and counters in the PLC, signal voltage levels, and choices of outputs are all important factors.

1.5.4 Distributed Control

For complex processes with a large number of I/O variables (e.g., a chemical plant, a nuclear power plant), systems with components located at large distances apart, and systems that have various and stringent operating requirements (e.g., the space station), centralized direct digital control is quite difficult to implement. In distributed control, the control functions are distributed, both geographically and functionally. Some form of distributed control is appropriate in large systems such as manufacturing workcells, factories, intelligent transportation systems, and multicomponent process plants. A distributed control system (DCS) will have many users who would need to use the resources simultaneously and, perhaps, would wish to communicate with each other as well. Also, the plant will need access to shared and public resources and means of remote monitoring and supervision. Furthermore, different types of devices from a variety of suppliers with different specifications, data types, and levels may have to be interconnected. A communication

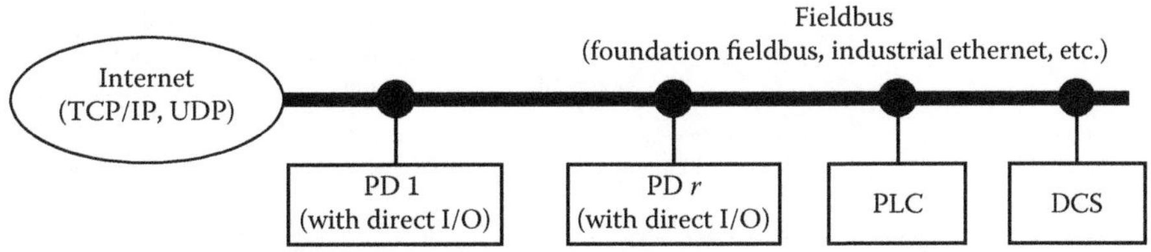

FIGURE 1.7
A networked industrial plant. PD, process device; PLC, programmable logic controller; DCS, distributed control system (supervisory controller).

network with switching nodes and multiple routes is needed for this purpose. This is essentially a networked control systems (NCS).

In order to achieve connectivity between different types of devices having different origins, it is desirable to use a standardized bus that is supported by all major suppliers of the needed devices. The Foundation Fieldbus or Industrial Ethernet may be adopted for this purpose. Fieldbus is a standardized bus for a plant, which may consist of an interconnected system of devices. It provides connectivity between different types of devices having different origins. Also, it provides access to shared and public resources. Furthermore, it can provide means of remote monitoring and supervision.

A suitable architecture for networking an industrial plant is shown in Figure 1.7. The industrial plant in this case consists of many "process devices" (PDs), one or more PLCs, and a DSC or a supervisory controller. The PDs will have direct I/O with their own components while possessing connectivity through the plant network. Similarly, a PLC may have direct connectivity with a group of devices as well as networked connectivity with other devices. The DSC will supervise, manage, coordinate, and control the overall plant.

1.5.4.1 Hierarchical Control

A popular distributed control architecture is provided by hierarchical control. Here, control functions are distributed functionally in different hierarchical levels (layers). The distribution of control may be done both geographically and functionally. A hierarchical structure can facilitate efficient control and communication in a complex control system.

Consider a three-level hierarchy. Management decisions, supervisory control, and coordination between plants in the overall facility may be provided by the supervisory control computer, which is at the highest level (level 3) of the hierarchy. The next lower level (intermediate level) generates control settings (or reference inputs) for each control region (subsystem) in the corresponding plant. Set points and reference signals are inputs to the direct digital controllers (DDC), which control each control region. The computers in the hierarchical system communicate using a suitable communication network. Information transfer in both directions (up and down) should be possible for best performance and flexibility. In master–slave distributed control, only downloading of information is available.

As an illustration, a three-level hierarchy of an intelligent mechatronic system (IMS) is shown in Figure 1.8. The bottom level consists of electromechanical components with component-level sensing. Furthermore, actuation and direct feedback control are carried out at this level. The intermediate level uses intelligent preprocessors for abstraction of the information generated by the component-level sensors. The sensors and their intelligent preprocessors together perform tasks of intelligent sensing. State of performance

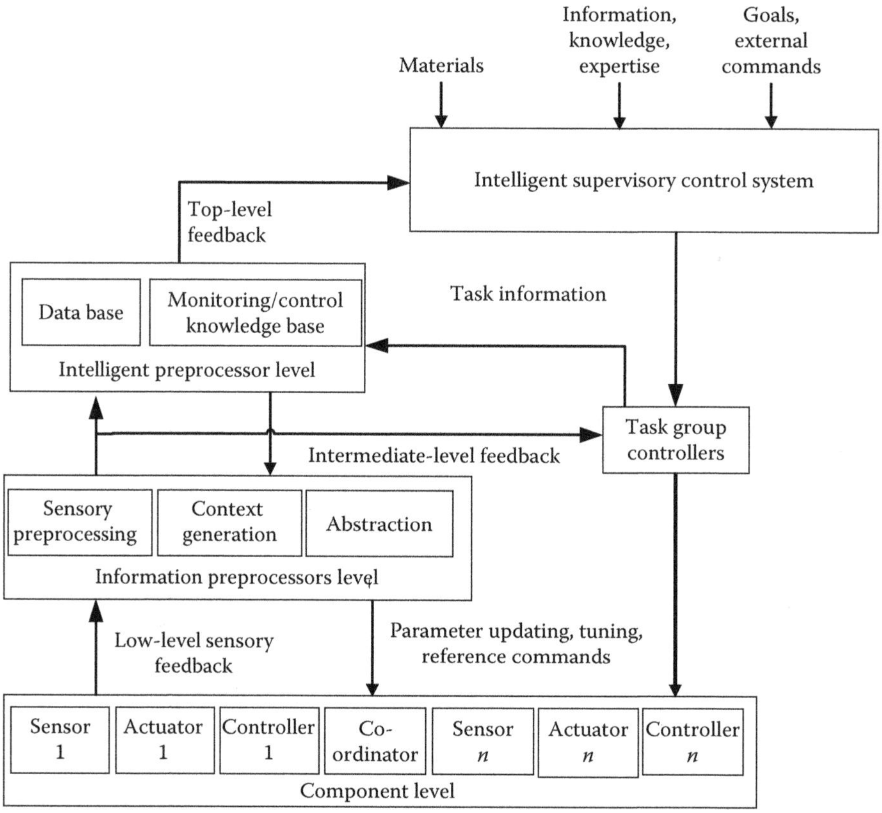

FIGURE 1.8
A hierarchical control/communications structure for an intelligent mechatronic system.

of the system components may be evaluated by this means, and component tuning and component-group control may be carried out as a result. The top level of the hierarchy performs task-level activities including planning, scheduling, monitoring of the system performance, and overall supervisory control. Resources such as materials and expertise may be provided at this level, and a human–machine interface would be available. Knowledge-based decision making is carried out at both intermediate and top levels. The resolution of the information that is involved will generally decrease as the hierarchical level increases, while the level of "intelligence" that would be needed in decision making will increase.

Within the overall system, the communication protocol provides a standard interface between various components such as sensors, actuators, signal conditioners, and controllers and also with the system environment. The protocol will not only allow highly flexible implementations, but will also enable the system to use distributed intelligence to perform preprocessing and information understanding. The communication protocol should be based on an application-level standard. In essence, it should outline what components can communicate with each other and with the environment, without defining the physical data link and network levels. The communication protocol should allow for different component types and different data abstractions to be interchanged within the same framework. It should also allow for information from geographically removed locations to be communicated to the control and communication system of the IMS.

1.6 Instrumentation Process

In some situations, each function or operation within an engineering system can be associated with one or more physical devices, components, or pieces of equipment, and in other situations, one hardware unit may accomplish several of the system functions. In the present context, by instrumentation we mean the identification of these instruments or hardware components with respect to their functions, operation, parameters, ratings, and interaction with each other and the proper selection, interfacing, and tuning of these components for a given application—in short, instrumenting the system. By design, we mean the process of selecting suitable equipment to accomplish various functions in an engineering system, developing the system architecture, matching and interfacing these devices, and selecting the parameter values, depending on the system characteristics, to achieve the desired objectives of the overall system (i.e., to meet design specifications), preferably in an optimal manner and according to some performance criterion. It follows that design may be included as an instrumentation objective. In particular, there can be many designs that meet a given set of performance requirements. Identification of key design parameters, modeling of various components, and analysis are often useful in the design process. Modeling (both analytical and experimental) is important in analyzing, designing, and evaluating a system.

Identification of the hardware components (perhaps commercially available off-the-shelf items) for various functions (e.g., sensing, actuation, control) is one of the first steps in the instrumentation of an engineering system. For example, in process control applications off-the-shelf analog, PID controllers may be used. These controllers for process control applications traditionally have knobs or dials for control parameter settings, that is, proportional band or gain, reset rate (in repeats of the proportional action per unit time), and rate time constant. The operating bandwidth (operating frequency range) of these control devices is specified. Various control modes—on/off, proportional, integral, and derivative or combinations—are provided by the same control box.

Actuating devices (i.e., actuators) include stepper motors, DC motors, AC motors, solenoids, valves, pumps, heaters/coolers, and relays, which are also commercially available to various specifications. An actuator may be directly connected to the driven load, and this is known as the direct-drive arrangement. More commonly, however, a transmission device (gearbox, harmonic drive, lead screw and nut, etc.) may be needed to convert the actuator motion into a desired load motion and for proper matching of the actuator with the driven load. Potentiometers, differential transformers, resolvers, synchros, gyros, strain gauges, tachometers, piezoelectric devices, fluid flow sensors, pressure gauges, thermocouples, thermistors, and RTDs are examples of sensors (see Chapters 8 through 10) that are used to measure process response for monitoring its performance and possibly for control.

An important factor that we must consider in any practical engineering system is random errors or noise. Noise may represent actual contamination of signals and measurement errors or the presence of other unknowns, uncertainties, and errors, such as parameter variations, modeling errors, external disturbances, and model errors. Such random factors may be removed through a process of "estimation" (see Chapter 7) such as least-squares estimation (LSE), maximum likelihood estimation (MLE), and various types of Kalman filtering (KF) including extended and unscented methods (EKF and UKF). Prior to estimation, noise may be removed through direct filtering using tracking filters, low-pass filters, high-pass filters, band-pass filters, band-reject filters or notch filters, and so on (see Chapter 3). Of course, by selecting proper and accurate sensors and sensing procedures, it may be possible to avoid at least some of the noise and signal uncertainties. Furthermore, weak signals have to be amplified, and the form of the signal might have to be modified at various points of interaction. Charge amplifiers, lock-in amplifiers, power amplifiers, switching amplifiers, linear amplifiers, and pulse-width modulated amplifiers are devices of direct signal conditioning and modification used in engineering systems (see Chapter 3). Additional components, such as power supplies and surge-protection units, are often needed in the operation of the system. Relays and other switching and transmission devices and modulators and demodulators may also be needed.

1.6.1 Instrumentation Steps

Instrumentation of an engineering system will primarily involve the selection and integration of proper sensors, actuators, controllers, and signal modification/interface hardware and software and the integration of the entire system, so as to meet a set of performance specifications. Of course, the steps involved in instrumentation will depend on the specific engineering system and performance requirements. But as a general guideline, some basic steps can be stated. They involve understanding the system that is instrumented. This may involve the development of a model (particularly, one that can be used for computer simulation—a computer model). Next, design/performance specifications have to be established for the system. Selecting and sizing the sensors, transducers, actuators, drive systems, controllers, signal conditioning, and interface hardware and software that will meet the overall performance specifications for the system constitute the next major step. After iterations of simulation, evaluation, and modification of the instrumentation choices, the final selection is made. The ultimate test for the validity of the instrumentation comes after integration of the selected components into the system and operating the integrated system. The main steps of instrumentation are as follows:

1. Study the plant (engineering process) that is to be instrumented. The purpose of the plant, how it operates, its important inputs and outputs (responses), and other relevant variables (state variables) including undesirable inputs and disturbances and parameters should be identified.
2. Separate the plant into its main subsystems (this may be done, e.g., based on the physical domains of the subsystems—mechanical, electrical/electronic, fluid, thermal, etc.) and formulate the physical equations for the processes of the subsystems. A computer-simulation model may be developed using these equations. The plant may be one that already exists or a conceptual plant that needs to be developed and instrumented, as long as the plant can be described in sufficient detail and be modeled.
3. Indicate the operating requirements (performance specifications) for the plant (i.e., how the plant should behave in carrying out its intended tasks) under proper control. We may use any available information on such requirements as accuracy, resolution, speed, linearity, stability, and operating bandwidth for this purpose.
4. Identify any constraints related to cost, size, weight, environment (e.g., operating temperature, humidity, dust-free or clean room conditions, lighting, washdown needs), and so on.
5. Select the type and the nature of the sensors/transducers, actuators, and signal-conditioning devices (including interfacing and data acquisition hardware and

software, filters, amplifiers, modulators, ADC, DAC) that are necessary for the operation and control of the plant. For sensors and actuators, establish the associated ratings and specifications (signal levels, bandwidths, accuracy, resolution, dynamic range, etc.). For actuators, establish the associated ratings and specifications (e.g., power, torque, speed, temperature, and pressure characteristics, including curves and numerical values). Identify possible manufacturers/vendors for the components, give the model numbers, and so on.

6. Establish the architecture of the overall integrated system together with appropriate controllers and/or control schemes. Modify the original computer model to accommodate the new instrumentation that has been integrated into the system.
7. Carry out computer simulations and make modifications to the instrumentation until the system performance meets the specifications. An optimization scheme (e.g., one that uses an "MDQ" as the performance measure) may be employed in this exercise.
8. Once the computer analysis provides acceptable results, we may proceed to the acquisition and integration of the actual components. In some situations, off-the-shelf components may not be available. Then, they have to be designed and developed.

1.6.2 Application Examples

We now present five examples of engineering systems that benefit from proper sensors, sensor systems, and related instrumentation.

1.6.2.1 Networked Application

A machine that we developed for head removal of salmon is shown in Figure 1.9. The conveyor, driven by an AC induction motor, indexes the fish in an intermittent manner. Image of each fish, obtained using a charge-coupled device (CCD) camera, is processed to determine the geometric features, which in turn establish the proper cutting location. A two-axis hydraulic drive then positions the cutter accordingly, and the cutting blade is operated using a pneumatic actuator. Position sensing of the hydraulic manipulator is carried out using linear magnetostrictive displacement transducers (see Chapter 10), which have a resolution of 0.025 mm when used with a 12-bit ADC. A set of six gauge pressure transducers is installed to measure the fluid pressure in the head and rod sides of each hydraulic cylinder and also in the supply lines. A high-level imaging system determines the cutting quality, according to which adjustments are made online, to the operating conditions of the machine and the parameters of the control system to improve the process performance. The control system has a hierarchical structure with conventional direct control at the component level (low level) and an intelligent monitoring and supervisory control system at an upper level.

The primary vision module of the machine is responsible for fast and accurate detection of the gill position of a fish, on the basis of an image of the fish that is captured by the primary CCD camera. This module is located in the machine host and comprises a CCD camera for image acquisition, an ultrasonic sensor for thickness measurement of fish, a trigger switch for detecting a fish on the conveyor, an image-processing board for frame grabbing and image analysis, and an I/O card for data acquisition and digital

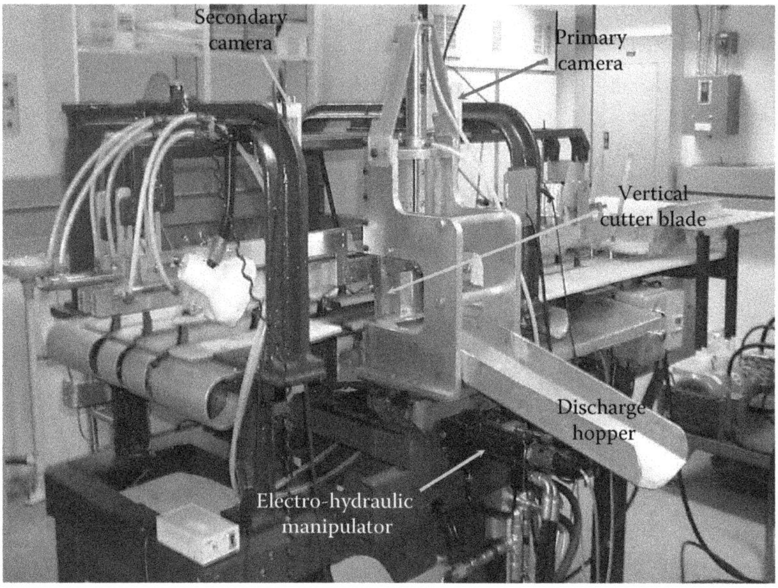

FIGURE 1.9
An automated fish-cutting machine.

data communication with the control computer of the electrohydraulic manipulator. This vision module is capable of reliably detecting and computing the cutting location of a fish in approximately 300–400 ms. The secondary vision module is responsible for acquisition and processing of visual information pertaining to the quality of processed fish that goes through the cutter assembly. This module functions as an *intelligent sensor* in providing high-level information feedback into the control computer. The hardware associated with this module are a CCD camera at the exit end for capturing images of processed fish and an image-processing board for visual data analysis. The CCD camera acquires images of processed fish under the direct control of the host computer, which determines the proper instance to trigger the camera by timing the duration of the cutting operation. The image is then transferred to the image buffer in the image-processing board for further processing. In this situation, however, image processing is accomplished to extract high-level information, such as the quality of processed fish.

With the objective of monitoring and controlling industrial processes from remote locations, we have developed a universal network architecture, for both hardware and software. The developed infrastructure is designed to perform optimally with Fast Ethernet backbone where each network device needs only a low-cost network interface card. Figure 1.10 shows a simplified hardware architecture, which networks two machines (a fish-processing machine and an industrial robot). Each machine is directly connected to its individual control server, which handles networked communication between the process and the web server, data acquisition, sending of control signals to the process, and the execution of low-level control laws. The control server of the fish-processing machine contains one or more data acquisition boards, which have ADC, DAC, digital I/O (see Chapter 4), and frame grabbers for image processing (see Chapter 11).

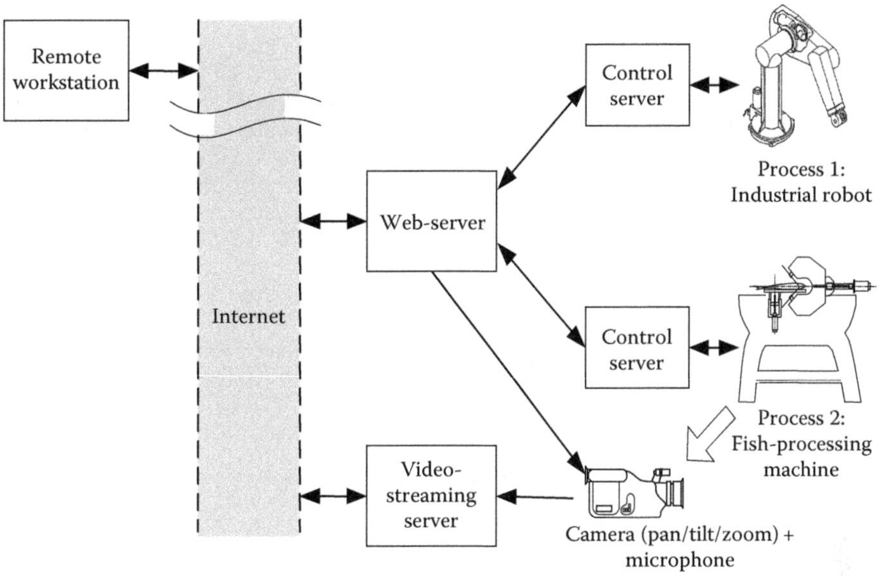

FIGURE 1.10
Hardware architecture of the networked system.

Video cameras and microphones are placed at strategic locations to capture live audio and video signals, allowing the remote user to view and listen to a process facility, and to communicate with local personnel. The camera has built-in pan, tilt, and 21× zoom capabilities, which can be controlled through a standard communication protocol from the host computer. Multiple cameras can be connected in a daisy-chained manner to the video-streaming server. For capturing and encoding the audio–video (AV) feed from the camera, a PCI board is installed in the video-streaming server. It can capture video signals at a standard frame resolution of 640 × 480 pixels at 30 frames per second (fps), with a hardware compression that significantly reduces computational overheads of the video-streaming server. In the present implementation, each of the AV capture boards can support only one AV input. Hence, multiple boards have to be installed.

1.6.2.2 Telemedicine System

A system of telemedicine is being developed by us. It employs the following approach: advanced sensing, signal processing, and public telecommunication are used for clinical monitoring of the patients located in their own community and for transmitting only the pertinent information, online, to a medical professional at a hospital at distance. The medical professional interacts with the patient remotely, through audio and video links, and simultaneously examines the data transmitted by the monitoring system, and does medical assessment, diagnosis, and prescription. The medical professional may consult with other professionals, online, and may also use other available resources in arriving at the diagnosis and prescription. The use of human medical professionals to perform health assessment, diagnosis, and prescription is far more desirable than the popular approach to telemedicine and telehealth where an automated system may provide medical advice based on the input generated by the patient, which is known to be biased and prone

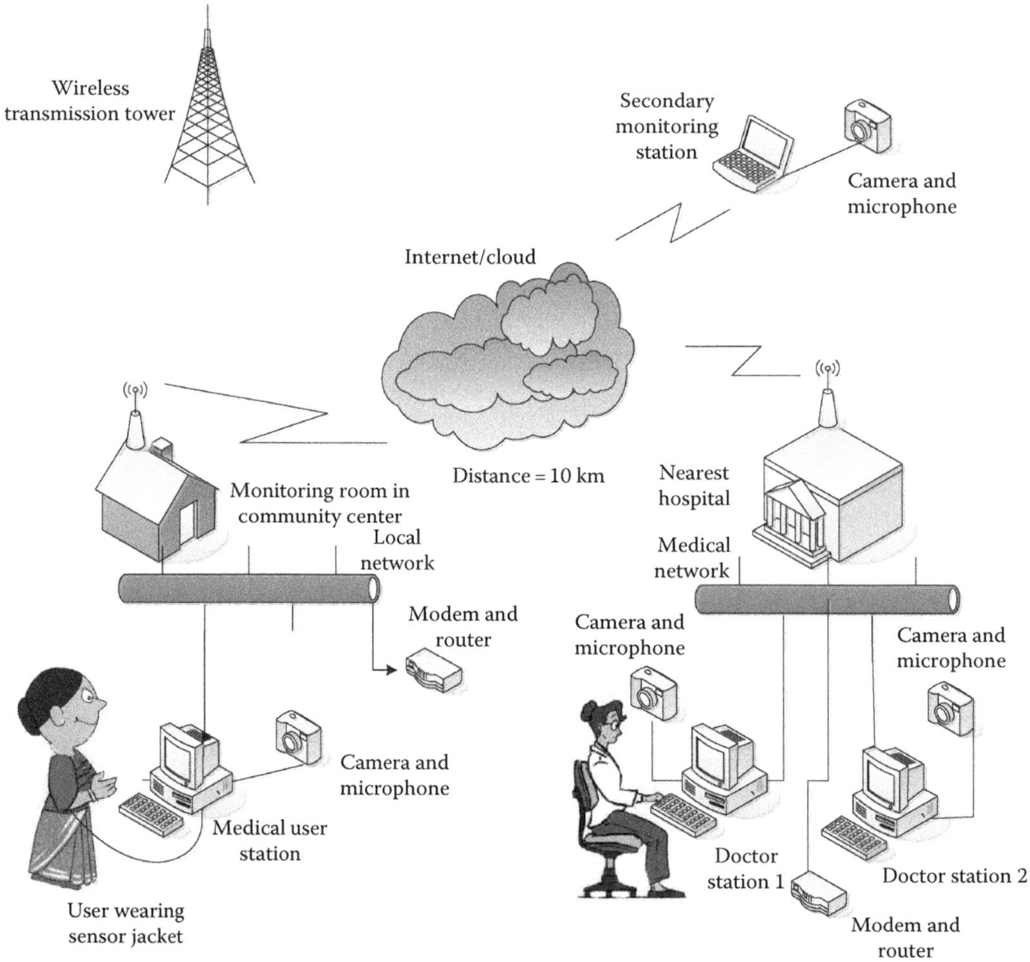

FIGURE 1.11
Structure of the telemedicine system.

to error. A schematic diagram of the proposed system is shown in Figure 1.11. The relevant issues of development and instrumentation of the system are listed as follows:

- Integrated mechatronic design of the sensor jacket to be worn by the subject for online health monitoring
- Selection of the embedded sensors and hardware, particularly with respect to their type, size, and features for acquiring the vital information from the subject
- Sensor location and configuration in the jacket to improve/optimize the process of data acquisition
- Power requirements and flexibility
- Signal processing and communication hardware on the sensor jacket
- Software for signal processing, artifact removal, data reduction, interpretation, and representation within the host computer at the patient end, for transmission to the doctor's computer

- Graphical user interface (GUI) at both ends (patient location and doctor location)
- Assistive methodologies for fast and accurate communication of information from the host computer at the patient end to the doctor

In the design of the sensor jacket, a mechatronic (optimal) approach that employs an MDQ as the performance function is utilized. The design indices in the MDQ include such aspects as component location, accuracy, speed, size, complexity, maintainability, design life, reliability, robustness, fault tolerance, reconfigurability, flexibility, cost, user-friendliness, and performance expectations. Parameters such as sensor location and configuration in the jacket are decided so as to improve/optimize the process of data acquisition, body conformability, weight, robustness, cost, and so on.

Selection of the sensors and associated hardware, particularly with respect to their type, size, and features to match the performance specifications of the system (as established in the mechatronic design) is an important aspect of the development of the sensor jacket. Pertinent sensors for the jacket are

- Standard electrocardiogram (ECG or EKG) sensors (skin/chest electrodes)
- Blood pressure sensors (arm cuff–based monitor)
- Temperature sensors (temperature probe or skin patch)
- Respiratory sensors (piezoelectric/piezoresistive sensors)
- Electromyogram (skin electrodes)
- Oximetry sensor
- Electrical stethoscope (neck and lung)
- Pure light ear clip sensor
- Circular stretch sensor

Some of the commercially available pertinent sensors and their key features are given as follows:

Digital stethoscope: (Agilent Technologies; 4.5 V DC, 1 mA)

- Captures sounds from the heart and lungs.
- Signals have to be amplified before acquisition by computer.
- Eight levels of sound amplification.
- Active noise filtering
- Mode selection: standard diaphragm and bell modes; extended diaphragm mode to hear high-frequency sounds (e.g., produced by mechanical heart valve prostheses)

Digital ECG recorder: (Fukuda Denshi, 12-Lead Digital ECG Unit, 100–240 V/50–60 Hz AC adapter)

- Captures full electrocardiogram and forms a data file
- Built-in software to process and interpret the signals (to assist diagnosis of some types of heart problems by doctor)
- Channel (lead) selection feature (to output different types of processed information)

Imaging, blood pressure, temperature, and blood oxygen sensing:

- Medical CCD camera (AMD Telemedicine, 110–220 V AC, 50–60 Hz or 12 V DC), with built-in illumination source.
- Digital blood pressure monitor (Bios Diagnostics or Omron, 110–230 V AC adapter, PC connectivity) provides blood pressure and pulse rate; cuff is inflated by pressing a button.
- Digital ear thermometer (Becton Dickinson and Co./Advanced Monitors Corp.).
- Pulse oximeter (Devon Medical Products; mounting on fingertip or earlobe is typical; forehead and chest models are available as well).

Note: Blood pressure and temperature readings may be wirelessly transmitted to patient-end computer by embedding low-power miniature transceivers into the sensors.

Sensor power supply capabilities: The following off-the-shelf sensors have built-in AC adapters (100–240 V universal, 50–60 Hz):

- ECG unit
- Medical CCD camera
- Blood pressure monitor

Stethoscope, thermometer, and pulse oximeter are typically powered by disposable batteries.

Other types of sensors, particularly, wearable ambulatory sensors/monitors may be integrated as well. The accessories required for the jacket include

- Complete low-power integrated analog front end for ECG applications
- One-piece ECG cable with lead wire
- Yokemate LWS® 3-lead universal adapter
- Dry electrode
- AMC&E reusable DIN connector lead-wire, 3-lead, snap connection
- Step-down converter with bypass mode for ultralow-power wireless applications
- Needle to clip converter
- De2 development and education board
- Arduino microcontroller (MCU)
- BLE 4.0 module
- Soft potentiometer
- Wearable kit (textile push button, conductive thread, etc.)
- Pressure vest

A graphic representation of the sensor jacket is given in Figure 1.12.

1.6.2.3 Home Care Robotic System

One of the ways that can reduce the spending on healthcare of the aging population is to exploit the recent technological advancements in sensor systems, robotics, and information and communication technologies (ICT) for providing high-quality supportive

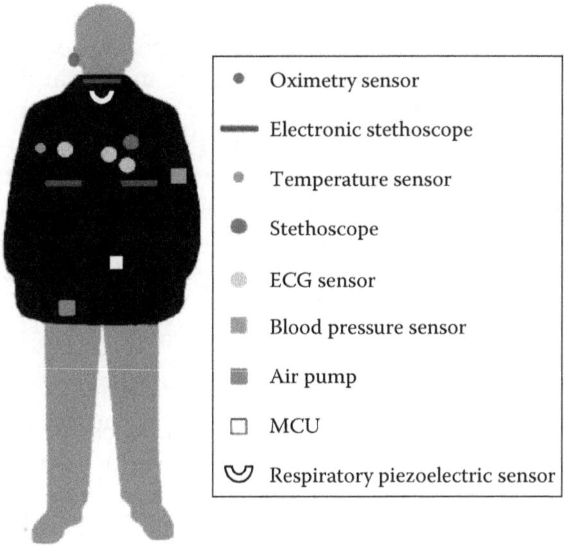

FIGURE 1.12
Graphic representation of the telemedicine sensor jacket.

environments to people in their own homes. A robotic home care environment may have autonomous robots, which can be augmented with haptic teleoperation capability comprising a haptic-assisted remotely controlled robot to monitor and assist individuals within the home environment (see Figure 1.13). Specifically, the system will have the capability for two modes of operation: (a) more autonomous operation for 24 hour routine basic care (mobility, bathing, dressing, toileting, meal preparation, providing medicine, monitoring and seeking external help, etc.) and (b) remote monitoring and haptic teleoperation in emergency situations (until regular help arrives—ambulance, paramedics, police, firefighters, etc.). Haptic teleoperation will incorporate advanced sensing and actuation capabilities of a robot together with dexterity and cognitive skills of a human.

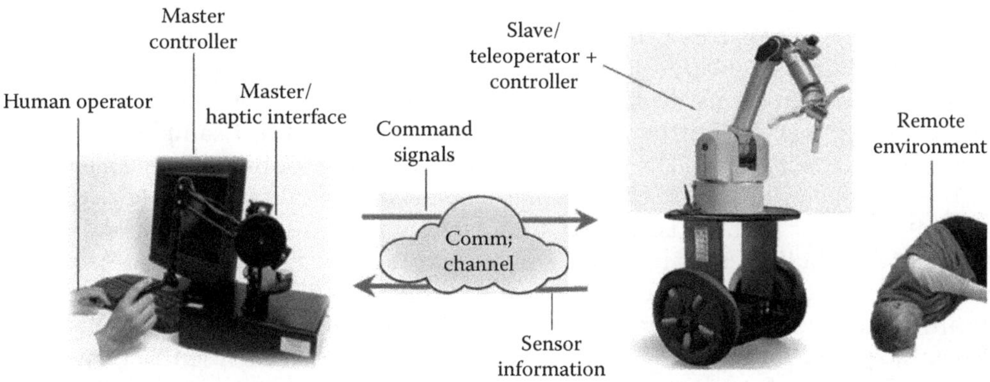

FIGURE 1.13
Haptic teleoperation of a home care robotic system.

Specifically, haptics incorporate the capability of "feeling" the conditions (e.g., forces of interaction) at the remote site (home) by the human operator of the master manipulator. In addition to the sensors in the robots (including the master and slave units for teleoperation), additional sensor systems are necessary for monitoring the environment, which is dynamic, unstructured, and not fully known. Pertinent sensors for this application include optical encoders for motion sensing and torque sensors for the joints of the robots and the master manipulator; tactile sensors for the robotic fingers; laser and ultrasonic rangefinders for the mobile platforms; optical encoders for the wheels of the mobile platforms; cameras for the mobile platforms and the work environment; and basic sensors for monitoring human health (e.g., as in telemedicine).

1.6.2.4 Automated Spatiotemporal Monitoring of Water Quality

Common sources of water include surface water (from catchments, lakes, reservoirs, rivers, waterways, etc.), groundwater (from wells, springs, etc.), ocean water (requires desalination), and rainwater (requires collection and retention). Water is essential for human survival (for consumption, hygiene, irrigation, power, etc.). In developing nations, approximately 80% of diseases are water related. Treated wastewater is used by many (e.g., approximately 74% of Americans, 57% of Canadians, 86.5% of Germans, and 99% of Swedes). Groundwater is used by many for domestic use (e.g., 26% of Canadians). About 70% of water usage is for agriculture.

Water contains naturally occurring substances (e.g., bicarbonates, sulfates, sodium, chlorides, calcium, magnesium, and potassium) and harmful contaminants (microbes and dissolved chemicals). Monitoring of the quality of water sources will not only benefit public health but also the ecosystem. In particular, water quality can determine the ecological balance among species. The types of water quality parameters can be categorized as physical (color, density, sediments, etc.), chemical (dissolved inorganic material), and biological (dissolved organic material, bacteria).

Commonly monitored water quality parameters are pH value, temperature, conductivity, oxidation–reduction potential, and dissolved oxygen. The manual approach is widely used to monitor the water quality by sensing these parameters. The process involves collection of water samples in the field and then testing them in laboratory. There are challenges to this practice including unavailability of proper sensors, cost of data loggers (sometimes, approximately US$7000 for a sensor), unreliability of data obtained through field tests, calibration requirements and difficulties infield, and low-density deployment of monitoring sites. In particular, the manual approach is time consuming and labor intensive, not convenient in relation with taking heavy and complex equipment to the field, may create safety problems for field workers, and is typically a low-resolution monitoring (time/space) process. Furthermore, transportation of field test samples to laboratory may disturb their properties. This will cause inaccurate and nonrepresentative test results. Also, this approach generally leads to delays in analysis and taking action (e.g., generating advisories). In view of these issues, an automated approach to spatiotemporal (i.e., both geographic distribution and time variation) monitoring of water quality is desired.

In the automated approach, the quality of water coming from various sources is automatically measured using several SNs that are distributed over a large geographic area. The data acquired from an SN may be locally conditioned and then transmitted to a mother node, then to a base station, and finally to a central assessment unit (CAU) using ICT. Spatiotemporal of the quality of water may be monitored and assessed, and appropriate actions can be taken in this manner.

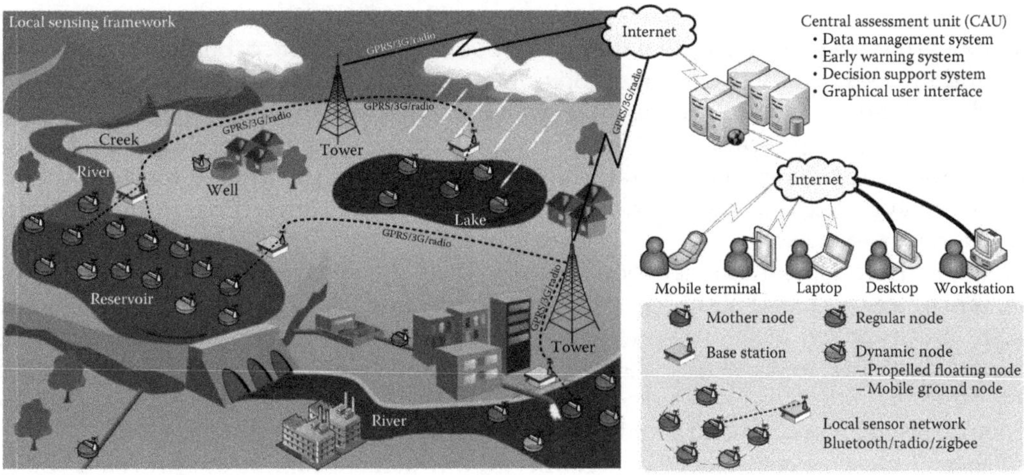

FIGURE 1.14
The structure of a multisensor system for water quality monitoring.

The general framework of the system is shown in Figure 1.14. In this platform, an SN may contain multiple sensor, for example, those for temperature, pH value, turbidity, dissolved oxygen, electrical conductivity, and oxidation–reduction potential. These sensors are commercially available and may be used directly or with some modification (e.g., for interfacing for automatic data transmission; sealing for proper protection from environmental factors). A microcontroller system is used for sensor data acquisition at each SN. Most nodes are stationary (static), while some dynamic (geographically movable) nodes may be implemented. The dynamic nodes may be in the form of propelled nodes in water or nodes mounted on mobile robotic devices that can navigate on ground. Low-level processing of sensory data (e.g., filtering, amplification, and digital representation) is carried out by the microcontroller before transmitting the data through a radio transceiver to a more powerful mother node.

Processed data from the SNs are transmitted to a base station. There, the received data are subjected to high-level processing and data compression and are subsequently transmitted to a central computer called CAU. The CAU is a knowledge-based decision making unit, which assess the data from different geographic locations in a temporal manner (i.e., spatiotemporal). Based on the assessment, the CAU provides advisories, alarms, trends, and other useful information of water quality at various locations and also provides the rationale and explanations for these decisions. Policy decisions can be made based on the outcomes. Furthermore, the CAU can "optimize" the operation of the ICT platform in order to make the system outcomes more accurate, uniform, and effective. It can map out regions having water quality problems, at different levels (based on a *water quality index*).

1.7 Organization of the Book

The book consists of 12 chapters. The chapters are devoted to presenting the fundamentals, analytical concepts, modeling and design issues, technical details, applications of sensors, sensor systems, and interfacing and signal modification within the framework of engineering system instrumentation. The book uniformly incorporates the underlying fundamentals as analytical methods, modeling approaches, component selection procedures, and design techniques in a systematic manner throughout the main chapters. The practical application of the concepts, approaches, and tools presented in the introductory chapters is demonstrated through numerous illustrative examples and a comprehensive set of case studies. Suggestions are made concerning student projects and laboratory exercises.

Chapter 1 introduces the subject of instrumentation of an engineering system using sensors, sensor systems, and signal modification hardware. The relevance of modeling and design in the context of instrumentation is indicated. Common control system architectures are described and the role played by sensors and sensor systems in these architectures is highlighted. This introductory chapter sets the tone for the study, which spans the remaining 11 chapters. Several problems are presented as student exercises.

Chapter 2 presents component interconnection, which is a significant unifying subject within engineering system instrumentation. Impedance considerations of component interconnection and matching are studied. The subject is addressed in the context of both electrical systems and mechanical systems. Several problems are presented as student exercises.

Chapter 3 addresses signal conditioning, particularly through amplifiers and filters. Several problems are presented as student exercises. Starting with the basic building block of an operational amplifier, various important types of amplifiers and filters are studied, particularly with respect to their modeling, analysis, performance specification, and application. Both analog filters and digital filters are considered. Several problems are presented as student exercises.

Chapter 4 presents the subject of signal conversion. Modulators and demodulators, ADC, DAC, bridge circuits, and other common signal conversion techniques and devices are discussed. Several problems are presented as student exercises.

Chapter 5 covers performance analysis of a device, component, or instrument within an engineering system. Methods of performance specification are addressed, in both time domain and frequency domain. Common instrument ratings that are used in industry and generally in engineering practice are discussed. Related analytical methods are given. Several problems are presented as student exercises.

Chapter 6 studies some important issues of device bandwidth, data sampling, and error considerations in the context of sensor systems and instrumentation. Instrument bandwidth considerations are highlighted, and a design approach based on component bandwidth is presented. Errors in digital devices, particularly resulting from signal sampling, are discussed from analytical and practical points of view. Representation, combination, propagation, and analysis of device error are studied. Application of the concept of sensitivity in error combination is discussed. Several problems are presented as student exercises.

Chapter 7 concerns the estimation of parameters and signals through measured data. The role of estimation in sensing is introduced. The concepts of model error and measurement error are discussed. Handling of randomness in error (mean, variance, or covariance) is studied. The following approaches are presented and illustrated using examples: LSE, MLE, and four versions of Kalman filter (scalar static Kalman filter; linear multivariable dynamic Kalman filter; extended Kalman filter, which is applicable in nonlinear situations; and unscented Kalman filter, which is also applicable in nonlinear situations and has advantages over the extended Kalman filter, because it directly takes into account the propagation of random characteristics through system nonlinearities). Several problems are presented as student exercises.

Chapter 8 presents important types, characteristics, and operating principles of analog sensors for motion measurement. Particular attention is given to sensors that are commonly used in control engineering practice. Considered motion sensors include potentiometers, differential transformers, resolvers, permanent-magnet transducers, tachometers, eddy-current transducers, variable-capacitance transducers, piezoelectric sensors, and gyroscopic and Coriolis sensors. Analytical basis, selection criteria, and application areas are indicated. Several problems are presented as student exercises.

Chapter 9 presents important types, characteristics, and operating principles of effort sensors. The treatment begins with a study of strain gauges. Force, torque, and tactile sensors are studied. The concepts of impedance sensing are treated. Automatic (self-) compensation for temperature is discussed. Analytical basis, selection criteria, and application areas are indicated. Several problems are presented as student exercises.

Chapter 10 discusses several miscellaneous types of sensors that are used in engineering applications. They include fiber-optic, laser-based, ultrasonic, magnetostrictive, acoustic emission, and thermofluid sensors and sensors that are useful in monitoring the quality of water (e.g., pH value, dissolved oxygen, and oxidation–reduction potential). Several problems are presented as student exercises.

Chapter 11 studies common types of digital transducers, digital camera, and some other innovative and advanced sensing technologies. Unlike analog sensors, digital transducers generate pulses, counts, or digital outputs. These devices have clear advantages, particularly when used in computer-based digital systems. They do possess quantization errors, which are unavoidable in digital representation of an analog quantity. Related issues of accuracy and resolution are addressed. Digital camera as a practical sensor and image acquisition through it are discussed. Other innovative sensor technologies including Hall-effect sensors are studied. Some applications of advanced sensing are given. Several problems are presented as student exercises.

Chapter 12 studies the topics of MEMS sensors and multisensor systems. Basic principles of MEMS sensors and their characteristics, fabrication, rating parameters, and application are discussed. Technologies of networked sensing and localization are presented. Sensor fusion through Bayesian approach, Kalman filter, and neural networks is studied. Several applications of advanced sensing are given. Several problems are presented as student exercises.

Several laboratory exercises on the subject of sensor systems are suggested in Appendix A. Typical student projects and case studies are indicated in Appendix B. Some basics of probability and statistics are presented in Appendix C. Reliability considerations and associated probability models of multicomponent systems are outlined in Appendix D.

Summary Sheet

Sensor: Measures (senses) unknown signals and parameters of a plant and its environment (sensors are needed to monitor and "learn" about the system).

Useful in: Process monitoring; product quality assessment; fault prediction, detection, and diagnosis; warning generation; surveillance; controlling a system.

Sensor system: May mean (1) multiple sensors or sensor/data fusion (one sensor may not be adequate for the particular application) or (2) sensor and its accessories (signal processing, data acquisition, display, etc.).

MEMSs: Use microminiature sensors and actuators.
Their scientific principles are often the same as those of their "macro" counterparts (e.g., piezoelectric, capacitive, electromagnetic, and piezoresistive principles).

Benefits of MEMS devices: Small size and light weight (negligible loading errors), high speed (high bandwidth), and convenient mass production (low cost).

Controller: Generates control signals according to which the plant (and the control devices) is driven.

Instrumentation process: Identify instrumentation components (type, functions, operation, interaction, etc.), address component interfacing (interconnection), and decide parameter values (component sizing, system tuning, accuracy, etc.) to meet performance requirements (specifications).

- Instrumentation should be considered as an integral part of design.
- Design enables us to build a system that meets the performance requirements—starting, perhaps, with a few basic components (sensors, actuators, controllers, signal modification devices, etc.).

Human sensory system (five senses): Sight (visual), hearing (auditory), touch (tactile), smell (olfactory), and taste (flavor).
Note: Other types of sensory features (e.g., sense of balance, pressure, temperature, pain, and motion) will involve the basic five senses, simultaneously through the CNS.

Human sensory process: Stimulus (e.g., light for vision, sound waves for hearing) is received at receptor; dendrites of neurons convert stimulus energy into electromechanical impulses; axons of neurons conduct action potentials into the CNS of the brain. Brain interprets these potentials to create the corresponding sensory perception.

Mechatronics: Synergistic application of mechanics, electronics, controls, and computer engineering in the development of electromechanical products and systems, through an integrated design approach; study of mechatronic engineering: modeling, design, development, integration, instrumentation, control, testing, operation, and maintenance.

- Consider "instrumentation" as an integral part of design.
- Perform "optimal" and "concurrent" instrumentation (consider all aspects and components of instrumentation simultaneously) with regard to sensors, actuators, interface hardware, and controllers.

Benefits of mechatronic design and instrumentation: Optimality and better component matching; increased efficiency; cost effectiveness; ease of system integration, compatibility and ease of cooperation with other systems; improved controllability; increased reliability; increased product life.

Bottlenecks for mechatronic instrumentation: For existing processes, the level flexibility for modification (dictated by the mechatronic approach) will be limited; available components (sensors, actuators, controllers, accessories) are limited and may not be fully compatible.

Feedback control: Control signal is determined according to plant response.

Feedforward control (closed-loop control): Control signal is determined according to plant excitation (input) or a model of the plant.

Open-loop control: No feedback of measured responses.

Digital control: Controller is a digital computer.

Advantages of digital control: Less susceptible to noise or parameter variation because data can be represented, generated, transmitted, and processed as binary words; high accuracy and speed (hardware implementation is faster than software implementation); can handle repetitive tasks well, through programming; complex control laws and signal-conditioning algorithms can be programmed; high reliability, by minimizing analog hardware components and decentralization using dedicated microprocessors for control; large amounts of data can be stored using compact, high-density data storage methods; data can be stored or maintained for very long periods of time without drift or getting affected by adverse environmental conditions; fast data transmission over long distances without excessive dynamic delays and attenuation, as in analog; easy and fast data retrieval capabilities; uses low operational voltages (e.g., 0–12 V DC) and cost effective.

Control performance characteristics: Stability (asymptotic, BIBO, etc.); speed of response or bandwidth; sensitivity; robustness; accuracy; and cross-sensitivity or dynamic coupling.

PLC: Concerns discrete-state control (or discrete-event control). Sequence a task consisting of many discrete operations and involving several devices that need to be carried out in a particular order. Used to connect input devices (e.g., switches) to output devices (e.g., valves) at high speed at appropriate times, governed by a program (ladder logic). Can perform simpler computations and control tasks (e.g., PID control).

Distributed control: Control functions are distributed, both geographically.

NCS: Control system components (sensors, actuators, controllers, etc.) are networked together, locally or at distance.

Hierarchical control: Control functions are distributed functionally in different hierarchical levels (layers).

Instrumentation procedure: (1) Study the instrumented plant (purpose, how it operates, important variables and parameters). (2) Separate the plant into its main subsystems (e.g., according to domains—mechanical, electrical/electronic, fluid, thermal) and formulate physical equations for the processes of the subsystems. Develop a computer-simulation model. (3) Indicate operating requirements (performance specifications) for the plant. (4) Identify constraints related to cost, size, weight, and environment (e.g., operating temperature, humidity, dust-free or clean room conditions, lighting, washdown needs). (5) Select type and nature of sensors/transducers, actuators, signal-conditioning devices (including interfacing and data acquisition hardware and software, filters, amplifiers, modulators, ADC, DAC, etc.). Establish the associated ratings and specifications (signal levels, bandwidths, accuracy, resolution, dynamic range, power, torque, speed, temperature, pressure characteristics, etc.). Identify manufacturers/vendors for the components (model numbers, etc.). (6) Establish system architecture (includes controllers and/or control schemes). Revise the original computer model as necessary. (7) Carry out computer simulations. Make modifications to instrumentation until the system performance meets the specifications (a mechatronic optimization scheme may be used). (8) Once acceptable results are achieved, acquire and integrate the actual components. Some new developments may be needed.

Problems

1.1 What are open-loop control systems and what are feedback control systems? Give one example of each.

A simple mass–spring–damper system (simple oscillator) is excited by an external force *f*(*t*). Its displacement response y (see Figure P1.1a) is given by the differential equation $m\ddot{y} + b\dot{y} + ky = f(t)$. A block diagram representation of this system is shown in Figure P1.1b. Is this a feedback control system? Explain and justify your answer.

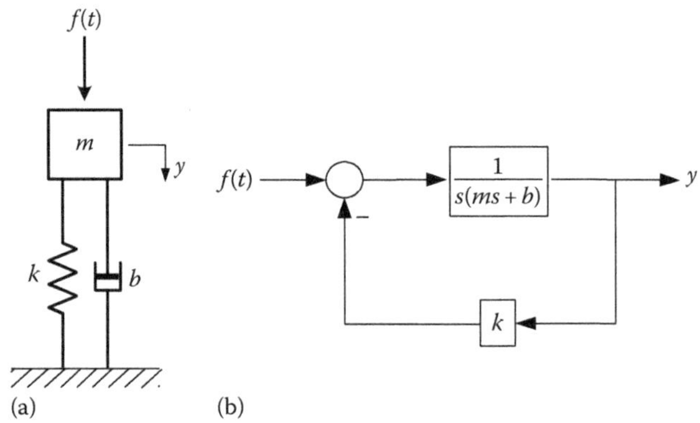

FIGURE P1.1
(a) A mechanical system representing a simple oscillator. (b) A block diagram representation of the simple oscillator.

1.2 You are asked to design a control system to turn on lights in an art gallery at night, provided there are people inside the gallery. Explain a suitable control system, identifying the open-loop and feedback functions, if any, and describe the control system components.

1.3 Into what classification of control system components—actuators, signal modification devices, controllers, and measuring devices—would you put the following devices: stepper motor, proportional plus integration circuit, power amplifier, ADC, DAC, optical incremental encoder, process computer, FFT analyzer, digital signal processor.

1.4 (a) Discuss possible sources of error that can make either open-loop control or feedforward control meaningless in some applications.
(b) How would you correct the situation?

1.5 Compare analog control and direct digital control for motion control in high-speed applications of industrial manipulators. Give some advantages and disadvantages of each control method for this application.

1.6 A soft drink bottling plant uses an automated bottle-filling system. Describe the operation of such a system, indicating various components in the control system and their functions. Typical components would include a conveyor belt, a motor for the conveyor, with start/stop controls; a measuring cylinder, with an inlet valve, an exit valve, and level sensors; valve actuators; and an alignment sensor for the bottle and the measuring cylinder.

1.7 Consider the natural gas home heating system shown in Figure 1.5. Describe the functions of various components in the system and classify them into the following functional groups: controller, actuator, sensor, and signal modification device. Explain the operation of the overall system and suggest possible improvements to obtain more stable and accurate temperature control.

1.8 In each of the following examples, indicate at least one (unknown) input that should be measured and used for feedforward control to improve the accuracy of the control system.

(a) A servo system for positioning a mechanical load. The servo motor is a field-controlled DC motor, with position feedback using the pulse count of an optical encoder and velocity feedback using the pulse rate from the encoder.
(b) An electric heating system for a pipeline carrying liquid. The exit temperature of the liquid is measured using a thermocouple and is used to adjust the power of the heater.
(c) A room heating system. Room temperature is measured and compared with the set point. If it is low, the valve of a steam radiator is opened; if it is high, the valve is shut.
(d) An assembly robot that grips a delicate part to pick it up without damaging the part.
(e) A welding robot that tracks the seam of a part to be welded.

1.9 A typical input variable is identified for each of the following examples of dynamic systems. Give at least one output variable for each system.

(a) Human body: neuroelectric pulses
(b) Company: information
(c) Power plant: fuel rate
(d) Automobile: steering wheel movement
(e) Robot: voltage to joint motor

1.10 Hierarchical control has been applied in many industries, including steel mills, oil refineries, chemical plants, glass works, and automated manufacturing. Most applications have been limited to two or three levels of hierarchy, however. The lower levels usually consist of tight servo loops, with bandwidths in the order of 1 kHz. The upper levels typically control production planning and scheduling events measured in units of days or weeks.

A five-level hierarchy for a flexible manufacturing facility is as follows: The lowest level (level 1) handles servo control of robotic manipulator joints and machine tool degrees of freedom. The second level performs activities such as coordinate transformation in machine tools, which are required in generating control commands for various servo loops. The third level converts task commands into motion trajectories (of manipulator end effector, machine tool bit, etc.) expressed in world coordinates. The fourth level converts complex and general task commands into simple task commands. The top level (level 5) performs supervisory control tasks for various machine tools and material-handling devices, including coordination, scheduling, and definition of basic moves. Suppose that this facility is used as a flexible manufacturing workcell for turbine blade production. Estimate the event duration at the lowest level and the control bandwidth (in hertz) at the highest level for this type of application.

1.11 According to some observers in the process control industry, early brands of analog control hardware had a product life of about 20 years. New hardware controllers can become obsolete in a couple of years, even before their development costs are recovered. As a control instrumentation engineer responsible for developing an off-the-shelf process controller, what features would you incorporate into the controller to correct this problem to a great extent?

1.12 The PLC is a sequential control device, which can sequentially and repeatedly activate a series of output devices (e.g., motors, valves, alarms, and signal lights) on the basis of the states of a series of input devices (e.g., switches, two-state sensors). Show how a programmable controller and a vision system consisting of a digital camera and a simple image processor (say, with an edge-detection algorithm) could be used for sorting fruits on the basis of quality and size for packaging and pricing.

1.13 Measuring devices (sensors, transducers) are useful for measuring the outputs of a process for feedback control. Give other situations in which signal measurement would be important. List at least five different sensors used in an automobile engine.

1.14 One way to classify controllers is to separately consider their sophistication and physical complexity. For instance, we can use an x–y plane with the x-axis denoting the physical complexity and the y-axis denoting the controller sophistication. In this graphical representation, simple open-loop on/off controllers (say, opening and closing a valve) would have a very low controller sophistication value, and an artificial intelligence–based intelligent controller would have a high controller sophistication value. Moreover, a passive device is considered to have less physical complexity than an active device. Hence, a passive spring-operated device (e.g., a relief valve) would occupy a position very close to the origin of the x–y plane, and an intelligent machine (e.g., sophisticated robot) would occupy a position diagonally far from the origin. Consider five control devices of your choice. Mark the locations that you expect them to occupy (in relative terms) on this classification plane.

1.15 A dental hygienist assures a patient that they have nothing to worry about the x-rays taken of the mouth as everything is "digital" now. Critically discuss the hygienist's statement and how it should be interpreted.

1.16 You are an engineer who has been assigned the task of designing and instrumenting a practical system. In the project report, you have to describe the steps of establishing the design/performance specifications for the system and selecting and sizing sensors, transducers, actuators, drive systems, controllers, and signal-conditioning and interface hardware and software for the instrumentation and component integration of this system. Keeping this in mind, write a project proposal giving the following information:

1. Select a process (plant) as the system to be developed. Describe the plant indicating the purpose of the plant, how the plant operates, what is the system boundary (physical or imaginary), what are the important inputs (e.g., voltages, torques, heat transfer rates, flow rates) and response variables (e.g., displacements, velocities, temperatures, pressures, currents, voltages), and what are important plant parameters (e.g., mass, stiffness, resistance, inductance, thermal conductivity, fluid capacity). You may use sketches.
2. Indicate the performance requirements (or operating specifications) for the plant (i.e., how the plant should behave in normal operation). You may use any available

information on such requirements as accuracy, resolution, speed, linearity, stability, and operating bandwidth.

3. Give any constraints related to cost, size, weight, environment (e.g., operating temperature, humidity, dust-free or clean room conditions, lighting, and wash-down needs), and so on.
4. Indicate the type and the nature of the sensors and transducers present in the plant and what additional sensors and transducers might be needed to properly operate and control the system.
5. Indicate the type and the nature of the actuators and drive systems present in the plant and which of these actuators have to be controlled. If you need to add new actuators (including control actuators) and drive systems, indicate such requirements in detail.
6. Mention what types of signal modification and interfacing hardware would be needed (i.e., filters, amplifiers, modulators, demodulators, ADC, DAC, and other data acquisition and control needs). Describe the purpose of these devices. Indicate any software (e.g., driver software) that may be needed along with this hardware.
7. Indicate the nature and operation of the controllers in the system. State whether these controllers are adequate for your system. If you intend to add new controllers, briefly give their nature, characteristics, objectives, and so on (e.g., analog, digital, linear, nonlinear, hardware, software, control bandwidth).
8. Describe how the users and operators interact with the system and the nature of the user interface requirements (e.g., GUI).

The following plants or systems may be considered:

1. A hybrid electric vehicle
2. A household robot
3. A smart camera
4. A smart airbag system for an automobile
5. Rover mobile robot for Mars exploration, developed by NASA
6. An AGV for a manufacturing plant
7. A flight simulator
8. A hard disk drive for a personal computer
9. A packaging and labeling system for a grocery item
10. A vibration testing system (electrodynamic or hydraulic)
11. An active orthotic device to be worn by a person to assist a disabled or weak hand (which has some sensation, but not fully functional)

2

Component Interconnection

Chapter Objectives

- Introduction to Component Interconnection
- Impedance Matching in Component Interconnection
- Impedance Matching Methods
- Maximum Power Transfer
- Power Transfer at Maximum Efficiency
- Reflection Prevention in Signal Transmission
- Loading Reduction
- Impedance Matching in Mechanical Systems

2.1 Introduction

An engineering system typically consists of a wide variety of components that are interconnected to perform the intended functions. When two components are interconnected, signals (and power) will flow between them, and as the two components interact (i.e., dynamically coupled), their signals (responses) will vary with time, depending on the dynamics of both components. When two devices are interfaced, it is essential to guarantee that the resulting changes to the dynamic conditions of the system will not lead to unacceptable performance. It is clear that considerations of component *interconnection* and *interface* between the connected components are important in *instrumentation* of an engineering system.

2.1.1 Component Interconnection

Engineering systems are typically multidomain (mixed) systems, which consist of more than one type of components that are interconnected. This is particularly true with mechatronic systems, which employ an integrated and concurrent optimal approach in their design and development. Commonly, mechanical (including fluid and thermal), electrical, electronic, and computer hardware are integrated to form practical applications. When components are interconnected, the behavior of the individual components in the integrated system can deviate significantly from that when each component operates independently. It follows that component interconnection is an important consideration in the design and instrumentation (and overall development) of an engineering system.

The nature and type of the signals that are present at the interface of the interconnected components depend on the nature and type of the components. For example, when a motor is coupled with a load through a gear (transmission) unit, mechanical power flows at the interfaces of these components. In that case, the power that is transmitted is of the same type (mechanical) and we are particularly interested in the associated signals of angular velocity and torque. Similarly, when a motor is connected to its electronic drive system (e.g., the electrical dive circuit may be connected to a stator or rotor or both of a DC motor depending on the type of motor), there is conversion of electrical power of the drive circuit into mechanical power of the rotor. Their interface may be represented by an electromechanical transformer, as shown in Figure 2.1. On one side, we have voltage and current as the power signals and on the other side, we have angular velocity and torque as the power signals. *Note*: In both examples, there will be energy dissipation (wastage) on both sides, and hence, the energy conversion will not take place at 100% efficiency.

Generally, when two components are interconnected, dynamic interactions (dynamic coupling) take place between them and hence, the conditions of either component will be different from what they were before the connection. It is clear that the interconnected components should be properly *matched* in order for the resulting system to operate in the desired manner. For example, in the case of a motor and its electronic drive system, maximum efficiency may be a primary objective. Then, the dynamic interaction between the two components will be significant. In contrast, in the case of a sensor and a monitored object, it is important that the dynamic conditions of the object would not be altered by the sensor (i.e., the *loading* of the object by the sensor should be negligible; e.g., with regard to a motion sensor, both electrical loading and mechanical loading should be negligible). In other words, dynamic interaction between the sensor and the monitored object should be negligible while maintaining the ability to accurately measure the required quantity.

Component interface plays an important role in the proper operation of the interconnected system. Specifically, the interface has to be designed, developed, or selected depending on the specific function of the interconnected system. Matching of components in a multicomponent system should be done carefully to improve the system performance and accuracy. In this context, impedance considerations are important because *impedance*

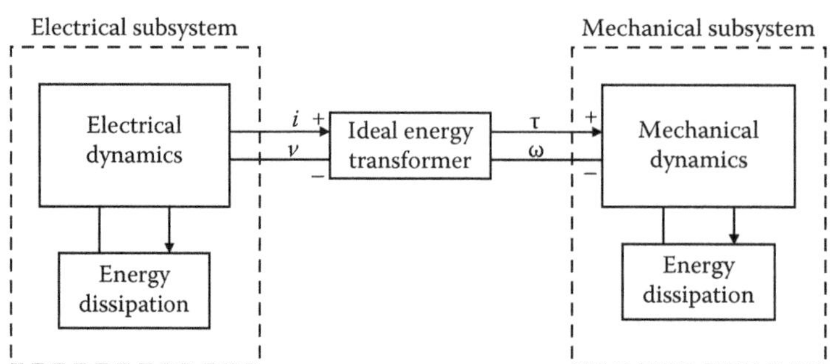

FIGURE 2.1
A model for mixed-domain (electromechanical) component interconnection.

matching is necessary to realize the best performance from the interconnected system, depending on its functional objective.

The following considerations are relevant in component interconnection:

1. Characteristics of the interconnected components (e.g., domain of the component [mechanical, electrical/electronic, thermal], type of the component [actuator, sensor, drive circuit, controller, mounting or housing])
2. Purpose of the interconnected system (e.g., drive a load, measure a signal, communicate information, minimize noise and disturbances—mechanical shock and vibration in particular)
3. Signal/power levels of operation

2.1.2 Chapter Objectives

This chapter addresses interconnection of components such as sensors, data acquisition (DAQ) hardware, signal-conditioning circuitry, actuators, power transmission devices, and mounting mechanics in an engineering system. The basic concepts of impedance and component matching are studied. Desirable impedance characteristics for interconnected components, depending on the purpose and application, are discussed. Several methods of impedance matching are presented and their applications are indicated. Impedance matching in both electrical and mechanical devices is addressed. Reliability considerations and associated probability models of multicomponent systems are outlined in Appendix D.

Discussions and developments given here may be rather general in some situations. However, the concepts presented here are applicable to many types of components in an engineering system (multidomain).

2.2 Impedance

2.2.1 Definition of Impedance

Impedance can be interpreted in the traditional electrical sense as "generalized resistance." It may be interpreted in the mechanical sense as well or in a general sense with regard to other domains (e.g., fluid, thermal) as well depending on the type of signals involved. For example, a voltmeter can modify the currents (and voltages) in a circuit, and this concerns *electrical resistance* of a DC circuit or more generally, *electrical impedance*, when AC circuits are considered. As another example, a heavy accelerometer will introduce an additional dynamic (mechanical) load, which will modify the actual acceleration at the monitored location. This concerns *mechanical impedance*. As a third example, a thermocouple junction can modify the temperature that is measured as a result of the heat transfer into the junction. This concerns *thermal impedance*. Similarly, we can define impedance for *fluid* systems, *magnetic* systems (*reluctance*), and so on. In general, impedance is defined as

$$\text{Impedance} = \frac{\text{Across variable}}{\text{Through variable}}. \tag{2.1}$$

The across variable is measured across the two ends (ports) of a component and the through variable transmits unaltered through the component. Examples of across variables are voltage, velocity, temperature, and pressure. Examples of through variables are current, force, heat transfer rate, and fluid flow rate. Even though electrical impedance is defined as voltage/current, which is consistent with the definition (2.1), mechanical impedance, historically, has been defined as force/velocity, which is the inverse of the definition (2.1). It is the "mobility" that is defined as velocity/force, and mobility should be interpreted as impedance in the general sense (i.e., generalized impedance), in our analysis.

2.2.2 Importance of Impedance Matching in Component Interconnection

When two electrical components are interconnected, current (and energy) flows between the two components and changes the original (unconnected) conditions. This is known as the (electrical) loading effect, and it has to be minimized. In practical situations, adequate power and current would be needed for signal communication, conditioning, display, and so on. Such requirements of instrumentation can be accommodated through proper matching of component impedances. In general, when components such as sensors and transducers, power sources, control hardware, DAQ boards, process (i.e., plant) equipment, and signal-conditioning hardware and cables are interconnected, it is necessary to match the impedances properly at each interface to realize their rated performance levels. This matching should be done according to the purpose of the interconnected system. Several categories of impedance matching are given in the following:

1. *Source and load matching for maximum power transfer*: In a drive system, an important objective may be to maximize the power transmitted from the power source to the actuator or the load. In that case, dynamic interactions between the interconnected components will be significant. Proper impedance matching can achieve the requirement of maximum power transfer.
2. *Power transfer at maximum efficiency*: Achieving maximum efficiency in power transfer is different from achieving maximum power simply because maximum efficiency is not achieved at maximum power transfer. The load impedance can be properly chosen to achieve high efficiency.
3. *Reflection prevention in signal transmission*: When two components are connected by a cable (e.g., coaxial cable) with characteristic impedance (e.g., 50 or 75 Ω for a coaxial cable), due to the impedance difference at the two ends (due to the impedances of the connected components), there will be signal reflection (similar to elastic wave reflection due to density difference in two media). The reflected signals (echoes) will cause additional power dissipation, drop in signal strength, and signal distortion, all of which are undesirable. The end impedances and the characteristic impedance of the cable have to be matched in order to avoid signal reflection.
4. *Loading reduction*: When two components are interconnected, in some applications, it is required that the output component does not load the input component. For example, in a sensing process, the sensor should not alter the conditions of the sensed object. In other words, the measuring instrument should not distort the signal that is measured. Quite simply, the sensor should not "load" the measured object. As another example, in a signal acquisition system of a sensor, the signal acquisition hardware should not distort the acquired signal from the sensor (i.e., the signal acquisition system, which may have such functions as filtering

and amplification, should not load the sensor). As a third example, in a regulated power source, the load that is connected to the power source should not considerably change the output voltage of the power source. The impedances can be chosen to reduce loading effects. In this case, impedance matching is called impedance bridging or voltage bridging. An impedance transformer would be required to achieve proper impedance matching for loading reduction.

Another adverse effect of improper impedance consideration is inadequate output signal levels, which will deteriorate such functions as signal processing and transmission, component driving, and actuation of a final control element or plant. In the context of sensor–transducer technology, it should be noted here that many types of transducers (e.g., piezoelectric accelerometers, impedance heads, and microphones) have high-output impedances in the order of 1000 MΩ (megaohm; $1\ \text{M}\Omega = 1 \times 10^6\ \Omega$). These devices generate low-output signals, and they would require conditioning to step up the signal level. Impedance-matching amplifiers (or impedance transformers) or impedance-bridging devices, which have high-input impedances and low-output impedances (a few ohms), are used for this purpose (e.g., charge amplifiers are used in conjunction with piezoelectric sensors). A device with a high-input impedance has the further advantages that it will consume less *power* (i.e., v^2/R is low) particularly from the input device to which it is connected, for a given input voltage, and furthermore, the power transfer will take place at higher *efficiency*. The fact that an output device having low input impedance extracts a high level of power from its input device may also be interpreted as the reason for loading error in the input device. In that situation, there will be significant dynamic interaction between the two interconnected devices (input device and output device).

2.3 Impedance-Matching Methods

In instrumentation procedures, component interconnection is done in order to achieve some specific purposes. The impedances of the connected components should be matched in order to improve the system performance with regard to these functional purposes (objectives). We will consider the following categories of objectives of impedance matching now:

1. Maximum power transfer from a source to a load
2. Power transfer at maximum efficiency
3. Reflection prevention in signal transmission
4. Loading reduction

2.3.1 Maximum Power Transfer

There are applications where an objective is to absorb maximum power from a source. The *internal impedance of the source* Z_s has to be matched with the impedance of the load Z_l in order to maximize the load power. If there are other components in the interconnected system apart from the source and the load, we can use the same approach simply by taking as the source impedance the Thevenin equivalent impedance of the circuit without including the load. The approach can be presented first by considering a purely resistive (DC) example.

Example 2.1

Consider a DC power supply of voltage v_s and internal (output) impedance (resistance) R_s. It is used to power a load of resistance R_l, as shown in Figure 2.2. What should be the relationship between R_s and R_l, if the objective is to maximize the power absorbed by the load?

Solution

The current through the circuit is

$$i = \frac{v_s}{R_l + R_s}.$$

Accordingly, the voltage across the load is

$$v_l = iR_l = \frac{v_s R_l}{R_l + R_s}.$$

The power absorbed by the load is

$$p_l = iv_l = \frac{v_s^2 R_l}{[R_l + R_s]^2}. \quad \text{(i)}$$

For maximum power, we need

$$\frac{dp_l}{dR_l} = 0. \quad \text{(ii)}$$

We differentiate the RHS expression of Equation i with respect to R_l and equate the result to 0, to satisfy (ii). This gives the requirement for maximum power as

$$R_l = R_s.$$

It follows that the requirement for maximum power transfer is that the load resistance must be equal to the source resistance.

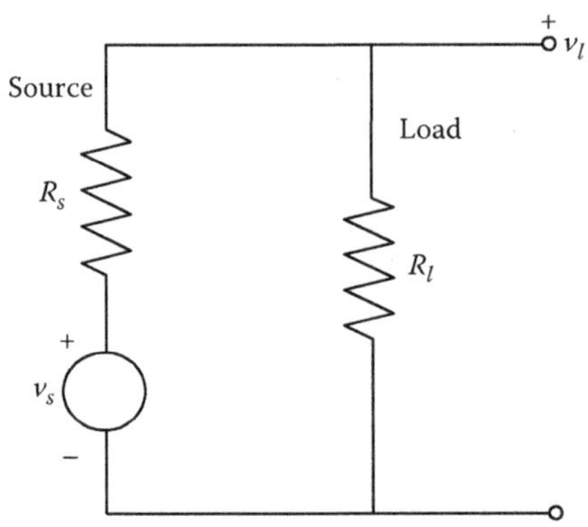

FIGURE 2.2
A DC circuit with a source and a load.

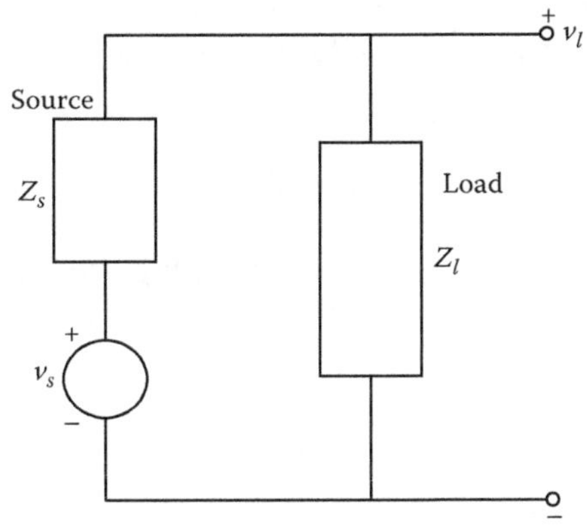

FIGURE 2.3
An impedance (AC) circuit with a source and a load.

The result obtained in Example 2.1 can be easily extended to the general case of impedance, which concerns AC circuits, which have both resistance and reactance (reactance is caused by inductance and capacitance). This situation is shown in Figure 2.3, where v_s is the source voltage, Z_s is the source impedance, and Z_l is the load impedance. If there are components other than the source and the load, they can be integrated with the source. Then, Z_l represents the equivalent Thevenin impedance of those components (excluding the load). If those components have sources as well, then v_s represents the equivalent source voltage of the Thevenin equivalent circuit.

In an AC circuit, the quantities have both magnitude and phase angle, and mathematically, they are represented by complex quantities (with a real part and an imaginary part). Using their magnitudes, the magnitude of the current through the circuit is $|I| = |V_s|/|Z_l+Z_s|$.

Power absorbed by the load is the resistive power and is given by

$$p_l = I_{rms}^2 R_l = \frac{1}{2}|I|^2 R_l = \frac{1}{2}\frac{|V_s|^2}{|Z_l+Z_s|^2}R_l = \frac{1}{2}\frac{|V_s|^2}{(R_l+R_s)^2+(X_l+X_s)^2}R_l,$$

where

"*rms*" denotes the root-mean-square value (for a sine signal, it is $1/\sqrt{2}$ of the amplitude)
R is the resistive (real) part of an impedance
X is the reactive (imaginary) part of an impedance

It is clear from the last term of this equation that one requirement for maximizing power is that the reactance contribution in the denominator is a minimum (i.e., zero, because it is a square). Hence, we need

$$X_l = -X_s. \tag{2.2a}$$

Once this condition is satisfied, the expression for load power as given earlier is identical to that of the purely resistive case, which was solved in Example 2.1. Hence, for load power maximization, we also need

$$R_l = R_s. \tag{2.2b}$$

By combining the requirements (2.2a) and (2.2b), it is seen that the overall requirement for maximizing the load power is that the load impedance must be the *complex conjugate* of the source impedance:

$$Z_l = Z_s^*. \tag{2.2}$$

This is known as "conjugate matching." By substituting the impedance-matching requirement (2.2) into the load power expression, we have the maximum power as

$$p_{l\max} = \frac{|V_s|^2}{8R_s} = \frac{|V_s|^2}{8R_l}. \tag{2.3}$$

2.3.2 Power Transfer at Maximum Efficiency

The efficiency of power absorption by the load is given by the fraction of the absorbed power from the total power:

$$\eta = \frac{1/2 \times |I|^2 R_l}{1/2 \times |I|^2 (R_l + R_s)} = \frac{R_l}{(R_l + R_s)}. \tag{2.4}$$

It is seen that the efficiency is a maximum when the load resistance is a maximum (or load impedance is a maximum). Hence, for increasing the efficiency of load power absorption, the load efficiency has to be increased. In theory, we get 100% efficiency when the load impedance is infinite.

Clearly, the condition for maximum efficiency is quite different from the condition for maximum power (Equation 2.2). In fact, by substituting Equation 2.2b into 2.3, we see that at maximum power, the efficiency is 50%. This is indeed a condition of rather poor efficiency.

2.3.3 Reflection Prevention in Signal Transmission

When an electric signal encounters an abrupt change in impedance, part of the signal will be reflected back. The *reflection coefficient* Γ is given by the ratio of the reflected signal voltage v_r to the incident signal voltage v_i:

$$\Gamma = \frac{v_r}{v_i}. \tag{2.5}$$

If a signal transmitted through an impedance Z_c abruptly encounters a terminating impedance Z_l, the corresponding reflection coefficient is

$$\Gamma = \left|\frac{Z_l - Z_c}{Z_l + Z_c}\right|. \tag{2.6}$$

A reflecting signal results in signal deterioration (both magnitude and phase angle) and dissipation (power loss) both of which are undesirable. It follows that, ideally, we like to have $\Gamma = 0$.

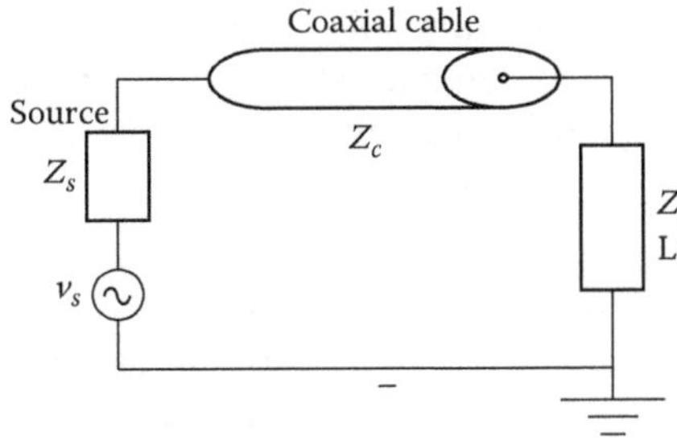

FIGURE 2.4
A source and a load connected by a cable.

Consider a source of internal impedance Z_s connected to a load of impedance Z_l through a cable of characteristic impedance Z_c (e.g., 50 Ω) as shown in Figure 2.4.

To avoid signal reflection at either termination point (load and source), we must have the impedance-matching condition:

$$Z_s = Z_c = Z_l. \tag{2.7}$$

If the impedance matching is not present, it can be achieved with a grounding impedance (or an *impedance-matching pad*). An example is shown in Figure 2.5. In this example, suppose that $Z_s \neq Z_c$. Then, the grounding impedance Z_g has to be placed such that the following condition is satisfied:

$$\frac{1}{Z_s} + \frac{1}{Z_g} = \frac{1}{Z_c}. \tag{2.8}$$

This arrangement provides an equivalent terminating impedance of Z_c at the source end.

In a short cable, the reflected signal travels to its terminals very quickly and decays fast. Hence, signal reflection is not important in short cables. The principle of signal reflection may be used in practical applications. For example, since damage in a cable will result in an abrupt change in the impedance, there will be signal reflection at the damaged location. Then by determining the time it takes for a voltage pulse to be reflected back to the source, the distance to the damaged location can be determined. This is the principle behind a *reflectometer*, which is used to detect damages in cables.

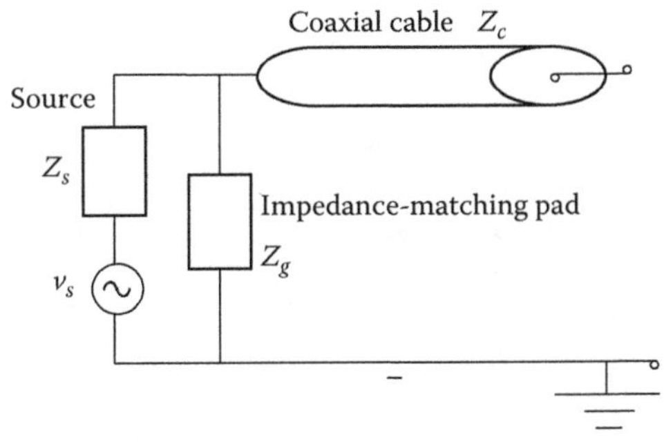

FIGURE 2.5
Application of an impedance-matching pad.

Signal reflection is not limited to metal cables (copper, aluminum, etc.). For example, it can be observed in optical fibers and in acoustic signals that encounter a change in the acoustic impedance in the transmission medium.

2.3.4 Loading Reduction

An adverse effect of improper impedance conditions is the *loading* effects, which distort signals. The resulting error can far exceed other types of error such as measurement error, sensor error, noise, and input disturbances. Loading can occur in any physical domain such as electrical and mechanical. *Electrical loading* errors result from connecting an output unit such as a measuring device or signal acquisition hardware that has low *input impedance* to an input device such as a signal source or a sensor with low to moderate impedance. *Mechanical loading* errors can result in an input device (e.g., an actuator) because of inertia, friction, and other resistive forces generated by an output component connected to it (e.g., a gear transmission, a mechanical load).

In engineering systems, loading errors can appear as phase distortions as well. Digital hardware also can produce loading errors. For example, analog-to-digital conversion hardware in a DAQ board can load the amplifier output from a strain-gauge bridge circuit, thereby affecting the digitized data (see Chapter 4).

2.3.4.1 Cascade Connection of Devices

To obtain a model for loading distortion and a method for reducing loading effects, we now consider cascade connection of two-port electrical devices. A model for a two-port electrical device is shown in Figure 2.6a. It shows in particular the input impedance Z_i and the output impedance Z_o of the device. They are defined in the following texts.

Input impedance: Input impedance Z_i is defined as the ratio of the rated input voltage to the corresponding current through the input terminals while the output terminals are maintained in open circuit.

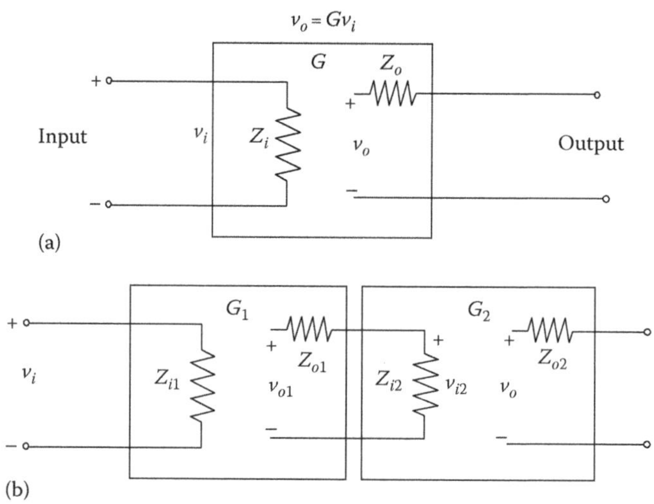

FIGURE 2.6
(a) Schematic representation of input impedance and output impedance and (b) cascade connection of two two-port devices.

Output impedance: The output impedance Z_o is defined as the ratio of the open-circuit (i.e., no load) voltage at the output port to the short-circuit current at the output port. Open-circuit voltage at output is the output voltage present when there is no current flowing at the output port. This is the case if the output port is not connected to a load (impedance). As soon as a load is connected at the output of the device, a current will flow through it, and the output voltage will drop to a value less than that of the open-circuit voltage. To measure the open-circuit voltage, the rated input voltage is applied at the input port and maintained constant, and the output voltage is measured using a voltmeter that has a high (input) impedance. To measure the short-circuit current, a very low-impedance ammeter is connected at the output port.

These definitions are given with reference to an electrical device. However, a generalization to mechanical devices is possible by interpreting voltage and velocity as "across variables" and current and force as "through variables," as noted earlier. Then, mechanical mobility should be used in place of electrical impedance, in the associated analysis. Similar generalization is possible for other physical domains as well.

It is seen that the input impedance Z_i and the output impedance Z_o as represented in Figure 2.6a agree with their definitions given earlier. Note that v_o is the open-circuit output voltage. When a load is connected at the output port, the voltage across the load will be different from v_o. This is caused by the presence of a current through Z_o. In the frequency domain, v_i and v_o are represented by their respective *Fourier spectra* (or in the complex form with a real part and an imaginary part or a magnitude and a phase). The corresponding transfer relation can be expressed in terms of the complex frequency response (transfer) function $G(j\omega)$ under open-circuit (no load) conditions:

$$v_o = Gv_i. \tag{2.9}$$

Next, consider two devices connected in cascade, as shown in Figure 2.6b. It can be easily verified that the following relations hold:

$$v_{o1} = G_1 v_i; \quad v_{i2} = \frac{Z_{i2}}{Z_{o1} + Z_{i2}} v_{o1}; \quad \text{and} \quad v_o = G_2 v_{i2}.$$

These relations can be combined to give the overall input–output relation:

$$v_o = \frac{Z_{i2}}{Z_{o1} + Z_{i2}} G_2 G_1 v_i.$$

We observe from this result that the overall frequency transfer function differs from the ideally expected product (G_2G_1) by the factor

$$\frac{Z_{i2}}{Z_{o1} + Z_{i2}} = \frac{1}{(Z_{o1}/Z_{i2}) + 1}. \tag{2.10}$$

Note from this equation that cascading has distorted the frequency response characteristics of the two devices, and this represents the *loading error*. The loading error becomes insignificant when $Z_{o1}/Z_{i2} \ll 1$. From this observation, it can be concluded that when two components are interconnected (cascaded), in order to reduce the loading error, the input impedance of the second device (output device) should be much larger than the output impedance of the first device (input device).

Example 2.2

A lag network used as the compensation element of a control system is shown in Figure 2.7a. Show that its transfer function is given by $v_o/v_i = Z_2/(R_1 + Z_2)$ where $Z_2 = R_2 + (1/Cs)$.

What are the input and output impedances of this circuit?

Moreover, if two such lag circuits are cascaded as shown in Figure 2.7b, what is the overall transfer function? How would you bring this transfer function close to the ideal result: $\{Z_2/(R_1 + Z_2)\}^2$?

Solution

To solve this problem, first note that in Figure 2.7a, the voltage drop across the element $R_2 + 1/(Cs)$ is

$$v_o = \frac{(R_2 + (1/Cs))}{\{R_1 + R_2 + (1/Cs)\}} v_i.$$

Hence,

$$\frac{v_o}{v_i} = \frac{Z_2}{R_1 + Z_2}. \tag{i}$$

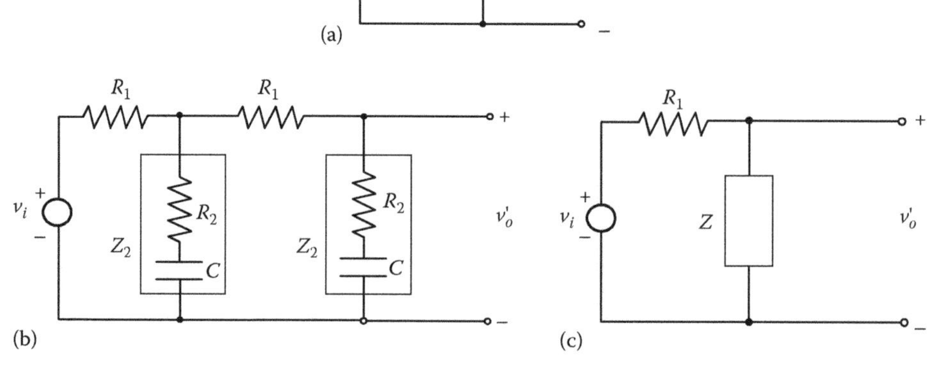

FIGURE 2.7
(a) A single-circuit module, (b) cascade connection of two modules, and (c) an equivalent circuit for (b).

Now, the input impedance Z_i is derived by using the input current $i = v_i/(R_1 + Z_2)$; thus,

$$Z_i = \frac{v_i}{i} = R_1 + Z_2. \tag{ii}$$

The output impedance Z_o is derived by using the short-circuit current $i_{sc} = v_i/R_1$; thus,

$$Z_o = \frac{v_o}{i_{sc}} = \frac{Z_2/(R_1+Z_2)v_i}{v_i/R_1} = \frac{R_1 Z_2}{R_1+Z_2}. \tag{iii}$$

Next, consider the equivalent circuit shown in Figure 2.7c. Since Z is formed by connecting Z_2 and $(R_1 + Z_2)$ in parallel, we have

$$\frac{1}{Z} = \frac{1}{Z_2} + \frac{1}{R_1+Z_2}. \tag{iv}$$

The voltage drop across Z is

$$v_o' = \frac{Z}{R_1+Z} v_i. \tag{v}$$

Now, apply the single-circuit module result (i) to the second circuit stage in Figure 2.7b. We get $v_o = (Z_2/(R_1+Z_2))v_o'$. Substituting this into Equation v, we get $v_o = (Z_2/(R_1+Z_2))(Z/(R_1+Z))v_i$. The overall transfer function for the cascaded circuit is

$$G = \frac{v_o}{v_i} = \frac{Z_2}{(R_1+Z_2)}\frac{Z}{(R_1+Z)} = \frac{Z_2}{(R_1+Z_2)}\frac{1}{(R_1/Z+1)}.$$

Now, substituting Equation iv for $1/Z$, we get

$$G = \left[\frac{Z_2}{(R_1+Z_2)}\right]^2 \frac{(R_1+Z_2)}{Z_2}\frac{1}{\left[R_1(1/Z_2+1/(R_1+Z_2))+1\right]}$$

$$= \left[\frac{Z_2}{(R_1+Z_2)}\right]^2 \frac{(R_1+Z_2)^2}{\left[R_1(R_1+Z_2+Z_2)+Z_2(R_1+Z_2)\right]}$$

$$= \left[\frac{Z_2}{(R_1+Z_2)}\right]^2 \frac{(R_1+Z_2)^2}{\left[(R_1+Z_2)^2+R_1Z_2\right]}$$

$$\rightarrow G = \left[\frac{Z_2}{R_1+Z_2}\right]^2 \frac{1}{[1+R_1Z_2/(R_1+Z_2)^2]}.$$

We observe that the ideal transfer function is approached by making $R_1Z_2/(R_1+Z_2)^2$ small compared with unity.

2.3.4.2 Impedance Matching for Loading Reduction

From the analysis given in the preceding section, it is clear that to reduce loading errors, the signal-conditioning circuitry should have a considerably large input impedance in comparison with the output impedance of the sensor–transducer unit. The problem is quite critical in measuring devices such as piezoelectric sensors, which have very high-output impedances. In such cases, the input impedance of the signal-conditioning unit might be inadequate to reduce loading effects; also, the output signal level of these high-impedance sensors is quite low for signal transmission, processing, actuation, and control. The solution for this problem is to introduce several stages of amplifier circuitry between the output of the first hardware unit (e.g., sensor) and the input of the second hardware unit (e.g., DAQ unit). The first stage of such an interfacing device is typically an impedance-matching amplifier (or impedance transformer) that has high-input impedance, low-output impedance, and almost unity gain. This is known as "impedance bridging." The last stage is typically a stable high-gain amplifier stage to step up the signal level. Impedance-matching amplifiers are, in fact, operational amplifiers (op-amps) with feedback.

In conclusion, we make the following comments:

1. When connecting a device to a signal source, loading problems can be reduced by making sure that the device has a high-input impedance. Unfortunately, this will also reduce the level (amplitude, power) of the signal received by the device. A stage of signal amplification may be needed then, while maintaining the required impedance levels at the output.
2. A high-impedance device may reflect back some harmonics of the source signal, as we noted under signal reflection. As presented there, a termination resistance (impedance pad) might be connected in parallel with the device to reduce this problem of signal reflection. In many DAQ systems, output impedance of the output amplifier is made equal to the transmission line impedance (characteristic impedance).
3. When maximum power amplification is desired, conjugate matching is recommended. In this case, input and output impedances of the matching amplifier are made equal to the complex conjugates of the source and load impedances, respectively.

2.3.5 Impedance Matching in Mechanical Systems

The concepts of impedance matching can be extended to mechanical systems and to mixed systems (e.g., electromechanical systems or mechatronic systems) in a straightforward manner. The procedure follows from the familiar electromechanical analogies. Two specific applications are in (1) shock and vibration isolation and (2) transmission systems (gears). These two applications are discussed as follows.

2.3.5.1 Vibration Isolation

A good example of component interconnection in mechanical systems is vibration isolation. Proper operation of engineering systems such as delicate instruments, computer hardware, machine tools, and vehicles is hampered due to shock and vibration. The purpose of vibration isolation is to "isolate" such devices from undesirable vibration and shock disturbances that come from its environment (including the supporting structure or road). This is achieved by connecting a *vibration isolator* or *shock mount* in between them.

2.3.5.1.1 Force Isolation and Motion Isolation

External disturbance can reach a system in the form of force or motion input, and depending on that, force isolation (related to force transmissibility) or motion isolation (related to motion transmissibility) would be applicable in the design of the isolator. Luckily, the isolator design is quite similar for the two situations.

In force isolation, vibration forces that would be ordinarily transmitted directly from a source to a supporting structure (isolated system) are filtered out by an isolator through its flexibility (spring) and dissipation (damping) so that part of the force is routed through an inertial path. In motion isolation, vibration motions that are applied to a system (e.g., vehicle) by a moving platform are absorbed by an isolator through its flexibility and dissipation so that the motion that is transmitted to the system of interest is weakened. The design problem in both cases is to select applicable parameters for the isolator so that the vibrations entering the system of interest are below the specified values within a frequency band of interest (the operating frequency range). This design problem is essentially a situation of "mechanical impedance matching," because impedance parameters (mechanical) of the isolator are chosen depending on the impedance parameters of the isolator.

Note: As indicated before, generalized impedance (across variable/through variable) corresponds to mechanical *mobility*, which is the inverse of what is traditionally called mechanical impedance.

Force isolation and motion isolation are represented in Figure 2.8. In Figure 2.8a, vibration force at the source is $f(t)$. In view of the isolator, the source (with mechanical impedance Z_m) is made to move at the same speed as the isolator (with mechanical impedance Z_s). This is a parallel connection of impedances. Hence, the force $f(t)$ is split so that part of it is taken up by the inertial path (broken line) of Z_m, and only the remainder (f_s) is transmitted through Z_s to the supporting structure, which is the isolated system. Force transmissibility is

$$T_f = \frac{f_s}{f} = \frac{Z_s}{Z_m + Z_s}. \tag{2.11}$$

In Figure 2.8b, vibration motion $v(t)$ of the source is applied through an isolator (with impedance Z_s and mobility M_s) to the isolated system (with mechanical impedance Z_m and mobility M_m). The resulting force is assumed to transmit directly from the isolator to the isolated system, and hence, these two units are connected in series. Consequently, we have the motion transmissibility:

$$T_m = \frac{v_m}{v} = \frac{M_m}{M_m + M_s} = \frac{Z_s}{Z_s + Z_m}. \tag{2.12}$$

It is noticed that, according to these two models, we have

$$T_f = T_m = T. \tag{2.13}$$

As a result, usually, both types of isolators can be designed in the same manner using a common transmissibility function T.

Simple examples of force isolation and motion isolation are shown in Figure 2.8c and d. First, we obtain the transmissibility (force transmissibility) function for system in Figure 2.8c. Then, in view of Equation 2.13, the motion transmissibility of the system

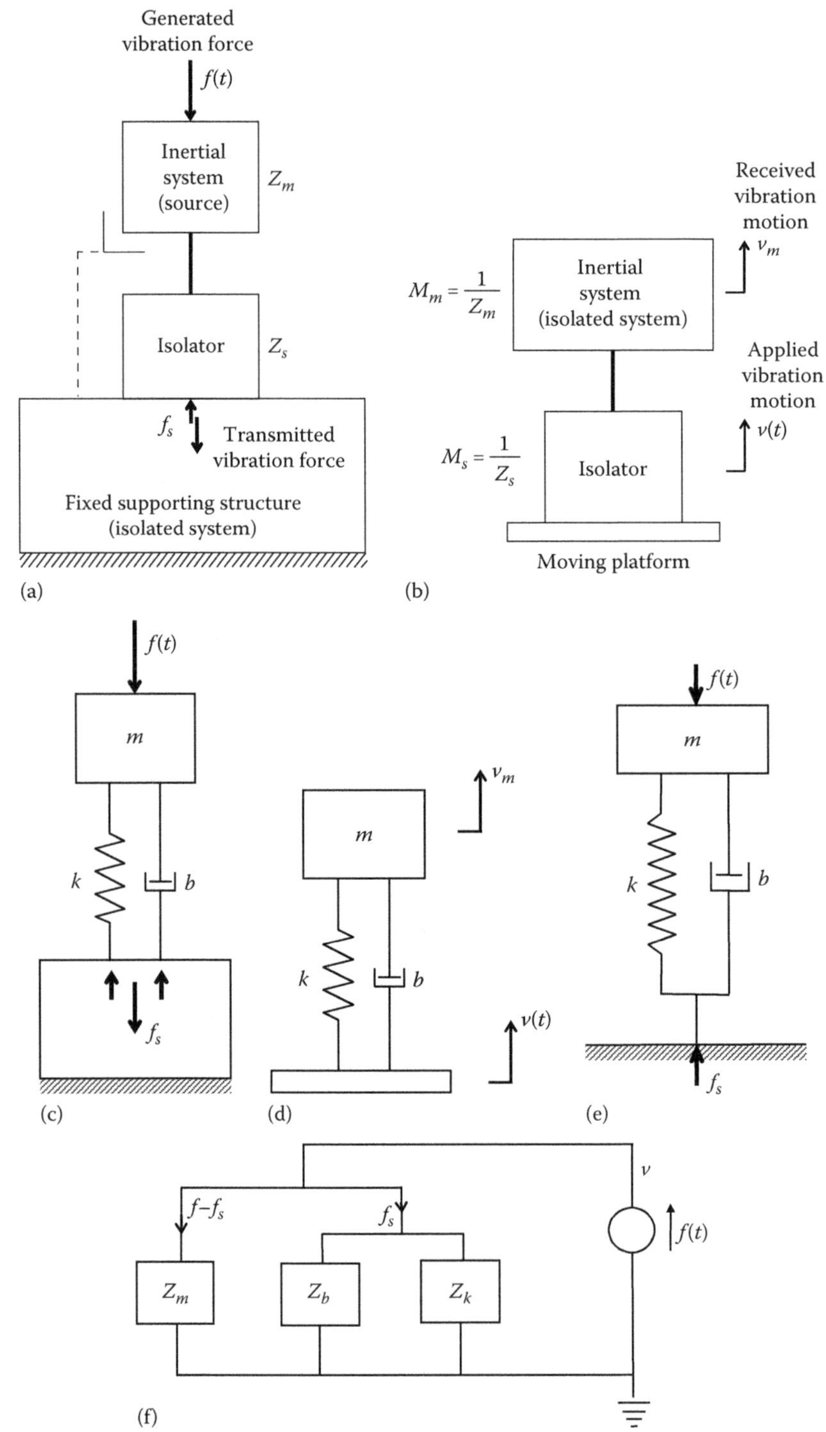

FIGURE 2.8
(a) Force isolation, (b) motion isolation, (c) force isolation example, (d) motion isolation example, (e) simplified model of a machine tool and its supporting structure, (f) and mechanical impedance circuit of the force isolation problem.

in Figure 2.8d is equal to the same expression. First, consider the force transmissibility problem of Figure 2.8c, which is shown again in Figure 2.8e. This may represent a simplified model of a machine tool and its supporting structure.

Clearly, the elements m, b, and f are in parallel, since they have a common velocity v across them. Hence, its mechanical–impedance circuit is shown in Figure 2.8f. In this circuit, the impedances of the basic elements are $Z_m = mj\omega$, $Z_b = b$, and $Z_k = k/j\omega$, for mass (m), spring (k), and viscous damper (b), respectively (see Table 2.1). Substitute the element impedances into the force transmissibility expression (which is obtained by using the laws of element interconnection shown in Figure 2.9 and the circuit in Figure 2.8f):

$$T_f = \frac{F_s}{F} = \frac{F_s/V}{F/V} = \frac{Z_s}{Z_s + Z_0} = \frac{Z_b + Z_k}{Z_m + Z_b + Z_k}. \tag{i}$$

TABLE 2.1

Mechanical Impedance and Mobility of Basic Mechanical Elements

Element	Time-Domain Model	Impedance	Mobility (Generalized Impedance)
Mass, m	$m\frac{dv}{dt} = f$	$Z_m = ms$	$M_m = \frac{1}{ms}$
Spring, k	$\frac{df}{dt} = kv$	$Z_k = \frac{k}{s}$	$M_k = \frac{s}{k}$
Damper, b	$f = bv$	$Z_b = b$	$M_b = \frac{1}{b}$

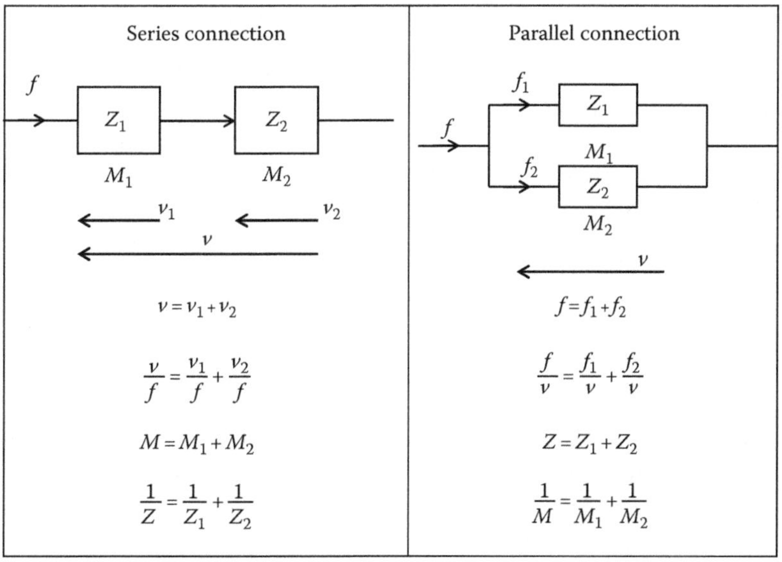

FIGURE 2.9
Interconnection laws for mechanical impedance and mobility.

We get

$$T_f = \frac{b+(k/j\omega)}{mj\omega+b+(k/j\omega)} = \frac{bj\omega+k}{-\omega^2 m+bj\omega+k} = \frac{j\omega b/m+k/m}{-\omega^2+j\omega b/m+k/m}. \tag{2.14a}$$

This expression is obtained by dividing the numerator and the denominator by m. Now, use the fact that $k/m=\omega_n^2$ and $b/m=2\zeta\omega_n$ (or $\omega_n=\sqrt{k/m}$ = undamped natural frequency of the system and $\zeta=b/2\sqrt{km}$ = damping ratio of the system) and divide (2.14a) throughout by ω_n^2. We get

$$T_f = \frac{\omega_n^2+2\zeta\omega_n j\omega}{\omega_n^2-\omega^2+2\zeta\omega_n j\omega} = \frac{1+2\zeta rj}{1-r^2+2\zeta rj}, \tag{2.14b}$$

where the nondimensional excitation frequency is defined as $r=\omega/\omega_n$.

The transmissibility function has a phase angle and magnitude. In practical applications of vibration isolation, it is the level of attenuation of the vibration excitation that is of primary importance, rather than the phase difference between the vibration excitation and the response. Accordingly, we use the transmissibility magnitude

$$|T_f| = \sqrt{\frac{1+4\zeta^2r^2}{(1-r^2)^2+4\zeta^2r^2}}. \tag{2.15a}$$

To determine the peak point of $|T_f|$, differentiate the expression within the square root sign in (2.15a) and equate to zero:

$$\frac{[(1-r^2)^2+2\zeta^2r^2]8\zeta^2r-[1+4\zeta^2r^2][2(1-r^2)(-2r)+8\zeta^2r]}{[(1-r^2)^2+4\zeta^2r^2]^2}=0.$$

Hence, $4r\{[(1-r^2)^2+2\zeta^2r^2]2\zeta^2+[1+4\zeta^2r^2][(1-r^2)-2\zeta^2]\}=0$. This simplifies to $r(2\zeta^2r^4+r^2-1)=0$. Its roots are $r=0$ and $r^2=(-1\pm\sqrt{1+8\zeta^2})/4\zeta^2$.

The root $r=0$ corresponds to the initial stationary point (at zero frequency). That does not represent a peak. Taking only the positive root for r^2 and then its positive square root, the peak point of the transmissibility magnitude is given by

$$r = \frac{\left[\sqrt{1+8\zeta^2}-1\right]^{1/2}}{2\zeta}. \tag{ii}$$

For small ζ, Taylor series expansion gives $\sqrt{1+8\zeta^2}\approx 1+(1/2)\times 8\zeta^2=1+4\zeta^2$. With this approximation, Equation ii evaluates to 1. Hence, for small damping, the transmissibility magnitude will have a peak at $r=1$, and from Equation 2.15a, its value is $|T_f|\approx(\sqrt{1+4\zeta^2})/2\zeta\approx(1+(1/2)\times 4\zeta^2)/2\zeta$ or

$$|T_f| \approx \frac{1}{2\zeta}+\zeta \approx \frac{1}{2\zeta}. \tag{2.16}$$

The five curves of $|T_f|$ versus r for $\zeta=0$, 0.3, 0.7, 1.0, and 2.0 are shown in Figure 2.10a. These curves use the exact expression (2.15a) and can be generated using the following MATLAB® program:

```
clear;
zeta=[0.0 0.3 0.7 1.0 2.0];
for j=1:5
for i=1:1201
  r(i)=(i-1)/200;
  T(i,j)=sqrt((1+4*zeta(j)^2*r(i)^2)/((1-r(i)^2)^2+4*zeta(j)^2*r(i)^2));
end
plot(r,T(:,1),r,T(:,2),r,T(:,3),r,T(:,4),r,T(:,5));
```

From the transmissibility curves in Figure 2.10a, we observe the following:

1. There is always a nonzero frequency value at which the transmissibility magnitude will peak. This is the resonance.
2. For small ζ, the peak transmissibility magnitude is obtained at approximately $r=1$. As ζ increases, this peak point shifts to the left (i.e., a lower value for peak frequency).
3. The peak magnitude decreases as ζ increases.
4. All the transmissibility curves pass through the magnitude value 1.0 at the same frequency $r=\sqrt{2}$.
5. The region of isolation (i.e., $|T_f|<1$) is given by $r>\sqrt{2}$. In this region, $|T_f|$ increases with ζ.
6. In the isolation region, the transmissibility magnitude decreases as r increases.

In two particular situations, from the transmissibility curves, we observe the following:

For $|T_f|<1.05$, $r>\sqrt{2}$ for all ζ.
For $|T_f|<0.5$, $r>(1.73, 1.964, 2.871, 3.77, 7.075)$ for $\zeta=(0.0, 0.3, 0.7, 1.0, 2.0)$, respectively.

Next, suppose that the device in Figure 2.8e has a primary, undamped natural frequency of 6 Hz and a damping ratio of 0.2. Suppose, for proper operation, it is required that the system achieves a force transmissibility magnitude of less than 0.5 for operating frequency values greater than 12 Hz. We need

$$\sqrt{\frac{1+4\zeta^2r^2}{(1-r^2)^2+4\zeta^2r^2}}<\frac{1}{2}\rightarrow 4+16\zeta^2r^2<(1-r^2)^2+4\zeta^2r^2\rightarrow r^4-2r^2-12\zeta^2r^2-3>0.$$

For $\zeta=0.2$ and $r=12/6=2$, this expression computes to $2^4-2\times2^2-12\times(0.2)^2\times2^2-3=3.08>0$.

Hence, the requirement is met. In fact, since, for $r=2$, the expression becomes $2^4-2\times2^2-12\times2^2\zeta^2-3=5-48\zeta^2$, it follows that the requirement would be met for $5-48\zeta^2>0\rightarrow\zeta<\sqrt{5/48}=0.32$. If the requirement was not met (say, $\zeta=0.4$), an option would be to reduce damping.

In design problems of vibration isolator, what is normally specified is the *percentage isolation*, as given by

$$I=[1-|T|]\times100\%. \tag{2.17}$$

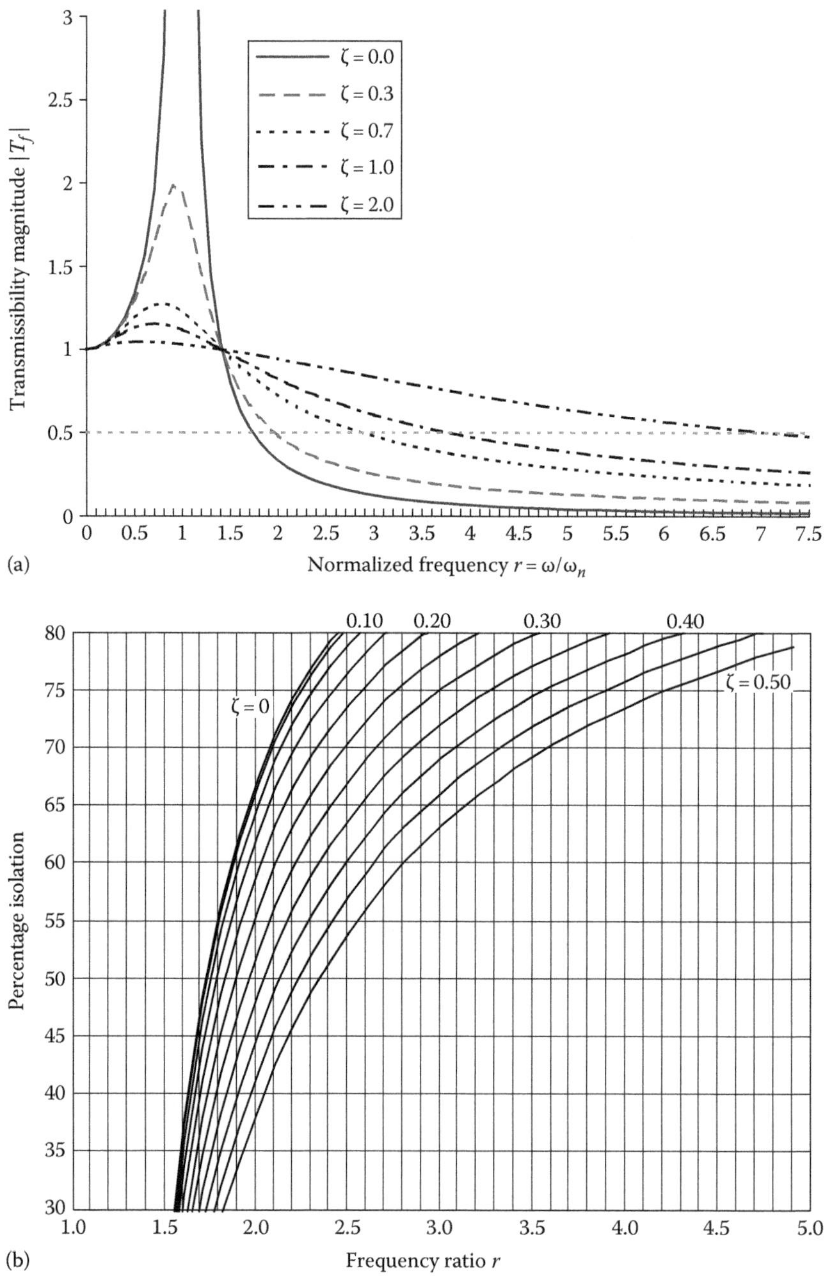

FIGURE 2.10
(a) Transmissibility curves for a simple oscillator model. (b) Curves of vibration isolation.

For the result in Equation 2.15, this corresponds to

$$I = \left[1 - \sqrt{\frac{1+4\zeta^2 r^2}{(r^2-1)^2+4\zeta^2 r^2}}\right] \times 100. \tag{2.18}$$

The isolation curves given by Equation 2.18 are plotted in Figure 2.10b. These curves are useful in the design of vibration isolators.

Note: The model in Figure 2.8 is not restricted to sinusoidal vibrations. Any general vibration excitation may be represented by a Fourier spectrum, which is a function of frequency ω. Then, the response vibration spectrum is obtained by multiplying the excitation spectrum by the transmissibility function T. The associated design problem involves the selection of the isolator impedance parameters k and b to meet the specifications of isolation.

Example 2.3

A machine tool, sketched in Figure 2.11a, weighs 1000 kg and normally operates in the speed range of 300–1200 rpm. A set of spring mounts has to be placed beneath the base of the machine so as to achieve a vibration isolation level of at least 70%. A commercially available spring mount has the load–deflection characteristic shown in Figure 2.11b. It is recommended that an appropriate number of these mounts be used, along with an inertia block, if necessary. The damping constant of each mount is 1.56×10^3 Nm^{-1} s. Design a vibration isolation system for the machine. Specifically, decide upon the number of spring mounts that are needed and the mass of the inertia block that should be added.

Solution

First, we assume zero damping (since, in practice, the level of damping in a system of this type is small) and design an isolator (spring mount and inertia block) for a level of isolation greater than the required 70%. Then, we will check for the case of damped isolator to see whether the required 70% level isolation is achieved.

For the undamped case, Equation 2.15a becomes

$$|T| = \frac{1}{r^2-1}. \tag{2.15b}$$

Note: We have used the case $r>1$, since the isolation region corresponds to $r>\sqrt{2}$.

For the percentage isolation, assume the conservative value $I=80\% \Rightarrow |T|=0.2$. Using Equation 2.15b, we have

$$r^2 = \frac{1}{|T|}+1 = \frac{1}{0.2}+1 = 6.0 = \frac{\omega^2}{\omega_n^2} = \frac{m\omega^2}{k}.$$

The lowest operating speed (frequency) is the most significant (critical) one (because it corresponds to the lowest isolation, as is clearly seen from Figure 2.10a). Hence,

$$\omega = \frac{300}{60} \times 2\pi = 10\pi \,\text{rad/s}.$$

From the load–deflection curve of a spring mount (Figure 2.11b), we have

$$\text{Mount stiffness} = \frac{3000^{-1}}{6\times 10^{-2}} = 50\times 10^3 \text{ N/m}.$$

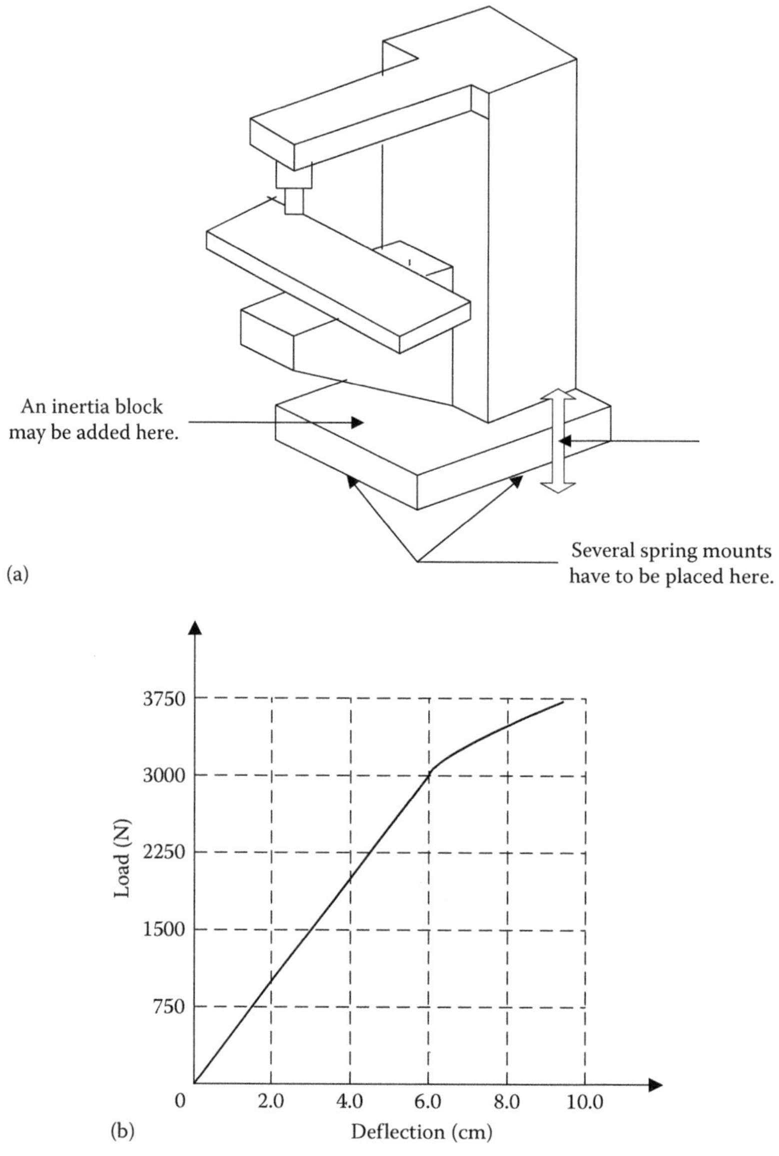

FIGURE 2.11
(a) A machine tool. (b) Load–deflection characteristic of a spring mount.

For a stable base, we will use four mounts. Then, $k = 4 \times 50 \times 10^3$ N/m
Hence,

$$\frac{m \times (10\pi)^2}{4 \times 50 \times 10^3} = 6.0 \rightarrow m = 1.216 \times 10^3 \text{ kg}.$$

Note that, according to this result, an inertia block of mass 216 kg has to be added.

Now, we must check whether the required level of vibration would be achieved in the damped case:

$$\text{Damping ratio } \zeta = \frac{b}{2\sqrt{km}} = \frac{4 \times 1.56 \times 10^3}{2\sqrt{4 \times 50 \times 10^3 \times 1.216 \times 10^3}} = 0.2.$$

Substitute in the damped isolator Equation 2.15a:

$$|T| = \sqrt{\frac{1 + 4\zeta^2 r^2}{(r^2 - 1)^2 + 4\zeta^2 r^2}} \quad \text{with } r^2 = 6.$$

We have

$$|T| = \sqrt{\frac{1 + 4 \times (0.2)^2 \times 6}{(6-1)^2 + 4 \times 0.2^2 \times 6}} = 0.27.$$

This corresponds to an isolation level of 73%, which is better than the required 70%.

2.3.5.1.2 Mechanical Transmission

Another application of component interconnection and impedance matching in mechanical system concerns speed transmission (gears, harmonic drives, lead-screw-nut devices, belt drives, rack-and-pinion devices, etc.). For a specific application, consider a mechanical load driven by a motor. Often, direct driving is not practical owing to the limitations of the speed–torque characteristics of the available motors. By including a suitable gear transmission between the motor and the load, it is possible to modify the speed–torque characteristics of the drive system as felt by the load. This is a process of component interconnection, interfacing, and impedance matching of mechanical systems. We will illustrate the application of these concepts using an example.

Example 2.4

Consider the mechanical system where a torque source (motor) of torque T and moment of inertia J_m is used to drive a purely inertial load of moment of inertia J_L, as shown in Figure 2.12a. What is the resulting angular acceleration $\ddot{\theta}$ of the system? Neglect the flexibility of the connecting shaft.

Now, suppose that the load is connected to the same torque source through an ideal (loss-free) gear of motor-to-load speed ratio r:1, as shown in Figure 2.12b. What is the resulting acceleration $\ddot{\theta}_g$ of the load?

Obtain an expression for the normalized load acceleration $a = \ddot{\theta}_g / \ddot{\theta}$ in terms of r and $p = J_L/J_m$. Sketch a versus r for $p = 0.1$, 1.0, and 10.0. Determine the value of r in terms of p that will maximize the load acceleration a.

Comment on the results obtained in this problem.

Solution

For the unit without the gear transmission, Newton's second law gives $(J_m + J_L)\ddot{\theta} = T$.

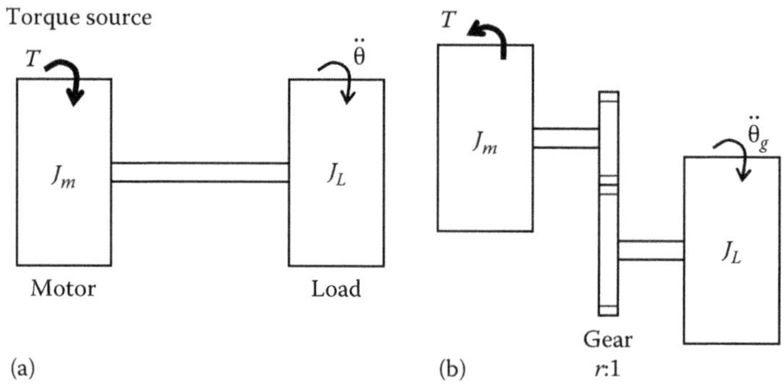

FIGURE 2.12
An inertial load driven by a motor: (a) without a gear transmission and (b) with a gear transmission.

Hence,

$$\ddot{\theta} = \frac{T}{J_m + J_L}. \tag{i}$$

For the unit with the gear transmission, see the free-body diagram shown in Figure 2.13, in the case of a loss-free (i.e., 100% efficient) gear transmission. Newton's second law gives

$$J_m r\ddot{\theta}_g = T - \frac{T_g}{r} \tag{ii}$$

and

$$J_L\ddot{\theta}_g = T_g. \tag{iii}$$

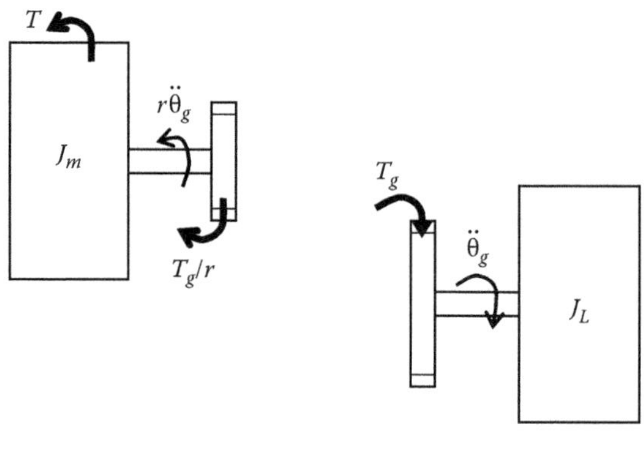

FIGURE 2.13
Free-body diagram.

where T_g is the gear torque on the load inertia. By eliminating T_g in Equations ii and iii, we get

$$\ddot{\theta}_g = \frac{rT}{(r^2 J_m + J_L)}. \tag{iv}$$

Divide Equation iv by Equation i:

$$\frac{\ddot{\theta}_g}{\ddot{\theta}} = a = \frac{r(J_m + J_L)}{(r^2 J_m + J_L)} = \frac{r(1 + J_L/J_m)}{(r^2 + J_L/J_m)}.$$

We get the *transmitted acceleration ratio*:

$$a = \frac{r(1+p)}{(r^2 + p)}, \tag{v}$$

where $p = J_L/J_m$.

From this, note that for $r = 0$, we have $a = 0$, and for $r \to \infty$, we have $a \to 0$. The peak value of a is obtained through differentiation:

$$\frac{\partial a}{\partial r} = \frac{(1+p)[(r^2+p) - r \times 2r]}{(r^2+p)^2} = 0.$$

We get, by taking the positive root,

$$r_p = \sqrt{p}, \tag{vi}$$

where r_p is the value of r corresponding to the peak of a. The peak value of a is obtained by substituting Equation vi in Equation v. Thus,

$$a_p = \frac{1+p}{2\sqrt{p}}. \tag{vii}$$

Also, note from Equation v that when $r = 1$, we have $a = r = 1$ for any value of p. Hence, all curves in Equation v will pass through the same point (1,1).

The relation (v) is sketched in Figure 2.14 for $p = 0.1$, 1.0, and 10.0. The peak values are tabulated in Table 2.2.

Note from Figure 2.14 that the transmission speed ratio can be chosen, depending on the inertia ratio, to maximize the load acceleration. Hence, the associated *impedance matching* (design) problem is as follows: For a specified (required) peak acceleration ratio a_p, select r and p using $a_p = (1+p)/(2\sqrt{p})$ and $r_p = \sqrt{p}$.
In particular, we can state the following:

1. When $J_L = J_m$, pick a direct-drive system (no gear transmission; i.e., $r = 1$).
2. When $J_L < J_m$, pick a speed-up gear at the peak value of $r = \sqrt{J_L/J_m}$.
3. When $J_L > J_m$, pick a speed-down gear at the peak value of r.

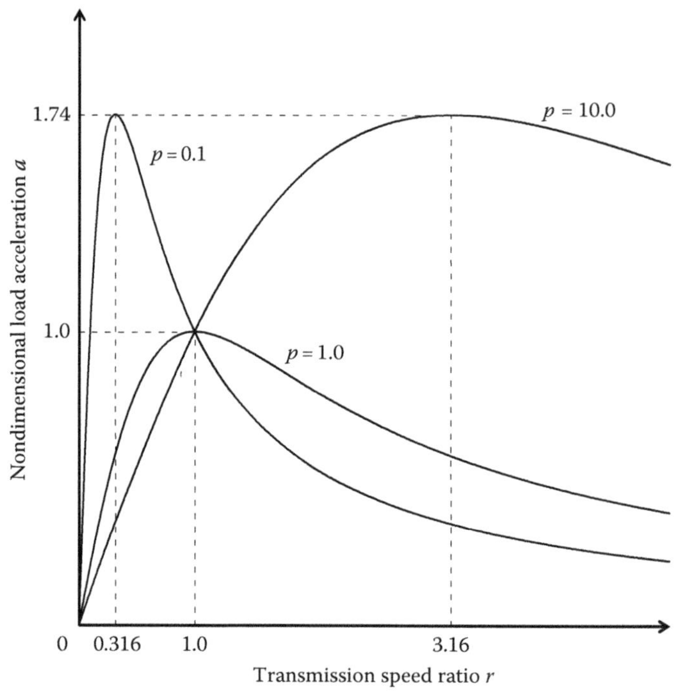

FIGURE 2.14
Normalized acceleration versus speed ratio.

TABLE 2.2
Peak Performance of Transmission System

p	r_p	a_p
0.1	0.316	1.74
1.0	1.0	1.0
10.0	3.16	1.74

Summary Sheet

Component interconnection: Due to dynamic interactions (coupling), the conditions in the components change after interconnection. Effects: change in signals, loading, etc.

Loading: The signals of the input component change undesirably when an output component is connected. Example of electrical loading: sensor that generates an electrical signal (input device) and signal acquisitioning hardware (output device). Example of mechanical loading: sensed object (input device) and a heavy sensor mounted on it (output device).

Component interconnection considerations: Impedance matching, signal conversion, and signal conditioning.

Impedance: Across-variable to through-variable; applicable in multiple domains (electrical, mechanical, thermal, fluid, etc.). *Note*: Electrical impedance is analogous to mechanical mobility (inverse of mechanical impedance).

Impedance matching: Match the impedances of the interconnected devices and add compensating impedances if necessary.

Impedance-matching categories: (1) Source and load matching for maximum power transfer, (2) power transfer at maximum efficiency, (3) reflection prevention in signal transmission, and (4) loading reduction.

Maximum power transfer: Uses conjugate matching $Z_l = Z_s^*$; $p_{l\max} = |V_s|^2/8R_s = |V_s|^2/8R_l$.

Power transfer at maximum efficiency: Efficiency $\eta = R_l/(R_l + R_s)$. Increased load impedance to increase efficiency.

Reflection coefficient: Reflected signal voltage/incident signal voltage; $\Gamma = |(Z_l - Z_c)/(Z_l + Z_c)|$, Z_i and Z_c are component impedances.

Reflection prevention: Compensates for change in impedance (e.g., use impedance pad) such that $(1/Z_i) + (1/Z_g) = 1/Z_c$.

Input impedance Z_i: Rated input voltage/corresponding current through input terminals; output terminals in open circuit.

Output impedance Z_o: Open-circuit (i.e., no load) voltage/short-circuit current, at output port.

Cascade connection: $v_o = 1/(Z_{o1}/Z_{i2} + 1)G_2G_1v_i$.

Loading error reduction: Make $Z_{o1}/Z_{i2} \ll 1$.

Impedance matching in mechanical systems: Examples vibration isolation and speed conversion (gears, etc.).

Force transmissibility T_f: Transmitted force/applied force.

Motion transmissibility T_m: Transmitted motion/applied motion.

Transmissibility: $T = T_f = T_m = (Z_s/(Z_s + Z_m)) = (M_m/(M_m + M_s))$.

Simple oscillator model:

$$T_f = \frac{\omega_n^2 + 2\zeta\omega_n j\omega}{\omega_n^2 - \omega^2 + 2\zeta\omega_n j\omega} = \frac{1 + 2\zeta rj}{1 - r^2 + 2\zeta rj};$$

$r = \omega/\omega_n$; $\omega_n = \sqrt{k/m}$ = undamped natural frequency; $\zeta = b/(2\sqrt{km})$ = damping ratio.

Transmissibility magnitude:

$$|T| = \sqrt{\frac{1 + 4\zeta^2 r^2}{(1 - r^2)^2 + 4\zeta^2 r^2}} \approx \frac{1}{r^2 - 1} \quad \text{for low } \zeta.$$

Vibration isolation: $I = [1 - |T|] \times 100\%$.

Isolator design: Specified I. Increase it by 10%; at lowest operating frequency (speed), determine r using approximate (low damping) formula for I; hence, $\omega_n = \sqrt{k/m}$. Pick k

(vibration mount stiffness) and m (inertia block). Check for the damped case using the full formula for I.

Mechanical transmission: Transmitted acceleration ratio $a = r(1+p)/(r^2+p)$; motor-to-load speed ratio r:1; inertia ratio $p = J_L/J_m$; peak a, $a_p = (1+p)/2\sqrt{p}$ occurs at $r_p = \sqrt{p}$.

Design problem: Select r and p for a specified (required) a_p.

Problems

2.1 (a) Define electrical impedance and mechanical impedance. (b) Identify a defect in this nomenclature in relation to the force–current analogy. (c) What improvements would you suggest? (d) What roles do input impedance and output impedance play in relation to the accuracy of a measuring device?

2.2 List four reasons why impedance matching is important in component interconnection.

2.3 What is meant by loading error in a signal measurement? Also, suppose that a piezoelectric sensor of output impedance Z_s is connected to a voltage-follower amplifier of input impedance Z_i, as shown in Figure P2.3. The sensor signal is v_i volts, and the amplifier output is v_o volts. The amplifier output is connected to a device with very high-input impedance. Plot to scale the signal ratio v_o/v_i against the impedance ratio Z_i/Z_s for values of the impedance ratio in the range 0.1–10.

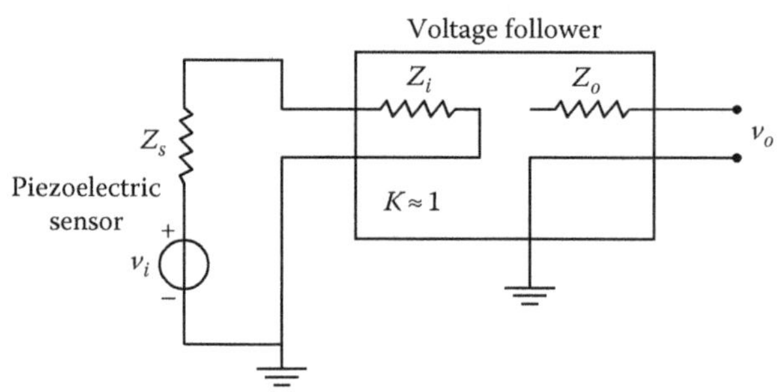

FIGURE P2.3
System with a piezoelectric sensor.

2.4 Thevenin's theorem states that with respect to the characteristics at an output port, an unknown subsystem consisting of linear passive elements and ideal source elements may be represented by a single across-variable (voltage) source v_{eq} connected in series with a single impedance Z_{eq}. This is illustrated in Figure P2.4a and b. Note in Figure P2.4b that v_{eq} is equal to the open-circuit across variable v_{oc} at the output port, because the current through Z_{eq} is zero. Consider the circuit shown in Figure P2.4c. Determine the equivalent source voltage v_{eq} and the equivalent series impedance Z_{eq}, in the frequency domain, for this circuit.

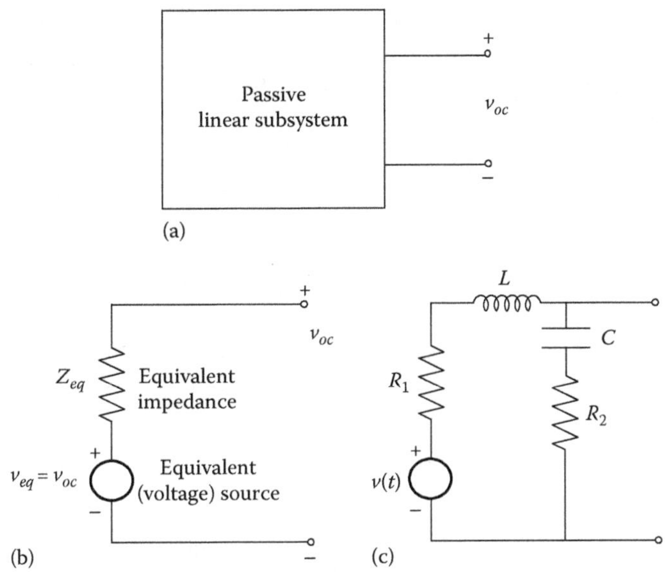

FIGURE P2.4
Illustration of Thevenin's theorem. (a) Unknown linear subsystem. (b) Equivalent representation. (c) Example.

2.5 For the circuit with resistive source and load circuit, shown in Figure 2.2, plot the curves of (a) load power efficiency and (b) ratio of (load power)/(maximum load power) against the ratio (load resistance)/(source resistance). Comment on the results.

2.6 Explain why a voltmeter should have a high resistance and an ammeter should have a very low resistance. What are some of the design implications of these general requirements for the two types of measuring instruments, particularly with respect to instrument sensitivity, speed of response, and robustness? Use a classical moving-coil galvanometer as the model for your discussion. *Note*: Galvanometers are currently not used in measuring electrical signals. Instead they are used in positioning and motion control applications.

2.7 Indicate a suitable impedance for the connected component in the following two applications:

(a) A pH sensor of output impedance 10 MΩ is connected to a conditioning amplifier.
(b) A power amplifier of output impedance 0.1 Ω is connected to a passive speaker.

In each case, estimate a possible percentage error in the transmitted signal.

2.8 A two-port nonlinear device is shown schematically in Figure P2.8. The transfer relations under static equilibrium (i.e., steady state) conditions are given by

$$v_o = F_1(f_o, f_i),$$

$$v_i = F_2(f_o, f_i),$$

where

v denotes an across variable
f denotes a through variable
o and i are the subscripts that represent the output port and the input port, respectively

Obtain expressions for the input impedance and the output impedance of the system in the neighborhood of an operating point, under static conditions, in terms of partial derivatives of the functions F_1 and F_2. Explain how these impedances could be determined experimentally.

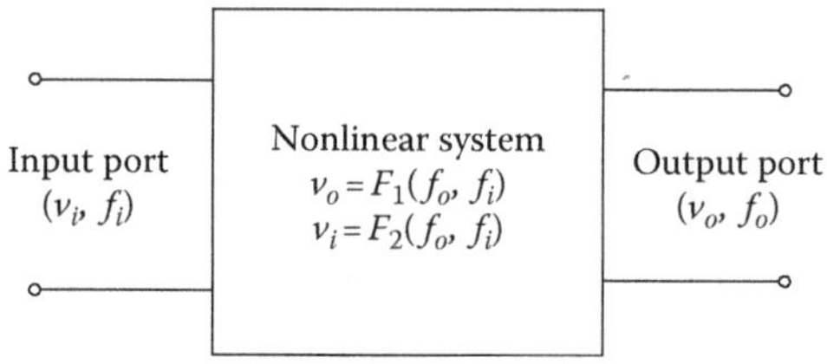

FIGURE P2.8
Impedance characteristics of a nonlinear system.

2.9 A signal is transmitted through a cable of impedance Z_c and transmitted through an antenna of impedance Z_l (see Figure P2.9).

(a) Show that $v_t = (2Z_l/(Z_l + Z_c))v_i$, where v_i is the voltage of the incident signal at the cable–antenna interface and v_t is the voltage of the signal that is transmitted from the cable to the antenna.

(b) What is the required relationship between Z_l and Z_c for proper impedance matching in this example?

(c) One method of impedance matching in this application is by using an impedance pad at the antenna connection. Suggest another method.

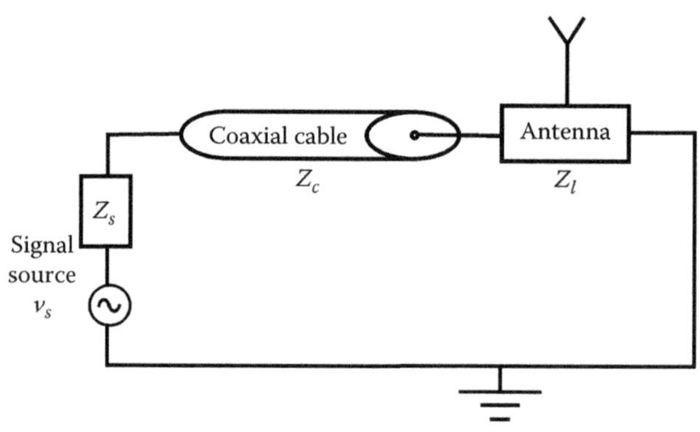

FIGURE P2.9
Impedance matching of transmission cable and antenna.

2.10 A machine of mass m has a rotating device, which generates a harmonic forcing excitation $f(t)$ in the vertical direction. The machine is mounted on the factory floor using a vibration isolator of stiffness k and damping constant b. The harmonic component of the force that is transmitted to the floor, due to the forcing excitation, is $f_s(t)$. A simplified model of the system is shown in Figure P2.10. The corresponding force transmissibility magnitude $|T_f|$ from f to f_s is given by

$$\left|T_f\right| = \sqrt{\frac{1+4\zeta^2r^2}{(1-r^2)^2+4\zeta^2r^2}},$$

where
$r = \omega/\omega_n$
ζ is the damping ratio
ω_n is the undamped natural frequency of the system
ω is the excitation frequency (of $f(t)$)

Suppose that $m = 100$ kg and $k = 1.0 \times 10^6$ N/m. Also, the frequency of the excitation force $f(t)$ in the operating range of the machine is known to be 200 rad/s or higher. Determine the damping constant b of the vibration isolator so that the force transmissibility magnitude does not exceed 0.5.

Using MATLAB, plot the resulting transmissibility function and verify that the design requirements are met.

Note: 2.0 = 6 dB; $\sqrt{2}$ = 3 dB; $1/\sqrt{2}$ = −3 dB; and 0.5 = −6 dB.

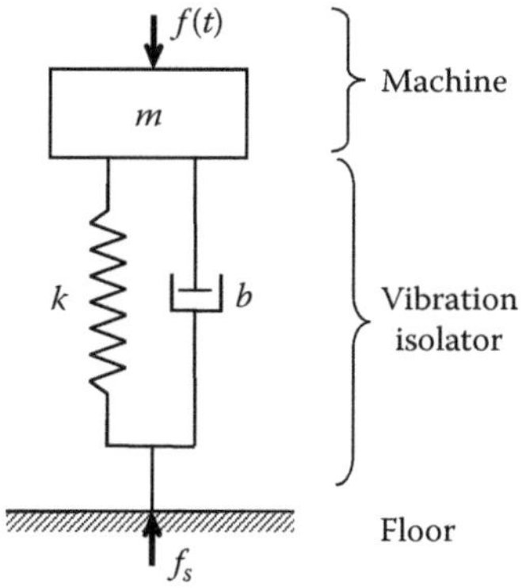

FIGURE P2.10
Model of a machine mounted on a vibration isolator.

3

Amplifiers and Filters

Chapter Objectives

- Signal Modification and Conditioning
- Operational Amplifier
- Amplifier Performance Ratings
- Voltage, Current, and Power Amplifiers
- Instrumentation Amplifiers
- Noise and Ground Loops
- Analog Filters
- Passive Filters and Active Filters
- Low-Pass Filters
- High-Pass Filters
- Band-Pass Filters
- Band-Reject Filters
- Digital Filters

3.1 Signal Modification and Conditioning

An engineering system typically consists of a wide variety of components that are interconnected to perform the intended functions. When two devices are interfaced, it is essential to guarantee that the resulting changes to the dynamic conditions will not lead to unacceptable performance. In particular, a signal leaving one device and entering the other must do so at proper signal levels (i.e., the values of voltage, current, speed, force, and power), in the proper form (i.e., electrical, mechanical, analog, digital, modulated, demodulated) and without distortion, as required by the specific application. In particular, disturbance and noise have to be eliminated. It is clear that considerations of *signal conditioning* are important in *instrumentation* of an engineering system. Signal amplification and filtering are applicable in this context, and this is the focus of this chapter. Signal conditioning is a particular case of signal modification.

3.1.1 Signal Conditioning

Signal conditioning and signal processing fall within the subject of signal modification, which plays a crucial role in component interconnection or interfacing and proper operation of the system. Specifically,

1. For proper operation of a system, the operating signals may have to be modified (with regard to power, type, magnitude, etc.). Amplifiers are useful here.
2. In view of noise, disturbances and other errors, and system requirements, the signals have to be conditioned. Filtering and amplification are relevant in this context.

Signal conditioning includes amplification and analog and digital filtering. For example, for transmission, a signal should be properly modified by amplification, filtering, and so on so that the signal–noise ratio of the transmitted signal is sufficiently large at the receiver. In general, signal conditioning is important in the context of component interconnection and integration due to the presence of noise and unknown/unwanted disturbances and errors in the associated signals. Hence, signal conditioning is an important subject in the study of instrumentation.

3.1.2 Chapter Objectives

This chapter addresses signal-conditioning circuitry, particularly related to amplification and filtering. This chapter describes pertinent signal-conditioning operations. The operational amplifier (op-amp) is introduced as a basic element in signal-conditioning and impedance-matching circuitry for electronic systems. Various types of signal-conditioning devices, particularly common types of amplifiers and filters, are analyzed, their properties are given, pertinent performance ratings are presented, and the applications of amplifiers and filters are indicated. Specific hardware components and designs are considered as examples of signal conditioning.

3.2 Amplifiers

Voltages, velocities, pressures, and temperatures are "across variables" since they are present across an element. Currents, forces, fluid flow rates, and heat transfer rates are "through variables" since they transmit through an element, unaltered. The level of an electrical signal can be represented by variables such as voltage, current, and power. Analogous across variables, through variables, and power variables can be defined for other types of signals (e.g., mechanical variables such as velocity, force, and power) as well. Signal levels at various interface locations of components in an engineering system have to be properly adjusted for a satisfactory performance of these components and of the overall system. For example, input to an actuator should possess adequate power to drive the actuator and its load. A signal should maintain its signal level above some threshold during transmission so that errors due to signal weakening would not be excessive. Signals applied to digital devices must remain within the specified logic levels. Many types of sensors produce weak signals, which have to be upgraded before they could be fed into a monitoring system, data processor, controller, or data logger.

Signal amplification concerns proper adjustment of the signal level for performing a specific task. Amplifiers are used to accomplish signal amplification. An amplifier is an *active* device, as it needs an external power source to operate. Even though various active circuits, amplifiers in particular, are commonly produced in the monolithic form using an original integrated circuit (IC) layout to accomplish a particular amplification task, it is convenient to study their performance using discrete circuit models, with the op-amp as the basic building block. Of course, op-amps are themselves available as monolithic IC packages. They are widely used as the basic building blocks in producing other types of amplifiers and numerous other hardware and, in turn, for modeling and analyzing these various kinds of amplifiers and devices. For these reasons, our discussion on amplifiers will build upon the op-amp.

3.2.1 Operational Amplifier

The origin of the op-amp dates back to the 1940s when the vacuum tube op-amp was introduced. Op-amp got its name because originally it was used almost exclusively to perform mathematical operations, for example, in analog computers. Subsequently, in the 1950s, the transistorized op-amp was developed. It used discrete elements such as bipolar junction transistors and resistors. Still it was too large, relatively slow, consumed too much power, and was too expensive for widespread use in general applications. This situation changed in the late 1960s when the IC op-amp was developed in the monolithic form as a single IC chip. Today, the IC op-amp, which consists of a large number of circuit elements on a substrate of typically a single silicon crystal (the monolithic form), is a valuable component in virtually all electronic signal modification devices. Bipolar complementary metal–oxide–semiconductor op-amps in various plastic packages and pin configurations are commonly available.

An op-amp could be manufactured in the discrete-element form using, say, 10 bipolar junction transistors and as many discrete resistors or alternatively (and preferably) in the modern monolithic form as an IC chip that may be equivalent to over 100 discrete elements. In any form, the device has an *input impedance* Z_i, an *output impedance* Z_o, and an *open-loop gain* K. Hence, a schematic model for an op-amp can be given as in Figure 3.1a. Op-amp packages are available in several forms. Very common is the eight-pin dual in-line package or V package, as shown in Figure 3.1b. The assignment of the pins (i.e., pin configuration or pin out) is as shown in the figure, which should be compared with Figure 3.1a. Note the counterclockwise numbering sequence starting with the top left pin next to the semicircular notch (or dot). This convention of numbering is standard for any type of IC package, not just op-amp packages. Other packages include 8-pin metal-can package or T package, which has a circular shape instead of the rectangular shape of the previous package, and the 14-pin rectangular "quad" package, which contains four op-amps (with a total of 8 input pins, 4 output pins, and 2 power supply pins). The conventional symbol of an op-amp is shown in Figure 3.1c. Typically, there are five terminals (pins or lead connections) to an op-amp. Specifically, there are two input ("differential") leads (a positive or noninverting lead with voltage v_{ip} and a negative or inverting lead with voltage v_{in}), an output lead (voltage v_o), and two bipolar power supply leads ($+v_s$ or v_{CC} or collector supply and $-v_s$ or v_{EE} or emitter supply). A "chip select" pin may be available in some op-amps. The supply voltage can be as low as 2.7 V and as high as ±22 V and quiescent current of about 250 μA. Normally, some of the pins may not be connected, for example, pins 1, 5, and 8 in Figure 3.1b.

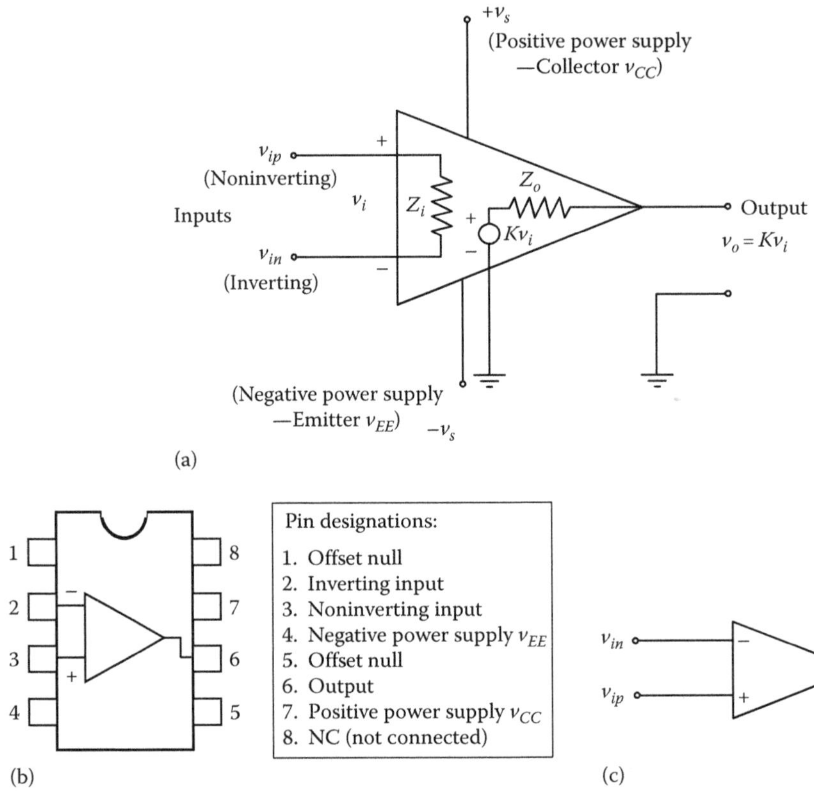

FIGURE 3.1
Operational amplifier. (a) A schematic model, (b) eight-pin dual in-line package (DIP), and (c) conventional circuit symbol.

Note: IC packages with multiple op-amps and correspondingly more pins (e.g., quad package with four op-amps and 14 pins: 8 differential input pins, 4 differential output pins, and 2 power supply pins) are commercially available.

Differential input voltage: From Figure 3.1a, under open-loop (i.e., no feedback) conditions, we have

$$v_o = Kv_i \tag{3.1}$$

Here, the input voltage v_i is the differential input voltage, which is defined as the algebraic difference between the voltages at the positive and negative leads of the op-amp. Thus,

$$v_i = v_{ip} - v_{in} \tag{3.2}$$

The open-loop voltage gain K is very high (10^5–10^9) for a typical op-amp. Furthermore, the input impedance Z_i could be as high as 10 MΩ (typical is 2 MΩ), and the output impedance is low, of the order 10 Ω, and may reach about 75 Ω for some op-amps. Since v_o is typically 1–15 V, from Equation 3.1, it follows that $v_i \cong 0$ since K is very large. Hence, from Equation 3.2, we have $v_{ip} \cong v_{in}$. In other words, the voltages at the two input leads are nearly equal. Now, if we apply a large voltage differential v_i (say, 10 V) at the input, then according to Equation 3.1, the output voltage should be extremely high. This never happens in practice, however, since the device *saturates* quickly beyond moderate output voltages (of the order 15 V).

From Equations 3.1 and 3.2, it is clear that if the negative input lead is grounded (i.e., $v_{in} = 0$), then

$$v_o = Kv_{ip} \tag{3.3a}$$

and if the positive input lead is grounded (i.e., $v_{ip} = 0$),

$$v_o = -Kv_{in} \tag{3.3b}$$

This is the reason why v_{ip} is termed *noninverting* input and v_{in} is termed *inverting* input.

Example 3.1

Consider an op-amp with an open-loop gain of 1×10^5. If the saturation voltage is 15 V, determine the output voltage in the following cases:

(a) 5 μV at the positive lead and 2 μV at the negative lead
(b) −5 μV at the positive lead and 2 μV at the negative lead
(c) 5 μV at the positive lead and −2 μV at the negative lead
(d) −5 μV at the positive lead and −2 μV at the negative lead
(e) 1 V at the positive lead and the grounded negative lead
(f) 1 V at the negative lead and the grounded positive lead

Solution

This problem can be solved using Equations 3.1 and 3.2. The results are given in Table 3.1. Note that in the last two cases the output will saturate and Equation 3.1 will no longer hold.

Field-effect transistors (FETs) and metal–oxide–semiconductor FETs (MOSFETs), for instance, are commonly used in the IC form of an op-amp. The MOSFET type has advantages over many other types, for example, higher input impedance and more stable output (almost equal to the power supply voltage) at saturation, making the MOSFET op-amps preferable over bipolar junction transistor op-amps in many applications.

TABLE 3.1

Solution to Example 3.1

v_{ip}	v_{in}	v_i	v_o
5 μV	2 μV	3 μV	0.3 V
−5 μV	2 μV	−7 μV	−0.7 V
5 μV	−2 μV	7 μV	0.7 V
−5 μV	−2 μV	−3 μV	−0.3 V
1 V	0	1 V	15 V
0	1 V	−1 V	−15 V

In analyzing op-amp circuits under unsaturated conditions, we use the following two characteristics of an op-amp:

1. Voltages of the two input leads should be (almost) equal.
2. Currents through each of the two input leads should be (almost) zero.

As explained earlier, the first property is credited to high open-loop gain and the second property to high input impedance in an op-amp. We will repeatedly use these two properties to obtain input–output equations for amplifier systems and other electronic devices that use op-amps.

3.2.2 Amplifier Performance Ratings

Many factors affect the performance of an amplifier, particularly an op-amp. For good performance, we need to consider such factors as

1. Stability
2. Speed of response (bandwidth, slew rate)
3. Input impedance and output impedance

The level of stability of an amplifier, in the conventional sense, is governed by the dynamics of the amplifier circuitry and may be represented by a time constant. In this context, if the negative feedback loop of an amplifier has a unity gain and a phase shift of 2π, then *sustained oscillation* will be generated. This is a condition of instability (or marginal stability). Another important consideration for an amplifier is the parameter variation due to aging, temperature, and other environmental factors. Parameter variation is also classified as a stability issue, in the context of devices such as amplifiers, because it pertains to the steadiness of the response when the input is maintained steady. Of particular importance in this context is the *temperature drift*. This may be specified as the drift in the output signal per unity change in temperature. Temperature drift also depends on the offset voltage of the amplifier (e.g., 3.6 μV/°C per 1.0 mV of offset voltage). Temperature drift can be reduced through compensating devices and by reducing the current draw in the amplifier circuit.

The speed of response of an amplifier dictates the ability of the amplifier to faithfully respond to transient inputs. In particular, we seek high speed of response. Conventional time-domain parameters such as rise time may be used to represent this. Alternatively, in the frequency domain, speed of response may be represented by a *bandwidth* parameter. For example, the frequency range over which the frequency response function is considered constant (flat) may be taken as a measure of bandwidth. Since there is some nonlinearity in any amplifier, bandwidth can depend on the signal level itself. Specifically, small-signal bandwidth refers to the bandwidth that is determined using small input signal amplitudes.

With regard to op-amps, another measure of the speed of response is the *slew rate*, which is defined as the largest possible rate of change of the amplifier output for a particular frequency of operation. Since for a given input amplitude, the output amplitude depends on the amplifier gain, slew rate is usually defined for unity gain.

Ideally, for a linear device, the frequency response function (transfer function) does not depend on the output amplitude (i.e., the product of the DC gain and the input amplitude). But for a device that has a limited slew rate, the bandwidth (or the maximum operating frequency at which output distortions may be neglected) will depend on the output amplitude.

The larger the output amplitude, the smaller the bandwidth for a given slew rate limit. A bandwidth parameter that is usually specified for a commercial op-amp is the gain-bandwidth product (GBP or GBWP). This is the product of the open-loop gain and the bandwidth of the op-amp. For example, for an op-amp with GBP = 15 MHz and an open-loop gain of 100 dB (i.e., 10^5), the bandwidth is $15 \times 10^6/10^5$ Hz = 150 Hz. Clearly, this bandwidth value is rather low. Since the gain of an op-amp with feedback is significantly lower than 100 dB, its effective bandwidth is much higher than that of an open-loop op-amp.

As discussed, generally, we wish to have a high input impedance and low output impedance. These requirements are generally satisfied by an op-amp.

Example 3.2

Obtain a relationship between the slew rate and the bandwidth for a slew rate–limited device. An amplifier has a slew rate of 1 V/μs. Determine the bandwidth of this amplifier when operating at an output amplitude of 5 V.

Solution

Clearly, the amplitude of the rate of change signal divided by the amplitude of the output signal gives an estimate of the output frequency. Consider a sinusoidal output voltage given by

$$v_o = a \sin 2\pi ft \tag{3.4}$$

The rate of change of output is $dv_o/dt = 2\pi fa \cos 2\pi ft$. Hence, the maximum rate of change of output is $2\pi fa$. Since this corresponds to the slew rate when f is the maximum allowable frequency, we have

$$s = 2\pi f_b a, \tag{3.5}$$

where
s is the slew rate
f_b is the bandwidth
a is the output amplitude

Now, with $s = 1$ V/μs and $a = 5$ V, we get $f_b = (1/2\pi) \times (1/(1 \times 10^{-8})) \times (1/5)$ Hz = 31.8 kHz.

3.2.2.1 Sources of Error in Op-Amps

Stability problems and frequency response errors are prevalent in the open-loop form of an op-amp. These problems can be eliminated using feedback, as will be discussed, because the effect of the open-loop transfer function on the closed-loop transfer function is negligible if the open-loop gain is very large, which is the case for an op-amp.

Unmodeled signals can be a major source of amplifier error, and these signals include

1. Bias currents
2. Offset signals
3. Common-mode output voltage
4. Internal noise

In analyzing op-amps, we assume that the current through the input leads is zero. This is not strictly true because bias currents for the transistors within the amplifier circuit have to flow through these leads. As a result, the output signal of the amplifier will deviate slightly from the ideal value.

Another assumption that we make in analyzing op-amps is that the voltage is equal at the two input leads. In practice, however, offset currents and voltages are present at the input leads, due to minute discrepancies inherent to the internal circuits within an op-amp.

The common-mode input voltage is the voltage at the input leads of the op-amp. Since any practical amplifier has some imbalances in the internal circuitry (e.g., the gain with respect to one input lead may not be equal to the gain with respect to the other input lead, and furthermore, bias signals are needed for operation of the internal circuitry), there will be an error voltage at the output, which depends on the common-mode input.

The three types of unmodeled signals mentioned earlier can be considered as noise. In addition, there are other types of noise signals that degrade the performance of an amplifier. For example, ground-loop noise can enter the output signal. Furthermore, stray capacitances and other types of unmodeled circuit effects can generate internal noise. Usually in amplifier analysis, unmodeled signals (including noise) can be represented by a noise voltage source at one of the input leads. The effects of unmodeled signals can be reduced by using suitably connected compensating circuitry, including variable resistors that can be adjusted to eliminate the effect of unmodeled signals at the amplifier output.

Useful terminology concerning the performance of an op-amp is summarized in the following:

Bandwidth: Operating frequency range of an op-amp (e.g., 56 MHz).

GBWP: The product of DC gain and bandwidth. It is expressed in MHz (because gain has no units). Typically, as the gain increases, the bandwidth decreases.

Slew rate: Maximum possible rate of change of the output voltage, without significantly distorting the output. It is expressed in V/μs (e.g., 160 V/μs). A high slew rate is desirable.

Differential input impedance: Impedance *between* the two inputs (noninverting and inverting) of the amplifier.

Common-mode input impedance: Impedance from each input to ground.

Offset-null pin: This is also called "balance" pin. It is used to remove the offset voltage. Offset voltage is the output voltage (due to imperfection in the op-amp) that appears when the noninverting and inverting voltages are equal (ideally, the corresponding output voltage should be zero). Some op-amps have a way to automatically remove this offset.

Input offset voltage: The voltage difference that is needed at the two input terminals (for imperfect op-amps) to make the output voltage zero.

Equation 3.1 applies for a perfect op-amp, and the gain K given there is the differential gain. Strictly, for an op-amp that is not perfect, the correct version of Equation 3.1 is

$$v_o = K_d(v_{ip} - v_{in}) + K_{cm} \times \frac{1}{2}(v_{ip} + v_{in}) \tag{3.6}$$

where

K_d is the differential gain

K_{cm} is the common-mode gain

Common-mode voltage: This is the average voltage at the two input leads, as given by $(1/2)(v_{ip} + v_{in})$ in Equation 3.6. In a good op-amp, this should not be amplified and should not be present at the output. It should be rejected, which can be realized by having a very low common-mode gain compared to the differential gain.

Common-mode rejection ratio (*CMRR*): This is the ratio of the differential gain to the common-mode gain (K_d/K_{cm}), expressed in decibels (i.e., 20 $\log_{10}$ (K_d/K_{cm}) dB). It should be rather high to assure proper rejection of the common-mode voltage (e.g., 113 dB).

Quiescent current: Current drawn by the op-amp from its power source when there is no load (i.e., output is in open circuit) and there are no signals at the input leads.

Some useful information about op-amps is summarized in Box 3.1.

3.2.2.2 Use of Feedback in Op-Amps

Op-amp is a very versatile device, primarily owing to its very high input impedance, low output impedance, and very high gain. However, it cannot be used as a practical amplifier without modification because it is not very stable in the form shown in Figure 3.1. The two main factors that contribute to this problem are *frequency response* and *drift*. Stated in another way, op-amp gain K is very high, and furthermore, it does not remain constant. It can vary with the frequency of the input signal (i.e., the frequency response function of an op-amp is not flat in the operating range) and also with time (i.e., there is drift). Because the gain is very high, a moderate input signal will saturate the op-amp. The frequency response problem arises because of circuit dynamics of an op-amp. This problem is usually not severe unless the device is operated at very high frequencies. The drift problem arises as a result of the sensitivity of gain K to environmental factors such as temperature, light, humidity, and vibration and also as a result of the variation of K due to aging. Drift in an op-amp can be significant, and steps should be taken to eliminate that problem.

It is virtually impossible to avoid the drift in gain and the frequency response error in an op-amp. However, an ingenious way has been found to remove the effect of these two problems at the amplifier output. Since gain K is very large, by using feedback, we can virtually eliminate its effect at the amplifier output. This closed-loop form of an op-amp has the advantage that the characteristics and the accuracy of the output of the overall circuit depend on the passive components (e.g., resistors and capacitors) in it, which can be provided at high precision, and not on the parameter values of the op-amp itself. The closed-loop form is preferred in almost every application; in particular, voltage follower and charge amplifier are devices that use the properties of high Z_i, low Z_o, and high K of an op-amp along with feedback through a high-precision resistor, to eliminate errors due to large and variable K. In summary, op-amp is not very useful in its open-loop form, particularly because gain K is very large and variable. However, it is because K is very large that the mentioned problems can be removed by using *feedback*. It is this closed-loop form that is commonly used in practical applications of an op-amp.

In addition to the large and unsteady nature of gain, there are other sources of error that contribute to the less-than-ideal performance of an op-amp circuit. As mentioned earlier, noteworthy are

BOX 3.1 OPERATIONAL AMPLIFIERS

IDEAL OP-AMP PROPERTIES

- Infinite open-loop differential gain
- Infinite input impedance
- Zero output impedance
- Infinite bandwidth
- Zero output for zero differential input

IDEAL ANALYSIS ASSUMPTIONS

- Voltages at the two input leads are equal.
- Current through either input lead is zero.

DEFINITIONS

- Open-loop gain $= \left|\dfrac{\text{Output voltage}}{\text{Voltage difference at input leads}}\right|$, with no feedback.
- Input impedance $= \dfrac{\text{Voltage between an input lead and ground}}{\text{Current through that lead}}$, with other input lead grounded and the output in open circuit.
- Output impedance $= \dfrac{\text{Voltage between output lead and ground in open circuit}}{\text{Current through that lead}}$, with normal input conditions.
- *Bandwidth*: Frequency range in which the frequency response is flat (gain is constant).
- GBWP = Open-loop gain × bandwidth at that gain.
- *Input bias current*: Average DC current through one input lead.
- *Input offset current*: Difference in the two input bias currents.
- *Differential input voltage*: Voltage at one input lead with the other grounded when the output voltage is zero.
- Common-mode gain $= \dfrac{\text{Output voltage when input leads are at the same voltage}}{\text{Common input voltage}}$
- CMRR $= \dfrac{\text{Open-loop differential gain}}{\text{Common-mode gain}}$
- *Slew rate*: Rate of change of output of a unity-gain op-amp, for a step input.

1. The offset current present at the input leads due to bias currents that are needed to operate the solid-state circuitry
2. The offset voltage that might be present at the output even when the input leads are open
3. The unequal gains corresponding to the two input leads (i.e., the inverting gain not equal to the noninverting gain)
4. Noise and environmental effects (thermal drift, etc.)

Such problems can produce nonlinear behavior in op-amp circuits. They can be reduced, however, by proper circuit design and through the use of compensating circuit elements.

3.2.3 Voltage, Current, and Power Amplifiers

Any type of amplifier can be constructed from scratch in the monolithic form as an IC chip or in the discrete form as a circuit containing several discrete elements such as discrete bipolar junction transistors or discrete FETs, discrete diodes, and discrete resistors. But almost all types of amplifiers can also be built using the op-amp as the basic building block. Since we are already familiar with op-amps and since op-amps are extensively used in electronic amplifier circuitry, we will use the latter approach, which uses discrete op-amps for building general amplifiers. Also, modeling, analysis, and design of a general amplifier may be performed on this basis.

If an electronic amplifier performs a voltage amplification function, it is termed a *voltage amplifier*. These amplifiers are so common that the term *amplifier* is often used to denote a voltage amplifier. A voltage amplifier can be modeled as

$$v_o = K_v v_i \tag{3.7}$$

where

v_o is the output voltage
v_i is the input voltage
K_v is the *voltage gain*

Voltage amplifiers are commonly used to achieve voltage compatibility (or *level shifting*) in circuits.

Similarly, *current amplifiers* are used to achieve current compatibility in electronic circuits. A current amplifier may be modeled by

$$i_o = K_i i_i \tag{3.8}$$

where

i_o is the output current
i_i is the input current
K_i is the *current gain*

A *voltage follower* has a unity gain; $K_v = 1$. Hence, it may be considered as a current amplifier. Besides, it provides impedance compatibility and acts as a buffer between a low-current (high impedance) output device (signal source or the device that provides the signal) and an input device (signal receiver or the device that receives the signal) of high current (low impedance) that are interconnected. Hence, the name *buffer amplifier* or *impedance transformer* is sometimes used for a current amplifier with unity voltage gain.

If the objective of signal amplification is to upgrade the associated power level, then a *power amplifier* should be used for that purpose. A simple model for a power amplifier is

$$p_o = K_p p_i \tag{3.9}$$

where
p_o is the output power
p_i is the input power
K_p is the *power gain*

It is easy to see from Equations 3.7 through 3.9 that

$$K_p = K_v K_i \tag{3.10}$$

All three types of amplification could be achieved simultaneously from the same amplifier. Furthermore, a current amplifier with unity voltage gain (e.g., a voltage follower) is a power amplifier. Usually, voltage amplifiers and current amplifiers are used in the first stages of a signal path (e.g., sensing, data acquisition, and signal generation), where signal levels and power levels are relatively low, while power amplifiers are typically used in the final stages (e.g., final control, actuation, recording, long-distance communication, display), where high signal levels and power levels are usually required.

In deriving the equations for any op-amp implementation of a practical device, we use the following two main properties of the op-amp:

1. Voltages at the two input leads (inverting and noninverting) are equal (due to high differential gain).
2. Currents at each input lead are zero (due to high input impedance).

We will use these characteristics repeatedly in the following derivations of the equations for practical amplifiers.

Figure 3.2a gives an op-amp circuit for a voltage amplifier. Note the feedback resistor R_f that serves the purposes of stabilizing the op-amp and providing an accurate voltage gain. The positive lead is grounded, and the input voltage is applied to the negative lead, through an accurately known resistor R, whose value is chosen as needed. The output is fed back to the negative lead through the feedback resistor R_f, whose value is also precisely chosen as needed. To determine the voltage gain, remember that the voltages at the two input leads of an op-amp should be equal (in the ideal case). Since the +ve lead is grounded, the voltage at point A is also zero. Next, recall that, ideally, the current through the input lead of an op-amp is zero, and write the current balance equation for the node point A:

$$\frac{v_i}{R} + \frac{v_o}{R_f} = 0$$

This gives the amplifier equation (3.7)

$$v_o = \left(1 + \frac{R_f}{R}\right) v_i \tag{3.11}$$

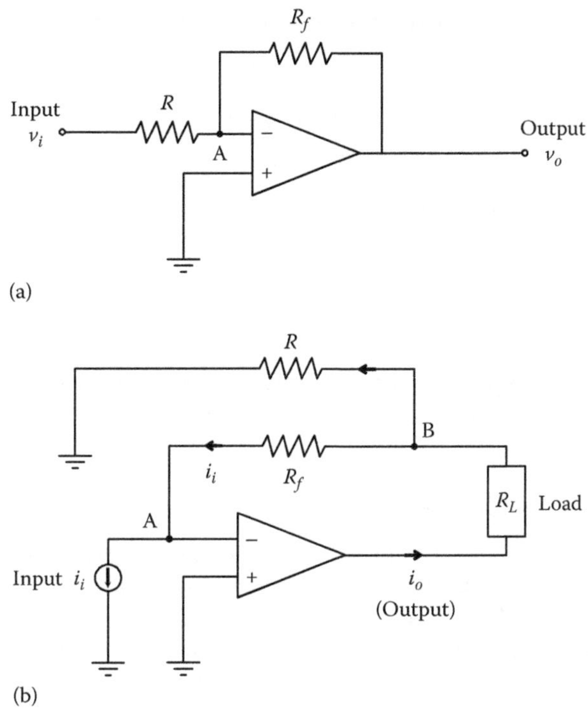

FIGURE 3.2
(a) A voltage amplifier and (b) a current amplifier.

Hence, the voltage gain is given by

$$K_v = -\frac{R_f}{R} \tag{3.12}$$

Note: We can disregard the –ve sign in the gain because it can be changed by simply reversing the terminals of the input to the application. Also, note that K_v depends on R and R_f and not on the op-amp gain. Hence, the voltage gain can be accurately determined by selecting the two passive elements (resistors) R and R_f precisely. Also, the output voltage has the same sign as the input voltage. Hence, this is a *noninverting amplifier*. If the voltages are of the opposite sign, we have an *inverting amplifier*.

A current amplifier is shown in Figure 3.2b. The input current i_i is applied to the negative lead of the op-amp as shown, and the positive lead is grounded. There is a feedback resistor R_f connected to the negative lead through the load R_L. The resistor R_f provides a path for the input current since the op-amp takes in virtually zero current. There is a second resistor R through which the output is grounded. This resistor is needed for current amplification. To analyze the amplifier, use the fact that the voltage at point A (i.e., at the negative lead) should be zero because the positive lead of the op-amp is grounded (zero voltage). Furthermore, the entire input current i_i passes through the resistor R_f as shown. Hence, the voltage at point B is $R_f i_i$. Consequently, current through the resistor R is $R_f i_i/R$, which is positive in the direction shown. It follows that the output current i_o is given by $i_o = i_i + (R_f/R)i_i$ or

$$i_o = \left(1 + \frac{R_f}{R}\right) i_i \tag{3.13}$$

It follows that the current gain of the amplifier is

$$K_i = 1 + \frac{R_f}{R} \tag{3.14}$$

As before, the amplifier gain can be accurately set using the high-precision resistors R and R_f. These are called "gain-setting resistors" of the amplifier.

3.2.4 Instrumentation Amplifiers

An instrumentation amplifier is typically a special-purpose voltage amplifier that is intended for instrumentation applications. An important characteristic of an instrumentation amplifier is the adjustable gain capability. The gain value can be adjusted manually in most instrumentation amplifiers. In more sophisticated instrumentation amplifiers, the gain is programmable and can be set by means of digital logic (e.g., through a control computer). Instrumentation amplifiers are normally used with low-voltage signals. Application examples of instrumentation amplifiers include the following:

1. Applications that need the difference of two signals, for example, control hardware such as the comparator, which generates the "control error" signal (i.e., difference between a reference signal and the output sensor signal). This error signal is used in feedback, for controlling the system.
2. Removing the common noise component (e.g., 60 Hz line noise from an AC power source) in two signals (when the same noise appears in both signals), by taking their difference. As desired in practical applications, a common noise component can be removed in this manner.
3. If the noise or the nonlinear component in a signal can be directly measured (e.g., at the source), it can be subtracted from the signal.
4. Amplifier that is used for producing the output from a bridge circuit (i.e., bridge amplifier).
5. Amplifiers used with various sensors and transducers.

3.2.4.1 Differential Amplifier

Usually, an instrumentation amplifier is also a *differential amplifier* (sometimes termed *difference amplifier*). It generates the difference between two signals, which has many uses as mentioned in the context of instrumentation amplifiers. Ground-loop noise can be a serious problem in single-ended amplifiers. Ground-loop noise can be effectively eliminated using a differential amplifier because noise loops are formed with both inputs of an amplifier, and hence, these noise signals are subtracted at the amplifier output. Since the noise level is almost the same for both inputs, it is canceled out. Any other noise (e.g., 60 Hz line

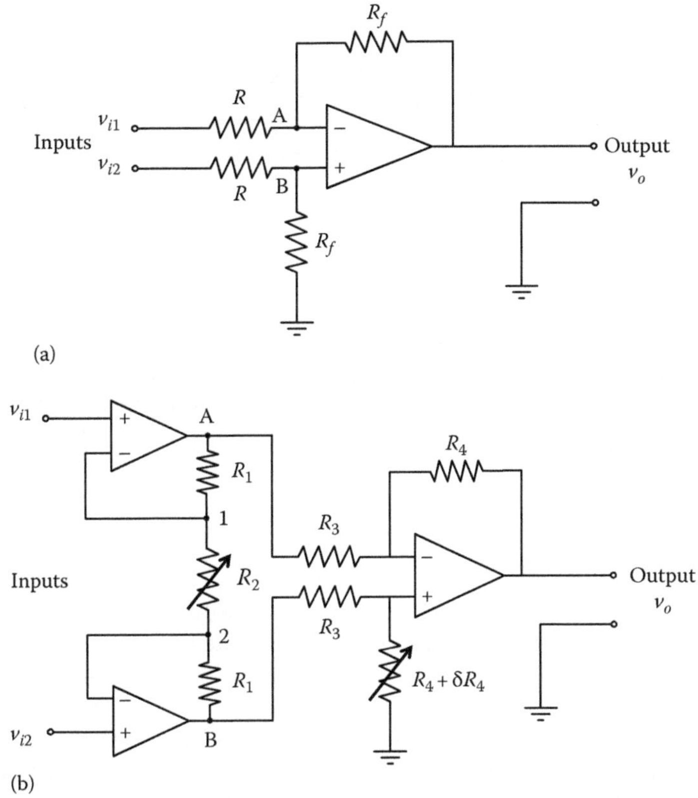

FIGURE 3.3
(a) A basic differential amplifier and (b) a basic instrumentation amplifier.

noise) that might enter both inputs with the same intensity will also be canceled out at the output of a differential amplifier.

In a differential amplifier both input leads are used for signal input, whereas in a single-ended amplifier one of the leads is grounded and only one lead is used for signal input.

A basic differential amplifier that uses a single op-amp is shown in Figure 3.3a. The input–output equation for this amplifier can be obtained in the usual manner. For instance, since the current through an op-amp is negligible, the current balance at point B gives

$$\frac{v_{i2} - v_B}{R} = \frac{v_B}{R_f} \tag{i}$$

where v_B is the voltage at *B*. Similarly, current balance at point A gives

$$\frac{v_o - v_A}{R_f} = \frac{v_A - v_{i1}}{R} \tag{ii}$$

Now, to eliminate v_A and v_B from Equations i and ii, we use the op-amp property

$$v_A = v_B \tag{iii}$$

This gives

$$\frac{v_{i2}}{(1+R/R_f)} = \frac{(v_o R/R_f + v_{i1})}{(1+R/R_f)}$$

or

$$v_o = \frac{R_f}{R}(v_{i2} - v_{i1}) \tag{3.15}$$

Two things are clear from this equation. First, the amplifier output is proportional to the difference and not the absolute value of the two inputs v_{i1} and v_{i2}. Second, the voltage gain of the amplifier is R_f/R. This is known as the "differential gain." It is clear that the differential gain can be accurately set by using the high-precision resistors R and R_f.

3.2.4.2 Instrumentation Amplifier

The basic differential amplifier, shown in Figure 3.3a and discussed earlier, is an important component of an instrumentation amplifier. In addition, an instrumentation amplifier should possess the capability of adjustable gain. Furthermore, it is desirable to have a very high input impedance and very low output impedance at each input lead. It is desirable for an instrumentation amplifier to possess a higher and more stable gain and also a higher input impedance than that of a basic differential amplifier. An instrumentation amplifier that possesses these basic requirements may be fabricated in the monolithic IC form as a single package. Alternatively, it may be built using three differential amplifiers and high-precision resistors, as shown in Figure 3.3b. The amplifier gain can be adjusted using the fine-tunable resistor R_2. Impedance requirements are provided by two voltage follower–type amplifiers, one for each input, as shown. The variable resistance δR_4 is necessary to compensate for errors due to unequal common-mode gain. Let us first consider this aspect and then obtain an equation for the instrumentation amplifier.

3.2.4.3 Common Mode

Now, we extend the discussion on this topic to differential amplifiers. When $v_{i1} = v_{i2}$, ideally, the output voltage v_o should be zero. In other words, ideally, any common-mode signals are rejected by a differential amplifier. But since commercial op-amps are not ideal and since they usually do not have exactly identical gains with respect to the two input leads, the output voltage v_o will not be zero when the two inputs are identical. The associated common-mode error can be compensated for by providing a variable resistor with fine resolution at one of the two input leads of the differential amplifier. Hence, in Figure 3.3b, to compensate for the common-mode error (i.e., to achieve a satisfactory level of common-mode rejection), first, the two inputs are made equal and then δR_4 is varied carefully until the output voltage level is sufficiently small (minimum). Usually, δR_4 that is required to achieve this compensation is small compared with the nominal feedback resistance R_4.

The CMRR of a differential amplifier is defined as

$$\text{CMRR} = \frac{Kv_{cm}}{v_{ocm}} \tag{3.16}$$

where

K is the gain of the differential amplifier (i.e., differential gain)
v_{cm} is the common-mode voltage (i.e., voltage common to both input leads)
v_{ocm} is the common-mode output voltage (i.e., output voltage due to common-mode input voltage)

Ideally, $v_{ocm} = 0$ and CMRR should be infinity. It follows that the larger the CMRR, the better the performance of a differential amplifier performance.

Since ideally $\delta R_4 = 0$, we can neglect δR_4 in the derivation of the instrumentation amplifier equation. Now, note from a basic property of an op-amp with no saturation (specifically, the voltages at the two input leads have to be almost identical) that in Figure 3.3b, the voltage at point 2 should be v_{i2} and the voltage at point 1 should be v_{i1}. Next, we use the property that the current through each input lead of an op-amp is negligible. Accordingly, current through the circuit path B → 2 → 1 → A has to be the same. This gives the current continuity equations: $((v_B - v_{i2})/R_1) = ((v_{i2} - v_{i1})/R_2) = ((v_{i1} - v_A)/R_1)$, where v_A and v_B are the voltages at points A and B, respectively. Hence, we get the two equations: $v_B = v_{i2} + (R_1/R_2)(v_{i2} - v_{i1})$ and $v_A = v_{i1} - (R_1/R_2)(v_{i2} - v_{i1})$. Now, by subtracting the second of these two equations from the first, we have the equation for the first stage of the instrumentation amplifier:

$$v_B - v_A = \left(1 + \frac{2R_1}{R_2}\right)(v_{i2} - v_{i1}) \tag{3.17a}$$

Next, from the previous result (3.15) for a differential amplifier, we have (with $\delta R_4 = 0$)

$$v_o = \frac{R_4}{R_3}(v_B - v_A) \tag{3.17b}$$

Equations 3.17a and b provide the equations for an instrumentation amplifier. Only the resistor R_2 is varied to adjust the gain (differential gain) of the amplifier. In Figure 3.3b, the two input op-amps (the voltage follower op-amps) do not have to be exactly identical as long as the resistors R_1 and R_2 are chosen accurately. This is so because the op-amp parameters such as open-loop gain and input impedance do not enter into the amplifier equations, provided that their values are sufficiently high, as noted earlier.

3.2.4.4 Charge Amplifier

An important category of instrumentation amplifiers is the charge amplifier. It is primarily used to condition the signals from high-impedance sensors such as piezoelectric sensors. It uses an op-amp with a feedback capacitor to provide signal conditioning for high-impedance devices. The charge amplifier will be discussed in detail in a future chapter, under piezoelectric accelerometers.

3.2.4.5 AC-Coupled Amplifiers

In some applications, it is necessary to restrict the DC component of a signal and admit only the AC component. Also, it is important to remove biases and offsets (DC). The DC component of a signal can be blocked off by connecting the signal through a capacitor (*Note*: The impedance of a capacitor is $1/(j\omega C)$, and hence, at zero frequency, there will be an infinite impedance). If the input lead of a device has a series capacitor, we say that the input is *AC coupled* and if the output lead has a series capacitor, then the output is AC coupled. Typically, an AC-coupled amplifier has a series capacitor both at the input lead and at the output lead. Hence, its frequency response function will have a high-pass characteristic; in particular, the DC components will be filtered out. Errors due to bias currents and offset signals are negligible for an AC-coupled amplifier. Furthermore, in an AC-coupled amplifier, stability problems are not very serious.

3.2.5 Noise and Ground Loops

In instruments that handle low-level signals (e.g., sensors such as accelerometers, signal-conditioning circuitry such as strain-gauge bridges, and sophisticated and delicate electronic components such as computer disk drives and automobile control modules), electrical noise can cause excessive error, unless proper corrective actions are taken. One form of noise is caused by fluctuating magnetic fields due to nearby AC power lines or electric machinery. This is commonly known as "electromagnetic interference" (EMI). This problem can be avoided by removing the source of EMI so that fluctuating external magnetic fields and currents are not present near the affected instrument. Another solution would be to use fiber-optic (optically coupled) signal transmission so that there is no noise conduction along with the transmitted signal from the source to the subject instrument. In the case of hard-wired transmission, if the two signal leads (positive and negative or hot and neutral) are twisted or if shielded cables are used, the induced noise voltages become equal in the two leads, which cancel each other.

Proper grounding practices are important to mitigate unnecessary electrical noise problems and, more importantly, to avoid electrical safety hazards. A standard single-phase AC outlet (120 V, 60 Hz) has three terminals, one carrying power (*hot*), the second *neutral*, and the third connected to earth *ground* (which is maintained at zero potential rather uniformly from point to point in the power network). Correspondingly, the power plug of an instrument should have three prongs. The shorter flat prong is connected to a black wire (hot), and the longer flat prong is connected to a white wire (neutral). The round prong is connected to a green wire (ground), which at the other end is connected to the chassis (or casing) of the instrument (chassis ground). By grounding the chassis in this manner, the instrument housing is maintained at zero potential, even in the presence of a fault in the power circuit (e.g., a leakage or a short). The power circuitry of an instrument also has a local ground (signal ground), with reference to which its power signal is measured. This is a sufficiently thick conductor within the instrument, and it provides a common and uniform reference of 0 V. Consider the sensor signal-conditioning example shown in Figure 3.4. The DC power supply can provide both positive (+) and negative (−) outputs. Its zero-voltage reference is denoted by common ground (COM), and it is the signal ground

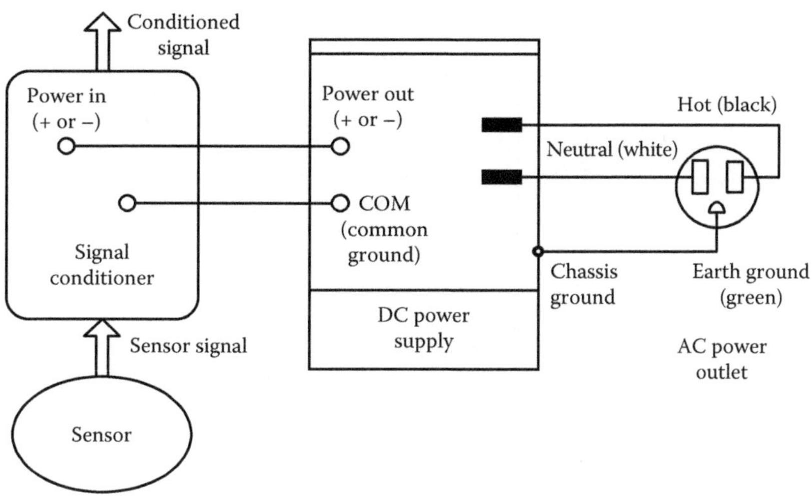

FIGURE 3.4
An example of grounding of instruments.

of the device. The COM of the DC power supply is not connected to the chassis ground but connected to the earth ground through the round prong of the power plug of the power supply. This is necessary to avoid the danger of an electric shock. Note that the COM of the power supply is connected to the signal ground of the signal-conditioning module. In this manner, a common 0 V reference is provided for the DC voltage that is supplied to the signal-conditioning module.

3.2.5.1 Ground-Loop Noise

The main cause of electrical noise is the ground loops that are created due to improper grounding of instruments. If two interconnected instruments are grounded at two separate locations that are far apart (multiple grounding), ground-loop noise can enter the signal leads because of the possible potential difference between the two ground points. The reason is that ground itself is not generally a uniform-potential medium (the difference can be about 100 mV), and a nonzero (and finite) impedance may exist from point to point within this medium. This is, in fact, the case with a typical ground medium such as a COM wire. An example is shown schematically in Figure 3.5a. In this example, the two leads of a sensor are directly connected to a signal-conditioning device such as an amplifier, with one of its input leads (+) grounded (at point B). The 0 V reference lead of the sensor is grounded through its housing to the earth ground (at point A). In this manner, both devices (sensor and amplifier in this example) are ground referenced (i.e., connected to the building ground, which is the ground of the three-pin wall sockets in the building). Because of nonuniform ground potentials, the two ground points A and B are subjected to a potential difference v_g. This will create a ground loop with the common reference lead, which interconnects the two devices. The solution to this problem is to isolate (i.e., provide an infinite impedance to) either one of the two devices. This is called "floating." Figure 3.5b shows *internal isolation* of the sensor. External isolation, by insulating the housing of the sensor, will also remove the ground loop. Floating off the COM of a power supply (see Figure 3.4) is another approach to eliminating ground loops. Specifically, COM is not connected to earth ground.

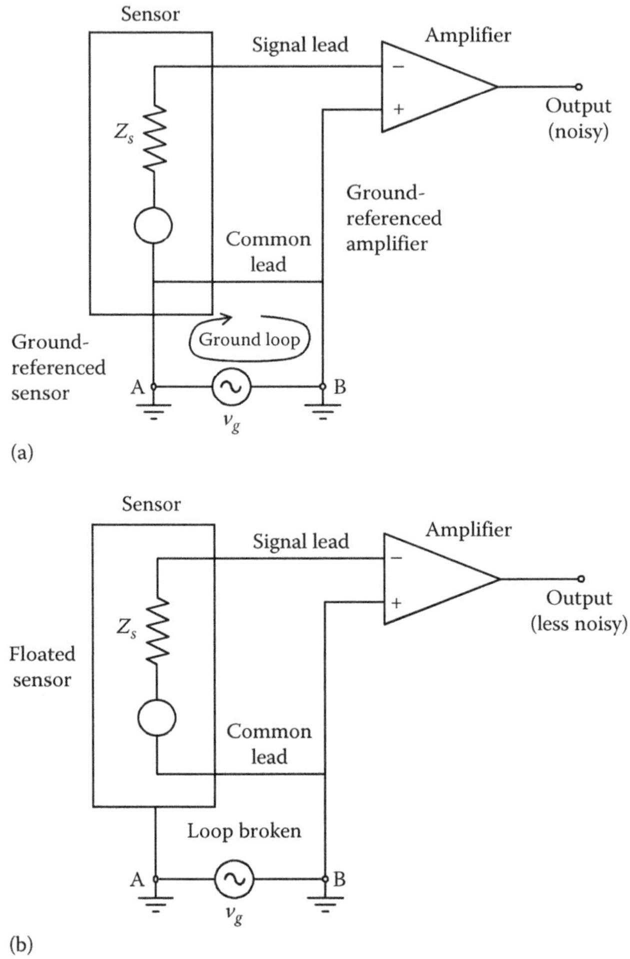

FIGURE 3.5
(a) Illustration of a ground loop and (b) device isolation to eliminate ground loops (an example of internal isolation).

3.3 Analog Filters

A filter is a device that allows through only the desirable part of a signal while rejecting the unwanted part. Unwanted signals can seriously degrade the performance of an engineering system. External disturbances, error components in excitations, and noise generated internally within system components and instrumentation are such spurious signals, which may be removed by a filter. In addition, a filter is capable of shaping a signal in a desired manner.

In typical applications of acquisition and processing of signals in an engineering system, the filtering task would involve the removal of signal components in a specific frequency range. In this context, we can identify the following four broad categories of filters:

1. Low-pass filters
2. High-pass filters
3. Band-pass filters
4. Band-reject (or notch) filters

The ideal frequency response characteristic of each of these four types of filters is shown in Figure 3.6. Only the magnitude of the frequency response function (magnitude of the frequency transfer function) is shown. It is understood, however, that the phase distortion of the input signal should also be small within the *passband* (the allowed frequency range). Practical filters are less than ideal. Their frequency response functions do not exhibit sharp cutoffs as in Figure 3.6, and furthermore, some phase distortion will be unavoidable.

A special type of band-pass filter that is widely used in acquisition and monitoring of response signals (e.g., in product dynamic testing) is *tracking filter.* This is simply a band-pass filter with a narrow passband that is variable (tunable). Specifically, the center frequency (midvalue) of the passband is varied, usually by coupling it to the frequency of a carrier signal (e.g., drive signal). In this manner, signals whose frequency varies with some basic variable in the system (e.g., rotor speed, frequency of a harmonic excitation signal,

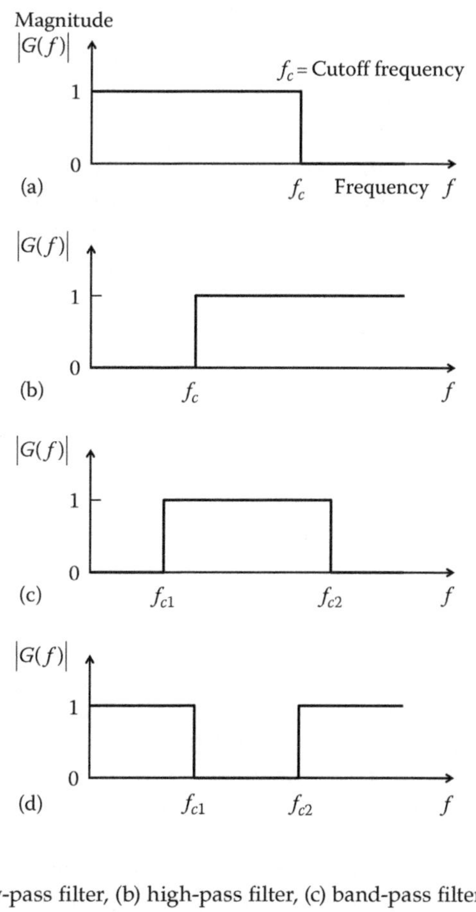

FIGURE 3.6
Ideal filter characteristics: (a) low-pass filter, (b) high-pass filter, (c) band-pass filter, and (d) band-reject (notch) filter.

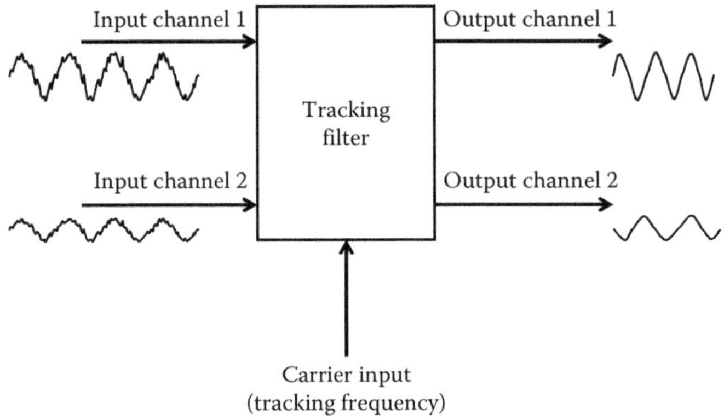

FIGURE 3.7
Schematic representation of a two-channel tracking filter.

frequency of a sweep oscillator) can be accurately tracked in the presence of noise and other signals. The inputs to a tracking filter are the signal that is tracked and the variable tracking frequency (carrier input). A typical tracking filter that can simultaneously track two signals is schematically shown in Figure 3.7.

Filtering can be achieved by digital filters as well as analog filters. Before digital signal processing became efficient and economical, analog filters were exclusively used for signal filtering and are still widely used. An analog filter is typically an active filter containing active components such as transistors or op-amps. In an analog filter, the input signal is passed through an analog circuit. Dynamics of the circuit will determine which (desired) signal components would be allowed through and which (unwanted) signal components would be rejected. Earlier versions of analog filters employed discrete circuit elements such as discrete transistors, capacitors, resistors, and even discrete inductors. Since inductors have several shortcomings such as susceptibility to electromagnetic noise, unknown resistance effects, and large size, today they are rarely used in filter circuits. Furthermore, due to well-known advantages of IC devices, today analog filters in the form of monolithic IC chips are extensively used in modem applications and are preferred over discrete-element filters. Digital filters, which employ digital signal processing to achieve filtering, are also widely used today.

3.3.1 Passive Filters and Active Filters

Passive analog filters employ analog circuits containing passive elements such as resistors and capacitors (and sometimes inductors) only. An external power source is not needed in a passive filter. Active analog filters employ active elements and components such as transistors and op-amps in addition to passive elements. Since external power is needed for the operation of the active elements and components, an active filter is characterized by the need for an external power supply. Active filters are widely available in a monolithic IC package and are usually preferred over passive filters.

Advantages of active filters include the following:

1. Loading effects and interaction with other components are negligible because active filters can provide a very high input impedance and a very low output impedance.

2. They can be used with low signal levels because both signal amplification and filtering can be provided by the same active circuit.
3. They are widely available in a low-cost and compact IC form.
4. They can be easily integrated with digital devices.
5. They are less susceptible to noise from EMI.

Commonly mentioned disadvantages of active filters are the following:

1. They need an external power supply.
2. They are susceptible to saturation-type nonlinearity at high signal levels.
3. They can introduce many types of internal noise and unmodeled signal errors (offset, bias signals, etc.).

Note that advantages and disadvantages of passive filters can be directly inferred from the disadvantages and advantages of active filters, as listed here.

3.3.1.1 Number of Poles

Analog filters are dynamic systems, and they can be represented by transfer functions, assuming linear dynamics. Number of poles of a filter is the number of poles in the associated transfer function. This is also equal to the *order* of the characteristic polynomial of the filter transfer function (i.e., order of the filter). *Note*: Poles (or eigenvalues) are the roots of the characteristic equation.

In our discussion, we show simplified versions of filters, typically consisting of a single filter stage. Performance of such a basic filter can be improved at the expense of circuit complexity (and increased pole count). Basic op-amp circuits are given here for active filters. More complex devices are commercially available, but our primary purpose is to illustrate the underlying principles rather than to provide complete descriptions and data sheets for commercial filters.

3.3.2 Low-Pass Filters

The purpose of a low-pass filter is to allow through all signal components below a certain (cutoff) frequency and block off all signal components above that cutoff. Analog low-pass filters are widely used as *anti-aliasing filters* in digital signal processing (see Chapter 6). An error known as "aliasing" enters the digitally processed results of a signal if the original signal has frequency components above half the sampling frequency (half the sampling frequency is called the Nyquist frequency). Hence, aliasing distortion can be eliminated if the signal is filtered using a low-pass filter with its cutoff set at the Nyquist frequency, before sampling and digital processing. This is one of the numerous applications of analog low-pass filters. Another typical application would be to eliminate high-frequency noise in a sensed signal.

A single-pole, active, low-pass filter is shown in Figure 3.8a. If two active filter stages similar to Figure 3.8a are connected together, loading errors will be negligible because the op-amp with feedback (i.e., a voltage follower) introduces a high input impedance and low output impedance, while maintaining the voltage gain at unity. With similar reasoning, it can be concluded that an active filter has the desirable property of very low interaction with any other connected component.

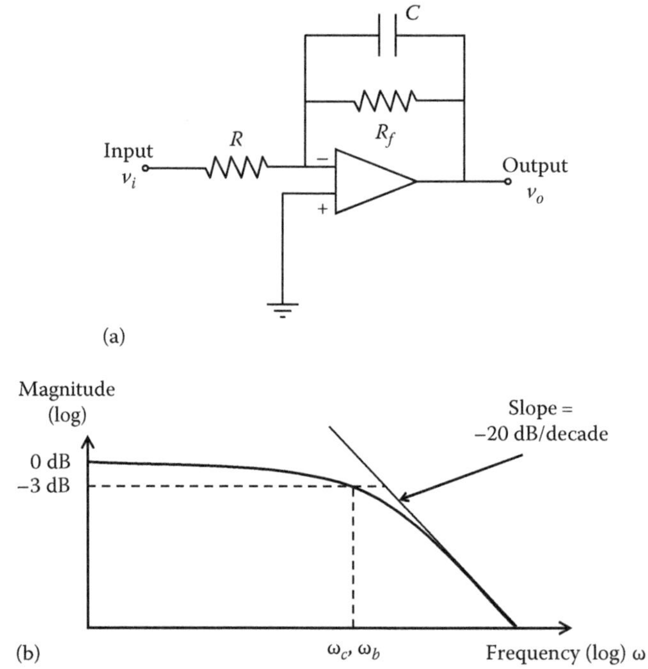

FIGURE 3.8
(a) A single-pole, active, low-pass filter and (b) the frequency response characteristic.

To obtain the filter equation for Figure 3.8a, we write the current balance at the inverting input lead of the op-amp (current into the op-amp is equal to 0; voltage there is equal to 0 in view of grounding of the inverting input lead): $(v_i/R) + (v_o/R_f) + C_f\,(dv_o/dt) = 0$. We get

$$\tau \frac{dv_o}{dt} + v_o = -kv_i \tag{3.18}$$

where the filter *time constant* is

$$\tau = R_f C_f \tag{3.19a}$$

Now, from Equation 3.17, it follows that the filter transfer function is

$$\frac{v_o}{v_i} = G(s) = -\frac{k}{(\tau s + 1)} \tag{3.20a}$$

Filter gain

$$k = \frac{R_f}{R} \tag{3.19b}$$

From this transfer function, it is clear that an analog low-pass filter is essentially a lag circuit (i.e., it provides a phase lag).

The frequency response function corresponding to Equation 3.20a is obtained by setting $s = j\omega$; thus,

$$G(j\omega) = -\frac{k}{(\tau j\omega + 1)} \tag{3.20b}$$

This gives the response of the filter when a sinusoidal signal of frequency ω is applied. The magnitude $|G(j\omega)|$ of the frequency transfer function gives the signal amplification, and the phase angle $\angle G(j\omega)$ gives the phase lead of the output signal with respect to the input. The magnitude curve (Bode magnitude curve), normalized by dividing by the DC gain k, is shown in Figure 3.8b. Note from Equation 3.20b that for small frequencies (i.e., $\omega \ll 1/\tau$) the magnitude (normalized) is approximately unity. Hence, $1/\tau$ can be considered the cutoff frequency ω_c:

$$\omega_c = \frac{1}{\tau} \tag{3.21}$$

Example 3.3

Show that the cutoff frequency given by Equation 3.21 is also the half-power bandwidth for the low-pass filter. Show that for frequencies much larger than this, the filter transfer function on the Bode magnitude plane (i.e., log magnitude vs. log frequency) can be approximated by a straight line with slope –20 dB/decade. This slope is known as the "roll-off rate."

Solution

Using the normalized transfer function ($k = 1$), the frequency corresponding to half power (or $1/\sqrt{2}$ magnitude) is given by $1/|\tau j\omega + 1| = 1/\sqrt{2}$. By cross multiplying, squaring, and simplifying the equation, we get $\tau^2\omega^2 = 1$. Hence, the half-power bandwidth is

$$\omega_b = \frac{1}{\tau} \tag{3.22}$$

This is identical to the cutoff frequency given by Equation 3.21.

Now for $\omega \gg 1/\tau$ (i.e., $\omega\tau \gg 1$), normalized Equation 3.19b can be approximated by $G(j\omega) = 1/\tau j\omega$. This has the magnitude $|G(j\omega)| = 1/\tau\omega$.

Converting to the log scale, we get

$$\log_{10}|G(j\omega)| = -\log_{10}\omega - \log_{10}\tau.$$

It follows that the $\log_{10}$ (magnitude) versus $\log_{10}$ (frequency) curve is a straight line with slope –1. In other words, when the frequency increases by a factor of 10 (i.e., a decade), the $\log_{10}$ magnitude decreases by unity (i.e., by 20 dB). Hence, the roll-off rate is –20 dB/decade. These observations are shown in Figure 3.8b. An amplitude change by a factor of $\sqrt{2}$ (or power by a factor of 2) corresponds to 3 dB. Hence, when the DC (zero-frequency magnitude) value is unity (0 dB), the half-power magnitude is –3 dB.

The *cutoff frequency* and the *roll-off rate* are the two main design specifications for a low-pass filter. Ideally, we would like a low-pass filter magnitude curve to be flat up to the required passband limit (cutoff frequency) and then roll off very rapidly. The low-pass filter shown in Figure 3.8 only approximately meets these requirements. In particular, the roll-off rate is not large enough. We would prefer a roll-off rate of at least −40 dB/decade or even −60 dB/decade, in practical filters. This can be realized by using a high-order filter (i.e., a filter with many poles). Low-pass Butterworth filter is of this type and is widely used.

3.3.2.1 Low-Pass Butterworth Filter

A low-pass Butterworth filter with two poles can provide a roll-off rate of −40 dB/decade, and one with three poles can provide a roll-off rate of −60 dB/decade. Furthermore, the steeper the roll-off slope, the flatter the filter magnitude curve within the passband.

A two-pole, low-pass Butterworth filter is shown in Figure 3.9. We could construct a two-pole filter simply by connecting together two single-pole stages of the type shown in Figure 3.8a. Then, we would require two op-amps, whereas the circuit shown in Figure 3.9 achieves the same objective by using only one op-amp (i.e., at a lower cost).

Example 3.4

Show that the op-amp circuit in Figure 3.9 is a low-pass filter with two poles. What is the transfer function of the filter? Estimate the cutoff frequency under suitable conditions. Show that the roll-off rate is −40 dB/decade.

Solution

To obtain the filter equation, we write the current balance equations first. Specifically, the sum of the currents through R_1 and C_1 passes through R_2. The same current has to pass through C_2 because the current through the op-amp lead is zero (a property of an op-amp). Hence,

$$\frac{v_i - v_A}{R_1} + C_1\frac{d}{dt}(v_o - v_A) = \frac{v_A - v_B}{R_2} = C_2\frac{dv_B}{dt} \tag{i}$$

Also, the current through the feedback resistor R_f goes entirely through the ground resistor R because the current through the second op-amp lead is also zero. Hence, this path of voltage division gives

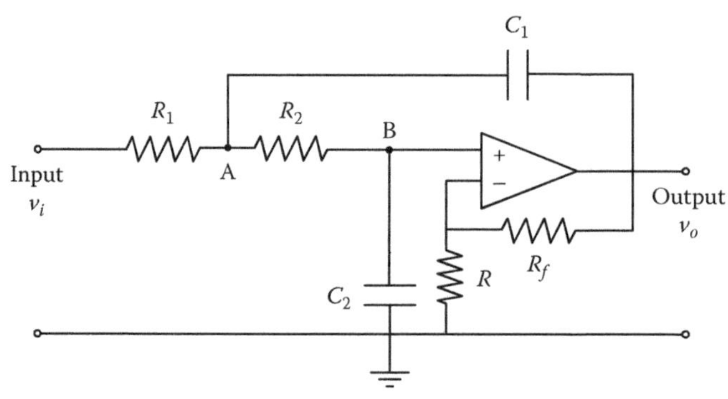

FIGURE 3.9
A two-pole low-pass Butterworth filter.

$$v_B = kv_o \tag{ii}$$

where

$$k = \frac{R}{R_f} \tag{3.23}$$

From Equations i and ii, we get

$$\frac{v_i - v_A}{R_1} + C_1\frac{dv_o}{dt} - C_1\frac{dv_A}{dt} = C_2 k\frac{dv_o}{dt} \tag{iii}$$

$$\frac{v_A - kv_o}{R_2} = C_2 k\frac{dv_o}{dt} \tag{iv}$$

Now, we define the following constants:

$$\tau_1 = R_1C_1, \quad \tau_2 = R_2C_2, \quad \tau_3 = R_1C_2 \tag{3.24}$$

Eliminate v_A by substituting Equation iv into Equation iii and introduce the Laplace variable s. We get the filter transfer function

$$\frac{v_o}{v_i} = \frac{1}{k\left[\tau_1\tau_2 s^2 + ((1-1/k)\tau_1 + \tau_2 + \tau_3)s + 1\right]} = \frac{\omega_n^2}{k\left[s^2 + 2\zeta\omega_n s + \omega_n^2\right]} \tag{3.25a}$$

This second-order transfer function becomes oscillatory if the poles are complex, that is, if $((1-1/k)\tau_1 + \tau_2 + \tau_3)^2 < 4\tau_1\tau_2$. Ideally, we would like to have a zero-resonant frequency, which corresponds to a damping ratio value $\zeta = 1/\sqrt{2}$.

Undamped natural frequency:

$$\omega_n = \frac{1}{\sqrt{\tau_1\tau_2}} \tag{3.26a}$$

Damping ratio:

$$\zeta = \frac{(1-1/k)\tau_1 + \tau_2 + \tau_3}{2\sqrt{\tau_1\tau_2}} \tag{3.27}$$

Resonant frequency:

$$\omega_r = \sqrt{1-2\zeta^2}\,\omega_n \tag{3.26b}$$

For a low-pass filter, ideal conditions correspond to $\omega_r = 0$ (i.e., no resonant peak, giving a wider flat region) that occurs when $\zeta = 1/\sqrt{2}$. For this optimal case, from Equations 3.27 and 3.28, we get

$$\left(\left(1-\frac{1}{k}\right)\tau_1 + \tau_2 + \tau_3\right)^2 = 2\tau_1\tau_2 \tag{3.28}$$

The frequency response function of the filter is (see Equation 3.25a)

$$G(j\omega) = \frac{\omega_n^2}{k\left[\omega_n^2 - \omega^2 + 2j\zeta\omega_n\omega\right]} \tag{3.25b}$$

For convenience, we normalize this transfer function using k (i.e., set k =1). Now, for $\omega \ll \omega_n$, the filter frequency response is flat with a unity gain. For $\omega \gg \omega_n$, the filter frequency response can be approximated by $G(j\omega) = -\omega_n^2/\omega^2$.

In a log (magnitude) versus log (frequency) scale, this function is a straight line with slope of −2. Hence, when the frequency increases by a factor of 10 (i.e., one decade), the $\log_{10}$ (magnitude) drops by two units (i.e., by 40 dB). In other words, the roll-off rate is −40 dB/decade.

Filter cutoff frequency: This is the frequency up to which low-pass filtering is valid. For the ideal filter (i.e., $\zeta = 1/\sqrt{2}$), ω_n can be taken as the cutoff frequency. Hence,

$$\omega_c = \omega_n = \frac{1}{\sqrt{\tau_1 \tau_2}} \tag{3.29}$$

It can be easily verified using Equation 3.25b that when $\zeta = 1/\sqrt{2}$, this frequency is identical to the half-power bandwidth (i.e., the frequency at which the transfer-function magnitude becomes $1/\sqrt{2}$, where the normalized DC value is 1.0).

Note: If two single-pole stages (of the type shown in Figure 3.8a) are cascaded, the resulting two-pole filter has an overdamped (i.e., nonoscillatory) transfer function ($\zeta > 1$), and it is not possible to achieve $\zeta = 1/\sqrt{2}$ unlike the present case. Furthermore, a three-pole low-pass Butterworth filter can be obtained by cascading the two-pole unit shown in Figure 3.9 with a single-pole unit shown in Figure 3.8a. Higher-order low-pass Butterworth filters can be obtained in a similar manner by cascading an appropriate set of basic filter units.

It is clear that the transfer function for an ideal two-pole (i.e., second order) low-pass Butterworth filter of cutoff frequency ω_c is

$$\frac{v_o}{v_i} = \frac{\omega_c^2}{\left[s^2 + \sqrt{2}\omega_c s + \omega_c^2\right]} = \frac{(\omega_c/\omega_o)^2}{\left[(s/\omega_o)^2 + \sqrt{2}(\omega_c/\omega_o)(s/\omega_o) + (\omega_c/\omega_o)^2\right]} \tag{3.30}$$

The second transfer function in Equation 3.30 is the normalized form, which is used in MATLAB®, with a normalizing frequency so that $0 < \omega_c/\omega_o < 1$. Then, once a normalized filter transfer function is determined using MATLAB, it can be scaled to any other frequency using a proper scaling frequency ω_o.

Example 3.5

Determine a second-order low-pass Butterworth filter of cutoff frequency $\omega_c = 1/\sqrt{2}$ rad/s. Verify the result using MATLAB.

Plot the magnitude of the filter transfer function.

How will the filter for 10 times this cutoff frequency be obtained using this result?

Next, obtain a four-pole (fourth order) Butterworth filter with the same cutoff frequency ($\omega_c = 1/\sqrt{2}$ rad/s) and compare the two results.

Solution

By directly substituting into Equation 3.30, we get the filter transfer function as

$$\frac{v_o}{v_i} = \frac{0.5}{\left[s^2 + s + 0.5\right]}$$

The corresponding MATLAB command is

```
>> [b,a] = butter(n,Wn,'s')
```

where
 n is the filter order
 Wn is the cutoff frequency
 b is the numerator coefficient vector of the transfer function
 a is the denominator coefficient vector of the transfer function

We get the following result:

```
>> [b,a]=butter(2,1/sqrt(2),'s')
b =
          0          0      0.5000
a =
  1.0000    1.0000    0.5000
```

This agrees with the analytical result. We can plot the magnitude of the frequency response function of this filter (in linear scale for frequency) using MATLAB, as follows:

```
>> w=linspace(0.005,0.705,142);
>> h = freqs(b,a,w);
>> plot(w,abs(h),'-')
```

The result is shown (solid curve) in Figure 3.10a.

From a normalized filter result (for any filter order n), we can obtain the filter transfer function corresponding to any other cutoff frequency and the same filter order in a straightforward manner. We simply change the polynomial coefficients (both numerator and denominator) of the normalized result as follows:

$$s^0 \text{ coefficient: Multiply by } r^n$$

$$s^1 \text{ coefficient: Multiply by } r^{n-1}$$

where r is the multiplication factor for changing the cutoff frequency.

Now, with $r = 10$, which corresponds to a cutoff frequency of $10/\sqrt{2}$ rad/s, we have the optimal filter transfer function

$$\frac{v_o}{v_i} = \frac{0.5 \times 100}{[s^2 + s \times 10 + 0.5 \times 100]} = \frac{50}{[s^2 + 10s + 50]}$$

Next, let us use a four-pole Butterworth filter to design a better low-pass filter for the same example and compare the two results. We use the following MATLAB commands for this purpose:

```
>> [b2,a2]=butter(4,1/sqrt(2),'s')
b2 =
          0          0          0          0      0.2500
a2 =
     1.0000     1.8478     1.7071   0.9239     0.2500
>> w=linspace(0.005,0.705,142);
>> h2 = freqs(b2,a2,w);
>> plot(w,abs(h2),'-')
```

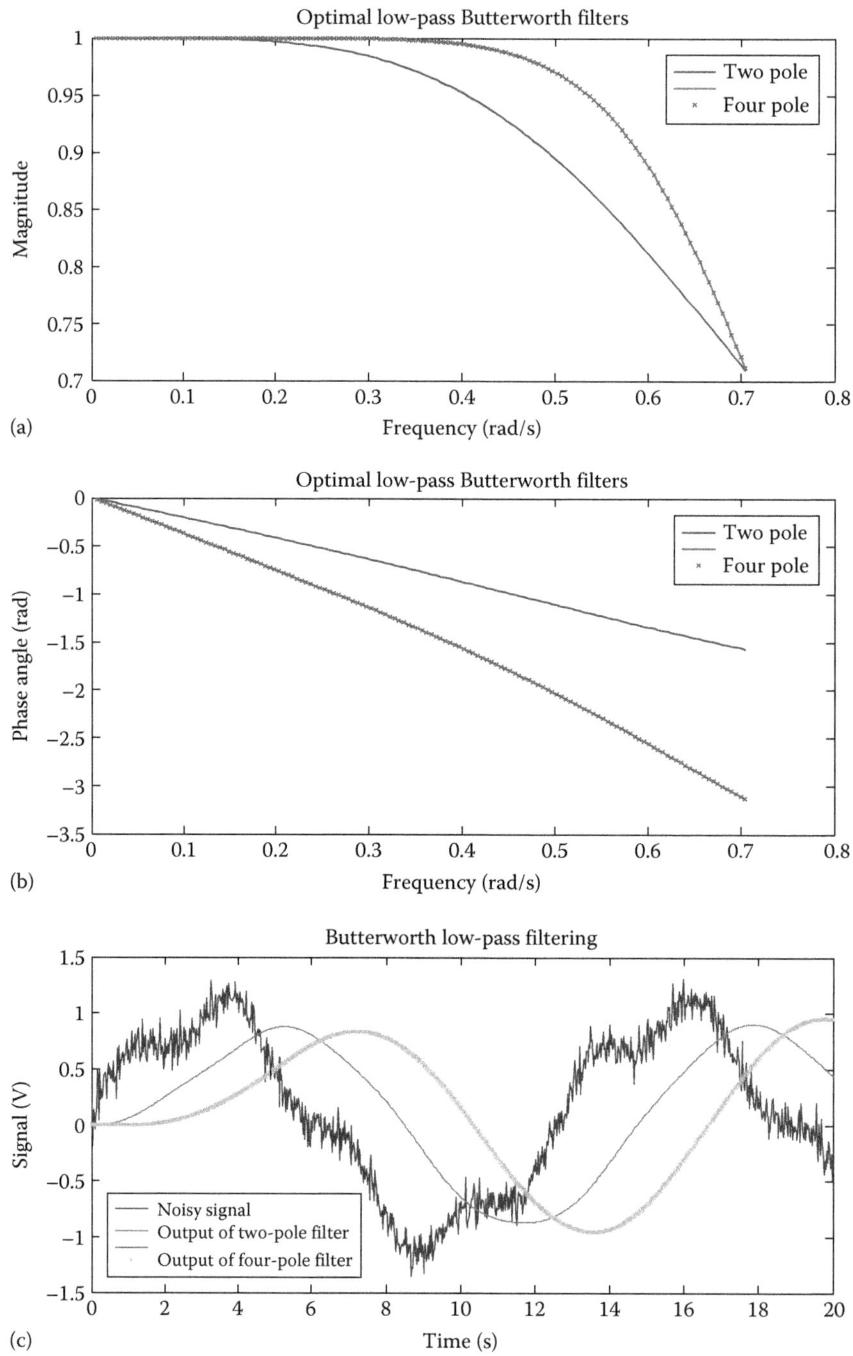

FIGURE 3.10
Optimal low-pass Butterworth filtering: (a) filter magnitudes, (b) filter phase angles, and (c) time signals.

The magnitude of the frequency response function of the four-pole Butterworth filter is shown by "x" in Figure 3.10a. It is seen that the flatness of the passband has improved considerably. In particular, the two-pole filter is quite flat up to about 0.2 rad/s, while the four-pole filter is flat up to about 0.4 rad/s.

Phase distortion: A filter has obvious benefits, but we normally achieve these while sacrificing something, with respect to signal distortion. There are two types of distortions that enter into the signal (when removing the undesirable components): (1) the signal magnitude (amplitude) will be distorted and (2) signal phase angle will be distorted (a phase lag will be introduced). We have already observed the magnitude distortion, which can be significant when the frequency is more than half the cutoff frequency. We have been able to improve the magnitude distortion by increasing the number of filter poles. Now let us consider the phase distortion, using the same example.

The phase angle curves (in radians) for the two filter are obtained using the following MATLAB command:

```
>> plot(w,angle(h),'-', w,angle(h2),'-',w,angle(h2),'x')
```

The results are shown in Figure 3.10b by a solid curve for the two-pole filter and a curve with "x" for the four-pole filter. It is seen that the phase distortion is quite significant. With regard to the phase distortion, however, the two-pole filter is better than the four-pole filter. In particular, at the cutoff frequency, the phase lag of the two-pole filter is $\pi/2$, while that of the four-pole filter is π.

Suppose that a sinusoidal signal with random noise is generated, as shown in Figure 3.10c, using the MATLAB script:

```
% Low-pass filter data
t=0:0.02:20.0;
u=sin(0.5*t)+0.2*sin(2*t);
for i=1:1001
u(i)=u(i)+normrnd(0.0,0.1); % Gaussian random noise
end
```

This signal is separately applied to the two-pole filter and the four-pole filter, using the following MATLAB commands:

```
>> y1=lsim(b,a,u,t);
>> y2=lsim(b2,a2,u,t);
```

Next, the input (noisy) signal and the filter outputs are plotted using

```
>> plot(t,u,'-',t,y1,'-',t,y2,'-',t,y2,'x')
```

The plots are shown in Figure 3.10c. The following observations can be made:

1. Both filters are equally effective in removing the noise.
2. The two-pole filter introduces slightly more amplitude distortion.
3. The four-pole filter introduces more phase distortion.

3.3.3 High-Pass Filters

Ideally, a high-pass filter allows through it all signal components above a certain frequency (*cutoff frequency*) and blocks off all signal components below that frequency. A single-pole, high-pass filter is shown in Figure 3.11a. As for the low-pass filter that was discussed earlier, an active filter is desired, however, because of its many advantages, including negligible

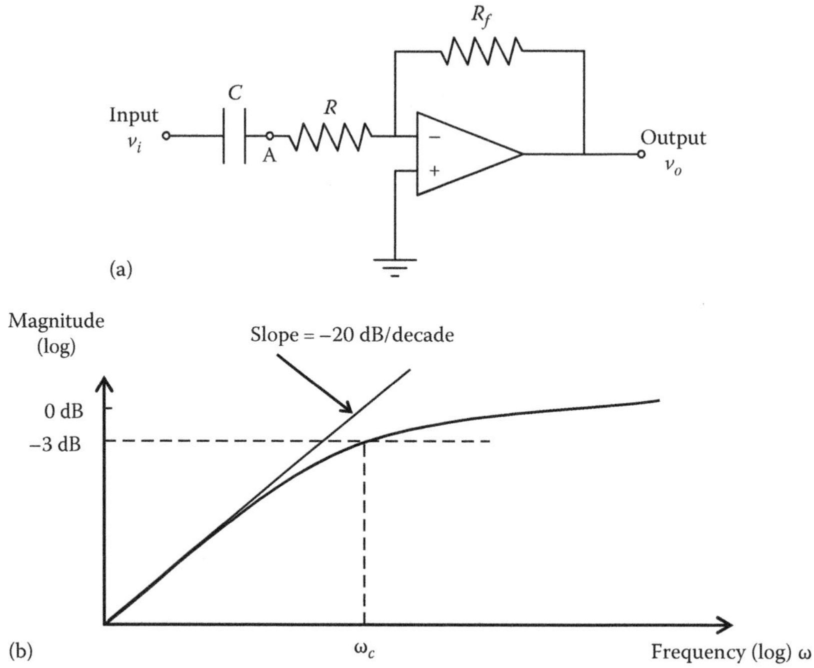

FIGURE 3.11
(a) A single-pole high-pass filter and (b) frequency response characteristic.

loading error due to high input impedance and low output impedance of the op-amp voltage follower that is present in this circuit.

The filter equation is obtained by noting that the current through the path C–R–R_f is the same (since no current can flow into the op-amp lead). Let v_A equal to voltage at A. We have $C(d/dt)(v_i - v_A) = v_A/R = -v_o/R_f$. By eliminating v_A in these two equations, we get

$$\tau\frac{d(v_i + kv_o)}{dt} = -v_o$$

which can be written as

$$\tau\frac{dv_i}{dt} = -\left(k\tau\frac{dv_o}{dt} + v_o\right) \tag{3.31}$$

where the filter time constant

$$\tau = RC \tag{3.32}$$

For convenience (without loss of generality), take $k = 1$ (i.e., $R = R_f$). Then, introducing the Laplace variable s, the filter transfer function is written as

$$\frac{v_o}{v_i} = G(s) = \frac{\tau s}{(\tau s + 1)} \tag{3.33a}$$

This corresponds to a *lead circuit* (i.e., an overall phase lead is provided by this transfer function). The corresponding frequency response function is

$$G(j\omega) = \frac{\tau j\omega}{(\tau j\omega + 1)} \tag{3.33b}$$

Since its magnitude is zero for $\omega \ll 1/\tau$, and it is unity for $\omega \gg 1/\tau$, the cutoff frequency is

$$\omega_c = \frac{1}{\tau} \tag{3.34}$$

An ideal high-pass filter will allow through it all signals above this cutoff frequency, undistorted, and will completely block off all signals below the cutoff. However, the actual behavior of the basic high-pass filter shown in Figure 3.11 is not perfect, as observed from the frequency response characteristic shown in Figure 3.11b. It can be easily verified that the half-power bandwidth of the basic high-pass filter is equal to the cutoff frequency given by Equation 3.33b, as in the case of the basic low-pass filter. The roll-up slope of the single-pole high-pass filter is 20 dB/decade. Steeper slopes are desirable. Multiple-pole, high-pass Butterworth filters can be constructed to give steeper roll-up slopes and reasonably flat passband magnitude characteristics.

3.3.4 Band-Pass Filters

An ideal band-pass filter passes all signal components within a finite frequency band and blocks off all signal components outside that band. The lower frequency limit of the passband is called the "lower cutoff frequency" (ω_{c1}), and the upper frequency limit of the band is called the "upper cutoff frequency" (ω_{c2}). The most straightforward way to form a band-pass filter is to cascade a high-pass filter of cutoff frequency ω_{c1} with a low-pass filter of cutoff frequency ω_{c2}. We will do this by connecting a passive low-pass stage to the high-pass filter of Figure 3.11. This arrangement is shown in Figure 3.12. Even though now there will be a load current at the output of the high-pass filter, it should be clear from the derivation of its equations, as given earlier, that the filter equation will be the same.

To obtain the filter equation, first consider the high-pass portion of the circuit shown in Figure 3.12a. From the previously obtained result (3.31a) for the high-pass filter, we have

$$\frac{v_{o1}}{v_i} = \frac{\tau s}{(\tau s + 1)} \tag{i}$$

where v_{o1} is the output of the high-pass stage.

The equation for the passive, low-pass stage is simply (output is in open circuit)

$$(v_{o1} - v_o)/R_2 = C_2(dv_o/dt)$$

This gives the transfer function for the low-pass stage as

$$\frac{v_o}{v_{o1}} = \frac{1}{(\tau_2 s + 1)} \tag{ii}$$

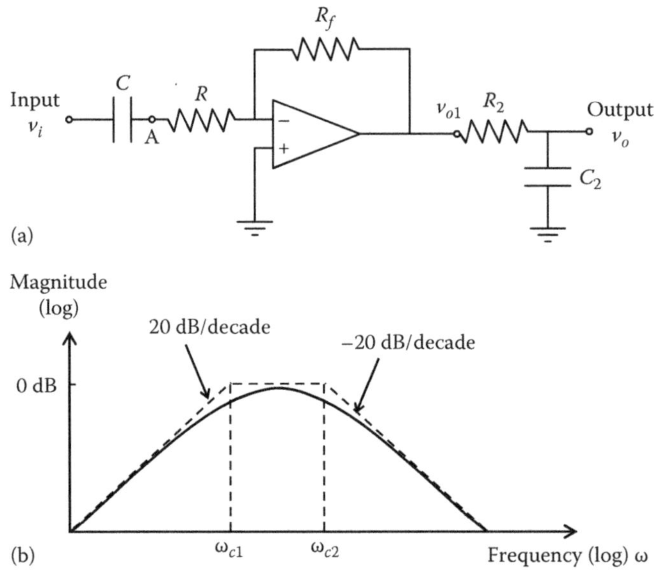

FIGURE 3.12
(a) An active band-pass filter and (b) frequency response characteristic.

Then, by combining the results (i) and (ii), we get the transfer function of the band-pass filter as

$$\frac{v_o}{v_i} = \frac{\tau s}{(\tau s + 1)(\tau_2 s + 1)} \tag{3.35}$$

where

$$\tau_2 = R_2 C_2 \tag{3.36}$$

The cutoff frequencies are $\omega_{c1} = (1/\tau)$ and $\omega_{c2} = (1/\tau_2)$. These are indicated in the frequency characteristic of Figure 3.12b. It can be verified that, for this basic band-pass filter, the roll-up slope is +20 dB/decade and the roll-down slope is −20 dB/decade. These slopes are not sufficient in many applications. Furthermore, the flatness of the frequency response within the passband of the basic filter is not adequate. More complex (higher order) band-pass filters with sharper cutoffs and flatter passbands are commercially available.

3.3.4.1 Resonance-Type Band-Pass Filters

There are many applications where a filter with a very narrow passband is required. The tracking filter mentioned at the beginning of the section on analog filters is one such application. A filter circuit with a sharp resonance can serve as a narrowband filter. A cascaded RC circuit does not provide an oscillatory response (because the filter poles are all real), and hence, it does not form a resonance-type filter. On the other hand, the circuit shown in Figure 3.13a will produce the desired effect.

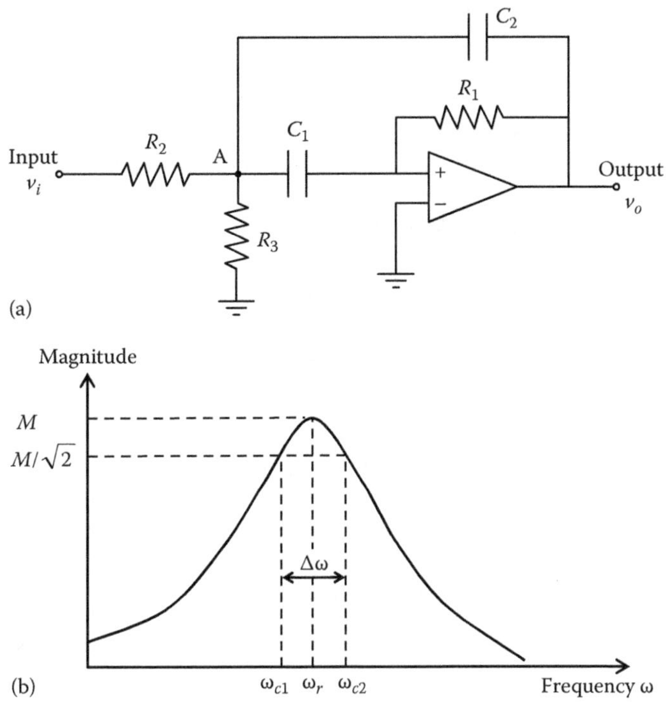

FIGURE 3.13
(a) A resonance-type narrow band-pass filter and (b) frequency response characteristic.

To obtain the filter equation, first write the current summation at the –ve lead of the op-amp:

$$\frac{v_o}{R_1} + C_1\frac{dv_A}{dt} = 0 \tag{i}$$

Next, write the current summation at A:

$$\frac{v_i - v_A}{R_2} + C_2\frac{d}{dt}(v_o - v_A) = \frac{v_A}{R_3} - \frac{v_o}{R_1} \tag{ii}$$

$$\text{(i): } v_o + \tau_1 s v_A = 0 \tag{iii}$$

$$\text{(ii): } v_i - v_A + \tau_2 s(v_o - v_A) = k_2 v_A - k_1 v_o \tag{iv}$$

Eliminate v_A from (iii) and (iv): $[\tau_1\tau_2 s^2 + (k_1\tau_1 + \tau_2)s + 1 + k_2]v_o = -\tau_1 s v_i$.

The filter transfer function is

$$\frac{v_o}{v_i} = G(s) = -\frac{\tau_1 s}{\left[\tau_1\tau_2 s^2 + (k_1\tau_1 + \tau_2)s + 1 + k_2\right]} \tag{3.37}$$

where

$$\tau_1 = R_1C_1,\ \tau_2 = R_2C_2,\ k_1 = \frac{R_2}{R_1},\ k_2 = \frac{R_2}{R_3}$$

It can be shown that the characteristic equation can possess complex roots (i.e., complex poles).

Example 3.6

Verify that the band-pass filter shown in Figure 3.13a can have a frequency response with a resonant peak, as shown in Figure 3.13b. Verify that the half-power bandwidth $\Delta\omega$ of the filter is given by $2\zeta\omega_r$ at low damping values (*Note*: ζ is the damping ratio and ω_r is the resonant frequency).

Solution

We may verify that the transfer function given by Equation 3.37 can have a resonant peak by showing that its characteristic equation can have complex roots. For example, if we use the parameter values $C_1 = 2$, $C_2 = 1$, $R_1 = 1$, $R_2 = 2$, and $R_3 = 1$, we have $\tau_1 = 2$, $\tau_2 = 2$, $k_1 = 2$, and $k_1 = 2$. The corresponding characteristic equation is $4s^2 + 6s + 3 = 0$, which has the roots $-(3/4) \pm j(\sqrt{3}/4)$. Clearly, the poles are complex.

To obtain an expression for the half-power bandwidth of the filter, note that the filter transfer function may be written as

$$G(s) = \frac{ks}{(s^2 + 2\zeta\omega_n s + \omega_n^2)} \tag{3.38a}$$

where

ω_n is the undamped natural frequency
ζ is the damping ratio
k is a gain parameter

The frequency response function is given by

$$G(j\omega) = \frac{k\,j\omega}{\left[\omega_n^2 - \omega^2 + 2j\zeta\omega_n\omega\right]} \tag{3.38b}$$

For low damping, the resonant frequency $\omega_r \cong \omega_n$. The corresponding peak magnitude M is obtained by substituting $\omega = \omega_n$ in Equation 3.38b and taking the transfer-function magnitude. Thus,

$$M = \frac{k}{2\zeta\omega_n} \tag{3.39}$$

At half-power frequencies, we have

$$\left|G(j\omega)\right| = \frac{M}{\sqrt{2}} \quad \text{or} \quad \frac{k\omega}{\sqrt{(\omega_n^2 - \omega^2)^2 + 4\zeta^2\omega_n^2\omega^2}} = \frac{k}{2\sqrt{2}\zeta\omega_n}$$

This gives

$$(\omega_n^2 - \omega^2)^2 = 4\zeta^2\omega_n^2\omega^2 \tag{3.40}$$

The positive roots of Equation 3.36 provide the passband frequencies ω_{c1} and ω_{c2}. The roots are given by $\omega_n^2 - \omega^2 = \pm 2\zeta\omega_n\omega$. Hence, the two roots ω_{c1} and ω_{c2} satisfy the following two equations: $\omega_{c1}^2 + 2\zeta\omega_n\omega_{c1} - \omega_n^2 = 0$ and $\omega_{c2}^2 - 2\zeta\omega_n\omega_{c2} - \omega_n^2 = 0$.

Accordingly, by solving these two quadratic equations and selecting the appropriate sign, we get

$$\omega_{c1} = -\zeta\omega_n + \sqrt{\omega_n^2 + \zeta^2\omega_n^2} \tag{3.41}$$

and

$$\omega_{c2} = \zeta\omega_n + \sqrt{\omega_n^2 + \zeta^2\omega_n^2} \tag{3.42}$$

The half-power bandwidth is

$$\Delta\omega = \omega_{c2} - \omega_{c1} = 2\zeta\omega_n \tag{3.43}$$

Now, since $\omega_n \cong \omega_r$ for low ζ, we have

$$\Delta\omega = 2\zeta\omega_r \tag{3.44}$$

A notable shortcoming of a resonance-type filter is that the frequency response within the bandwidth (passband) is not flat. Hence, quite nonuniform signal attenuation will take place inside the passband.

3.3.5 Band-Reject Filters

Band-reject filters or notch filters are commonly used to filter out a narrowband of noise components from a signal. For example, 60 Hz line noise in a signal can be eliminated by using a notch filter with a notch frequency of 60 Hz.

An active circuit that can serve as a notch filter is shown in Figure 3.14a. This is known as the "twin T circuit" because its geometric configuration resembles two T-shaped circuits connected together.

To obtain the filter equation, note that the voltage at point P is $-v_o$ because of unity gain (because of the equal resistances R_f) of the voltage follower and since the input lead to the op-amp is grounded. Now, we write the current balance at nodes A and B:

$$\frac{v_i - v_B}{R} = 2C\frac{dv_B}{dt} + \frac{v_B + v_o}{R} \quad ; \quad C\frac{d}{dt}(v_i - v_A) = \frac{v_A}{R/2} + C\frac{d}{dt}(v_A + v_o)$$

Next, since the current through the positive lead of the op-amp is zero, we have the current balance at node P as

$$\frac{v_B + v_o}{R} + C\frac{d}{dt}(v_A + v_o) = \frac{-v_o}{R_f}$$

These three equations are written in the Laplace form

$$v_i = 2(\tau s + 1)v_B + v_o \tag{i}$$

$$\tau s v_i = 2(\tau s + 1)v_A + \tau s v_o \tag{ii}$$

$$v_B + (\tau s + 1 + k)v_o + \tau s v_A = 0 \tag{iii}$$

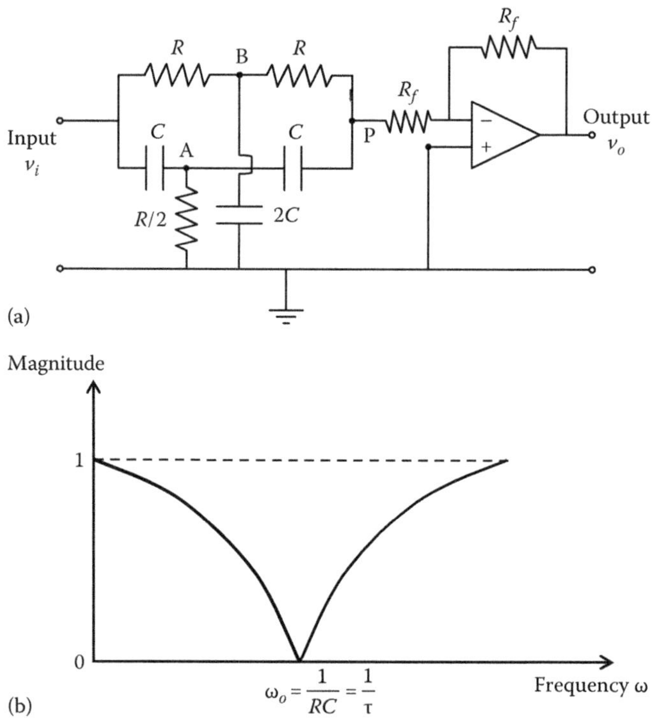

FIGURE 3.14
(a) A twin T filter circuit and (b) frequency response characteristic.

where

$$\tau = RC \quad \text{and} \quad k = \frac{R}{R_f} \tag{3.45}$$

Finally, by eliminating v_A and v_B in Equations i through iii, we get

$$\frac{v_o}{v_i} = G(s) = -\frac{(\tau^2 s^2 + 1)}{[\tau^2 s^2 + (4+k)\tau s + 1 + 2k]} \tag{3.46a}$$

The frequency response function of the filter (with $s = j\omega$) is

$$G(j\omega) = \frac{(1 - \tau^2\omega^2)}{[1 - \tau^2\omega^2 + (4+k)j\tau\omega]} \tag{3.46b}$$

The magnitude of this function becomes zero at the frequency

$$\omega_o = \frac{1}{\tau} \tag{3.47}$$

This is known as the "notch frequency." The magnitude of the frequency response function of the notch filter is sketched in Figure 3.14b. It is noticed that any signal component at frequency ω_o will be completely eliminated by the notch filter. Sharp *roll down* and *roll up* are needed to allow the other (desirable) signal components through without too much attenuation.

While the previous three types of filters achieve their frequency response characteristics through the poles of the filter transfer function, a notch filter achieves its frequency response characteristic through its zeroes (roots of the numerator polynomial equation).

Some useful information about filters is summarized in Box 3.2.

BOX 3.2 FILTERS

ACTIVE FILTERS (NEED EXTERNAL POWER)

Advantages

- Smaller loading errors and low interaction (have high input impedance and low output impedance and, hence, do not affect the input circuit conditions, output signals, and other components)
- Lower cost
- Better accuracy

PASSIVE FILTERS (NO EXTERNAL POWER, USE PASSIVE ELEMENTS)

Advantages

- Passive filters can be used at very high frequencies (e.g., radio frequency).
- A power supply is not needed.

FILTER TYPES

- *Low pass*: Allows frequency components up to the cutoff and rejects the higher-frequency components
- *High pass*: Rejects frequency components up to the cutoff and allows the higher-frequency components
- *Band pass*: Allows frequency components within an interval and rejects the rest
- *Notch (or band reject)*: Rejects frequency components within an interval (usually, a narrowband) and allows the rest

DEFINITIONS

- *Filter order*: Number of poles in the filter circuit or transfer function
- *Anti-aliasing filter*: Low-pass filter with cutoff at less than half the sampling rate (i.e., at less than Nyquist frequency) for digital processing
- *Butterworth filter*: A high-order filter with a flat passband
- *Chebyshev filter*: An optimal filter with uniform ripples in the passband
- *Sallen–Key filter*: An active filter whose output is in phase with the input

3.3.6 Digital Filters

In analog filtering, the filter is a physical dynamic system, typically an electric circuit. The signal to be filtered is applied as the input to this dynamic system. The output of the dynamic system is the filtered signal. In essence, any physical dynamic system can be interpreted as an analog filter.

An analog filter can be represented by a differential equation with respect to time. It takes an analog input signal $u(t)$, which is defined continuously in time t and generates an analog output $y(t)$. A digital filter is a device that accepts a sequence of discrete input values (say, sampled from an analog signal at sampling period Δt), represented by

$$\{u_k\} = \{u_0, u_1, u_2, \ldots\} \tag{3.48}$$

and generates a sequence of discrete output values

$$\{y_k\} = \{y_0, y_1, y_2, \ldots\} \tag{3.49}$$

It follows that a digital filter is a discrete-time system, and it can be represented by a *difference equation*.

An nth-order linear difference equation can be written in the form

$$a_0 y_k + a_1 y_{k-1} + \cdots + a_n y_{k-n} = b_0 u_k + b_1 u_{k-1} + \cdots + b_m u_{k-m} \tag{3.50}$$

This is a recursive algorithm, in the sense that it generates one value of the output sequence using the previous values of the output sequence and all the values of the input sequence up to the present time point. Digital filters that are represented in this manner are termed *recursive digital filters*. There are filters that employ digital processing where a block (a collection of samples) of the input sequence is converted by a one-shot computation into a block of the output sequence. They are not recursive filters. Nonrecursive filters usually employ digital Fourier analysis, the fast Fourier transform algorithm in particular.

3.3.6.1 Software Implementation and Hardware Implementation

In digital filters, signal filtering is accomplished through digital processing of the input signal. The sequence of input data (usually obtained by sampling and digitizing the corresponding analog signal) is processed according to the recursive algorithm of the particular digital filter. This generates the output sequence. The resulting digital output can be converted into an analog signal using a digital-to-analog converter, if so desired (see Chapter 4).

A recursive digital filter is an implementation of a recursive algorithm that governs the particular filtering scheme (e.g., low pass, high pass, band pass, and band reject). The filter algorithm can be implemented either in software or in hardware. In the software implementation, the filter algorithm is programmed into a digital computer. The processor (e.g., microprocessor or DSP) of the computer can process an input data sequence according to the run-time filter program that is stored in the memory (in machine code) to generate the filtered output sequence.

In the software approach, the filter algorithm is programmed and executed in a digital computer. Alternatively, a hardware digital filter can be implemented in an IC chip, using logic elements to carry out the filtering scheme.

The software implementation of digital filters has the advantage of flexibility; specifically, the filter algorithm can be easily modified by changing the software program that is stored in the computer. If, on the other hand, a large number of filters of a particular (fixed) structure are commercially needed, then it would be more economical to design the filter as an IC, which can be mass produced. In this manner, very-low-cost digital filters can be produced. A hardware filter can operate at a much faster speed in comparison to a software filter because in the former case, processing takes place automatically through logic circuitry in the filter chip without using a software program and various data items that are stored in the computer memory. The main disadvantage of a hardware filter is that its algorithm and parameter values cannot be modified, and the filter is dedicated to perform a fixed function.

Summary Sheet

Component interconnection considerations: Impedance matching, signal conversion, and signal conditioning.

Operational amplifier (Op-amp): High input impedance (~2 MΩ), low output impedance (~10 Ω), and very high open-loop voltage (differential) gain (10^5–10^9).

Op-amp equation: $v_o = K_d(v_{ip} - v_{in}) + K_{cm} \times (1/2)(v_{ip} + v_{in})$; noninverting lead voltage, inverting lead voltage v_{in}, K_d = differential gain, K_{cm} = common-mode gain.

Common-mode voltage: $(1/2)(v_{ip} + v_{in})$.

Common-mode rejection ratio (CMRR): K_d/K_{cm}.

Bandwidth: Operating frequency range (e.g., 56 MHz).

Slew rate: Maximum possible rate of change of output voltage, without significantly distorting the output (e.g., 160 V/μs).

Quiescent current: Current drawn by op-amp when output is in open-circuit and no input signals.

Two assumptions for op-amp circuit equations: (1) Voltages at input leads (inverting and noninverting) are equal (due to high differential gain) and (2) currents at an input lead are zero (due to high input impedance).

Instrumentation amplifiers: Take and amplify the difference between two input signals; have op-amp (high input impedance) at each input; gain is adjustable; have tuning capability (error correction).

Applications of instrumentation amplifier: Control hardware such as the comparator, removing common noise (e.g., 60 Hz line noise from the ac power source) from two signals, removing measurable noise or nonlinear component, amplifiers for bridge circuits, amplifiers for sensors and transducers.

Grounding and isolation: Avoiding transmission of electrical noise and harmful signals into instruments.

Filters: Low pass, high pass, band pass, and band reject (including notch).

Analog filters: Use analog circuitry (with elements like op-amp, resistors, and capacitors). Filtering is done by circuit dynamics.

Passive filters: Use passive elements; power source not needed.

Active filters: Use external power source, use op-amps, high input impedance, cheaper, smaller, more accurate, and more efficient.

Filter parameters: Passband = frequency band of allowed signals; cutoff frequency = end frequencies of the passband (determined by filter time constants); poles are = roots of denominator equation (characteristic equation) of the filter transfer function.

Low-pass filter: $G(s) = k/(\tau s + 1)$ (one pole), cutoff frequency = half-power bandwidth = $\omega_c = 1/\tau$, and roll-off rate = −20 dB/decade; $G(s) = \omega_n^2/[s^2 + 2\zeta\omega_n s + \omega_n^2]$ (two poles), cutoff = $\omega_c = \omega_n$, roll-off rate = −40 dB/decade, and optimal filter $\rightarrow \zeta = 1/\sqrt{2} \rightarrow \omega_n$ = half-power bandwidth.

Note: This two-pole filter is better than two cascaded one-pole filters because it requires one op-amp and it is optimal ($\zeta = 1/\sqrt{2}$).

High-pass filter: $G(s) = \tau s/(\tau s + 1)$ (one pole), cutoff frequency $\omega_c = 1/\tau$, and roll-up slope = 20 dB/decade.

Band-pass filter: $G(s) = \tau s/((\tau s + 1)(\tau_2 s + 1))$, cutoff frequencies $\omega_{c1} = 1/\tau$, $\omega_{c2} = 1/\tau_2$, roll-up slope = +20 dB/decade, roll-down slope = −20 dB/decade; $G(s) = \omega_n^2/[s^2 + 2\zeta\omega_n s + \omega_n^2]$ with $\zeta < 1/\sqrt{2} \rightarrow$ resonance-type band-pass filter with half-power bandwidth $\Delta\omega = 2\zeta\omega_n$.

Band-reject (notch) filter: $G(s) = (\tau^2 s^2 + 1)/[\tau^2 s^2 + (4 + k)\tau s + 1 + 2k]$ and notch frequency $\omega_o = 1/\tau$.

Digital filters: Used digital processing for the filtering action; filter model is a difference equation $a_0 y_k + a_1 y_{k-1} + \cdots + a_n y_{k-n} = b_0 u_k + b_1 u_{k-1} + \cdots + b_m u_{k-m}$ or Z transfer function; both software implementation (in a computer program) and hardware implementation (in fixed logic hardware) are possible.

Problems

3.1 In the context of motion sensing, define the following terms:

(a) Mechanical loading
(b) Electrical loading

Explain how these loading effects can be reduced. The following table gives ideal values for some parameters of an op-amp. Give typical, practical values for these parameters (e.g., output impedance of 50 Ω).

Parameter	Ideal Value	Typical Value
Input impedance	Infinity	—
Output impedance	Zero	50 Ω
Gain	Infinity	—
Bandwidth	Infinity	—

Note: Under ideal conditions, inverting lead voltage is equal to the noninverting lead voltage (i.e., offset voltage is zero).

3.2 Usually, an op-amp circuit is analyzed by making use of the following two assumptions:

1. The potential at the positive input lead is equal to the potential at the negative input lead.
2. The current through each of the two input leads is zero.

Explain why these assumptions are valid under unsaturated conditions of an op-amp.

(a) An amateur electronics enthusiast to a circuit connects an op-amp without a feedback element. Even when there is no signal applied to the op-amp, the output was found to oscillate between +12 and −12 V once the power supply is turned on. Give a reason for this behavior.

(b) An op-amp has an open-loop gain of 5×10^5 and a saturated output of ±14 V. If the noninverting input is −1 μV and the inverting input is +0.5 μV, what is the output? If the inverting input is 5 μV and the noninverting input is grounded, what is the output?

3.3 Define the following terms in connection with an op-amp:

(a) Offset current
(b) Offset voltage (at input and output)
(c) Unequal gains
(d) Slew rate

Give typical values for these parameters. The open-loop gain and the input impedance of an op-amp are known to vary with frequency and are known to drift (with time) as well. Still, the op-amp circuits are known to behave very accurately. What is the main reason for this?

3.4 (a) What is a voltage follower? Give a practical use of a voltage follower.

(b) Consider the amplifier circuit shown in Figure P3.4. Determine an expression for the voltage gain K_v of the amplifier in terms of the resistances R and R_f. Is this an inverting amplifier or a noninverting amplifier?

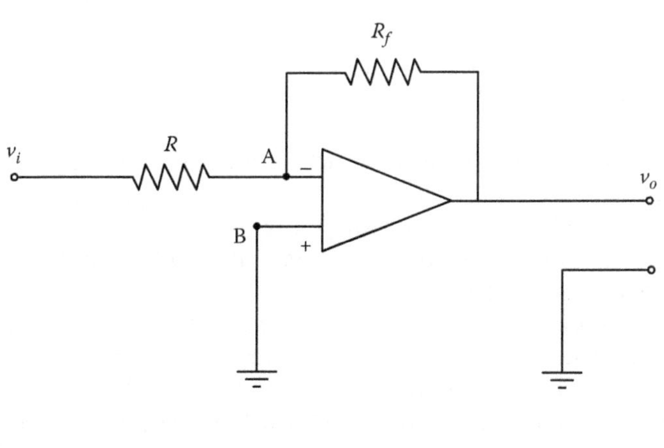

FIGURE P3.4
An amplifier circuit.

3.5 The speed of response of an amplifier may be represented using the three parameters: bandwidth, rise time, and slew rate. For an idealized linear model (transfer function), it can be verified that the rise time and the bandwidth are independent of the size of the input and the DC gain of the system. Since the size of the output (under steady conditions) may be expressed as the product of the input size and the DC gain, it is seen that the rise time and the bandwidth are independent of the amplitude of the output, for a linear model.
Discuss how slew rate is related to bandwidth and rise time of a practical amplifier. Usually, amplifiers have a limiting slew rate value. Show that the bandwidth decreases with the output amplitude in this case.

A voltage follower has a slew rate of 0.5 V/μs. If a sinusoidal voltage of amplitude 2.5 V is applied to this amplifier, estimate the operating bandwidth. If, instead, a step input of magnitude 5 V is applied, estimate the time required for the output to reach 5 V.

3.6 Define the terms

(a) Common-mode voltage
(b) Common-mode gain
(c) CMRR

What is a typical value for the CMRR of an op-amp? Figure P3.6 shows a differential amplifier circuit with a flying capacitor. The switch pairs A and B are turned on and off alternately during operation. For example, first the switches denoted by A are turned on (closed) with the switches B off (open). Next, the switches A are opened and the switches B are closed. Explain why this arrangement provides good common-mode rejection characteristics.

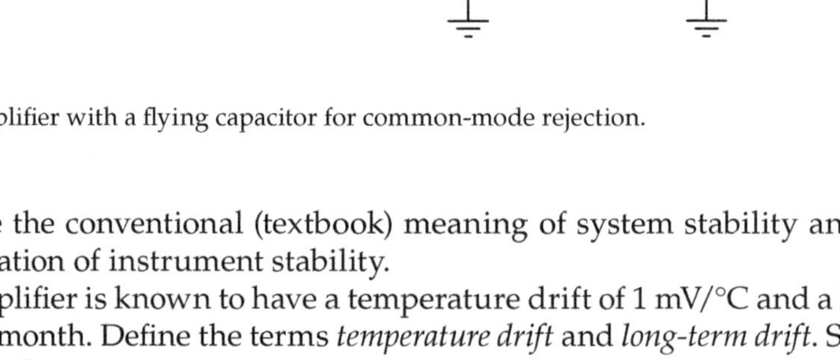

FIGURE P3.6
A differential amplifier with a flying capacitor for common-mode rejection.

3.7 Compare the conventional (textbook) meaning of system stability and the practical interpretation of instrument stability.

An amplifier is known to have a temperature drift of 1 mV/°C and a long-term drift of 25 μV/month. Define the terms *temperature drift* and *long-term drift*. Suggest ways to reduce drift in an instrument.

3.8 Electrical isolation of a device (or circuit) from another device (or circuit) is very useful in the engineering practice. An isolation amplifier may be used to achieve this. It provides a transmission link, which is almost one way and avoids loading problems. In this manner, damage in one component due to increase in signal levels in the other components (perhaps due to short circuits, malfunctions, noise, high common-mode signals, etc.) could be reduced. An isolation amplifier can be constructed from a transformer and a demodulator with other auxiliary components such as filters and amplifiers. Draw a suitable schematic diagram for an isolation amplifier and explain the operation of this device.

3.9 A newspaper report has described a death by electrocution of a person while using a cell phone and a laptop computer. According to the report, the person was using both devices while they were being charged (see Figure P3.9). In particular, the person was wearing headphones, which were connected to the laptop. Burns were found on the ears and the chest of the person. While it was alleged that the cause was the faulty cell phone charger sending a high-voltage electrical pulse into the body, this suspicion cannot be conclusive, which should be clear from Figure P3.9. Discuss possible causes of this electrocution.

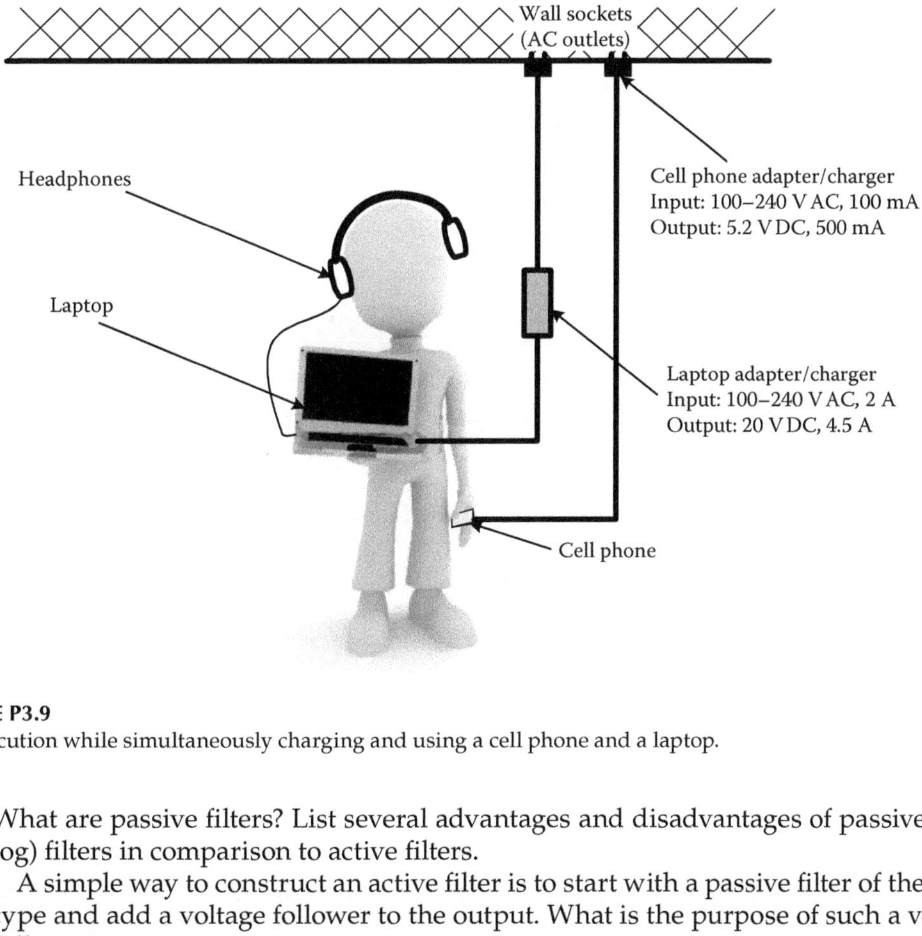

FIGURE P3.9
Electrocution while simultaneously charging and using a cell phone and a laptop.

3.10 What are passive filters? List several advantages and disadvantages of passive (analog) filters in comparison to active filters.

A simple way to construct an active filter is to start with a passive filter of the same type and add a voltage follower to the output. What is the purpose of such a voltage follower?

3.11 Give one application each for the following types of analog filters:

(a) Low-pass filter
(b) High-pass filter
(c) Band-pass filter
(d) Notch filter

Suppose that several single-pole active filter stages are cascaded. Is it possible for the overall (cascaded) filter to possess a resonant peak? Explain.

3.12 Butterworth filter is said to have a maximally flat magnitude. Explain what is meant by this. Give another characteristic that is desired from a practical filter.

3.13 An active filter circuit is given in Figure P3.13.

(a) Obtain the filter transfer function. What is the order of the filter?
(b) Sketch the magnitude of the frequency transfer function. What type of filter does it represent?
(c) Estimate the cutoff frequency and the roll-off slope of the filter.

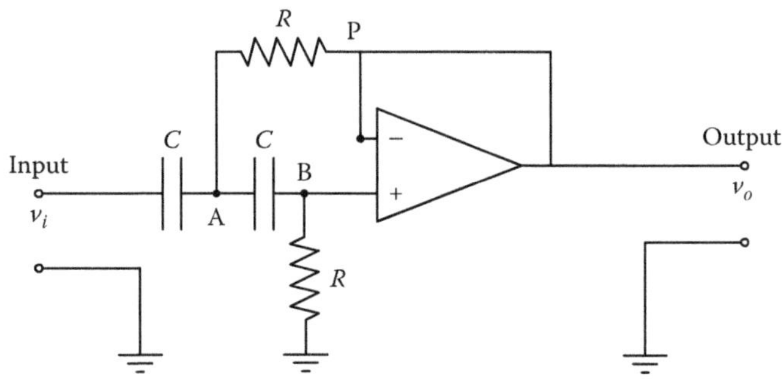

FIGURE P3.13
An active filter circuit.

3.14 Select a set of sensors and identify the type of noise that may be present in the measurement from those sensors. Indicate what type of filtering may be used for filtering out that noise.

3.15 Generate a noisy signal (501 points sampled at sampling periods of 0.02 s), as shown in Figure P3.15, using the MATLAB script:

```
% Filter input data
t=0:0.02:10.0;
u=sin(t)+0.2*sin(10*t);
for i=1:501
u(i)=u(i)+normrnd(0.0,0.1); % normal random noise
end
% plot the results
plot(t,u,'-')
```

(a) Identify some characteristics of this signal (assuming that you did not generate the signal and it was given to you without any description).
(b) Use a four-pole Butterworth low-pass filter with cutoff frequency at 2.0 rad/s and obtain the filtered signal. Describe the nature of this signal.
(c) Use a four-pole Butterworth band-pass filter with the passband: [9.9, 10.1], [9.0, 11.0], and [8.0, 12.0] in rad/s and obtain the filtered signals. Discuss these results.

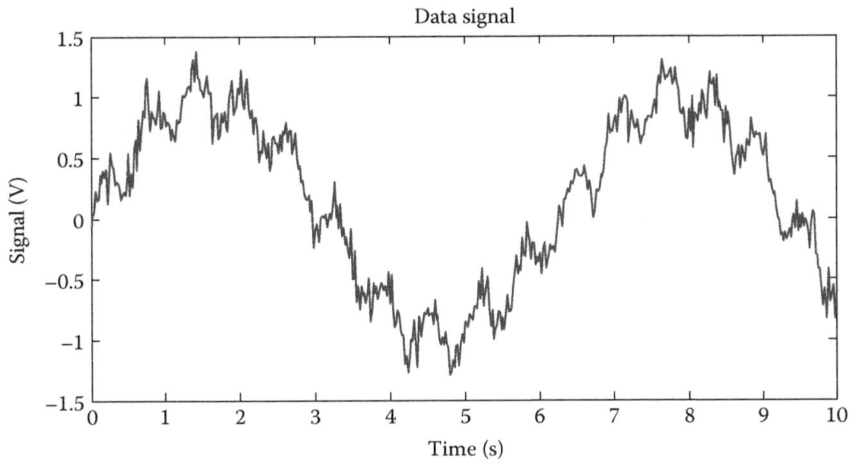

FIGURE P3.15
A noisy signal.

3.16 Single-chip amplifiers with built-in compensation and filtering circuits are becoming popular for signal-conditioning tasks in engineering applications, particularly those associated with data acquisition, machine health monitoring, and control. Signal processing such as integration that would be needed to convert, say, an accelerometer into a velocity sensor could also be accomplished in the analog form using an IC chip. What are the advantages of such signal-modification chips in comparison with the conventional analog signal-conditioning hardware that employs discrete circuit elements and separate components to accomplish various signal-conditioning tasks?

4

Signal Conversion Methods

Chapter Objectives

- Signal Modification and Signal Conversion
- Modulators and Demodulators
- Data Acquisition (DAQ) Hardware
- Digital-to-Analog Conversion (DAC)
- Analog-to-Digital Conversion (ADC)
- Sample-and-Hold Hardware
- Multiplexers
- Bridge Circuits
- Linearizing Devices
- Phase Shifters
- Voltage-to-Frequency Converter (VFC)
- Frequency-to-Voltage Converter (FVC)
- Voltage-to-Current Converter (VCC)
- Peak-Hold Circuit

4.1 Signal Modification and Conversion

Many types of signals can be present in an engineering system. Furthermore, depending on the purpose of the system, it may be required to convert signals into different types and forms. Specifically, a signal leaving one device in a system and entering another device should do so at proper signal levels (i.e., the values of voltage, current, speed, force, and power) and in the proper form (i.e., electrical, mechanical, analog, digital, modulated, demodulated), as required by the specific application. In previous chapters, we have considered several relevant issues such as component *interconnection*, *loading*, and *signal conditioning*, which are all important in *instrumentation* of an engineering system. In this chapter, we will address signal conversion, which is a category of signal modification.

4.1.1 Signal Conversion

Signal modification concerns changing the signal in some manner. Amplification and filtering, which have been studied in Chapter 3, are operations of signal modification. Another category of signal modification involves converting the signal from one type to another. The subject of signal conversion is studied in this chapter. It can be crucial in the development and operation of practical engineering systems.

The operations of signal conversion include analog-to-digital conversion (ADC), digital-to-analog conversion (DAC), voltage-to-frequency conversion, and frequency-to-voltage conversion, amplitude modulation (AM), frequency modulation (FM), phase modulation (PM), pulse-width modulation (PWM), pulse-frequency modulation (PFM), pulse-code modulation (PCM), and demodulation (i.e., the reverse process of modulation). In addition, many other types of useful signal modification operations can be identified. For example, a sample-and-hold (S/H) circuit is used in digital data acquisition (DAQ) systems. Devices such as analog and digital multiplexers (MUXs) and bridge circuits are needed in many applications of sensing and DAQ. Phase shifting, curve shaping, offsetting, and linearization can also be classified as signal conversion.

When an analog signal is acquired by a digital computer for processing, first it has to be digitized. Similarly, when a digital signal from a digital device is used in an analog application (e.g., in driving a device such as a motor), it has to be converted into an analog form. Particularly for transmission, a signal may have to be modulated, digitized, and so on so that the SNR ratio of the transmitted signal is sufficiently large at the receiver and the received signal is in the desired form. Depending on the application, the signal parameter of significance for operating a particular device may be frequency, voltage, or current. Then, if the signal is in a form that is different from what is needed, it has to be converted into the proper form through some means. The significance of signal modification is clear from these observations, and it is an important topic in the study of instrumentation.

4.1.2 Chapter Objectives

This chapter studies several practical operations of signal conversion. Various types of signal conversion devices such as modulators, demodulators, bridge circuits, analog-to-digital converters (ADCs), digital-to-analog converters (DACs), S/H hardware, MUX, bridge circuits, linearizing devices, phase shifter, voltage-to-frequency converters (VFCs), frequency-to-voltage converters (FVCs), voltage-to-current converter (VCC), and peak-hold circuit are discussed. Specific hardware components, designs, and their key parameters and specifications are considered as examples.

4.2 Modulators and Demodulators

Sometimes, signals are deliberately converted into a different form to maintain their authenticity/accuracy during generation, transmission, conditioning, and processing. A specific example is signal modulation where the data signal, known as the "modulating signal," is used to vary (*modulate*) a property (such as amplitude or frequency) of a *carrier signal*. In this manner, the carrier signal is modulated by the data signal. It is this "modulated carrier signal" that is used for subsequent handling (transmission, processing, etc.). After transmitting or conditioning a modulated signal, typically, the data signal has to be recovered by removing the carrier signal. This process is known as "demodulation" or "discrimination."

A variety of modulation techniques exist, and several other types of signal modification (e.g., digitizing) can also be classified as signal modulation even though they may not be commonly termed as such. The following four types of modulation are illustrated in Figure 4.1:

1. AM
2. Frequency modulation (FM)
3. Pulse-width modulation (PWM)
4. PFM

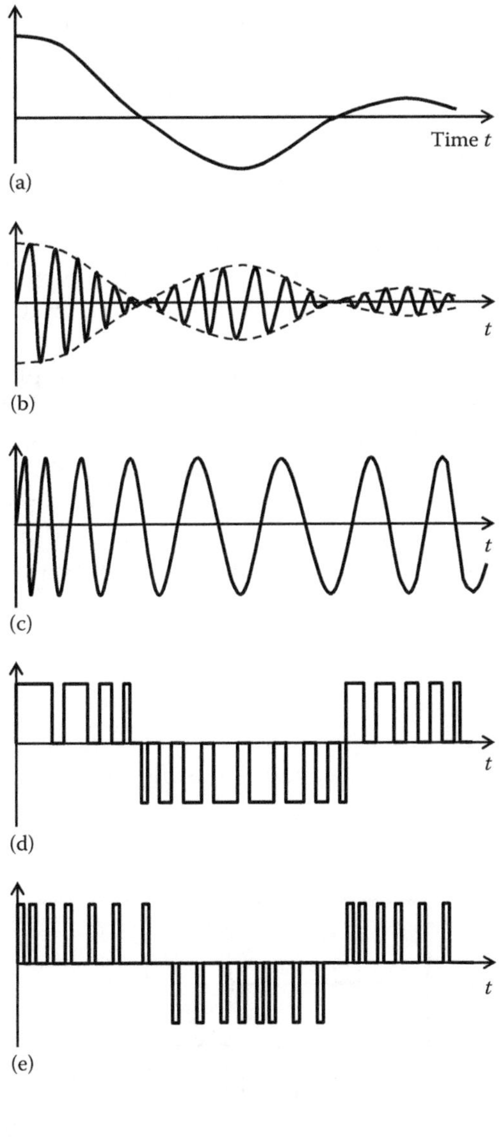

FIGURE 4.1
(a) Modulating signal (data signal), (b) amplitude-modulated signal, (c) frequency-modulated signal, (d) pulse-width-modulated signal, and (e) pulse-frequency-modulated signal.

In AM, the amplitude of a periodic carrier signal is varied according to the amplitude of the data signal (modulating signal) while keeping the frequency of the carrier signal (carrier frequency) constant. Suppose that the transient signal shown in Figure 4.1a is the modulating signal and a high-frequency sinusoidal signal is used as the carrier signal. The resulting amplitude-modulated signal is shown in Figure 4.1b. AM is used in telecommunication, transmission of radio and TV signals, instrumentation, and signal conditioning. The underlying principle is particularly useful in applications such as sensing and instrumentation of engineering systems and fault detection and diagnosis in rotating machinery.

In frequency modulation (FM), the frequency of the carrier signal is varied in proportion to the amplitude of the data signal (modulating signal) while keeping the amplitude of the carrier signal constant. Suppose that the data signal shown in Figure 4.1a is used to frequency modulate a sinusoidal carrier signal. The modulated result will appear as in Figure 4.1c. Since in FM, the information is carried as frequency rather than amplitude, any noise that might alter the signal amplitude would have virtually no effect on the transmitted data. Hence, FM is less susceptible to noise than AM. Furthermore, since in FM the carrier amplitude is kept constant, signal weakening and noise effects that are unavoidable in long-distance data communication/transmission will have less effect than in the case of AM, particularly if the data signal level is low in the beginning. However, more sophisticated techniques and hardware are needed for signal recovery (demodulation) in FM transmission because FM demodulation involves frequency discrimination rather than amplitude detection. FM is also widely used in radio transmission and in data recording and replay.

In pulse-width modulation (PWM), the carrier signal is a pulse sequence of constant amplitude. The pulse width is changed in proportion to the amplitude of the data signal while keeping the pulse spacing (pulse period) constant. This is illustrated in Figure 4.1d. Suppose that the high level of the PWM signal corresponds to the "on" condition of a circuit and the low level corresponds to the "off" condition. Then, as shown in Figure 4.2, the pulse width is equal to the on time ΔT of the circuit within each signal cycle period T. The duty cycle of the PWM is defined as the percentage on time in a pulse period and is given by

$$\text{Duty cycle} = \frac{\Delta T}{T} \times 100\% \tag{4.1}$$

PWM signals are extensively used for controlling electric motors and other mechanical devices such as valves (hydraulic and pneumatic) and machine tools. Note that in a given

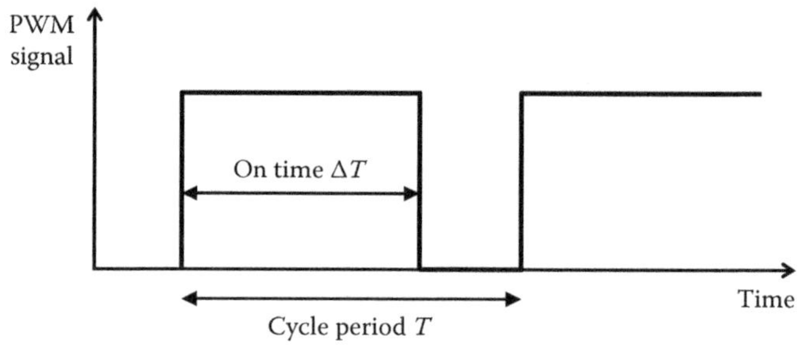

FIGURE 4.2
Duty cycle of a PWM signal.

(short) time interval, the average value of the PWM signal is an estimate of the average value of the data signal in that period. Hence, PWM signals can be used directly in controlling a process, without demodulating it. Advantages of PWM include better energy efficiency (less dissipation) and better performance in nonlinear devices. For example, a device may stick at low speeds due to Coulomb friction. This can be avoided by using a PWM signal with an amplitude that is sufficient to overcome friction, while maintaining the required average control signal, which might be very small.

In PFM, as well, the carrier signal is a pulse sequence of constant amplitude. In this method, it is the frequency (or period) of the pulses that is changed in proportion to the value of the data signal, while keeping the pulse-width constant. PFM has the same advantages as those of ordinary FM. Further advantages exist due to the fact that electronic circuits (digital circuits in particular) can handle pulses very efficiently. Furthermore, pulse detection is not susceptible to noise because it involves distinguishing between the presence and the absence of a pulse, rather than accurate determination of the pulse amplitude (or width). PFM may be used in place of PWM in most applications, with better results.

Another type of modulation is phase modulation (PM). In this method, the phase angle of the carrier signal is varied in proportion to the amplitude of the data signal. Conversion of discrete (sampled) data into the digital (binary) form is also considered a form of modulation. In fact, this is termed PCM. In PCM, each discrete data sample is represented by a binary number containing a fixed number of binary digits (bits). Since each digit in the binary number can take only two values, 0 or 1, it can be represented by the absence or the presence of a voltage pulse. Hence, each data sample can be transmitted using a set of pulses. This is known as "encoding." At the receiver, the pulses have to be interpreted (or decoded) to determine the data value. As with any other pulse technique, PCM is quite immune to noise because decoding involves detection of the presence or absence of a pulse, rather than determination of the exact magnitude of the pulse signal level. Also, since pulse amplitude is constant, long-distance signal transmission (of the digital data) can be accomplished without the danger of considerable signal weakening and associated distortion. Of course, there will be some error introduced by the digitization process itself, which is governed by the finite word size (or dynamic range) of the binary data element. This is known as the "quantization error" and is unavoidable in signal digitization.

In any type of signal modulation, it is essential to preserve the algebraic sign of the modulating signal (data). Different types of modulators handle this in different ways. For example, in PCM, an extra sign bit is added to represent the sign of the transmitted data sample. In AM and FM, a phase-sensitive demodulator is used to extract the original (modulating) signal with the correct algebraic sign. In AM and FM, a sign change in the modulating signal can be represented by a 180° phase change in the modulated signal. This is not quite noticeable in Figure 4.1b and c. In PWM and PFM, a sign change in the modulating signal can be represented by changing the sign of the pulses, as shown in Figure 4.1d and e. In PM, a positive range of phase angles (say 0 to π) could be assigned for the positive values of the data signal, and a negative range of phase angles (say $-\pi$ to 0) could be assigned for the negative values of the signal.

4.2.1 Amplitude Modulation

AM can naturally and inherently enter into many physical phenomena. More important, perhaps, is the deliberate (artificial or practical) use of AM to facilitate data transmission and signal conditioning. Let us first examine the mathematics of AM.

AM is achieved by multiplying the data signal (modulating signal) $x(t)$ by a high-frequency (periodic) carrier signal $x_c(t)$. Hence, amplitude-modulated signal $x_a(t)$ is given by

$$x_a(t) = x(t)x_c(t) \tag{4.2}$$

The carrier can be any periodic signal such as harmonic (sinusoidal), square wave, or triangular. The main requirement is that the fundamental frequency of the carrier signal (carrier frequency) f_c be significantly large (say, by a factor of 5 or 10) than the highest frequency of interest (bandwidth) of the data signal (also see Chapter 6 for further treatment of signal bandwidth). Analysis can be simplified by assuming a sinusoidal carrier frequency, however. Thus,

$$x_c(t) = a_c \cos 2\pi f_c t \tag{4.3}$$

4.2.1.1 Analog, Discrete, and Digital AM

In analog AM, what is transmitted is the analog signal $x_a(t) = x(t)x_c(t)$, which is continuous in time. Less dissipative and more efficient is the discrete AM, also called pulse AM (PAM). Here, the modulated analog signal $x_a(t)$ is sampled, and the resulting discrete values (or pulses), whose magnitude is the signal magnitude, are transmitted. According to *Shannon's sampling theorem* (see Chapter 6), the signal has to be sampled at a minimum rate of twice the maximum frequency of interest (which is the carrier frequency) in the signal. In PAM, what is transmitted are signal magnitudes, not their digital representations. Hence, they are still prone to noise. An AM method that is immune to noise during transmission is digital AM or pulse-coded AM (PCM). Here, the data samples are first digitized (represented as a digital word), and the corresponding bits are transmitted. In fact, in PCM, the actual modulation (i.e., the product operation in Equation 4.2) may be performed digitally with digitized carrier and data (modulating) signals and then the digital (coded) data are transmitted. This method is more efficient (with regard to power loss, etc.) and far more immune to noise during transmission than analog AM and PAM.

4.2.1.2 Modulation Theorem

Known also as the "frequency-shifting theorem," this relates the fact that if a signal is multiplied by a sinusoidal signal, the Fourier spectrum of the product signal is simply the Fourier spectrum of the original signal shifted through the frequency of the multiplied sinusoidal signal. In other words, the Fourier spectrum $X_a(f)$ of the amplitude-modulated signal $x_a(t)$ can be obtained from the Fourier spectrum $X(f)$ of the original data signal $x(t)$, simply by shifting the spectrum of the original signal through the carrier frequency f_c. It is this shifted spectrum that is transmitted.

To mathematically explain the modulation theorem, we use the definition of the Fourier integral transform to get $X_a(f) = a_c \int_{-\infty}^{\infty} x(t)\cos 2\pi f_c t \exp(-j2\pi f t)dt$. Next, since $\cos 2\pi f_c t = (1/2)[\exp(j2\pi f_c t) + \exp(-j2\pi f_c t)]$, we have

$$X_a(f) = \frac{1}{2}a_c \int_{-\infty}^{\infty} x(t)\exp[-j2\pi(f - f_c)t]dt + \frac{1}{2}a_c \int_{-\infty}^{\infty} x(t)\exp[-j2\pi(f + f_c)t]dt$$

or

$$X_a(f) = \frac{1}{2}a_c[X(f - f_c) + X(f + f_c)] \tag{4.4}$$

This equation is the mathematical statement of the modulation theorem. It is schematically illustrated in Figure 4.3. Consider a transient signal $x(t)$ with a (continuous) Fourier spectrum $X(f)$, whose magnitude $|X(f)|$ is as shown in Figure 4.3a. If this signal is used to amplitude modulate a high-frequency sinusoidal carrier signal of frequency f_c, the resulting modulated signal $x_a(t)$ and the magnitude of its Fourier spectrum are as shown in Figure 4.3b. As given by Equation 4.4, the magnitude has been multiplied by $a_c/2$.

Note: In this schematic example, it is assumed that the data signal is band limited, with bandwidth f_b. Of course, the modulation theorem is not limited to band-limited signals, but for practical reasons, we need to have some upper limit on the useful frequency of the data signal. Also, for practical reasons (not for the theorem itself), the carrier frequency f_c should be several times larger than f_b so that there is a reasonably wide frequency band from 0 to $(f_c - f_b)$, within which the magnitude of the modulated signal is virtually zero. The significance of this should be clear when we discuss the applications of AM.

Figure 4.3 shows only the magnitude of the frequency spectra. However, every Fourier spectrum has a phase angle spectrum as well. This is not shown for the sake of brevity. But, clearly, the phase angle spectrum is also similarly affected (frequency shifted) by AM.

4.2.1.3 Side Frequencies and Sidebands

The modulation theorem, as described earlier, assumes transient data signals with associated continuous Fourier spectra. The same ideas are equally applicable to periodic signals (with discrete spectra). Periodic signals represent merely a special case of what was discussed earlier and can be analyzed directly by using the Fourier integral transform. Then, however, we have to cope with impulsive spectral lines (for discrete spectra). Alternatively, Fourier series expansion may be employed, thereby avoiding the introduction of impulsive discrete spectra into the analysis. As shown in Figure 4.3c and d, however, no analysis is actually needed for the case of periodic signals because the final answer can be deduced from the results for a transient signal. Specifically, in the Fourier series expansion of the data signal, each frequency component f_o of amplitude $a/2$ will be shifted by $\pm f_c$ to the two new frequency locations $f_c + f_o$ and $-f_c + f_o$ with an associated amplitude $aa_c/4$. The negative frequency component $-f_o$ should also be considered in the same way, as illustrated in Figure 4.3d. The modulated signal does not have a spectral component at the carrier frequency f_c but rather, on each side of it, at $f_c \pm f_o$. Hence, these spectral components are termed *side frequencies*. When a band of side frequencies is present, as in Figure 4.3b, it is termed a *sideband*. Side frequencies are very useful in fault detection and diagnosis of rotating machinery.

4.2.2 Application of Amplitude Modulation

The main hardware component of an amplitude modulator is an analog multiplier. It is commercially available in the monolithic IC form. Alternatively, it can be assembled using IC operational amplifiers (op-amps) and various discrete circuit elements.

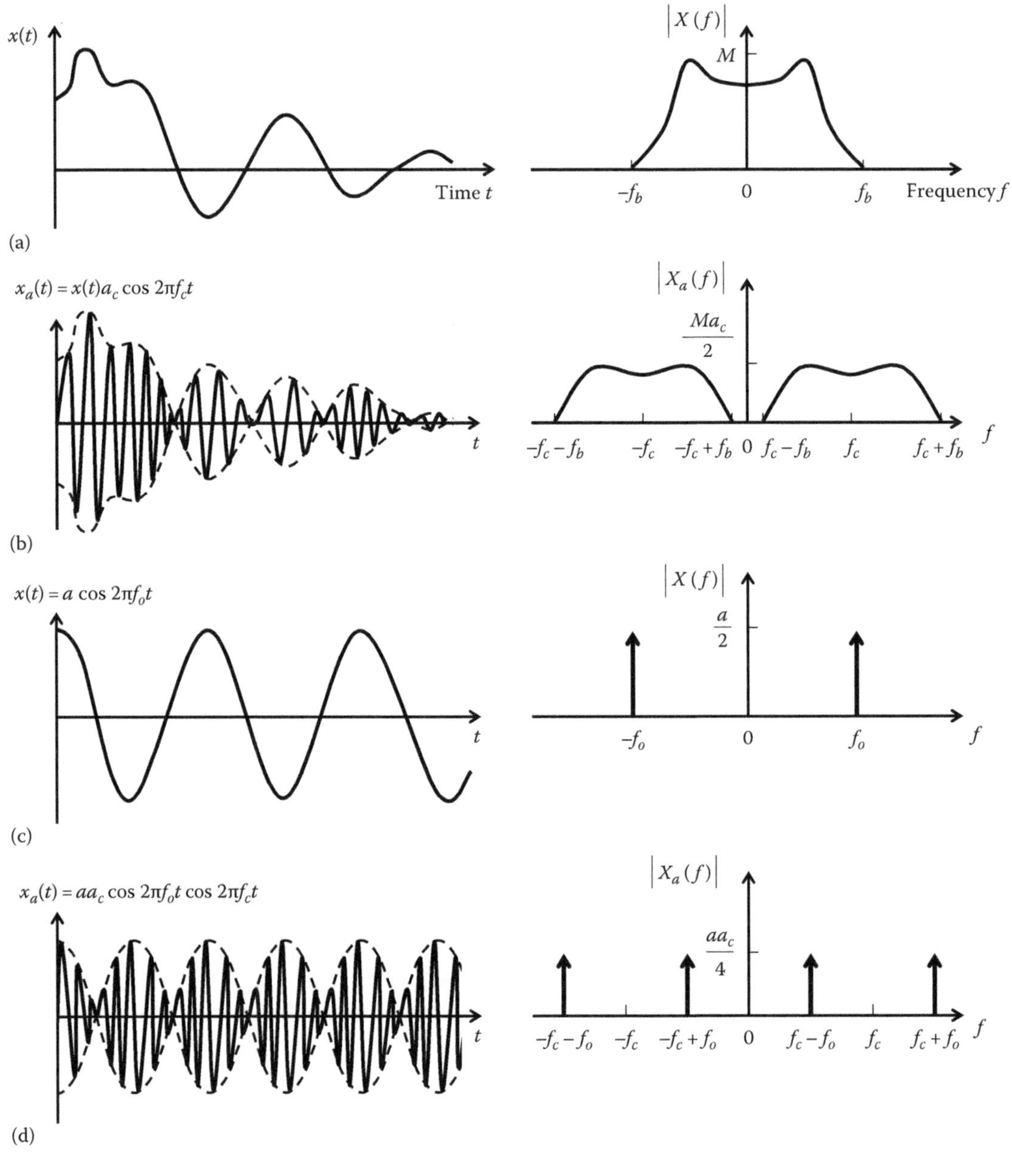

FIGURE 4.3
Illustration of the modulation theorem: (a) a transient data signal and its Fourier spectrum magnitude, (b) amplitude-modulated signal and its Fourier spectrum magnitude, (c) a sinusoidal data signal, and (d) amplitude modulation by a sinusoidal signal.

Schematic representation of an amplitude modulator is shown in Figure 4.4. In practice, to achieve satisfactory modulation, other components such as signal preamplifiers and filters will be needed.

There are many applications of AM. In some applications, modulation is performed intentionally. In others, modulation occurs naturally as a consequence of the physical

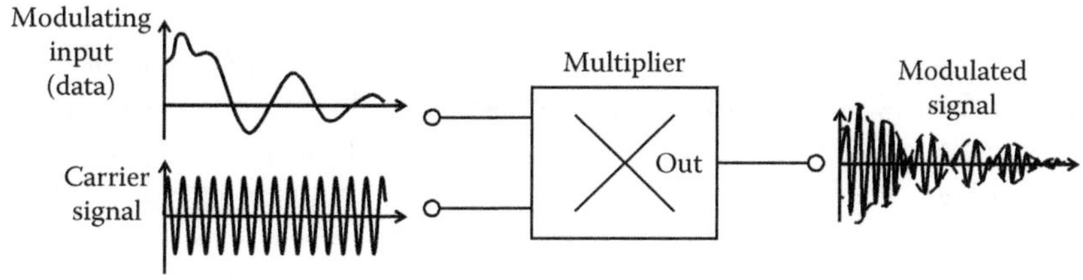

FIGURE 4.4
Representation of an amplitude modulator.

process, and the resulting signal is used to meet a practical objective. Typical applications of AM include the following:

1. Conditioning of general signals (including DC, transient, and low frequency) by exploiting the advantages of AC signal-conditioning hardware
2. Making low-frequency signals immune to low-frequency noise
3. Transmission of general signals (DC, low frequency, etc.) by exploiting the advantages of AC signal transmission
4. Transmission of low-level signals under noisy conditions
5. Transmission of several signals simultaneously through the same medium (e.g., same telephone line, same transmission antenna)
6. Fault detection and diagnosis of rotating machinery

The role of AM in many of these applications should be obvious if one understands the frequency-shifting property of AM. Several other types of applications are also feasible due to the fact that the power of the carrier signal can be increased somewhat arbitrarily, irrespective of the power level of the data (modulating) signal. Let us discuss, one by one, the listed six categories of applications.

1. *Signal conditioning*: AC signal-conditioning devices such as AC amplifiers are known to be more stable than their DC counterparts. In particular, drift (instability) problems are not as severe and nonlinearity effects are lower in AC signal-conditioning devices. Hence, instead of conditioning a DC signal using DC hardware, we can first use the signal to modulate a high-frequency carrier signal. Then, the resulting high-frequency-modulated signal (AC) can be conditioned more effectively using AC hardware.
2. *Noise immunity*: The frequency-shifting property of AM can be exploited in making low-frequency signals immune to low-frequency noise. Note from Figure 4.3 that using AM, the low-frequency spectrum of the modulating signal can be shifted out into a very high-frequency region, by choosing a carrier frequency f_c that is sufficiently large. Then, any low-frequency noise (within the band 0 to $f_c - f_b$) would not distort the spectrum of the modulated signal. Hence, this noise could be removed by a high-pass filter (with cutoff at $f_c - f_b$) so that it would not affect the data. Finally, the original data signal can be recovered using demodulation. Since the frequency of a noise component can very well be within the bandwidth f_b of the data signal, if AM was not employed, the noise would directly distort the data signal.

3. *AC signal transmission*: Transmission of AC signals is more efficient than the transmission of DC signals. Advantages of AC transmission include lower energy dissipation problems. As a result, a modulated signal can be transmitted over long distances more effectively than could the original data signal alone. Furthermore, the transmission of low-frequency (large wavelength) signals requires large antennas. Hence, when AM is employed (with an associated reduction in signal wavelength), the size of the broadcast antenna can be effectively reduced.
4. *Weak signal transmission*: Transmission of weak signals over long distances is not desirable because further signal weakening and corruption by noise can produce disastrous results. Even if the power of the data signal is low, by increasing the power of the carrier signal to a sufficiently high level, the strength of the resulting modulated signal can be elevated to an adequate level, for long-distance transmission.
5. *Simultaneous signal transmission*: It is not possible to transmit two or more signals in the same frequency range simultaneously using a single telephone line. This problem can be resolved by using carrier signals with significantly different carrier frequencies to amplitude modulate the data signals. By picking the carrier frequencies sufficiently farther apart, the spectra of the modulated signals can be made nonoverlapping, thereby making simultaneous transmission possible. Similarly, with AM, simultaneous broadcasting by several radio (AM) broadcast stations in the same broadcast area has become possible.
6. *Fault detection and diagnosis*: *A manifestation of AM that is* particularly useful in the practice of electromechanical systems is in the fault detection and diagnosis of rotating machinery. In this method, modulation is not deliberately introduced, but rather results from the dynamics of the machine. Flaws and faults in a rotating machine are known to produce periodic forcing signals at frequencies higher than, and that are typically integer multiples of, the rotating speed of the machine. For example, backlash in a gear pair will generate forces at the tooth-meshing frequency (equal to the product: number of teeth × gear-rotating speed). Flaws in roller bearings can generate forcing signals at frequencies proportional to the product of the rotating speed and the number of rollers in the bearing race. Similarly, blade passing in turbines and compressors and eccentricity and unbalance in a rotor can generate forcing components at frequencies that are integer multiples of the rotating speed. The resulting system response (e.g., acceleration in the housing) is clearly an amplitude-modulated signal, where the rotating response of the machine modulates the high-frequency forcing response. This can be confirmed experimentally through Fourier analysis (fast Fourier transform [FFT]) of the resulting response signals. For a gearbox, for example, it will be noticed that, instead of getting a spectral peak at the gear-tooth-meshing frequency, two sidebands are produced around that frequency. Faults can be detected by monitoring the evolution of these sidebands. Furthermore, since sidebands are the result of modulation of a specific forcing phenomenon (e.g., gear-tooth meshing, bearing-roller hammer, turbine-blade passing, unbalance, eccentricity, misalignment), one can trace the source of a particular fault (i.e., diagnose the fault) by studying the Fourier spectrum of the measured response.

AM is an integral part of many types of sensors. In these sensors, a high-frequency carrier signal (typically the AC excitation in a primary winding) is modulated by the motion that is sensed. Actual motion signal can be recovered by demodulating the output. Examples of sensors that generate modulated outputs are differential transformers (linear-variable differential transducer or transformer [LVDT], RVDT), magnetic-induction proximity sensors, eddy current proximity sensors, AC tachometers, and strain-gauge devices that use AC bridge circuits. These are discussed in Chapters 8 and 9. Signal conditioning and transmission are facilitated by AM in these cases. The signal has to be demodulated at the end, for most practical purposes such as analysis and recording.

4.2.3 Demodulation

Demodulation or discrimination or detection is the process of extracting the original data signal from a modulated signal. In general, demodulation has to be *phase sensitive* in the sense that the algebraic sign of the data signal should be preserved and determined by the demodulation process. In *full-wave demodulation*, an output is generated continuously. In *half-wave demodulation*, no output is generated for every alternate half period of the carrier signal.

A simple and straightforward method of demodulation is by detection of the envelope of the modulated signal. For this method to be feasible, the carrier signal must be quite powerful (i.e., signal level has to be high) and the carrier frequency also should be very high. An alternative method of demodulation, which generally provides more reliable results, involves a further step of modulation performed on the already modulated signal, followed by low-pass filtering. This method can be explained by referring to Figure 4.3.

Consider the amplitude-modulated signal $x_a(t)$ shown in Figure 4.3b. Multiply this signal by the scaled sinusoidal carrier signal $2/a_c \cos 2\pi f_c t$. We get

$$\tilde{x}(t) = \frac{2}{a_c} x_a(t) \cos 2\pi f_c t \tag{4.5}$$

Now, by applying the modulation theorem (Equation 4.4) to Equation 4.5, we get the Fourier spectrum of $\tilde{x}(t)$ as

$$\tilde{X}(f) = \frac{1}{2}\frac{2}{a_c}\left[\frac{1}{2}a_c\{X(f-2f_c)+X(f)\}+\frac{1}{2}a_c\{X(f)+X(f+2f_c)\}\right]$$

or

$$\tilde{X}(f) = X(f) + \frac{1}{2}X(f-2f_c) + \frac{1}{2}X(f+2f_c) \tag{4.6}$$

The magnitude of this spectrum is shown in Figure 4.5a. Observe that we have recovered the spectrum $X(f)$ of the original data signal, but there exist two sidebands at locations far removed (centered at $\pm 2f_c$) from the bandwidth of the original signal. We can conveniently low-pass filter the signal $\tilde{x}(t)$ using a filter with cutoff at f_b to recover the original data signal. A schematic representation of this method of amplitude demodulation is shown in Figure 4.5b.

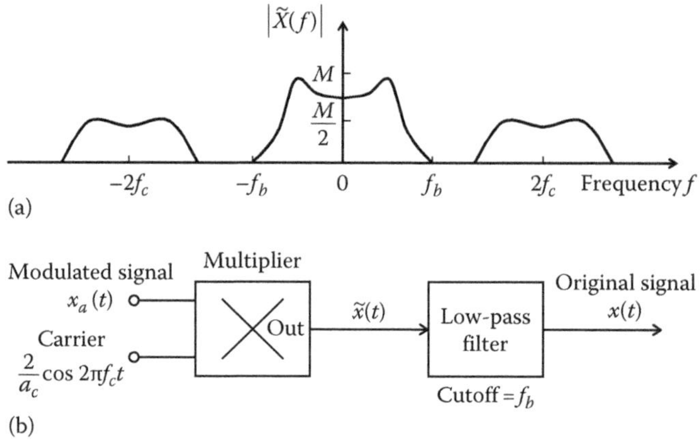

FIGURE 4.5
Amplitude demodulation: (a) spectrum of the signal after the second modulation and (b) demodulation schematic diagram (modulation + filtering).

4.2.3.1 Advantages and Disadvantages of AM

The main advantage of AM is the use of a carrier signal (of higher power and higher frequency) to "carry" the information of the data signal (modulating signal). The data are transmitted at a much higher frequency (in the form of sidebands) than that of the data signal and are recovered (through demodulating) at the receiver end. Also, the modulation process is quite simple (multiplication of two signals). However, there are several disadvantages of AM. They include the following:

1. Since an analog signal of high power and high frequency is transmitted, power loss during transmission can be high. Hence, it is somewhat wasteful and not quite efficient.
2. Since the amplitude of the transmitted signal varies with that of the data signal, it is prone to noise (at low signal-to-noise ratio [SNR]), when the signal level is low.
3. AM signal uses up more bandwidth since the carrier signal has to be transmitted as well as the data signal.

The key disadvantages of AM can be overcome by using digital AM (or PCM) or other methods of modulation such as FM and PWM where the modulated signal has a constant amplitude (and also digital methods can be used with added advantages).

4.2.3.2 Double-Sideband Suppressed Carrier

The AM given by Equation 4.2: $x_a(t) = x(t)x_c(t)$ is called "suppressed carrier" AM or double-sideband suppressed carrier (DSBSC) AM. As shown in Figure 4.3b, its spectrum comprises the two sidebands, which are the frequency-shifted spectra of the data signal (modulating signal). Since it is these two sidebands that are transmitted, it is rather efficient with respect to signal power. Often, however, AM is represented by

$$x_a(t) = x_c(t) + x(t)x_c(t) = (1 + x(t))x_c(t) \tag{4.7}$$

Here, the carrier signal is added to the product signal so that the product signal rides on the carrier signal. This overall modulated signal has more power. Then, a *modulation index* is defined as

$$\text{Modulation index} = \frac{\text{Amplitude of data signal}}{\text{Amplitude of carrier signal}} \tag{4.8}$$

Clearly, Equations 4.2 and 4.7 carry the same information content. So, in theory, they are equivalent. In particular, at high levels of modulation index, the two modulated signals are quite similar, as shown in Figure 4.6. However, the nature and the power content of the two types of modulated signals are different. When power efficient modulation is important, the AM given by Equation 4.2 is suitable. When high-power AM is desired, the AM given by Equation 4.7 is preferred.

4.2.3.3 Analog AM Hardware

The most critical component in analog AM is the analog multiplier, where the data signal and the carrier signal are multiplied. Analog hardware multipliers are commercially available. For example, an analog multiplier that can multiply two analog signals and add to the product another signal (say, add the carrier), which is exactly the AM operation given by Equation 4.7 in the frequency range DC to 2 GHz, is available as an IC package. The feature of product scaling [called gain scaling] is available as well, which corresponds to setting the modulation index.

Note: The miniature [3 mm] analog multiplier [or amplitude modulator] IC package ADL5391 from analog devices has 16 leads corresponding to 3 differential signal inputs [6], differential output [2], DC supply voltage leads [3] for 4.5–5.5 V, device common leads [2], scaling input [1], chip enable [1], and DC reference output [1].

There are some drawbacks of analog multiplication. It is a nonlinear operation with corresponding disadvantages. The noise in either signal will affect the product. Furthermore, the effect on the phase angle is much more complex.

4.3 Data Acquisition Hardware

Engineering systems use digital DAQ for a variety of purposes such as process condition monitoring and performance evaluation, fault detection and diagnosis, product quality assessment, dynamic testing, system identification (i.e., experimental modeling), and process control. A typical DAQ system consists of the following key components:

1. Sensors and transducers (to measure the variables in the process that is monitored)
2. Signal conditioning (filtering and amplification of the sensed signals)
3. DAQ hardware (to receive different types of monitored signals and make them available to the bus of a computer. [*Note*: Some signal conditioning is typically built into the DAQ unit itself.])
4. Computer (personal computer [PC], laptop, microcontroller, microprocessor, etc., to process the acquired signals so as to achieve the end objective of the DAQ system)

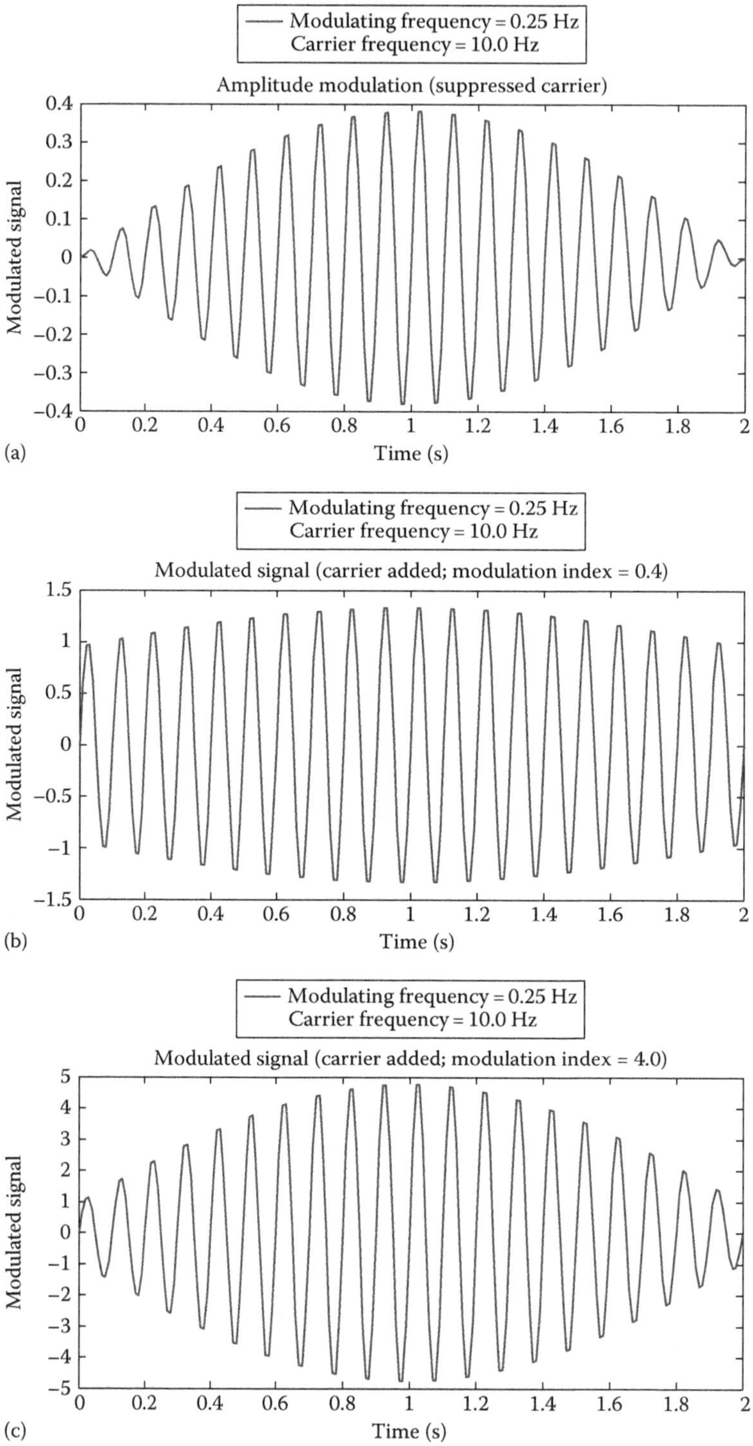

FIGURE 4.6
(a) Modulated signal with suppressed carrier, (b) modulated signal with added carrier at modulation index = 0.4, and (c) modulated signal with added carrier at modulation index = 4.0.

5. Power supply (external signal conditioning and active sensors will need power; DAQ power typically comes from the computer)
6. Software (driver software to operate the DAQ hardware for properly acquiring the sensed data; application software for use by the computer to process the data for the end objective)

Consider the process monitoring and control system shown in Figure 4.7. Typically, the measured variables (responses or outputs, inputs) of a physical system (process, plant, machine) are available in the analog form, which are signals that are continuous in time. Furthermore, typically, the drive signals (or control inputs) for a physical system have to be provided in the analog form. These signals may have to be filtered to remove the undesirable components and amplified to bring the signals to proper levels for further use. Filtering and amplification have been studied in Chapter 3. A digital computer is an integral component of a typical engineering system and may take the form of a PC, laptop, or one or more general-purpose microprocessors with powerful processing capability or more specific microcontrollers with extensive input–output (I/O) capabilities. For additional processing power, coprocessors such as digital signal processors may be incorporated. In the system, the digital computer will perform tasks such as signal processing, data analysis and reduction, parameter estimation and model identification, diagnosis, performance analysis, decision-making, tuning, and control. Essentially, the computer performs the end objective of the monitoring and DAQ process.

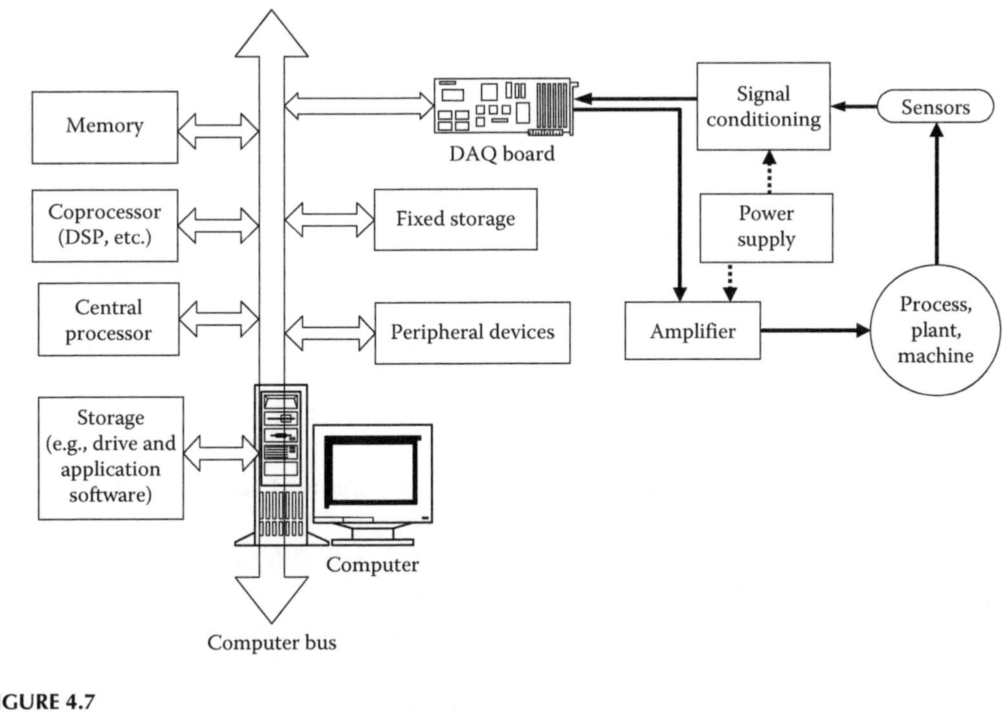

FIGURE 4.7
Components of a process monitoring and control system.

Computer Architecture and Hardware

The computer uses a bus (e.g., peripheral component interconnect [PCI] bus) to transfer data between components in the computer. In a typical PC, the DAQ card goes into an expansion slot of the computer. The power for DAQ comes from the PC itself. The operation of the DAQ is managed by "driver software," which is provided by the DAQ supplier and is stored in the computer. The driver software has to be compatible with the operating system (e.g., Windows, Mac OS, Linux) of the computer. This software manages accessing data from the DAQ and making them available to the computer for further processing. This further processing is done by "application software," which may be programmed using such tools as MATLAB® and LabVIEW or using high-level programming languages such as C and C++. This software not only will process the acquired data to achieve the end objective (performance analysis, diagnosis, model identification, control, etc.) but also may be used to develop a suitable graphical user interface for the monitoring system.

Motherboard: The motherboard (or main board or system board) of a computer represents interconnected key hardware components of a computer. External devices and I/O ports are also connected to the motherboard through a computer bus. Various IC packages and other hardware devices are mounted on the motherboard, which is located in the computer housing. Other devices (various cards including DAQ) are mounted in the expansion slots of the computer housing. A typical architecture of a computer motherboard is shown in Figure 4.8a. It shows the main components such as the central processing unit (CPU), memory, and clock; expansion slots for hardware such as DAQ, network card, video card, storage, sound card, and memory expansion card; and I/O ports for peripheral devices and communication, such as monitor, keyboard, mouse, printer, scanner, external storage, and local area network.

Here are some acronyms that are used in the context of computer hardware, operation, and communication:

SCSI: *small computer system interface*. Standards and protocols for connecting and transferring data between computers and peripheral devices such as hard drives, CD drives, and scanners.

EISA: *extended industry standard architecture*. A bus standard for PCs.

PCI bus: A popular bus of a PC for connecting hardware devices in it and data transfer between them.

Internal bus: A bus for connecting the internal hardware of a computer. Also known as system bus and front-side bus.

External bus: A bus for connecting external hardware to a computer. Also known as expansion bus.

USB: *universal serial bus*. A computer bus for connection and communication with peripheral devices.

FIFO: *first in, first out*. A method for arranging data in a buffer or stack, where the oldest data (bottom of the stack) are processed first.

DMA: *direct memory access*. Capability where a hardware component in the computer can directly access the computer memory (without going through the CPU).

RS-232: A standard for serial communication of data.

RS-422: Extends the range of RS-232 connections.

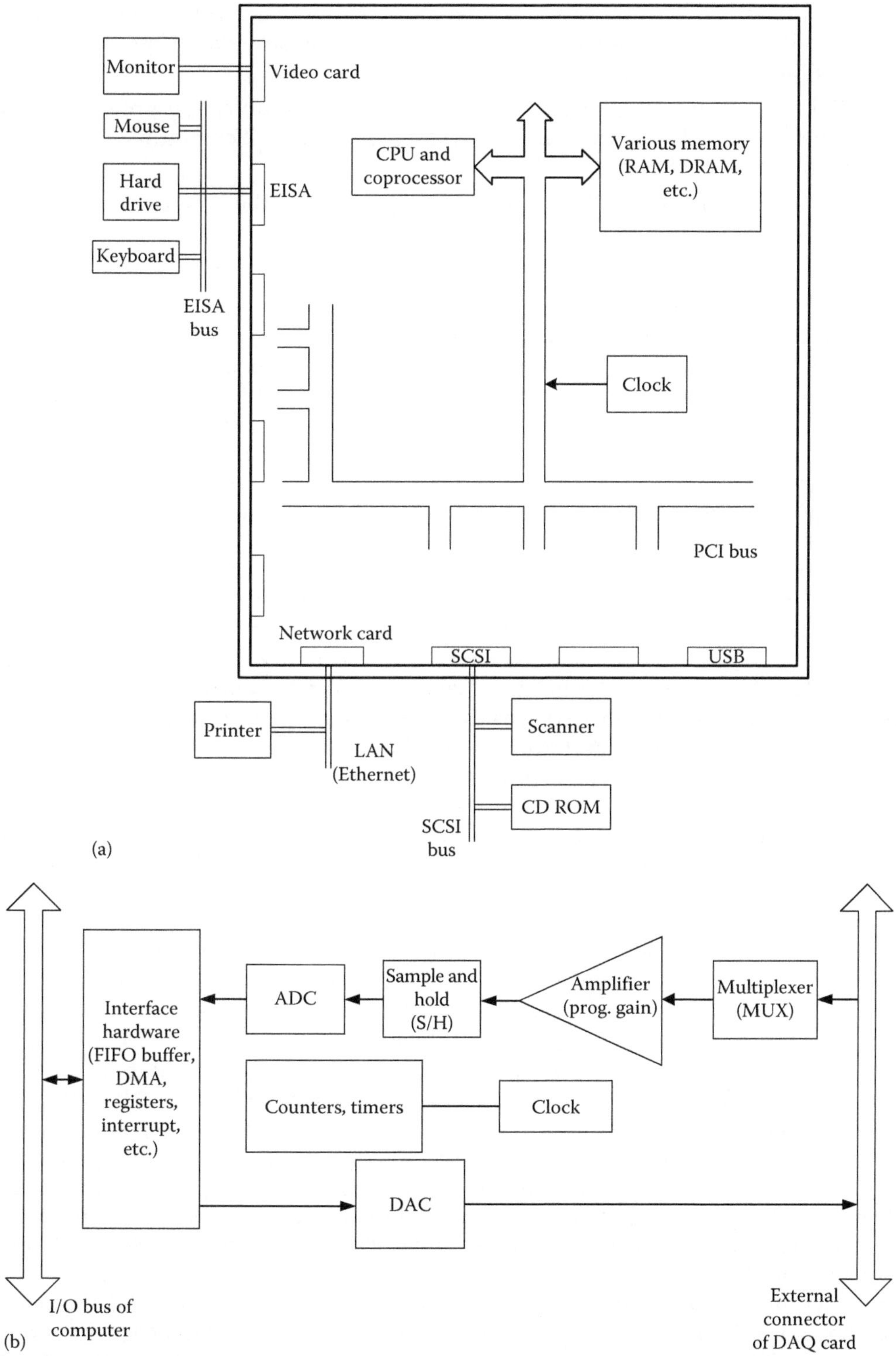

FIGURE 4.8
(a) Hardware components of a computer. (b) Main components of a data acquisition card of a computer.

UART: *universal asynchronous receiver/transmitter.* A hardware component that converts data between parallel and serial forms for transmission. Commonly used with RS-232 and RS-422.

TCP/IP: *transmission control protocol* (*TCP*) is a core communication protocol of the Internet protocol (IP) suite. This is a protocol for network communication. More reliable at the expense of speed.

UDP: *user datagram protocol.* A communication protocol in the IP suite. Faster, at the expense of reliability.

Data Acquisition and Analog-to-Digital Conversion

Inputs to a digital device (typically, a computer or a microcontroller) and outputs from a digital device are necessarily present in the digital form. Hence, when a digital device is interfaced with an analog device (e.g., sensor, actuator), the interface hardware and associated driver software have to perform several important functions. Two of the most important components of interface hardware are *digital-to-analog converter* (DAC) and *analog-to-digital converter* (ADC). An analog signal has to be converted into the digital form, using an ADC, according to an appropriate code, before it is read by a digital processor. For this, the analog signal is first "sampled" into a sequence of discrete values, and each discrete value is converted into the digital form. During this conversion, the sampled discrete value has to be maintained constant by means of S/H hardware. If multiple signals (from multiple sensors) are acquired simultaneously, an analog or digital MUX may have to be used to read the multiple signals sequentially by the computer. On the other hand, a digital output from a computer has to be converted into the analog form, using a DAC, for feeding into an analog device such as drive amplifier, actuator or analog recording, or display unit. DAC, ADC, S/H, and MUX are studied in this section.

Both ADC and DAC are elements or components in a typical DAQ card (or I/O board, or I/O board, or DAQ and control card or DAC). Complete DAQ cards and associated driver software are available from such companies as the National Instruments, ADLINK, Agilent, Precision MicroDynamics, and Keithley Instruments (MetraByte). A DAQ card can be directly plugged into an expansion slot of a PC and automatically linked with the bus of the PC. Its operation is managed by the driver software, which has to be stored in the computer. Powerful microcontroller units (e.g., Intel Galileo) have DAQ functions and hardware already integrated into them (e.g., 14 digital I/O pins 6 of which are for pulse-width-modulated outputs; 6 analog inputs with a built-in analog-to-digital converter).

The main components of a DAQ card are shown in Figure 4.8b. The MUX selects the appropriate input channel for the incoming analog data. The signal is amplified by a programmable amplifier before ADC. As discussed in a later section, the S/H samples the analog signal and maintains its value at the sampled level until the conversion by the ADC is completed. The FIFO element stores the ADC output until it is accessed by the computer for digital processing. The DAQ card can provide an analog output through the DAC. Furthermore, a typical DAQ card can provide digital outputs as well. An encoder (i.e., a pulse-generating position sensors) can be directly interfaced to the DAQ card, for use in motion control applications. Specifications of a typical DAQ card are given in Box 4.1. Many of the indicated parameters are discussed in this chapter. Others are either self-explanatory or discussed elsewhere in this book. Particular note should be made about the sampling rate. This is the rate at which an analog input signal is sampled by the ADC. The Nyquist frequency (or the bandwidth limit) of the sampled data would be half this number (e.g., for a sampling rate of 100 kS/s, it is 50 kHz). When multiplexing is used (i.e., several

BOX 4.1 TYPICAL SPECIFICATIONS OF A PLUG-IN DATA ACQUISITION CARD

Number of analog input channels = 2–16 single ended or 1–8 differential
Analog input ranges = ±5; 0–10; ±10; 0–24 V
Buffer size = 512–2048 samples
Input gain ranges (programmable) = 1, 2, 5, 10, 20, 50, 100
Sampling rate for A/D conversion = 10k samples/s (100 kHz) to 1 MS/s
Word size (resolution) of ADC = 12 bits, 16 bits
Number of D/A output channels = 1–4
Word size (resolution) of DAC = 12 bits
Ranges of analog output = 0–10 V (unipolar mode); ±10 V (bipolar mode)
Number of digital input lines = 12
Low voltage of input logic = 0.8 V (maximum)
High voltage of input logic = 2.0 V (minimum)
Number of digital output lines = 12
Low voltage of output logic = 0.45 V (maximum)
High voltage of output logic = 2.4 V (minimum)
Number of counters/timers = 3
Resolution of a counter/timer = 16 bits
Input impedance = 2.4 kΩ at 0.5 W
Output impedance = 75 Ω

input channels are read at the same time), the effective sampling rate for each channel will be reduced by a factor equal to the number of channels. For example, if 16 channels are sampled simultaneously, the effective sampling rate will be 100 kHz/16 = 6.25 kS/s, giving a Nyquist frequency of 3.125 kHz.

Since DAC and ADC play important functions in engineering applications of monitoring, they are discussed now. DACs are simpler and less expensive than ADCs. Furthermore, some types of ADCs employ a DAC to perform their function. For these reasons, we will discuss DAC before ADC.

4.3.1 Digital-to-Analog Converter

The function of a digital-to-analog converter (DAC or D/A or D2A) is to convert a digital word stored in its data register (called DAC register), typically in the straight binary form, into an analog value (voltage or current). In this manner, a sequence of digital data can be converted into an analog signal. Some form of interpolation (or "reconstruction filter") has to be used to connect and smooth the resulting discrete analog values, for forming the analog signal. Typically, the data in the DAC register are arriving from the data bus of the computer to which the DAC is connected (e.g., the DAC located in the DAQ card of the computer).

Each binary digit (bit) of information in the DAC register may be present as a state of a bistable (two stage) logic element, which can generate a voltage pulse or a voltage level to represent that bit. For example, the off state of a bistable logic element or absence of a voltage pulse or low level of a voltage signal or no change in a voltage level can represent binary 0. Conversely, the on state of a bistable device or presence of a voltage pulse or high level of a voltage signal or change in a voltage level will represent binary 1. The combination of these bits, which form the digital word in the DAC register, will correspond to a numerical value of the analog output signal. Then, the purpose of the DAC is to generate an output voltage (signal level) that has this numerical value and maintain the value until the next digital word in the arriving digital data sequence is converted into the analog form. Since a voltage output cannot be arbitrarily large or small for practical reasons, some form of scaling would have to be done in the DAC process. This scale will depend on the reference voltage v_{ref} that is used in the particular DAC circuit.

A typical DAC unit is an active circuit in the form of an IC chip. It may consist of a data register (digital circuits), solid-state switching elements, resistors, and op-amps powered by an external power supply (possibly that of the host computer), which can provide the reference voltage for the DAC. The reference voltage will determine the maximum value of the DAC output (full-scale voltage). As noted earlier, the IC chip that represents the DAC is usually one of the many components mounted on a printed circuit board, which is the DAQ card (or I/O card or interface board or DAQ and control board). This card is plugged into a slot of the host PC (see Figures 4.7 and 4.8).

DAC operation: The typical operation of the DAC chip is based on turning on and off of semiconductor switches (e.g., complementary metal–oxide–semiconductor [CMOS] switches) at proper times, as governed by some logic dependent on the digital data value. This switching will determine the output of an op-amp circuit, which is the analog output of the DAC. There are many types and forms of DAC circuits. The form will depend mainly on the DAC method, manufacturer, and requirements of the user or of the particular application. Most types of DAC are variations of two basic types: the weighted type (or summer type or adder type) and the ladder type. The latter type of DAC is more desirable and more power efficient even though the former type could be somewhat simpler and less expensive. Another straightforward and simpler (but possibly less accurate) method uses a pulse-width modulator (PWM) chip. Two representative DAC methods are outlined next.

4.3.1.1 Ladder (or R–2R) DAC

A DAC that uses an R–$2R$ ladder circuit is known as a ladder DAC or R–$2R$ DAC. This circuit uses only two types of resistors, one with resistance R and the other with $2R$. Hence, the precision of the resistors is not as stringent as what is needed for the weighted-resistor DAC. Schematic representation of an R–$2R$ ladder DAC is shown in Figure 4.9. Switching of each element occurs depending on the corresponding bit value (0 or 1) of the digital word. The sum of the corresponding voltage values is generated by the op-amp, which is the analog output.

To obtain the I/O equation for the ladder DAC, suppose that the voltage output from the solid-state switch associated with the bit b_i of the digital word is v_i. Furthermore, suppose that $\tilde{v}_i$ is the voltage at node i of the ladder circuit, as shown in Figure 4.9. Now, writing the current summation at node i, we get

$$\frac{v_i-\tilde{v}_i}{2R}+\frac{\tilde{v}_{i+1}-\tilde{v}_i}{R}+\frac{\tilde{v}_{i-1}-\tilde{v}_i}{R}=0 \quad \text{or} \quad \frac{1}{2}v_i=\frac{5}{2}\tilde{v}_i-\tilde{v}_{i-1}-\tilde{v}_{i+1}, \quad \text{for } i=1,2,\ldots,n-2 \tag{i}$$

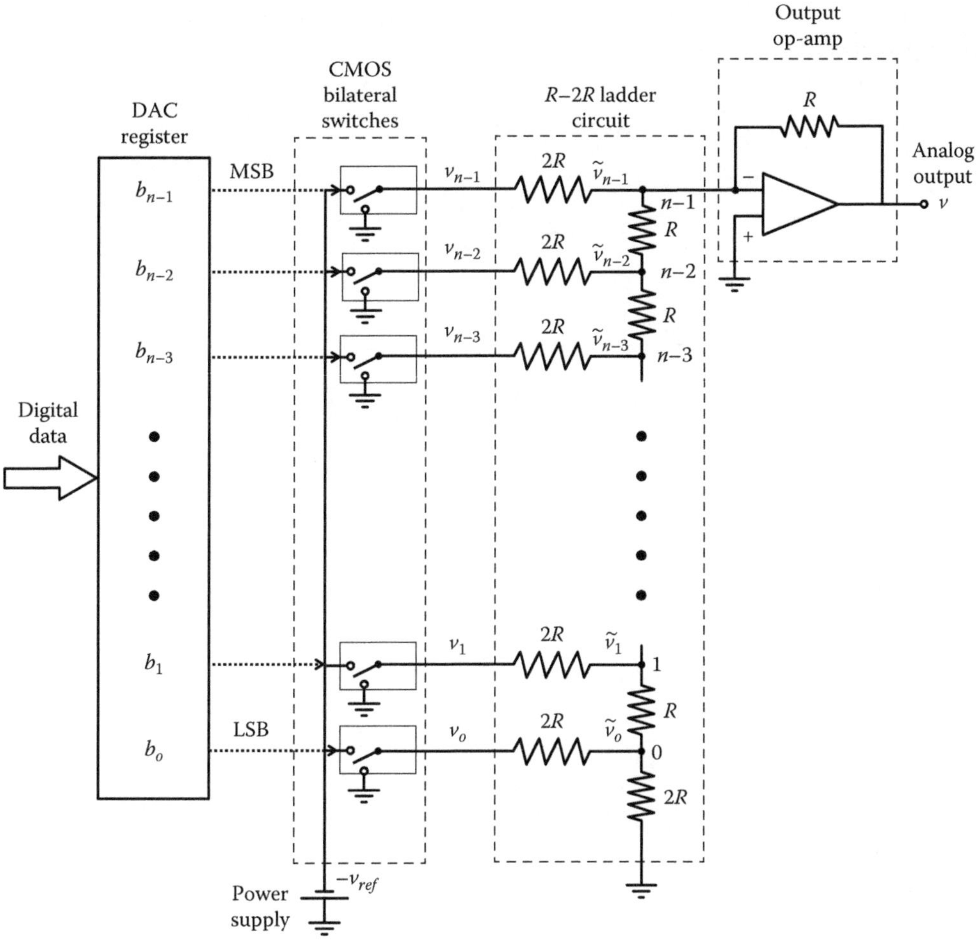

FIGURE 4.9
Circuit of ladder DAC.

This is valid for all nodes, except for nodes 0 and $n-1$. It is seen that the current summation for node 0 gives

$$\frac{v_0 - \tilde{v}_0}{2R} + \frac{\tilde{v}_1 - \tilde{v}_0}{R} + \frac{0 - \tilde{v}_0}{2R} = 0 \quad \text{or} \quad \frac{1}{2}v_0 = 2\tilde{v}_0 - \tilde{v}_1 \tag{ii}$$

The current summation for node $n-1$ gives

$$\frac{v_{n-1} - \tilde{v}_{n-1}}{2R} + \frac{v - \tilde{v}_{n-1}}{R} + \frac{\tilde{v}_{n-2} - \tilde{v}_{n-1}}{R} = 0$$

Now, since the positive lead of the op-amp is grounded, we have $\tilde{v}_{n-1} = 0$. Hence,

$$\frac{1}{2}v_{n-1} = -\tilde{v}_{n-2} - v \tag{iii}$$

Next, by using Equations i through iii, along with the fact that $\tilde{v}_{n-1} = 0$, we can write the following series of equations:

$$\frac{1}{2}v_{n-1} = -\tilde{v}_{n-2} - v$$

$$\frac{1}{2^2}v_{n-2} = \frac{1}{2}\frac{5}{2}\tilde{v}_{n-2} - \frac{1}{2}\tilde{v}_{n-3}$$

$$\frac{1}{2^3}v_{n-3} = \frac{1}{2^2}\frac{5}{2}\tilde{v}_{n-3} - \frac{1}{2^2}\tilde{v}_{n-4} - \frac{1}{2^2}\tilde{v}_{n-2}$$

$$\frac{1}{2^{n-1}}v_1 = \frac{1}{2^{n-2}}\frac{5}{2}\tilde{v}_1 - \frac{1}{2^{n-2}}\tilde{v}_0 - \frac{1}{2^{n-2}}\tilde{v}_2$$

$$\frac{1}{2^2}v_0 = \frac{1}{2^{n-1}}2\tilde{v}_0 - \frac{1}{2^{n-1}}\tilde{v}_1 \tag{iv}$$

If we sum these n equations, first denoting

$$S = \frac{1}{2^2}\tilde{v}_{n-2} + \frac{1}{2^3}\tilde{v}_{n-3} + \cdots + \frac{1}{2^{n-1}}\tilde{v}_1$$

we get

$$\frac{1}{2}v_{n-1} + \frac{1}{2^2}v_{n-2} + \cdots + \frac{1}{2^n}v_0 = 5S - 4S - S + \frac{1}{2^{n-1}}2\tilde{v}_0 - \frac{1}{2^{n-2}}\tilde{v}_0 - v = -v$$

Finally, since $v_i = -b_i v_{ref}$, the analog output is

$$v = \left[\frac{1}{2}b_{n-1} + \frac{1}{2^2}b_{n-2} + \cdots + \frac{1}{2^n}b_0\right]v_{ref} \tag{4.9}$$

Hence, the analog output is proportional to the value D of the digital word. Furthermore, the full-scale value (FSV) of the ladder DAC is given by

$$\text{FSV} = \left(1 - \frac{1}{2^n}\right)v_{ref} \tag{4.10}$$

Note: The same results are obtained for a weighted-resistor (adder) DAC.

4.3.1.2 PWM DAC

As noted earlier, in pulse-width modulation (PWM), the pulse width is varied (modulated) in a pulse sequence of fixed amplitude. Consider the pulse signal shown in Figure 4.10a. Here,

T is the pulse period

p is the fraction of the period in which the pulse is on

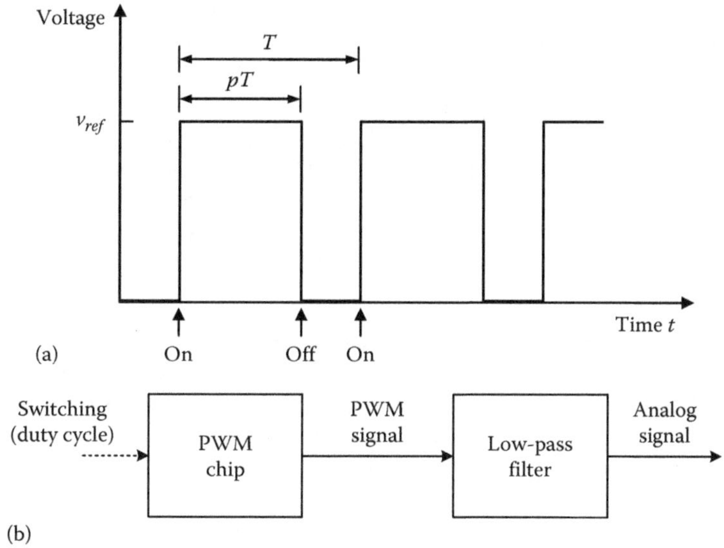

FIGURE 4.10
(a) Duty cycle of a PWM signal and (b) operation of a PWM DAC.

When expressed as a percentage (i.e., $100p$), p represents the *duty cycle* of the pulse. Hence, the two extremes of the modulation are duty cycle of 0% where the pulse is fully off and 100% where the pulse is fully on during the entire period. Also, it is clear that the average value (i.e., the DC value) of the pulse is pv_{ref}. It follows that when the duty cycle varies from 0% to 100%, the DC value of the pulse signal varies in proportion, from 0 to v_{ref}. This principle is used in a DAC that uses a PWM chip. Specifically, the PWM signal is generated by switching the PWM on for a time period that is proportional to the value of the digital word. The resulting signal is low-pass filtered with a very low-frequency cutoff, as shown in Figure 4.10b. The magnitude of the resulting analog signal is almost equal to the DC value of the PWM, which is pv_{ref}. In this manner, an analog output in the range 0 to v_{ref} is obtained in proportion to the value of the digital word in the DAC register.

4.3.1.3 DAC Error Sources

For a given digital word, the analog output voltage from a DAC would not be exactly equal to what is given by the analytical formulas (e.g., Equation 4.9). The difference between the actual output and the ideal output is the error. The DAC error could be normalized with respect to the FSV.

There are many causes for DAC error. Typical error sources include parametric uncertainties and variations, circuit time constants, switching errors, and variations and noise in the reference voltage. Several types of error sources and representations of a DAC are given in the following.

1. *Code ambiguity*: In many digital codes (e.g., in the straight binary code), incrementing a number by a least significant bit (LSB) will involve more than one bit switching. If the speed of switching from 0 to 1 is different from that for 1 to 0 and if switching pulses are not simultaneously applied to the switching circuit, the

switching of the bits will not take place simultaneously. For example, in a four-bit DAC, incrementing from decimal 2 to decimal 4 will involve changing the digital word from 0011 to 0100. This requires two bit switchings from 1 to 0 and one bit switching from 0 to 1. If 1 to 0 switching is faster than the 0 to 1 switching, then an intermediate value given by 0000 (decimal zero) will be generated, with a corresponding analog output. Hence, there will be a momentary code ambiguity and associated error in the DAC signal. This problem can be reduced (and eliminated in the case of single-bit increments) if a gray code is used to represent the digital data. Improving the switching circuitry will also help reduce this error.

2. *Settling time*: The circuit hardware in a DAC unit will have some dynamics, with associated time constants and perhaps oscillations (underdamped response). Hence, the output voltage cannot instantaneously settle to its ideal value upon switching. The time required for the analog output to settle within a certain band (say ±2% of the final value or ±1/2 resolution), following the application of the digital data, is termed *settling time*. Naturally, settling time should be smaller for better (faster and more accurate) performance. As a rule of the thumb, the settling time should be less than half the data arrival period. *Note*: The data arrival time is equal to the time interval between the arrivals of two successive data values and is equal to the inverse of the data arrival rate.
3. *Glitches*: Switching of a circuit will involve sudden changes in magnetic flux due to changes in current. This will induce voltages, which will produce unwanted signal components. In a DAC circuit, these induced voltages due to rapid switching can cause signal spikes, which will appear at the output. At low conversion rates, the error due to these noise signals is not significant.
4. *Parametric errors*: The resistor elements in a DAC may not be very precise, particularly when resistors within a wide range of magnitudes are employed, as in the case of weighted-resistor DAC. These errors appear at the analog output. Furthermore, aging and environmental changes (primarily, change in temperature) will change the values of circuit parameters, resistance in particular. This will also result in DAC error. These types of error due to the imprecision of circuit parameters and the variations of parameter values are termed *parametric errors*. The effects of such errors can be reduced by several ways including the use of compensation hardware (and perhaps software) and directly by using precise and robust circuit components and employing good manufacturing practices.
5. *Reference voltage variations*: Since the analog output of a DAC is proportional to the reference voltage v_{ref}, any variation in the voltage supply will directly appear as an error. This problem can be overcome by using a stabilized voltage source with sufficiently low-output impedance.
6. *Monotonicity*: Clearly, the output of a DAC should change by its resolution ($\Delta y = v_{ref}/2^n$) for each step of one LSB increment in the digital value. This ideal behavior might not exist in some practical DACs due to such errors as those mentioned earlier. At least, the analog output should not decrease as the value of the digital input increases. This is known as the monotonicity requirement, and it should be met by a practical DAC.
7. *Nonlinearity*: Suppose that the digital input to a DAC is varied from [0 0 … 0] to [1 1 … 1] in steps of one LSB. Ideally, the analog output should increase in constant jumps of $\Delta y = v_{ref}/2^n$, giving a staircase-shaped analog output. If we draw

BOX 4.2 RATINGS OF A COMMERCIAL DAC CHIP

Number of DAC channels = 2–16 single ended or 1–8 differential

Resolution = 16-bit

Offset error = ±2 mV (max)

Current settling time = 1 μs

Slew rate = 5 V/μs

Power dissipation = 20 mW

Single power supply = 5–15 V DC

Maximum sample rate = 1.6×10^9 samples/s (1.6 GSPS)

Size = 25 mm × 6 mm

the best linear fit for this ideally monotonic staircase response, it will have a slope equal to the resolution/bit. This slope is known as the ideal scale factor. Nonlinearity of a DAC is measured by the largest deviation of the DAC output from this best linear fit. *Note*: In the ideal case, the nonlinearity is limited to half the resolution ($(1/2)\Delta y$).

One cause of nonlinearity is faulty bit transitions. Another cause is circuit nonlinearity in the conventional sense. Specifically, due to nonlinearities in circuit elements such as op-amps and resistors, the analog output will not be proportional to the value of the digital word as dictated by the bit switchings (faulty or not). This latter type of nonlinearity can be accounted for by using calibration.

Multiple DACs in a single package are commercially available, for example, a package of 16 DACs each of 16-bit resolution and independently software programmable or pin configurable in the output voltage range ±10 V, with internal 16:1 analog multiplexing (MUX). Typical ratings of a commercial DAC chip are given in Box 4.2.

4.3.2 Analog-to-Digital Converter

The measured variables of an engineering system are typically continuous in time; they are analog signals. However, common applications that use these signals, such as performance monitoring, fault diagnosis, and control, will require digital processing of these signals. Hence, the analog signals have to be sampled at discrete time points, and the sample values have to be represented in the digital form (according to a suitable code) to be read into a digital system such as a computer or a microcontroller. An IC device called analog-to-digital converter (ADC, A/D, or A2D) is used to accomplish this. A feedback control system scenario, which involves both DAC and ADC, is shown in Figure 4.7.

Sampling an analog signal into a sequence of discrete values introduces *aliasing distortion* (see Chapter 6). This error can be reduced by increasing the sampling rate and also by using an antialiasing filter. Nevertheless, according to Shannon's sampling theorem, the frequency spectrum of the analog signal beyond half the sampling frequency (i.e., Nyquist frequency) is completely lost due to data sampling. Furthermore, in representing a discrete signal value by a digital value (say in straight binary, two's complement binary, or gray code), an error called "quantization error" is introduced due to the finite bit length of the digital word. This is also the *resolution* of the ADC. DACs and ADCs are usually situated on the same DAQ card (see Figure 4.8b). But the ADC process is more complex and time-consuming than the DAC process. Furthermore, many types of ADCs use DACs to accomplish the ADC. Hence, ADCs are usually more costly, and their conversion rate is usually slower in comparison to DACs.

Many types of ADCs are commercially available. However, their principle of operation may be classified into two primary methods:

1. Uses an internal DAC and comparator hardware (the analog value is compared with the DAC output, and the DAC input is incremented until a match is achieved).
2. Analog value is represented by a count (digital) in proportion. The full count corresponds to the FSV of the ADC.

Two common types of ADC are discussed now. Some other (related) versions are considered as end-of-chapter problems.

4.3.2.1 Successive Approximation ADC

This ADC uses an internal DAC and a comparator. The DAC input starts with the most significant bit (MSB) and changed. The DAC output is compared with the analog data, until a match is found. It is very fast and is suitable for high-speed applications. The speed of conversion depends on the number of bits in the output register of the ADC but is virtually independent of the nature of the analog input signal.

A schematic diagram for a successive approximation ADC is shown in Figure 4.11. *Note*: DAC is an integral component of this ADC. The sampled analog signal (from the S/H

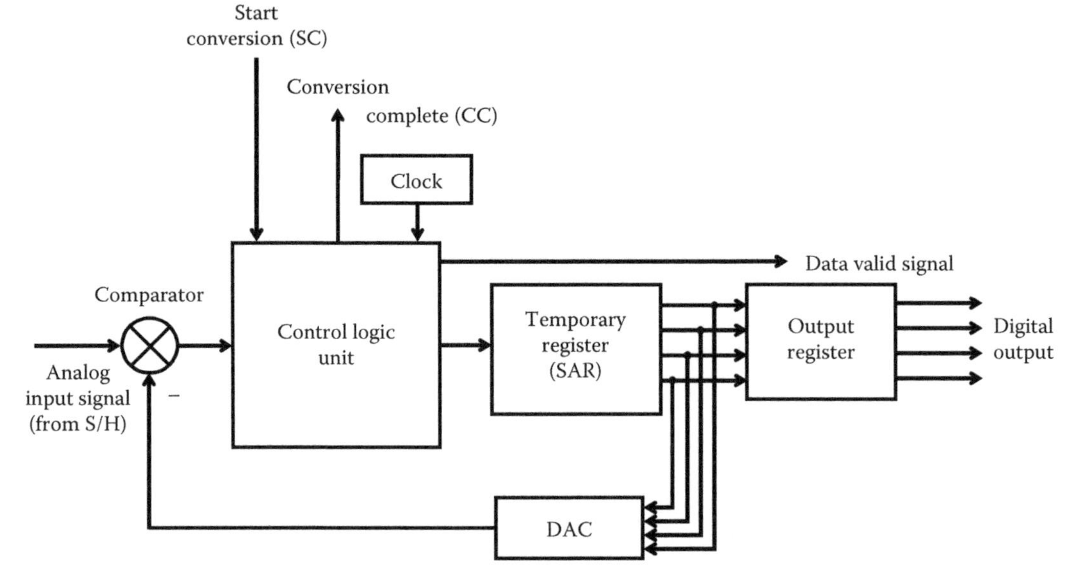

FIGURE 4.11
Successive approximation ADC.

device of the DAQ) is applied to a comparator (a differential amplifier). Simultaneously, a start conversion (SC) control pulse is sent into the control logic unit by the external device (perhaps a microcontroller) that controls the operation of the ADC. Then, no new data will be accepted by the ADC until a conversion complete (CC) pulse is sent out by the control logic unit. Initially, the registers are cleared so that they contain all zero bits. Then, the ADC is ready for its first conversion approximation.

The first approximation begins with a clock pulse. Then, the control logic unit sets the MSB of the temporary register (DAC control register or successive approximation register [SAR]) to one, all the remaining bits in that register being zero (*Note*: This corresponds to half the FSV of the ADC. For example, if FSV = 12 V, the DAC output now is 6 V, which is compared with the analog value, which could be anything in the range 0–12 V, such as 8.2, 4.9 V). This digital word in the temporary register is supplied to the DAC. The analog output of the DAC is (half the FSV now) subtracted from the analog input (sampled data), using the comparator. If the comparator output is >0, the control logic unit will keep the MSB of the temporary register at binary 1 and will proceed to the next approximation. If the comparator output is <0, the control logic unit will change the MSB to binary 0 before proceeding to the next approximation.

The second approximation will start at another clock pulse. This approximation will consider the second MSB of the temporary register. As earlier, this bit is set to 1 and the comparison is made. If the comparator output is >0, this bit is left at value 1 and the third MSB is considered. If the comparator output is <0, the bit value will be changed to 0 before proceeding to the third MSB.

In this manner, all bits in the temporary register are set successively starting from the MSB and ending with the LSB. The contents of the temporary register (SAR) are then transferred to the output register, and a data valid signal is sent by the control logic unit, signaling the digital processor (computer) to read the contents of the output register of ADC. The computer will not read the register if a data valid signal is not present. Next, a CC pulse is sent out by the control logic unit, and the temporary register is cleared. The ADC is now ready to accept another data sample for digital conversion. *Note*: The conversion process is essentially the same for every bit in the temporary register. Hence, the total conversion time is approximately n times the conversion time for one bit. Typically, one bit conversion can be completed within one clock period.

Signal value and sign: It should be clear that if the maximum value of an analog input signal exceeds the FSV of a DAC, then the excess signal value cannot be converted by the ADC. The excess value will directly contribute to the error in the digital output of the ADC. Hence, this situation should be avoided, either by properly scaling the analog input or by properly selecting the reference voltage for the internal DAC unit. Thus far, we have assumed that the value of the analog input signal is positive. If the value is negative, the sign has to be accounted for in the ADC, by some means. For example, the sign of the signal can be detected initially from the sign of the comparator output, when all bits are zero. If the sign is negative, then the same A/D conversion process, as for a positive signal, is carried out after switching the polarity of the comparator. Finally, the sign is correctly represented in the digital output (e.g., by the two's complement representation for negative quantities). Another approach to account for signed (bipolar) input signals is to offset the signal by a sufficiently large constant voltage, such that the analog input is always positive. After the conversion, the digital number corresponding to this offset is subtracted from the converted data in the output register to obtain the correct digital output. Then, we may assume that the analog input signal is positive.

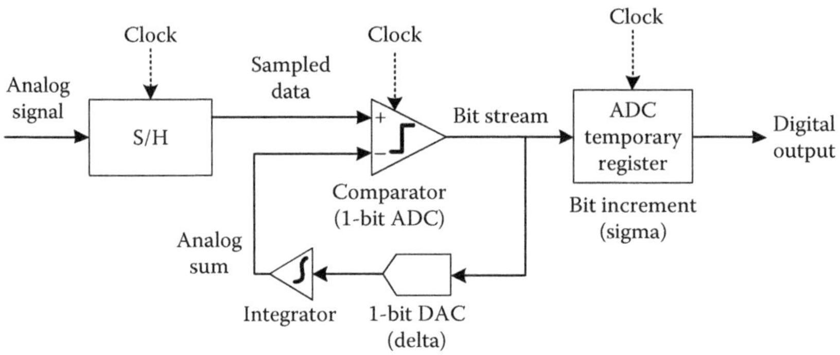

FIGURE 4.12
The principle of a delta–sigma ADC.

4.3.2.2 Delta–Sigma ADC

Also known as a ΔΣ ADC or delta–sigma ADC, this popular variety of ADC has a relatively low cost, low bandwidth, and high resolution. The basic principle of the device is shown in Figure 4.12. The sampled analog data value is compared with the integrated (summed) output of a 1-bit DAC. If the difference is positive, the comparator (1-bit ADC) generates a "1" bit. Otherwise, it generates a "0" bit and the conversion, as stored in the temporary registered, is complete and available for reading by the computer.

Each "1" bit from the comparator increments the digital value in the temporary register by 1 bit. Hence, the digital word in this register increases from 0, one bit at a time, until the value equals the analog sampled value (up to the quantization error). The term *delta* is used because the digital value in the register is incremented by 1 bit at a time, corresponding to the output of the 1-bit ADC, and also because the compared analog value is incremented 1 bit at a time, corresponding to the output of the 1-bit DAC (Greek letter delta is normally used to denote a small increment). The word *sigma* is used because the delta values are added to form the digital output (Greek, uppercase sigma is normally used to denote summation).

There are several other variations of delta–sigma ADC. In them, typically, the integrator is located in the forward path of the feedback loop, after taking the difference between the sampled data value and the output of the 1-bit DAC. Then, the 1-bit ADC generates a bit stream whose bit density represents the sampled data value.

4.3.2.3 ADC Performance Characteristics

For an ADC that uses an internal DAC, the same error sources that were discussed previously for DACs will apply. The code ambiguity is not a problem when only one bit is converted at a time and also because the entire record in the temporary ADC register is transferred instantaneously to the output register. In an ADC that uses an internal DAC, however, ambiguity in the DAC register can cause error.

Conversion time is a major consideration because this is much larger for an ADC. In addition to resolution and dynamic range, quantization error will be applicable to an ADC. These considerations, which govern the performance of an ADC, are outlined next.

4.3.2.3.1 *Resolution and Quantization Error*

The number of bits n in an ADC register determines the resolution and the dynamic range of an ADC. For an n-bit ADC, the size of the output register is n bits. Hence, the smallest possible increment of the digital output is one LSB. The change in the analog input that results in a change of one LSB at the output is the resolution of the ADC. For the unipolar (unsigned) case, the available range of the digital outputs is from 0 to $2^n - 1$. This represents the dynamic range. It follows that, as for a DAC, the dynamic range of an n-bit ADC is given by the ratio

$$\text{DR} = 2^n - 1 \tag{4.11}$$

or in decibels

$$\text{DR} = 20\log_{10}(2^n - 1)\ \text{dB} \tag{4.12}$$

Note: The resolution improves with n.

The FSV of an ADC is the value of the analog input that corresponds to the maximum digital output.

Suppose that an analog signal within the dynamic range of a particular ADC is converted by that ADC. Since the analog input (sampled value) has an infinitesimal resolution and the digital representation has a finite resolution (one LSB), an error is introduced into the process of ADC. This is known as the "quantization error." A digital number undergoes successive increments in constant steps of 1 LSB. If an analog value falls at an intermediate point within a step of single LSB, a quantization error is caused as a result. Rounding off the digital output can be accomplished as follows: The magnitude of the error when quantized up is compared with that when quantized down, say, using two hold elements and a differential amplifier. Then, we retain the digital value corresponding to the lower error magnitude. If the analog value is below the 1/2 LSB mark, then the corresponding digital value is represented by the value at the beginning of the step. If the analog value is above the 1/2 LSB mark, then the corresponding digital value is the value at the end of the step. It follows that with this type of rounding off, the quantization error does not exceed 1/2 LSB.

4.3.2.3.2 *Monotonicity, Nonlinearity, and Offset Error*

Considerations of monotonicity and nonlinearity are important for an ADC as well as for a DAC. For an ADC, the input is an analog signal and the output is digital. Disregarding quantization error, the digital output of an ADC will increase in constant steps in the shape of an ideal staircase function, when the analog input is increased from 0 in steps of the device resolution (δy). This is the monotonic case. The best straight-line fit to this curve has a slope equal to $1/\delta y$ (LSB/volts). This is the ideal gain or ideal scale factor. Still there will be an offset error of 1/2 LSB because the best linear fit will not pass through the origin. Adjustments can be made for this offset error.

Incorrect bit transitions can take place in an ADC, due to various errors that might be present and also possibly due to circuit malfunctions. The best linear fit under such faulty conditions will have a slope that is different from the ideal gain. The difference is the *gain error*. Nonlinearity is the maximum deviation of the output from the best linear fit. It is clear that with perfect bit transitions, in the ideal case, a nonlinearity of 1/2 LSB would be present. Nonlinearities larger than this would result due to incorrect bit transitions. As in the case of a DAC, another source of nonlinearity in an ADC is circuit nonlinearities, which would deform the analog input signal before converting it into the digital form.

4.3.2.3.3 ADC Conversion Rate

It is clear that ADC is much slower than DAC. The conversion time is a very important factor, particularly in real-time applications, because the rate at which the conversion can take place governs many aspects of DAQ. For example, the data sampling rate has to synchronize with the ADC conversion rate. This, in turn, will determine the Nyquist frequency (half the sampling rate), which corresponds to the bandwidth of the sampled signal, and is the maximum value of useful frequency that is retained as a result of sampling. Furthermore, the sampling rate will dictate the requirements of storage and memory. Another important consideration related to the conversion rate of an ADC is the fact that a signal sample has to be maintained at the same value during the entire process of conversion into the digital form. This would require a *hold circuit*, and this circuit should be able to perform accurately at the largest possible conversion time for the particular ADC device.

The time needed for a sampled analog input to be converted into the digital form will depend on the type of ADC. Usually, in a comparison-type ADC (which uses an internal DAC), each bit transition will take place in one clock period Δt. Also, in an integrating (dual slope) ADC, each clock count will need a time of Δt. On this basis, the conversion time for a successive approximation ADC may be estimated as follows:

For an n-bit ADC, n comparisons are needed. Hence, the conversion time is given by

$$t_c = n \cdot \Delta t \tag{4.13}$$

where Δt is the clock period.

Note: For this ADC, t_c does not depend on the signal level (analog input).

The total time taken to convert an analog signal will depend on other factors besides the time taken for the conversion of sampled data into digital form. For example, in multiple-channel DAQ (multiplexing), the time taken to select the channels has to be counted in. Furthermore, the time needed to sample the data and the time needed to transfer the converted digital data into the output register have to be included. In fact, the conversion rate for an ADC is the inverse of this overall time needed for a conversion cycle. Typically, however, the conversion rate depends primarily on the bit conversion time, in the case of a comparison-type ADC, and on the integration time, in the case of an integration-type ADC. A typical time period for a comparison step or counting step in an ADC is $\Delta t = 5$ μs. Hence, for an 8-bit successive approximation ADC, the conversion time is 40 μs. The corresponding sampling rate would be of the order of (less than) $1/40 \times 10^{-6} = 25 \times 10^3$ samples/s (or 25 kHz). The maximum conversion rate for an 8-bit counter ADC would be about $5 \times (2^8 - 1) = 1275$ μs. The corresponding sampling rate would be of the order of 780 samples/s. Note that this is considerably slow. The maximum conversion time for a dual-slope ADC would likely be larger (i.e., slower rate).

Ratings of a commercial ADC package are given in Box 4.3.

4.3.3 Sample-and-Hold Hardware

Typical applications of DAQ use analog signals, which have to be converted into the digital form using an ADC for subsequent processing. The analog input to an ADC can be very

BOX 4.3 RATINGS OF A COMMERCIAL ADC PACKAGE

Number of analog input channels = 6 (bipolar)
(6 independent ADCs)
Resolution = 16 bits
Sampling rate = 250 kSPS
SNR = 88 dB
Voltage ranges = ±5 or ±10 V
Power = 140 mW (with 5 V supply)

transient, and furthermore, the process of ADC itself is not instantaneous (ADC time will be many times larger than the DAC time). Specifically, the incoming analog signal might be changing at a rate higher than the ADC conversion rate. Then, the input signal value will vary during the conversion period, and there will be an ambiguity as to what analog input value actually corresponds to a particular digital output value. Hence, it is necessary to sample the analog input signal and maintain the input to the ADC at this sampled value, until the conversion process is completed. In other words, since we are typically dealing with analog signals that can vary at a high speed, it would be necessary to S/H the input signal during each ADC cycle. Each data sample must be generated and captured by the S/H circuit on the issue of the SC control signal, and the captured voltage level has to be maintained constant until a CC control signal is issued by the ADC unit.

The main element in an S/H circuit is the holding capacitor. A schematic diagram of an S/H chip is shown in Figure 4.13. The analog input signal is supplied through a voltage follower to a solid-state switch. The switch typically uses a field-effect transistor (FET), such as the metal–oxide–semiconductor FET (MOSFET). The switch is closed in response to a sample pulse and is opened in response to a hold pulse. Both control pulses are generated by the control logic unit of the ADC. During the time interval between these two pulses, the holding capacitor is charged to the voltage of the sampled input. This capacitor voltage is then supplied to the ADC through a second voltage follower.

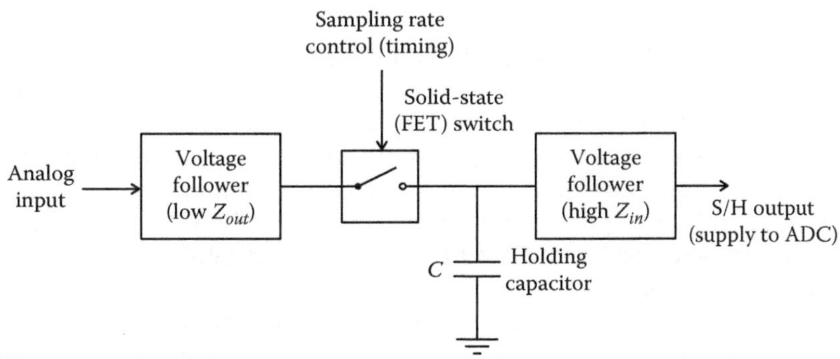

FIGURE 4.13
Circuit of a sample-and-hold chip.

The functions of the two voltage followers are explained now. When the FET switch is closed in response to a sample command (pulse), the capacitor has to be charged as quickly as possible. The associated time constant (charging time constant) τ_c is given by

$$\tau_c = R_s C \tag{4.14}$$

where
R_s is the source resistance
C is the capacitance of the holding capacitor

Since τ_c has to be very small for fast charging and since C is fixed by the holding requirements (typically, C is of the order of 100 pF, where 1 pF $= 1 \times 10^{-12}$ F), we need a very small source resistance. The requirement is met by the input voltage follower (which is known to have a very low-output impedance), thereby providing a very small R_s.

Next, once the FET switch is opened in response to a hold command (pulse), the capacitor should not discharge. This requirement is met due to the presence of the output voltage follower. Since the input impedance of a voltage follower is very high, the current through its leads would be almost zero. Because of this, the holding capacitor will have a virtually zero discharge rate under "hold" conditions. Furthermore, we like the output of this second voltage follower to be equal to the voltage of the capacitor. This condition is also satisfied due to the fact that a voltage follower has a unity gain. Hence, the sampling would be almost instantaneous, and the output of the S/H circuit would be maintained (almost) constant during the holding period, due to the presence of the two voltage followers. A typical S/H chip has 14 pins (8 pins for the two op-amps, 3 pins for the switch, 2 pins for DC power [bipolar], and 1 pin for ground [GND]). Also, acquisition time of 3 μs, a *droop rate* (the rate at which the voltage of the holding capacitor drops) of 1 mV/ms, and a maximum *offset error* of 3 mV are typical.

Note: The practical S/H circuits are zero-order hold devices, by definition.

4.3.4 Multiplexer

An MUX (sometimes called a scanner) is used to select one channel at a time from a bank of signal channels and connect it to a common hardware unit. In this manner, a costly and complex hardware unit (e.g., a computer or even a sophisticated ADC) can be time shared among several signal channels. Typically, channel selection is done in a sequential manner at a fixed channel-select rate.

There are two types of MUX: analog MUX and digital MUX. An *analog MUX* is used to scan a group of analog channels. Alternatively, a *digital MUX* is used to read one digital data channel at a time sequentially from a set of digital data channels.

Conversely, the process of distributing a single channel of data among several output channels is known as demultiplexing. A *demultiplexer* (or data distributor) performs the reverse function of an MUX. It may be used, for example, when the same (processed) signal from a computer is needed for several purposes (e.g., digital display, analog reading, digital plotting, and control).

Multiplexing used in short-distance signal transmission applications (e.g., data logging and process control) is usually time-division multiplexing. In this method, channel selection is made with respect to time. Hence, only one input channel is connected to the output channel of the MUX. This is the method described here. Another method of multiplexing, used particularly in long-distance transmission of several data signals, is known as frequency-division multiplexing. In this method, the input signals are modulated (e.g., by AM, as discussed previously) with carrier signals with different frequencies and are transmitted simultaneously through the same data channel. The signals are separated by demodulation at the receiving end.

4.3.4.1 Analog Multiplexers

Monitoring of an engineering system often requires the sensing of several process variables (mainly, responses or outputs). These signals have to be conditioned (e.g., by amplification and filtering) and modified in some manner (e.g., by ADC) before supplying to a common-purpose system such as a computer, microcontroller, or data logger. Usually, data-modification devices are costly. In particular, we have noted that ADCs are more expensive than DACs. An expensive option for interfacing several analog signals with a common-purpose system such as a computer would be to provide separate data-modification hardware for each signal channel. For example, multichannel DAQs with multiple ADCs are commercially available. This method has the advantage of high speed. An alternative, low-cost method is to use an analog MUX to select one signal channel at a time sequentially and connect it to a common signal modification hardware unit (consisting of amplifiers, filters, S/H, ADC, etc.). In this way, by time-sharing expensive hardware among many data channels, the DAQ speed is traded off to some extent for significant cost savings. Because very high speeds of channel selection are possible with solid-state switching (e.g., solid-state speeds of the order of 100 MHz or channel-switch time of 10 ns), the speed reduction due to multiplexing is not a significant drawback in most applications. On the other hand, since the cost of hardware components such as ADC is declining due to rapid advances in solid-state technologies, cost reductions attainable through the use of multiplexing might not be substantial in some applications. Hence, some economic evaluation and engineering judgment would be needed when deciding on the use of signal multiplexing for a particular DAQ, monitoring, or control application.

A schematic diagram of an analog MUX is shown in Figure 4.14. The figure represents the general case of N input channels and one output channel. This is called an $N \times 1$ (or N:1) analog MUX. Each input channel is connected to the output through a solid-state switch, typically an FET or CMOS switch. One switch is closed (turned on) at a time. A switch is selected by a digital word, which contains the corresponding channel address. Note that an n-bit address can assume 2^n digital values in the range 0 to $2^n - 1$. Hence, an MUX with an n-bit address can handle $N = 2^n$ channels. The channel selection may be done by the computer that acquires the data (e.g., an external microcontroller), which places the address of the channel on the address bus and simultaneously sends a control signal to the MUX to enable the MUX. The address decoder decodes the address and activates the corresponding solid-state switch. In this manner, channel selection can be done in an arbitrary order and with arbitrary timing, controlled by the computer or microcontroller. In simple versions of MUX, the channel selection is made in a fixed order at a fixed speed, however.

MUX pinout: For example, an 8:1 multiplexer (8 channels of data in and 1 channel of data out) will have the following 16 pins: 8 input pins, 1 output pin, 3 channel-select pins (for the 8 input channels), 1 control (enable) pin, 2 (bipolar) supply voltage pins, and 1 GND pin. Sometimes, there may be a no-function (or not connected) pin, an additional GND pin, and so on.

Typically, the output of an analog MUX chip is connected to an S/H chip and an ADC chip. A voltage follower may be provided at both input and output of an MUX to reduce

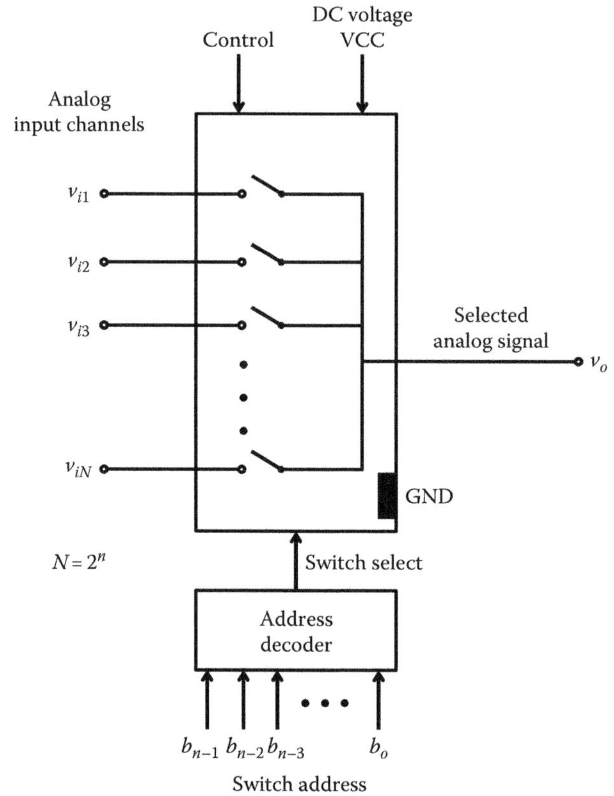

FIGURE 4.14
An N-channel analog multiplexer.

loading problems. A differential amplifier (or instrumentation amplifier) may be used as well at the MUX output to amplify the signal while reducing noise problems, particularly rejecting common-mode interference, as discussed in Chapter 3. The channel-select speed has to be synchronized with the sampling and ADC speeds for each signal channel. The MUX speed is not a major limitation because very high speeds (solid-state speeds of 100 MHz or channel switching times of 10 ns) are available with solid-state switching.

4.3.4.2 Digital Multiplexers

Sometimes, it is required to select one data word at a time from a set of digital data words, to be fed into a common device. For example, the set of data may be the outputs from a bank of digital transducers (e.g., shaft encoders that measure angular motions; see Chapter 11) or outputs from a set of ADCs that are connected to a series of analog signal channels. Then, a particular digital output (data word) can be read by a computer by using standard techniques of addressing and data transfer.

A digital multiplexing (or logic multiplexing) configuration is shown in Figure 4.15. The N registers of the MUX hold a set of N data words. The contents of each register may correspond to a measured variable and may change rapidly. The registers may represent

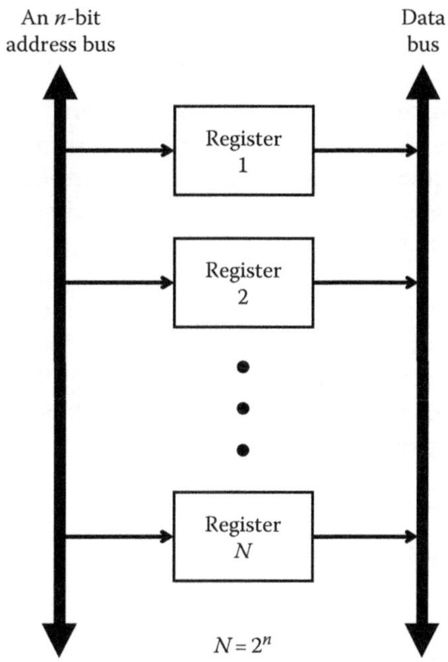

FIGURE 4.15
An $N \times 1$ digital multiplexer.

separate hardware devices (e.g., output registers of a bank of ADCs) or may represent locations in a computer memory to which data are transferred (read in) regularly. Each register has a unique binary address. As in the case of analog MUX, an n-bit address can select (address) 2^n registers. Hence, the number of registers will be given by $N = 2^n$, as earlier. When the address of the register to be selected is placed on the address bus, it enables the particular register. This causes the contents of that register to be placed on the data bus. Now the data bus is read by the device (e.g., computer), which is time shared among the N data registers. Placing a different address on the address bus will result in selecting another register and reading the contents of that register, as earlier.

Digital multiplexing is usually faster than analog multiplexing and has the usual advantages of digital devices, for example, high accuracy, better noise immunity, robustness (no drift and errors due to parameter variations), long-distance data transmission capability without associated errors due to signal weakening, and capability to handle very large numbers of data channels. Furthermore, a digital MUX can be modified using software, usually without the need for hardware changes. If, however, instead of using an analog MUX followed by a single ADC, a separate ADC is used for each analog signal channel and then digital multiplexing is used, it is quite possible for the digital multiplexing approach to be more costly. If, on the other hand, the measurements are already available in the digital form (for instance, as encoder outputs of displacement measurement), then digital multiplexing tends to be rather cost-effective and more desirable.

Transfer of a digital word from a single data source (e.g., a data bus) into several data registers, which are to be accessed independently, may be interpreted as digital demultiplexing. This is also a straightforward process of digital data transfer and reading.

4.4 Bridge Circuits

A bridge circuit is used to make some form of measurement. Typical measurements include change in resistance, change in inductance, change in capacitance (or, generally, change in impedance), oscillating frequency, or some variable (stimulus) that causes these changes. The typical output of a bridge is a voltage. In this sense, a bridge circuit may be considered as a device of signal conversion.

A full bridge is a circuit with four arms connected in a lattice form. Four nodes are formed in this manner. Two opposite nodes are used for powering (voltage or current supply) the bridge, and the remaining two opposite nodes provide the bridge output. *Note*: It is these two output nodes that "bridge" the circuit, giving its name. A bridge is said to be *balanced* when its output voltage is zero. There are two basic methods of making a measurement using a bridge circuit:

1. Bridge-balance method
2. Imbalance output method

In the bridge-balance method, we start with a balanced bridge. When making a measurement, the balance of the bridge will be upset due to the associated variation. As a result, a nonzero output voltage will be produced. The bridge can be balanced again by varying one of the arms of the bridge (assuming, of course, that some means is provided for fine adjustments that may be required). In this method, the change in the bridge element (resistance or impedance) that is required to restore the balance is the "measurement." A bridge can be precisely balanced using a servo device to change the bridge parameter, where an error feedback mechanism (a servo) is employed to make the bridge output zero.

In the imbalance output method as well, we start with a balanced bridge. As before, the balance of the bridge will be upset as a result of a change in the variable that is measured. Now, instead of balancing the bridge again, the output voltage of the bridge due to the resulted imbalance is measured and used as the bridge measurement.

There are many types of bridge circuits. If the supply to the bridge is DC, then we have a *DC bridge*. Similarly, an *AC bridge* uses an AC supply. A *resistance bridge* has only resistance elements in its four arms, and it is typically a DC bridge. An *impedance bridge* has impedance elements consisting of resistors, capacitors, and inductors in one or more of its arms. This is necessarily an AC bridge. If the bridge excitation is a constant-voltage supply, we have a *constant-voltage bridge*. If the bridge supply is a constant-current source, we get a *constant-current bridge*.

4.4.1 Wheatstone Bridge

Wheatstone bridge is a resistance bridge with a constant DC voltage supply (i.e., it is a constant-voltage resistance bridge). A Wheatstone bridge is particularly useful in strain-gauge measurements and, consequently, in force, torque, and tactile sensors that employ strain-gauge techniques (see Chapter 9). Since a Wheatstone bridge is used primarily in the measurement of small changes in resistance, it can be used in other types of sensing applications as well. For example, in resistance temperature detectors (RTDs), the change in resistance in a metallic (e.g., platinum) element, as caused by a change in temperature, is measured using a bridge circuit (see Chapter 10).

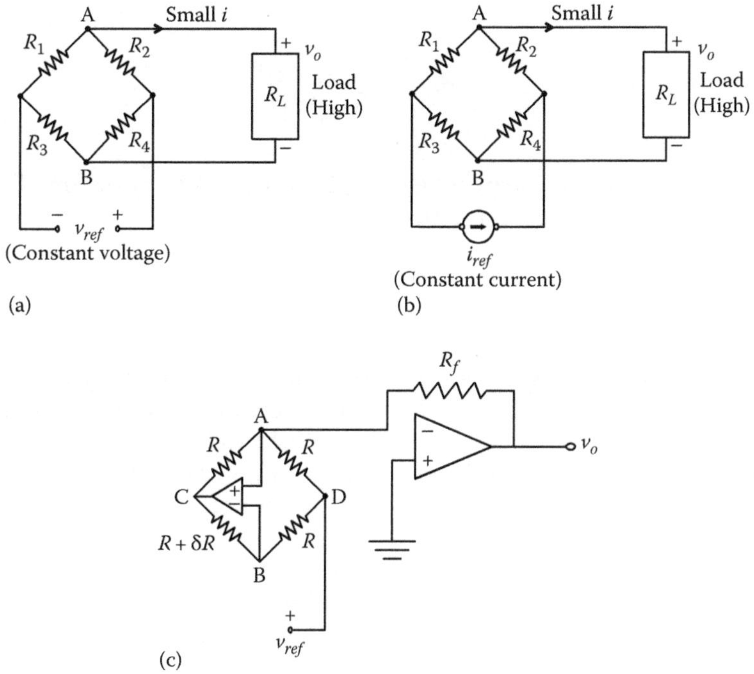

FIGURE 4.16
(a) Wheatstone bridge (constant-voltage resistance bridge), (b) constant-current resistance bridge, and (c) a linearized bridge.

Note: The temperature coefficient of resistance is positive for a typical metal (i.e., the resistance increases with temperature). For platinum, this value (change in resistance per unit resistance per unit change in temperature) is about 0.00385/°C.

Consider the Wheatstone bridge circuit shown in Figure 4.16a. Assuming that the bridge output is open circuit (i.e., has a very high load resistance), the output v_o may be expressed as

$$v_o = v_A - v_B = \frac{R_1 v_{ref}}{(R_1 + R_2)} - \frac{R_3 v_{ref}}{(R_3 + R_4)} = \frac{(R_1 R_4 - R_2 R_3)}{(R_1 + R_2)(R_3 + R_4)} v_{ref} \tag{4.15}$$

For a balanced bridge, the numerator of the RHS expression of Equation 4.15 must vanish. Hence, the condition for bridge balance is

$$\frac{R_1}{R_2} = \frac{R_3}{R_4} \tag{4.16}$$

Suppose that at first $R_1 = R_2 = R_3 = R_4 = R$. Then, according to this equation, the bridge is balanced. Now, increase R_1 by δR. For example, R_1 may represent the only active strain gauge, while the remaining three elements in the bridge are identical dummy elements. In view of Equation 4.15, the change in the bridge output due to the change δR is given by

$$\delta v_o = \frac{\left[(R + \delta R)R - R^2\right]}{(R + \delta R + R)(R + R)} v_{ref} - 0$$

or

$$\frac{\delta v_o}{v_{ref}} = \frac{\delta R/R}{(4+2\delta R/R)} \tag{4.17}$$

Observe that the output is nonlinear in $\delta R/R$. If, however, $\delta R/R$ is assumed small in comparison to 2, we have the linearized relationship

$$\frac{\delta v_o}{v_{ref}} = \frac{\delta R}{4R} \tag{4.18}$$

The factor 1/4 on the RHS of this equation represents the *sensitivity* of the bridge, as it gives the change in the bridge output for a given change in the active resistance, while the other parameters are kept fixed. Strictly speaking, the bridge sensitivity is given by $\delta v_o/\delta R$, which is equal to $v_{ref}/(4R)$ whose nondimensional parameter is 1/4.

The error due to linearization, which is a measure of nonlinearity, may be given as the percentage

$$N_P = 100\left(1 - \frac{\text{Linearized output}}{\text{Actual output}}\right)\% \tag{4.19}$$

Hence, from Equations 4.17 and 4.18, we have

$$N_P = 50\frac{\delta R}{R}\% \tag{4.20}$$

Example 4.1

Suppose that in Figure 4.16a, at first $R_1 = R_2 = R_3 = R_4 = R$. Now, increase R_1 by δR and decrease R_2 by δR. This will represent two active elements that act in reverse, as in the case of two strain-gauge elements mounted on the top and the bottom surfaces of a beam in bending. Show that the bridge output is linear in δR in this case.

Solution

From Equation 4.15, we get $\delta v_o = ([(R+\delta R)R - R^2]/((R+\delta R+R-\delta R)(R+R)))v_{ref} - 0$. This simplifies to $(\delta v_o/v_{ref}) = (\delta R/4R)$, which is linear.

Similarly, it can be shown using Equation 4.15 that the pair of changes: $R_3 \rightarrow R+\delta R$ and $R_4 \rightarrow R - \delta R$ will result in a linear relation for the bridge output.

4.4.2 Constant-Current Bridge

When large resistance variations δR are required for a measurement, the Wheatstone bridge may not be satisfactory due to its nonlinearity, as indicated by Equation 4.17. The constant-current bridge is less nonlinear and is preferred in such applications. However, it needs a current-regulated power supply, which is typically more costly than a voltage-regulated power supply.

As shown in Figure 4.16b, the constant-current bridge uses a constant-current supply i_{ref}, instead of a constant-voltage supply. The output equation for a constant-current bridge can be determined from Equation 4.15, simply by substituting the voltage at the current source. Suppose that this voltage is v_{ref}, with the polarity shown in Figure 4.16a. Now, since the load current is assumed small (i.e., a high-impedance load), the current through R_2 is equal to the current through R_1 and is given by $v_{ref}/(R_1+R_2)$. Similarly, current through R_4 and R_3 is given by $v_{ref}/(R_3+R_4)$. Accordingly, by current summation, we get

$$i_{ref} = \frac{v_{ref}}{(R_1+R_2)} + \frac{v_{ref}}{(R_3+R_4)}$$

or

$$v_{ref} = \frac{(R_1+R_2)(R_3+R_4)}{(R_1+R_2+R_3+R_4)} i_{ref} \tag{4.21}$$

This result may be directly obtained from the equivalent resistance of the bridge, as seen by the current source. Substituting Equation 4.21 into Equation 4.15, we have the output equation for the constant-current bridge as

$$v_o = \frac{(R_1R_4 - R_2R_3)}{(R_1+R_2+R_3+R_4)} i_{ref} \tag{4.22}$$

Note from Equation 4.22 that for this bridge as well, the bridge-balance requirement (i.e., $v_o=0$) is given by Equation 4.16.

To estimate the nonlinearity of a constant-current bridge, we start with the balanced condition: $R_1=R_2=R_3=R_4=R$ and change R_1 by δR while keeping the remaining resistors inactive. Again, R_1 will represent the active element (sensing element) of the bridge and may correspond to an active strain gauge. The change in output δv_o is given by

$$\delta v_o = \frac{[(R+\delta R)R - R^2]}{(R+\delta R+R+R+R)} i_{ref} - 0$$

or

$$\frac{\delta v_o}{R i_{ref}} = \frac{\delta R/R}{(4+\delta R/R)} \tag{4.23}$$

By comparing the denominator on the RHS of this equation with Equation 4.17, we observe that the constant-current bridge is less nonlinear. Specifically, using the definition given by Equation 4.19, the percentage nonlinearity may be expressed as

$$N_p = 25\frac{\delta R}{R}\% \tag{4.24}$$

It is noted that the nonlinearity is halved by using a constant-current excitation, instead of a constant-voltage excitation.

Example 4.2

Suppose that in the constant-current bridge circuit shown in Figure 4.16b, at first $R_1=R_2=R_3=R_4=R$. Assume that R_1 and R_4 represent strain gauges mounted on the same side of a rod in tension. Due to the tension, R_1 increases by δR and R_4 also increases by δR. Derive an expression for the bridge output (normalized) in this case and show that it is linear. What would be the result in this example if R_2 and R_3 represented the active tensile strain gauges?

Solution

From Equation 4.22, we get $\delta v_o=([(R+\delta R)(R+\delta R)-R^2]/(R+\delta R+R+R+R+\delta R))i_{ref}-0$. By simplifying and canceling the common term in the numerator and the denominator, we get the linear relation

$$\frac{\delta v_o}{Ri_{ref}}=\frac{\delta R/R}{2} \tag{4.25}$$

If R_2 and R_3 are the active elements, it is clear from Equation 4.22 that we get the same linear result, except for a sign change. Specifically,

$$\frac{\delta v_o}{Ri_{ref}}=-\frac{\delta R/R}{2} \tag{4.26}$$

4.4.3 Hardware Linearization of Bridge Outputs

From the foregoing developments and as illustrated in the examples, it should be clear that the output of a resistance bridge is not linear in general, with respect to the change in resistance of the active elements. Particular arrangements of the active elements, however, can result in a linear output. It is seen from Equations 4.15 and 4.22 that, when there is only one active element, the bridge output is nonlinear. Such a nonlinear bridge can be linearized using hardware, particularly op-amp elements. To illustrate this approach, consider a constant-voltage resistance bridge. We modify it by connecting two op-amp elements, as shown in Figure 4.16c. The output amplifier has a feedback resistor R_f. The output equation for this circuit can be obtained by using the properties of an op-amp, in the usual manner. In particular, the potentials at the two input leads must be equal and the current through these leads must be zero. From the first property, it follows that the potentials at both nodes A and B are zero. Let the potential at node C be denoted by v. Now, use the second property and write current summations at nodes A and B.

$$\text{Node A}: \quad \frac{v}{R}+\frac{v_{ref}}{R}+\frac{v_o}{R_f}=0 \tag{i}$$

$$\text{Node B}: \quad \frac{v_{ref}}{R}+\frac{v}{R+\delta R}=0 \tag{ii}$$

Substitute Equation ii into Equation i to eliminate v and simplify to get the linear result

$$\frac{\delta v_o}{v_{ref}}=\frac{R_f}{R}\frac{\delta R}{R} \tag{4.27}$$

Compare this result with Equation 4.17 for the original bridge with a single active element. Note that, when $\delta R = 0$, from Equation ii, we get $v = -v_{ref}$ and from Equation i we get $v_o = 0$. Hence, when we start with a balanced bridge, v_o and δv_o mean the same thing, as used in Equation 4.27.

4.4.3.1 Bridge Amplifiers

The output signal from a resistance bridge is usually very small in comparison to the reference signal, and it has to be amplified to increase its voltage level to a useful value (e.g., for use in system monitoring, data logging, or control). A bridge amplifier is used for this purpose. This is typically an instrumentation amplifier, which is essentially a sophisticated differential amplifier (see Chapter 3). The bridge amplifier is modeled as a simple gain K_a, which multiplies the bridge output. Typical characteristics are

- Gain up to 1000 (adjustable)
- Low drift
- Wide operating range (±200 μV to ±5 V, adjustable in steps)
- Supply (DC) voltage equal to ±10 V (the same source supplies the bridge circuit as well)
- Max current equal to 30 mA
- A high input impedance (2 MΩ) so that the bridge output would not be loaded
- Multiple channels (for simultaneous use with multiple bridges)
- Low-pass filter cutoff up to 2 kHz (selectable)
- Common-mode rejection ratio (CMRR) 100 dB at 50 Hz

4.4.4 Half-Bridge Circuits

A half bridge may be used in some applications that require a bridge circuit. A half bridge has only two arms, and the output is tapped from the midpoint of these two arms.

Note: The half-bridge circuit is somewhat similar to a potentiometer circuit or a voltage divider. In some half-bridge circuits, there may exist a third arm that is connected across the entire span of these two arms. Still, the output is tapped at the common node of the first two arms, while the other output lead is at an end node of an arm.

The ends of the two arms are excited by two voltages, one of which is positive and the other negative (i.e., bipolar supply voltage). Initially, the two arms have equal resistances so that nominally the bridge output is zero. One of the arms has the active element. Its change in resistance results in a nonzero output voltage.

A half-bridge amplifier consisting of a resistance half bridge and an output amplifier is shown in Figure 4.17. The two bridge arms have resistances R_1 and R_2, and the output amplifier uses a feedback resistance R_f. To get the output equation, we use the two basic facts for an unsaturated op-amp (see Chapter 3): the voltages at the two input leads are equal (due to high gain), and the current in either lead is zero (due to high input impedance). Hence, the voltage at node A is zero, and the current balance equation at node A is given by

$$\frac{v_{ref}}{R_1} + \frac{(-v_{ref})}{R_2} + \frac{v_o}{R_f} = 0$$

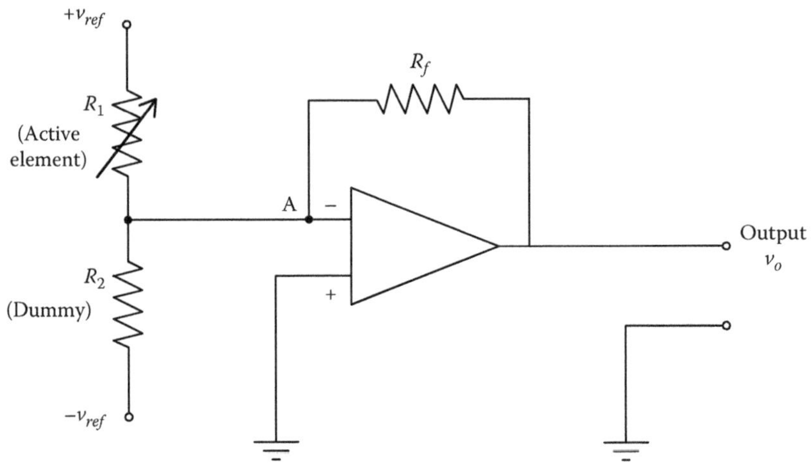

FIGURE 4.17
A half bridge with an output amplifier.

This gives

$$v_o = R_f\left(\frac{1}{R_2}-\frac{1}{R_1}\right)v_{ref} \tag{4.28}$$

Now, suppose that initially $R_1=R_2=R$, and the active element R_1 changes by δR. The corresponding change in the output is

$$\delta v_o = R_f\left(\frac{1}{R}-\frac{1}{R+\delta R}\right)v_{ref}-0$$

or

$$\frac{\delta v_o}{v_{ref}} = \frac{R_f}{R}\frac{\delta R/R}{(1+\delta R/R)} \tag{4.29}$$

Note that R_f/R is the amplifier gain. Now in view of Equation 4.19, the percentage nonlinearity of the half-bridge circuit is

$$N_p = 100\frac{\delta R}{R}\% \tag{4.30}$$

It follows that the nonlinearity of a half-bridge circuit is worse than that for a Wheatstone bridge.

4.4.5 Impedance Bridges

An impedance bridge is an alternating-current (AC) bridge. It contains general impedance elements Z_1, Z_2, Z_3, and Z_4 in its four arms, as shown in Figure 4.18a. The bridge is powered by an AC supply voltage v_{ref}.

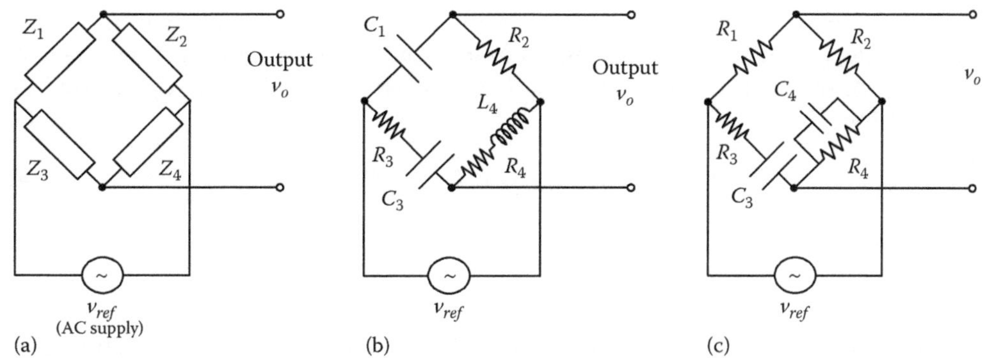

FIGURE 4.18
(a) General impedance bridge, (b) Owen bridge, and (c) Wien-bridge oscillator.

Note: v_{ref} would represent a carrier signal, and the output voltage v_o would have to be demodulated if a transient signal representative of the variation in one of the bridge elements was needed.

Impedance bridges may be used, for example, to measure capacitances in capacitive sensors and changes of inductance in variable-inductance sensors and eddy current sensors. Also, impedance bridges can be used as oscillator circuits. An oscillator circuit may serve as a constant-frequency source of a signal generator (e.g., in product dynamic testing or in generating carrier signals), or it may be used to determine an unknown circuit parameter by measuring the oscillating frequency.

Analyzing by using frequency-domain concepts, the frequency spectrum of the impedance bridge output is given by

$$v_o(\omega) = \frac{(Z_1Z_4 - Z_2Z_3)}{(Z_1 + Z_2)(Z_3 + Z_4)} v_{ref}(\omega) \tag{4.31}$$

This reduces to Equation 4.15 in the DC case of a Wheatstone bridge. The balanced condition is given by

$$\frac{Z_1}{Z_2} = \frac{Z_3}{Z_4} \tag{4.32}$$

This equation is used to measure an unknown circuit parameter in the bridge. Let us consider two particular impedance bridges.

4.4.5.1 Owen Bridge

The Owen bridge is shown in Figure 4.18b. It may be used, for example, to measure both inductance L_4 and capacitance C_3, by the bridge-balance method. To derive the necessary equations, note that the voltage–current relation for an inductor is $v = L(di/dt)$, and for a capacitor, it is $i = C(dv/dt)$. It follows that the voltage–current transfer function (in the Laplace domain) for an inductor is $(v(s)/i(s)) = Ls$, and for a capacitor, it is $(v(s))/(i(s)) = 1/Cs$. Accordingly, the impedances of an inductor element and a capacitor element at frequency ω are $Z_L = j\omega L$ and $Z_c = 1/(j\omega C)$, respectively. By applying these results for the Owen bridge, we get $Z_1 = 1/(j\omega C_1)$, $Z_2 = R_2$, $Z_3 = R_3 + 1/(j\omega C_3)$, $Z_4 = R_4 + j\omega L_4$, where ω is the excitation frequency. Now, for a balanced bridge, from Equation 4.32, we have $1/(j\omega C_1)(R_4+j\omega L_4)=R_2(R_3+1/(j\omega C_3))$. By equating separately the real parts and the imaginary parts of this equation, we get the two equations: $(L_4/C_1)=R_2R_3$ and $R_4/C_1=R_2/C_3$. Hence, we get the final results:

$$L_4 = C_1R_2R_3 \tag{4.33}$$

$$C_3 = C_1\frac{R_2}{R_4} \tag{4.34}$$

It follows that L_4 and C_3 can be determined with the knowledge of C_1, R_2, R_3, and R_4 under balanced conditions. For example, with fixed C_1 and R_2, an adjustable R_3 may be used to measure the variable L_4, and an adjustable R_4 may be used to measure the variable C_3.

4.4.5.2 Wien-Bridge Oscillator

Now, consider the Wien-bridge oscillator shown in Figure 4.18c. For this circuit, we have $Z_1=R_1$, $Z_2=R_2$, $Z_3=R_3+(1/(j\omega C_3))$, $1/Z_4=1/R_4+j\omega C_4$. Hence, from Equation 4.32, the bridge-balance requirement is

$$\frac{R_1}{R_2} = \left(R_3 + \frac{1}{j\omega C_4}\right)\left(\frac{1}{R_4} + j\omega C_4\right)$$

By equating the real parts, we get

$$\frac{R_1}{R_2} = \frac{R_3}{R_4} + \frac{C_4}{C_3} \tag{4.35}$$

Next, by equating the imaginary parts, we get $0=\omega C_4R_3-(1/\omega C_3R_4)$. Hence,

$$\omega = \frac{1}{\sqrt{C_3C_4R_3R_4}} \tag{4.36}$$

This confirms that the circuit is an oscillator whose natural frequency is given by this equation, under balanced conditions. If the frequency of the supply is equal to the natural frequency of the circuit, large-amplitude oscillations will take place. The circuit can be used to measure an unknown resistance (e.g., in strain-gauge devices) by first measuring the frequency of the bridge signals at resonance (natural frequency). Alternatively, an oscillator that is excited at its natural frequency can be used as an accurate source of periodic signals (i.e., a signal generator).

4.5 Linearizing Devices

Nonlinearity is present in any physical device to varying degrees. If the level of nonlinearity in a system (component, device, or equipment) can be neglected without exceeding the error tolerance, then for practical purposes the system can be assumed linear.

In general, a linear system is one that can be expressed by a *linear analytical model* (e.g., a set of linear differential equations or linear algebraic equations). Furthermore, the *principle of superposition* holds for linear systems. Specifically, according to this principle, if the system response to an input u_1 is y_1 and the response to another input u_2 is y_2, then the response to $a_1u_1 + a_2u_2$ would be $a_1y_1 + a_2y_2$, for any arbitrary a_1 and a_2.

4.5.1 Nature of Nonlinearities

Nonlinearities in a system can appear in two forms:

1. Dynamic manifestation of nonlinearities
2. Static manifestation of nonlinearities

In many applications, the useful operating region of a system can exceed the frequency range where the frequency response function is flat. The operating response of such a system is said to be dynamic. Examples include a typical control system (e.g., automobile, aircraft, milling machine, robot), sensor (e.g., LVDT), actuator (e.g., hydraulic motor), and controller (e.g., proportional–integral–derivative control hardware). Nonlinearities of such systems can manifest themselves in a static form (where the output is not proportional to the input) or in a dynamic form such as the jump phenomenon (also known as the fold catastrophe), limit cycles, and frequency creation. The use of a nonlinear calibration curve (e.g., change in the scale from linear to logarithmic) may be adequate to remove a static nonlinearity. Design changes, extensive adjustments, or reduction of the operating signal levels and bandwidths may be necessary in general to reduce or eliminate dynamic manifestations of nonlinearity. In many instances, such changes might not be practical, and we may have to somehow cope with the presence of the nonlinearities under dynamic conditions. Design changes for reducing nonlinearities might involve replacing conventional gear drives by devices such as harmonic drives to reduce backlash, replacing nonlinear actuators by linear actuators, and using components that have negligible nonlinear (e.g., Coulomb) friction and that make small motion excursions.

A wide majority of sensors, transducers, and signal modification devices are expected to operate in the flat region of their frequency response function. The I/O relation of these types of devices, in the operating range, is expressed (modeled) as a static curve rather than a differential equation. Nonlinearities in these devices will manifest themselves in the static operating curve in many forms. These manifestations include saturation, hysteresis, and offset.

4.5.1.1 Linearization Methods

In one category of systems (e.g., plants, actuators, and compensators), if a nonlinearity is exhibited in the dynamic form, proper modeling and control practices should be employed to avoid unsatisfactory degradation of the system performance. In another category of systems (e.g., sensors, transducers, and signal modification devices), if nonlinearities are exhibited in the static operating curve, again the overall performance of the system will be degraded. Hence, it is important to linearize the output of such devices. Note that in dynamic manifestations it is not possible to realistically linearize the output because the response is generated in the dynamic form. The solution in that case may include the following: (1) reduce the operating range, (2) minimize the nonlinearities in the system by design modifications and adjustments, (3) apply feedback linearization where the system response and a nonlinear analytical model are used to estimate the nonlinearity and feed it back into the system to remove (compensate for) that nonlinearity, and (4) take the nonlinearities into account in system modeling and control. In this section, we will not address dynamic nonlinearities. Instead, we will consider static linearization of devices, whose operating characteristics can be expressed by static I/O curves.

Linearization of a static device may be done as well by making design changes and adjustments, as in the case of dynamic devices. But since the response is static and since we normally deal with an available device (i.e., a fixed design) whose internal hardware cannot be modified, we will only consider the ways of linearizing the I/O characteristic by modifying the output itself.

Static linearization of a device can be made in several ways such as

1. Linearization using nonlinear calibration curves, nonlinear transformations, and associated software
2. Linearization using digital (logic) hardware
3. Linearization using analog hardware

In the software approach to linearization, the output of the device is read into a digital processor with software-programmable memory, and the output is modified (recalibrated) according to the program instructions so that the I/O relation is linear. An example is the use of log scale to linearize nonlinear expressions. In the logic hardware approach, the output is read by a device with fixed logic circuitry for processing (modifying) the data. The resulting output is also digital. In the analog hardware approach, a linearizing analog circuit is directly connected at the output of the device so that the output signal (analog) of the linearizing device is proportional to the input of the original device. An example of this type of (analog) linearization is the linearized bridge as discussed previously (see Figure 4.16c). These three approaches of linearization are discussed next, while heavily emphasizing the analog-circuit approach.

Hysteresis-type static nonlinearity characteristics have the property that the I/O curve is not one to one. In other words, one input value may correspond to more than one (static) output value, and one output value may correspond to more than one input value. Disregard this type of nonlinearities, let us concentrate on the linearization of a device with a single-valued static response curve that is not a straight line. An example of a typical nonlinear I/O characteristic is shown in Figure 4.19a. Strictly speaking, a straight-line characteristic with a simple offset, as shown in Figure 4.19b, is also a nonlinearity. In particular, note that superposition does not hold for an I/O characteristic of this type, given by

$$y = ku + c \tag{4.37}$$

It is very easy, however, to linearize such a device because a simple addition of a DC component will convert the characteristic into the linear form, given by

$$y = ku \tag{4.38}$$

This method of linearization is known as "offsetting." Linearization is more difficult in the general case where the characteristic curve might be much more complex.

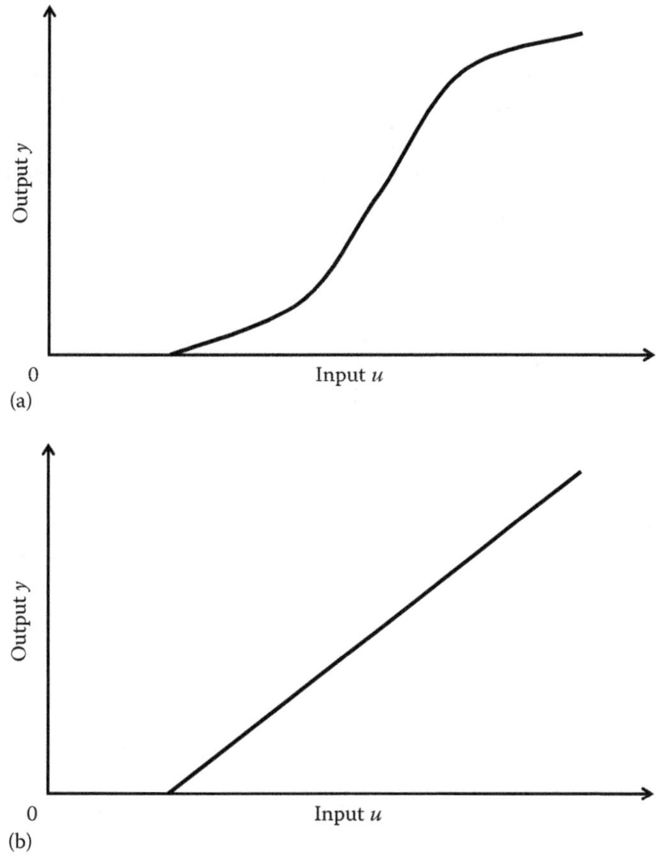

FIGURE 4.19
(a) A general static nonlinear characteristic and (b) an offset nonlinearity.

4.5.1.2 Linearization by Software

If the nonlinear relationship (static) between the input and the output of a nonlinear device is known, the input can be computed for a known value of the output. In the software approach of linearization, a processor and memory that can be programmed using software (i.e., a digital computer) are used to compute the input using output data. Three approaches can be used. They are

1. Variable transformation
2. Equation inversion
3. Table lookup

In the first two methods, the nonlinear characteristic of the device is known in the analytic (equation) form (i.e., an algebraic model)

$$y = f(u) \tag{4.39}$$

where
u is the device input
y is the device output

In the first method, the variables u and y are transformed into two new variables u' and y' (e.g., log of variable) such that the relationship between u' and y' is linear. Then, linear methodologies can be used to analyze the transformed data of u' and y' without sacrificing any accuracy. Since the transformation relation is known, the results can be transformed back into the domain of u and y, if necessary.

In the second method, assuming that Equation 4.39 is a one-to-one relationship, a unique inverse can be determined, as given by the equation

$$u = f^{-1}(y) \tag{4.40}$$

This equation may be programmed as a computation algorithm into the read-and-write memory (RAM) of the computer. When the output values y are supplied to the computer, the processor will compute the corresponding input values u using the instructions (executable program) stored in the RAM.

In the third method (table lookup), a sufficiently large number of pairs of values (y, u) are stored in the memory of the computer in the form of a table of ordered pairs. These values should cover the entire operating range of the device. Then, when a value for y is entered into the computer, the processor scans the stored data to check whether that value is present. If so, the corresponding value of u will be read, and this is the linearized output. If the value of y is not present in the data table, then the processor will interpolate the data in the vicinity of the value and will compute the corresponding output. In the linear interpolation method, the neighborhood of the data table where the y value falls is fitted with a straight line and the corresponding u value is computed using this straight line. Higher-order interpolations use nonlinear interpolation curves such as quadratic and cubic polynomial equations (splines).

The variable transformation method and the equation inversion method are usually more accurate than the table lookup method. Furthermore, the first two methods do not need excessive memory for data storage. But they are relatively slow because data are transferred, transformed, and processed within the computer using program instructions, which are stored in the memory and which typically have to be accessed in a sequential manner. The table lookup method is faster. Since accuracy depends on the amount of stored data values, this is a memory-intensive method. For better accuracy, more data should be stored. But since the entire data table has to be scanned to check for a given data value, this increase in accuracy is derived at the expense of speed as well as memory requirements.

4.5.1.3 Linearization by Logic Hardware

The software approach of linearization is flexible because the linearization algorithm can be modified (e.g., improved, changed) simply by modifying the program stored in the computer memory. Furthermore, highly complex nonlinearities can be handled by the software method. As mentioned earlier, the method is relatively slow, however.

In the logic hardware method of linearization, the linearization algorithm is permanently implemented in the IC form using appropriate digital logic circuitry for data processing and memory elements (e.g., flip–flops). Note that the algorithm and the numerical values of the model parameters (not the input values) cannot be modified without redesigning the IC chip, because a hardware digital device typically does not have programmable memory. Furthermore, it will be difficult to implement very complex linearization algorithms by this method, and unless the chips are mass produced for an extensive commercial market, the initial chip development cost will make the production of linearizing chips economically infeasible. In bulk production, however, the per unit cost will be very small. Since both the access of stored program instructions and extensive data manipulation are not involved, the hardware method of linearization can be substantially faster than the software method.

A digital linearizing unit with a processor and a read-only memory (ROM), whose program cannot be modified, also lacks the flexibility of a programmable software device. Hence, such a ROM-based device also falls into the category of hardware logic devices. The use of microcontroller for static linearization is a software approach, yet it is quite economical and can be made quite fast with modern microcontroller technologies.

4.5.2 Analog Linearizing Hardware

The following three types of analog linearizing hardware are discussed now:

1. Offsetting circuitry
2. Circuitry that provides a proportional output
3. Curve shapers

An offset is a nonlinearity that can be easily removed using an analog device. This is accomplished by simply adding a DC offset of equal value to the response, in the opposite direction. Deliberate addition of an offset in this manner is known as "offsetting." The associated removal of original offset is known as "offset compensation." There are many applications of offsetting. Unwanted offsets such as those present in the results of ADC and DAC can be removed by analog offsetting. Constant (DC) error components, such as steady-state errors in dynamic systems due to load changes, gain changes, and other disturbances, can be eliminated as well by offsetting. Common-mode error signals in amplifiers and other analog devices can also be removed by offsetting. In measurement circuitry such as potentiometer (ballast) circuits, where the actual measurement signal is a small change δv_o of a steady output signal v_o, the measurement can be completely masked by noise. To overcome this problem, first, the output should be offset by $-v_o$ so that the net output is δv_o and not $v_o+\delta v_o$. Subsequently, this output can be conditioned through filtering and amplification. Another application of offsetting is the additive change of the scale of a measurement from a relative scale to an absolute scale (e.g., in the case of velocity). In summary, some applications of offsetting are the following:

1. Removal of unwanted offsets and DC components in signals (e.g., in ADC, DAC, signal integration)
2. Removal of steady-state error components in dynamic system responses (e.g., due to load changes and gain changes in Type 0 dynamic systems) (*Note*: Type 0 systems are open-loop systems that do not possess free integrators.)
3. Rejection of common-mode levels (e.g., in amplifiers and filters; see Chapter 3)
4. Error reduction when a measurement is an increment of a large steady output level (e.g., in ballast—potentiometer circuits for strain-gauge and RTD sensors; see Chapters 9 and 10)
5. Scale changes in an additive manner (e.g., conversion from relative to absolute units or from absolute to relative units)

We can remove unwanted offsets in a simple manner as discussed earlier. Let us now consider more complex nonlinear responses that are nonlinear, in the sense that the I/O curve is not a straight line. Analog circuitry can be used to linearize this type of responses as well. The employed linearizing circuit will generally depend on the particular device and the nature of its nonlinearity. Hence, often linearizing circuits of this type have to be discussed with respect to a particular application. For example, such linearization circuits are useful in a transverse-displacement capacitive sensor. Several useful circuits are described here.

Consider the type of linearization that is known as "curve shaping." A curve shaper is a linear device whose gain (output–input) can be adjusted so that response curves with different slopes can be obtained. Suppose that a nonlinear device with an irregular (nonlinear) I/O characteristic is to be linearized. First, we apply the operating input simultaneously to both the device and the curve shaper and then adjust the gain of the curve shaper such that its output closely matches that of the actual device in a small range of operation. Now, the output of the curve shaper can be utilized for any task that requires the device output. The advantage here is that linear assumptions are valid with the curve shaper (and hence, linear methodologies are applicable), which is not the case for the actual device. When the operating range changes, the curve shaper has to be adjusted to the new range. Comparison (recalibration) of the curve shaper and the nonlinear device can be done off line, and once a set of gain values corresponding to a set of operating ranges is determined in this manner for the curve shaper, it is possible to completely replace the nonlinear device by the curve shaper. Then, the gain of the curve shaper can be adjusted, depending on the actual operating range during system operation. This is known as "gain scheduling." *Note*: In this manner, we can replace a nonlinear device by a linear device (curve shaper) within a multicomponent system without greatly sacrificing the accuracy of the overall system.

4.5.2.1 Offsetting Circuitry

Common-mode outputs and offsets in amplifiers and other analog devices can be minimized by including a compensating resistor, which will provide fine adjustments at one of the input leads (see Chapter 3). Also, the larger the magnitude of the feedback signal in a control system, the smaller the steady-state error. Hence, steady-state offsets can be reduced by decreasing the feedback resistance (thereby increasing the feedback signal). Furthermore, since a ballast (potentiometer) circuit provides an output of $v_o + \delta v_o$ and a bridge circuit provides an output of δv_o, the use of a bridge circuit can be interpreted as an offset compensation method.

The most straightforward way of offsetting a nonlinear device is by using a differential amplifier (or a summing amplifier; see Chapter 3) to subtract (or add) a DC voltage to the output of the device. The DC level has to be variable so that various levels of offset can be provided with the same circuit. This is accomplished by using an adjustable resistance at the DC input lead of the amplifier.

An op-amp circuit that can be used for offsetting is shown in Figure 4.20. Since the input v_i is connected to the negative lead of the op-amp, we have an inverting amplifier, and the input signal will appear in the output v_o with its sign reversed. This is also a summing amplifier because two signals can be added together by this circuit. If the input v_i is connected to the positive lead of the op-amp, we will have a noninverting amplifier.

The DC voltage v_{ref} provides the offsetting voltage. The compensating resistor R_c is variable so that different values of offset can be compensated for using the same circuit. To obtain the circuit equation, we write the current balance equation for node A, using the

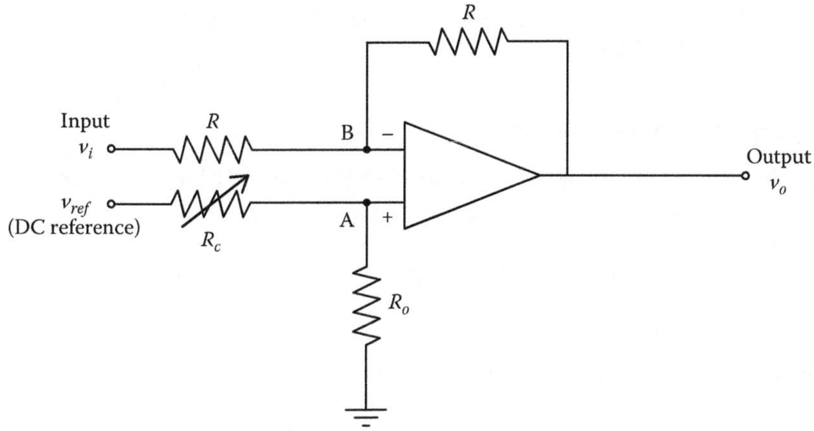

FIGURE 4.20
An inverting amplifier circuit for offset compensation.

usual assumption that the current through an input lead is zero for an op-amp (because of its very high input impedance):

$$\frac{v_{ref} - v_A}{R_c} = \frac{v_A}{R_o}$$

or

$$v_A = \frac{R_o}{(R_o + R_c)} v_{ref} \tag{i}$$

Similarly, the current balance at node B gives

$$\frac{v_i - v_B}{R} + \frac{v_o - v_B}{R} = 0$$

or

$$v_o = -v_i + 2v_B \tag{ii}$$

Since $v_A = v_B$ for the op-amp (because of its very high open-loop gain), we can substitute Equation i into Equation ii to get

$$v_o = -v_i + \frac{2R_o}{(R_o + R_c)} v_{ref} \tag{4.41}$$

Note the sign reversal of v_i at the output (because this is an inverting amplifier). This sign reversal is not a problem because the polarity can be reversed at the input or the output when connecting this circuit to other circuitry, thereby recovering the original sign. The important result here is the presence of a constant offset term on the RHS of

Equation 4.41. This term can be adjusted by picking the proper value for R_c so as to compensate for a given offset in v_i.

4.5.2.2 Proportional-Output Hardware

An op-amp circuit may be employed to linearize the output of a capacitive transverse-displacement sensor. When studying bridge circuits, in this chapter, we have noted that in the constant-voltage and constant-current resistance bridges and in the constant-voltage half bridge, the relation between the bridge output δv_o and the measurand (i.e., the measured quantity, which is the change in resistance in the active element) is nonlinear in general. The lowest nonlinearity is provided by the constant-current bridge and the highest is by the half bridge. As δR is small compared with R, the nonlinear relations can be linearized without introducing large errors. However, the linear relations are inexact and are not suitable if δR cannot be neglected in comparison to R. Then, the use of a linearizing circuit would be more appropriate.

One way to obtain a proportional output from a Wheatstone bridge is to feed back a suitable factor of the bridge output into the bridge supply v_{ref}. This approach has been illustrated previously (see Figure 4.16c). Another way is to use the op-amp circuit shown in Figure 4.21. This should be compared with the Wheatstone bridge shown in Figure 4.16a. In Figure 4.21, R_1 represents the only active element (e.g., an active strain gauge).

First, let us show that the output equation for the circuit in Figure 4.21 is quite similar to Equation 4.15. Using the fact that the current through an input lead of an unsaturated op-amp can be neglected, we have the current balance equations for nodes A and B: $((v_{ref}-v_A)/R_4)=(v_A/R_2)$ and $((v_{ref}-v_B)/R_3)+((v_o-v_B)/R_1)=0$. Hence, $v_A=(R_2/(R_2+R_4))v_{ref}$ and $v_B=(R_1v_{ref}+R_3v_o)/(R_1+R_3)$. Now using the fact that $v_A=v_B$ for an op-amp, we get $((R_1v_{ref}+R_3v_o)/(R_1+R_3))=(R_2/(R_2+R_4))v_{ref}$. Accordingly, we have the circuit output equation

$$v_o=\frac{(R_2R_3-R_1R_4)}{R_3(R_2+R_4)}v_{ref} \tag{4.42}$$

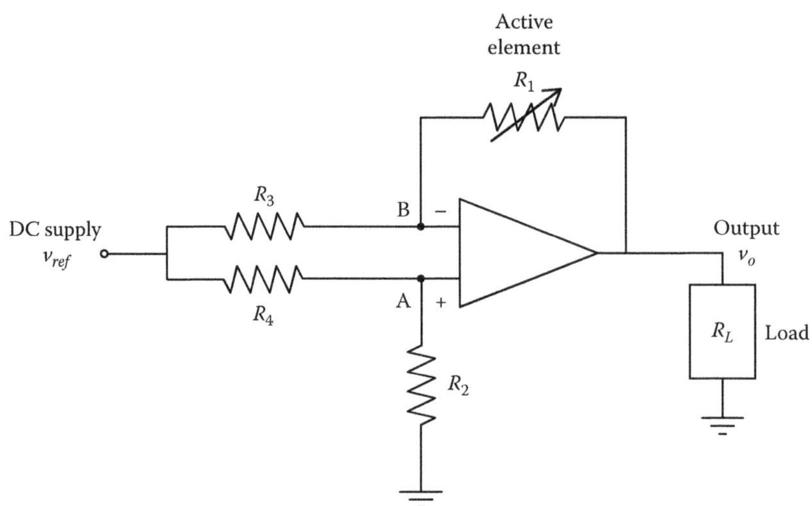

FIGURE 4.21
A proportional-output circuit for an active resistance element (strain gauge).

This relation is quite similar to the Wheatstone bridge equation (4.15). Again, the balance condition (i.e., $v_o = 0$) is given by Equation 4.16.

Suppose that $R_1 = R_2 = R_3 = R_4 = R$ in the beginning (hence, the circuit is balanced) so that $v_o = 0$. Next, suppose that the active resistance R_1 is changed by δR (say, due to a change in strain in the strain gauge R_1). Then, using Equation 4.42, we can write an expression for the resulting change in the circuit output as

$$\delta v_o = \frac{[R^2 - R(R + \delta R)]}{R(R + R)} v_{ref} - 0$$

or

$$\frac{\delta v_o}{v_{ref}} = -\frac{1}{2}\frac{\delta R}{R} \tag{4.43}$$

By comparing this result with Equation 4.17, we observe that the circuit output δv_o is proportional to the measurand δR. Furthermore, note that the sensitivity (1/2) of the circuit in Figure 4.21 is double that of a Wheatstone bridge (1/4) with one active element, which is a further advantage of the proportional-output circuit. The sign reversal is not a drawback because it can be removed by reversing the load polarity.

4.5.2.3 Curve-Shaping Hardware

A curve shaper can be interpreted as an amplifier whose gain is adjustable. A typical arrangement for a curve-shaping circuit is shown in Figure 4.22. The feedback resistance R_f is adjustable by some means. For example, a switching circuit with a bank of resistors (say, connected in parallel through solid-state switches as in the case of a weighted-resistor DAC) can be used to switch the feedback resistance to the required value. Automatic switching can be realized by using Zener diodes as well, which will start conducting at certain voltage levels. In both cases (i.e., external switching by switching pulses and automatic switching using Zener diodes), amplifier gain is variable in discrete steps. Alternatively, a potentiometer may be used to provide the resistance R_f so that the gain can be continuously adjusted (manually or automatically).

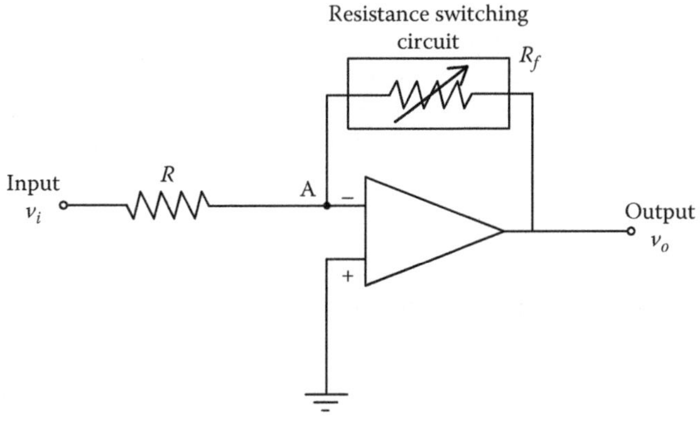

FIGURE 4.22
A curve-shaping circuit.

The output equation for the curve-shaping circuit shown in Figure 4.22 is obtained by writing the current balance at node A, noting that $v_A = 0$:

$$\frac{v_i}{R} + \frac{v_o}{R_f} = 0$$

which gives

$$v_o = -\frac{R_f}{R} v_i \tag{4.44}$$

It is seen that the gain (R_f/R) of the amplifier can be adjusted by changing R_f.

4.6 Miscellaneous Signal Conversion Hardware

In addition to the signal conversion devices discussed thus far in this chapter, there are many other types of circuitry that are used for signal conversion and related tasks. Examples are phase shifters, VFCs, FVCs, voltage-to-current converts (VCC), and peak-hold circuits. The objective of this section is to briefly discuss several of such miscellaneous circuits and components that are useful in the instrumentation, monitoring, and control of engineering systems.

4.6.1 Phase Shifters

A phase shifter changes the *phase angle* of a signal. Consider a sinusoidal signal given by

$$v = v_a \sin(\omega t + \phi) \tag{4.45}$$

It has the following three representative parameters:

1. Amplitude, v_a
2. Frequency, ω
3. Phase angle, ϕ

The phase angle represents the time reference (i.e., starting point) of the signal. It is a particularly important consideration when two or more signal components are compared and also when different time instants of a signal (generally not sinusoidal) are compared. The Fourier spectrum of a signal is presented as both its amplitude (magnitude) and the phase angle with respect to frequency.

4.6.1.1 Applications

Phase-shifting circuits have many applications. The applications can be classified into general two types:

1. Detecting the phase angle of a signal (typically, by shifting the phase angle and comparing with a reference signal)
2. Shifting the phase angle of a signal for subsequent use in the application

As a specific application, phase lead or lag of two quadrature signals (i.e., out of phase by 90°) generated by a digital transducer corresponds to the direction of motion. In this context, phase determination is used in establishing the direction of motion.

Another application of phase angle determination is in "system identification" where the goal is to obtain an *experimental model* of a system. When a signal passes through a system, the phase angle of the signal changes due to the system dynamics. Consequently, the phase change provides very useful information not only about the output signal but also about the dynamic characteristics of the system. Specifically, for a linear constant-parameter system, this phase shift is equal to the phase angle of the frequency response function (i.e., frequency transfer function) of the system at that particular frequency. This phase-shifting behavior is, of course, not limited to electrical systems and is equally exhibited by other types of systems, including mechanical systems, thermal systems, fluid systems, and multidomain (mixed) systems. The phase shift between two signals can be determined by converting the signals into the electrical form (using suitable transducers) and shifting the phase angle of one signal through known amounts using a phase-shifting circuit until the two signals are in phase.

A further application of phase shifters is in signal demodulation. For example, as noted earlier in this chapter, one method of amplitude demodulation involves processing the modulated signal together with the carrier signal. This, however, requires the modulated signal and the carrier signal to be in phase. But, usually, since the modulated signal has already transmitted through hardware with impedance characteristics, its phase angle would have changed. Then, it is necessary to shift the phase angle of the carrier until the two signals are in phase so that demodulation can be performed accurately. Hence, phase shifters are used in demodulating, for example, the outputs of LVDT displacement sensors (see Chapter 8).

Phase shifters are also used in signal communication (e.g., phase modulation in digital communication and modems) and transmission (e.g., antennas that do not require reorientation).

4.6.1.2 Analog Phase Shift Hardware

A phase-shifter circuit, ideally, should not change the signal amplitude while changing the phase angle by a required amount. However, practical analog phase shifters may introduce some degree of amplitude distortion (with respect to frequency) as well. A simple phase-shifter circuit can be constructed using resistor (R) and capacitor (C) elements. A resistor or a capacitor of such an RC circuit is made fine adjustable so as to realize variable phase shifting.

An op-amp-based phase-shifter circuit is shown in Figure 4.23. We can show that this circuit provides a phase shift without distorting the signal amplitude. The circuit equation is obtained by writing the current balance equations at nodes A and B, as usual, noting that the current through the op-amp leads can be neglected due to the high input impedance. Thus,

$$\frac{v_i - v_A}{R_c} = C\frac{dv_A}{dt} \quad \text{and} \quad \frac{v_i - v_B}{R} + \frac{v_o - v_B}{R} = 0$$

On simplifying and introducing the Laplace variable s, we get

$$v_i = (\tau s + 1)v_A \tag{i}$$

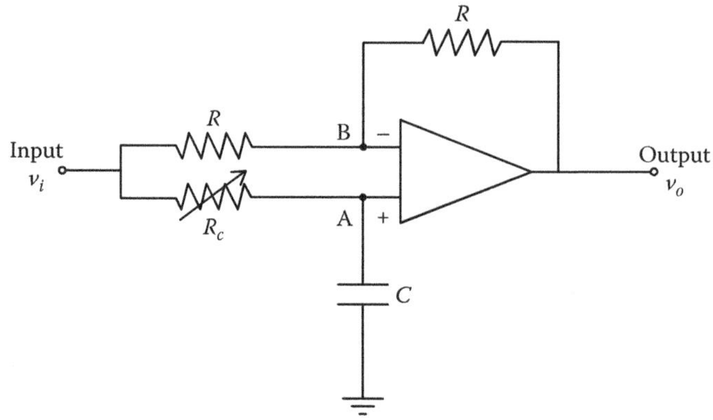

FIGURE 4.23
A phase-shifter circuit.

and

$$v_B = \frac{1}{2}(v_i + v_o) \tag{ii}$$

where the circuit time constant τ is given by $\tau = R_cC$. Since $v_A = v_B$, as a result of very high gain in the op-amp, we have by substituting Equation ii into Equation i $v_i = (1/2)(\tau s + 1)(v_i + v_o)$. It follows that the transfer function $G(s)$ of the circuit is given by

$$\frac{v_o}{v_i} = G(s) = \frac{(1 - \tau s)}{(1 + \tau s)} \tag{4.46}$$

It is seen that the magnitude of the frequency response function $G(j\omega)$ is $|G(j\omega)| = \sqrt{1+\tau^2\omega^2}/\sqrt{1+\tau^2\omega^2}$ or

$$|G(j\omega)| = 1 \tag{4.47}$$

and the phase angle of $G(j\omega)$ is $\angle G(j\omega) = -\tan^{-1}\tau\omega - \tan^{-1}\tau\omega$ or

$$\angle G(j\omega) = -2\tan^{-1}\tau\omega = -2\tan^{-1}R_cC\omega \tag{4.48}$$

As needed, the transfer-function magnitude is unity, indicating that the circuit does not distort the signal amplitude over the entire bandwidth. Equation 4.48 gives the phase lead of the output v_o with respect to the input v_i. Note that this angle is negative, indicating that actually a phase "lag" is introduced by the circuit, as expected. The phase shift can be adjusted by varying the resistance R_c.

4.6.1.3 Digital Phase Shifter

In a digital phase shifter, a digital hardware processor is used to phase shift an incoming sequence of data bits. Digital phase shifters in the form of monolithic IC chips (e.g., a 4 mm package of GaAs 6-bit digital phase shifter with an integrated CMOS driver, operating frequency range 3.5 GHz, phase shift range 360°, max error 1°, phase shift step 6°, supply voltage ±8 V DC) for frequency shifting of radio frequency signals are commercially available. Their applications include satellite communication, antennas, and active phased array radars. For example, as the digital signal sequences are transmitted and received, the phase change of the received signal is used to determine the distance between the transmitter and the receiver (or the time of flight of the data). Measurement of object deformation using laser holographic interferometry and phase shifting of holographic data frames (software based) has been reported. Another application is in 3D measurement that uses stereo images and phase-shifted fringe patterns.

4.6.2 Voltage-to-Frequency Converter

A VFC generates a periodic output signal whose frequency is proportional to the level of an input voltage. Since such an oscillator generates a periodic output according to the voltage input, it is also called a "voltage-controlled oscillator" (VCO). Furthermore, since a frequency can be counted and represented as a digital word, a VFC can serve as an *analog-to-digital converter* (ADC) as well. Furthermore, a VFC is essentially a *frequency modulator* (FM) as well. The voltage input to the VFC may correspond to a transducer signal (e.g., strain gauge, temperature sensor, accelerometer).

A common type of VFC circuit uses a capacitor. The time needed to charge the capacitor to a fixed voltage level depends on (inversely proportional to) the charging voltage. Suppose that this voltage is governed by the input voltage. Then, if the capacitor is made to periodically charge and discharge, we have an output whose frequency (inverse of the charge–discharge period) is proportional to the charging voltage. The output amplitude will be given by the fixed voltage level to which the capacitor is charged in each cycle. Consequently, we have a signal with a fixed amplitude and a frequency that depends on the charging voltage (input).

A VFC (or VCO) circuit is shown in Figure 4.24a. The voltage-sensitive switch closes when the voltage across it exceeds a reference level v_s and it opens again when the voltage across it falls below a lower limit $v_o(0)$. In a discrete element circuit, a programmable unijunction transistor may be used as a switching device. However, modern VFCs are available in the monolithic form as IC chips.

Note that the polarity of the input voltage v_i is reversed. Suppose that the switch is open. Then, current balance at node A of the op-amp circuit is given by $(v_i/R)=C(dv_o/dt)$. As usual, v_A, the voltage at the positive lead, is 0 because the op-amp has a very high gain, and the current through the op-amp leads is 0 because the op-amp has a very high input impedance. The capacitor charging equation can be integrated for a given value of v_i. This gives $v_o(t)=(1/RC)v_i t+v_o(0)$. The switch will close when the voltage across the capacitor $v_o(t)$ equals the reference level v_s.

Then, the capacitor will be immediately discharged through the closed switch. Hence, the capacitor charging time T is given by

$$v_s = \frac{1}{RC} v_i T + v_o(0)$$

Accordingly,

$$T = \frac{RC}{v_i}(v_s - v_o(0)) \tag{4.49}$$

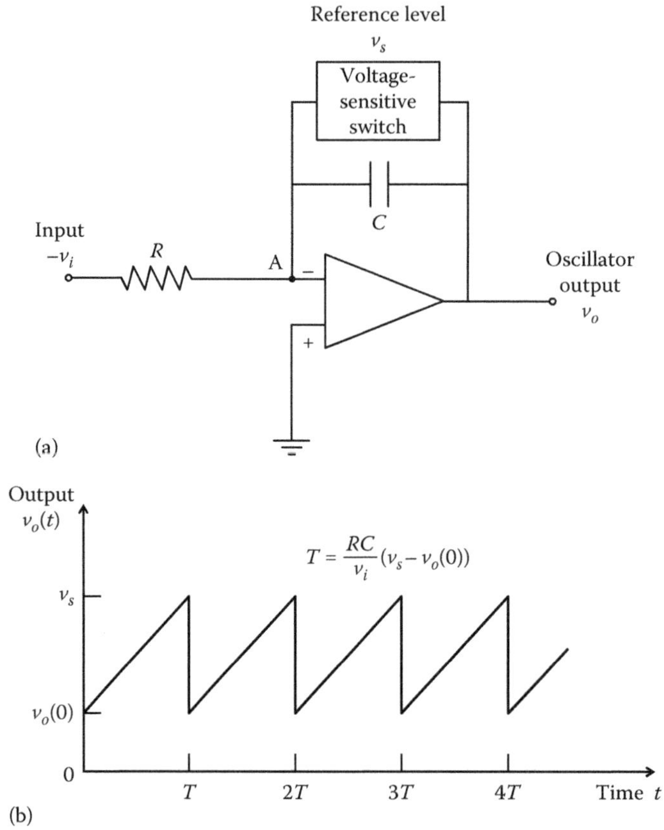

FIGURE 4.24
A voltage-to-frequency converter or voltage-controlled oscillator. (a) Circuit and (b) output signal.

The switch opens again when the voltage across the capacitor drops to $v_o(0)$, and the capacitor will begin to charge again from $v_o(0)$ up to v_s. This cycle of charging and instantaneous discharging will repeat periodically. The corresponding output signal is as shown in Figure 4.24b. This is a periodic (saw tooth) wave with period T. The frequency of oscillation ($1/T$) of the output is given by

$$f = \frac{v_i}{RC(v_s - v_o(0))} \tag{4.50}$$

It is seen that the oscillator frequency is proportional to the input voltage v_i. The oscillator amplitude is v_s, which is fixed.

4.6.2.1 Applications

VFCs have many applications. One application is in ADC. In a VFC-type ADC, the analog signal is converted into an oscillating signal using a VFC. Then, the oscillator frequency is measured using a digital counter. This count, which is available in the digital form, is representative of the input analog signal level. Another application is in digital voltmeters.

Here, the same method as for ADC is used. Specifically, the voltage is converted into an oscillator signal, and its frequency is measured using a digital counter. The count can be scaled and displayed to provide the voltage measurement. A direct application of VCO is apparent from the fact that VFC is actually a frequency modulator (FM), providing a signal whose frequency is proportional to the input (modulating) signal. Hence, VFC is useful in applications that require FM. Also, a VFC can function as a signal (wave) generator for variable-frequency applications, for example, excitation inputs for shakers in product dynamic testing, excitations for frequency-controlled motors, and pulse signals for drive circuits of stepping motors (incremental actuators).

4.6.2.2 Voltage-to-Frequency Converter Chips

VFCs are commonly available in the monolithic form as IC chips. Typically, the timing resistor (e.g., 1 kΩ) and capacitor (e.g., 390 pF) have to be externally connected to the chip. The key pins are bipolar supply DC voltage (2 pins), input voltage signal (1 pin), output frequency signal (1 pin), GND (1 pin), external resistor pin (1 pin), external capacitor pin (1 pin), and logic common pin for connection with other logic devices and is connected to GND or negative supply (1 pin).

Typical parameters are as follows:

Operating frequency range (linear): 1 Hz to 250 kHz

Supply voltage: ±5 to ±20 V DC

Power consumption: 10 mW

Pin structure: 8PDIP (8-pin dual inline)

Size: 10 mm × 6 mm package

Amplification: With a signal amplifier

Input impedance: 250 MΩ

4.6.3 Frequency-to-Voltage Converter

An FVC generates an output voltage whose level is proportional to the frequency of its input signal. One way to obtain an FVC is to use a digital counter to count the signal frequency and then use a DAC to obtain a voltage proportional to the frequency. A schematic representation of this type of FVC is shown in Figure 4.25a.

An alternative FVC circuit is schematically shown in Figure 4.25b. In this method, the frequency signal is supplied to a comparator along with a threshold voltage level. The sign of the comparator output will depend on whether the input signal level is larger or smaller than the threshold level. The first sign change (negative to positive) in the comparator output is used to trigger a switching circuit that will respond by connecting a capacitor to a fixed charging voltage. This will charge the capacitor. The next sign change (positive to negative) of the comparator output will cause the switching circuit to short the capacitor, thereby instantaneously discharging it. This charging–discharging process will be repeated in response to the oscillator input. The voltage level to which the capacitor is charged each time will depend on the switching period (charging voltage is fixed), which is in turn governed by the frequency of the input signal. Hence, the output voltage of the capacitor circuit will be representative of the frequency of the input signal. Since the output is not steady due to the ramp-like charging curve and instantaneous discharge, a

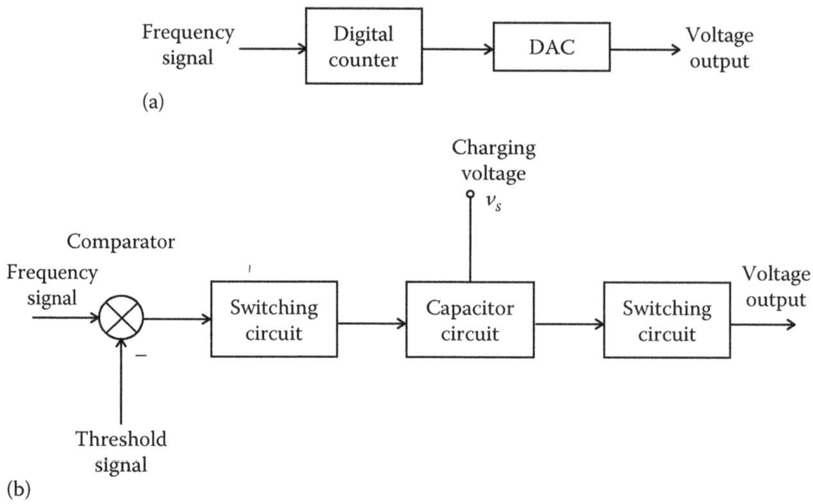

FIGURE 4.25
Frequency-to-voltage converter. (a) Digital counter method and (b) capacitor charging method.

smoothing circuit is provided at the output to remove the resulting noise ripples. It should be clear that the circuitry for this second approach to FVC is similar to that for VFC. In fact, the same IC chip is commercially available for both VFC and FVC.

Applications of FVC include demodulation of frequency-modulated signals, frequency measurement in control applications, digital-to-analog conversion, and conversion of pulse outputs in some types of sensors and transducers into analog voltage signals (e.g., output generation for digital tachometers).

4.6.4 Voltage-to-Current Converter

Measurement and feedback signals are usually transmitted as current levels in the range of 4–20 mA rather than as voltage levels. This is particularly useful when the measurement site is not close to the monitoring room. Since the measurement itself is usually available as a voltage, it has to be converted into current by using a VCC. For example, pressure transmitters and temperature transmitters in systems of operability testing (testing of proper system operation) provide current outputs that are proportional to the measured values of pressure and temperature. Furthermore, the torque of a motor corresponds to a current that generates the torque-producing magnetic field. Hence, torque control of a motor can be achieved indirectly through current control. Voltage-controlled current sources are useful in driving and testing of devices. The usefulness of a VCC is clear from these observations.

In signal transmission through a cable, there are advantages to transmitting the current rather than the voltage. The voltage level will drop due to the resistance in the transmission path, but the current through a conductor will remain unchanged unless the conductor is branched. Hence, current signals are less likely to acquire errors due to signal weakening. Another advantage of using current, instead of voltage as the measurement signal, is that the same signal can be used to operate several devices in series (e.g., a display, a recorder, and a signal processor simultaneously), again without causing errors due to signal weakening by the power lost at each device, because the same current is applied

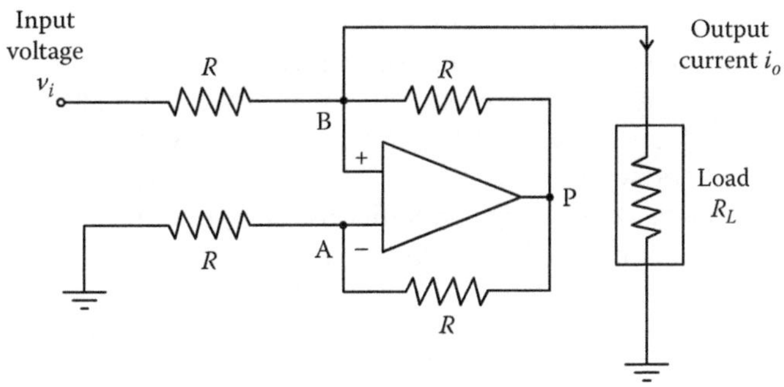

FIGURE 4.26
A voltage-to-current converter.

to all devices. A VCC should provide a current proportional to an input voltage, without being affected by the load resistance to which the current is supplied.

An op-amp-based VCC circuit is shown in Figure 4.26. Using the fact that the currents through the input leads of an unsaturated op-amp can be neglected (due to its very high input impedance), we write the current summation equations for the two nodes A and B: $(v_A/R) = ((v_P - v_A)/R)$ and $((v_i - v_B)/R) + ((v_P - v_B)/R) = i_o$. Accordingly, we have

$$2v_A = v_P \tag{i}$$

and

$$v_i - 2v_B + v_P = Ri_o \tag{ii}$$

Now using the fact that $v_A = v_B$ for the op-amp (due to its very high gain), we substitute Equation i into Equation ii to obtain

$$i_o = \frac{v_i}{R} \tag{4.51}$$

where
i_o is the output current
v_i is the input voltage

It follows that the output current is proportional to the input voltage, irrespective of the value of the load resistance R_L, as required for a VCC.

Commercially, VCCs are available as multipin IC chips. Some parameters of a VCC chip are as follows: it has an operating voltage range of 0–40 V and an operating current range of 0–40 mA, and it uses an external resistor.

4.6.5 Peak-Hold Circuit

The purpose of an analog peak-hold device is to receive an analog signal and hold its maximum value in a storage capacitor, until a larger value is received. Unlike an S/H circuit, which holds every sampled value of a signal, a peak-hold circuit holds only the largest value reached by the signal during the monitored period. Peak holding is useful in a variety of applications. For example, using this device, a signal envelope can be generated, which represents the extreme or most severe values of a signal. A peak-hold circuit may be used as a signal shaping device as well. As another application, in signal processing for shock and vibration studies of dynamic systems, what is known as "response spectra" (e.g., shock response spectrum) are determined by using a response spectrum analyzer, which exploits peak holding. Specifically, suppose that a signal is applied to a simple oscillator (a single-degree-of-freedom second-order system with no zeros) and the peak value of the response (output) is determined. A plot of the peak output as a function of the natural frequency of the oscillator, for a specified damping ratio, is known as the response spectrum of the signal for that damping ratio. Peak detection is also useful in machine monitoring and alarm-generating systems. In short, when just one representative value of a signal is needed in a particular application, the peak value is most likely a good choice.

Digital peak detector: Peak detection of a signal can be conveniently done using digital processing. For example, the signal is sampled and the previous sample value is replaced by the present sample value, if and only if the latter is larger than the former. In this manner, the peak value of the signal is retained by sampling and then retaining the larger value. Note that usually the time instant at which the peak occurs is not recorded.

Analog peak detector: Peak detection can be done by using analog circuitry as well. This is in fact the basis of analog spectrum analyzers. A peak-holding circuit is shown in Figure 4.27. The circuit consists of two voltage followers. The first voltage follower has a diode at its output that is forward biased by the positive output of the voltage follower and reverse biased by a low-leakage capacitor, as shown. The second voltage follower presents to the circuit output the peak voltage that is held by the capacitor, at a low-output impedance without loading the previous circuit stage (capacitor and first voltage follower). To explain the operation of the circuit, suppose that the input voltage v_i is larger than the voltage up to which capacitor is charged (v). Since the voltage at the positive lead of the op-amp is v_i and the voltage at the negative lead is v, the first op-amp will be saturated. Since the differential input to the op-amp is positive under these conditions, the op-amp output will be positive. The output will charge the capacitor until the capacitor voltage v equals the input voltage v_i. This voltage (call it v_o) is in turn supplied to the second voltage follower, which presents the same value to its output (*Note*: The gain of a voltage follower is 1), but at a very low impedance level. The op-amp output remains at the saturated value only for a

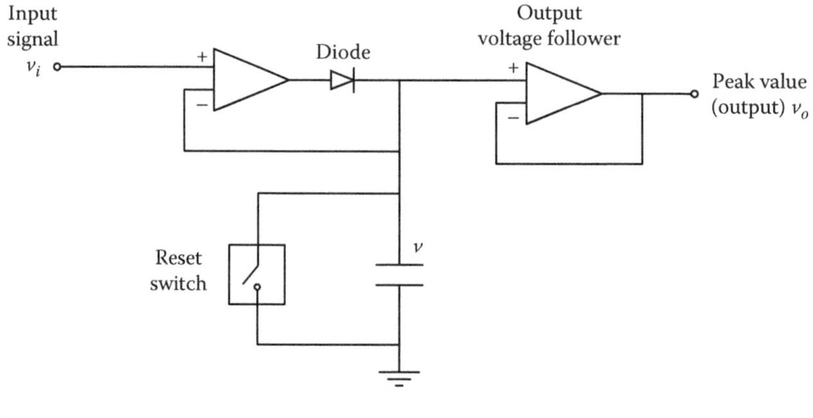

FIGURE 4.27
A peak-holding circuit.

very short time (the time taken by the capacitor to charge). Now, suppose that v_i is smaller than v. Then, the differential input of the op-amp will be negative, and the op-amp output will be saturated at the negative saturation level. This will reverse bias the diode. Hence, the output of the first op-amp will be in open circuit, and as a result, the voltage supplied to the output voltage follower will still be the capacitor voltage and not the output of the first op-amp. It follows that the voltage level of the capacitor (and hence the output of the second voltage follower) will always be the peak value of the input signal. The circuit can be reset by discharging the capacitor through a solid-state switch that is activated by an external pulse.

Commercial analog peak detectors are available in the monolithic form as IC chips. The hold and reset modes can be digitally selected through a pin. Both positive and negative peaks can be detected since both polarities are available in the op-amp. Typical parameters are as follows: input voltage range = ±10 V, CMRR = 90 dB, slew rate = 0.5 V/μs, and bandwidth = 0.5 MHz.

Summary Sheet

Modulation: A data signal (modulating signal) modulates a property of a carrier signal. The data signal is recovered through demodulation (discrimination). *Examples*: amplitude modulation (AM), frequency modulation (FM), pulse-width modulation (PWM), pulse-frequency modulation (PFM), phase modulation (PM), and pulse code modulation (PCM).

Amplitude modulation: Modulated signal $x_a(t) = x(t)x_c(t)$; data signal (modulating signal) = $x(t)$, carrier signal $x_c(t) = a_c \cos 2\pi f_c t$, frequency $= f_c$; available as monolithic IC.

Modulation theorem (frequency-shifting theorem): Signal with Fourier spectrum $X(f)$ is multiplied by carrier signal $x_c(t) = a_c \cos 2\pi f_c t$. Spectrum of resulting signal: $X_a(f) = (1/2) a_c[X(f - f_c) + X(f + f_c)]$.

Sidebands: A frequency component (sinusoidal), when multiplied by a carrier (sinusoidal), is shifted to either side of the carrier frequency through component frequency → an entire spectral band is shifted to the sides of the carrier. These are the two sidebands.

Amplitude demodulation: Multiply the AM signal by $2/a_c \cos 2\pi f_c t \rightarrow \tilde{X}(f) = X(f) + (1/2)X(f - 2f_c) + (1/2)X(f + 2f_c)$. Filter (low pass) out the two sidebands.

Carrier-added AM: $x_a(t) = x_c(t) + x(t)x_c(t)$, modulated signal rides on carrier → more power. Spectrum = carrier spectrum + sidebands.

Double sideband suppressed carrier (DSBSC): This is the conventional AM $x_a(t) = x(t)x_c(t)$ → no carrier, only sidebands → more efficient transmission (less power).

DAQ card: Data acquisition card. Input/output (I/O) card. Has multichannel DC, DAC, S/H, MUX, filtering, amplification, etc. Fits into an expansion slot of computer.

DAC: Digital-to-analog converter. (1) Weighted type (or summer type or adder type) DAC: one-step weighted adding of bit values using resistors. (2) Ladder type DAC: ladder-like recursive adding of bit values using resistors. (3) Pulse-width modulator (PWM) type DAC: uses PWM chip to switch on time of pulse according to digital input. Low-pass filter is the resulting PWM output.

DAC error sources: Code ambiguity, settling time, glitches, parametric errors, reference voltage variations, monotonicity, and nonlinearity.

ADC: Analog-to-digital converter. Methods of operation: (1) Using an internal DAC and comparator, analog value is compared with the DAC output, and DAC input is incremented until matched). (2) Analog value is represented by a digital count in proportion. Full count corresponds (→) to the FSV of ADC.

Delta–sigma ADC: ΔΣ ADC. Sampled analog value is compared with integrated (summed) output of a 1-bit DAC. If difference > 0, comparator (1-bit ADC) generates a "1" bit. Otherwise, it generates a "0" bit and conversion is complete. Sigma → summing; delta → bit increment.

ADC performance characteristics: Resolution (1 LSB), quantization error (1/2 LSB), monotonicity (output should increase/decrease as the input increases/decreases), ADC conversion rate, nonlinearity, and offset error (1/2 LSB). LSB = least significant bit.

Sample and hold (S/H): Sampled value is held constant until it is converted into the digital form. A capacitor is used.

Multiplexer (MUX): Both analog and digital. Selects one data channel at a time and connects it to a common hardware unit. In the analog case, CMOS switching is used. In the digital case, addressing of data register is used.

Bridge circuit: Has four arms with impedances Z_i, $i = 1, \ldots, 4$. Two opposite nodes are used for powering the bridge. The other two nodes form the output (which "bridges" the circuit).

Bridge output:

$$v_o = v_A - v_B = \frac{Z_1 v_{ref}}{(Z_1 + Z_2)} - \frac{Z_3 v_{ref}}{(Z_3 + Z_4)} = \frac{(Z_1 Z_4 - Z_2 Z_3)}{(Z_1 + Z_2)(Z_3 + Z_4)} v_{ref};$$ v_{ref} is the bridge excitation voltage.

Balanced bridge: Output = 0. $(Z_1/Z_2) = (Z_3/Z_4)$.

Wheatstone bridge: Constant-voltage resistance bridge. $Z_i \equiv R_i$. With equal R_i and one active element (changed by δR), bridge output: $(\delta v_o/v_{ref}) = ((\delta R/R)/(4 + 2\delta R/R))$.

Constant-current bridge: Has constant-current excitation i_{ref}. Bridge output: $v_o = ((R_1R_4 - R_2R_3)/(R_1 + R_2 + R_3 + R_4))i_{ref}$.

With equal R_i and one active element (changed by δR), bridge output: $(\delta v_o/Ri_{ref}) = ((\delta R/R)/(4 + \delta R/R))$ (less nonlinear than Wheatstone).

Hardware linearization of bridge: Equal bridge resistors R. Connect the input nodes to a third node using op-amp. Remaining node is connected to an output amplifier (feedback resistance R_f). Output (for one active element with increment δR): $(\delta v_o/v_{ref}) = (R_f/R)(\delta R/R)$ (linear).

Half bridge: Has only two arms. End nodes are excited by two voltages. Output is tapped from the midpoint of the two arms. With equal resistors R, output amplifier (feedback resistance R_f), and one active element: $(\delta v_o/v_{ref}) = (R_f/R)((\delta R/R)/(1 + \delta R/R))$ (mostly nonlinear).

Owen bridge: An impedance bridge. $Z_1 = (1/j\omega C_1)$, $Z_2 = R_2$, $Z_3 = R_3 + (1/j\omega C_3)$, $Z_4 = R_4 + j\omega L_4$; ω = excitation frequency. For a balanced bridge: $L_4 = C_1R_2R_3$, $C_3 = C_1(R_2/R_4)$ (can be used to measure capacitor and inductor simultaneously).

Wien-bridge oscillator: An impedance bridge. $Z_1 = R_1$, $Z_2 = R_2$, $Z_3 = R_3 + (1/j\omega C_3)$, $(1/Z_4) = (1/R_4) + j\omega C_4$. For a balanced bridge: $\omega = 1/\sqrt{C_3 C_4 R_3 R_4}$ (can serve as an oscillator or frequency sensor).

Linearizing devices (analog, digital, software): Offsetting, proportional output, curve shaping.

Miscellaneous signal modification devices: The types of miscellaneous signal modification devices are phase shifter, VFC (periodically charge a capacitor to input voltage and discharge and can serve as ADC or FM), FVC (uses a digital counter or VFC and signal comparison), voltage-to-current convert (VCC) (useful in sensing and signal transmission through cables), and peak-hold circuits (useful in signal enveloping and device testing). All are available in monolithic form as IC chips.

Problems

4.1 LVDT is a displacement sensor, which is commonly used in monitoring and control of mechanical devices and systems. Consider a digital control loop that uses an LVDT to measure and control the motion of a machine. Typically, the LVDT is energized by a DC power supply. An oscillator provides an excitation signal in the kilohertz range to the primary windings of the LVDT. The secondary winding segments are connected in series opposition. An AC amplifier, demodulator, low-pass filter, amplifier, and ADC are used in the monitoring path. Figure P4.1 shows the various hardware components in the control loop. Indicate the functions of these components.

At null position of the LVDT stroke, there was a residual voltage. A compensating resistor was used to eliminate this voltage. Indicate the connections for this compensating resistor.

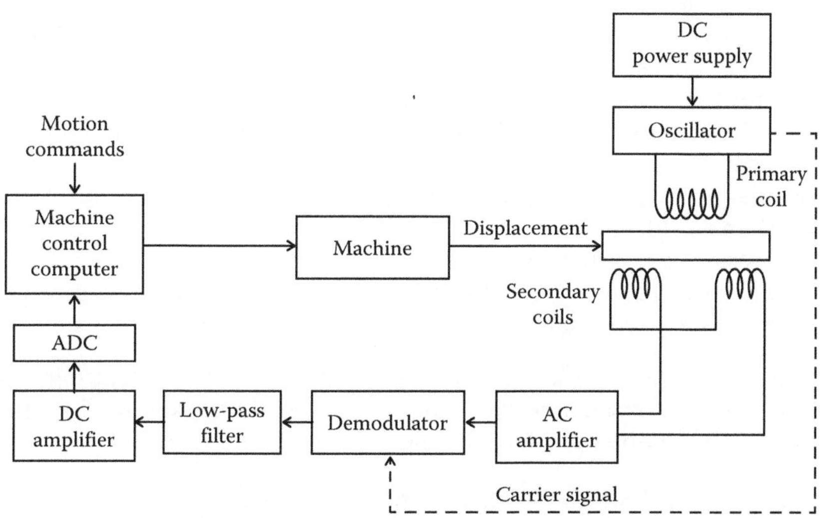

FIGURE P4.1
Components of a linear-variable differential transformer–based machine control loop.

4.2 Today, digital image sensors are used in many industrial tasks including process monitoring and control and product quality assessment. There are two main types of image sensors: charge-coupled device (CCD) and CMOS, depending on the sensing element. Both devices receive light from a monitored object and generate electrical charges, which are amplified and converted into voltages for subsequent analog-to-digital conversion (in the case of digital image sensor as opposed to an analog image sensor, which provides an analog "video" signal) and image processing. The steps of doing this are different in the two cases, but the end result of object image is essentially the same. The image sensor provides image frames, which are acquired by a frame grabber in a computer (with the necessary software). The results from image processing in the computer are used to determine the necessary information for subsequent actions. This is the software approach of image processing. The need for very large data-handling rates is a limitation on a real-time controller that uses software-based image processing.

A CCD camera has an image plate consisting of a matrix of MOSFET elements. The electrical charge that is held by each MOSFET element is proportional to the intensity of light falling on the element. The output circuit of the camera has a charge amplifier–like device (capacitive coupled), which is supplied by each MOSFET element. The MOSFET element that is to be connected to the output circuit at a given instant is determined by the control logic, which systematically scans the matrix of MOSFET elements. The capacitor circuit provides a voltage that is proportional to the charge in each MOSFET element.

An image may be divided into pixels (or picture elements) for representation and subsequent processing. A pixel has a well-defined coordinate location in the picture frame, relative to some reference coordinate frame. In a CCD sensor, the number of pixels per image frame is equal to the number of CCD elements in the image plate. The information carried by a pixel (in addition to its location) is the photointensity (or gray level) at the image location. This number has to be expressed in the digital form (using a certain number of bits) for digital image processing.

(a) Draw a schematic diagram for an industrial process that uses a CCD sensor and a computer to monitor an object and based on that carry out mechanical actions (e.g., object movement). Indicate the necessary signal modification operations at various stages in the monitoring and action loop, showing filters, amplifiers, ADC, and DAC as necessary. *Note*: There are many ways to link a digital image sensor to a computer. Details of such hardware and associated software are not needed here.

(b) Consider an image frame of the size 488×380 pixels. The refresh rate of the picture frame is 30 frames/s. If 8 bits are needed to represent the gray level of each pixel, what is the associated data (bits/s or baud) rate?

(c) Discuss whether you prefer hardware-based image processing or programmable-software-based image processing in this application.

4.3 What is meant by each of the following terms: modulation, modulating signal, carrier signal, modulated signal, and demodulation? Explain the following types of signal modulation giving an application for each case:

(a) AM, (b) FM, (c) PM, (d) PWM, (e) PFM, and (f) PCM

How could the sign of the modulating signal be accounted for during demodulation in each of these types of modulation?

4.4 Give two situations where AM is intentionally introduced and, in each situation, explain how AM is beneficial. Also, describe two devices where AM might be naturally present. Could the fact that AM is present be exploited to our advantage in these two natural situations as well? Explain.

4.5 The monitoring system for a ball bearing of a rotating machine is schematically shown in Figure P4.5a. It consists of an accelerometer to measure the bearing vibration and an FFT analyzer to compute the Fourier spectrum of the response signal. This spectrum is examined over a period of one month after installation of the rotating machine to detect any degradation in the bearing performance. An interested segment of the Fourier spectrum can be examined at high resolution by using the zoom analysis capability of the FFT analyzer. The magnitude of the original spectrum and that of the current spectrum (determined 1 month later), in the same zoom region, are shown in Figure P4.5b.

(a) Estimate the operating speed of the rotating machine and the number of balls in the bearing.
(b) Do you suspect any bearing problems?

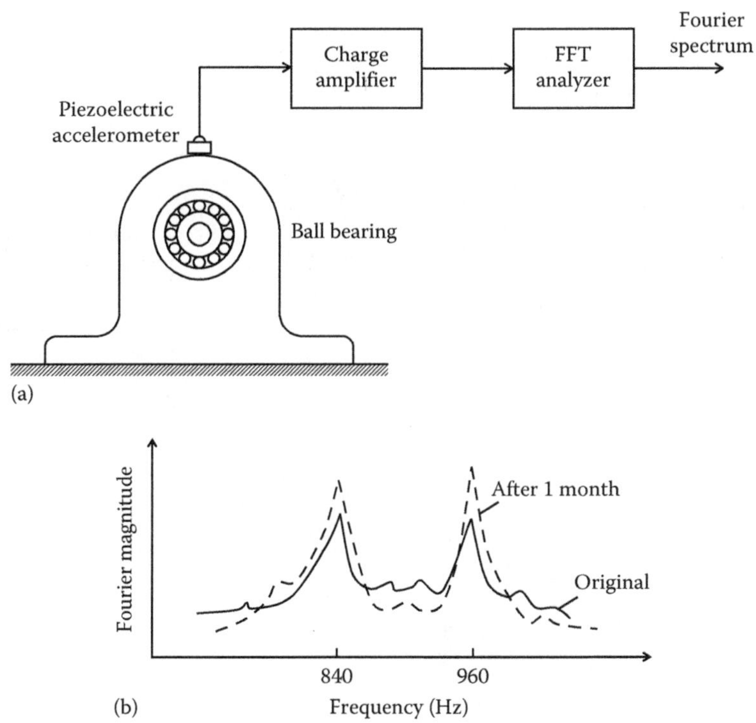

FIGURE P4.5
(a) A monitoring system for a ball bearing and (b) a zoomed Fourier spectrum.

4.6 Explain the following terms:

(a) Phase-sensitive demodulation
(b) Half-wave demodulation
(c) Full-wave demodulation

When vibrations in rotating machinery such as gearboxes, bearings, turbines, and compressors are monitored, it is observed that a peak of the spectral magnitude curve does not usually occur at the frequency corresponding to the forcing function (e.g., tooth meshing, ball or roller hammer, blade passing). Instead, two peaks occur on the two sides of this frequency. Explain the reason for this fact.

4.7 An 8-bit analog-to-digital converter (ADC) has a maximum analog input (FSV) of 10 V. What is the resolution and what is the quantization error of the ADC?

4.8 A schematic representation of a weighted-resistor DAC (or summer DAC or adder DAC) is shown in Figure P4.8. This is a general n-bit DAC, and n is equal to the number of bits in the output register. The binary word in the register is $w = [b_{n-1}b_{n-1}b_{n-3}\ldots b_1b_0]$, where b_i is the bit in the ith position and it can take the value 0 or 1, depending on the value of the digital output.

(a) Obtain an equation for the analog output in terms of the digital input.
(b) What is the full-scale value?
(c) Give a drawback of this DAC over the ladder DAC.

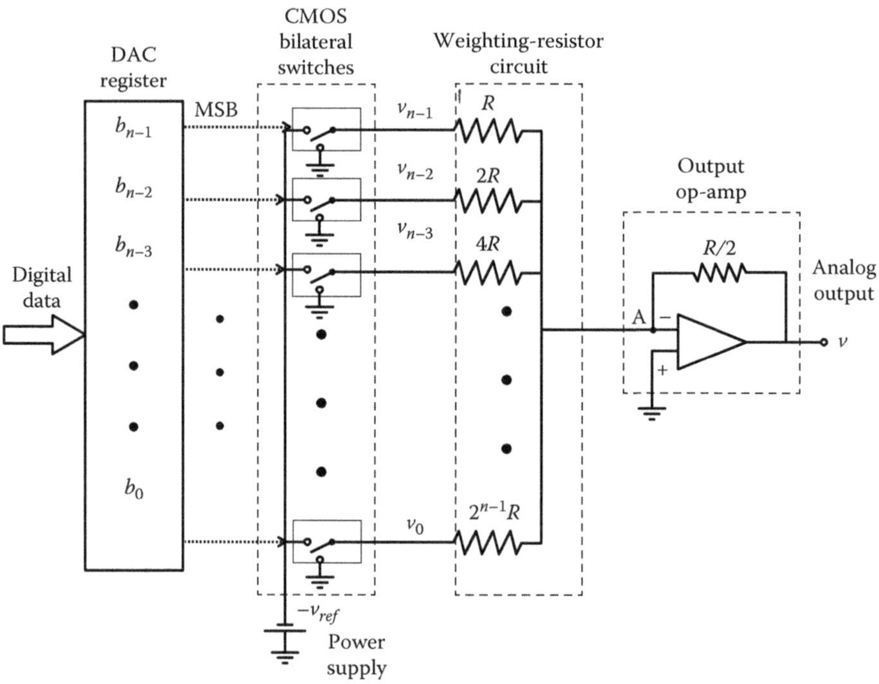

FIGURE P4.8
Weighted-resistor (adder) DAC.

4.9 Define the following terms in relation to an analog-to-digital converter (ADC):

(a) Resolution
(b) Dynamic range
(c) FSV
(d) Quantization error

4.10 Describe the operation of the following types of ADC:

(a) Dual-slope ADC (integrating ADC)
(b) Counter ADC

4.11 Estimate the conversion times for an n-bit dual-slope (integrating) ADC and counter ADC. Compare these estimates with that for a successive approximation ADC.

4.12 Briefly describe the operation of the following types of analog-to-digital converters:

(a) Direct-conversion ADC (flash ADC)
(b) Ramp-compare ADC
(c) Wilkinson ADC
(d) Delta-encoded ADC (counter-ramp ADC)
(e) Pipeline ADC (subranging quantizer)
(f) ADC with intermediate FM stage

4.13 Compare the three types of bridge circuits: constant-voltage bridge, constant-current bridge, and half bridge, in terms of nonlinearity, effect of change in temperature, and cost.

Obtain an expression for the percentage error in a half-bridge circuit output due to an error δv_{ref} in the voltage supply v_{ref}. Compute the percentage error in the output if voltage supply has a 1% error.

4.14 Suppose that in the constant-voltage (Wheatstone) bridge circuit shown in Figure 4.16a we have $R_1 = R_2 = R_3 = R_4 = R$. Let R_1 represent a strain gauge mounted on the tensile side of a bending beam element and R_3 represent another strain gauge mounted on the compressive side of the bending beam. Due to bending, R_1 increases by δR and R_3 decreases by δR. Derive an expression for the bridge output in this case, and show that it is nonlinear. What would be the result if instead R_2 represents the tensile strain gauge and R_4 represents the compressive strain gauge?

4.15 Suppose that in the constant-current bridge circuit shown in Figure 4.16b we have $R_1 = R_2 = R_3 = R_4 = R$. Assume that R_1 and R_2 represent strain gauges mounted on a rotating shaft, at right angles, and symmetrically about the axis of rotation. Also, in this configuration and in a particular direction of rotation of the shaft, suppose that R_1 increases by δR and R_2 decreases by δR. Derive an expression for the bridge output (normalized) in this case, and show that it is linear. What would be the result if R_4 and R_3 were to represent the active strain gauges in this example, with the former element in tension and the latter in compression?

4.16 Consider the constant-voltage bridge shown in Figure 4.16a. The output Equation 4.15 can be expressed as $v_o = ((R_1/R_2 - R_3/R_4)/((R_1/R_2 + 1)(R_3/R_4 + 1)))v_{ref}$. Now, suppose that the bridge is balanced, with the resistors set according to $(R_1/R_2) = (R_3/R_4) = p$. Then, if the active element R_1 increases by δR_1, show that the resulting output of the bridge is given by

$$\delta v_o = \frac{p\delta r}{\left[p(1+\delta r)+1(p+1)\right]} v_{ref}$$

where $\delta r = \delta R_1/R_1$, which is the fractional change in resistance in the active element.

For a given δr, it should be clear that δv_o represents the sensitivity of the bridge. For what value of the resistance ratio p would the bridge sensitivity be a maximum? Show that this ratio is almost equal to 1.

4.17 The Maxwell bridge circuit is shown in Figure P4.17. Obtain the conditions for a balanced Maxwell bridge in terms of the circuit parameters R_1, R_2, R_3, R_4, C_1, and L_4. Explain how this circuit could be used to measure variations in both C_1 and L_4.

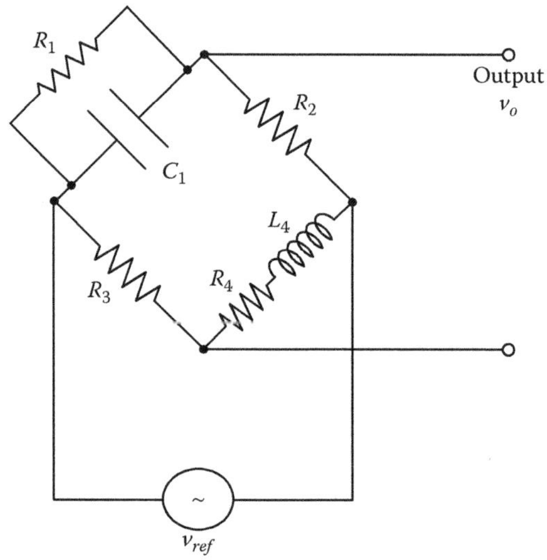

FIGURE P4.17
Maxwell bridge.

4.18 The standard LVDT arrangement has a primary coil and two secondary coil segments connected in series opposition. Alternatively, some LVDTs use a bridge circuit to produce their output. An example of a half-bridge circuit for an LVDT is shown in Figure P4.18. Explain the operation of this arrangement. Extend this idea to a full impedance bridge, for LVDT measurement.

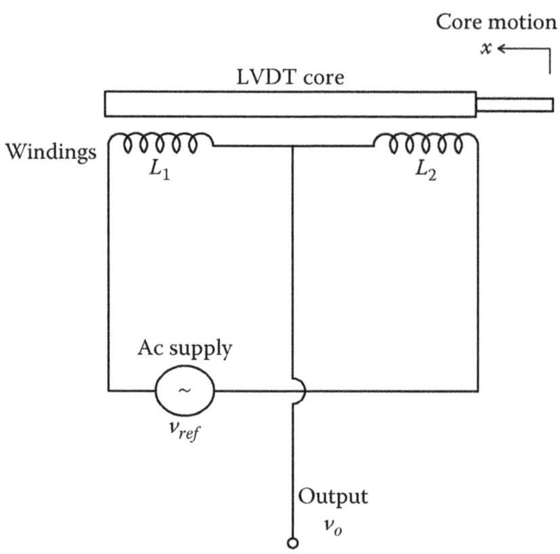

FIGURE P4.18
A half-bridge circuit for a linear-variable differential transformer.

4.19 The output of a Wheatstone bridge is nonlinear with respect to the variations in a bridge resistance. This nonlinearity is negligible for small changes in resistance. For large variations in resistance, however, some method of calibration or linearization should be employed. One way to linearize the bridge output is to use positive feedback of the output voltage signal into the bridge supply using a feedback op-amp. Consider the Wheatstone bridge circuit shown in Figure 4.16a. Initially, the bridge is balanced with $R_1 = R_2 = R_3 = R_4 = R$. Then, the resistor R_1 is varied to $R + \delta R$. Suppose that the bridge output δv_o is fed back (positive) with a gain of 2 into the bridge supply v_{ref}. Show that this will linearize the bridge equation.

4.20 Compare the potentiometer (ballast) circuit with the Wheatstone bridge circuit for strain-gauge measurements, with respect to the following considerations:

(a) Sensitivity to the measured strain
(b) Error due to ambient effects (e.g., temperature changes)
(c) SNR of the output voltage
(d) Circuit complexity and cost
(e) Linearity

4.21 In the strain-gauge bridge shown in Figure 4.16a, suppose that the load current i is not negligible. Derive an expression for the output voltage v_o in terms of R_1, R_2, R_3, R_4, R_L, and v_{ref}. Initially, the bridge was balanced, with equal resistances in the four arms. Then, one of the resistances (say, R_1) was increased by 1%. Plot to scale the ratio (actual output from the bridge)/(output under open-circuit, or infinite-load-impedance, conditions) as a function of the nondimensionalized load resistance R_L/R in the range 0.1 to 10.0, where R is the initial resistance in each arm of the bridge.

4.22 Consider the strain-gauge bridge shown in Figure 4.16a. Initially, the bridge is balanced, with $R_1 = R_2 = R$. (*Note*: R_3 may not be equal to R_1.) Then, R_1 is changed by δR. Assuming that the load current is negligible, derive an expression for the percentage error as a result of neglecting the second-order and higher-order terms in δR. If $\delta R/R = 0.05$, estimate this nonlinearity error.

4.23 What is meant by the term *bridge sensitivity*? Describe methods of increasing the bridge sensitivity. Assuming that the load resistance is very high in comparison with the arm resistances in the strain-gauge bridge shown in Figure 4.16a, obtain an expression for the power dissipation p in terms of the bridge resistances and the supply voltage. Discuss how the limitation on power dissipation can affect the bridge sensitivity.

4.24 Consider a standard bridge circuit (Figure 4.16a) where R_1 as the only active gauge and $R_3 = R_4$. Obtain an expression for R_1 in terms of R_2, v_o, and v_{ref}. Show that when $R_1 = R_2$, we get $v_o = 0$—a balanced bridge—as required. Note that the equation for R_1, assuming that v_o is measured using a high-impedance sensor, can be used to detect large resistance changes in R_1. Now, suppose that the active gauge R_1 is connected to the bridge using a long, twisted wire pair, with each wire having a resistance of R_c. The bridge circuit has to be modified as in Figure P4.24 in this case.

Show that the equation of the modified bridge is given by

$$R_1 = R_2\left[\frac{v_{ref} + 2v_o}{v_{ref} - 2v_o}\right] + 4R_c\frac{v_o}{[v_{ref} - 2v_o]}$$

Obtain an expression for the fractional error in the R_1 measurement due to cable resistance R_c. Show that this error can be decreased by increasing R_2 and v_{ref}.

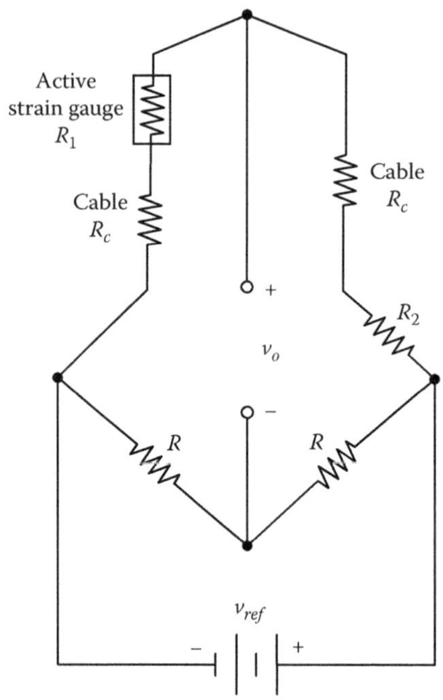

FIGURE P4.24
The influence of cable resistance on bridge measurements.

4.25 A furnace used in a chemical process is controlled in the following manner. The furnace is turned on in the beginning of the process. When the temperature within the furnace reaches a certain threshold value T_o, the (temperature) × (time) product is measured in the units of Celsius minutes. When this product reaches a specified value, the furnace is turned off. The available hardware includes an RTD (a temperature sensor using change in resistance); a differential amplifier; a diode circuit, which does not conduct when the input voltage is negative and conducts with a current proportional to the input voltage when the input is positive; a current-to-voltage converter circuit; a VFC; a counter; and an on/off control unit. Draw a block diagram for this control system and explain its operation. Clearly identify the signal modification operations in this control system, indicating the purpose of each operation.

4.26 Typically, when a digital transducer is employed to generate the feedback signal for an analog controller, a DAC would be needed to convert the digital output from the transducer into a continuous (analog) signal. Similarly, when a digital controller is used to drive an analog process, a DAC has to be used to convert the digital output from the controller into the analog drive signal. There exist ways, however, to eliminate the need for a DAC in these two types of situations.

(a) Show how a shaft encoder and an FVC can replace an analog tachometer in an analog speed-control loop.
(b) Show how a digital controller with PWM can be employed to drive a DC motor without the use of a DAC.

4.27 The noise in an electrical circuit can depend on the nature of the coupling mechanism. In particular, the following types of coupling are available:

(a) Conductive coupling
(b) Inductive coupling
(c) Capacitive coupling
(d) Optical coupling

Compare these four types of coupling with respect to the nature and level of noise that is fed through or eliminated in each case. Discuss ways to reduce the noise that is fed through in each type of coupling.

The noise due to variations in ambient light can be a major problem in optically coupled systems. Briefly discuss a method that could be used in an optically coupled device to make the device immune to variations in the ambient light level.

4.28 What are the advantages of using optical coupling in electrical circuits? For optical coupling, diodes that emit infrared radiation are often preferred over light-emitting diodes that emit visible light. What are the reasons behind this? Discuss why pulse-modulated light (or pulse-modulated radiation) is used in many types of optical systems. List several advantages and disadvantages of laser-based optical systems.

Young's modulus of a material of known density can be determined by measuring the frequency of the fundamental mode of transverse vibration of a uniform cantilever beam specimen of the material. A photosensor and a timer can be used for this measurement. Describe an experimental setup for this method of determining the modulus of elasticity.

4.29 For an engineering application of your choice, complete the following table. *Note*: You may use an online search to obtain the necessary information.

Item	Information
Parameters or variables that have to be measured in your application	
Nature of the information (parameters and variables) needed in the particular application (analog, digital, modulated, demodulated, power level, bandwidth, accuracy, etc.)	
List of sensors needed for the application.	
Signal provided by each sensor (type [analog, digital, modulated, etc.], power level, frequency range, etc.)	
Errors present in the sensor output (SNR, etc.)	
Type of signal conditioning or conversion needed for the sensors (filtering, amplification, modulation, demodulation, ADC, DAC, VFC, FVC, etc.)	
Any other comments	

5

Performance Specification and Rating Parameters

Chapter Objectives

- Performance Specification
- Parameters for Performance Specification
- Dynamic Reference Models
- Time-Domain Specifications
- Stability and Speed of Response
- Frequency-Domain Specifications
- Linearity
- Instrument Ratings
- Rating Parameters

5.1 Performance Specification

An engineering system consists of an integration of several components such as sensors, transducers, signal-conditioning and modification devices, controllers, and a variety of other electronic and digital hardware. The performance and realization of intended purpose of the overall system depend on the performance of the individual components and how the components are interconnected. All devices that assist in the intended functions of an engineering system can be interpreted as components of the system. Activities related to *system instrumentation* and operation such as prescription of the components for the system, selection of available components for a particular application, design of new components, analysis and evaluation of the system performance, making adjustments (parametric, structural, etc.) to the system, control of the system, and monitoring and fault/malfunction diagnosis should rely heavily on performance specifications. The *performance requirements* have to be specified or established based on the functional needs and associated constraints of the overall system. Then, these *specifications* have to be correlated and established in terms of the *rating parameters* (*performance parameters*) of the individual components. Some performance parameters are found in the product data sheets, which can be obtained from the manufacturer or vendor. For new developments of products, the required performance specifications have to be developed by the product development team (engineers, etc.) in consultation with the users, regulatory agencies, vendors, and so on.

5.1.1 Parameters for Performance Specification

In this chapter, we study ratings and parameters for performance specification of components in an engineering system, which in turn determine the performance of the overall system. Typically, the performance of an engineering system is specified under three important types of performance measures:

1. Speed of performance
2. Stability
3. Accuracy

Performance parameters in all three types are discussed in this chapter. As expected, due to the dynamic interactions in an engineering system, there is some degree of interrelation among parameters of these three types.

Two categories of parameters are found in the performance specification of components in an engineering system:

1. Parameters used in engineering practice (e.g., parameters listed commercially in the component data sheets)
2. Parameters defined using engineering theoretical considerations and a *reference model*, either in the time domain or in the frequency domain

Instrument ratings for commercial products (Category 1) are often developed on the basis of the analytical engineering parameters (Category 2). However, the nomenclature and the definitions used in Category 1 may not be quite identical or consistent with the precise analytical definitions used in Category 2, for reasons of convention, common usage, and the history of engineering practice. Nevertheless, both categories of performance parameters are equally important in the instrumentation practice and are addressed in this chapter. Specifically, this chapter addresses the basis (analytical basis, practical reasons, rationale, etc.) of performance representation and specification for an engineering system and the parameters used for that purpose. Even though sensors and associated hardware are particularly emphasized in this chapter, the procedures are generally applicable to a variety of components in an engineering system since these components can be represented by similar "dynamic models," which are used in the development of the parameters for performance specification. Of course, the performance of not just the components but of the entire system can also be represented and specified in the same way.

A great majority of instrument ratings provided by manufacturers (or parameters provided in commercial instrument data sheets, which come under Category 1) are in the form of *static parameters*. In engineering applications, however, *dynamic performance specifications* are also very important, and they primarily come under Category 2. Both static and dynamic characteristics of instruments and relevant parameters are discussed in this chapter.

A *sensor* detects (feels) the quantity that is measured (*measurand*). The *transducer* converts the detected measurand into a convenient form for subsequent use (monitoring, diagnosis, control, actuation, prediction, recording, etc.). The transducer input signal may be filtered, amplified, and suitably modified as needed for its subsequent use. The components used for all these purposes may be addressed in the same manner in the context of performance specification and rating parameters. Of course, the primary end goal of instrumentation is to achieve the desired performance from the overall integrated system. Performance of the individual components is critical in this regard because the overall performance of the system depends on the performance of the individual components and how the components are interconnected (and matched) in the system.

For performance specification in the analytical domain (i.e., Category 2 as indicated before), two types of dynamic models are used:

1. Differential-equation models in the time domain
2. Transfer-function models in the frequency domain

Specifically, the parameters for performance specification are commonly developed using these two types of dynamic models.

Models are quite useful in representing, analyzing, designing, evaluating, operating, and controlling such devices as sensors, transducers, controllers, actuators, and interface hardware (including signal-conditioning and modification devices). In the time domain, performance parameters such as rise time, peak time, settling time, and % overshot may be specified. Alternatively, in the frequency domain, parameters such as bandwidth, static gain, resonant frequency, magnitude at resonance, impedances, gain margin, and phase margin may be specified. These various parameters of performance specification will be discussed in this chapter. In particular, bandwidth plays an important role in specifying and characterizing many components of an engineering system. Notably, the useful frequency range, operating bandwidth, and control bandwidth are important considerations. In Chapter 6, we will study in detail several important issues related to system bandwidth.

5.1.1.1 Performance Specification in Design and Control

As observed in the previous chapters, *instrumentation* is relevant in both *design* and *control*. Instrumentation completes the design of a system. Control helps achieve performance requirements, and in some sense, control compensates for design shortcomings. Particularly in the context of *mechatronics*, both instrumentation and control should be considered concurrently within the mechatronic design problem, which involves *integrated multidomain optimal design*. It is clear that performance specifications are directly related to design specifications. Both instrumentation and control help in achieving these specifications.

It will be clear in the present study that control specifications are rather similar to the specifications for instrumentation and design. Specifically, a particular rating parameter such as "sensitivity" may be adapted to achieve some performance objective through control as well as design and instrumentation.

5.1.1.2 Perfect Measurement Device

Measuring devices or sensor systems, which include sensors and related hardware, are an important category of components or subsystems in the instrumentation of an engineering system. Their performance may be specified with reference to a *perfect measuring device*. A perfect measuring device can be defined as one that possesses the following main characteristics:

1. The output of the measuring device instantly reaches the measured value (*fast response, zero error*).
2. Transducer output is sufficiently large (*high gain, low output impedance, high sensitivity*).
3. The device output remains at the measured value (*zero error*, without *drifting* or getting affected by environmental effects and other undesirable disturbances and noise) unless the measurand (i.e., what is measured) itself changes (*stability and robustness*).
4. The output signal level of the transducer varies in proportion to the signal level of the measurand (*static linearity*).
5. The connection of a measuring device does not distort the measurand itself (*loading* effects are absent and *impedances* are matched; see Chapter 2).
6. Power consumption is small (high input impedance; see Chapter 2).

All these properties are based on dynamic characteristics and, therefore, can be explained in terms of dynamic behavior of the measuring device. In particular, items 1 through 4 can be specified in terms of the device response, either in the *time domain* or in the *frequency domain*. Items 2, 5, and 6 can be specified using the impedance characteristics of the device. First, we discuss the response characteristics that are important in the performance specification of a component of an engineering system.

5.1.2 Dynamic Reference Models

As noted earlier, in engineering applications, both *static parameters* and *dynamic parameters* are used in performance specification. Dynamic performance parameters for a device concern dynamics of the device. For example, perfect requirements are never precisely realized for a sensor perhaps in view of the sensor dynamics. For instance, the sensor will have a delay in providing its final reading due to the sensor dynamics (time constant).

Dynamic performance parameters are established with respect to a dynamic model, which represents the dynamics of the considered component or subsystem (e.g., sensor, sensor system). It may not be a complete and precise model of the device but rather a model representing the performance specifications. Hence, it is a *reference model*. However, the dynamics of the reference model has to be related to the dynamics of the actual device (or a precise model of it). Two types of dynamic models are used:

1. Differential-equation models in the time domain
2. Transfer-function models in the frequency domain

Time-domain models can be converted into transfer-function (i.e., frequency domain) models and vice versa, by means of a simple operation (i.e., replacing the time-derivative operation d/dt by the Laplace variable s and vice versa). However, for practical reasons of significance of the performance parameters in the two domains, it is important to consider models in both domains.

The widely used reference models for a component are

1. First-order model
2. Second-order (simple oscillator) model

Both first-order and second-order models have to be considered because a complete second-order model cannot be constructed by cascading two first-order models. Notably, the cascading first-order models will always result in an overdamped model, which cannot represent oscillations that commonly and naturally occur in the device dynamics.

5.1.2.1 First-Order Model

A first-order linear dynamic system is given by (in time domain)

$$\tau\dot{y} + y = ku \tag{5.1a}$$

where
 u is the input
 y is the output
 τ is the time constant
 k is the direct current (DC) gains

The corresponding transfer-function model is

$$\frac{Y(s)}{U(s)} = H(s) = \frac{k}{\tau s + 1} \tag{5.1b}$$

Suppose that the system starts from $y(0) = y_0$ and a step input of magnitude A is applied at that initial state. The corresponding response is

$$y_{\text{step}} = y_0 e^{-t/\tau} + Ak(1 - e^{-t/\tau}) \tag{5.2}$$

This response is sketched in Figure 5.1a. The first term on the RHS of Equation 5.2 is the "free response" and the second term is the "forced response." It should be clear that a significant parameter of performance specification using a first-order model is the time constant τ.

Note 1: It is clear from Equation 5.2 that if a line is drawn at $t = 0$, with its slope equal to the initial slope of the response (i.e., tangential line with slope $= (Ak - y_0)/\tau$), it will reach the final (steady state) value (Ak) at the time $t = \tau$. This is another interpretation of the time constant, as shown in Figure 5.1a.

Note 2: It can be shown that (see later) the half-power bandwidth $= 1/\tau$.

It is clear that only two performance parameters can be specified by using a first-order reference model (time constant τ and DC gain k). The time constant represents both speed and stability in this case. In fact, time constant is the only performance parameter for a first-order system since the transfer function can be normalized by using gain $k = 1$. The gain can be adjusted as appropriate (physically using an amplifier or computationally through a simple multiplication of the response by a constant value).

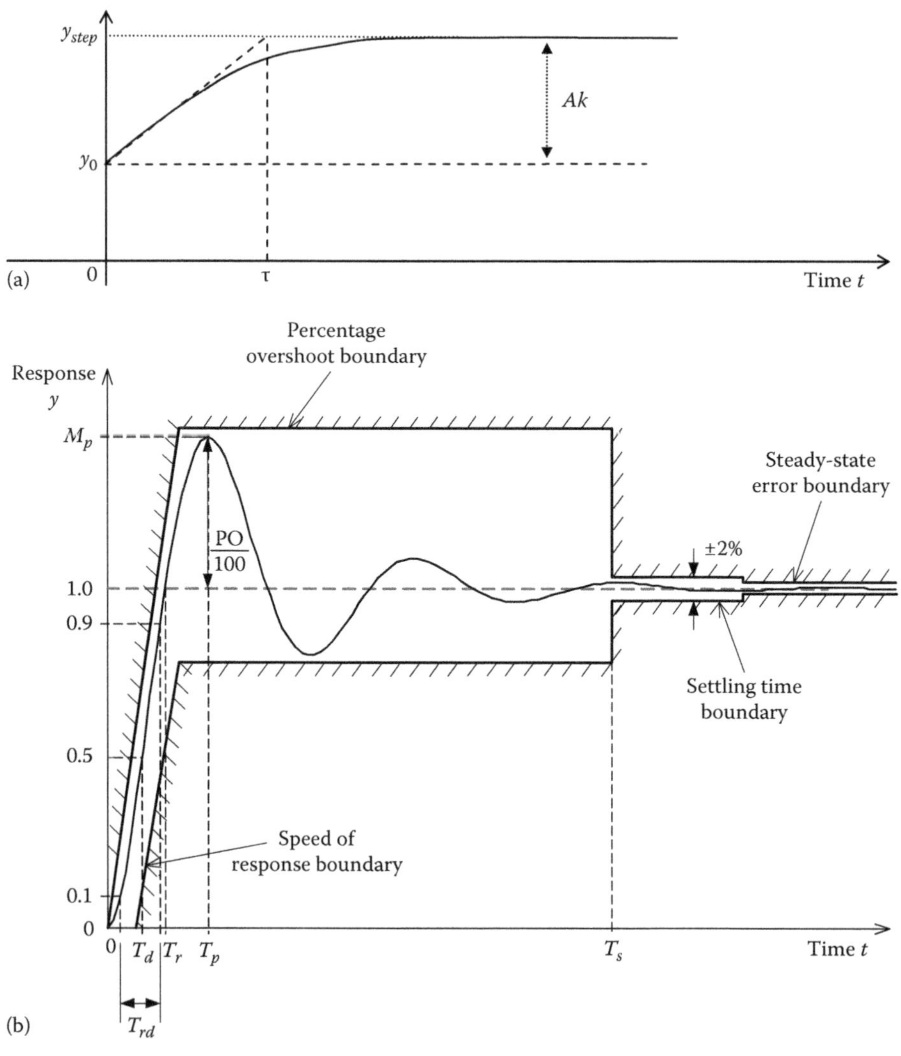

FIGURE 5.1
(a) Performance parameters based on a first-order response. (b) Response parameters for time-domain specification of performance.

5.1.2.1.1 Simple Oscillator Model

The simple oscillator is a versatile model, which can represent the performance of a variety of devices, and specifies the desired performance. Depending on the level of damping that is present, both oscillatory and nonoscillatory behavior can be represented by this model. The model can be expressed as

$$\ddot{y} + 2\zeta\omega_n \dot{y} + \omega_n^2 y = \omega_n^2 u(t) \tag{5.3a}$$

where

ω_n is the undamped natural frequency
ζ is the damping ratio

The corresponding transfer-function model is

$$\frac{Y(s)}{U(s)} = H(s) = \left[\frac{\omega_n^2}{s^2 + 2\zeta\omega_n s + \omega_n^2}\right] \tag{5.3b}$$

Note: We have normalized the model by making the static gain equal to 1. However, we can simply add a gain k to the numerator, as in Equation 5.1b, if necessary.

The damped natural frequency is given by

$$\omega_d = \sqrt{1-\zeta^2}\,\omega_n \tag{5.4}$$

The actual (damped) system executes free (natural) oscillations at this frequency. The response of the system to a unit-step excitation, with zero initial conditions, is known to be

$$y_{step} = 1 - \frac{1}{\sqrt{1-\zeta^2}} e^{-\zeta\omega_n t} \sin(\omega_d t + \phi) \tag{5.5}$$

where ϕ is the phase angle in the response and is given by

$$\cos\phi = \zeta \tag{5.6}$$

5.2 Time-Domain Specifications

As noted earlier, even though specific reference may be made to sensors, transducers, and encompassing measuring devices and sensor systems, the concepts and specifications presented here are applicable to a variety of other types of components in an engineering system. Figure 5.1b shows a typical step response in the dominant mode of a device. Note that the curve is normalized with respect to the steady-state value. We have identified several parameters that are useful for the time-domain specification of performance of the device. Some important parameters for performance specification in the time domain, using the simple oscillator model given by Equation 5.3 and its step response (5.5), are listed in Table 5.1. Definitions of these parameters are given next.

Rise time: This is the time taken for the response to pass through the steady-state value for the first time. In an overdamped system, the response is nonoscillatory; consequently, there is no overshoot. Hence, this definition of rise time is not valid for all systems. To avoid such problems, rise time may be defined as the time taken by the response to pass 90% of the steady-state value for the first time. Furthermore, rise time may be measured from 10% of the steady-state value in order to leave out start-up irregularities and time lags that might be present in a system. A modified rise time (T_{rd}) may be defined in this manner (see Figure 5.1b). An alternative definition of rise time, particularly suitable for nonoscillatory

TABLE 5.1

Time-Domain Performance Parameters Using the Simple Oscillator Model

Performance Parameter	Expression
Rise time	$T_r = \frac{\pi - \phi}{\omega_d}$ with $\cos\phi = \zeta$
Peak time	$T_p = \frac{\pi}{\omega_d}$
Peak value	$M_p = 1 - e^{-\pi\zeta/\sqrt{1-\zeta^2}}$
Percentage overshoot	$\text{PO} = 100e^{-\pi\zeta/\sqrt{1-\zeta^2}}$
Time constant	$\tau = \frac{1}{\zeta\omega_n}$
Settling time (2%)	$T_s = -\frac{\ln[0.02\sqrt{1-\zeta^2}]}{\zeta\omega_n} \approx 4\tau = \frac{4}{\zeta\omega_n}$

responses, is the reciprocal of the slope of the step response curve at 50% of the steady-state value, multiplied by the steady-state value. In process control terminology, this is called the "cycle time." No matter what definition is used, rise time represents the speed of response of a device—a small rise time indicates a fast response.

Delay time: This is usually defined as the time taken by the response to reach 50% of the steady-state value for the first time. This parameter is also a measure of the *speed of response.*

Peak time: The time at the first peak of the device response is the peak time. This parameter also represents the *speed of response* of the device.

Settling time: This is the time taken for the device response to settle down within a certain percentage (typically ±2%) of the steady-state value. This parameter is related to the degree of damping present in the device as well as the degree of *stability.*

Note: According to the simple oscillator model, the settling time (at low damping) is almost equal to four times the time constant. As a specific use of this fact, consider a sensing process. Since for a sensor, the sensing time for a data value should be greater than its settling time (so that the data will not have errors from the sensor dynamics), the sensor should be more than 4 times (preferably 10 times) faster than the fastest signal component (determined by its frequency) that needs to be accurately measured.

Percentage overshoot (*PO*): This is defined using the normalized-to-unity step response curve as

$$\text{PO} = 100(M_p - 1)\% \tag{5.7}$$

where M_p is the peak value. PO is a measure of damping or *relative stability* in the device.

Steady-state error: This is the deviation of the actual steady-state value of the device response from the desired final value. Steady-state error may be expressed as a percentage with respect to the (desired) steady-state value. In a device output, the steady-state error manifests itself as an offset. This is a *systematic* (*deterministic*) *error*. It can be normally corrected by recalibration. In servo-controlled devices, steady-state error can be reduced by increasing loop gain or by introducing lag compensation. Steady-state error can be completely eliminated using the *integral control* (*reset*) action (or by adding an integrator into the system dynamics).

For the best performance of an output device (e.g., sensor–transducer unit), we should aim to have the values of all the foregoing parameters as small as possible. In actual practice, however, it might be difficult to meet all the specifications, particularly when conflicting requirements are present. For instance, T_r can be decreased by increasing the dominant natural frequency ω_n of the device. This, however, will increase the PO and sometimes the T_s. On the other hand, the PO and T_s can be decreased by increasing the device damping, but it has the undesirable effect of increasing T_r.

Example 5.1

In a particular application, the fastest component of a signal that needs to be accurately measured is 100 Hz. Estimate an upper limit for the time constant of a sensor that may be employed for this application.

Solution

Fastest signal component = $(100 \times 2\pi)$ rad/s

To make the sensor 10 times faster than the fastest signal component, we need

$$\frac{1}{\tau} = 10 \times (100 \times 2\pi) \text{ rad/s} \rightarrow \tau = \frac{1}{10 \times (100 \times 2\pi)} \text{ s} = 159\ \mu\text{s}$$

where τ is the time constant of the sensor.

5.2.1 Stability and Speed of Response

The free response of a device can provide valuable information concerning the natural characteristics of the device. The free (unforced) excitation may be obtained, for example, by giving an initial-condition excitation to the device and then allowing it to respond freely. The two important characteristics that can be determined in this manner are

1. Stability
2. Speed of response

The stability of a dynamic system implies that the response will not grow without bounds when the excitation force itself is finite. The speed of response of a system indicates how fast the system responds to an excitation (input). It is also a measure of how fast the free response (1) rises or falls if the system is oscillatory (i.e., underdamped) or (2) decays if the system is nonoscillatory (i.e., overdamped). It follows that the two characteristics, stability and speed of response, are not completely independent. In particular, for nonoscillatory systems these two properties are very closely related.

The level of stability of a linear dynamic system depends on the real parts of the *eigenvalues* (or *poles*), which are the roots of the *characteristic equation*. (*Note*: Characteristic polynomial is the denominator of the system transfer function.) Specifically, if all the roots have real parts that are negative, then the system is stable. Additionally, the more negative the real part of a pole, the faster the decay of the free response component corresponding to that pole. The inverse of the negative real part is the *time constant*. Hence, the smaller the time constant, the faster the decay of the corresponding free response and, hence, the higher the level of stability associated with that pole. We can summarize these observations as follows:

Level of stability	Depends on the decay rate of free response (and hence on time constants or real parts of poles)
Speed of response	Depends on the natural frequency and damping for oscillatory systems and the decay rate for nonoscillatory systems
Time constant	Determines the system stability and the decay rate of free response (and speed of response as well in nonoscillatory systems)

Example 5.2

An automobile weighs 1000 kg. The equivalent stiffness at each wheel, including the suspension system, is approximately 60.0×10^3 N/m. If the suspension is designed for a percentage overshoot of 1%, estimate the damping constant that is needed at each wheel.

Solution

For a quick estimate use a simple oscillator (quarter vehicle) model, which is of the form

$$m\ddot{y} + b\dot{y} + ky = ku(t) \tag{i}$$

where

m is the equivalent mass equal to 250 kg
b is the equivalent damping constant (to be determined)
k is the equivalent stiffness equal to 60.0×10^3 N/m and u is the displacement excitation at the wheel

By comparing Equation i with Equation 5.3a, we get

$$\zeta = \frac{b}{2\sqrt{km}} \tag{ii}$$

Note: The equivalent mass at each wheel is taken as one-fourth of the total mass.

For a PO of 1%, from Table 5.1, we have $1.0 = 100\exp(-(\pi\zeta/\sqrt{1-\zeta^2}))$.

This gives $\zeta = 0.83$. Substituting values in Equation ii, we get $0.83 = b/(2\sqrt{60 \times 10^3 \times 250.0})$ or $b = 6.43 \times 10^3$ N/m/s.

With respect to time-domain specifications of a device such as a transducer, it is desirable to have a very small rise time, and a very small settling time in comparison with the time constants of the system whose response is measured, and a low percentage overshoot. These conflicting requirements will lead to a fast, stable, and steady response.

Example 5.3

Consider an underdamped system and an overdamped system with the same undamped natural frequency but with damping ratios ζ_u and ζ_o, respectively. Show that the underdamped system is more stable and faster than the overdamped system if and only if

$$\zeta_o > \frac{\zeta_u^2 + 1}{2\zeta_u}$$

where $\zeta_o > 1 > \zeta_u > 0$ by definition.

Solution

Use the simple oscillator model (Equation 5.3). The characteristic equation is

$$\lambda^2 + 2\zeta\omega_n\lambda + \omega_n^2 = 0 \tag{5.8}$$

The eigenvalues (poles) are

$$\lambda = -\zeta\omega_n \pm \sqrt{\zeta^2 - 1}\,\omega_n \tag{5.9}$$

To be more stable, we should locate the underdamped pole farther away from the origin than the dominant overdamped pole. Specifically, we must have

$$\zeta_u\omega_n > \zeta_o\omega_n - \sqrt{\zeta_o^2 - 1}\,\omega_n$$

This gives

$$\zeta_o > \frac{\zeta_u^2 + 1}{2\zeta_u} \tag{5.10}$$

The corresponding region is shown as the shaded area in Figure 5.2.

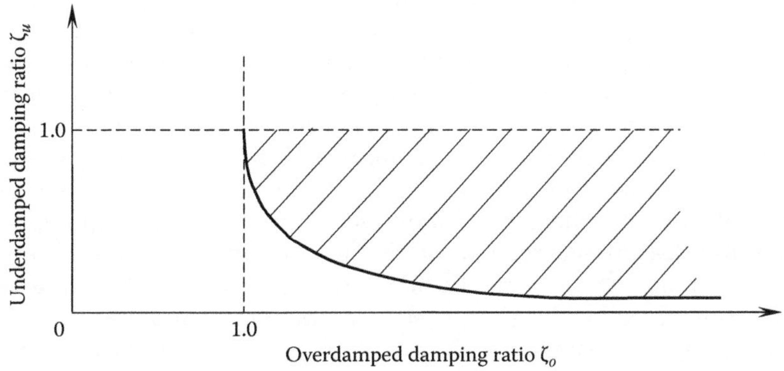

FIGURE 5.2
Region (shaded) where underdamped system is faster and more stable than the corresponding overdamped system.

Note: Then, the underdamped response not only decays faster (more stable) but also has a faster speed of response (due to its oscillation).

This result indicates that high damping does not necessarily mean increased stability. To explain this result further, consider an undamped ($\zeta = 0$) simple oscillator of natural frequency ω_n. Now, let us add damping and increase ζ gradually from 0 to 1. Then, the complex conjugate poles $-\zeta\omega_n \pm j\omega_d$ will move away from the imaginary axis as ζ increases (because $\zeta\omega_n$ increases), and hence, the level of stability will increase. When ζ reaches the value 1 (i.e., *critical damping*), we get two identical and real poles at $-\omega_n$. When ζ is increased beyond 1, the poles will be real and unequal, with one pole having a magnitude smaller than ω_n and the other having a magnitude larger than ω_n. The former (which is closer to the "origin" of zero value) is the dominant pole, and it will determine both the stability and speed of response of the resulting overdamped system. Specifically, as ζ increases beyond 1, the two poles will branch out from the location $-\omega_n$, one moving toward the origin (becoming less stable) and the other moving away from the origin. It is now clear that as ζ is increased beyond the point of critical damping, the system becomes less stable. Specifically, for a given value of $\zeta_u < 1.0$, there is a value of $\zeta_o > 1$, governed by Equation 5.10, above which the overdamped system is less stable and slower than the underdamped system.

5.3 Frequency-Domain Specifications

Figure 5.3 shows a representative *frequency transfer function* (FTF) (often termed *frequency response function* [FRF]) of a device. This constitutes the plots of *gain* (FRF magnitude) and *phase angle*, using frequency as the independent variable. This pair of plots is commonly known as the "Bode diagram," particularly when the magnitude axis is calibrated in *decibels* (dB) and the frequency axis in a log scale such as *octaves* (i.e., changing by a factor of 2) or *decades* (i.e., changing by a factor of 10). Experimental determination of these curves can be accomplished either by applying a harmonic excitation and noting the amplitude amplification and the phase lead in the response signal at steady state or by Fourier analysis of the excitation and response signals for either transient or random excitations. Experimental determination of the transfer functions is known as "system identification" in the frequency domain.

Transfer functions provide complete information regarding the system response (of a linear system) to a sinusoidal excitation. Since any time signal can be decomposed into sinusoidal components through Fourier transformation, it is clear that the response of a linear system to an arbitrary input excitation can also be determined using the transfer-function information for that system. In this sense, transfer functions are frequency-domain models, which can completely describe a linear system. In fact, a linear time-domain model with constant coefficients can be transformed into a transfer function and vice versa. Hence, a time-domain model and a frequency-domain model are completely equivalent and interchangeable. For this reason, one may argue that it is redundant to use both time-domain specifications and frequency-domain specifications, as they carry the same information. Often, however, both specifications are used simultaneously, because this can provide a better (and more practical) insight into the system performance. In fact, the physical interpretation of some performance parameters (e.g., bandwidth and resonance) is more convenient in the frequency domain, and for some other parameters (e.g., speed of response and stability), it is more convenient in the time domain. In particular, frequency-domain

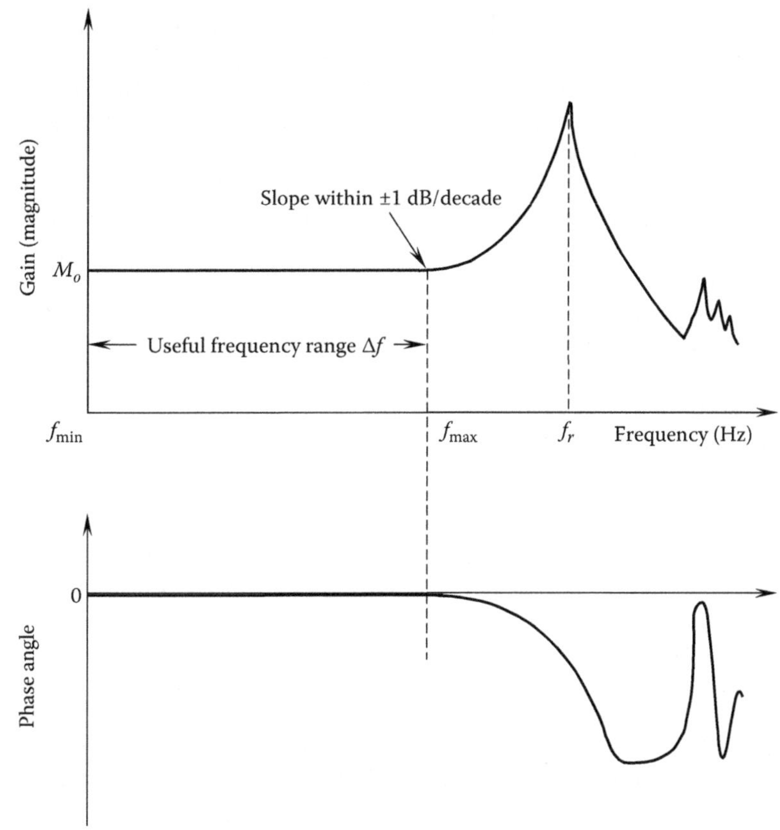

FIGURE 5.3
Response parameters for frequency-domain specification of performance. *Mo*, static gain (DC Gain); *fr*, resonant frequency.

parameters are more suitable in representing some characteristics of a system under harmonic (sinusoidal) excitation.

Some useful parameters for performance specification of a device, in the frequency domain, are the following:

- Useful frequency range (*operating interval*)
- Bandwidth (*speed of response*)
- Static gain (*steady-state performance*)
- Resonant frequency (*speed of response* and *critical frequency region*)
- Magnitude at resonance (*stability*)
- Input impedance (*loading, efficiency, interconnectability, maximum power transfer, signal reflection*)
- Output impedance (*loading, efficiency, interconnectability, maximum power transfer, signal level*)
- Gain margin (*stability*)
- Phase margin (*stability*)

The first three items are discussed in detail in this chapter and is also indicated in Figure 5.3. Resonant frequency corresponds to an excitation frequency where the response magnitude peaks. The dominant resonant frequency typically is the lowest resonant frequency, which usually also has the largest peak magnitude. It is shown as f_r in Figure 5.3. The term *magnitude at resonance* is self-explanatory and is the peak magnitude mentioned earlier and shown in Figure 5.3. Resonant frequency is a measure of speed of response and bandwidth and is also a frequency that should be avoided during normal operation and whenever possible. This is particularly true for devices that have poor stability (e.g., low damping). Specifically, a high magnitude at resonance is an indication of poor stability. Input impedance and output impedance are discussed in Chapter 2. Bandwidth considerations are studied in detail in Chapter 6.

5.3.1 Gain Margin and Phase Margin

Gain and phase margins are measures of *stability* of a device. To define these two parameters, consider the feedback system of Figure 5.4a. The *forward transfer function* of the system is $G(s)$ and the *feedback transfer function* is $H(s)$, with $s = j\omega$. These transfer functions are frequency-domain representations of the overall system, which may include a variety of components such as the plant, sensors, transducers, actuators, controllers and interfacing, and signal modification devices.

The Bode diagram of the system constitutes the magnitude and phase lead plots of the *loop transfer function* $G(j\omega)H(j\omega)$ as a function of frequency. This is sketched in Figure 5.4b.

Suppose that at a particular frequency ω the forward transfer function $G(j\omega)$ provides a phase lag of ϕ_1 and the feedback transfer function $H(j\omega)$ provides a phase lag of ϕ_2.

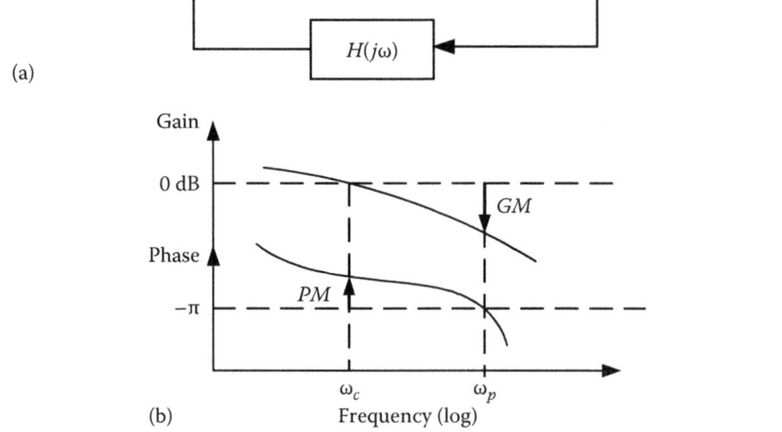

FIGURE 5.4
Illustration of gain and phase margins. (a) A feedback system and (b) Bode diagram.

Now, in view of the negative feedback, the feedback signal undergoes a further phase lag of π: Hence,

Total phase lag in the loop $= \phi + \pi$

where

Phase lag of $GH = \phi_1 + \phi_2 = \phi$

It follows that when the overall phase lag of the *loop transfer function* $GH(j\omega)$ is equal to π, the loop phase lag becomes 2π, which means that if a signal of frequency ω travels through the system loop, it will not experience a net phase lag. Additionally, if at this particular frequency, the loop gain $|GH(j\omega)|$ is unity, a sinusoidal signal with this frequency will be able to repeatedly travel through the loop without ever changing its phase or altering its magnitude, even in the absence of any external excitation input. This corresponds to a condition of *marginally stability*.

If, on the other hand, the loop gain $|GH(j\omega)| > 1$ at this frequency while the loop phase lag is π, the signal magnitude will monotonically grow as the signal travels through the loop. This is an unstable situation. On the other hand, if the loop gain is <1 at this frequency while the loop phase lag is π, the signal magnitude will monotonically decay as the signal cycles through the loop. This is a stable situation. In summary,

1. If $|GH(j\omega)| = 1$ when $\angle GH(j\omega) = -\pi$, the system is marginally stable.
2. If $|GH(j\omega)| > 1$ when $\angle GH(j\omega) = -\pi$, the system is unstable.
3. If $|GH(j\omega)| < 1$ when $\angle GH(j\omega) = -\pi$, the system is stable.

It follows that the margin of smallness of $|GH(j\omega)|$ when compared to 1 at the frequency ω, where $\angle GH(j\omega) = -\pi$, provides a measure of stability. This is termed *gain margin* (see Figure 5.4b). Similarly, at the frequency ω, where $|GH(j\omega)| = 1$, the amount (margin) of phase lag that can be added to the system so as to make the loop phase lag equal to π is a measure of stability. This amount is termed *phase margin* (see Figure 5.4b).

In terms of frequency-domain specifications, a device such as a transducer, an amplifier, or a sensor system should have a wide useful frequency range. For this, it must have a high fundamental natural frequency (about 5–10 times the maximum frequency of the operating range) and a somewhat low damping ratio (slightly < 1).

5.3.2 Simple Oscillator Model in Frequency Domain

The transfer function $H(s)$ for a simple oscillator is given by Equation 5.3b.

The frequency transfer function $H(j\omega)$ is defined as $H(s)|_{s=j\omega}$, where ω is the excitation frequency.

$$H(j\omega) = \left[\frac{\omega_n^2}{\omega_n^2 - \omega^2 + 2j\zeta\omega_n\omega}\right] \tag{5.3c}$$

Note that $H(j\omega)$ is a complex function in ω. We have

Gain $= |H(j\omega)| =$ magnitude of $H(j\omega)$

Phase lead $= \angle H(j\omega) =$ phase angle of $H(j\omega)$

These represent *amplitude gain* and *phase lead* of the output (response) when a sine input signal (excitation) of frequency ω is applied to the system.

Resonant frequency ω_r corresponds to the excitation frequency when the amplitude gain is a maximum and is given by

$$\omega_r = \sqrt{1-2\zeta^2}\,\omega_n \tag{5.11}$$

This expression is valid for $\zeta \le 1/\sqrt{2}$. It can be shown that

$$\text{Gain} = \frac{1}{2\zeta}; \quad \text{Phase lead} = -\frac{\pi}{2} \text{ at } \omega = \omega_n \tag{5.12}$$

This concept is used to measure the damping in a device, in addition to specifying the performance in the frequency domain. Frequency-domain concepts are discussed further under bandwidth considerations in Chapter 6.

Note: The first-order model given by Equation 5.1a or 5.1b has the frequency response function

$$H(j\omega) = \frac{k}{1+j\tau\omega} \tag{5.1c}$$

It is clear that this model has only one performance parameter (time constant τ) since the gain k can be normalized to 1 (and adjusted physically by an amplifier or computationally by simply multiplying the response by a constant). Furthermore, it cannot represent an underdamped system and hence a condition of resonance. In particular, a simple oscillator model cannot be obtained by cascading two first-order models.

5.4 Linearity

In a theoretical and dynamic sense, a device is considered linear if it can be modeled by a set of linear differential equations, with time t as the independent variable (or by a set of transfer functions, with frequency ω as the independent variable). A useful property of a linear system is that the *principle of superposition* is applicable.

In particular, if the input u_1 generates the output y_1 and if the input u_2 generates the output y_2, then the input $a_1u_1+a_2u_2$ will generate the output $a_1y_1+a_2y_2$, for any a_1 and a_2.

A property of a nonlinear system is that its stability may depend on the system inputs and/or initial conditions. Nonlinear devices are often analyzed using linear techniques by considering small excursions about an operating point. This "local linearization" is accomplished by introducing incremental variables for the inputs and outputs. If one increment can cover with sufficient accuracy the entire operating range of a device, it is an indication that the device is linear. If the input–output relations are nonlinear algebraic equations, they represent a *static nonlinearity*. Such a situation can be handled simply by using nonlinear calibration curves, which linearize the device without introducing nonlinearity errors. If, on the other hand, the input–output relations are nonlinear differential equations, analysis usually becomes more complex. This situation represents a *dynamic nonlinearity*. The transfer-function representation of an instrument implicitly assumes linearity.

According to industrial and commercial terminology, a linear measuring instrument provides a *measured value* that varies linearly with the value of the *measurand*—the variable that is measured. This is consistent with the definition of static linearity and is appropriate because for those commercial devices it is typically required that the operating range is outside the region where the dynamics of the device appreciably affects the device output. All physical devices are nonlinear to some degree. This stems due to deviation from the ideal behavior because of causes such as electrical and magnetic saturation, deviation from Hooke's law in elastic elements, Coulomb friction, creep at joints, aerodynamic damping, backlash in gears and other loose components, and component wear-out.

Nonlinearities in devices are often manifested as some peculiar characteristics. In particular, the following properties are important in detecting nonlinear behavior in devices.

Saturation: Nonlinear devices may exhibit saturation (see Figure 5.5a). This may be the result of causes such as magnetic saturation, which is common in magnetic-induction devices and transformer-like devices (e.g., differential transformers), electronic saturation (e.g., in an amplifier circuit), plasticity in mechanical components, and nonlinear springs.

Dead zone: A dead zone is a region in which a device would not respond to an excitation. Stiction in mechanical devices with Coulomb friction is a good example. Because of stiction, a component will not move until the applied force reaches a certain minimum value. Once the motion is initiated, subsequent behavior can be either linear or nonlinear. Another example is the backlash in loose components such as gear wheel pairs. Bias signal in electronic devices is a third example. In them, until the bias signal reaches a specific level, the circuit action will not take place. A dead zone with subsequent linear behavior is shown in Figure 5.5b.

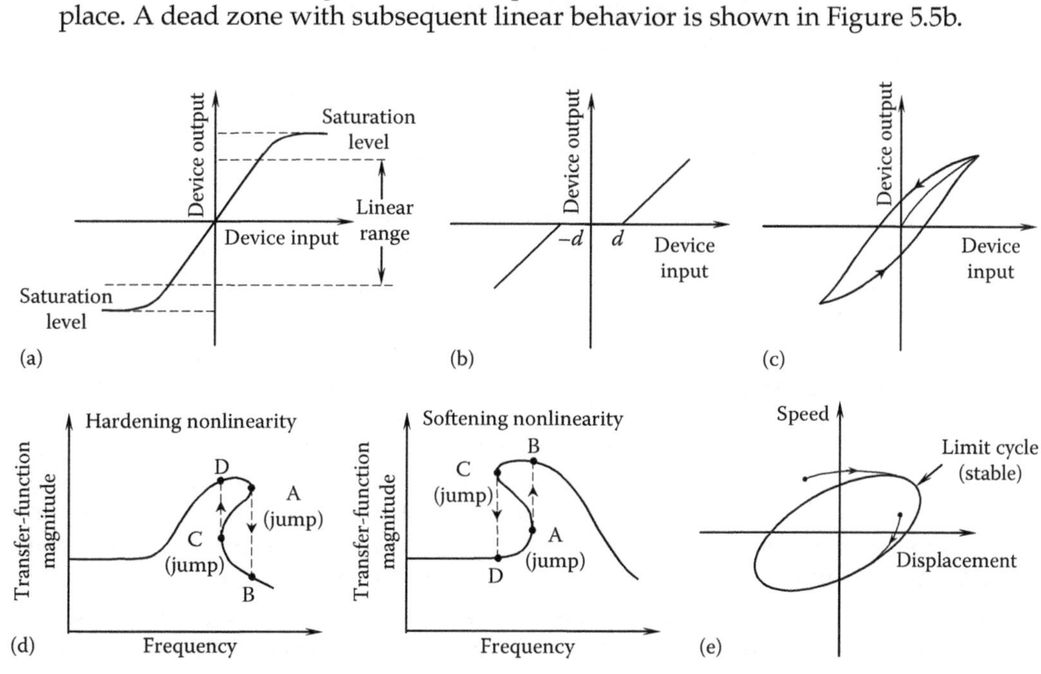

FIGURE 5.5
Common manifestations of nonlinearity in devices: (a) saturation, (b) dead zone, (c) hysteresis, (d) the jump phenomenon, and (e) limit-cycle response.

Hysteresis: Nonlinear devices may produce hysteresis. In hysteresis, the input–output curve changes depending on the direction of the input (see Figure 5.5c), resulting in a hysteresis loop. This behavior is common in loose components such as gears, which have backlash; in components with nonlinear damping, such as Coulomb friction; and in magnetic devices with ferromagnetic media and various dissipative mechanisms (e.g., eddy current dissipation). For example, consider a coil wrapped around a ferromagnetic core. If a DC current is passed through the coil, a magnetic field is generated. As the current is increased from zero, the field strength will also increase. Now, if the current is decreased back to zero, the field strength will not return to zero because of residual magnetism in the ferromagnetic core. A negative current has to be applied to demagnetize the core. It follows that the field strength versus current curve looks somewhat like Figure 5.5c. This is magnetic hysteresis.

Linear viscous damping also exhibits a hysteresis loop in its force–displacement curve. This is a property of any mechanical component that dissipates energy (the area within the hysteresis loop gives the energy dissipated in one cycle of motion). In general, if force depends on displacement (as in the case of a spring) and velocity (as in the case of a damping element), the value of force at a given value of displacement will change with the direction of the velocity. In particular, the force when the component is moving in one direction (say positive velocity) will be different from the force at the same location when the component is moving in the opposite direction (negative velocity), thereby giving a hysteresis loop in the force–displacement plane. If the relationship of displacement and velocity to force is linear (as in viscous damping), the hysteresis effect is linear. If, on the other hand, the relationship is nonlinear (as in Coulomb damping and aerodynamic damping), the resulting hysteresis is nonlinear.

Jump phenomenon: Some nonlinear devices exhibit an instability known as the jump phenomenon (or *fold catastrophe*) in the frequency response (transfer) function curve. This is shown in Figure 5.5d for both *hardening* devices and *softening* devices. With increasing frequency, the jump occurs from A to B, and with decreasing frequency, it occurs from C to D. Furthermore, the transfer function itself may change with the magnitude of input excitation in the case of nonlinear devices.

Limit cycles: Nonlinear devices may produce limit cycles. An example is given in Figure 5.5e on the phase plane (2D) of velocity versus displacement. A limit cycle is a closed trajectory in the state space that corresponds to sustained oscillations at a specific frequency and amplitude, without decay or growth. The amplitude of these oscillations is independent of the initial location from which the response started. In addition, an external input is not needed to sustain a limit-cycle oscillation. In the case of a *stable limit cycle*, the response will move onto the limit cycle irrespective of the location in the neighborhood of the limit cycle from which the response was initiated (see Figure 5.5e). In the case of an *unstable limit cycle*, the response will move away from it with the slightest disturbance.

Frequency creation: A linear device when excited by a sinusoidal signal will generate, at steady state, a response at the same frequency as the excitation. On the other hand, at steady state, a nonlinear device may create frequencies that are not present in the excitation signals. These frequencies might be *harmonics* (integer multiples of the excitation frequency), *subharmonics* (integer fractions of the excitation frequency), or *nonharmonics* (usually, rational fractions of the excitation frequency).

Example 5.4

Consider a nonlinear device modeled by the differential equation $\{dy/dt\}^{1/2} = u(t)$, where $u(t)$ is the input and y is the output. Show that this device creates frequency components that are different from the excitation frequencies.

Solution

First, note that the response of the system is given by $y = \int_0^t u^2(t)\,dt + y(0)$.

Now, for an input given by $u(t) = a_1 \sin \omega_1 t + a_2 \sin \omega_2 t$, straightforward integration using properties of trigonometric functions gives the following response:

$$y = \left(a_1^2 + a_2^2\right)\frac{t}{2} - \frac{a_1^2}{4\omega_1}\sin 2\omega_1 t - \frac{a_2^2}{4\omega_2}\sin 2\omega_2 t$$

$$+ \frac{a_1 a_2}{2(\omega_1 - \omega_2)}\sin(\omega_1 - \omega_2)t - \frac{a_1 a_2}{2(\omega_1 + \omega_2)}\sin(\omega_1 + \omega_2)t - y(0)$$

It is seen that the discrete frequency components $2\omega_1$, $2\omega_2$, $(\omega_1 - \omega_2)$, and $(\omega_1 + \omega_2)$ are created. Additionally, there is a continuous spectrum that is contributed by the linear function of t that is present in the response.

Nonlinear systems can be analyzed in the frequency domain using the *describing function* approach. As observed earlier, when a harmonic input (at a specific frequency) is applied to a nonlinear device, the resulting output at steady state will have a component at this fundamental frequency and also components at other frequencies (as a result of frequency creation by the nonlinear device), typically harmonics. The response may be represented by a *Fourier series*, which has frequency components that are integer multiples of the input frequency. The describing function approach neglects all the higher harmonics in the response and retains only the fundamental component. This output component, when divided by the input, produces the describing function of the device. This is similar to the transfer function of a linear device, but unlike for a linear device, the gain and the phase shift will be dependent on the input amplitude. Details of the describing function approach can be found in textbooks on nonlinear control theory.

5.4.1 Linearization

A popular method of linearization of a nonlinear device considers the local behavior over a small operating range. This local linearization is straightforward but is not generally applicable due to such reasons as follows:

1. The operating conditions can change considerably and a single local slope may not be valid over the entire range.
2. The local slope may not exist or insignificant compared to O(2) terms of the Taylor series expansion (e.g., Coulomb friction).
3. In some nonlinear systems, the use of local slopes (e.g., negative damping in a control law) may lead to undesirable consequences (e.g., instability).

Several other methods are available to reduce or eliminate nonlinear behavior in devices. They include the use of *calibration curves* (in the static case), *linearizing elements* (e.g., resistors

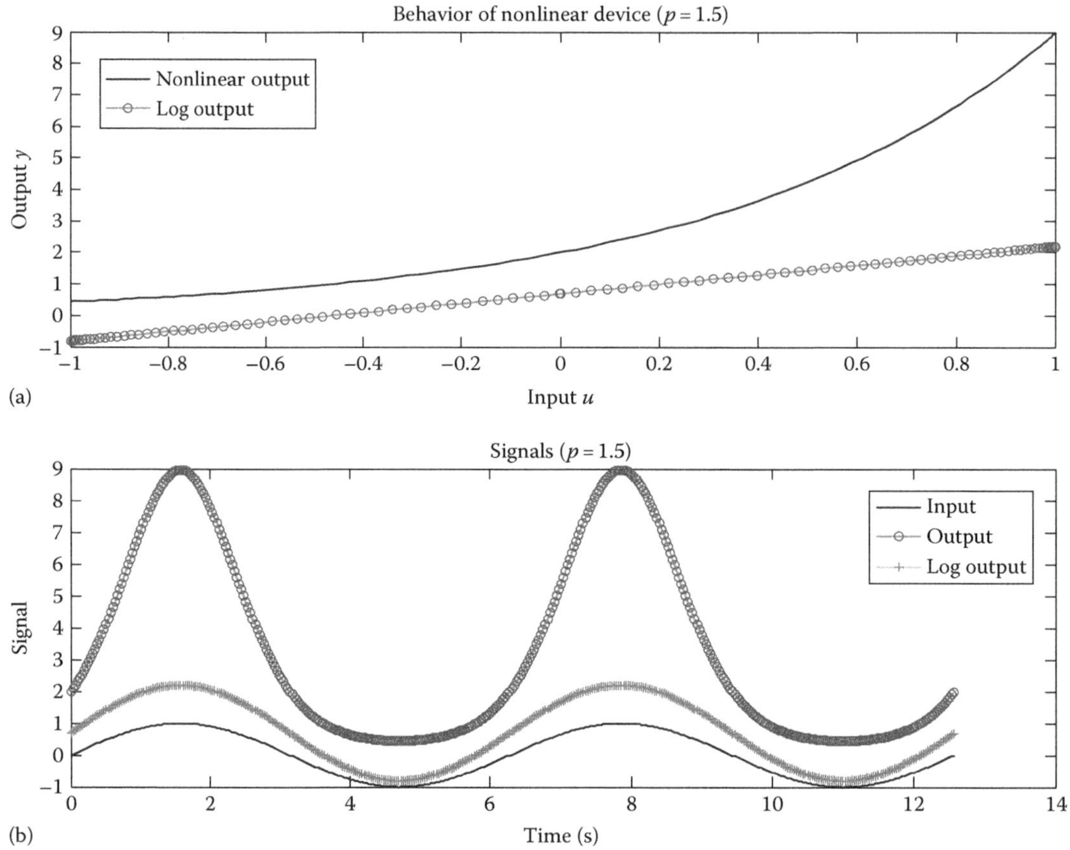

FIGURE 5.6
Sine response of a nonlinear device; (a) input–output behavior and (b) signals.

and amplifiers in bridge circuits) to neutralize the nonlinear effects, and nonlinear feedback (*feedback linearization*).

A significant consequence of a static nonlinearity is the distortion of the output. This can be linearized by recalibration or rescaling. For example, suppose that the input (u)–output (y) behavior of a device is given by $y = ke^{pu}$. It is clear that a sinusoidal input $u = u_0 \sin\omega t$ will be far from sinusoidal at the output.

Clearly, we can "transform" the problem as $\log(y) = pu + \log(k)$. Hence, the input–output relationship can be accurately linearized by simply using a log scale for the output and also adding a constant offset of $-\log(k)$. In this recalibrated form, the output for a sinusoidal input will be purely sinusoidal.

To illustrate this, let us use the parameter values: $k = 2.0$, $p = 1.5$, $u_0 = 2.0$, and $\omega = 1.0$. We use the following MATLAB® function to determine the input–output behavior (Figure 5.6a) and the corresponding two signals (Figure 5.6b):

```
% Response of nonlinear device
u0=2.0;k=2.0;p=1.5;
t=0:0.01*pi:4*pi;
u=sin(t);
y=k*exp(p*u);
```

```
y2=log(y);
% plot the results
plot(u,y,'-',u,y2,'-',u,y2,'o')
plot(t,u,'-',t,y,'-',t,y,'o',t,y2,'-',t,y2,'+')
```

It is seen that the actual nonlinear function considerably distorts the sine signal while using a log scale for the output can conveniently and accurately linearize the behavior, giving an undistorted output. Furthermore, with the log output shown in Figure 5.6a, we can extract the two parameters p and k from the slope and the y-intercept of the linear curve. Specifically, $p = \text{slope} = 3.0/2.0 = 1.5$; $\text{Log } k = 0.7 \rightarrow k = 2.0$.

For mitigating nonlinearity problems, it is a good practice to take such precautions as the following:

1. Avoid operating the device over a wide range of signal levels (inputs).
2. Avoid operation over a wide frequency band.
3. Use devices that do not generate large mechanical motions.
4. Minimize Coulomb friction and stiction (e.g., use proper lubrication).
5. Avoid loose joints and gear coupling (i.e., use *direct-drive* mechanisms).
6. Minimize environmental influences.
7. Minimize sensitivity to undesirable influences.
8. Minimize wear and tear.

5.5 Instrument Ratings

Instrument manufacturers do not usually provide complete dynamic information for their products. In most cases, for practical engineering systems and complex instruments, it is unrealistic to expect complete dynamic models (in the time domain or the frequency domain) and accurate parameter values for the models. Performance characteristics provided by manufacturers and vendors are primarily static parameters. Known as instrument ratings, these are available as parameter values, tables, charts, calibration curves, and empirical equations. Dynamic characteristics such as transfer functions (e.g., transmissibility curves expressed with respect to excitation frequency) might also be provided for more sophisticated instruments, but the available dynamic information is never complete. The rationale for this is that under normal operating conditions of a typical device (e.g., sensor, amplifier, data acquisition hardware), the device dynamics should have a minimal effect on its output. Nevertheless, some information on the dynamics of the device (e.g., time constants, bandwidth) would be useful in selecting the operating conditions and components for a practical application.

Definitions of rating parameters that are used by manufacturers and vendors of instruments are in some cases not the same as the analytical definitions used in textbooks. This is particularly true in relation to the terms *linearity* and *stability*. Still, instrument ratings provided by manufacturers and vendors are very useful in the selection, installation and interconnection, operation, control, and maintenance of components in an engineering system. Let us examine key performance parameters.

5.5.1 Rating Parameters

Typical rating parameters provided by instrument manufacturers and vendors (in their data sheets) are the following:

1. Sensitivity and sensitivity error
2. Signal-to-noise ratio (SNR)
3. Dynamic range (DR)
4. Resolution
5. Offset or bias
6. Linearity
7. Zero drift, full-scale drift, and calibration drift (stability)
8. Useful frequency range
9. Bandwidth
10. Input and output impedances

We have already discussed the meaning and significance of some of these terms. Bandwidth considerations are further examined in Chapter 6. In this section, we examine the conventional definitions for rating parameters, as given by instrument manufacturers and vendors.

5.5.2 Sensitivity

The *sensitivity* of a device (e.g., transducer) is measured by the magnitude (peak, root-mean-square [rms] value, etc.) of the output signal corresponding to a unit input (e.g., measurand). This may be expressed as the ratio of incremental output and incremental input (e.g., slope of input–output curve of the device) or, analytically, as the corresponding partial derivative of the input–output relationship. It is clear that sensitivity is also the *gain* of the device. In the case of vectorial or tensorial signals (e.g., displacement, velocity, acceleration, strain, force), the direction of sensitivity should be specified.

A countless number of factors (including environment) can affect the output of a device such as sensor. Then, the important objectives of instrumentation with respect to sensitivity would be the following:

1. Select a reasonable number of factors that have noteworthy sensitivity levels on the device output.
2. Determine the sensitivities (say, relative sensitivities—nondimensional) of the selected factors.
3. Maximize the sensitivity to desirable factors (e.g., the measured quantity).
4. Minimize the sensitivity to undesirable factors (e.g., thermal effects on a strain reading) or cross sensitivity.

We will revisit the subject of sensitivity in Chapter 6, under Section 6.5.

Cross sensitivity: This is the sensitivity along directions that are orthogonal to the primary direction of sensitivity. It is normally expressed as a percentage of direct sensitivity. High direct sensitivity and low cross sensitivity are desirable in any input–output device (e.g., measuring instrument). Sensitivity to parameter changes and noise has to be small in any device, however, and this is an indication of its *robustness*. On the other hand, in *adaptive control* and *self-tuning control*, the sensitivity of the system to control parameters has to be sufficiently high. Often, sensitivity and robustness are conflicting requirements.

5.5.2.1 Sensitivity in Digital Devices

Digital devices generate digital outputs. They include devices that generate pulses or counts or those with built-in analog-to-digital converters (ADCs). Sensitivity of any digital device can be represented in the usual manner. Specifically,

$$\text{Sensitivity} = \frac{\text{Digital output}}{\text{Corresponding input}} \tag{5.13a}$$

Most commonly, the input is analog, but digital inputs can be accommodated in the same definition.

An n-bit device can represent 2^n values including 0. Then, the maximum possible value (unsigned) is 2^n–1. To represent signed values, we need to assign one bit as the sign bit. Then, an n-bit device can represent 2^n–1 positive values (including zero) and the same number of corresponding negative values. Another way of interpreting a digital output is as a count. Indeed, the actual output of the device may be a count (of pulses or events). Then, an n-bit device can represent a maximum of 2^n counts (because 0 and the sign are not relevant now). If we use this latter approach, the digital sensitivity of a device may be expressed as (for an n-bit device)

$$\frac{2^n}{(\text{Full-scale input})} \text{ in “counts per unit input”} \tag{5.13b}$$

Sometimes, a sensitivity may be expressed with respect to more than one input variable. For example, suppose that a potentiometric displacement sensor gives an output of 1.5 V for a displacement of 5 cm. If the power supply of the potentiometer (or its reference voltage) is 10 V, then the sensitivity of the device may be given as 1.5/5.0/10.0 V/cm/V = 30.0 mV/cm/V. Some examples of sensor sensitivities are given in Table 5.2.

TABLE 5.2

Sensitivities of Some Practical Sensing Devices

Sensor	Sensitivity
Blood pressure sensor	10 mV/V/mm Hg
Capacitive displacement sensor	10.0 V/mm
Charge sensitivity of piezoelectric (PZT) accelerometer	110 pC/N (picocoulomb per newton)
Current sensor	2.0 V/A
DC tachometer	5 ± 10% for 1000 rpm
Fluid pressure sensor	80 mV/kPa
Light sensor (digital output with ADC)	50 counts/lux
Strain gauge (gauge factor)	150 $\Delta R/R$/strain (dimensionless)
Temperature sensor (thermistor)	5 mV/K

Example 5.5

A photovoltaic light sensor can detect a maximum of 20 lux of light and generate a corresponding voltage of 5.0 V. The device has an 8-bit ADC, which gives its maximum count for the full-scale input of 5.0 V. What is the overall sensitivity of the device?

Solution

The maximum count of the ADC = 2^8 = 256 counts

This corresponds to 5.0 V into the ADC, which is the sensor output for the maximum possible light level of 20 lux. Hence, the overall sensitivity of the device is

$$\frac{256}{20.0}\text{counts/lux} = 12.8\text{ counts/lux}$$

Note: The sensitivity of the ADC alone is 256/5.0 counts/V = 51.2 counts/V

5.5.2.2 Sensitivity Error

The rated sensitivity of a device, as given in its data sheet, may not be accurate. The difference between the rated sensitivity and the actual sensitivity is called the sensitivity error.

The sensitivity, which is the slope of the input–output curve of a device, may not be accurate for reasons such as

1. Effect of cross sensitivities of undesirable inputs.
2. Drifting due to wear and environmental effects.
3. Dependence on the value of the input. This means the slope changes with the input value, and it is a reflection of the *nonlinearity* in the device.
4. Local slope of the input–output curve (*local sensitivity*) may not be defined or may be insignificant (compared to O(2) terms).

Concerning item 4 in the list, the local slope (derivative) may be

1. Zero (as in saturation or Coulomb friction)
2. Infinity (as in Coulomb friction)
3. Less significant than the higher derivatives (i.e., O(2) terms of the Taylor series expansion cannot be neglected)

Then, either a local sensitivity cannot be defined or a local sensitivity is not significant. In such situations, a global sensitivity (i.e., [overall or full-scale output]/corresponding input]) may be used. Errors in sensitivity may be represented using a ±range of variation of the average sensitivity, which corresponds to the difference between the minimum and maximum values within which the actual sensitivity may vary. This total variation of the sensitivity (max–min) is a measure of the static *nonlinearity* of the device.

As mentioned earlier, the sensitivity in instrumentation may be handled as either a design issue or a control issue. In particular, the main objective being maximization of the sensitivity to desirable factors and minimization of the sensitivity for undesirable factors, it may be achieved through both design and control. Once a system is designed for the sensitivity objective, it may be further improved or specific sensitivity objectives may be achieved through control. This issue is discussed next.

5.5.2.3 Sensitivity Considerations in Control

The accuracy of a control system is affected by parameter changes in the system components and by the influence of external disturbances. Furthermore, some types of control (e.g., adaptive control, self-tuning control) depend on the sensitivity of the system to control parameters. It follows that analyzing the sensitivity of a feedback control system to parameter changes and to external disturbances is important.

Consider the block diagram of a typical feedback control system, shown in Figure 5.7a. In the usual notation, we have

$G_p(s)$ as the plant (or controlled system) transfer function
$G_c(s)$ as the controller (including compensator and other hardware) transfer function
$H(s)$ as the feedback (including measurement system) transfer function
u as the system command, y as the system output, and u_d as the external disturbance input

For linear systems, the *principle of superposition* applies. In particular, if we know the outputs corresponding to two inputs when applied separately, the output when both inputs are applied simultaneously is given by the sum of the individual outputs. We use this principle here.

First, set $u_d = 0$. Then, it is straightforward to obtain the input–output relationship:

$$y = \left[\frac{G_c G_p}{1 + G_c G_p H}\right] u \tag{i}$$

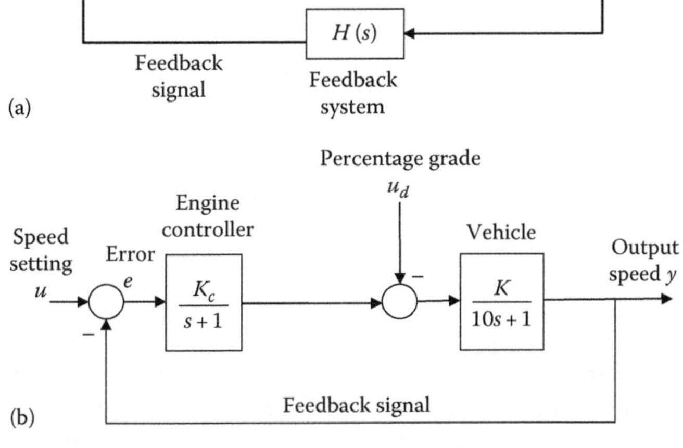

FIGURE 5.7
(a) Block diagram representation of a feedback control system. (b) A cruise control system.

Next, set u equal to 0. Then, we obtain the following input–output relationship:

$$y = \left[\frac{G_p}{1+G_cG_pH}\right]u_d \tag{ii}$$

By applying the principle of superposition on (i) and (ii), we obtain the overall input–output relationship:

$$y = \left[\frac{G_cG_p}{1+G_cG_pH}\right]u + \left[\frac{G_p}{1+G_cG_pH}\right]u_d \tag{5.14}$$

The closed-loop transfer function $\tilde{G}$ is given by y/u, with $u_d=0$; thus,

$$\tilde{G} = \frac{G_cG_p}{\left[1+G_cG_pH\right]} \tag{5.15}$$

System sensitivity to parameter change: The sensitivity of the system to a change in some parameter k may be expressed as the ratio of the change in the system output to the change in the parameter: $\Delta y/\Delta k$. In the *nondimensional form*, this sensitivity is given by $S_k=(k/y)(\Delta y/\Delta k)$.

Note: The nondimensional form of sensitivity is generally suitable because it enables a fair comparison of different sensitivities (a different dimension or scale will change a sensitivity value for the same condition).

Since $y=\tilde{G}u$, with $u_d=0$, it follows that for a given input u, $(\Delta y/y)=(\Delta\tilde{G}/\tilde{G})$. Consequently, Equation 5.16 may be expressed as $S_k=(k/\tilde{G})(\Delta\tilde{G}/\Delta k)$ or in the limit

$$S_k = \frac{k}{\tilde{G}}\frac{\partial\tilde{G}}{\partial k} \tag{5.16}$$

Now, by applying Equation 5.16 to 5.15, we are able to determine expressions for the control system sensitivity to changes in various components in the control system. Specifically, by straightforward partial differentiation of Equation 5.15, separately with respect to G_p, G_c, and H, respectively, we get

$$S_{Gp} = \frac{1}{[1+G_cG_pH]} \tag{5.17}$$

$$S_{Gc} = \frac{1}{[1+G_cG_pH]} \tag{5.18}$$

$$S_H = -\frac{G_cG_pH}{[1+G_cG_pH]} \tag{5.19}$$

It is clear from these three relations that as the static gain (or DC gain) of the loop (i.e., G_cG_pH, with $s=0$) is increased, the sensitivity of the control system to changes in the plant and the controller decreases, but the sensitivity to changes in the feedback (measurement) system approaches (negative) unity. Furthermore, it is clear from Equation 5.14 that the effect of the disturbance input can be reduced by increasing the static gain of G_cH.

By combining these observations, the following design criterion concerning sensitivity can be stipulated for a feedback control system:

1. Make the measurement system (H) robust, stable, and very accurate.
2. Increase the loop gain (i.e., gain of G_cG_pH) to reduce the sensitivity of the control system to changes in the plant and controller.
3. Increase the gain of G_cH to reduce the influence of external disturbances.

In practical situations, the plant G_p is usually fixed and cannot be modified. Furthermore, once a suitable and accurate measurement system is chosen, H is essentially fixed. Hence, most of the design freedom is available with respect to G_c only. It is virtually impossible to achieve all the design requirements simply by increasing the gain of G_c. The dynamics (i.e., the entire transfer function) of G_c (not just the gain value at $s=0$) also have to be properly designed in order to obtain the desired performance in a control system.

Example 5.6

Consider the cruise control system given by the block diagram in Figure 5.7b. The vehicle travels up a constant incline with constant speed setting from the cruise controller.

(a) For a speed setting of $u = u_o$ and a constant road inclination of $u_d = u_{do}$, derive an expression for the steady-state values, y_{ss} of the speed and e_{ss} of the speed error. Express your answers in terms of K, K_c, u_o, and u_{do}.
(b) At what minimum percentage grade would the vehicle stall? Use steady-state conditions, and express your answer in terms of the speed setting u_o and controller gain K_c.
(c) Suggest a way to reduce e_{ss}.
(d) If $u_o = 4$ and $u_{do} = 2$ and $K = 2$, determine the value of K_c such that $e_{ss} = 0.1$.

Solution

(a) For $u_d = 0$: $y = \dfrac{\dfrac{K_cK}{(s+1)(10s+1)}}{\left[1+\dfrac{K_cK}{(s+1)(10s+1)}\right]}u = \dfrac{K_cK}{[(s+1)(10s+1)+K_cK]}u$

For $u = 0$: $y = \dfrac{\dfrac{K}{(10s+1)}}{\left[1+\dfrac{K_vK}{(s+1)(10s+1)}\right]}(-u_d) = -\dfrac{K(s+1)}{[(s+1)(10s+1)+K_cK]}u_d$

Hence, with both u and u_d present, using the principle of superposition (linear system), we have

$$y = \frac{K_cK}{[(s+1)(10s+1)+K_cK]}u - \frac{K(s+1)}{[(s+1)(10s+1)+K_cK]}u_d \quad \text{(i)}$$

If the inputs are constant at steady state, the corresponding steady-state output does not depend on the nature of the input during transition to the steady state. Hence, in this problem what matters is the fact that the inputs and the output are constant at steady state. Hence, without loss of generality, we can assume the inputs to be step functions.

Note: Even if we assume a different starting shape for the inputs, we should get the same answer for the steady-state output, for the same steady-state input values. But the mathematics of getting that answer would be more complex.

Now, we use the following *final value theorem* at steady state:

$$y_{ss} = \lim_{s\to 0}\left[\frac{K_cK}{[(s+1)(10s+1)+K_cK]}\cdot\frac{u_o}{s}\cdot s - \frac{K(s+1)}{[(s+1)(10s+1)+K_cK]}\cdot\frac{u_{do}}{s}s\right]$$

or

$$y_{ss} = \frac{K_cK}{(1+K_cK)}u_o - \frac{K}{(1+K_cK)}u_{do} \quad \text{(ii)}$$

Hence, the steady-state error is

$$e_{ss} = u_o - y_{ss} = u_o - \frac{K_cK}{(1+K_cK)}u_o + \frac{K}{(1+K_cK)}u_{do}$$

or

$$e_{ss} = \frac{1}{(1+K_cK)}u_o + \frac{K}{(1+K_cK)}u_{do} \quad \text{(iii)}$$

(b) Stalling condition is $y_{ss}=0$. Hence, from (ii) we get

$$u_{do} = K_cu_o$$

(c) Since K is usually fixed (a plant parameter) and cannot be adjusted, we should increase K_c to reduce e_{ss}.

(d) Substituting $u_o=4,\ u_{do}=2,\ K=2,\ e_{ss}=0.1$ into (iii), we get

$$0.1 = \frac{1}{(1+2K_c)}\times 4 + \frac{2}{(1+2K_c)}\times 2 \to 1+2K_c = 80 \to K_c = 39.5$$

Sensitivity-based control: Sometimes, sensitivity is used to determine the control law for a system. Adaptive control and self-tuning control are examples of this, where the parameters of the controller are changed (adapted, tuned) depending on the performance requirements. The sensitivity of the controller parameters to the system performance plays a direct role in the control scheme. This procedure primarily uses locally linearized models (i.e., local sensitivities).

For some nonlinear systems, however, the use of local sensitivities can lead to undesirable results. For example, in nonlinear damping, the local slope may be negative, which corresponds to a negative damping constant and will generate positive (unstable) poles (to illustrate this point, consider the simple oscillator with linear damping, make the damping constant negative, and find the corresponding poles). This corresponds to an unstable system. As a specific example, consider the Stribeck friction model as shown in Figure 5.8. Regions 1 and 2 have negative slopes, and they correspond to unstable behavior, whereas viscous damping (Region 3) corresponds to stable behavior.

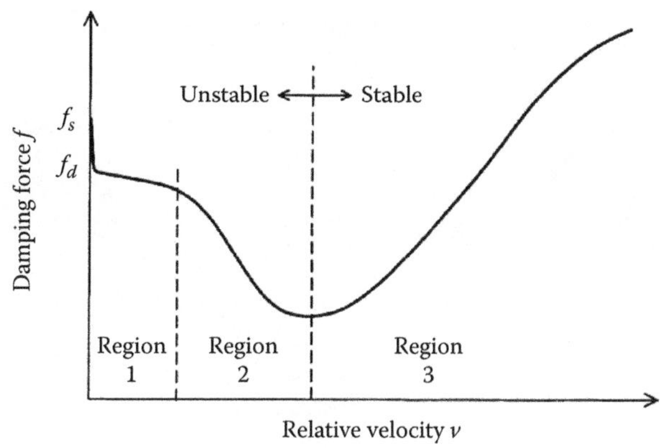

FIGURE 5.8
Stribeck friction.

SNR: The SNR is the ratio of the signal magnitude to the noise magnitude, expressed in dB. We have

$$SNR = 10\log_{10}\left(\frac{P_{signal}}{P_{noise}}\right) = 20\log_{10}\left(\frac{M_{signal}}{M_{noise}}\right) \tag{5.20}$$

where
- *P* denotes signal power
- *M* denotes signal magnitude

For each sinusoidal Fourier signal component, *P* is proportional to M^2, and this is consistent with the two expressions given in (5.20). Also, the signal value that is commonly used in the SNR is the rms value.

As an example, consider the noisy signal, true (noise-free) signal, and the noise shown in Figure 5.9. These were generated using the following MATLAB script:

```
% Signal-to-noise ratio
for i=1:501
n(i)=normrnd(0.0,0.1); % nrandom noise
end
t=0:0.02:10.0;
u=sin(t);
u2=0.2*sin(50*t);
n=n+u2;
un=u+n;
% rms of signal
sigrms=std(u)
% rms of noise
noirms=std(n)
SNR=20*log10(sigrms/noirms)
% plot the results
plot(t,u,'-',t,un,'-',t,n,'o',t,n,'-',t,n,'x')
```

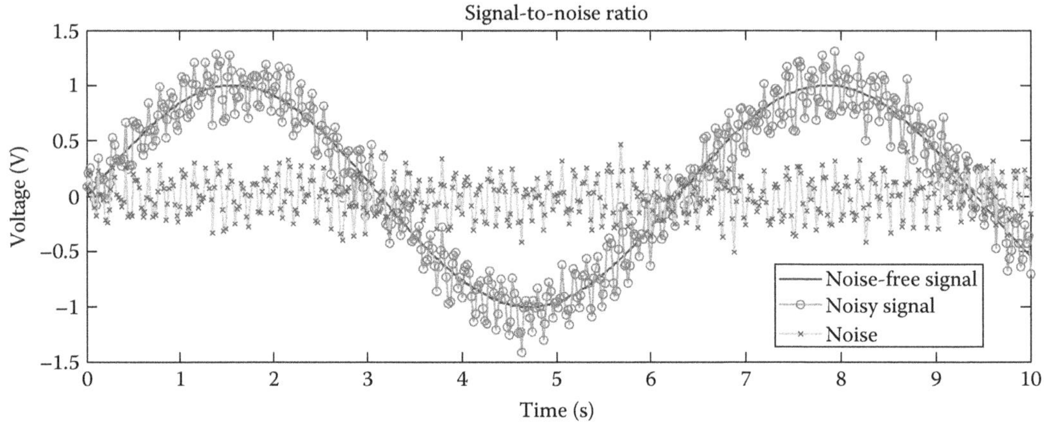

FIGURE 5.9
A noise-free signal and noise.

Using MATLAB, the rms values of the signal and the noise and the corresponding SNR (in dB) were found to be

```
sigrms =
    0.6663
noirms =
    0.1769
SNR =
    11.5168
```

Note: In this example, the signal and noise have zero mean.

As a rule of thumb, an SNR value of 10 dB or more would be satisfactory. A value of 3 dB (half power for the noise) or less is not acceptable. Of course in such calculations, typically only the noisy signal is known (by measurement), and the true (noise-free) signal is not accurately known. Strictly, in the SNR computation, the signal value (rms) has to be for the noise-free signal. Then, a procedure would be to filter the noisy signal and remove its noise as much as possible, compute the rms value of the filtered signal, obtain the noise signal by subtracting the filtered signal from the noisy signal, and calculate the rms value of the noise.

SNR may be interpreted from the viewpoint of sensitivity as well. Then, it will represent the ratio of the sensitivities to the desirable signals and undesirable signals (noise).

DR: The DR or simply "range" of an instrument is determined by the allowed lower and upper limits of its output (response) while maintaining a required level of output accuracy. This range is usually expressed as a ratio (e.g., a log value in *decibels* or dB). In many situations, the lower limit of the DR is equal to the resolution of the device. Hence, the DR (ratio) is usually expressed as [range of operation]/[resolution] in dB. We have

$$\text{Dynamic range} = 20\log_{10}\left[\frac{\text{Range of operation}}{\text{Resolution}}\right] = \frac{y_{\max} - y_{\min}}{\delta y} \tag{5.21a}$$

Note: $y_{\min}$ can be zero, positive, or negative.

Resolution: The resolution of an input–output instrument is the smallest change in a signal (input) that can be detected and accurately presented (as output) by the instrument. The instrument may be an input–output device such as a sensor, transducer, signal conversion hardware, or sensor system. It is usually expressed as a percentage of the maximum range of the instrument or as the inverse of the DR ratio. It follows that DR and resolution are closely related.

DR and resolution of a digital device: The meaning of DR (and resolution) can easily be extended to cover digital instruments. The instrument may be a digital device in its own right such as one that generates pulses and counts or an analog device with an ADC. Still, the real resolution will be some analog value δy, depending on the particular application. For example, δy may represent an output signal increment of 0.0025 V of a transducer (e.g., bridge output of a strain gauge).

For an n-bit digital device, resolution is the change in the analog output in proper units corresponding to one increment in the least significant bit (LSB). Hence,

$$\text{Resolution} = 1 \text{ last significant bit} = \delta y$$

Since an n-bit word can represent a combination of 2^n values, if the smallest value is denoted by y_{min}, the largest value is $y_{max} = y_{min} + (2^n - 1)\,\delta y$. Hence,

$$\text{Range} = y_{max} - y_{min} = (2^n - 1)\,\delta y$$

Then, for an n-bit device (say, a device with n-bit ADC), the DR is

$$DR = \frac{y_{max} - y_{min}}{\delta y} = \frac{(2^n - 1)\delta y}{\delta y} = 2^n - 1 \tag{5.21b}$$

Note: This needs to be expressed in dB.

The result given by Equation 5.21b does not necessarily mean that the overall DR of the device depends only on its digital component. We obtained the result (5.21b), which depends only on the number of bits (n) of the device because we have correlated the analog quantities of the device directly to digital quantities (specifically, digital increment of "1" to the analog quantity δy and digital range (2^n–1) to the analog range y_{max}–y_{min}). However, in practice, when several devices (both analog and digital) are interconnected, we need to consider their individual DR values and use the most critical one (i.e., smallest value) as the DR of the overall system.

Example 5.7

Consider an instrument that has a 12-bit ADC. Estimate the DR of this instrument.

Solution

In this example, DR is determined (primarily) by the word size of the ADC. Each bit can take the binary value 0 or 1. Since the resolution is given by the smallest possible increment, that is, a change by the LSB, it is clear that digital resolution is equal to 1. The largest value represented by a 12-bit word corresponds to the case when all 12 bits are unity. This value is decimal $2^{12} - 1$. The smallest value (when all 12 bits are zero) is zero. Hence, according to Equation 5.21b, the dynamic range of the instrument is given by

$$20\log_{10}\left[\frac{2^{12}-1}{1}\right] = 72 \text{ dB}.$$

Offset (bias): The offset in an output can create difficulties in instrumentation practice. Particularly important consideration is the *zero offset*, which is the output of the device when the input is zero. If a sensor, for example, has a zero offset, the control action that is generated using it can be incorrect. Particularly, if there is an offset in an error signal, it can lead to incorrect actions (because corrective actions would be made even when there is no error). As another example, under balanced conditions, a bridge circuit should generate a zero output. If the output of a balanced bridge is not zero, it has to be compensated for so as to remove the offset. A further example is a differential amplifier whose output has to be zero when the two input signals are equal.

A known offset can be corrected by several methods including

1. Recalibration of the device
2. Programming a digital output (i.e., subtract the offset)
3. Using analog hardware for offsetting at the device output

Linearity: This topic has been addressed already, but some important concepts are mentioned now. Linearity is determined by the calibration curve of an instrument. The curve of the output value (e.g., peak or rms value) versus input value under static (or steady state) conditions within the DR of the instrument is known as the "static calibration curve." Its closeness to a straight line measures the degree of linearity of the instrument. Manufacturers provide this information as the maximum deviation of the calibration curve either from the *least-squares straight-line fit* (also see Chapter 7) of the calibration curve or from some other reference straight line. If the least-squares fit is used as the reference straight line, the maximum deviation is called "independent linearity" (more correctly, independent nonlinearity, because the larger the deviation, the greater the nonlinearity). Nonlinearity may be expressed as a percentage of either the actual reading at an operating point or the full-scale reading or as the maximum variation of the sensitivity as a percentage of a reference sensitivity.

Zero drift and full-scale drift: Zero drift is defined as the drift from the null reading of an instrument when the input is maintained steady for a long period. In this context, the input is kept at zero or any other level (if there is a zero offset) that corresponds to the null reading of the instrument. Similarly, *full-scale drift* is defined with respect to the full-scale reading (i.e., the input is maintained at the full-scale value). In the instrumentation practice, drift is a consideration of *stability*. This interpretation, however, is not identical to the standard textbook definitions of stability. Usual causes of drift include instrument instability (e.g., instability in amplifiers), ambient changes (e.g., changes in temperature, pressure, magnetic fields, humidity, and vibration level), changes in power supply (e.g., changes in reference DC voltage or alternating current [AC] line voltage), and parameter changes in an instrument (because of aging, wear and tear, nonlinearities, etc.). Drift due to parameter changes caused by nonlinearities, environmental effects, etc., is known as "parametric drift," "sensitivity drift," or "scale-factor drift." For example, a change in spring stiffness or electrical resistance due to changes in ambient temperature will result in a parametric drift. Parametric drift generally depends on the input level, while zero drift is assumed to be the same at any input level if the other conditions are kept constant. For example, a change in the reading caused by thermal expansion of the readout mechanism because of changes in ambient temperature is considered a zero drift. Drift in electronic devices can be reduced by using AC circuitry rather than DC circuitry. For example, AC-coupled amplifiers have fewer drift problems than DC amplifiers. Intermittent checking for instrument response level with zero input is a popular way to calibrate for zero drift. In digital devices, for example, this can be done automatically from time to time between sample points (hold periods) or at other times when the input signal can be bypassed without affecting the system operation. The calibration curve of a device can change with time due to changes in the device as mentioned here. This is called "calibration drift." Recalibration can overcome the calibration drift.

Useful frequency range: This corresponds to a flat *gain curve* and a zero *phase curve* in the frequency response characteristics (*FTF* or *FRF*) of an instrument. The upper frequency in this band should be typically less than half (say, one-fifth) of the dominant resonant frequency of the instrument. This is a measure of the instrument bandwidth.

Bandwidth: The bandwidth of an instrument determines the maximum speed or frequency at which the instrument is capable of operating. High bandwidth implies faster *speed of response* (the speed at which an instrument reacts to an input signal). Bandwidth is determined by the dominant natural frequency ω_n or the dominant resonant frequency ω_r of the device. (*Note*: For low damping, ω_r is approximately equal to ω_n, as we have seen from the expressions for the simple oscillator model). It is inversely proportional to *rise time* and the *dominant time constant*. Half-power bandwidth is also a useful parameter (see Chapter 6). Instrument bandwidth has to be several times greater than the maximum frequency of interest in the input signals. For example, the bandwidth of a measuring device is important particularly when measuring transient signals. The *sensor bandwidth* should be several times larger than the frequency of the fastest signal component that should be accurately measured; note further that bandwidth is directly related to the *useful frequency range*. Chapter 6 addresses the subject of bandwidth in detail.

Summary Sheet

Performance specifications: Parameters that are used to indicate the rated or expected performance of a device.

Categories of performance specification: Error, speed of performance, and stability.

Types of performance parameters: (1) Parameters used in engineering practice (provided by device manufacturers and vendors) and (2) parameters defined using engineering theoretical considerations (model based).

Models used for performance specification: (1) Differential-equation models (time domain) and (2) transfer-function models (frequency domain) are two types of dynamic models that are used for performance specification in the analytical domain.

Perfect measurement device : The main characteristics of a perfect measuring device are (1) its output instantly reaches the measured value (*fast response, zero error*), (2) the output level is high (*high gain, low output impedance, high sensitivity*), (3) the output remains steady at measured value when the measurand is steady (zero error, *no drift or environmental effects and noise, stable, and robust*), (4) the output varies in proportion to the measurand (*static linearity*), (5) the device does not distort the measurand (no *loading* effects, matched *impedances*), and (6) it acquires low power consumption (high input impedance).

First-order model: $\tau\dot{y} + y = ku \rightarrow Y(s)/U(s) = H(s) = \dfrac{k}{\tau s + 1}$, $y_{\text{step}} = y_0 e^{-t/\tau} + Ak(1 - e^{-t/\tau})$; initial slope = $(Ak-y_0)/\tau$, steady-state value = Ak, half-power bandwidth = $1/\tau$, input = u, output=y, time constant = τ, and gain = k; tangential line at $y_{step}(0)$ will reach steady-state value at $t=\tau$ (another interpretation of time constant).

Note: τ is the key performance parameter here.

Simple oscillator model: $\ddot{y} + 2\zeta\omega_n\dot{y} + \omega_n^2 y = \omega_n^2 u(t) \rightarrow$

$$\frac{Y(s)}{U(s)} = H(s) = \left[\frac{\omega_n^2}{s^2 + 2\zeta\omega_n s + \omega_n^2}\right]; \quad y_{step} = 1 - \frac{1}{\sqrt{1-\zeta^2}} e^{-\zeta\omega_n t} \sin(\omega_d t + \phi);$$

$$\cos\phi = \zeta, \quad \omega_d = \sqrt{1-\zeta^2}\,\omega_n, \quad \omega_r = \sqrt{1-2\zeta^2}\,\omega_n$$

Time-domain performance parameters: Rise time $T_r = (\pi - \phi)/\omega_d$ with $\cos\phi = \zeta$ (speed); peak time $T_p = \pi/\omega_d$ (speed); peak value $M_p = 1 - e^{-\pi\zeta/\sqrt{1-\zeta^2}}$ (stability); percentage overshoot (PO) $\text{PO} = 100e^{-\pi\zeta/\sqrt{1-\zeta^2}}$ (stability); *time constant* $\tau = 1/\zeta\omega_n$ (speed and stability); and settling time (2%)

$$T_s = -\frac{\ln\left[0.02\sqrt{1-\zeta^2}\right]}{\zeta\omega_n} \approx 4\tau = \frac{4}{\zeta\omega_n} \text{(stability)}.$$

Steady-state error: Important performance parameter, cannot be represented by simple oscillator model, and can be included through an offset term at output; integral control or loop gain will decrease this error.

Level of stability: Depends on decay rate of free response (and hence on time constants or real parts of poles).

Speed of response: Depends on natural frequency and damping for oscillatory systems and decay rate for nonoscillatory systems.

Time constant: Determines system stability and decay rate of free response (and speed of response as well in nonoscillatory systems).

Underdamped is more stable than overdamped: For the same natural frequency, if damping ratio $\zeta_o > (\zeta_u^2 + 1)/2\zeta_u$; u—underdamped, o—overdamped.

Frequency-domain specifications: Useful frequency range (*operating interval*); bandwidth (*speed of response*); static gain (*steady-state performance*); resonant frequency (*speed* and *critical frequency region*); magnitude at resonance (*stability*); input impedance (*loading, efficiency, interconnectability, maximum power transfer, signal reflection*); output impedance (*loading, efficiency, interconnectability, maximum power transfer, signal level*); gain margin (*stability*); and phase margin (*stability*).

Manifestations of nonlinearity: Saturation, dead zone, hysteresis, jump phenomenon, limit cycles, and frequency creation.

Linearization methods: Use of nonlinear calibration curve (*in static case*), linearizing elements (*e.g., resistors and amplifiers in bridge circuits*) to neutralize nonlinear effects, nonlinear feedback (feedback linearization), and local linearization (*local slope*).

Drawbacks of local linearization: Local slope (1) will change with operating conditions (nonlinear), (2) may not exist or insignificant compared to O(2) terms of Taylor series (e.g., Coulomb friction), and (3) may result in instability (e.g., negative damping in a control law).

Rating parameters (commercial, in device data sheets): Sensitivity and sensitivity error; signal-to-noise ratio (SNR); dynamic range (DR); resolution; offset or bias; linearity; zero drift, full-scale drift, and calibration drift (stability); useful frequency range; bandwidth; and input and output impedances.

Sensitivity: Device output–input; goal maximize for desirable inputs and minimize for undesirable inputs.

Handling sensitivity: Select a reasonable number of factors having noteworthy sensitivity, determine their sensitivities (say, relative sensitivities), maximize the sensitivity to desirable factors (e.g., the measured quantity), and minimize the sensitivity to undesirable factors (e.g., thermal effects on a strain reading) or cross sensitivity.

Cross sensitivity: Sensitivity along directions orthogonal to primary direction of sensitivity (expressed as a % of direct sensitivity).

Sensitivity in digital devices: Digital Output/Corresponding Input = 2^n/[full-scale input] in "counts per unit input" for n-bit device.

Sensitivity error: [Rated sensitivity] – [Actual sensitivity]. Causes of sensitivity error: effect of cross sensitivities of undesirable inputs; drifting due to wear, environmental effects, etc.; dependence on the value of the input (i.e., slope changes with input → *nonlinear* device); and local slope (*local sensitivity*) possibly not defined or insignificant compared to higher-order terms.

Sensitivity considerations in control: Input (including disturbance)–output relation:

$$y = \left[\frac{G_c G_p}{1 + G_c G_p H}\right] u + \left[\frac{G_p}{1 + G_c G_p H}\right] u_d$$

$G_p(s)$ = plant (or controlled system) transfer function, $G_c(s)$ = controller (including compensator and other hardware) transfer function, and $H(s)$ = feedback (including measurement system) transfer function. Sensitivities to $G_p(s)$, $G_c(s)$, and $H(s)$:

$$S_{Gp} = \frac{1}{[1 + G_c G_p H]}, \quad S_{Gc} = \frac{1}{[1 + G_c G_p H]}, \quad S_H = -\frac{G_c G_p H}{[1 + G_c G_p H]}$$

Sensitivity-based design strategy: (1) Make the measurement system (H) robust, stable, and very accurate; (2) increase the loop gain (i.e., gain of $G_c G_p H$) to reduce the sensitivity of the control system to changes in the plant and controller; and (3) increase the gain of $G_c H$ to reduce the influence of external disturbances.

Singal-to-noise ratio (SNR): [Signal magnitude (rms)]/[Noise magnitude (rms)] in dB:

$$SNR = 10\log_{10}\left(\frac{P_{signal}}{P_{noise}}\right) = 20\log_{10}\left(\frac{M_{signal}}{M_{noise}}\right)$$

P = signal power
M = signal magnitude

$$P \propto M^2$$

Another interpretation: SNR = [signal sensitivity]/[noise sensitivity].

Rule of thumb: SNR ≥ 10 dB is good; SNR ≤ 3 dB (half power for noise) is bad.

DR: Or "range" of an instrument = allowed lower to upper range of output, while maintaining a required level of output accuracy. Expressed as a ratio (dB). Typically, lower limit = the instrument resolution → DR = (Range of operation)/(Resolution) in dB

$$DR = \frac{y_{\max} - y_{\min}}{\delta y}$$

For a digital (n bit) device, $DR = ((2^n - 1)\delta y)/\delta y = (2^n - 1)$.

Resolution: Smallest change in a signal (input) that can be detected and presented (as output) accurately by the instrument, for example, sensor, transducer, signal conversion hardware (e.g., ADC), and sensor system. Usually expressed as a % maximum range or inverse of DR.

Offset (bias): *Zero offset* = device output when input = 0 (e.g., output of imperfect bridge under balanced conditions; output of imperfect amplifier when the two input signals are equal). Methods of correction (when offset is known): (1) recalibration, (2) programming a digital output (i.e., subtract the offset), and (3) using analog hardware for offsetting at device output.

Linearity: Curve of output (peak or rms) versus input value under static (or steady-state) conditions within DR of instrument → *static calibration curve*. Its closeness to average straight line measures degree of linearity. If *least-squares fit* is used as the reference, max deviation is called "independent linearity" (more correctly, independent nonlinearity). Nonlinearity is expressed as (1) % of actual reading at operating point or full-scale reading and (2) max variation of sensitivity/reference sensitivity as a %.

Zero drift: Drift from null reading of instrument when input is maintained steady for a long period.

Full-scale drift: Drift from full-scale reading when input is maintained at full-scale value.

Parametric drift: Drift in device parameter values (due to environmental effects, etc.).

Sensitivity drift: Drift in sensitivity of a parameter.

Scale-factor drift: Drift in scale factor of output. These are closely related.

Calibration drift: Drift in calibration curve of device (due to mentioned reasons).

Useful frequency range: Corresponds to flat *gain curve* and a zero *phase curve* in frequency response characteristics (FTF or *FRF*) of instrument. Upper frequency in this range <0.5 (say, one-fifth) × dominant resonant frequency ← measure of instrument bandwidth.

Bandwidth: Interpretations: (1) speed of response of device, (2) pass band of filter, (3) operating frequency range of device, (4) uncertainty in frequency content of signal, and (5) information capacity of communication network.

Problems

5.1 What do you consider a perfect measuring device? Suppose that you are asked to develop an analog device for measuring angular position in an application related to control of a kinematic linkage system (e.g., a robotic manipulator). What instrument ratings (or specifications) would you consider crucial in this application? Discuss their significance.

5.2 List and explain some time-domain parameters and frequency-domain parameters that can be used to predominantly represent the (a) speed of response and (b) degree of stability of a control system. In addition, briefly discuss any conflicts that can arise in specifying these parameters.

5.3 A tactile (distributed touch) sensor (see Chapter 9) of the gripper of a robotic manipulator consists of a matrix of piezoelectric sensor elements placed 2 mm apart. Each element generates an electric charge when it is strained by an external load. Sensor elements are multiplexed at very high speed in order to avoid charge leakage and to read all data channels using a single high-performance charge amplifier. Load distribution on the surface of the tactile sensor is determined from the charge amplifier readings, since the multiplexing sequence is known. Each sensor element can read a maximum load of 50 N and can detect load changes in the order of 0.01 N.

(a) What is the spatial resolution of the tactile sensor?
(b) What is the load resolution (in N/m^2) of the tactile sensor?
(c) What is the dynamic range?

5.4 A useful rating parameter for a robotic tool is *dexterity*. Though not complete, an appropriate analytical definition for the dexterity of a device is

$$\text{Dexterity} = \frac{\text{Number of degrees of freedom}}{\text{Motion resolution}}$$

where the number of degrees of freedom is equal to the number of independent variables that is required to completely define an arbitrary position increment of the tool (i.e., for an arbitrary change in its kinematic configuration).

(a) Explain the physical significance of dexterity and give an example of a device for which the specification of dexterity would be very important.

(b) The power rating of a tool may be defined as the product of maximum force that can be applied by it in a controlled manner and the corresponding maximum speed. Discuss why the power rating of a manipulating device is usually related to the dexterity of the device. Sketch a typical curve of power versus dexterity.

5.5 The resolution of a feedback sensor (or the resolution of a response measurement used in feedback) has a direct effect on the accuracy that is achievable in a control system. This is true because the controller cannot correct a deviation of the response from the desired value (set point) unless the response sensor can detect that change. It follows that the resolution of a feedback sensor will govern the minimum (best) possible deviation band (about the desired value) of the system response, under feedback control. An angular position servo uses a resolver (see Chapter 8) as its feedback sensor. If peak-to-peak oscillations of the servo load (the plant) under steady-state conditions have to be limited to no more than 2°, what is the worst tolerable resolution of the resolver?

Note: In practice, the feedback sensor should have a resolution better (smaller) than this worst value.

5.6 Consider a simple mechanical dynamic device (single degree of freedom) that has low damping. An approximate design relationship between the two performance parameters T_r and f_b may be given as $T_r f_b = k$, where T_r is the rise time in nanoseconds (ns) and f_b is the bandwidth in megahertz (MHz). Estimate a suitable value for k.

5.7 List several response characteristics of nonlinear dynamic systems that are not generally exhibited by linear dynamic systems. Additionally, determine the response y of the nonlinear system $[dy/dt]^{1/3}=u(t)$ when excited by the input $u(t)=a_1 \sin\omega_1 t+a_2 \sin\omega_2 t$. What characteristic of a nonlinear system does this result illustrate?

5.8 Consider the "static" (or "algebraic") system represented by $y=pu^2+c$. Sketch this input–output relationship for $p=1$ and $c=0.2$. What is the response of this device on application of the sine input $u=\sin t$? How would you linearize this device without losing its accuracy and the operating range?

5.9 Consider a mechanical component whose response x is governed by the relationship

$$f = f(x,\dot{x})$$

where

f denotes applied (input) force
$\dot{x}$ denotes velocity

Three special cases are

(a) Linear spring: $f=kx$
(b) Linear spring with viscous (linear) damping: $f = kx + b\dot{x}$
(c) Linear spring with Coulomb friction: $f = kx + f_c \sin(\dot{x})$

Suppose that a harmonic excitation of the form $f=f_o \sin\omega t$ is applied in each case. Sketch the force–displacement curves for the three cases at steady state. Which of the three components exhibit hysteresis? Which components are nonlinear? Discuss your answers.

5.10 You are required to select a sensor for a position control application. List several important considerations that you have to take into account in this selection. Briefly indicate why each of them is important.

5.11 (a) Sketch (not to scale) the magnitude versus frequency curves of the following two transfer functions: (i) $G_i(s) = 1/(\tau_i s + 1)$ and (ii) $G_d(s) = 1/(1 + (1/\tau_d s))$

Explain why these transfer fractions may be used as an integrator, a low-pass filter, a differentiator, or a high-pass filter. In your magnitude versus frequency curves, indicate in which frequency bands these four respective realizations are feasible. You may make appropriate assumptions for the time constant parameters τ_i and τ_d.

(b) Active vibration isolators, known as electronic mounts, have been considered for high-end automobile engines. The purpose is to actively filter out the cyclic excitation forces generated by the internal-combustion engine before they would adversely vibrate the components such as seats, floor, and steering column, which come into contact with the vehicle occupants. Consider a four-stroke, four-cylinder engine. It is known that the excitation frequency on the engine mounts is twice the crank-shaft speed, as a result of the firing cycles of the cylinders. A schematic representation of an active engine mount is shown in Figure P5.11a. The crank-shaft speed is measured and supplied to the controller of a valve actuator. The servovalve of a hydraulic cylinder is operated on the basis of this measurement. The hydraulic cylinder functions as an active suspension with a variable (active) spring and a damper. A simplified model of the mechanical interactions is shown in Figure P5.11b.

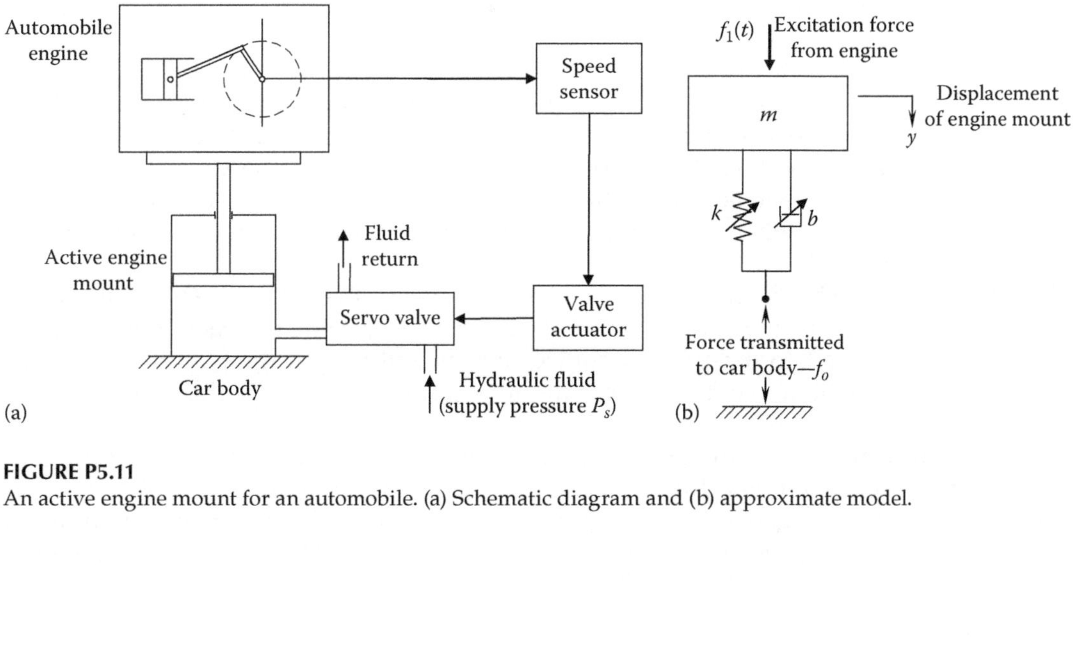

FIGURE P5.11
An active engine mount for an automobile. (a) Schematic diagram and (b) approximate model.

(i) Neglecting gravity forces (which cancel out because of the static spring force) show that a linear model for system dynamics may be expressed as

$$m\ddot{y} + b\dot{y} + ky = f_i$$

$$b\dot{y} + ky - f_o = 0$$

where
f_i is the excitation force from the engine
f_o is the force transmitted to the passenger compartment
y is the displacement of the engine mount with respect to a frame fixed to the passenger compartment
m is the mass of the engine unit
k is the equivalent stiffness of the active mount
b is the equivalent viscous damping constant of the active mount

(ii) Determine the transfer function (with the Laplace variable s) f_o/f_i for the system.

(iii) Sketch the magnitude versus frequency curve of the transfer function obtained in Part (ii) and show a suitable operating range for the active mount.

(iv) For a damping ratio $\zeta = 0.2$, what is the magnitude of the transfer function when the excitation frequency ω is 5 times the natural frequency ω_n of the suspension (engine mount) system?

(v) Suppose that the magnitude estimated in Part (iv) is satisfactory for the purpose of vibration isolation. If the engine speed varies from 600 to 1200 rpm, what is the range in which the spring stiffness k (N/m) should be varied by the control system in order to maintain this level of vibration isolation? Assume that the engine mass m is equal to 100 kg and the damping ratio is approximately constant at ζ is equal to 0.2.

5.12 (a) Briefly discuss any conflicts that can arise in specifying parameters that can be used to predominantly represent the speed of response and the degree of stability of a process (plant).

(b) Consider a measuring device that is connected to a plant for feedback control. Explain the significance of the following parameters in the performance of the feedback control system: (i) bandwidth, (ii) resolution, (iii) linearity, (iv) input impedance, and (v) output impedance of the measuring device.

5.13 (a) Explain why mechanical loading error due to tachometer inertia can be significantly higher when measuring transient speeds than when measuring constant speeds.

(b) A DC tachometer has an equivalent resistance $R_a = 20\ \Omega$ in its rotor windings. In a position plus velocity servo system, the tachometer signal is connected to a feedback control circuit with equivalent resistance 2 kΩ. Estimate the percentage error due to electrical loading of the tachometer at steady state.

(c) If the conditions were not steady, how would the electrical loading be affected in this application (Part (b))?

5.14 Consider a strain-gauge load cell (force sensor) in an active prosthetic application, where the sensing process is shown in Figure P5.14. The force f is sensed using the

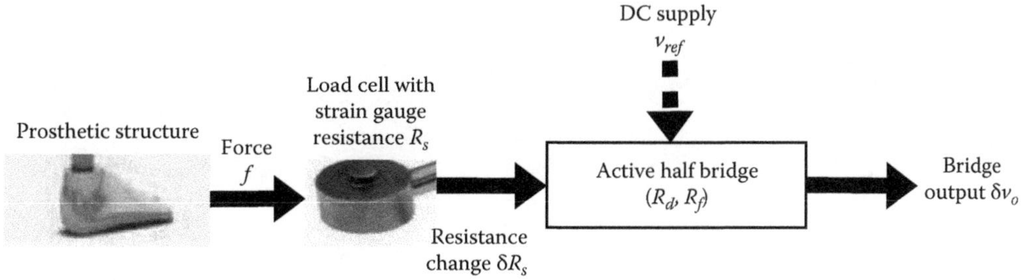

FIGURE P5.14
Prosthetic sensing through a strain-gauge load cell.

change δR_s in the strain-gauge resistance R_s. The active half bridge (Figure 4.17, with $R_1 \equiv R_s$ and $R_2 \equiv R_d$) is initially balanced (i.e., $R_d = R_s$) and generates an output v_o, which represents the measured force. The bridge output is given by $v_o/v_{ref} = (R_f/R)((\delta R_s/R)/(1 + \delta R_s/R))$ with $R_d = R_s = R$. The strain ε is related to the resistance change through $(\delta R_s)/R = S_s\varepsilon$ where S_s is the gauge factor (sensitivity of the strain gauge). The relationship between the force f, which is measured, to the strain may be taken as linear: $f = k\varepsilon$.

Characterize the following for this sensor:

(a) The *measurand* (primary input) and the *measurement* (required output y)
(b) Many possible secondary inputs that will affect the measurement y

Note: These may include undesirable input variables (e.g., noise, disturbance inputs), the parameters that can change due to environmental effects, and the needed inputs (e.g., power) whose changes/errors will affect the measurement in an undesirable manner.

(c) Full-scale range (input/output)
(d) Resolution
(e) Full-scale nonlinearity
(f) Signal-to-noise ratio
  (i) Pick several secondary inputs, which you feel are important with respect to their effect on the measurement. (*Note*: This may be done based on your knowledge, common sense, or simply a guess.)
  (ii) Develop an analytical relationship to represent how the errors in these secondary inputs will affect the measurement v_o. (*Note*: This may be done based on physical relations, available information, your knowledge, common sense, guess work, etc. It may be a nonlinear relationship.)
  (iii) Arrange the error sources in their order of significance.
  (iv) Discuss whether some of the considered input error terms can be neglected.

6

Bandwidth, Sampling, and Error Propagation

Chapter Objectives

- Bandwidth Analysis
- System Bandwidth and Control Bandwidth
- Sampling Theorem
- Aliasing Distortion
- Anti-Aliasing Filter
- Bandwidth Design of a Control System
- Instrument Error Considerations
- Error Representation
- Instrument Accuracy and Measurement Accuracy
- Accuracy and Precision
- Error Propagation and Combination
- Application of Sensitivity in Error Combination

6.1 Introduction

This chapter addresses several further issues of sensing instrumentation that affect the performance of an engineering system. Particularly, bandwidth considerations, data sampling, error and its representation, combination, and propagation are studied in some detail.

Bandwidth has different meanings depending on the particular context and application. For example, when studying the response of a device, the bandwidth relates to the fundamental resonant frequency and correspondingly to the *speed of response* of the device for a given excitation. In band-pass filters, the bandwidth refers to the frequency band (*passband*) of the signal components that are allowed through the filter, while the frequency components outside the band are rejected. With respect to measuring instruments such as sensor systems, bandwidth refers to the range frequencies within which the instrument measures a signal accurately (*operating frequency range*). As a particular note, if a signal passes through a band-pass filter, we know that its frequency content is within the bandwidth of the filter, but we cannot determine the actual frequency content of the signal on the basis of that observation. In this context, the bandwidth appears to represent a *frequency uncertainty* in the observation (i.e., the larger the bandwidth of the filter, the less certain is our knowledge about the actual frequency content of a signal that passes through the filter). In digital communication networks (e.g., the Internet), the bandwidth denotes the capacity (*information capacity*) of the network in terms of information rate (bits/s). In this chapter, we will address various interpretations of bandwidth. The present focus, however, is on instrument bandwidth and control bandwidth. Within that focus, bandwidth directly concerns some form of largest possible operating "speed" such as the speed of response of a dynamic system and the speed of control.

In any multicomponent system, the overall error depends on the component error. Component error degrades the performance of an engineering system. This is particularly true for sensors and transducers as their error is directly manifested as incorrectly known system variables and parameters within the system. Since error may be separated into a systematic (or deterministic) part and a random (or stochastic) part, statistical considerations are important in error analysis. The degree of seriousness of how a component error affects the overall system error concerns sensitivity. In particular, the sensitivity to desirable factors has to be maximized, while the sensitivity to undesirable factors has to be minimized. Since there may be vast number of factors that can affect the system performance, particularly its accuracy (which is the inverse of error), we need to find ways to select and incorporate an appropriate set of them in the instrumentation task. Sensitivity considerations are particularly applicable here. This chapter deals with pertinent considerations of representation, classification, combination, and propagation of error and sensitivity analysis.

6.2 Bandwidth Analysis

Bandwidth plays an important role in specifying and characterizing the components of an engineering system. In particular, useful frequency range, operating bandwidth, and control bandwidth are important considerations. In this section, we study several interpretations of bandwidth and some important issues related to these topics.

6.2.1 Bandwidth

The term *bandwidth* may take the following meanings:

1. Speed of response of a device
2. Passband of a filter
3. Operating frequency range of a device
4. Uncertainty in frequency content of a signal
5. Information capacity of a communication network

These various interpretations of the term *bandwidth* may be somewhat related even though they are not identical.

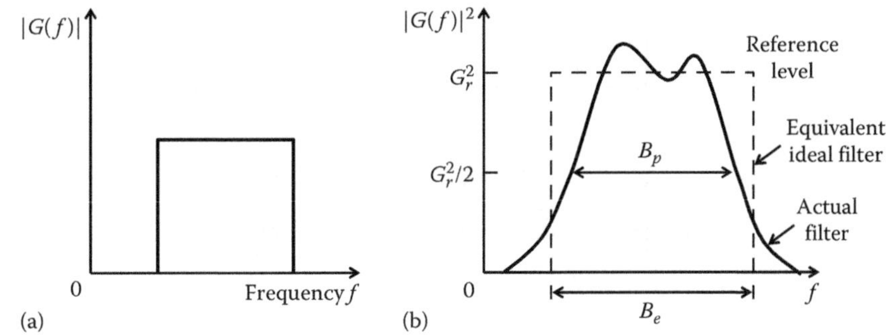

FIGURE 6.1
Characteristics of (a) an ideal band-pass filter and (b) a practical band-pass filter.

6.2.1.1 Transmission Level of a Band-Pass Filter

Practical filters can be interpreted as dynamic systems. In fact, all physical dynamic systems (e.g., electromechanical systems) are analog filters. It follows that the filter characteristic can be represented by the frequency transfer function (FTF) $G(f)$ of the filter. A magnitude-squared plot of such a filter transfer function is shown in Figure 6.1. In a logarithmic plot (e.g., in the Bode plot), the magnitude-squared curve is obtained by simply doubling the corresponding magnitude curve. Note that an actual filter transfer function (Figure 6.1b) is not quite flat like the ideal filter shown in Figure 6.1a. The reference level G_r is the average value of the transfer-function magnitude in the neighborhood of its peak.

6.2.1.2 Effective Noise Bandwidth

Effective noise bandwidth of a filter is equal to the bandwidth of an ideal filter that has the same reference level and that transmits the same amount of power from a white noise source. Recall that *white noise* has a constant (flat) power spectral density (PSD). Hence, for a noise source of unity PSD, the power transmitted by the practical filter is given by

$$\int_0^{\infty} |G(f)|^2 df$$

which, by definition, is equal to the power $G_r^2 B_e$ that is transmitted by the equivalent ideal filter. Hence, the effective noise bandwidth B_e is given by

$$B_e = \frac{\int_0^{\infty} |G(f)|^2 df}{G_r^2} \tag{6.1}$$

Note: The higher the B_e, the larger the frequency content uncertainty of the filtered signal (i.e., more unwanted signal components pass through).

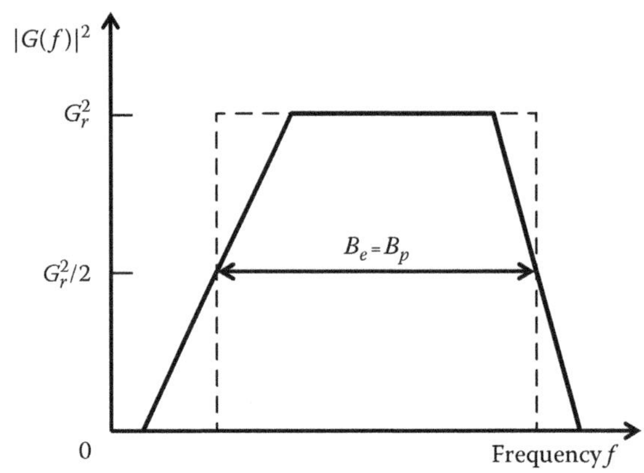

FIGURE 6.2
An idealized filter with linear segments.

6.2.1.3 Half-Power (or 3 dB) Bandwidth

Half of the power from a unity PSD noise source that is transmitted by the filter is $G_r^2B_e/2$. Since B_e is the width of the ideal band-pass filter, $G_r/\sqrt{2}$ is the *half-power level* (in amplitude). This is also known as a 3 dB level because $20\log_{10}\sqrt{2} = 10\log_{10}2 = 3$ dB. (*Note*: 3 dB refers to a power ratio of 2 or an amplitude ratio of $\sqrt{2}$. Hence, a 3 dB drop corresponds to a drop of power to half the original value. A 20 dB corresponds to an amplitude ratio of 10 or a power ratio of 100.) The 3 dB (or half-power) bandwidth is the width of the filter transfer function at the half-power level. This is denoted by B_p. For the ideal filter, this is at the $G_r/\sqrt{2}$ magnitude level. For the actual filter, equivalently, the half-power bandwidth B_p may be taken as the bandwidth at the same magnitude level $G_r/\sqrt{2}$, as shown in Figure 6.1b.

Note that B_e and B_p are different in general. In an approximate spectrum where the magnitude-squared filter characteristic has linear rising, falloff, and flat segments, these two bandwidths are equal (see Figure 6.2).

6.2.1.4 Fourier Analysis Bandwidth

In Fourier analysis, bandwidth is interpreted as the *frequency uncertainty* in the spectral results. In analytical results of *Fourier integral transform*, which assume that the entire signal is available for analysis, the spectrum is continuously defined over the entire frequency range $[-\infty, \infty]$ and the frequency increment δf is infinitesimally small ($\delta f \rightarrow 0$). There is no frequency uncertainty in this case, and the analysis bandwidth is infinitesimally narrow. In *digital Fourier analysis*, the discrete spectral lines are generated at frequency intervals of ΔF. Therefore, this finite frequency increment ΔF, which is the frequency uncertainty, is the analysis bandwidth B for this analysis (digital computation). It is known that $\Delta F = 1/T$, where T is the record length of the signal (or *window length* when a rectangular window is used to select the signal segment for analysis). It also follows that the minimum frequency that has a meaningful accuracy is the analysis bandwidth. This interpretation of the analysis bandwidth is further justified by noting the fact that harmonic components of frequency less than ΔF (or period greater than T) cannot be studied by observing a signal record of length less than T. The analysis bandwidth carries information regarding distinguishable minimum frequency separation in computed results. In this sense, bandwidth is directly related to the *frequency resolution* of the analyzed (computed) results. The *accuracy* of analysis (computation) increases by increasing the record length T (i.e., by decreasing the analysis bandwidth B).

When a time window other than the rectangular window is used to truncate a signal, then reshaping of the signal segment (data) occurs according to the shape of the window. This reshaping suppresses the *side lobes* of the Fourier spectrum of the original rectangular window and, hence, reduces the *frequency leakage* that arises from the truncation of the signal. At the same time, however, an error is introduced as a result of the information lost through data reshaping. This error is proportional to the bandwidth of the window itself. The *effective noise bandwidth* of a rectangular window is only slightly less than $1/T$, because the main lobe of its Fourier spectrum is nearly rectangular and a lobe has a width of $1/T$. Hence, for practical purposes, the effective noise bandwidth can be taken as the analysis bandwidth. Data truncation (i.e., multiplication by a window in the time domain) is equivalent to *convolution* of the Fourier spectrum of the signal with the Fourier spectrum of the window (in the frequency domain). Hence, the main lobe of the window spectrum uniformly affects all spectral lines in the discrete spectrum of the data signal. It follows that a window main lobe with a broader effective noise bandwidth introduces a larger error into the spectral results. Hence, in digital Fourier analysis, bandwidth is taken as the effective noise bandwidth of the time window that is employed.

6.2.1.5 Useful Frequency Range

This corresponds to the flat region (static region) in the gain curve and the zero-phase-lead region in the phase curve of a device (with respect to frequency). It is determined by the dominant (i.e., the lowest) resonant frequency f_r of the device. The upper frequency limit f_{max} in the useful frequency range is several times smaller than f_r for a typical input–output device (e.g., $f_{max}=0.25f_r$). The useful frequency range may also be determined by specifying the flatness of the static portion of the frequency response curve. For example, since a single pole or a single zero introduces a slope of approximately ±20 dB/decade to the Bode magnitude curve of the device, a slope within 5% of this value (i.e., ±1 dB/decade) may be considered flat for most practical purposes. For a measuring instrument, for example, operation in the useful frequency range implies that the significant frequency content of the measured signal is limited to this band. Then, faithful measurement and fast response are guaranteed, because dynamics of the measuring device will not corrupt the measurement.

6.2.1.6 Instrument Bandwidth

This is a measure of the *useful frequency range* of operation of an instrument. Furthermore, the larger the bandwidth of the device, the faster the speed of response. Unfortunately, the larger the bandwidth, the more susceptible the instrument will be to high-frequency noise as well as stability problems. Filtering will be needed to eliminate unwanted noise. Stability can be improved by dynamic compensation. Common definitions of instrument bandwidth include the frequency range over which the transfer-function magnitude is flat, the resonant frequency, and the frequency at which the transfer-function magnitude drops to $1/\sqrt{2}$ (or 70.7%) of the zero-frequency (or static) level. As noted earlier, the last definition corresponds to the *half-power bandwidth*, because a reduction of the amplitude level by a factor of $\sqrt{2}$ corresponds to a power drop by a factor of 2.

6.2.1.7 Control Bandwidth

This is used to specify the maximum possible *speed of control*. It is an important specification in both analog control and digital control. In digital control, the data sampling rate (in samples per second) has to be several times higher than the control bandwidth (in hertz or Hz) so that sufficient data would be available to compute the control action. Moreover, from *Shannon's sampling theorem*, control bandwidth is given by half the rate at which the control action is computed (see later under Section 6.3). The control bandwidth provides the frequency range within which a system can be controlled (assuming that all the devices in the system can operate within this bandwidth).

Example 6.1

Consider the speed control system that is schematically shown in Figure 6.3. Suppose that the plant and the controller together are approximated by the transfer function $G_p(s)=k/(\tau_p s+1)$, where τ_p is the plant time constant.

(a) Give an expression for the bandwidth ω_p of the plant, in the absence of feedback.
(b) If the feedback tachometer is ideal and is represented by a unity (negative) feedback, what is the bandwidth ω_c of the feedback control system?
(c) If the feedback tachometer can be represented by the transfer function $G_s(s)=1/(\tau_s s+1)$, where τ_s is the sensor time constant, explain why the bandwidth ω_{cs} of the feedback control system is given by the smaller quantity of $1/\tau_s$ and $(k+1)/(\tau_p+\tau_s)$. Assume that both τ_p and τ_s are sufficiently small.

Now suppose that approximately $\tau_p=0.016$ s. Estimate a sufficient bandwidth in hertz for the tachometer. Additionally, if $k=1$, estimate the overall bandwidth of the feedback control system.

If $k=49$, what would be the representative bandwidth of the feedback control system?

For the particular AC tachometer (with the bandwidth value as chosen in the present numerical example), what should be the frequency of the carrier signal? In addition, what should be the cutoff frequency of the low-pass filter that is used with its demodulator circuit?

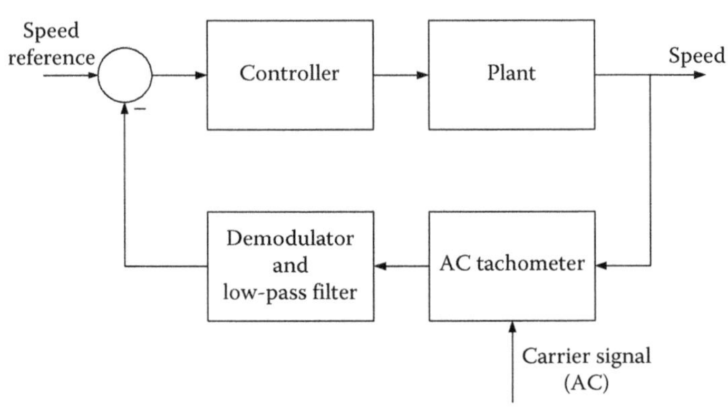

FIGURE 6.3
A speed control system.

Solution

(a) $G_p(s) = \dfrac{k}{(\tau_p s+1)} \rightarrow \omega_p = \dfrac{1}{\tau_p}$

(b) With unity feedback, the closed-loop transfer function is $G_c(s) = \dfrac{k/(\tau_p s+1)}{1+k/(\tau_p s+1)}$, which simplifies to $G_c(s) = \dfrac{k}{(\tau_p s+1+k)} \rightarrow \omega_c = \dfrac{1+k}{\tau_p}$.

Note: The bandwidth has increased.

(c) With feedback sensor of transfer function $G_s(s) = 1/(\tau_s s+1)$, the closed-loop transfer function is

$$G_{cs}(s) = \frac{k/(\tau_p s+1)}{1+k/\{(\tau_p s+1)(\tau_s s+1)\}} = \frac{k(\tau_s s+1)}{\tau_p\tau_s s^2+(\tau_p+\tau_s)s+1+k} \cong \frac{k(\tau_s s+1)}{(\tau_p+\tau_s)s+1+k}$$

Note: We have neglected $\tau_p\tau_s$ since both parameters in the product are relatively small.

Hence, to avoid the dynamic effect of the sensor (which has introduced a zero in $G_{cs}(s)$), we should limit the bandwidth to $1/\tau_s$.

Additionally, from the denominator of G_{cs}, it is seen that the closed-loop bandwidth is given by $1+k/(\tau_p+\tau_s)$.

Hence, for satisfactory performance, the bandwidth has to be limited to

$$\min\left[\frac{1}{\tau_s}, \frac{1+k}{(\tau_p+\tau_s)}\right].$$

With $\tau_p = 0.016$ s, we have

$$\omega_p = \frac{1}{0.016} = 62.5 \text{ rad/s} = 10.0 \text{ Hz}$$

Hence, pick a sensor bandwidth of 10 times this value $\rightarrow \omega_s = 100.0$ Hz $= 625.0$ rad/s.

Then $\tau_s = 1/\omega_s = 0.0016$ s. With $k=1$,

$$\frac{1+k}{(\tau_p+\tau_s)} = \frac{(1+1)}{(0.016+0.0016)} \text{rad/s} = 18.0 \text{ Hz}.$$

Also,

$$\frac{1}{\tau_s} = 100.0 \text{ Hz} \rightarrow \omega_{cs} \cong \min[100, 18.0]\text{Hz} = 18.0 \text{ Hz}$$

With $k=49$, $\dfrac{1+k}{(\tau_p+\tau_s)} = \dfrac{1+49}{(0.016+0.0016)}$ rad/s $= 450.0$ Hz, and as before, $1/\tau_s = 100.0$ Hz

$$\rightarrow \omega_{cs} = \min\,[100, 450.0] \text{ Hz} = 100.0 \text{ Hz}.$$

It follows that now the control system bandwidth has increased to about 100 Hz (possibly somewhat lower than 100 Hz).

For a sensor with 100 Hz bandwidth (see Sections 4.2.3 and 8.4.2),

$$\text{Carrier frequency} \cong 10 \times 100 \text{ Hz} = 1000.0 \text{ Hz}$$

$$\rightarrow 2 \times \text{carrier frequency} = 2000 \text{ Hz}$$

$$\rightarrow \text{Low-pass filter cutoff} = (1/10) \times 2000 \text{ Hz} = 200.0 \text{ Hz}$$

6.2.2 Static Gain

This is the gain (i.e., transfer-function magnitude) of a device (e.g., measuring instrument) within the useful (flat) range (or at very low frequencies) of the device. It is also termed *DC gain*. A high value for static gain results in a high-sensitivity device, which is a desirable characteristic. A high gain value increases the output level and can increase the speed of response and reduce the steady-state error in a feedback control system. But it has the undesirable effect of making the system less stable.

Example 6.2

A mechanical device for measuring angular velocity is shown in Figure 6.4. The main element of this tachometer is a rotary viscous damper (damping constant b) consisting of two cylinders. The outer cylinder carries a viscous fluid within which the inner cylinder rotates. The inner cylinder is connected to the shaft whose speed ω_i is to be measured. The outer cylinder is resisted by a linear torsional spring of stiffness k. The rotation θ_o of the outer cylinder is indicated by a pointer on a suitably calibrated scale. Neglecting the inertia of the moving parts, perform a bandwidth analysis for this device.

Solution

The damping torque is proportional to the relative velocity of the two cylinders and is resisted by the spring torque. The equation of motion is given by $b(\omega_i - \dot{\theta}_o) = k\theta_o$, or

$$b\dot{\theta}_o + k\theta_o = b\omega_i \tag{i}$$

The transfer function is determined by first replacing the time derivative by the Laplace operator s and taking the ratio output/input:

$$\frac{\theta_o}{\omega_i} = \frac{b}{[bs + k]} = \frac{b/k}{[(b/k)s + 1]} = \frac{k_g}{[\tau s + 1]} \tag{ii}$$

The static gain (i.e., DC gain; transfer-function magnitude at $s = 0$) is

$$k_g = \frac{b}{k} \tag{iii}$$

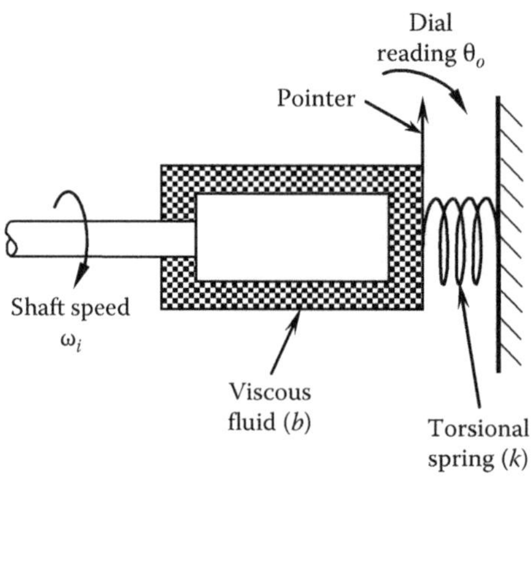

FIGURE 6.4
A mechanical tachometer.

and the time constant is

$$\tau = \frac{b}{k} \tag{iv}$$

It is seen that in this device, the static gain and the time constant are equal (and hence we have only one performance parameter). The design requirements of *speed* (which decreases with the time constant) and the *output level* (which increases with the static gain) become conflicting for this reason. On the one hand, we want to have a large static gain so that a sufficiently large reading is provided by the sensor. On the other hand, the time constant of the device must be small in order to obtain a quick reading that faithfully follows the measured variable (speed). A compromise must be reached here, depending on the specific design requirements. Alternatively, or in addition, a signal-conditioning device may be employed to amplify the sensor output.

Note: In this example, the speed and the *level of stability* are not conflicting (both improve when the time constant is decreased).

Now, let us examine the half-power bandwidth of the device. The FTF is

$$G(j\omega) = \frac{k_g}{\tau j\omega + 1} \tag{v}$$

By definition, the half-power bandwidth ω_b is given by

$$\frac{k_g}{|\tau j\omega_b + 1|} = \frac{k_g}{\sqrt{2}}.$$

Hence,

$$(\tau\omega_b)^2 + 1 = 2$$

As both τ and ω_b are positive, we have $\tau\omega_b = 1$, or

$$\tau = \frac{1}{\omega_b} \tag{vi}$$

Note that the bandwidth is inversely proportional to the time constant. This confirms our earlier assertion that bandwidth is a measure of the speed of response of a device.

6.3 Aliasing Distortion due to Signal Sampling

Aliasing distortion, also known as sampling error, arises due to sampling (discretization) of analog (continuous) data. Hence, it is an important source of error in a system that uses sampled data. Such systems include those that employ analog-to-digital converters (ADCs), which include digital devices or systems such as microcontrollers and various types of digital computers that are used in process monitoring control, and pulse-generating sensor–transducer devices such as shaft encoders (see Chapter 11). Sampling error may enter into computation in both time domain and frequency domain, depending on the domain in which the data are sampled.

6.3.1 Sampling Theorem

If a time signal $x(t)$ is sampled at equal steps of ΔT, information regarding its frequency spectrum $X(f)$ is lost for frequencies higher than $f_c = 1/(2\Delta T)$. This fact is known as "Shannon's sampling theorem," and the limiting (cutoff) frequency f_c in the spectrum (of the sampled data) is called the "Nyquist frequency," which is half the sampling frequency $f_s = 1/(\Delta T)$.

It can be shown that the *aliasing error* is caused by folding of the high-frequency segment of the frequency spectrum beyond the Nyquist frequency onto the low-frequency segment. This is illustrated in Figure 6.5. It should be clear that the aliasing error becomes more and more prominent for frequencies of the spectrum closer to the Nyquist frequency. In signal analysis, a sufficiently small sample step ΔT should be chosen in order to reduce the aliasing distortion in the frequency domain, depending on the highest frequency of interest in the analyzed signal. This, however, increases the number of data samples and hence the signal processing time and the computer storage requirements, which are undesirable particularly in real-time data processing (e.g., in process monitoring and control). It can also result in stability problems in numerical computations. The Nyquist sampling criterion requires that the sampling rate $(1/\Delta T)$ for a signal should be at least twice the highest frequency of interest. Instead of making the sampling rate very high, a moderate value that satisfies the Nyquist sampling criterion is used in practice, together with an *anti-aliasing filter* to remove the frequency components in the original signal that would distort the spectrum of the computed signal.

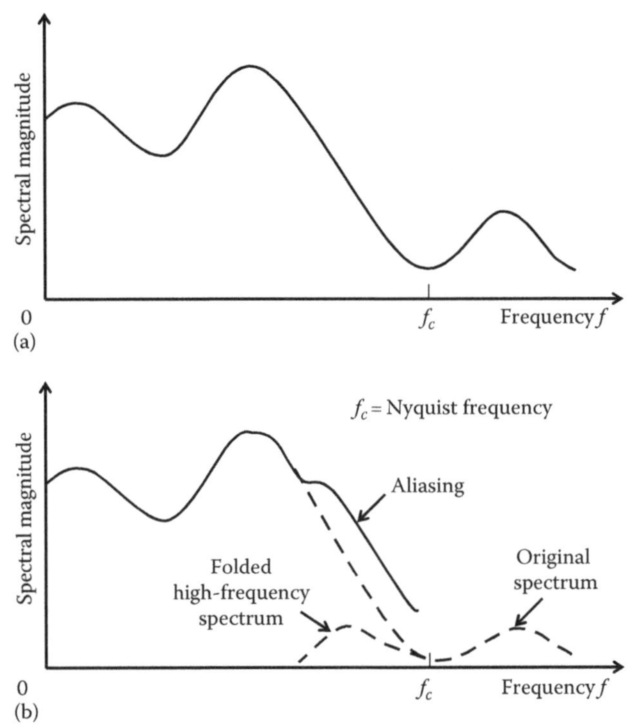

FIGURE 6.5
Aliasing distortion of a frequency spectrum: (a) original spectrum and (b) distorted spectrum due to aliasing.

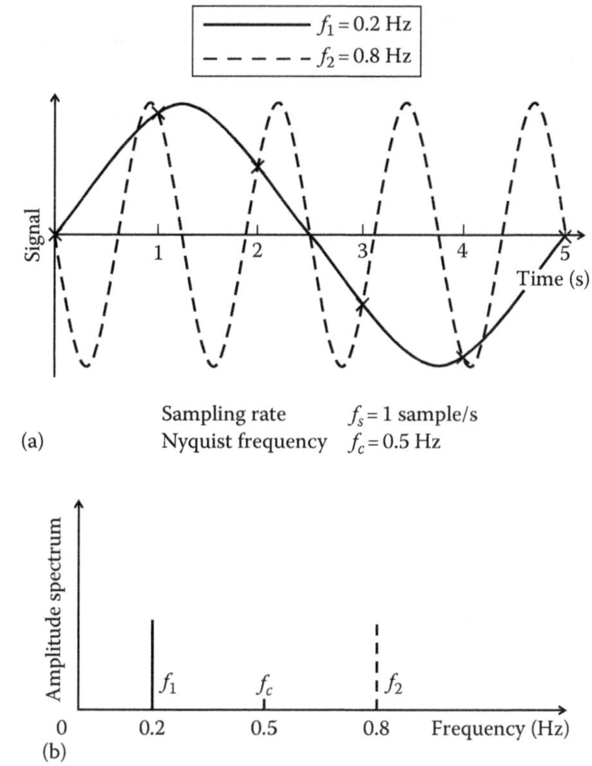

FIGURE 6.6
A simple illustration of aliasing: (a) two harmonic signals with identical sampled data and (b) frequency spectra of the two harmonic signals.

6.3.2 Simple Illustration of Aliasing

A simple illustration of aliasing is given in Figure 6.6. Here, two sinusoidal signals of frequency $f_1 = 0.2$ Hz and $f_2 = 0.8$ Hz are shown (Figure 6.6a). Suppose that the two signals are sampled at the rate of $f_s = 1$ sample/s. The corresponding Nyquist frequency is $f_c = 0.5$ Hz. It is seen that, at this sampling rate, the data samples from the two signals are identical. In other words, from the sampled data the high-frequency signal cannot be distinguished from the low-frequency signal. Hence, a high-frequency signal component of frequency 0.8 Hz will appear as a low-frequency signal component of frequency 0.2 Hz. This is indeed aliasing, as clear from the signal spectrum shown in Figure 6.6b. Specifically, the spectral segment of the signal beyond the Nyquist frequency (f_c) folds onto the low-frequency side due to data sampling and cannot be recovered.

6.3.3 Anti-Aliasing Filter

It should be clear from Figure 6.5 that if the original signal is low-pass filtered, using an ideal filter, at a cutoff frequency equal to the Nyquist frequency, then the aliasing distortion caused by sampling would not occur. A filter of this type is called an anti-aliasing filter. Analog hardware filters may be used for this purpose. In practice, however, it is not possible to achieve perfect filtering. Hence, some aliasing would remain even after using an anti-aliasing filter, further reducing the valid frequency range of the computed signal. Typically, the useful frequency limit is $f_c/1.28$, and the last 20% of the spectral points near the Nyquist frequency should be neglected. Hence, the filter cutoff frequency is chosen to be somewhat lower than the Nyquist frequency, for example, $f_c/1.28$ ($\cong 0.8f_c$). In this case, the computed spectrum is accurate up to the filter cutoff frequency $0.8\,f_c$ and not up to the Nyquist frequency f_c.

Example 6.3

Consider 1024 data points from a signal, sampled at 1 ms intervals.

$$\text{Sample rate } f_s = 1/0.001 \text{ samples/s} = 1000 \text{ Hz} = 1 \text{ kHz}$$

$$\text{Nyquist frequency} = 1000/2 \text{ Hz} = 500 \text{ Hz}$$

Because of aliasing, approximately 20% of the spectrum even in the theoretically useful range (i.e., spectrum beyond 400 Hz) will be distorted. Here, we may use an anti-aliasing filter with a cutoff at 400 Hz.

The digital Fourier transform computation of this set of 1024 time data points provides 1024 frequency spectrum data points (spectral lines) up to 1000 Hz. Half of this number of spectral lines is beyond the Nyquist frequency and will not give any new information about the signal spectrum.

$$\text{Spectral line separation} = 1000/1024 \text{ Hz} = 1 \text{ Hz (approx)}$$

We keep only the first 400 spectral lines as the useful spectrum.

Note: Almost 500 spectral lines may be retained if an accurate anti-aliasing filter with its cutoff frequency at 500 Hz is used.

Example 6.4

(a) Suppose that a sinusoidal signal of frequency f_1 Hz was sampled at the rate of f_s samples per second. Another sinusoidal signal of the same amplitude, but of a higher frequency f_2 Hz, was found to yield the same data when sampled at f_s. What is the likely analytical relationship between f_1, f_2, and f_s?

(b) Consider a plant of transfer function $G(s) = k/(1+\tau s)$. What is the static gain of this plant? Show that the magnitude of the transfer function reaches $1/\sqrt{2}$ of the static gain when the excitation frequency is $1/\tau$ rad/s. *Note*: The frequency $\omega_b = 1/\tau$ rad/s may be taken as the operating bandwidth of the plant.

(c) Consider a chip refiner that is used in the pulp and paper industry. The machine is used for mechanical pulping of wood chips. It has a fixed plate and a rotating plate, driven by an induction motor. The gap between the plates is sensed and is adjustable as well. As the plate rotates, the chips are ground into a pulp within the gap. A block diagram of the plate-positioning control system is shown in Figure 6.7.

Suppose that the torque sensor signal and the gap sensor signal are sampled at 100 and 200 Hz, respectively, into the digital controller, which takes 0.05 s to compute each positioning command for the servovalve. The time constant of the servovalve is $0.05/2\pi$ s, and for the mechanical load (plant), the time constant is $0.2/2\pi$ s. Estimate the control bandwidth and the operating bandwidth of the positioning system.

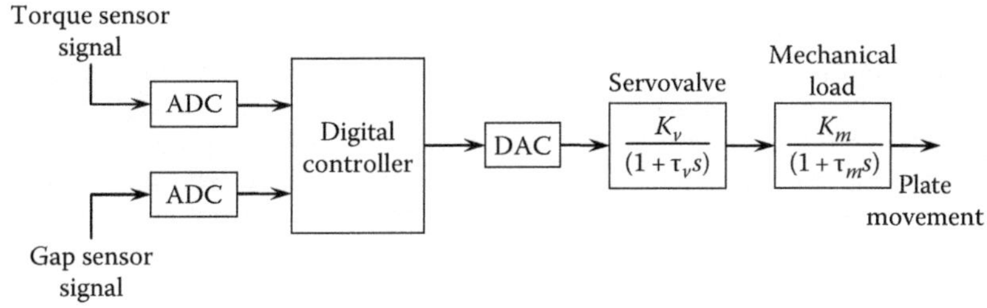

FIGURE 6.7
Block diagram of the plate-positioning control system for a chip refiner.

Solution

(a) It is likely that f_1 and f_2 are symmetrically located on either side of the Nyquist frequency f_c. Then, $f_2 - f_c = f_c - f_1$. This gives $f_2 = f_c + (f_c - f_1) = 2f_c - f_1$, or

$$f_2 + f_1 = f_s = 2f_c \tag{6.2}$$

(b) $G(j\omega) = \dfrac{k}{(1+\tau j\omega)}$ = frequency transfer function, where ω is in rad/s.

Static gain is the transfer-function magnitude at steady state (i.e., at zero frequency).

Hence, static gain $= G(0) = k$.

When $\omega = 1/\tau$, $G(j\omega) = k/(1+j) \rightarrow |G(j\omega)| = k/\sqrt{2}$ at this frequency.

This corresponds to the half-power bandwidth.

(c) Because of sampling, the torque signal has a bandwidth of $(1/2)\times 100$ Hz = 50 Hz, and the gap sensor signal has a bandwidth of $(1/2)\times 200 = 100$ Hz.

Control cycle time = 0.05 s, which generates control signals at a rate of $1/0.05 = 20$ Hz.

Since 20 Hz < min (50 Hz, 100 Hz), we have adequate bandwidth from the sampled sensor signals to accurately compute the control signal.

$$\text{Control bandwidth provided by the digital controller} = \frac{1}{2}\times 20\text{ Hz} = 10\text{ Hz}$$

(from Shannon's sampling theorem).

But the servovalve is also part of the controller (analog hardware).

$$\text{Its bandwidth} = \frac{1}{\tau_v}\text{rad/s} = \frac{1}{2\pi\tau_v}\text{Hz} = \frac{2\pi}{2\pi\times 0.05}\text{Hz} = 20\text{ Hz}.$$

Control bandwidth is limited by both digital control bandwidth (10 Hz) and analog hardware control bandwidth (20 Hz). Hence,

$$\text{Control bandwidth} = \min\,(10\text{ Hz}, 20\text{ Hz}) = 10\text{ Hz}$$

Bandwidth of the mechanical load (plant) is

$$\text{Bandwidth of the mechanical load (plant)} = \frac{1}{\tau_m}\text{rad/s} = \frac{1}{2\pi\tau_m}\text{Hz} = \frac{2\pi}{2\pi\times 0.2}\text{Hz} = 5\text{ Hz}.$$

Operating bandwidth is limited by both control bandwidth (10 Hz) and the plant bandwidth (5 Hz). Hence,

Operating bandwidth of the system = min (10 Hz, 5 Hz) = 5 Hz.

Example 6.5

Consider the digital control system for a mechanical position application, as schematically shown in Figure 6.8. The control computer generates a control signal according to an algorithm, on the basis of the desired position and actual position, as measured by an optical encoder (see Chapter 11). This digital signal is converted into the analog form using a digital-to-analog converter and is supplied to the drive amplifier. Accordingly, the current signals needed to energize the motor windings are generated by the amplifier. The inertial element, which needs to be properly moved to the correct location, is directly (and rigidly) linked to the motor rotor and is resisted by a spring and a damper, as shown.

Suppose that the combined transfer function of the drive amplifier and the electromagnetic circuit (torque generator) of the motor is given by

$$\frac{k_e}{\left(s^2 + 2\zeta_e\omega_e s + \omega_e^2\right)}$$

and the transfer function of the mechanical system including the inertia of the motor rotor is given by $k_m/(s^2 + 2\zeta_m\omega_m s + \omega_m^2)$. Here, k = equivalent gain, ζ = damping ratio, and ω = natural frequency. The subscripts $(\,)_e$ and $(\,)_m$ denote the electrical components and mechanical components, respectively. Moreover,

ΔT_c is the time taken to compute each control action
ΔT_p is the pulse period of the position sensing encoder

The following numerical values are

$$\omega_e = 1000\pi\,\text{rad/s},\ \zeta_e = 0.5,\ \omega_m = 100\pi\,\text{rad/s, and}\ \zeta_m = 0.3.$$

For the purpose of this example, you may neglect the loading effects and coupling effects that arise from component cascading, dynamic interaction, and signal feedback.

(a) Explain why the control bandwidth of this system cannot be much larger than 50 Hz.
(b) If ΔT_c = 0.02 s, estimate the control bandwidth of the system.
(c) Explain the significance of ΔT_p in this application. Why, typically, ΔT_p should not be greater than $0.5\Delta T_c$?
(d) Estimate the operating bandwidth of the positioning system, assuming that significant plant dynamics are to be avoided.
(e) If ω_m = 500π rad/s and ΔT_c = 0.02 s, with the remaining parameters kept as specified earlier, estimate the operating bandwidth of the system, again so as not to excite significant plant dynamics.

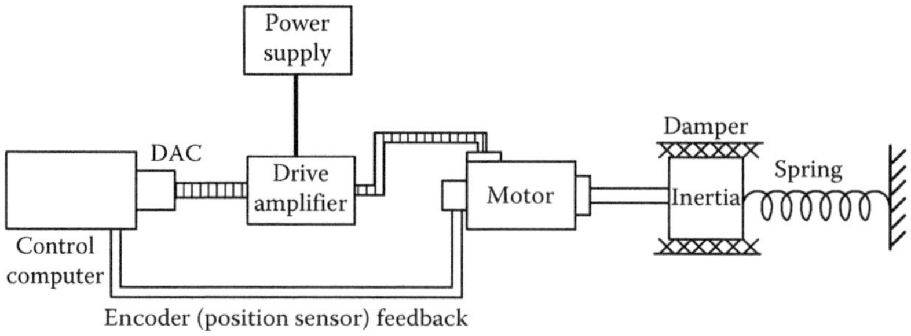

FIGURE 6.8
Digital control system for a mechanical positioning application.

Solution

(a) The drive system (electronic hardware) has a resonant frequency slightly less than 500 Hz. Hence, the flat region of the spectrum (i.e., operating region) of the drive system would be about 1/10 of this, that is, 50 Hz. This would limit the maximum useful spectral component of the drive signal to about 50 Hz. Hence, the control bandwidth (analog) would be limited by this value.

(b) Rate at which the digital control signal is generated = 1/0.02 Hz = 50 Hz. By Shannon's sampling theorem, the effective (useful) spectrum of the control signal is limited to $(1/2)\times 50$ Hz = 25 Hz. Even though the drive system can accommodate a bandwidth of about 50 Hz, the control bandwidth would be limited to 25 Hz, because of the lower bandwidth of the digital controller, in this case.

(c) Note that ΔT_p is the sampling period of the measurement signal (for feedback). Hence, its useful spectrum would be limited to $1/(2\Delta T_p)$, by Shannon's sampling theorem. Consequently, the feedback signal will not be able to provide any useful information of the process beyond the frequency $1/(2\Delta T_p)$. To generate a control signal at the rate of $1/\Delta T_c$ samples per second, the process information has to be provided at least up to $1/\Delta T_c$ Hz. To provide this information, we must have

$$\frac{1}{2\Delta T_p} \geq \frac{1}{\Delta T_c} \text{ or } \Delta T_p \leq 0.5\,\Delta T_c \tag{6.3}$$

Here, ΔT_c is the time taken to compute each control action.

This guarantees that at least two points of sampled data from the sensor are used for computing each control action.

(d) The resonant frequency of the plant (positioning system) is approximately (less than)

$$\frac{100\pi}{2\pi}\text{Hz} \approx 50\text{ Hz}.$$

At frequencies near this, the resonance will interfere with control and should be avoided if possible, unless the resonances (or modes) of the plant themselves need to be modified through control. At frequencies much larger than this, the process will not significantly respond to the control action and will not be of much use (the plant would be felt like a rigid wall). Hence, the operating bandwidth has to be sufficiently smaller than 50 Hz, say 25 Hz, in order to avoid significant plant dynamics.

Note: This is a matter of design judgment, based on the nature of the application (e.g., excavator, disk drive). Typically, however, one needs to control the plant dynamics. In that case, it is necessary to use the entire control bandwidth (i.e., maximum possible control speed) as the operating bandwidth. In the present case, even if the entire control BW (i.e., 25 Hz) is used as the operating BW, it still avoids (is much smaller than) the plant resonance.

(e) The plant resonance in this case is about $500\pi/(2\pi)$ Hz = 250 Hz. This limits the operating bandwidth to about $250\pi/(2\pi)$ Hz = 125 Hz, so as to avoid significant plant dynamics. But as determined earlier, the control bandwidth is about 25 Hz because $\Delta T_c = 0.02$ s. Hence, the operating bandwidth cannot be greater than this value and would be; 25 Hz.

Significance of Sensor Location: As a comment that is not related to the questions of this example, let us consider the location of the motion sensor (encoder). Since the encoder is integral with the motor, it measures the motion of the motor, not the driven load. When the flexibility of the shaft that connects the motor to the load is not negligible or if there is speed transmission unit between the motor and the load, the sensor does not read the actual motion of the load. If the corresponding error is not negligible, we must consider locating a motion sensor at the load. Then, however, we are moving the sensor away from the drive location. It is known that, in feedback control, moving the sensor location farther from the drive point can make the system less stable. In this manner, there can arise a trade-off between the level of stability and the accuracy of motion, which is governed by the location of the motion sensor.

6.3.4 Bandwidth Design of a Control System

Based on the concepts that have been introduced thus far, it is now possible to give a set of simple steps for designing a control system on the basis of bandwidth considerations.

- *Step 1*: Decide on the maximum frequency of operation (operating bandwidth BW_o) of the system based on the requirements of the particular application.
- *Step 2*: Select the process components (electromechanical) that have the capacity to properly operate at least up to the frequency BW_o in performing the required tasks.
- *Step 3*: Select feedback sensors with a flat frequency spectrum (sensor bandwidth or operating frequency range) greater than $4 \times BW_o$. *Note*: Typically, sensor bandwidth is not a limiting factor. So, we may even pick larger sensor BW.
- *Step 4*: Develop a digital controller with (a) sampling rate $> 4 \times BW_o$ for the feedback sensor signals (i.e., the useful spectrum sampled signal falls well within the sensor bandwidth and the process operating bandwidth by a factor of 2, in the present case) and (b) direct digital control cycle time (period) of $1/(2 \times BW_o)$. *Note*: Digital control actions are generated at a rate of $2 \times BW_o$, thereby providing a digital control bandwidth that matches the system operating bandwidth (BW_o).
- *Step 5*: Select the control drive system (analog devices such as interface analog hardware, filters, amplifiers, and actuators) that have a flat frequency spectrum of at least BW_o (preferably, $2 \times BW_o$) giving an analog (hardware) control bandwidth of at least BW_o.
- *Step 6*: Integrate the system and test the performance. If the performance specifications are not satisfied, make necessary adjustments and test again.

Clearly, a control system should not be designed using bandwidth considerations alone. Many other considerations that depend on the particular application, performance requirements, and constraints have to be considered, and appropriate control techniques have to be employed.

6.3.4.1 Comment about Control Cycle Time

In the engineering literature, it is often used that $\Delta T_c = \Delta T_p$, where ΔT_c is the control cycle time (period at which the digital control actions are generated) and ΔT_p is the period at which the feedback sensor signals are sampled (see Figure 6.9a).

This is acceptable in systems where the useful frequency range of the plant is considerably smaller than $1/\Delta T_p$ (and $1/\Delta T_c$). Then, the sampling rate $1/\Delta T_p$ of the feedback measurements

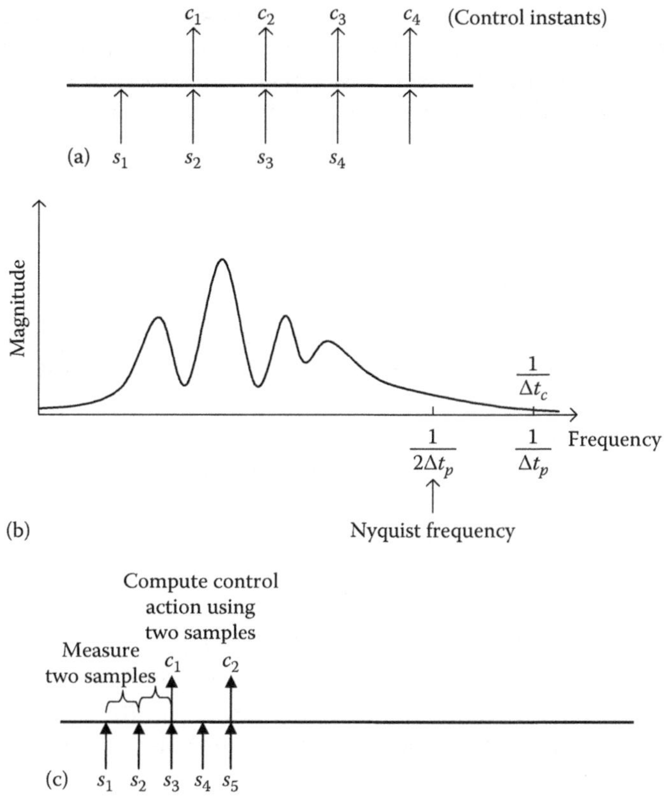

FIGURE 6.9
(a) Conventional sampling of feedback sensor signals for direct digital control. (b) Acceptable frequency characteristic of a plant for the top case. (c) Improved sampling criterion for feedback signals in direct digital control.

(and the Nyquist frequency $0.5/\Delta T_p$) will still be sufficiently larger than the useful frequency range (or the operating bandwidth) of the plant (see Figure 6.9b), and hence, the sampled data will accurately represent the plant response. But the bandwidth criterion presented earlier in this section satisfies $\Delta T_p \le 0.5\Delta T_c$ (see Equation 6.3). It can be argued that this is a more desirable choice. For example, in Figure 6.9c, the two previous measurement samples are used in computing each control action, and the control computation occupies twice the data sampling period. Here, the data sampling period is half the control cycle period, and for a specified control action frequency, the Nyquist frequency of the sampled feedback signals is double that of the previous, conventional approach. As a result, the sampled data will cover a larger (double) frequency range of the plant. Typically, in practice, having a higher sampling rate for the sensor is not a limitation because fast sensors and data acquisition systems with very fast sampling rates are commonly available. Of course, a third option would be to still use the two previous data samples to compute a control action, but do the computation faster, in one data sample period (rather than two). This option will use one previously used data sample and one new data sample (unlike in the previous option, which uses two new data samples). Hence, this third option will require increased processing power as well as a larger buffer for storing the sampled data for control action computation.

Note: One may argue that in some control schemes (e.g., proportional control), you need only one data sample to compute the control action. However, even in such control schemes, having more than one latest data samples can improve the measurement accuracy (e.g., simply by taking the local average, to reduce random noise).

6.4 Instrument Error Considerations

Analysis of error in an instrument or a multicomponent engineering system is a challenging task. Difficulties arise for many reasons, particularly the following:

1. The true value is usually unknown.
2. The instrument reading may contain random error (error of the measuring system, including sensor error, and other random errors that enter into the engineering system, including external disturbance inputs) that cannot be determined exactly.
3. The error may be a complex (i.e., not simple) function of many variables (input variables and state variables or response variables).
4. The monitored engineering system may be made up of many components that have complex interrelations (dynamic coupling, multiple degree-of-freedom responses, nonlinearities, etc.), and each component may contribute to the overall error.

The first item is a philosophical issue that would lead to an argument similar to the chicken-and-egg controversy. For instance, if the true value is known, there is no need to measure it; and if the true value is unknown, it is impossible to determine exactly how inaccurate a particular reading is. In fact, this situation can be addressed to some extent by using statistical representation of error, probability theory, and estimation (see Chapter 4), which takes us to the second item listed. The third and fourth items may be addressed by methods of *error combination* in multivariable systems and by *error propagation* in complex multicomponent systems. It is not feasible here to provide a full treatment of all these topics. Only an introduction to a useful analytical technique is given, using illustrative examples.

6.4.1 Error Representation

The error in a measurement (data), measurement error, and instrument (say, measuring instrument) error are not the same thing. The error in a piece of data is the difference between its value and the true (correct) value. It can come from many sources even before measuring it. Measurement errors enter through the process of measurement including the device (sensor) that is used for the measurement. Instrument error is the error in the particular instrument. These concepts should be clear from the following discussion.

In general, the error in an instrument reading is a random variable. Regardless of what different factors contribute to this error, it is defined as

$$\text{Error} = (\text{Instrument reading}) - (\text{True value})$$

Randomness associated with a *measurand* (the quantity to be measured) can be interpreted in two ways. First, since the true value of the measurand is a fixed quantity, randomness can be interpreted as the randomness in error that is usually originating from the random factors in instrument response. Second, looking at the issue in a more practical manner, error analysis can be interpreted as an *estimation problem* in which the objective is to estimate the true value of a measurand from a known set of readings. In this latter point of view, *estimated value* itself becomes a random variable. No matter what approach is used, the same statistical concepts may be used in representing error.

6.4.1.1 Instrument Accuracy and Measurement Accuracy

Various instrument ratings as discussed earlier determine the overall *accuracy* of an instrument. *Instrument accuracy* is represented by the worst accuracy level generated by the instrument within its dynamic range in a normal operating environment. Instrument accuracy depends not only on the physical hardware of the instrument but also on its calibration, actual operating conditions (power, signal levels, load, speed, etc.; environmental conditions, etc.), design operating conditions (operating conditions for which the instrument is designed for: normal, steady operating conditions; extreme transient conditions, such as emergency start-up; and shutdown conditions), shortcomings of how the instrument is set up, other components and systems to which the instrument is connected, undesirable external inputs and disturbances, and so on.

Accuracy can be assigned either to a particular reading or to an instrument. *Measurement accuracy* determines the closeness of the measured value (*measurement*) to the true value (*measurand*). It depends not only on the instrument accuracy but also on how the measurement process is conducted, how the measured data are presented (communicated, displayed, stored, etc.), and so on.

Note: The error in a measurement depends not only on the measuring device and how the measurement was conducted but also on other factors, particularly errors that enter into the measurand before it is sensed by the measuring device. These can include noise, disturbances, and other errors in the system that generates the measurand.

Measurement error is defined as

$$\text{Error} = (\text{Measured value}) - (\text{True value}) \tag{6.4}$$

The correction, which is the negative of error, is defined as

$$\text{Correction} = (\text{True value}) - (\text{Measured value}) \tag{6.5}$$

Each of these can also be expressed as a percentage of the true value. The accuracy of an instrument may be determined by measuring a parameter whose true value is known, near the extremes of the dynamic range of instrument, under certain operating conditions. For this purpose, standard parameters or signals than can be generated at very high levels of accuracy would be needed. The National Institute for Standards and Testing or National Research Council is usually responsible for generating these standards. Nevertheless, accuracy and error values cannot be determined to 100% exactness in typical applications, because the true value is not known to begin with. In a given situation, we can only make estimates for accuracy, by using the ratings provided by the instrument manufacturer or by analyzing data from previous measurements and models.

In general, causes of error in an engineering system (having interconnected and interacting multiple components) include instrument instability, external noise (disturbances), poor calibration, inaccurately generated information (e.g., inaccurate sensors, poor analytical models, inaccurate control laws), parameter changes (e.g., as a result of environmental changes, aging, and wear-out), unknown nonlinearities, and improper use of the instruments (shortcomings in the system setup, improper and extreme operating conditions, etc.).

6.4.1.2 Accuracy and Precision

Errors can be classified as *deterministic* (or *systematic*) and *random* (or *stochastic*). Deterministic errors are those caused by well-defined factors, including known nonlinearities and offsets in readings. These usually can be accounted for by proper calibration, testing, analysis and computational practices, and compensating hardware. Error ratings and calibration charts are commonly used to remove systematic errors from instrument readings. Random errors are caused by uncertain factors entering into instrument response. These include device noise, line noise, random inputs, and effects of unknown random variations in the operating environment. A statistical analysis using sufficiently large amounts of data is necessary to estimate random errors. The results are usually expressed as a *mean error*, which is the systematic part of random error, and a *standard deviation* or *confidence interval* for instrument response.

6.4.1.2.1 Precision

Precision is not synonymous with accuracy. Reproducibility (or repeatability) of an instrument reading determines the precision of an instrument. An instrument that has a high offset error might be able to generate a response at high precision, even though this output is clearly inaccurate. For example, consider a timing device (clock) that very accurately indicates time increments (say, up to the nearest nanosecond). If the reference time (starting time) is set incorrectly, the time readings will be in error, even though the clock has a very high precision.

Instrument error may be represented by a random variable that has a mean value μ_e and a standard deviation σ_e. If the standard deviation is zero, the variable is considered deterministic for most practical purposes. In that case, the error is said to be deterministic or *repeatable*. Otherwise, the error is said to be random. The precision of an instrument is determined by the standard deviation of the error in the instrument response. Readings of an instrument may have a large mean value of error (e.g., large offset), but if the standard deviation is small, the instrument has high precision. Hence, a quantitative definition for precision would be

$$\text{Precision} = \frac{(\text{Measurement range})}{\sigma_e} \tag{6.6}$$

Lack of precision originates from random causes and poor construction practices. It cannot be compensated for by recalibration, just as the precision of a clock cannot be improved by resetting the time. On the other hand, accuracy can be improved by recalibration. Repeatable (deterministic) accuracy is inversely proportional to the magnitude of the mean error μ_e.

Note: A device with low systematic (deterministic) error may not be precise if it has high zero-mean random error.

Matching instrument ratings with specifications is very important in selecting instruments for an engineering application. Several additional considerations should be looked into as well. These include geometric limitations (size, shape, etc.), environmental conditions (e.g., chemical reactions including corrosion, extreme temperatures, light, dirt accumulation, humidity, electromagnetic fields, radioactive environments, shock and vibration), power requirements, operational simplicity, availability, past record and reputation of the manufacturer and of the particular instrument, and cost-related economic aspects (initial cost, maintenance cost, cost of supplementary components such as signal-conditioning and processing devices, design life and associated frequency of replacement, and cost of disposal and replacement). Often, these considerations become the ultimate deciding factors in the selection process.

6.5 Error Propagation and Combination

System error is perhaps the most critical performance indicator in any engineering system. Analytically, error = actual value − correct (desired or ideal) value. Correction is the negative error, and it is the value that should be added to the actual value to obtain the correct value.

The overall error in a response variable (output) of a device or in an estimated parameter of a multicomponent dynamic system would depend on the errors present in (1) components (their variables and parameters) and how they interact and (2) measured variables or parameters (of individual components, etc.) that are used to compute (estimate) or determine the required quantity (variable or parameter value). Knowing how component errors are propagated within a multicomponent system and how the individual errors in the system variables and parameters contribute toward the overall error in a particular response variable or parameter would be important in estimating the error limits in a complex engineering system.

As an example, suppose that the output power of a gas turbine is computed by measuring the torque and the angular speed of the output shaft of the turbine and taking the product of the two measurements. Then, error margins in the two measured variables (torque and speed) will be directly combined into the error in the power computation. As another example, if the natural frequency of a vehicle suspension system is determined by measuring the parameters of mass and spring stiffness of the suspension, the natural frequency estimate will be directly affected by the possible errors in the mass and stiffness measurements. As a further example, in a robotic manipulator, the accuracy of the actual trajectory of the end effector will depend on the accuracy of the sensors and actuators at the manipulator joints and on the accuracy of the robot controller. Generalizing this idea, the overall error in an engineering system depends on the individual error levels in various components (sensors, actuators, controller hardware, filters, amplifiers, data acquisition devices, etc.) of the system and in the manner in which these components are physically interconnected and interacting.

Note that we are dealing with a generalized idea of error propagation, which considers the errors in system variables (e.g., input and output signals, such as velocity, force, torque, voltage, current, temperature, heat transfer rate, pressure, and fluid flow rate), system parameters (e.g., mass, stiffness, damping, capacitance, inductance, resistance, thermal conductivity, and viscosity), and system components (e.g., sensors, actuators, filters, amplifiers, interface hardware, control circuits, thermal conductors, and valves).

6.5.1 Application of Sensitivity in Error Combination

For the development of an analytical basis for error combination, we will use the familiar concepts of *sensitivity*. We have observed that sensitivity is applicable in several different practical situations; the following are examples:

1. It determines the output level and the gain of a component.
2. Variability of the sensitivity with the input level is an indication of nonlinearity in the device. The difference between the maximum and minimum sensitivities in the operating range is a measure of the device nonlinearity.
3. Signal-to-noise ratio may be interpreted as the ratio of the sensitivities to the desired signal and the undesirable signal.
4. Sensitivity in a control system may be used in design and control of the system, specifically, to compensate for disturbances and to determine the control signals and parameters. (*Note*: In adaptive control, e.g., the control parameter values are changed to achieve the control objectives.)

Now we specifically consider the application of sensitivity in error propagation and combination. To develop the necessary analytical basis for a device or system of interest, we start with a functional relationship of the form

$$y = f(x_1, x_2, \ldots, x_r) \tag{6.7}$$

Here, x_i are the independent system variables or parameters whose individual error components are propagated into a dependent variable (or parameter value) y, which may be the output of interest in the particular system. Determination of this functional relationship f is not always simple or straightforward, and in fact the relationship itself is a "model," which may be in error. In particular, this relationship depends on such considerations as

1. Characteristics and physics of the individual components
2. How the components are interconnected
3. Interactions (dynamic coupling) among the components
4. Inputs (desirable and undesirable) of the system

Since our objective is to make a reasonable estimate for the possible error in y due to the combined effect of the errors from x_i, an approximate functional relationship f would be adequate in most cases.

The error in a quantity (variable or parameter) changes that quantity. Hence, we will denote the error in a quantity by the differential of that quantity. Taking the differential of Equation 6.7, we get

$$\delta y = \frac{\partial f}{\partial x_1}\delta x_1 + \frac{\partial f}{\partial x_2}\delta x_2 + \cdots + \frac{\partial f}{\partial x_r}\delta x_r \tag{6.8}$$

Here, the differential terms $\delta(\cdot)$ represent errors, which are assumed "small" for analytical purposes. For those who are not familiar with differential calculus, Equation 6.8 should be interpreted as the first-order terms in a *Taylor series expansion* of Equation 6.7. The partial derivatives are evaluated at the *operating conditions* under which the error assessment is carried out. Now, rewriting Equation 6.8 in the fractional form, we get

$$\frac{\delta y}{y} = \sum_{i=1}^{r} \left[\frac{x_i}{y} \frac{\partial f}{\partial x_i} \frac{\delta x_i}{x_i} \right] \quad \text{or} \quad e_y = \sum_{i=1}^{r} \left[\frac{x_i}{y} \frac{\partial f}{\partial x_i} e_i \right] \tag{6.9}$$

where

$\delta y/y = e_y$ represents the overall (propagated) error

$\delta x_i/x_i = e_i$ represents the component error, where the errors are expressed as fractions

6.5.1.1 *Nondimensional Error Sensitivities*

The nondimensional, or fractional, representation of error in Equation 6.9 is quite appropriate. Each derivative $\partial f/\partial x_i$ represents the sensitivity of the error in x_i on the combined error in y. In our error analysis, we wish to retain the high-sensitivity factors and ignore the low-sensitivity factors. A comparison of the sensitivities for this purpose would not be realistic unless we use nondimensional forms for the sensitivities. Specifically, in Equation 6.9, the term $(x_i/y)(\partial f/\partial x_i)$ represents the nondimensional sensitivity of the error in x_i on the combined error in y. Hence, this term represents the degree of significance of the individual error component on the overall combined error.

Now, we will consider two ways to estimate the combined (propagated) error.

6.5.2 Absolute Error

Since error δx_i can be positive or negative, an upper bound for the overall error is obtained by summing the absolute values of the RHS terms in Equation 6.9. This estimate e_{ABS}, which is termed *absolute error*, is given by

$$e_{ABS} = \sum_{i=1}^{r} \left| \frac{x_i}{y} \frac{\partial f}{\partial x_i} \right| e_i \tag{6.10}$$

Note that the component error e_i and the absolute error e_{ABS} in this equation are always positive quantities. When specifying error, however, both positive and negative limits should be indicated or implied (e.g., $\pm e_{ABS}$, $\pm e_i$).

6.5.3 SRSS Error

Equation 6.10 provides a conservative (upper bound) estimate for the overall error. Since the estimate itself is not precise, it is often overly wasteful to use such a high conservatism. A nonconservative error estimate that is frequently used in practice is the *square root of sum of squares* (SRSS) error. As the name implies, this is given by

$$e_{SRSS} = \left[\sum_{i=1}^{r} \left(\frac{x_i}{y} \frac{\partial f}{\partial x_i} e_i \right)^2 \right]^{1/2} \tag{6.11}$$

This is not an upper bound estimate for error. In particular, $e_{SRSS} < e_{ABS}$ when more than one nonzero error contribution is present. The SRSS error relation is particularly suitable when component error is represented by the *standard deviation* of the associated variable or parameter value and when the corresponding error sources are independent. The underlying theoretical basis concerns independent random variables (see Appendix C). Now we present several examples of error propagation and combination.

6.5.4 Equal Contributions from Individual Errors

Using the absolute value method for error combination (e.g., Equation 6.10), we can determine the fractional error in each item x_i such that the contribution from each item to the overall error e_{ABS} would be the same. For equal error contribution from all r components, we must have

$$\left|\frac{x_1}{y}\frac{\partial f}{\partial x_1}\right|e_1 = \left|\frac{x_2}{y}\frac{\partial f}{\partial x_2}\right|e_2 = \cdots = \left|\frac{x_r}{y}\frac{\partial f}{\partial x_r}\right|e_r$$

Hence,

$$r\left|\frac{x_i}{y}\frac{\partial f}{\partial x_i}\right|e_i = e_{ABS}$$

Thus,

$$e_i = \frac{e_{ABS}}{\left(r\left|(x_i/y)(\partial f/\partial x_i)\right|\right)} \tag{6.12}$$

This represents the condition for equal error sensitivities.

The degree of importance of an error is determined by its nondimensional sensitivity $(x_i/y)(\partial f/\partial x_i)$. This and the result of Equation 6.12 are useful in the design of multicomponent systems and in the cost-effective selection of instrumentation for an application. In particular, using Equation 6.12, we can arrange the items x_i in their order of significance. For this, we rewrite Equation 6.12 as

$$e_i = \frac{K}{\left|x_i(\partial f/\partial x_i)\right|} \tag{6.13}$$

where K is a quantity that does not vary with x_i. It follows that for equal error contribution from all items, error in x_i should be inversely proportional to $|x_i(\partial f/\partial x_i)|$. In particular, the item with the largest $|x(\partial f/\partial x)|$ should be made most accurate. In this manner, the allowable relative accuracy for various components can be estimated. Since, in general, the most accurate device is also the costliest, instrumentation cost can be optimized if components are selected according to the required overall accuracy, using a criterion such as that implied by Equation 6.13. Hence, this result is useful in the design of multicomponent systems and in the cost-effective selection of instrumentation for a particular application.

Example 6.6

Figure 6.10 schematically shows an optical device for measuring displacement. This sensor is essentially an optical potentiometer (see Chapter 8). The potentiometer element is uniform and has a resistance R_c. A photoresistive layer is sandwiched between

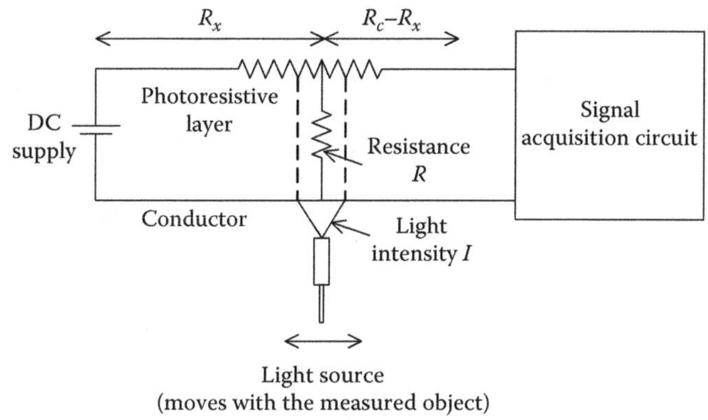

FIGURE 6.10
An optical displacement sensor.

this element and a perfect conductor of electricity. A light source that moves with the object whose displacement is to be measured, directs a beam of light whose intensity is I onto a narrow rectangular region of the photoresistive layer. As a result, this region becomes conductive with resistance R, which bridges the potentiometer element and the conductor, as shown.

Note: The output of the potentiometer directly depends on the bridging resistance R. Hence, this may be considered as the primary output of the device.

An empirical relation between R and I was found to be $\ln(R/R_o) = (I_o/I)^{1/4}$, where the resistance R is in kΩ and the light intensity I is expressed in watts per square meter (W/m²). The parameters R_o and I_o are empirical constants having the same units as R and I, respectively. These two parameters generally have some experimental error.

(a) Sketch the curve of R versus I and explain the significance of the parameters R_o and I_o.
(b) Using the absolute error method, show that the combined fractional error e_R in the bridging resistance R can be expressed as $e_R = e_{R_o} + (1/4)(I_o/I)^{1/4}[e_I + e_{I_o}]$, where e_{R_o}, e_I, and e_{I_o} are the fractional errors in R_o, I, and I_o, respectively.
(c) Suppose that the empirical error in the sensor model can be expressed as $e_{R_o} = \pm 0.01$ and $e_{I_o} = \pm 0.01$. Caused by the variations in the light source (due to power supply variations) and in ambient lighting conditions, the fractional error in I was found to be ±0.01. If the error e_R is to be maintained within ±0.02, at what light intensity level (I) should the light source operate? Assume that the empirical value of I_o is 2.0 W/m².
(d) Discuss the advantages and disadvantages of this device as a dynamic displacement sensor.

Solution

(a) We have $\ln(R/R_o) = (I_o/I)^{1/4}$. As $I \to \infty$, $\ln(R/R_o) \to 0$ or $R/R_o \to 1$. Hence, R_o represents the minimum resistance provided by the photoresistive bridge (which occurs at very high light intensity levels). When $I = I_o$, the bridge resistance R is calculated to be about $2.7R_o$. Hence, I_o represents a lower bound for the light intensity for satisfactory operation of the sensor. A suitable upper bound for the light intensity would be $10I_o$, for satisfactory operation. At this value, it can be computed that $R \approx 1.75R_o$. These characteristics are sketched in Figure 6.11.

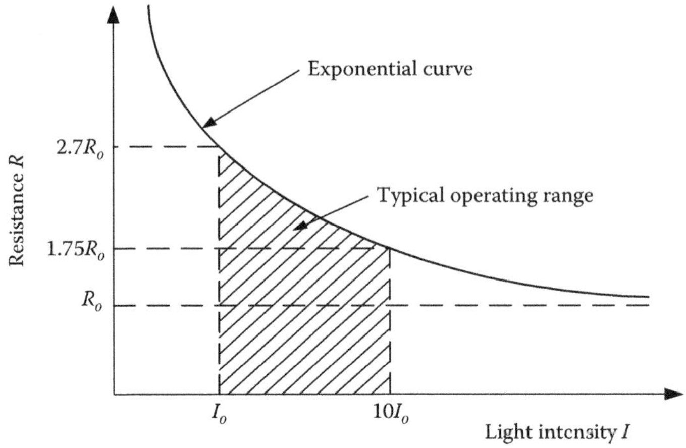

FIGURE 6.11
Characteristic curve of the sensor.

(b) First, we write $\ln R - \ln R_o = (I_o/I)^{1/4}$ and differentiate (take the differentials of the individual terms)

$$\frac{\delta R}{R} - \frac{\delta R_o}{R_o} = \frac{1}{4}\left(\frac{I_o}{I}\right)^{-3/4}\left[\frac{\delta I_o}{I} - \frac{I_o}{I^2}\delta I\right] = \frac{1}{4}\left(\frac{I_o}{I}\right)^{1/4}\left[\frac{\delta I_o}{I_o} - \frac{\delta I}{I}\right]$$

Hence, using the absolute method of error combination,

$$e_R = e_{Ro} + \frac{1}{4}\left(\frac{I_o}{I}\right)^{1/4}\left[e_{I_o} + e_I\right]$$

Note the use of the "+" sign instead of "−" since we employ the "absolute" method of error combination (i.e., positive magnitudes are used regardless of the actual algebraic sign).

(c) With the given numerical values, we have

$$0.02 = 0.01 + \frac{1}{4}\left(\frac{I_o}{I}\right)^{1/4}[0.01 + 0.01] \Rightarrow \left(\frac{I_o}{I}\right)^{1/4} = 2$$

$$\rightarrow I = \frac{1}{16}I_o = \frac{2.0}{16}\ \mathrm{W/m^2} = 0.125\ \mathrm{W/m^2}$$

Note: For larger values of I, the absolute error in R_o would be smaller. For example, for $I = 10I_o$ we have

$$e_R = 0.01 + \frac{1}{4}\left(\frac{1}{10}\right)^{1/4}[0.01 + 0.01] \approx 0.013$$

It is clear from this exercise that the operating conditions (e.g., I) can be properly chosen to obtain a desired level of accuracy. Also, we can determine the relative level of influence of various factors of error in the desired quantity (R).

(d) Advantages
- Noncontacting
- Small moving mass (low inertial loading)
- All advantages of a potentiometer (see Chapter 8)

Disadvantages
- Nonlinear and exponential variation of R
- Effect of ambient lighting
- Possible nonlinear behavior of the device (nonlinear input–output relation)
- Effect of variations in the power supply on the light source
- Effect of aging of the light source

Example 6.7

A schematic diagram of a chip refiner that is used in the pulp and paper industry is shown in Figure 6.12. This machine is used for mechanical pulping of wood chips. The refiner has one fixed disk and one rotating disk (typical diameter = 2 m). The plate is rotated by an AC induction motor. The plate separation (typical gap = 0.5 mm) is controlled by means of a hydraulic actuator (piston–cylinder unit with a servovalve). Wood chips are supplied to the "eye" of the refiner by a screw conveyor and are diluted with water. As the refiner plate rotates, the chips are ground into a pulp within the internal grooves of the plates. This is accompanied by the generation of steam due to the associated energy dissipation. The pulp is drawn and further processed for making paper.

An empirical formula relating the plate gap (h) and the motor torque (T) is given by $T = ah/(1 + bh^2)$, where the model parameters a and b are known to be positive.

(a) Sketch the curve T versus h. Express the maximum torque T_{max} and the plate gap (h_0) at this torque in terms of a and b only.

(b) Suppose that the motor torque is measured and the plate gap is adjusted by the hydraulic actuator according to the formula given previously. Show that the fractional error in h may be expressed as $e_h = \left[e_T + e_a + \frac{bh^2}{(1+bh^2)}e_b\right]\frac{(1+bh^2)}{|1-bh^2|}$ where e_T, e_a, and e_b are the fractional errors in T, a, and b, respectively, the latter two representing model error.

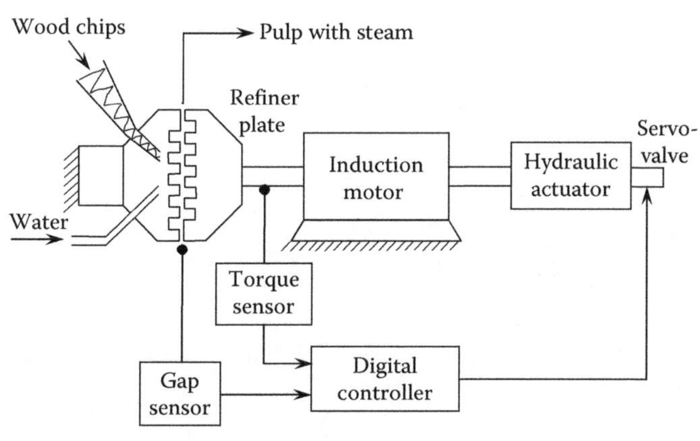

FIGURE 6.12
A single-disk chip refiner.

(c) The normal operating region of the refiner corresponds to $h>h_0$. The interval $0<h<h_0$ is known as the "pad collapse region" and should be avoided. If the operating value of the plate gap is $h=2/\sqrt{b}$ and if the error values are given as $e_T=\pm 0.05$, $e_a=\pm 0.02$, and $e_b=\pm 0.025$, compute the corresponding error in the plate gap estimate.

(d) Discuss why operation of the chip refiner at $h=1/\sqrt{b}$ is not desirable.

Solution

(a) See the sketch in Figure 6.13:

$$T=\frac{ah}{1+bh^2} \tag{i}$$

Differentiate Equation i with respect to h:

$$\frac{\partial T}{\partial h}=\frac{(1+bh^2)a-ah(2bh)}{(1+bh^2)^2}=0 \text{ at maximum } T$$

Hence, $1-bh^2=0 \rightarrow h_o=1/\sqrt{b}$. Substitute in (i):

$$T_{\max}=\frac{a}{2\sqrt{b}}$$

(b) The differential relation of Equation i is obtained by taking the differential of each term (i.e., slope times the increment). Thus,

$$\delta T=\frac{h}{(1+bh^2)}\delta a+\frac{\partial T}{\partial h}\delta h-\frac{ah}{(1+bh^2)^2}h^2\delta b$$

Substitute for $\partial T/\partial h$ from Part (a):

$$\delta T=\frac{h}{(1+bh^2)}\delta a+a\frac{(1-bh^2)}{(1+bh^2)^2}\delta h-\frac{ah^3}{(1+bh^2)^2}\delta b$$

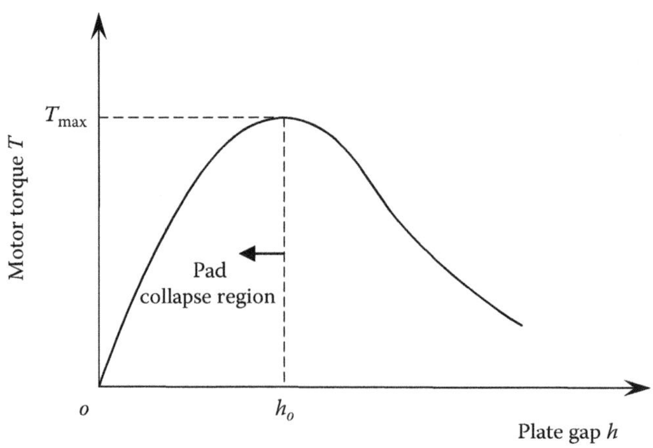

FIGURE 6.13
Characteristic curve of the chip refiner.

Divide throughout by Equation i:

$$\frac{\delta T}{T}=\frac{\delta a}{a}+\left[\frac{1-bh^2}{1+bh^2}\right]\frac{\delta h}{h}-\frac{bh^2}{(1+bh^2)}\frac{\delta b}{b}$$

Or

$$\frac{\delta h}{h}=\left[\frac{\delta T}{T}-\frac{\delta a}{a}+\frac{bh^2}{(1+bh^2)}\frac{\delta b}{b}\right]\left[\frac{1+bh^2}{1-bh^2}\right]$$

Now represent the fractional errors by the fractional deviations (differentials) and use the absolute value method of error combination. We have

$$e_h=\left[e_T+e_a+\frac{bh^2}{(1+bh^2)}e_b\right]\frac{(1+bh^2)}{|1-bh^2|} \tag{ii}$$

Note: The absolute values of the error terms are added. Hence, the minus sign in a term is ignored.

(c) With $h=2/\sqrt{b}$, we have $bh^2=4$. Substitute the given numerical values for fractional error in Equation ii:

$$e_h=\left[0.05+0.02+\frac{4}{5}\times 0.025\right]\frac{(1+4)}{|1-4|}=\pm 0.15$$

(d) When $h=1/\sqrt{b}$, we see from Equation ii that $e_h\to\infty$. In addition, from the curve in Part (a), we see that at this point the motor torque is not sensitive to changes in the plate gap. Hence, operation at this point is not appropriate and should be avoided.

The drawbacks of the sensitivity method of error propagation and combination include the following:

1. Sensitivities have to be determined either analytically using a model (which may be difficult or complex) or experimentally (which can be costly and time-consuming).
2. Sometimes, the sensitivities (local derivatives) may not exist (infinite).
3. System nonlinearities may result in sensitivities that vary with the operating conditions and/or local sensitivities that are insignificant in comparison to the contributions for the higher derivatives (i.e., O(2) terms of the Taylor series expansion).

Summary Sheet

Useful frequency range: The range of flat *gain curve* and zero-*phase curve* in frequency response characteristics (*frequency transfer function* [FTF] or *frequency response function* [*FRF*]) of instrument. Upper frequency in this range < 0.5 (say, one-fifth) × dominant resonant frequency ← measure of instrument bandwidth.

Bandwidth: Interpretations:

1. Speed of response of device
2. Passband of filter
3. Operating frequency range of device
4. Uncertainty in frequency content of signal
5. Information capacity of communication network

Effective noise bandwidth of filter (B_e): $B_e = \int_0^\infty |G(f)|^2 \, df / G_r^2$; G = filter FRF and G_r = average FRF magnitude near peak of FRF. *Note*: The higher the B_e, the larger the frequency content uncertainty of the filtered signal (i.e., more unwanted signal components pass through).

Half-power (or 3 dB) bandwidth (B_p): Width of the filter FRF at half-power (3 dB) level (i.e., at a drop of 3 dB from peak magnitude). For ideal (rectangular) equivalent filter, half power$=G_r^2 B_e/2$, which occurs at amplitude level $G_r/\sqrt{2}$. For the actual filter, $B_p \approx$ width of actual spectrum at this level ($B_p = B_e$ if filter power spectrum consists of linear segments).

Fourier analysis bandwidth: Frequency uncertainty in spectral results$=\Delta F = 1/T$; T = record length of signal (or window length for a rectangular window).

Control bandwidth : This is used to specify max possible *speed of control*. In digital control, it is given by half the rate at which the control action is computed (assuming that all the devices in the system can operate within this bandwidth). Output sensor data (response) sampling rate (samples/second) has to be several times higher than the control bandwidth so that sufficient data would be available to compute the control action. Typically, we need $\frac{1}{2\Delta T_p} \geq \frac{1}{\Delta T_c}$ or $\Delta T_p \leq 0.5\,\Delta T_c$, where ΔT_p is the sampling period of response measurement and ΔT_c is the time taken to compute each control action.

Static gain (DC gain): Gain (transfer-function magnitude) of device (e.g., measuring instrument) within useful (flat) range (or at very low frequencies). High static gain → high sensitivity → increases output level, increases speed of response, reduces steady-state error in a feedback control system, but makes it less stable.

Shannon's sampling theorem: Sampled data of a signal sampled at equal steps of ΔT have no information regarding signal spectrum beyond frequency $f_c = 1/(2\Delta T)$ = *Nyquist frequency*.

Aliasing error (distortion): Folding of high-frequency spectrum beyond Nyquist frequency onto low-frequency side, due to sampling → spectrum at frequency f_2 appears as spectrum at f_1 such that $f_2 + f_1 = f_s = 2f_c$; $f_s = 1/(\Delta T)$ = sampling rate.

Note: Increasing ΔT reduces aliasing.

Anti-aliasing filter: A low-pass filter with cutoff frequency at Nyquist frequency f_c to remove aliasing; a better cutoff frequency is $f_c/1.28$ ($\cong 0.8 f_c$).

Bandwidth design of a control system: The following are the simple steps for designing a control system:

Step 1: Decide on maximum frequency of operation (BW_o) based on application requirements.

Step 2: Select process components (electromechanical) that have the capacity to operate at least up to BW_o.

Step 3: Select feedback sensors with flat frequency spectrum (operating frequency range) $> 4 \times BW_o$.

Step 4: Develop digital controller with a sampling rate $> 4 \times BW_o$ for feedback sensor signals (within flat spectrum of sensors) and digital control cycle time (period) $1/(2 \times BW_o)$. *Note*: Digital control actions are generated at the rate of $2 \times BW_o$.

Step 5: Select control drive system hardware (interface analog hardware, filters, amplifiers, actuators, etc.) that have flat frequency spectrum of $\geq BW_o$.

Step 6: Integrate the system and test the performance. If the performance specifications are not satisfied, make necessary adjustments and test again.

Instrument accuracy: Represented by the worst accuracy level of instrument within its dynamic range in a specific operating environment. Depends on: the physical hardware, actual operating conditions (power, signal levels, load, speed, etc.; environmental conditions, etc.), design operating conditions (operating conditions for which the instrument is designed for: normal, steady operating conditions; extreme transient conditions, such as emergency start-up; and shutdown conditions), instrument setup shortcomings, other components and systems to which the instrument is connected, and so on.

Measurement accuracy: Closeness of measured value (*measurement*) to true value (*measurand*). Depends on: the instrument accuracy, how measurement process is conducted, how the measured data are presented (e.g., communicated, displayed, stored), and so on.

$$\text{Error} = (\text{Instrument reading}) - (\text{True value}); \quad \text{Correction} = -\text{Error}$$

Precision: Reproducibility (or repeatability) of an instrument reading (e.g., accurate clock with wrong time setting → precise, not accurate); precise → presented by low random error:

$$\text{Precision} = \frac{\text{Measurement range}}{\sigma_e}$$

σ_e = standard deviation of error.

Deterministic error: Repeatable (systematic) error; represented by mean error μ_e; can be corrected through recalibration.

Note: Device with low systematic error may not be precise if it has high zero-mean random error.

Difficulties in error analysis:

1. The true value is unknown.
2. Instrument reading may contain random error (error of the measuring system, including sensor error; other random errors that enter into the engineering system, including external disturbance inputs) that cannot be determined exactly.
3. Error may be a complex (i.e., not simple) function of many variables (input variables and state variables or response variables).
4. Monitored system may be multicomponent, having complex interrelations (dynamic coupling, multi-degree-of-freedom responses, nonlinearities, etc.), and each component may contribute to the overall error.

Application of sensitivity in error combination: Contribution of individual components or parameters to system output is expressed by the model $y=f(x_1,x_2,\ldots,x_r)$; x_i = independent system variables or parameter values whose errors are propagated into a dependent variable or output (or parameter value) y. This relationship depends on component characteristics, how the components are interconnected, interactions (dynamic coupling) among components, and inputs (desirable and undesirable) of the system.

$$\frac{\delta y}{y}=\sum_{i=1}^{r}\left[\frac{x_i}{y}\frac{\partial f}{\partial x_i}\frac{\delta x_i}{x_i}\right]\rightarrow e_y=\sum_{i=1}^{r}\left[\frac{x_i}{y}\frac{\partial f}{\partial x_i}e_i\right]$$

where

$\delta y/y=e_y$ is the overall (propagated) error

$\delta x_i/x_i=e_i$ is the component error

$(x_i/y)(\partial f/\partial x_i)$ is the nondimensional sensitivity of error in x_i on combined (propagated) error in y

Absolute error: $e_{ABS}=\sum_{i=1}^{r}\left|\frac{x_i}{y}\frac{\partial f}{\partial x_i}\right|e_i$

SRSS error: SRSS $e_{SRSS}=\left[\sum_{i=1}^{r}\left(\frac{x_i}{y}\frac{\partial f}{\partial x_i}e_i\right)^2\right]^{1/2}$

Equal contributions from individual errors: $\left|\frac{x_1}{y}\frac{\partial f}{\partial x_1}\right|e_1=\left|\frac{x_2}{y}\frac{\partial f}{\partial x_2}\right|e_2=\cdots=\left|\frac{x_r}{y}\frac{\partial f}{\partial x_r}\right|e_r\rightarrow$

$$e_i=\frac{e_{ABS}}{\left(r\left|\frac{x_i}{y}\frac{\partial f}{\partial x_i}\right|\right)}$$

Problems

6.1 Discuss how the accuracy of a sensory data acquisition system may be affected by

(a) Stability and bandwidth of amplifier
(b) Load impedance of the ADC

Moreover, what methods do you suggest to minimize problems associated with these parameters?

6.2 Consider the mechanical tachometer shown in Figure 6.4. Write expressions for sensitivity and bandwidth of the device. Using the example, show that the two performance ratings, sensitivity and bandwidth, generally conflict. Discuss ways to improve the sensitivity of this mechanical tachometer.

6.3 What is an anti-aliasing filter? In a particular application, the sensor signal is sampled at f_s Hz. Suggest a suitable cutoff frequency for an anti-aliasing filter to be used in this application.

6.4 Suppose that the frequency range of interest in a particular signal is 0–200 Hz. Estimate the sampling rate (digitization speed) for the data and the cutoff frequency for the anti-aliasing (low-pass) filter.

6.5 (a) Consider a multi-degree-of-freedom robotic arm with flexible joints and links. The purpose of the manipulator is to accurately place a payload. Suppose that the second natural frequency (i.e., the natural frequency of the second flexible mode) of bending of the robot, in the plane of its motion, is more than four times the first natural frequency. Discuss pertinent issues of sensing and control (e.g., types and locations of the sensors, types of control, operating bandwidth, control bandwidth, sampling rate for sensed information) if the primary frequency of the payload motion is

(i) One-tenth of the first natural frequency of the robot
(ii) Very close to the first natural frequency of the robot
(iii) Twice the first natural frequency of the robot

(b) A single-link space robot is shown in Figure P6.5. The link is assumed to be uniform with length 10 m and mass 400 kg. The total mass of the end effector and the payload is also 400 kg. The robot link is assumed to be flexible, although the other components are rigid. The modulus of rigidity of bending deflection of the link in the plane of robot motion is known to be $EI = 8.25 \times 10^9$ N·m^2. The primary natural frequency of bending motion of a uniform cantilever beam with an end mass is given by $\omega_1 = \lambda_1^2\sqrt{EI/m}$, where m is the mass per unit length and λ_1 is the mode shape parameter for mode 1. For [beam mass/end mass] = 1.0, it is known that $\lambda_1 l = 1.875$, where l is the beam length. Give a suitable operating bandwidth for the robot manipulation. Estimate a suitable sampling rate for the response measurements to be used in feedback control. What is the corresponding control bandwidth, assuming that the actuator and the signal-conditioning hardware can accommodate this bandwidth?

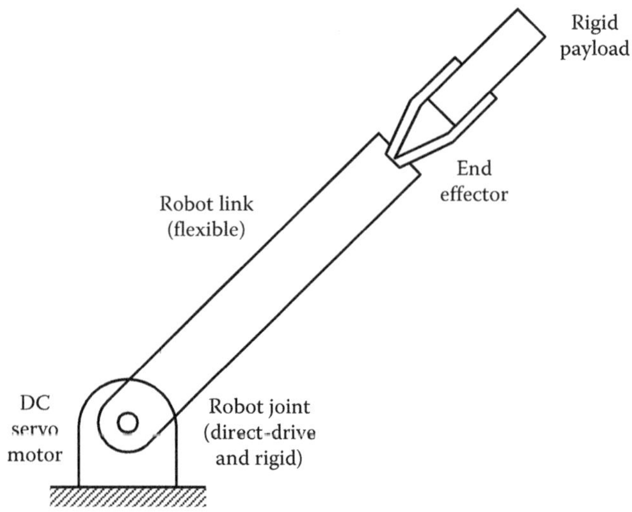

FIGURE P6.5
A single-link robotic manipulator.

6.6 (a) Define the following terms: sensor, transducer, actuator, controller, control system, operating bandwidth of a control system, control bandwidth, and Nyquist frequency.

(b) Choose a practical dynamic system that has at least one sensor, one actuator, and a feedback controller.

(i) Briefly describe the purpose and operation of each dynamic system.

(ii) For each system give a suitable value for the operating bandwidth, control bandwidth, operating frequency range of the sensor, and sampling rate for sensor signal for feedback control. Clearly justify the values that you have given.

6.7 Outline the following two approaches of control: (a) sensitivity-based control and (b) bandwidth-based control. Indicate shortcomings of each method and why either method alone may not be adequate to control an engineering system.

6.8 Discuss and contrast the following terms: (a) measurement accuracy, (b) instrument accuracy, (c) measurement error, and (d) precision.

In addition, for an analog sensor–transducer unit of your choice, identify and discuss various sources of error and ways to reduce or account for their influence.

6.9 Briefly explain what is meant by the terms *systematic error* and *random error* of a measuring device. What statistical parameters may be used to quantify these two types of error? State, giving an example, how *precision* is related to error.

6.10 Four sets of measurements were taken on the same response variable of a process using four different sensors. The true value of the response was known to be a constant. Suppose that the four sets of data are as shown in Figure P6.10a through d. Classify these data sets, and hence the corresponding sensors, as precise, offset-free, and accurate.

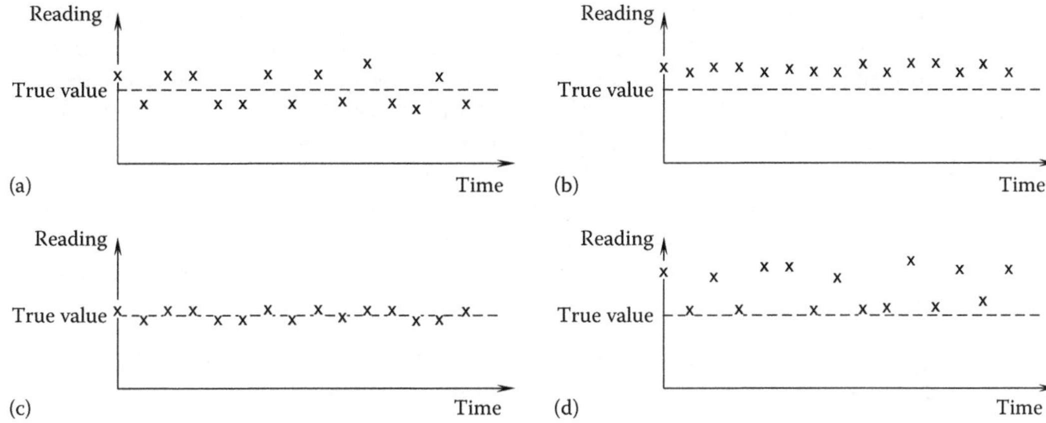

FIGURE P6.10
Four sets of measurements on the same response variable using different sensors.

6.11 The damping constant b of the mounting structure of a machine is determined experimentally. First, mass m of the structure is directly measured. Next, spring stiffness k is determined by applying a static load and measuring the resulting displacement. Finally, damping ratio ζ is determined using the logarithmic decrement method, by conducting an impact test and measuring the free response of the structure. A model of the structure is shown in Figure P6.11. Show that the damping constant is given by $b = 2\zeta\sqrt{km}$.

If the allowable levels of error in the measurements of k, m, and ζ are ±2%, ±1%, and ±6%, respectively, estimate a percentage absolute error limit for b.

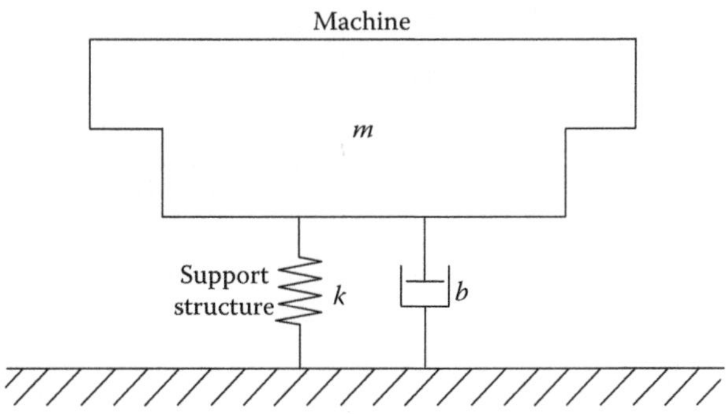

FIGURE P6.11
A model for the mounting structure of a machine.

6.12 Using the SRRSS method for error combination, determine the fractional error in each component x_i so that the contribution from each component to the overall error e_{SRSS} is the same.

6.13 A single-degree-of-freedom model of a robotic manipulator is shown in Figure P6.13a. The joint motor has rotor inertia J_m. It drives an inertial load that has moment of inertia J_l through a speed reducer of gear ratio 1:r (*Note*: $r<1$). The control scheme used in this system is the so-called feedforward control (strictly, "computed-torque control") method. Specifically, the motor torque T_m that is required to accelerate or decelerate the load is computed using a suitable dynamic model and a desired motion trajectory for the manipulator, and the motor windings are excited so as to generate that torque. A typical trajectory would consist of a constant angular acceleration segment followed by a constant angular velocity segment and finally a constant deceleration segment, as shown in Figure P6.13b.

(a) Neglecting friction (particularly bearing friction) and the inertia of the speed reducer shows that a dynamic model for torque computation during the accelerating and decelerating segments of the motion trajectory would be $T_m - (J_m + r^2 J_l)\ddot{\theta}_l/r$, where $\ddot{\theta}_l = \alpha_l$ is the angular acceleration of the load. Show that the overall system can be modeled as a single inertia element rotating at the motor speed. Using this result, discuss the effect of gearing on a mechanical drive.

(b) Given that $r=0.1$, $J_m=0.1$ kg·m^2, $J_l=1.0$ kg·m^2, and $\alpha_l=5.0$ rad/s^2, estimate the allowable error for these four quantities so that the combined error in the computed torque is limited to ±4% and that each of the four quantities contributes equally toward this error in the computed T_m. Use the absolute value method for error combination.

(c) Arrange the four quantities r, J_m, J_l, and α_l in the descending order of required accuracy, for the numerical values given in the problem.

(d) Suppose that $J_m = r^2 J_l$. Discuss the effect of error in r on the error in T_m.

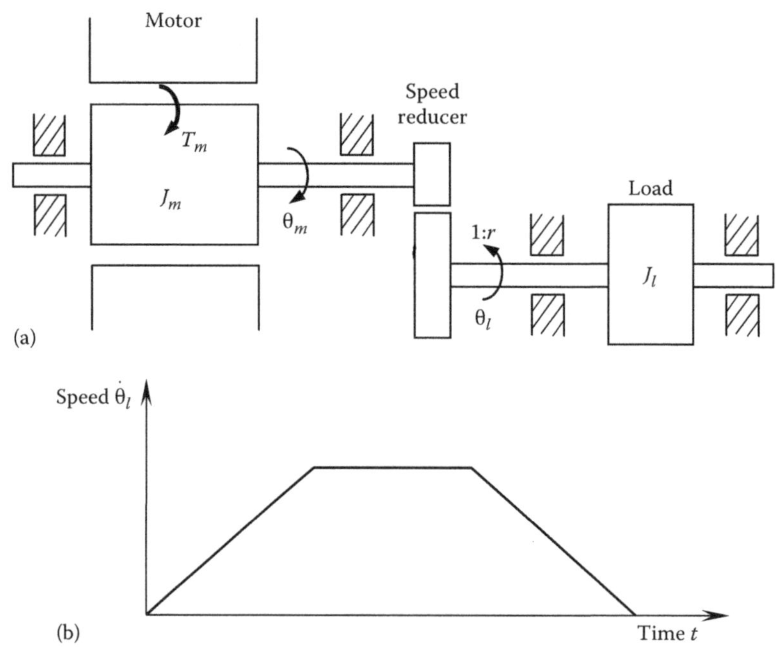

FIGURE P6.13
(a) A single degree-of-freedom model of a robotic manipulator. (b) A typical reference (desired) speed trajectory for computed-torque control.

6.14 An actuator (e.g., electric motor, hydraulic piston–cylinder mechanism, or ram) is used to drive a terminal device (e.g., gripper, hand, wrist with active remote center compliance) of a robotic manipulator. The terminal device functions as a force generator. A schematic diagram for the system is shown in Figure P6.14. Show that the displacement error e_x is related to the force error e_f through $e_f = \frac{x}{f}\frac{df}{dx}e_x$.

The actuator is known to be 100% accurate for practical purposes, but there is an initial position error δx_o (at $x = x_o$). Obtain a suitable transfer relation $f(x)$ for the terminal device so that the force error e_f remains constant throughout the dynamic range of the device.

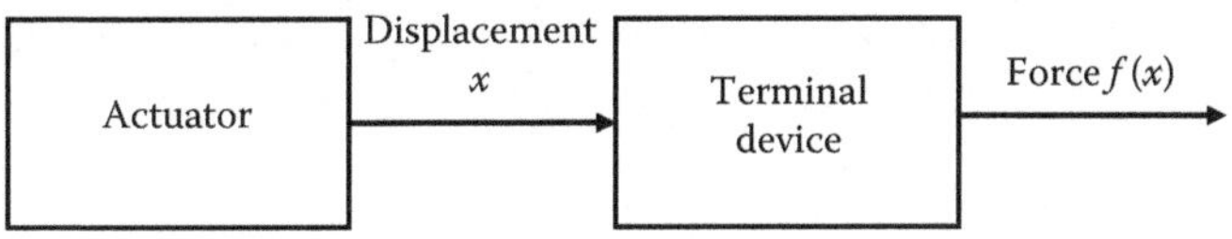

FIGURE P6.14
Block diagram of the terminal device of a robotic manipulator.

6.15 (a) Clearly explain why the "SRSS" method of error combination is preferred over the "absolute" method when the error parameters are assumed to be Gaussian (normal) and independent.

(b) Hydraulic pulse generators (HPGs) may be used in a variety of applications such as rock blasting, projectile driving, and seismic signal generation. In a typical HPG, water at very high pressure is supplied intermittently from an accumulator into the discharge gun through a high-speed control valve. The pulsating water jet is discharged through a shock tube and may be used, for example, for blasting granite. A model for an HPG was found to be $E = aV\left(b + \frac{c}{V^{1/3}}\right)$, where E is the hydraulic pulse energy (kJ), V is the volume of blast burden (m^3), and a, b, and c are model parameters whose values may be determined experimentally. Suppose that this model is used to estimate the blast volume of material (V) for a specific amount of pulse energy (E).

(i) Assuming that the estimation error values in the model parameters a, b, and c are independent and may be represented by appropriate standard deviations, obtain an equation relating these fractional errors e_a, e_b, and e_c to the fractional error e_v of the estimated blast volume.

(ii) Assuming that $a = 2175.0$, $b = 0.3$, and $c = 0.07$ with consistent units, show that a pulse energy of $E = 219.0$ kJ can blast a material volume of approximately 0.6^3 m^3. If $e_a = e_b = e_c = \pm 0.1$, estimate the fractional error e_v of this predicted volume.

6.16 The absolute method of error combination is suitable when the error contributions are additive (same sign). Under what circumstances would the SRSS method be more appropriate than the absolute method?

A simplified block diagram of a DC motor speed control system is shown in Figure P6.16. Show that in the Laplace domain, the fractional error e_y in the motor speed y is given by $e_y = -\frac{\tau s}{(\tau s + 1 + k)}e_\tau + \frac{(\tau s + 1)}{(\tau s + 1 + k)}e_k$, where e_τ is the fractional error in the time constant τ; e_k is the fractional error in the open-loop gain k, and the reference speed command u is assumed error-free. Express the absolute error combination relation for this system in the frequency domain ($s=j\omega$). Using it, show that

(a) At low frequencies, the contribution from the error in k will dominate and the error can be reduced by increasing the gain
(b) At high frequencies, k and τ will make equal contributions toward the speed error and the error cannot be reduced by increasing the gain

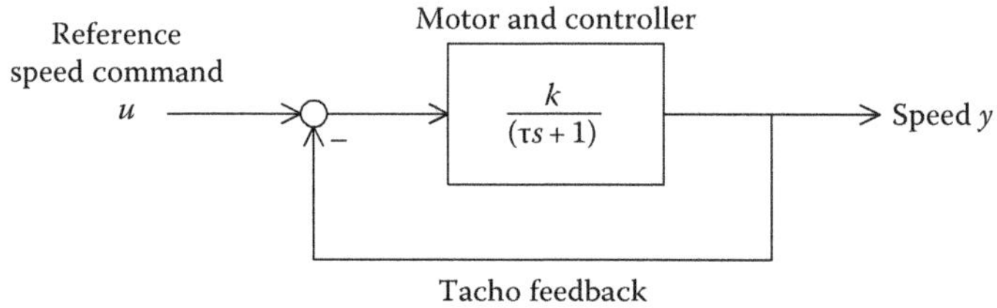

FIGURE P6.16
DC motor speed control system.

6.17 (a) Compare and contrast the "absolute error" method with the "SRSS" method in analyzing error combination of multicomponent systems. Indicate situations where one method is preferred over the other.
(b) Figure P6.17 shows a schematic diagram of a machine that is used to produce steel billets. The molten steel in the vessel (called tundish) is poured into the copper mold having a rectangular cross section. The mold has a steel jacket with channels to carry cooling water upward around the copper mold. The mold, which is properly lubricated, is oscillated using a shaker (electromechanical or hydraulic) to facilitate stripping of the solidified steel inside it. A set of power-driven friction rollers is used to provide the withdrawal force for delivering the solidified steel strand to the cutting station. A billet cutter (torch or shear type) is used to cut the strand into billets of appropriate length.

The quality of the steel billets produced by this machine is determined on the basis of several factors, which include various types of cracks, deformation problems such as rhomboidity, and oscillation marks. It is known that the quality can be improved through proper control of the following variables: Q is the coolant (water) flow rate, v is the speed of the steel strand, s is the stroke of the mold oscillations, and f is the cyclic frequency of the mold oscillations. Specifically, these variables are measured and transmitted to the central controller of the billet casting machine, which in turn generates proper control commands for the coolant valve controller, the drive controller of the withdrawal rollers, and the shaker controller.

A nondimensional quality index q has been expressed in terms of the measured variables as

$$q = \frac{\left[1 + \dfrac{s}{s_o}\sin\dfrac{\pi}{2}\left(\dfrac{f}{f_o + f}\right)\right]}{(1 + \beta v/Q)},$$

where s_o, f_o, and β are operating parameters of the control system and are exactly known. Under normal operating conditions, the following conditions are (approximately) satisfied: $Q \approx \beta v$, $f \approx f_o$, and $s \approx s_o$. *Note*: If the sensor readings are incorrect, the control system will not function properly, and the quality of the billets will deteriorate. It is proposed to use the "absolute error" method to determine the influence of the sensor errors on the billet quality.

(i) Obtain an expression for the quality deterioration δq in terms of the fractional errors $\delta v/v$, $\delta Q/Q$, $\delta s/s$, and $\delta f/f$ of the sensor readings.
(ii) If the sensor of the strand speed is known to have an error of ±1%, determine the allowable error percentages for the other three sensors so that there is equal contribution of error to the quality index from all four sensors, under normal operating conditions.

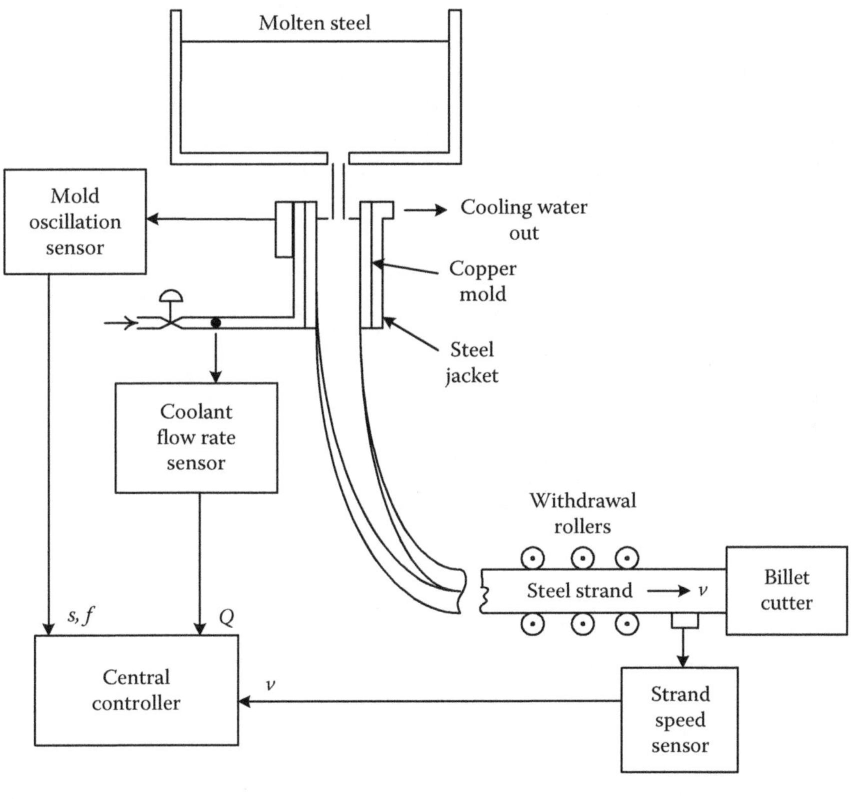

FIGURE P6.17
A steel-billet casting machine.

6.18 Consider the servo control system that is modeled as in Figure P6.16. Note that k is the equivalent gain and τ is the overall time constant of the motor and its controller.

(a) Obtain an expression for the closed-loop transfer function y/u.
(b) In the frequency domain, show that for equal contribution of parameter error toward the system response, we should have $(e_k/e_\tau) = \tau\omega/(\sqrt{\tau^2\omega^2+1})$, where

fractional errors (or variations) are, for the gain, $e_k = |\delta k/k|$ and, for the time constant, $e_\tau = |\delta\tau/\tau|$.

Using this relation, explain why at low frequencies the control system has a larger tolerance to error in τ than to that in k. Also, show that, at very high frequencies, the two error tolerance levels are almost equal.

6.19 Tension T at point P in a cable can be computed with the knowledge of the cable sag y, cable length s, cable weight w per unit length, and the minimum tension T_o at point O (see Figure P6.19). The applicable relationship is $1+\frac{w}{T_o}y=\sqrt{1+\frac{w^2}{T^2}s^2}$.

(a) For a particular arrangement, it is given that $T_o = 100$ lbf. The following parameter values were measured: $w = 1$1b/ft, $s = 10$ ft, $y = 0.412$ ft. Calculate the tension T.

(b) In addition, if the measurements y and s each have 1% error and the measurement w has 2% error in this example, estimate the percentage error in T.

(c) Now suppose that equal contributions to error in T are made by y, s, and w. What are the corresponding percentage error values for y, s, and w so that the overall error in T is equal to the value computed in the previous part of the problem? Which of the three quantities y, s, and w should be measured most accurately, according to the equal contribution criterion?

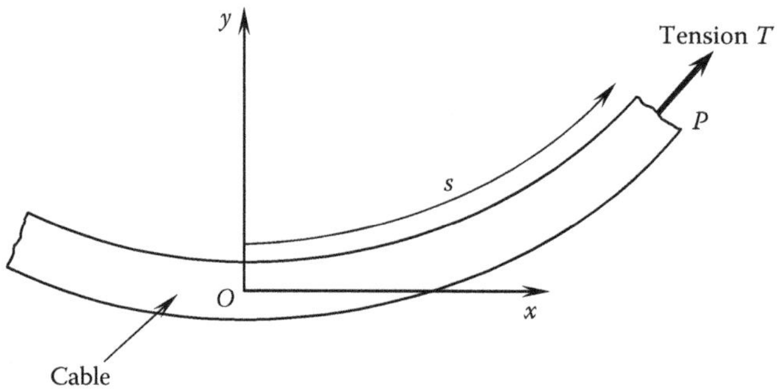

FIGURE P6.19
Cable tension example of error combination.

6.20 In Problem 6.19, suppose that the percentage error values specified are in fact standard deviations in the measurements of y, s, and w. Estimate the standard deviation in the estimated value of tension T.

6.21 A thermistor (a temperature sensor; see Chapter 10) has the empirical relation between its resistance R (in Ω) and the measured temperature T (in K), given by $R = R_o\exp\left[\beta\left(\frac{1}{T}-\frac{1}{T_o}\right)\right]$. The empirical parameter R_o is the resistance of the thermistor at the reference temperature T_o. Given are $R_o = 5000\,\Omega$ at $T_o = 298$ K (i.e., 25°C) and the "characteristic temperature" $\beta = 4200$ K. In a typical sensing procedure, the resistance R is measured, and the earlier equation (thermistor model) is used to compute the corresponding temperature T. There is measurement error in R and model error in R_o.

(a) Using the absolute error method, derive an equation for the combined fractional error e_T in the temperature measurement (estimation) in terms of the fractional errors e_R and e_{R_o} of R and R_o, respectively.
(b) Suppose that $e_{R_o} = \pm 0.02$ and at a temperature of 400 K, $e_R = \pm 0.01$. What is the fractional error e_T in the measured (estimated) temperature?
(c) Do you expect e_T to increase or decrease at higher temperatures? Why?

6.22 The quality control system in a steel rolling mill uses a proximity sensor to measure the thickness of the rolled steel (steel gage) at every 1 m along the sheet, and the mill controller adjustments are made on the basis of the last 20 measurements. Specifically, the controller is adjusted unless the probability that the mean thickness lies within ±1% of the sample mean exceeds 0.99. A typical set of 20 measurements in millimeters is as follows:

5.10	5.05	4.94	4.98	5.10	5.12	5.07	4.96	4.99	4.95
4.99	4.97	5.00	5.08	5.10	5.11	4.99	4.96	4.90	4.10

On the basis of these measurements, check whether adjustments would be made in the gage controller.

7

Estimation from Measurements

Chapter Objectives

- Sensing and Estimation
- Least-Squares Point Estimate
- Randomness in Data and Estimate
- Least-Squares Line Estimate
- Quality of Estimate
- Maximum Likelihood Estimation (MLE)
- Justification of MLE through Bayes' Theorem
- Recursive MLE
- Discrete MLE
- Scalar Static Kalman Filter
- Linear Multivariable Dynamic Kalman Filter
- State-Space Model
- Linear Kalman Filter Algorithm
- Extended Kalman Filter
- Unscented Kalman Filter

7.1 Sensing and Estimation

In the use of a sensor system, the measured quantity may be a constant parameter (e.g., moment of inertia of a link of a robotic arm), an average property of a batch of items (e.g., average internal diameter and its variance of a batch of ball bearings), a varying parameter (e.g., resistance of a strain gauge as the temperature changes), or a process variable (e.g., velocity of a vehicle). The sensor measurement may not provide the true value of the required parameter or variable for two main reasons:

1. What is measured may not be the required quantity and will have to be computed from the measured value (or values) using a suitable "model."
2. The sensor (or even the sensing process) is not perfect and will introduce "measurement error."

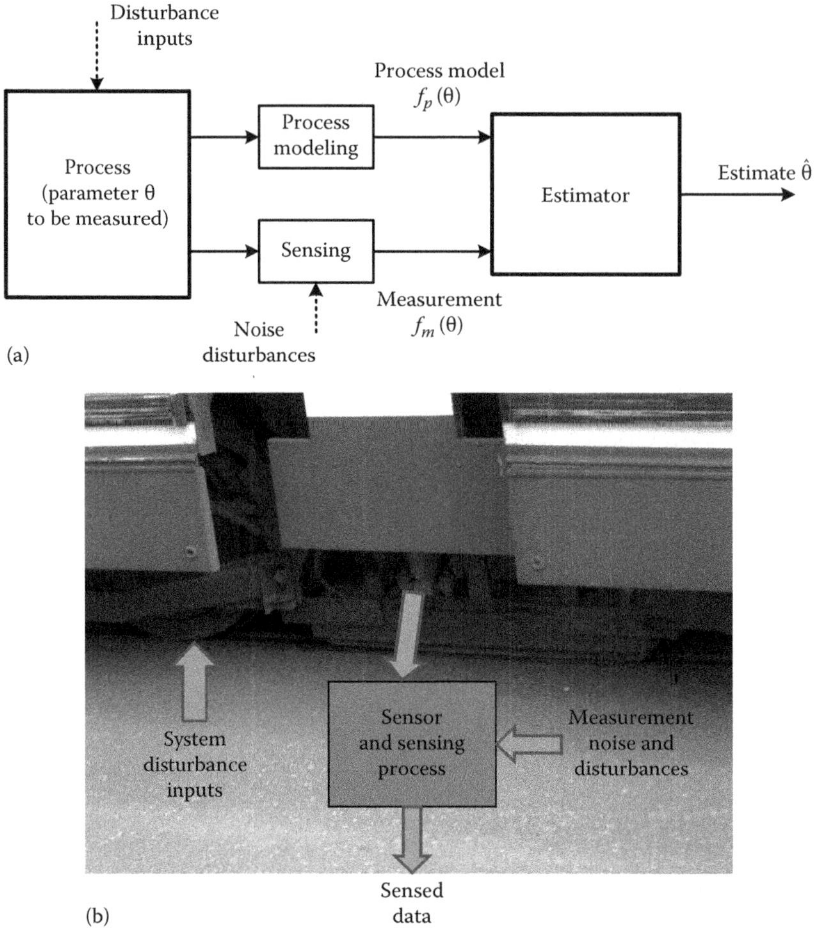

FIGURE 7.1
(a) The use of model error and measurement error in estimation and (b) an example of input disturbances and measurement error.

Hence, sensing may be viewed as a problem of estimation, where the "true value" of the measured quantity is "estimated" using the measured data. Clearly, two main categories of error, "model error" and "measurement error," enter into the process of estimation and will affect the accuracy of the result.

The underlying concept is schematically shown in Figure 7.1. The process model is shown here as a function $f_p(\theta)$ of the quantity θ, which is to be measured. In some situations, this function may be nonlinear and "dynamic" (i.e., not an algebraic function but a differential equation), and it may also include unknown and random effects. In other words, there can be model errors. *Note*: The errors in Category 1 also include undesirable and random inputs (disturbances) that enter into the dynamic system (process). This is because they also will affect the measured data and hence the value that is estimated using that data.

In Figure 7.1, the measurement is shown as a function $f_m(\theta)$ of the measured quantity θ. This function may also be nonlinear and "dynamic" (due to sensor dynamics) and may include unknown and random effects (e.g., sensor noise and errors in the measurement process) in general. In other words, the measurement may be neither direct nor exact. The estimator generates an estimate $\hat{\theta}$ of the measured quantity by using the available information (model $f_p(\theta)$ and measurement data $f_m(\theta)$) according to some method. Since the information will not be exact or complete, the estimate will not be precise. An optimal estimator will determine the "best" estimate according to some criterion (e.g., least-squares error).

Many methods of parameter estimation are used in sensing. Some methods use all the measured data simultaneously as a "batch" to estimate the required quantity. This is the "nonrecursive" approach. Other methods use the measured data as they are generated and "update" or "improve" the current result at each sensing step (in other words, the current estimate and the new data are used to compute a new estimate at each sensing step). This is the "recursive" approach. For example, Kalman filter is a recursive approach for parameter estimation, and it explicitly addresses both "model error" and "measurement error" to arrive at an estimate that is optimal (one that minimizes the squared error). Clearly, if the measured quantity itself changes with time, one has to use a recursive or time-varying approach of parameter estimation.

In this chapter, we will learn least-squares estimation (LSE), maximum likelihood estimation (MLE), and four versions of Kalman filter, namely, scalar static Kalman filter; linear multivariable dynamic Kalman filter; extended Kalman filter (EKF), which is applicable in nonlinear situations; and unscented Kalman filter (UKF), which is also applicable in nonlinear situations and has advantages over the EKF because it directly takes into account the propagation of random characteristics through system nonlinearities. Some basics of probability and statistics are presented in Appendix C. Reliability considerations and associated probability models of multicomponent systems are outlined in Appendix D.

7.2 Least-Squares Estimation

In the LSE, we estimate the unknown parameters by minimizing the sum of squared error (SSE) between the data and a model of the data. Hence, this is an "optimal" method of estimation. The unknown parameters are the model parameters. If the model is linear, we have linear LSE. If the model is nonlinear, we have nonlinear LSE.

7.2.1 Least-Squares Point Estimate

In least-squares point estimation, an unknown constant parameter is estimated using a batch of measurements (with error) of the parameter, such that the SSE between the data set and the estimated value is minimized. Suppose that the value of a constant parameter (e.g., mass) is to be estimated by using data from several repeated measurements. A reasonable estimate of the parameter would be the average value of the data set (a batch operation). It can be shown that this is also the optimal estimate in the sense of least squares.

Note: This approach is applicable to (a) repeated measurement of the same parameter of the same object (with measurement noise) and (b) measurement of a particular parameter in each object of a batch of objects that are nominally identical.

To show this, suppose that a constant parameter of unknown value m is repeatedly measured using a sensor (having some random error) N times, to generate the data set $\{Y_1, Y_2, \ldots, Y_N\}$. *Note*: As traditionally done, we use the "uppercase" Y to represent the data, in order to emphasize the fact that it contains "random" error. The SSE in the data set is

$$e = \sum_{i=1}^{N} (Y_i - m)^2 \tag{7.1}$$

To find the value of m (i.e., the estimate $\hat{m}$ of the unknown constant m) that would minimize the squared error (i.e., produce the "least-squares error"), we differentiate e with respect to (wrt) m and equate the result to zero:

$$\sum_{i=1}^{N} 2 \times (Y_i - m) \times (-1) = 0$$

We get

$$\hat{m} = \frac{1}{N} \sum_{i=1}^{N} Y_i \tag{7.2}$$

This result indicates that the least-squares point estimate (an optimal estimate) is the *sample mean* of the data.

Note: Here, the process "model" is simply an "identity" or "no change" operation (just "1" in the scalar case) since the model (static) is the constant parameter that we are interested in measuring.

Assuming that this model is correct, the only error that is present is the measurement error. Alternatively, both model error and measurement error may be integrated into a single error parameter, albeit without having the capability to distinguish between the two.

Note: The measurement error can come from both the sensor and the measurement process.

Clearly, all errors will affect the accuracy of the estimate given by Equation 7.2.

The estimation given by Equation 7.2 is a "batch" operation where the entire data set is used simultaneously. Hence, it has to be performed off-line. This operation may be converted into a "recursive" scheme, which can be executed online as the data are measured, as follows:

$$\begin{aligned} \hat{m}_1 &= Y_1 \\ \hat{m}_{i+1} &= \frac{1}{(i+1)} (i \times \hat{m}_i + Y_{i+1}), \quad i = 1, 2, \ldots \end{aligned} \tag{7.3}$$

We expect the accuracy of the estimate to increase as more data come in (assuming that the measured or estimated quantity is a constant and the error is random).

7.2.2 Randomness in Data and Estimate

In the previous case of point estimate, a nonrandom, constant quantity was estimated using direct measurements of that quantity. The process "model" was simply an "identity" or "no change" operation (just "1" in the scalar case) since the model (static) was the estimated constant parameter itself. Model error (or model randomness) was not explicitly considered. Of course, randomness in the measurement, through random errors in the sensor and the measurement process, affects the estimate. In the least-squares point estimate, we could not incorporate any knowledge of measurement randomness into the estimation process. Albeit, the effect of the measurement randomness (and model randomness) would reduce (average out) with the number of times the measurement is repeated. However, any constant error (bias, offset, or nonzero mean error) would not be eliminated or reduced.

Suppose that it is known that the sensor and the measuring process have random error, which is represented by a combined variance of σ_m^2.

Note: Zero-mean model error may be incorporated into this variance. Of course, once incorporated, the two components of error are indistinguishable.

Further, suppose that each measurement is independent of any other measurement in the data set $\{Y_1, Y_2, \ldots, Y_N\}$. More specifically, we assume that Y_i are "independent and identically distributed" (abbreviated by "iid") random variables (i.e., they have the same probability distribution). *Note*: Each measurement is a random variable, and hence, the estimate $\hat{m} = (1/N)\sum_{i=1}^{N} Y_i$ is also a random variable (because it is a function of the measured "random" data). Then, the variance of the estimate is

$$\text{Var}(\hat{m}) = \text{Var}\left[\frac{1}{N}(Y_1 + Y_2 + \cdots + Y_N)\right] = \frac{1}{N^2}\text{Var}(Y_1 + Y_2 + \cdots + Y_N)$$

$$= \frac{1}{N^2}[\text{Var}(Y_1) + \text{Var}(Y_2) + \cdots + \text{Var}(Y_N)] = \frac{N\sigma_m^2}{N^2}$$

Hence,

$$\text{Var}(\hat{m}) = \sigma_{\hat{m}}^2 = \frac{\sigma_m^2}{N} \tag{7.4}$$

or

$$\sigma_{\hat{m}} = \frac{\sigma_m}{\sqrt{N}} \tag{7.5}$$

This result confirms our previous statement that the randomness of the estimate decreases (and the precision improves) as the number of data items in the "measurement sample" is increased. Furthermore, as expected, it is clear from Equation 7.5 that the randomness of the estimate decreases (and the precision improves) as the precision of the measurement process (including the sensor) improves.

Example 7.1

A measuring instrument produces a random error whose standard deviation (std) is 1%. Assuming that each measurement is independent of another, how many measurements should be averaged in order to reduce the std of the error in a measured quantity to less than 0.05%?

Solution

Here, we use the fact that X_i are iid. Hence, from Equation 7.4 for the averaged measurement, $\text{Var}(\bar{X}) = \sigma^2/N$ and $\text{Std}(\bar{X}) = \sigma/\sqrt{N}$.

With $\sigma = 1\%$ and $\sigma/\sqrt{N} < 0.05\%$, we have

$$\frac{1}{\sqrt{N}} < 0.05 \rightarrow N > 400$$

Hence, we should average more than 400 measurements to achieve the specified accuracy.

7.2.2.1 Model Randomness and Measurement Randomness

In addition to the randomness in the measurement process (including sensor), the analytical representation of the measured (estimated) quantity itself contains some random component. Specifically, there is randomness in the model. The model may include the relationship between the measured quantity and the estimated quantity in addition to the analytical representation of the process (system) that generates the data.

To illustrate the related concepts, consider a high-precision manufacturing process where ball bearings are produced according to a tight tolerance (at the nanometer level). In particular, the "roundness" of the bearing balls is an important parameter of quality control of ball bearings. For this purpose, suppose that a sample of bearing balls of a specific nominal diameter (required size) is randomly chosen from a manufactured batch and the diameter of each ball is measured using a roundness probe.

A typical "roundness sensing" system is sketched in Figure 7.2. The sensing device has a turntable on which a ball would be mounted. A probe comes into contact with the ball, and a servomechanism in the device ensures that the contact with the ball is maintained continuously. As the turntable rotates, the probe moves according to the outer profile of the ball. The probe movement is sensed using a differential transformer (see Chapter 8) and is recorded. The maximum deviation of the ball diameter is measured in this manner. For such a sensing system, a typical sensor accuracy is ±25 nm and a typical sensor resolution is 5 nm.

In this measurement process, a sample set of balls is chosen randomly and each ball in the sample that is measured is also chosen randomly in sequence. Furthermore, due to

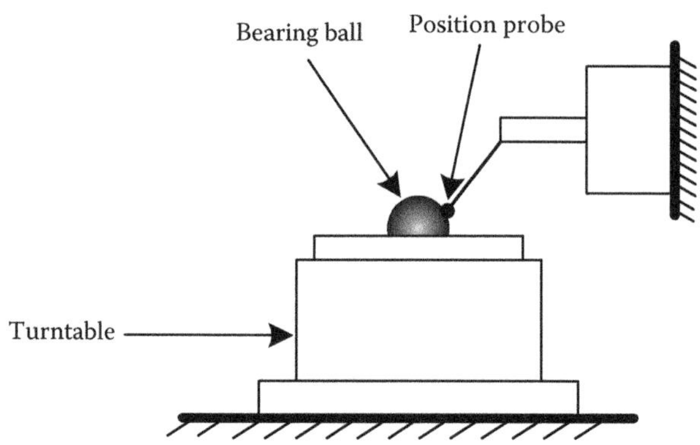

FIGURE 7.2
Roundness sensing setup.

the random effects in the manufacturing process, the actual ball diameter also changes randomly (albeit within some tolerance limit, which governs acceptable production quality). It follows that there is randomness in the data for each ball, due to the following causes:

1. Manufacturing process of each ball
2. Selection of the sample set of balls from the produced batch
3. Selection of a ball from the sample set
4. Ball mounting and probe contact in the roundness measuring device
5. Error in the probe (sensor)

Item 1 introduces randomness into the "process model" itself, and it directly affects the product quality. Items 2 through 5 all correspond to the randomness in the measurement process, which indirectly affects the product quality (by producing an inaccurate estimate of the actual roundness).

Consider a set of measurement data $\{Y_1, Y_2, \ldots, Y_N\}$. Since the goal of the measuring process in the present application is to determine the "quality" of the produced bearing balls, suppose that the model that is used for this purpose (to represent the product quality) consists of two quantities:

1. The *sample mean* is defined as

$$\bar{Y} = \frac{1}{N}\sum_{i=1}^{N} Y_i \tag{7.2}$$

2. The *sample variance* is defined as

$$S^2 = \frac{1}{(N-1)}\sum_{i=1}^{N}(Y_i - \bar{Y})^2 \tag{7.6}$$

These two quantities are known to be "unbiased estimates." Specifically, the expected value (E) of the sample mean equals the true mean, and the expected value of the sample variance equals the true variance. Here, assume that the measurements Y_i are iid random variables of mean value μ and variance σ^2.

Note: In the present test procedure, the iid assumption is quite valid because each measurement is taken without regard to any other measurement, and the nature of randomness of the measurements is quite similar.

Then, it can be shown that the expected values of the two estimates (Equations 7.2 and 7.6) are equal to the mean and variance of the measurements. That is,

$$E(\bar{Y}) = \mu \tag{7.7}$$

$$E(S^2) = \sigma^2 \tag{7.8}$$

As we observed earlier, in the present situation, the randomness of the measured data comes from both model randomness and measurement randomness (five types were listed earlier). However, in the present method of estimation, these various types of contributions are not treated separately and are represented in an integrated manner.

7.2.2.2 Randomness in Data and Estimates

It should be emphasized that the randomness is in the signal of the random process and not in the parameters that describe the random process (e.g., mean μ and std σ). Furthermore, the data measured from a random process are also random (i.e., if the measurement is repeated, it is unlikely that we will get exactly the same data), and hence, the "estimates" that are computed from the data (e.g., sample mean and sample std) are also random.

Example 7.2

A roundness sensor produces the following set of diameter measurements (in mm) for a sample of 10 bearing balls from a production batch: 5.01 5.01 5.02 4.95 4.98 4.99 4.99 5.01 5.02 4.99. To process the data, we define the MATLAB® function `Stat.m` using the script given by the following M-file:

```
x=[5.01 5.01 5.02 4.95 4.98 4.99 4.99, 5.01 5.02 4.99]
Sample_mean=mean (x) %Calculates sample mean of array x
Sample_variance=var(x) %Calculates sample variance of array x
```

Now, we can compute the sample mean and the sample variance (or mean squared error) of the data array x according to Equations 7.2 and 7.6 as follows:

```
>> Stat
x =
    5.0100    5.0100    5.0200    4.9500    4.9800    4.9900    4.9900
5.0100    5.0200    4.9900
Sample_mean =
    4.9970
Sample_variance =
   4.6778e-04
>>
```

We may treat the sample mean 4.997 mm as an estimate of the diameter of the batch of balls. The sample std 0.02163 mm is a measure of the dimensional accuracy of the batch of bearing balls.

Note: In this example, the error in the estimate results from the combined effect of both manufacturing process error and the measurement error.

7.2.3 Least-Squares Line Estimate

In the least-squares line estimation, a line (which can be linear or nonlinear) is fitted to the data so that the SSE between the data set and the line is minimized. In this case, the line is the "model" and is represented by more than one parameter. (*Note*: Two parameters are needed to represent a straight line, three parameters for a quadratic function, and so on.) Since many algebraic expressions become linear when plotted to a logarithmic scale, linear (straight line) fit becomes more accurate when log–log axes are used.

Clearly, linear least-squares fit is an estimation method, which "estimates" the two parameters of an input/output model (process model), the straight line. It fits a given set of data to a straight line such that the squared error is a minimum. The estimated straight line is known as the "linear regression line." In the context of sensing and instrumentation, it is also known as the "mean calibration curve." Instrument *linearity* may be represented by the largest deviation of the input/output data (or the actual calibration curve, which can be nonlinear) from the least-squares straight-line fit of the data (or the mean calibration curve). LSE comes under the general subject of "model identification," "system identification," or "experimental modeling," where a model (static or dynamic) is fitted to the data. Essentially, the parameters of the model are estimated. A treatment of estimating the parameters of a dynamic (nonalgebraic) model is beyond the scope of the present treatment.

Consider N pairs of data $\{(X_1, Y_1), (X_2, Y_2), \ldots, (X_N, Y_N)\}$ in which X denotes the *independent variable* (input variable) and Y denotes the *dependent variable* (output variable) of the "process" or the "system" that is being "identified." Suppose that the estimated linear regression (linear model) is given by

$$Y = mX + a \tag{7.9}$$

where
m is the *slope*
a is the *intercept* of the line

For the independent variable value X_i, the dependent variable value on the regression line is $(mX_i + a)$, but the measured value of the dependent variable is Y_i. The corresponding error (or *residual*) is $(Y_i - mX_i - a)$. Hence, the SSE for all the data points is

$$e = \sum_{i=1}^{N} (Y_i - mX_i - a)^2 \tag{7.10}$$

We have to minimize e with respect to the two parameters m and a. The required conditions are $\partial e/\partial m = 0$ and $\partial e/\partial a = 0$. By carrying out these differentiations in Equation 7.10, we get $\sum_{i=1}^{N} X_i(Y_i - mX_i - a) = 0$ and $\sum_{i=1}^{N} (Y_i - mX_i - a) = 0$.

Dividing these two equations by N and using the definition of sample mean, we get

$$\frac{1}{N}\sum X_iY_i - \frac{m}{N}\sum X_i^2 - a\bar{X} = 0 \tag{i}$$

$$\bar{Y} - m\bar{X} - a = 0 \tag{ii}$$

Solving these two simultaneous equations for m, we obtain

$$m = \frac{1/N\sum_{i=1}^{N} X_iY_i - \bar{X}\bar{Y}}{1/N\sum_{i=1}^{N} X_i^2 - \bar{X}^2} \tag{7.11}$$

The parameter a does not have to be explicitly expressed here, because by using Equations 7.9 and ii, we can eliminate a and express the linear regression line as

$$Y - \bar{Y} = m(X - \bar{X}) \tag{7.12}$$

Note, however, from Equation 7.4 that a is the Y-axis intercept (i.e., the value of Y when $X = 0$) and in view of Equation ii, it is given by

$$a = \bar{Y} - m\bar{X} \tag{7.13}$$

7.2.4 Quality of Estimate

The "quality" or "goodness" of an estimate depends on many factors such as

1. Accuracy of the data
2. Size of the data set
3. Method of estimation
4. Model used for estimation (e.g., linear fit, quadratic fit)
5. Number of estimated parameters

Some useful error statistics that indicate the "goodness" of a least-squares error fit are defined next.

SSE: This is the sum of the squares of error at each data point with respect to the corresponding point of the best fit. Specifically,

$$\text{SSE} = \sum_{i=1}^{N} (Y_i - \hat{Y}_i)^2 \tag{7.14}$$

where

Y_i is the measured data value

$\hat{Y}_i$ is the corresponding value as predicted by the line of best fit (i.e., line corresponding to least-squares error)

A value closer to 0 indicates that the model and the data have a better match (i.e., more accurate; the random error is smaller).

Mean square error (*MSE*): This is the "adjusted" average value of SSE and is given by

$$\text{MSE} = \frac{1}{(N-M)} \sum_{i=1}^{N} (Y_i - \hat{Y}_i)^2 \tag{7.15}$$

where M is the number of coefficients (of the fitted cure) that are estimated through curve fitting. The number $N - M$ is called the "residual degrees of freedom." *Note*: For a line fit, $M = 2$.

The rationale for this "adjusted average" should be clear. In particular, when the same number of data points is used to estimate more model coefficients, the estimation accuracy would be lower.

Root MSE (*RMSE*): This is the square root of MSE.

R-squared: This is also called the "coefficient of determination." It is defined as

$$R\text{-squared} = 1 - \frac{\text{SSE}}{\sum_{i=1}^{N} (Y_i - \bar{Y})^2} \tag{7.16}$$

where $\bar{Y}$ is the average value of all the data points.

Note: In Equation 7.16, SSE represents the deviation of the data from the model. The denominator represents the deviation of the data from its simple average (generally, the average is not a good model).

It follows that *R*-squared represents how well the data match the model. A value of *R*-square closer to 1 is desired since it indicates a better fit of the data with the model curve.

Adjusted R-squared: For a given set of data, when the number of coefficients in the fitted curve increases, the accuracy of the estimates decreases in general. This is taken into account in the adjusted *R*-squared, which is defined as

$$\text{Adjusted } R\text{-squared} = 1 - \frac{\text{MSE}}{\text{VAR}} \tag{7.17}$$

where

MSE is the mean square error, as given by Equation 7.15

VAR is the sample variance of the data, as given by Equation 7.6

Note: In the summations of Equations 7.14 through 7.17, we may include a weighting w_i for each data value Y_i. The weight may reflect such *a priori* considerations as the accuracy and the importance of a particular data value. In the formulas given earlier, we have assigned equal weighting to all data values (i.e., $w_i = 1$).

Example 7.3

Consider the capacitor circuit shown in Figure 7.3.

First, the capacitor is charged to voltage v_o using a constant DC voltage source (switch is in position 1), then it is discharged through a known resistance *R* (switch is in position 2). Voltage decay during discharge is measured at known time increments. Three separate tests are carried out. The measured data are given in Table 7.1.

If the resistance is accurately known to be 1000 Ω, estimate the capacitance *C* in microfarads (μF) and the source voltage v_o in volts (V).

Solution

To solve this problem, we assume the well-known expression for the free decay of voltage across a capacitor:

$$v(t) = v_o \exp\left[-\frac{t}{RC}\right] \tag{i}$$

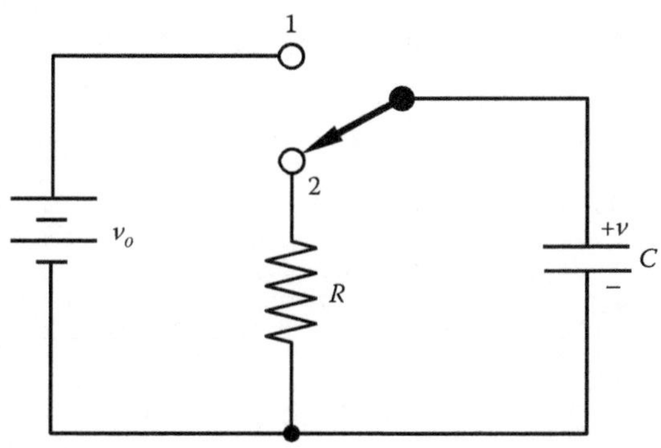

FIGURE 7.3
A circuit for the least-squares estimation of capacitance.

TABLE 7.1

Capacitor Discharge Data

Time t (s)	0.1	0.2	0.3	0.4	0.5
Voltage v (V)					
Test 1	7.3	2.8	1.0	0.4	0.1
Test 2	7.4	2.7	1.1	0.3	0.2
Test 3	7.3	2.6	1.0	0.4	0.1

Take the natural logarithm of Equation i:

$$\ln v = -\frac{t}{RC} + \ln v_o \tag{ii}$$

With $Y = \ln v$ and $X = t$, Equation ii represents a straight line with slope

$$m = -\frac{1}{RC} \tag{iii}$$

and the Y-axis intercept

$$a = \ln v_o \tag{iv}$$

Using all the data, the overall sample means and two useful summations can be computed:

$$\bar{X} = 0.3; \quad \bar{Y} = -0.01335; \quad \frac{1}{N}\sum X_i Y_i = -0.2067; \quad \text{and} \quad \frac{1}{N}\sum X_i^2 = 0.11$$

Now substitute these values in Equations 7.11 and 7.13. We get

$$m = -10.13 \quad \text{and} \quad a = 3.02565$$

Next, from Equation iii, with $R = 1000$, we have $C = 1/(10.13 \times 1000)$ F $= 98.72$ μF.

From Equation iv, $v_o = 20.61$ V.

Note: In this problem, the estimation error would be larger if we did not use log scaling for the linear fit.

We can obtain detailed results for this example by using linear least-squares error curve fitting in MATLAB. In particular, we may use the M-file:

```
x1=0.1:0.1:0.5; %one segment of x data as a row vector
x=[x1,x1,x1]; %assemble the three segments of x data as a row vector
x=x'; %convert x data to a column vector
y=[7.3,2.8,1,0.4,0.1,7.4,2.7,1.1,0.3,0.2,7.3,2.6,1.0,0.4,0.1]; %y data
as a row vector
y=log(y); %convert y data to log scale row vector
y=y'; %convert y data to a column vector
xlabel('Time (s)'), ylabel('Ln Voltage') %Label the axes
fit(x,y,'poly1')
```

MATLAB results:

```
Linear model Poly1:
     f(x) = p1*x + p2
```

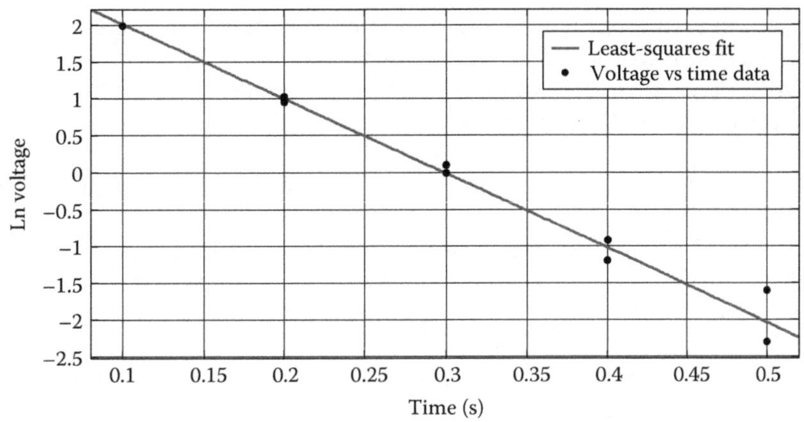

FIGURE 7.4
Linear least-squares error fit.

```
Coefficients (with 95% confidence bounds):
       p1 =      -10.13  (-10.82, -9.445)
       p2 =       3.027  (2.798, 3.255)

Goodness of fit:
  SSE: 0.3956
  R-square: 0.9873
  Adjusted R-square: 0.9863
  RMSE: 0.1744
```

The obtained linear least-squares error fit is shown in Figure 7.4.

As noted in the beginning of the present main section, least-squares curve fitting is not limited to linear (i.e., straight line) fit. The method can be extended to a polynomial fit of any order. For example, in *quadratic fit*, the data are fitted to a second-order (i.e., quadratic) polynomial. In that case, there are three unknown parameters, which would be determined by minimizing the SSE.

7.3 MLE

In the least-squares error method of estimation (LSE), we minimize the average squared error between the data and the estimate. Even though LSE is an "optimization" method, it is not necessarily the best method of estimation because it implicitly assumes that the model itself is very accurate (similar to the chicken-and-egg controversy) or it cannot separately account for model error and measurement error. For example, if the actual behavior of the data is nonlinear but the fitted model is linear, then LSE cannot properly compensate for that. Furthermore, if the data had a systematic error (an offset), the LSE method cannot remove that.

Another popular approach of parameter estimation is the maximum likelihood method. In the method of MLE, we maximize the likelihood of the estimated value, given the set of data that we have. In other words, we estimate the proper parameter value for the (random) process so that it would "most likely" generate the data set that we have. In computation, the MLE method is typically expressed as a recursive algorithm. A characteristic of recursive estimation is that it improves the estimation as more and more data values are supplied.

Note: MLE can estimate more than one parameter (i.e., a vector of parameters) simultaneously.

The MLE method assumes a particular probability distribution for the data, and it is the parameters of that probability function that are estimated by the method. If the probability distribution is assumed to be normal (i.e., Gaussian), then the results of the two methods LSE and MLE are equivalent, and they generate virtually the same results. This also implies that if the probability distribution of the data is known and is not Gaussian, then MLE is better than LSE.

Apart from parameter estimation, the MLE method is commonly used in *hypothesis testing* as well. The objective in that context is to select the most likely hypothesis (by maximizing the likelihood ratio) for the available data. Furthermore, since the MLE method is based on a probability distribution for the data, it can generate a "confidence interval" for the data.

7.3.1 Analytical Basis of MLE

Suppose that a process and a sensor with some randomness (possibly in both process and sensing) produce the following set of actual (numerical) data in a random manner: $\boldsymbol{y} = [y_1, y_2, \ldots, y_n]$. Also, suppose that the random process (actual process and sensing; e.g., the manufacturing process of a product and the measurement process of sensing the product dimensions) that generated the data is represented by a probability distribution function with the following parameters: $\boldsymbol{m} = [m_1, m_2, \ldots, m_r]$. These are the model parameters and are unknown. The goal of MLE is to estimate the model parameter value (vector) $\hat{\boldsymbol{m}}$ that "most likely" has produced the given data set $\boldsymbol{y}$. Note that the data set $\boldsymbol{y}$ is already generated and known, but the model parameters $\boldsymbol{m}$ are unknown.

The conditional probability density function (pdf) of data $\boldsymbol{y}$, conditioned upon the model parameters $\boldsymbol{m}$, is denoted by $f(\boldsymbol{y}|\boldsymbol{m})$. Since $\boldsymbol{y}$ is known and $\boldsymbol{m}$ is unknown, $f(\boldsymbol{y}|\boldsymbol{m})$ is a function of $\boldsymbol{m}$. In MLE, we estimate $\boldsymbol{m}$ by maximizing $f(\boldsymbol{y}|\boldsymbol{m})$. Specifically,

$$\hat{\boldsymbol{m}} = \underset{Max\, f(\boldsymbol{y}|\boldsymbol{m})}{\boldsymbol{m}} \tag{7.18}$$

Hence, the *likelihood function* $L(\boldsymbol{m}|\boldsymbol{y})$, which is maximized in order to estimate the unknown $\boldsymbol{m}$, given the data $\boldsymbol{y}$, is defined as

$$L(\boldsymbol{m} \mid \boldsymbol{y}) = f(\boldsymbol{y} \mid \boldsymbol{m}) \tag{7.19}$$

As it should be the case, the roles of $\boldsymbol{m}$ and $\boldsymbol{y}$ are reversed in L and f, even though the actual analytical functions of L and f are identical, as indicated by Equation 7.22. Notably, L represents the "likelihood" of a particular model parameter value, given a specific set of data (generated by the process that is represented by that model), while f represents the probability of achieving the specific data through a probability model having the particular parameter value.

Example 7.4

In the process of product quality monitoring, 10 items are randomly chosen from a batch of products and are individually and carefully tested. During testing, each product is either accepted (denoted by "A") or rejected (denoted by "R"). Suppose that such a procedure produced the data set $\boldsymbol{y}$ = [A, A, R, A, R, A, A, A, A, A]. Estimate the most likely probability $\hat{p}$ that a randomly selected product from the batch is acceptable (A).

Solution

Let p be the probability that a selected product from the batch is acceptable.

The probability of realizing the data set [A, A, R, A, R, A, A, A, A, A] is

$$\begin{aligned}\Pr(\boldsymbol{y} \mid \boldsymbol{m}) &= \Pr([A, A, R, A, R, A, A, A, A, A] \mid p)\\ &= p \times p \times (1-p) \times p \times (1-p) \times p \times p \times p \times p \times p = p^8(1-p)^2 \end{aligned} \quad \text{(i)}$$

Note: Pr(·) stands for "probability of."

Equation i is also the likelihood function

$$L(\boldsymbol{m} \mid \boldsymbol{y}) = L(p \mid [A, A, R, A, R, A, A, A, A, A]) = p^8(1-p)^2 \quad \text{(ii)}$$

We need to determine the value of p that maximizes this L. To obtain it, we may differentiate (ii) wrt p, equate to 0, and solve for p. Analytically easier way to achieve the same result is to use the fact that "log" is a monotonically increasing function and hence use log(L) as a *modified likelihood function* and maximize it. We have

$$\log L(p \mid [\mathrm{A, A, R, A, R, A, A, A, A, A}]) = 8\log p + 2\log(1-p)$$

Differentiate $8/p - (2/(1-p)) = 0 \rightarrow 8(1-p) - 2p = 8 - 10p = 0 \rightarrow \hat{p} = 0.8$.

Note: For the sake of completeness, we have to show that the second derivative of the likelihood function with respect to the estimated parameter (p) is negative (so that the likelihood function is a maximum, not a minimum). In the present example, the first derivative (of the log likelihood function) is $8/p-(2/(1-p))$. Its derivative is $-(8/p^2)-(2/(1-p)^2)$, which confirms that the turning point corresponds to a maximum (not a minimum).

7.3.2 Justification of MLE through Bayes' Theorem

We can justify the method of MLE by using Bayes' theorem as well. For this, we recall from the basic theory of probability that $\Pr(\boldsymbol{m},\boldsymbol{y}) = \Pr(\boldsymbol{m}|\boldsymbol{y})\Pr(\boldsymbol{y})$. This is called the "chain rule." Here, $\Pr(\boldsymbol{m},\boldsymbol{y})$ stands for the probability of the joint occurrence of $\boldsymbol{m}$ and $\boldsymbol{y}$. Similarly, $\Pr(\boldsymbol{m},\boldsymbol{y}) = \Pr(\boldsymbol{y}|\boldsymbol{m})\Pr(\boldsymbol{m})$. By equating these two results, we get $\Pr(\boldsymbol{m}|\boldsymbol{y})\Pr(\boldsymbol{y}) = \Pr(\boldsymbol{y}|\boldsymbol{m})\Pr(\boldsymbol{m})$ or

$$\Pr(\boldsymbol{m} \mid \boldsymbol{y}) = \frac{\Pr(\boldsymbol{y} \mid \boldsymbol{m})\Pr(\boldsymbol{m})}{\Pr(\boldsymbol{y})} \quad (7.20)$$

This result is a version of Bayes' theorem, and it can be used to rationalize the method of MLE. In particular, note that on the RHS of Equation 7.20, the denominator indicates the probability of obtaining a particular data set. Since in the MLE method, the data set is given, this denominator term has no effect and can be treated as a constant. In the numerator of the RHS, we have the term $\Pr(\boldsymbol{m})$. This is the "a priori probability" of $\boldsymbol{m}$ (i.e., the probability of $\boldsymbol{m}$ before the data set $\boldsymbol{y}$ is known). Once the data set is known, we have the "a posteriori probability" $\Pr(m|y)$. This represents the probability of occurrence of a particular value (estimate) of m once we have (know) the data set. From Equation 7.20, it is seen that this probability is directly proportional to $\Pr(y|m)$, which is also the likelihood function $L(m|y)$ as given by Equation 7.19. That is,

$$\Pr(m \mid y) \propto \Pr(m)L(m \mid y) \tag{7.21}$$

From Equation 7.21, it is clear that given an estimate of parameter m, the best update of the estimate (*a posteriori* estimate) is the one that maximizes the likelihood function $L(m|y)$.

7.3.3 MLE with Normal Distribution

If the data are iid and have the normal (i.e., Gaussian) distribution, the MLE results take a particularly convenient and familiar form.

Consider a random data set $\{Y_1, Y_2, \ldots, Y_N\}$. Suppose that the random variables Y_i are iid with $N(\mu, \sigma^2)$ (i.e., have normal distribution with mean μ and variance σ^2). The pdf of Y_i is

$$f(y_i) = \frac{1}{\sqrt{2\pi}\sigma} \exp\left[-\frac{(y_i - \mu)^2}{2\sigma^2}\right] \tag{7.22}$$

The joint pdf of the entire random data set, given the model parameters μ and σ, is

$$f(y_1, y_2, \ldots, y_N \mid \mu, \sigma) = \prod_{i=1}^{N} \frac{1}{\sqrt{2\pi}\sigma} \exp\left[-\frac{(y_i - \mu)^2}{2\sigma^2}\right] \tag{7.23}$$

Note: Since Y_i are iid, the joint pdf is the product of the individual pdfs.

This is indeed the likelihood function of the problem. Due to the exponential nature of this function, it is convenient to use its log version:

$$\log L = -N \log \sqrt{2\pi} - N \log \sigma - \frac{1}{2\sigma^2} \sum_{i=1}^{N} (y_i - \mu)^2$$

Differentiate wrt μ and σ, separately, and set to zero, to maximize the likelihood function:

$$\frac{1}{\sigma^2} \sum_{i=1}^{N} (y_i - \mu) = 0; \quad -\frac{N}{\sigma} + \frac{1}{\sigma^3} \sum_{i=1}^{N} (y_i - \mu)^2 = 0$$

Hence, we get the maximum likelihood estimates:

$$\hat{\mu} = \frac{1}{N} \sum_{i=1}^{N} y_i \tag{7.24}$$

$$\hat{\sigma}^2 = \frac{1}{N} \sum_{i=1}^{N} (y_i - \hat{\mu})^2 \tag{7.25}$$

Clearly, Equation 7.24 is identical to the least-squares estimate (see Equation 7.2), which is known to be an unbiased estimate for the mean. Equation 7.28, however, does not correspond to an unbiased estimate for the variance. The unbiased estimate for variance is the sample variance, as given by Equation 7.6, where the denominator is $N - 1$, not N.

Example 7.5

In this example, we will use MATLAB to determine the maximum likelihood estimate for the Gaussian probability density parameters corresponding to the data of Example 7.2. The MATLAB function `mle(x)` generates the maximum likelihood estimates of the mean and the std of a normal (Gaussian) distribution. We have

```
>> x=[5.01 5.01 5.02 4.95 4.98 4.99 4.99, 5.01 5.02 4.99];
>> Estimates=mle(x)
Estimates =
    4.9970    0.0205
```

Note: These values agree with the results obtained in Example 7.2. In particular, the MLE of the mean is the sample mean, 4.9970. The MLE of the std is 0.0205. Hence, the MLE of the variance is $(0.0205)^2 = 4.2025 \times 10^{-4}$. This value is almost 9/10 of the sample variance 4.6778×10^{-4}, which was obtained in Example 7.2.

7.3.4 Recursive Maximum Likelihood Estimation

We have seen that, in essence, MLE is an optimization problem. However, except for some special cases, the MLE optimization does not produce a convenient closed-form solution. A numerical solution of the recursive form is desirable in such situations. This recursive formulation may be expressed in the following general form:

$$\hat{m}_i = \hat{m}_{i-1} + u_{i-1} \tag{7.26}$$

where $\hat{m}$ represents the parameter that is estimated. In general, the recursive increment u depends on such factors as the data (measurements) of m, the model error variance, the measurement error variance, and even the present estimated value itself. We start with an initial guess $\hat{m}_0$ for the parameter. Then we update it in the increasing direction of the likelihood function L (*Note*: Typically, it is more convenient to use $\log L$ rather than L) step by step, until the required update is negligible (i.e., the maximum value of L is reached).

Clearly, the required direction of update is the gradient (slope) δ of L with respect to m. Specifically, we need

$$\hat{m}_i = \hat{m}_{i-1} + k_{i-1}\delta_{i-1} \tag{7.27}$$

where k_{i-1} is the step weighting factor of the update in the ith iteration. The gradient is given by

$$\delta = \frac{\partial L}{\partial m} \tag{7.28}$$

The weighting factor k also depends on such factors as the data (measurements) of m, the model error variance, the measurement error variance, and even the present estimated value itself. It needs to be adjusted in each step for these reasons and also because when the change in the gradient is smaller, then larger updates can be accommodated during iteration. Specifically, k should be inversely proportional to the second derivative of L wrt m. This second derivative is called the "Hessian."

Note: This second derivative is negative because the gradient (first derivative) decreases as we reach a maximum point.

Hence, the recursive formulation of the MLE may be expressed as

$$\hat{m}_i = \hat{m}_{i-1} - c\left(\frac{1}{(\partial^2 L/\partial m_{i-1}^2)}\right)\frac{\partial L}{\partial m_{i-1}} \tag{7.29}$$

We continue the iteration until the new estimate is equal to the old estimate (within a specified tolerance).

7.3.4.1 Recursive Gaussian Maximum Likelihood Estimation

Suppose that we are given measurements y of quantity m (which is estimated). The data y have both model error σ_m (also denoted by σ_v), which corresponds to the error in the model that is used to represent m and/or unknown disturbances that affect the true value of m, and measurement error σ_w. The stds σ_m and σ_w are also given. We need to determine an estimate $\hat{m}$ of m.

Assumptions: (1) Gaussian distributions and (2) zero-mean model error and zero-mean measurement error

Since the probability distributions are Gaussian, we know that

1. MLE estimate is the mean value given by the pdf $f(m|y)$
2. Estimation error is represented by the std as given by the pdf

To derive a recursive formula for MLE, we start with the Bayes' formula:

$$f(m\,|\,y) = \frac{f(m)f(y\,|\,m)}{f(y)} \tag{i}$$

where

$f(m)$ corresponds to the "a priori value" of estimated m

$f(m|y)$ corresponds to the "a posteriori value" of the estimate (i.e., the value that is computed, once the data y is applied)

Note: Then, in a recursive formulation, $f(m)$ corresponds to the previous estimate and $f(m|y)$ corresponds to the new estimate once new data are applied.

Normal (Gaussian) pdfs:

$$f(m) = \frac{1}{\sqrt{2\pi}\sigma_m} e^{-\frac{(m-\mu_m)^2}{2\sigma_m^2}} \tag{ii}$$

Once m is given (i.e., there is no randomness in it), the randomness in y comes only from that of the measurement error w. Hence, the $\mathrm{Var}(y|m) = \mathrm{Var}(w)$. Also, once m is given, the mean value of y is m (*Note*: Both model error and measurement error are zero mean). Hence, we have

$$f(y\,|\,m)=\frac{1}{\sqrt{2\pi}\sigma_w}e^{-\frac{(y-m)^2}{2\sigma_w^2}} \tag{iii}$$

Once y is known, $f(y)$ in the Bayes' formula (i) may be denoted by a constant parameter a. Then, by substituting Equations ii and iii in Equation ii, we get

$$f(m\,|\,y)=\frac{1}{a\sqrt{2\pi}\sigma_m\sigma_w}e^{-\frac{1}{2}\left(\frac{(m-\mu_m)^2}{\sigma_m^2}+\frac{(y-m)^2}{\sigma_w^2}\right)}=\frac{1}{\sqrt{2\pi}\sigma_{m|y}}e^{-\frac{1}{2}\left(\frac{(m-\mu_{m|y})^2}{\sigma_{m|y}^2}\right)}$$

where

$$\mu_{m|y}=\frac{\sigma_w^2}{\left(\sigma_m^2+\sigma_w^2\right)}\mu_m+\frac{\sigma_m^2}{\left(\sigma_m^2+\sigma_w^2\right)}y \tag{iv}$$

$$\frac{1}{\sigma_{m|y}^2}=\frac{1}{\sigma_m^2}+\frac{1}{\sigma_w^2} \tag{v}$$

Equations iv and v can be used to determine the estimate $\hat{m}$ and the estimation error variance σ_m^2, recursively, as

$$\hat{m}_i=\frac{\sigma_w^2}{\left(\sigma_{m_{i-1}}^2+\sigma_w^2\right)}\hat{m}_{i-1}+\frac{\sigma_{m_{i-1}}^2}{\left(\sigma_{m_{i-1}}^2+\sigma_w^2\right)}y_i$$

$$\frac{1}{\sigma_{m_i}^2}=\frac{1}{\sigma_{m_{i-1}}^2}+\frac{1}{\sigma_w^2}$$

It will be seen later that these results form the basis for the static Kalman filter.

7.3.5 Discrete MLE Example

Suppose that the estimated quantity m is discrete. Specifically, it takes one of a set of discrete values m_i, $i = 1, 2, \ldots, n$. It can be represented by the column vector

$$\boldsymbol{m}=[m_1,m_2,\ldots,m_n]^T$$

For example, each m_i may represent different states of distance of an object (near, far, very far, no object, etc.).

Also, suppose that the measurement/observation y is correspondingly discrete. It takes n discrete values y_i, $i = 1, 2, \ldots, n$. It can be represented by the column vector

$$\boldsymbol{y}=[y_1,y_2,\ldots,y_n]^T$$

A measurement or observation will produce these n discrete quantities with different probability values (the sum of these probabilities = 1). A special case would be to have a measurement/observation of just one of these n discrete quantities, at probability 1.

To perform the MLE, we need a likelihood matrix:

$$L(\boldsymbol{m} \mid \boldsymbol{y}) = P(\boldsymbol{y} \mid \boldsymbol{m}) = \begin{bmatrix} p_{11} & p_{12} & \cdots & p_{1n} \\ p_{21} & p_{22} & \cdots & p_{2n} \\ \cdot & \cdot & \cdot & \cdot \\ p_{n1} & p_{n2} & \cdots & p_{nn} \end{bmatrix} \tag{i}$$

Then, the probability vector of the estimate of $\boldsymbol{m}$, given the data $\boldsymbol{y}$, is given by (from the Bayes' formula)

$$P(\boldsymbol{m} \mid \boldsymbol{y}) = aP(\boldsymbol{y} \mid \boldsymbol{m})P(\boldsymbol{m}) \tag{ii}$$

where $P(\boldsymbol{m})$ is the probability vector of the prior estimate (i.e., a priori probability) of $\boldsymbol{m}$. Also, the parameter a has to be chosen such that the probability elements of the vector $P(\boldsymbol{m}|\boldsymbol{y})$ add to 1. Of course, the value of this parameter is irrelevant for the MLE because the approach prescribes the selection of the element of $\boldsymbol{m}$ that corresponds to the largest probability value in the probability vector $P(\boldsymbol{m}|\boldsymbol{y})$.

7.4 Scalar Static Kalman Filter

Kalman filter is a popular method of estimation using measured data. Even though it can be used to estimate parameters, as done with the methods of LSE and MLE, it is more commonly used for estimating variables in a dynamic system. For example, if the variables (some or all) that are needed to determine the "dynamic state" of a system (i.e., state variables) are not available for measurement, a Kalman filter can "estimate" them in an optimal manner (by minimizing the mean value—expected value of squared error of the estimate) using measurements of output variables. This estimation is possible in the presence of both model error (random) and measurement error (random). This ability to remove (or compensate for) random effects (noise and disturbance) is the reason why it is called a "filter" even though it is an "estimator" or an "observer."

Unlike the method of LSE (and in some cases, with MLE), Kalman filter is able to explicitly and separately accommodate model error and measurement error. This is an advantage when statistics of these two types of error are separately available.

As we will see, Kalman filter uses a "predictor–corrector" method where a model of the process is used to first "predict" the estimated quantity and its error covariance and then correct these two quantities by using the actual measured data (output). The "predictor" step generates the *a priori* estimate, and the corrector step generates the *a posteriori* estimate. Discrete-time Kalman filter uses a two-step recursive scheme to accomplish this.

In the present section, we will derive a Kalman filter formulation using Bayes' formula for a scalar (nonvector, with just one estimated parameter) static (nondynamic) problem. In Section 7.5, drawing inspiration from the scalar static results, a complete formulation of the discrete-time Kalman filter will be presented for a linear, multivariable (vector) dynamic problem. Subsequently, in Section 7.6, the *linear multivariable Kalman filter* will be extended to *nonlinear* problems using the *extended Kalman filter*. Finally, in Section 7.7, to overcome some shortcomings of the EKF in propagating random characteristics through a nonlinearity, an improved version called the "unscented Kalman filter" will be presented.

7.4.1 Concepts of Scalar Static Kalman Filter

Now, we present a Kalman filter–style recursive formulation for a scalar, static system. This formulation is inspired by MLE through the Bayes' formula (7.20). Here, we will explicitly and separately account for both *model error* and *measurement error*.

Estimation can be interpreted in general as "estimating" an unknown quantity $\boldsymbol{m}$ (a scalar or vector) using data $\boldsymbol{y}$ measured from the actual process in which the quantity $\boldsymbol{m}$ is embedded. Symbolically, we determine $\boldsymbol{m}|\boldsymbol{y}$, which denotes "$\boldsymbol{m}$ given $\boldsymbol{y}$." Since $\boldsymbol{m}$ is unknown and $\boldsymbol{y}$ can be affected by various random causes, $\boldsymbol{m}$ itself is a random quantity (before it is determined or estimated). Hence, once the probability distribution of $\boldsymbol{m}$ is obtained, it completely determines $\boldsymbol{m}$. In fact, it is the *mean value* (expected value) of $\boldsymbol{m}$ that is determined (with respect to some criteria, including optimization criteria). If $\boldsymbol{m}$ is not random, then the mean value alone is what is needed. If it is indeed random, other statistics (such as the variance) would be needed as well to characterize its randomness. If the probability distribution of $\boldsymbol{m}$ is Gaussian (i.e., normal), then the two parameters *mean* and *variance* alone will completely characterize $\boldsymbol{m}$. The mean value will give the estimated value, and the variance will represent the estimation error. This is the basis of the present approach. Specifically, we will present an algorithm to determine the pdf $f(m|y)$.

The present formulation is for a "static" system, and furthermore, the process (model) is the estimated quantity itself. In other words, the system model does not have any dynamics, and also it corresponds to the "identity" or "no change" operation of multiplying by "1." This means that the "predictor" step of the Kalman filter (which uses the model) is not needed, and only the "corrector" step (which uses the measured data to correct the estimate) is needed. Even though the process model is 1 (in the error-free case), some model error may be present and hence is explicitly accounted for in the present algorithm.

In deriving the present recursive algorithm, we make the following assumptions:

1. When there is no measurement error, the measured (estimated) parameter m and the measurement (data) y are linearly related through a known constant gain C, which is called the measurement gain (*Note*: C corresponds to the familiar "output formation matrix" or "measurement gain matrix" in the state-space formulation of a dynamic system. We will encounter this again in the multivariable dynamic formulation of a Kalman filter, which is addressed in a later section of this chapter).
2. The model error (e.g., manufacturing error of a product, how the product is mounted during actual use) and measurement error (e.g., sensor error, sensor usage error, signal acquisition error) are independent and Gaussian (i.e., normal) random variables.
3. The model error v has a zero mean and std σ_m. (*Note*: We may use the standard notation σ_v of Kalman filter to denote this std.)
4. The measurement error w has a zero mean and a std σ_w.

Note: If the mean value of the measurement error is not zero, it cannot be estimated by the present process of estimation. The measurement system should be recalibrated to remove that constant offset error in the readings. Also, we assume that the mean value of the model error is zero. If the mean value of the model error is nonzero and known *a priori*, we can simply adjust the estimated value using the mean value (a constant offset).

What is estimated is the mean value of the parameter. If the mean value of an error is nonzero and unknown, the estimation process cannot determine it. This can be further confirmed by using the maximum likelihood method with the following choice of likelihood function, which is consistent with the Kalman filter approach (minimization of error covariance). Specifically, as the "inverse" of the likelihood function, we choose

$$\frac{1}{L} = E[(m - \hat{m})^2 \mid y] \tag{7.30}$$

To minimize the RHS of Equation 7.30, we require

$$\frac{\partial E[(m - \hat{m})^2 \mid y]}{\partial \hat{m}} = 0 = 2E[(m - \hat{m}) \mid y]$$

The maximum likelihood estimate itself is a known quantity (once it is estimated using the given data y) and can be taken out of the "expectation" (mean) operation "E" on the RHS term earlier. Hence, the maximum likelihood estimate is

$$\hat{m} = E[m \mid y] \tag{7.31}$$

In other words, what is estimated is the mean value of m, given the data y.

In the present problem, the objective is to estimate m using measurements y. The measurement equation is

$$y = Cm + w \tag{7.32}$$

where C is the *measurement gain*, as noted earlier. The variables and parameters in this equation are defined in the list of assumptions given in the beginning.

Since w is zero mean, from Equation 7.32, we have the following relation between the mean values of m and y:

$$\mu_y = C\mu_m \tag{7.33}$$

Furthermore, since m and w are independent random variables, we have from Equation 7.32

$$\mathrm{Var}(y) = \mathrm{Var}(Cm) + \mathrm{Var}(w) = C^2\mathrm{Var}(m) + \mathrm{Var}(w)$$

or

$$\sigma_y^2 = C^2\sigma_m^2 + \sigma_w^2 \tag{7.34}$$

7.4.2 Use of Bayes' Formula

In deriving the algorithm of the scalar static Kalman filter, we now restate the Bayes' formula (7.20) using pdfs $f(\cdot)$ as

$$f(m \mid y) = \frac{f(m)f(y \mid m)}{f(y)} \tag{7.35}$$

Note: According to one of our assumptions, all the pdfs in this equation are Gaussian.

As recognized earlier, in Equation 7.35, $f(m)$ corresponds to the "a priori value" of the estimate and $f(m|y)$ corresponds to the "a posteriori value" of the estimate (i.e., the value that is computed, once the data y is applied). In a recursive formulation, $f(m)$ corresponds to the previous estimate and $f(m|y)$ corresponds to the new estimate once new data are applied.

We have the following normal (Gaussian) pdfs:

$$f(m) = \frac{1}{\sqrt{2\pi}\sigma_m} e^{-\frac{(m-\mu_m)^2}{2\sigma_m^2}}$$

and in view of Equation 7.33,

$$f(y) = \frac{1}{\sqrt{2\pi}\sigma_y} e^{-\frac{(y-C\mu_m)^2}{2\sigma_y^2}}$$

Once m is given (i.e., there is no randomness in it) the randomness in y comes only from that of w. Hence, the $\text{Var}(y|m) = \text{Var}(w)$. Also, once m is given, from Equation 7.32, the mean value of y is Cm (*Note*: w is zero mean). Hence, we have

$$f(y \mid m) = \frac{1}{\sqrt{2\pi}\sigma_w} e^{-\frac{(y-Cm)^2}{2\sigma_w^2}}$$

Substitute these three pdfs in Equation 7.35:

$$f(m \mid y) = \frac{1}{\sqrt{2\pi}} \frac{\sigma_y}{\sigma_m \sigma_w} e^{-\frac{1}{2}\left(\frac{(m-\mu_m)^2}{\sigma_m^2} + \frac{(y-Cm)^2}{\sigma_w^2} - \frac{(y-C\mu_m)^2}{\sigma_y^2}\right)}$$

$$= \frac{1}{\sqrt{2\pi}} \frac{\sigma_y}{\sigma_m \sigma_w} e^{-\frac{1}{2}\left(\frac{\sigma_y}{\sigma_m \sigma_w}\right)^2 \left(\frac{\sigma_w^2 (m-\mu_m)^2}{\sigma_y^2} + \frac{\sigma_m^2 (y-Cm)^2}{\sigma_y^2} - \frac{\sigma_m^2 \sigma_w^2 (y-C\mu_m)^2}{\sigma_y^4}\right)}$$

Now manipulate the exponent term in the final expression as follows:

$$\frac{\sigma_w^2 (m-\mu_m)^2}{\sigma_y^2} + \frac{\sigma_m^2 (y-Cm)^2}{\sigma_y^2} - \frac{\sigma_m^2 \sigma_w^2 (y-C\mu_m)^2}{\sigma_y^4}$$

$$= \frac{\sigma_w^2}{\sigma_y^2}\left(m^2 - 2m\mu_m + \mu_m^2\right) + \frac{\sigma_m^2}{\sigma_y^2}(y^2 - 2yCm + C^2 m^2) - \frac{\sigma_m^2 \sigma_w^2}{\sigma_y^4}\left(y^2 - 2yC\mu_m + C^2 \mu_m^2\right)$$

$$= m^2\left(\frac{\sigma_w^2}{\sigma_y^2} + \frac{C^2 \sigma_m^2}{\sigma_y^2}\right) - 2m\left(\frac{\sigma_w^2}{\sigma_y^2}\mu_m + \frac{Cy\sigma_m^2}{\sigma_y^2}\right) + \frac{\sigma_w^2}{\sigma_y^2}\mu_m^2 + \frac{\sigma_m^2}{\sigma_y^2} y^2 - \frac{\sigma_m^2 \sigma_w^2}{\sigma_y^4}\left(y^2 - 2yC\mu_m + C^2 \mu_m^2\right)$$

$$= m^2 - 2m(\mu_m + a) + (\mu_m + a)^2 = [m - (\mu_m + a)]^2$$

where

$$a = \frac{C\sigma_m^2}{\sigma_y^2}(y - C\mu_m)$$

We have

$$f(m \mid y) = \frac{1}{\sqrt{2\pi}}\frac{\sigma_y}{\sigma_m\sigma_w}e^{-\frac{1}{2}\left(\frac{\sigma_y}{\sigma_m\sigma_w}\right)^2([m-(\mu_m+a)]^2)} \tag{7.36}$$

It follows from Equation 7.36 that the mean value of m, given y, is

$$\mu_{m|y} = \mu_m + \frac{C\sigma_m^2}{\sigma_y^2}(y - C\mu_m) = \mu_m + K(y - C\mu_m) \tag{7.37}$$

and the variance of the estimate m, given y, is

$$\sigma_{m|y}^2 = \frac{\sigma_m^2\sigma_w^2}{\sigma_y^2} = \sigma_m^2\left(1 - C^2\frac{\sigma_m^2}{\sigma_y^2}\right) = \sigma_m^2(1 - CK) \tag{7.38}$$

where

$$K = \frac{C\sigma_m^2}{\sigma_y^2} = \frac{C\sigma_m^2}{C^2\sigma_m^2 + \sigma_w^2} \tag{7.39}$$

Note: To obtain the final result, Equation 7.38, we have substituted Equation 7.34.

7.4.3 Algorithm of Scalar Static Kalman Filter

As discussed earlier, according to Bayes' formula, μ_m is the *a priori* mean of m and $\mu_{m|y}$ is the *a posteriori* mean of m once the knowledge of data y is incorporated. Then, in a recursive scheme of Equation 7.37, μ_m is the prior estimate of m and $\mu_{m|y}$ is the new (updated or "corrected") estimate of m (after the knowledge of new data y is incorporated). Similarly, in a recursive scheme of Equation 7.38, σ_m is the prior estimate of the std of the estimation error and $\sigma_{m|y}$ is the new (updated or corrected) estimate of the std of the estimation error (after the knowledge of new data y is incorporated). Also, note that the estimated value of m is the mean value, as given by Equation 7.31. It follows that the recursive formulation of the present estimation (static Kalman filter) scheme is

$$\hat{m}_i = \hat{m}_{i-1} + K_i(y_i - C\hat{m}_{i-1}) \tag{7.40a}$$

$$\sigma_{m_i}^2 = \sigma_{m_{i-1}}^2(1 - CK_i) \tag{7.40b}$$

where

$$K_i = \frac{C\sigma_{m_{i-1}}^2}{C^2\sigma_{m_{i-1}}^2 + \sigma_w^2} \tag{7.40c}$$

The iteration process (7.40a) starts with the initial value $\hat{m}_0$ (which should be known). This is the expected value of m before any measurements are known. Typically, this is taken as the "nominal" or "ideal" value of the estimated quantity (in the absence of any error, e.g., as given in a product data sheet). The iteration process (7.40b) starts with the initial value $\sigma_{m0} = \sigma_m$, which is the std of the model error and is known (*Note*: In the standard Kalman filter notation, this std is denoted by σ_v). This initial choice is valid for estimation error because no measurements have been made/used yet (at $i = 0$).

Note: Equation 7.40c may be expressed as

$$K_i = \frac{C\sigma_{m_{i-1}}^2}{C^2\sigma_{m_{i-1}}^2 + \sigma_w^2} = \frac{1}{C} \times \frac{C^2\sigma_{m_{i-1}}^2}{C^2\sigma_{m_{i-1}}^2 + \sigma_w^2} = \frac{1}{C} \times k_i \tag{7.40d}$$

It is seen from Equation 7.40d that K_i is simply the scaling factor between the measurement and the measured quantity (i.e., $1/C$) weighted by k_i. The form of this weighting factor and the overall recursive scheme may be intuitively justified, as indicated now. For this, consider the two extreme cases:

1. Model error is negligible (i.e., $\sigma_m = 0$) in comparison to the measurement error (represented by σ_w).
2. Measurement error is negligible (i.e., $\sigma_w = 0$) in comparison to the model error (represented by σ_m).

In Case 1, from Equation 7.40d, it is seen that $k = 0$ and $K = 0$, and hence, $\hat{m}_i = \hat{m}_{i-1}$. This means the iteration will not rely on the new measurement and will retain the previous estimate. This is justifiable because the measurement process is not accurate while the model is accurate, in this case.

In Case 2, from Equation 7.40d, it is seen that $k = 1$ and $K = 1/C$, and hence, $\hat{m}_i = y_i/C$. This means that we completely rely on the new measurement and disregard the previous estimate. This is justifiable because the measurement process is very accurate now while the model is not.

For situations in between these two extremes, Equation 7.40d properly weights the estimate on the relative accuracies of the model and the measurement. Nevertheless, it is clear from Equation 7.40b that the estimation variance will steadily decrease with each additional recursion.

It will be seen in Section 7.5 that the recursive equations (7.40a through c) are quite similar to a recursive formulation of Kalman filter. However, as noted earlier, the present "static" scheme does not need the "predictor" step and uses only the "corrector" step of a Kalman filter. The present scheme is represented in Figure 7.5. The parameter K is the "Kalman gain."

Example 7.6

A structural monitoring system uses a semiconductor strain gauge to measure the strain at some critical location of a structure. The system consists of the strain-gauge, bridge circuit electronics, which converts the change in resistance of the strain gauge into a voltage proportionately, and a data acquisition and processing system, which records the voltage signal and digitally processes it as required (see Figure 7.6).

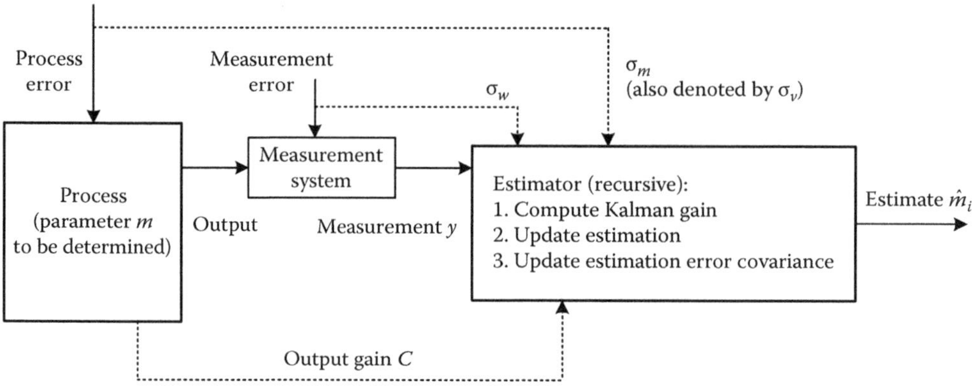

FIGURE 7.5
Static Kalman filter scheme (Bayes').

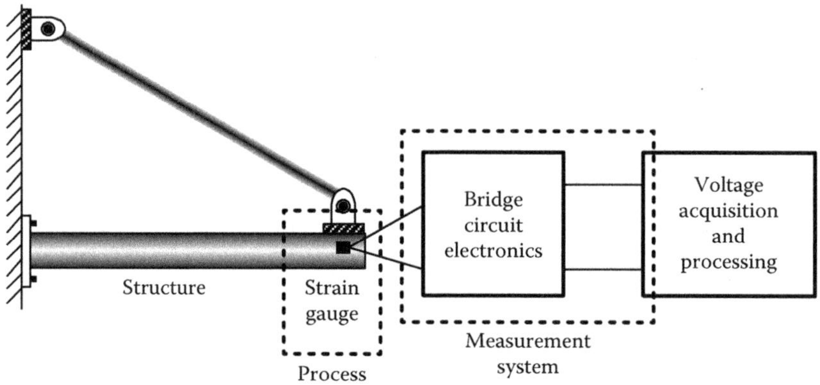

FIGURE 7.6
Structural monitoring using a strain-gauge sensor.

In this example, the strain that is measured by the strain gauge represents the "process." The strain gauge, bridge circuit and recording of its output voltage represent the measurement system. The following information is known:

Calibration constant of the monitoring system = 400 με/V

Note: 1 με = 1 microstrain = 1×10^{-6} strains

std of the process (model) error: $\sigma_m = 5.0$ με

std of the measurement error: $\sigma_w = 2.0$ mV

Measurements (V): [1.99, 2.10, 2.05, 1.98, 1.99, 2.08, 2.09, 2.10, 2.09, 2.11]

Note: This is a static (not dynamic) problem (with the error-free part of the model = 1).

$$C = \frac{1}{[400]}\ V/\mu\varepsilon$$

Case 1

Let us initialize the iteration at $\hat{m} = 800$ με (this seems a reasonable nominal value for the strain, in view of the given data and the calibration constant).

The following MATLAB program is used to get the recursive static Kalman filter results:

```
>> zw=2.0; %measurement std in mV
>> zw=zw*0.001; %change to V
>> zm1=5.0; % model standard deviation in microstrains
>> C=1/400.0; %measurement gain
>> y=[1.99,2.10,2.05,1.98,1.99,2.08,2.09,2.10,2.09,2.11]; %measurements
>> zw2=zw^2; %square
>> m(1)=800; %initialize
>> zm(1)=zm1; %initialize
>> zm2=zm1^2; %square
>> %iteration
>> for i=1:10
K=C*zm2/(C^2*zm2+zw2); %MLE update gain
m(i+1)=m(i)+K*(y(i)-C*m(i));
zm2=zm2*(1-C*K); %update model variance
zm(i+1)=sqrt(zm2); % updated model std
end
>> m
>> zm
>> %Results
m =
  Columns 1 through 10
  800.0000   796.0998   817.7725   818.5087   811.9237   808.7552
812.6129    815.9417   818.9394   820.8296
  Column 11
  823.1408
zm =
  Columns 1 through 10
    5.0000   0.7900   0.5621   0.4599   0.3987   0.3569   0.3259
0.3018   0.2824   0.2663
  Column 11
    0.2527
```

According to this result, we may use 823.14 με as the estimated strain. Note that the std of the estimate has improved (error has decreased) as the iteration progresses.

Case 2

Next let us change the starting value of the estimated parameter to 0 (instead of the more logical value of 800 με). We get the following MATLAB results:

```
m =
  Columns 1 through 10
         0   776.1310   807.6619   811.7398   806.8362   804.6800
809.2140  813.0266   816.3876   818.5605
  Column 11
  821.0980
zm =
  Columns 1 through 10
    5.0000   0.7900   0.5621   0.4599   0.3987   0.3569   0.3259
0.3018   0.2824   0.2663
  Column 11
    0.2527
```

It is seen that the final estimate is almost the same (821.10 με), with the same level of estimation error.

Case 3

Now let us increase the level of measurement error to $\sigma_w = 50.0$ mV (equivalent to 20.0 με) with the same starting value (0) for $\hat{m}$. We get the following MATLAB results:

```
m =
  Columns 1 through 10
         0    46.8235    90.8889    129.2632    162.4000    192.5714
221.6364   248.3478   273.0000   295.5200
  Column 11
  316.6154
zm =
  Columns 1 through 10
    5.0000    4.8507    4.7140    4.5883    4.4721    4.3644    4.2640
4.1703    4.0825    4.0000
  Column 11
    3.9223
```

The final estimate now is 316.62 με, which is clearly inaccurate and unacceptable. The reason is as follows. In this case, the measurement error is much greater than the model error, and hence, the estimation has placed less importance on the new measurements. Even though we started with a very poor guess (0) for the estimated value, the estimation did not improve rapidly for this reason. The estimated error std of the estimation was also high (3.922 compared to 0.2527 before) due to this reason. Had we proceeded further with more steps of iteration, the estimate would improve.

The results from these three cases are plotted in Figure 7.7. In particular, it is seen from Figure 7.7b that the std of the estimation gradually converges to a steady value.

Next, we use the conventional Gaussian MLE procedure where we estimate the parameters (mean and std) of a Gaussian distribution that would most likely generate the given measurement data y. We get the following MATLAB result:

```
>> mle_estimate=mle(y); %standard Gaussian mle estimate
>> mle_estimate=mle_estimate/C; %scale back to model parameter(strain)
>> mle_estimate
mle_estimate =
  823.2000   19.6611
```

This estimate is almost the same as what we obtained first with the static Kalman estimation. Also note that the std obtained from this procedure (19.661 με) is quite large because it relied entirely on the measurements and did not differentiate between model error and measurement error.

Finally, we directly compute the mean and std of the given measurement data using MATLAB. We have

```
>> mean(y)
ans =
  2.0580
>> var(y)
ans =
  0.0027
>> std(y)
ans =
  0.0518
```

When scaled back to microstrains, the mean value is 823.20 με. This is in fact the "least-squares" estimate and is also the estimate obtained from the standard MLE with Gaussian distribution. The std of the estimate is 0.0518 V (or 20.72 με), which is also comparable to what was obtained from the standard Gaussian MLE. The LSE method

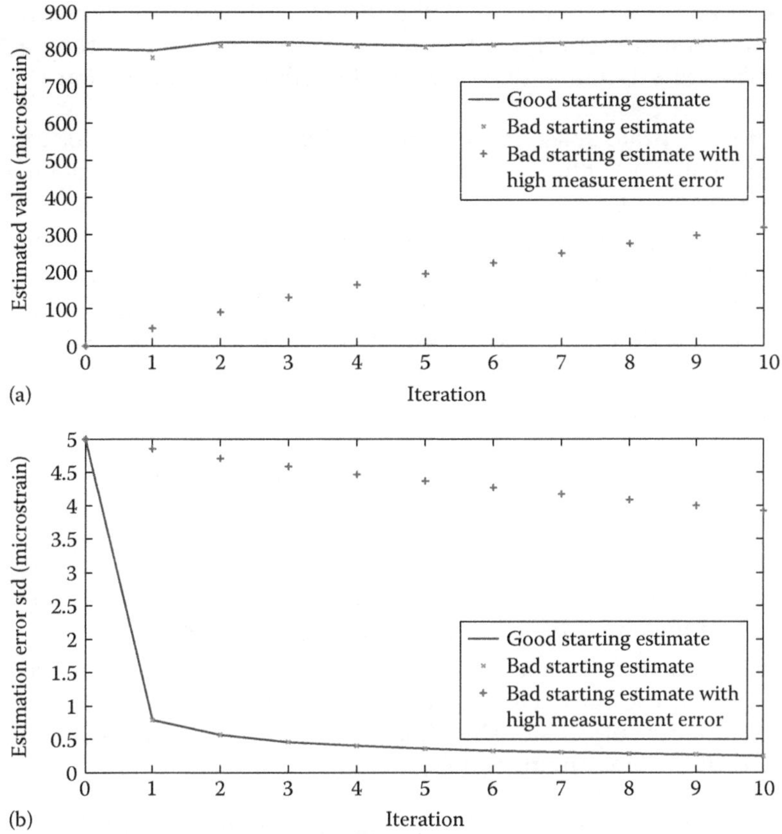

FIGURE 7.7
(a) Estimated strain and (b) standard deviation of the estimation error.

does not have the ability to indicate whether much of the error comes from the model or the measurement process. In general, when only a few data values are available, the Kalman filter–style recursive MLE is more desirable than the LSE, as we have seen from the earlier results in this example, since the LSE is simply the average value of the data. Furthermore, as expected, the LSE estimate is identical to the value obtained using the conventional MLE with Gaussian distribution.

7.5 Linear Multivariable Dynamic Kalman Filter

In the previous section, we presented a "static" Kalman filter where the process model was the estimated quantity itself. Since the model was an "identity" or "no-change" operation, in that case (model = 1 in the scalar situation), the "predictor" stage of the Kalman filter (which depends on the process model) was not needed. Only the "corrector" stage was needed. Now, we present the Kalman filter scheme for a linear multivariable dynamic system. This requires both the predictor stage (which uses the process model) and the corrector stage (which uses the measured data).

Kalman filter is an "optimal estimator." It performs estimation by minimizing the error covariance of the estimate. A "dynamic" Kalman filter is able to estimate (or observe) unknown variables of a dynamic system using the following information:

1. A linear model (state-space model) of the dynamic system including statistics (covariance matrix V) of the random disturbances that enter the system (including random model error).
2. A linear relationship between the measurable output variables of the dynamic system and the variables to be estimated. This relationship may be represented by an "output formation matrix" (also called "measurement gain matrix") C and also includes random noise (i.e., random measurement error) with covariance matrix W.
3. Output measurements.

As noted earlier, a general Kalman filter uses a "predictor–corrector" approach, which consists of the following two steps:

1. Predict the unknown variables and the associated error covariance matrix. This is the *a priori* estimate. This "predictor" step uses the process model and the covariance matrix of the input disturbances (including model error).
2. Correct the predicted variables and the associated error covariance matrix. This is the *a posteriori* estimate. This "corrector" step uses the measured data (output measurement), output relationship (measurement gain matrix), and the covariance matrix of the measurement noise.

Next, we present some preliminaries, which need to be understood before proceeding to the algorithm of the dynamic Kalman filter. They concern the model of the dynamic system, response of the system, discrete-time model, controllability (reachability), and observability (constructability).

7.5.1 State-Space Model

The standard linear dynamic Kalman filter uses a linear dynamic model, with constant coefficients, of the system that generates the variables to be estimated. The dynamic system has r inputs and m outputs. In particular, a linear time-invariant (i.e., constant coefficient) state-space model is used in this Kalman filter. Such a model may be expressed as given in the following.

State equation (coupled first-order linear differential equations) is

$$\dot{x} = Ax + Bu \tag{7.41}$$

and output equation (coupled algebraic equations) is

$$y = Cx + Du \tag{7.42}$$

where

$x = [x_1, x_2, \ldots, x_n]^T$ is the nth-order state vector
$u = [u_1, u_2, \ldots, u_r]^T$ is the rth-order input vector
$y = [y_1, y_2, \ldots, y_m]^T$ is the mth-order output vector

All these vectors are defined as column vectors.

Also,

A is the system matrix

B is the input distribution/gain matrix

C is the output/measurement formation/gain matrix

D is the input feedforward matrix

The state vector represents the "dynamic state" of the system. Its order (n) is the order (or the dynamic size) of the system. Some or all the state variables $\mathbf{x}$ may not be directly measurable (for various reasons), and it is the state variables that are estimated using the measured outputs and the Kalman filter. It is assumed that the outputs $\mathbf{y}$ are measurable (but may contain measurement noise). Also, the model itself may have errors, which may be assumed "additive" and hence represented as disturbance (random) inputs in Equation 7.41.

Note: Often, the matrix ***D*** is zero. It is nonzero only when the system has the feedforward character. An illustrative example is given next.

Example 7.7

The rigid output shaft of a diesel engine prime mover is running at known angular velocity $\Omega(t)$. It is connected through a friction clutch to a flexible shaft, which in turn drives a hydraulic pump (see Figure 7.8a). A linear model for this system is schematically shown in Figure 7.8b. The clutch is represented by a viscous rotatory damper of damping constant B_1 (units: torque/angular velocity). The stiffness of the flexible shaft is K (units: torque/rotation). The pump is represented by a wheel of moment of inertia J (units: torque/angular acceleration), and its fluid load by a viscous damper of damping constant B_2.

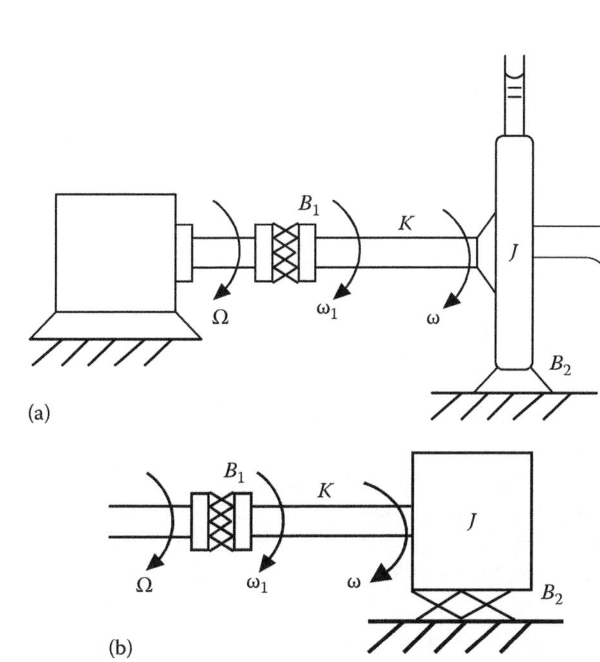

FIGURE 7.8
(a) Diesel engine–driven pump and (b) linear model.

We can write two state equations for this second-order system, relating the state variables T and ω to the input Ω where T is the torque in the flexible shaft and ω is the pump speed. We will consider two cases of system output:

(a) Output ω_1 is the angular speed at the left end of the shaft.
(b) Output ω is the pump speed.

The state equations are derived now.

The constitutive (physical) relation for shaft K is

$$\frac{dT}{dt} = K(\omega_1 - \omega) \tag{i}$$

The constitutive relation for damper B_1 is

$$T = B_1(\Omega - \omega_1) \tag{ii}$$

Substitute Equation ii into Equation i:

$$\frac{dT}{dt} = -\frac{K}{B_1}T - K\omega + K\Omega \tag{iii}$$

This is first state equation.

The constitutive equation for inertia J is

$$J\dot{\omega} = T - T_2 \tag{iv}$$

The constitutive relation for damper B_2 is

$$T_2 = B_2\omega \tag{v}$$

Substitute Equation v in Equation iv:

$$\frac{d\omega}{dt} = -\frac{B_2}{J}\omega + \frac{1}{J}T \tag{vi}$$

This is the second state equation.

The vector-matrix form of the state equations iii and vi is

$$\begin{bmatrix} \frac{dT}{dt} \\ \frac{d\omega}{dt} \end{bmatrix} = \begin{bmatrix} \frac{-K}{B_1} & -K \\ \frac{1}{J} & -\frac{B_2}{J} \end{bmatrix} \begin{bmatrix} T \\ \omega \end{bmatrix} + \begin{bmatrix} K \\ 0 \end{bmatrix} \Omega$$

with the state vector $x = [^T\ \omega]^T$ and the input $u = [\Omega]$.

Now we consider the two cases of output.

Case (a): The output equation is given by ii, which can be written as $\omega_1 = \Omega - (T/B_1)$.

The associated matrices of the model are

$$\boldsymbol{A} = \begin{bmatrix} \frac{-K}{B_1} & -K \\ \frac{1}{J} & -\frac{B_2}{J} \end{bmatrix}; \quad \boldsymbol{B} = \begin{bmatrix} K \\ 0 \end{bmatrix}; \quad \boldsymbol{C} = \begin{bmatrix} \frac{-1}{B_1} & 0 \end{bmatrix}; \quad \boldsymbol{D} = [1]$$

In this case, we notice direct "feedforward" of the input Ω into the output ω_1 through the clutch B_1.

Case (b): In this case, the output is simply the second state variable. Then we have the new matrices $\boldsymbol{C}$ and $\boldsymbol{D}$:

$$C = [0\,1]; \quad D = [0]$$

7.5.2 System Response

The system will respond to both initial condition $\mathbf{x}(0)$ and input $\boldsymbol{u}$. The overall response is given by

$$\boldsymbol{x}(t) = e^{At}\boldsymbol{x}(0) + \int_0^t e^{A(t-\tau)}\boldsymbol{B}(\tau)\boldsymbol{u}(\tau)d\tau = \Phi(t)\boldsymbol{x}(0) + \int_0^t \Phi(t-\tau)\boldsymbol{B}\boldsymbol{u}(\tau)d\tau \tag{7.43}$$

The matrix exponential is the *state transition matrix*:

$$\Phi(t) = e^{At} \tag{7.44}$$

According to Cayley–Hamilton theorem in linear algebra, the state transition matrix may be expressed as

$$\Phi(t) = e^{At} = \alpha_0\boldsymbol{I} + \alpha_1\boldsymbol{A} + \cdots + \alpha_{n-1}\boldsymbol{A}^{n-1} \tag{7.45}$$

where the coefficients α_j are exponential functions of the eigenvalues λ_i of $\boldsymbol{A}$. They can be determined as the solution of the simultaneous algebraic equations:

$$\begin{aligned} e^{\lambda_1 t} &= \alpha_0 + \alpha_1\lambda_1 + \cdots + \alpha_{n-1}\lambda_1^{n-1} \\ &\vdots \\ e^{\lambda_n t} &= \alpha_0 + \alpha_1\lambda_n + \cdots + \alpha_{n-1}\lambda_n^{n-1} \end{aligned} \tag{7.46}$$

7.5.3 Controllability and Observability

Controllability (*reachability*) assures that the state vector can be moved to any desired value (vector) in a finite time using the inputs. Observability (*constructibility*) assures that the state vector can be computed using the measured output. The following definitions formally present these two properties of a dynamic system.

Definition 7.1

A linear system is controllable at time t_0 if we can find an input $\boldsymbol{u}$ that transfers any arbitrary state $\boldsymbol{x}(t_0)$ to the origin ($\boldsymbol{x} = 0$) in some finite time t_1. If this is true for any t_0, then the system is said to be completely controllable.

A linear time-invariant system is

$$\text{Controllable (reachable) iff} \quad \text{Rank}\left[\boldsymbol{B} \,|\, \boldsymbol{AB} \,|\cdots|\, \boldsymbol{A}^{n-1}\boldsymbol{B}\right] = n \tag{7.47a}$$

For the dynamic system in Example 7.7, the *controllability matrix* is

$$[B, AB] = \begin{bmatrix} K & \frac{-K^2}{B_1} \\ 0 & \frac{K}{J} \end{bmatrix}$$

Since the rank of this matrix is 2 (because the determinant is nonzero), the system is controllable (i.e., reachable).

Definition 7.2

A linear system is observable (*constructible*) at time t_0 if we can completely determine the state $x(t_0)$ from the output measurements $\boldsymbol{y}$ over the duration $[t_0, t_1]$ for finite t_1. If this is true for any t_0, then the system is said to be completely observable.

A linear time-invariant system is

$$\text{Observable (constructible) iff} \quad \text{Rank}\left[C^T \mid A^T C^T \mid \cdots \mid A^{n-1} C^T\right] = n \tag{7.48a}$$

Note: Iff means "If and only if."

In the dynamic system of Example 7.7, for Part (a) the *observability matrix* is

$$[C^T, A^T C^T] = \begin{bmatrix} \frac{-1}{B_1} & \frac{-K}{B_1^2} \\ 0 & \frac{K}{B_1} \end{bmatrix}$$

Since the rank of this matrix is 2 (because the determinant is nonzero), the system is observable.

In the dynamic system of Example 7.7, for Part (b) the *observability matrix* is

$$[C^T, A^T C^T] = \begin{bmatrix} 0 & \frac{1}{J} \\ 1 & \frac{-B_2}{J} \end{bmatrix}$$

Again, since the rank of this matrix is 2 (because the determinant is nonzero), the system is observable.

7.5.4 Discrete-Time State-Space Model

Kalman filter equations in discrete time are based on the following discrete-time state-space model, which is equivalent to the continuous-time state-space model given by Equations 7.41 and 7.42:

$$\boldsymbol{x}_i = \boldsymbol{F}\boldsymbol{x}_{i-1} + \boldsymbol{G}\boldsymbol{u}_{i-1} + \boldsymbol{v}_{i-1} \tag{7.49}$$

$$\boldsymbol{y}_i = \boldsymbol{C}\boldsymbol{x}_i + \boldsymbol{D}\boldsymbol{u}_i + \boldsymbol{w}_i \tag{7.50}$$

The vectors v and w represent input disturbances (or model error, assumed "additive") and output (measurement) noise, respectively, which are assumed to be independent Gaussian white noise (i.e., zero-mean Gaussian random signals with a constant *power spectral density function*) whose covariance matrices are V and W.

Note: $V = E[vv^T]$ and $W = E[ww^T]$.

In the discrete model given by Equations 7.49 and 7.50, the signals are sampled at constant time periods (sampling period) T, and within each sampling period, the signal value is assumed constant (i.e., a *zero-order hold* is assumed). Hence, subscript i represents the signal value at the discrete-time point iT. Specifically, $x_i = x(iT)$.

The matrices F and G in the discrete model are related to the matrices A and B of the continuous model, because the discrete model is derived from the continuous model. The applicable relations are as follows:

$$F = \Phi(T) = e^{AT} \tag{7.51}$$

$$G = \int_0^T e^{A\tau} d\tau B = \int_0^T \Phi(\tau) d\tau B \approx T\Phi(T)B \tag{7.52}$$

where Φ is the state transition matrix.

The equation for discrete-time controllability is

$$\text{Controllable (reachable) iff} \quad \text{Rank}\left[G \,|\, FG \,|\cdots|\, F^{n-1}G\right] = n \tag{7.47b}$$

The equation for discrete-time observability is

$$\text{Observable (constructible) iff} \quad \text{Rank}\left[F^T \,|\, F^T C^T \,|\cdots|\, F^{n-1}C^T\right] = n \tag{7.48b}$$

7.5.5 Linear Kalman Filter Algorithm

The discrete-time equations of the standard linear, dynamic Kalman filter are presented next. The following assumptions are made in the derivation of these equations:

1. The dynamic system is linear and time invariant, and the associated model matrices (F, G, C, D) are known.
2. All outputs y are measurable (albeit with possible measurement noise).
3. The dynamic system is observable.
4. Input disturbances (or model error) and output (measurement) noise are additive, independent, and Gaussian white noise, with known covariance matrices V and W.

Typically, it is also assumed that in Equation 7.42 $D = 0$. Modifying the Kalman filter equations to include D is simple and straightforward.

Kalman filter is an "optimal" estimation method where the unknown variables (the state vector) are estimated by minimizing the error covariance of the estimate, given the measured data, as time goes to infinity (i.e., as we progress in the filter recursion). Specifically, at time step i, suppose that the actual state vector is x_i and its estimated value (using the measured data) is $\hat{x}_i$. The associated estimation error is

$$e_i = x_i - \hat{x}_i \tag{7.53}$$

Its covariance (the error covariance matrix) $\boldsymbol{P}_i$ is given by

$$\boldsymbol{P}_i = E[\boldsymbol{e}_i\boldsymbol{e}_i^T] \tag{7.54}$$

Kalman filter minimizes the trace of the error covariance matrix $\boldsymbol{P}_i$, as $t\rightarrow\infty$. Also, Kalman filter maintains that the error is zero mean, and hence,

$$E[\boldsymbol{x}_i] = \hat{\boldsymbol{x}}_i \tag{7.55}$$

As indicated earlier, Kalman filter follows a two-step process of "prediction" and "correction." In the prediction step, *a priori* values $\hat{\boldsymbol{x}}_i^-$ and $\boldsymbol{P}_i^-$, of the state estimate and estimation error covariance, respectively, are determined. (*Note*: The superscript "−" denotes the *a priori* estimates.) In the correction step, by using actual output measurements $\boldsymbol{y}_i$, these *a priori* estimates are "corrected" to the *a posteriori* estimates $\hat{\boldsymbol{x}}_i$ and $\boldsymbol{P}_i$. The associated equations are given as follows:

Predictor step (*a priori* estimation)

$$\hat{\boldsymbol{x}}_i^- = F\hat{\boldsymbol{x}}_{i-1} + \boldsymbol{G}\boldsymbol{u}_{i-1} \tag{7.56}$$

$$\boldsymbol{P}_i^- = \boldsymbol{F}\boldsymbol{P}_{i-1}\boldsymbol{F}^T + \boldsymbol{V} \tag{7.57}$$

Corrector step (*a posteriori* estimation)

$$\boldsymbol{K}_i = \boldsymbol{P}_i^-\boldsymbol{C}^T\left(\boldsymbol{C}\boldsymbol{P}_i^-\boldsymbol{C}^T + \boldsymbol{W}\right)^{-1} \tag{7.58}$$

$$\hat{\boldsymbol{x}}_i = \hat{\boldsymbol{x}}_i^- + \boldsymbol{K}_i\left(\boldsymbol{y}_i - \boldsymbol{C}\hat{\boldsymbol{x}}_i^-\right) \tag{7.59}$$

$$\boldsymbol{P}_i = (\boldsymbol{I} - \boldsymbol{K}_i\boldsymbol{C})\boldsymbol{P}_i^- \tag{7.60}$$

7.5.5.1 Initial Values of Recursion

The initial values $\hat{\boldsymbol{x}}_0$ and $\boldsymbol{P}_0$ of the estimated state vector and the error covariance matrix are needed to begin the Kalman filter recursion. These are chosen using prior information and/or intuition. In particular, we use

$$\boldsymbol{P}_0 = \boldsymbol{V}$$

The initial estimate $\hat{\boldsymbol{x}}_0$ is typically the initial value $\boldsymbol{x}(0)$ of the state vector, which should be known.

The computational procedure of a discrete Kalman filter is shown in Figure 7.9. Observe the prediction and correction stages of this "dynamic" Kalman filter. Compare this with Figure 7.5, for a "static" Kalman filter, which has only the correction stage. The matrix $\boldsymbol{K}_i$ is called the "Kalman gain." Note the similarity of the expression of this matrix to the scalar expression (7.40c) of the static scalar Kalman filter. Also, Equation 7.59 corresponds to Equation 7.40a, and Equation 7.60 corresponds to Equation 7.40b.

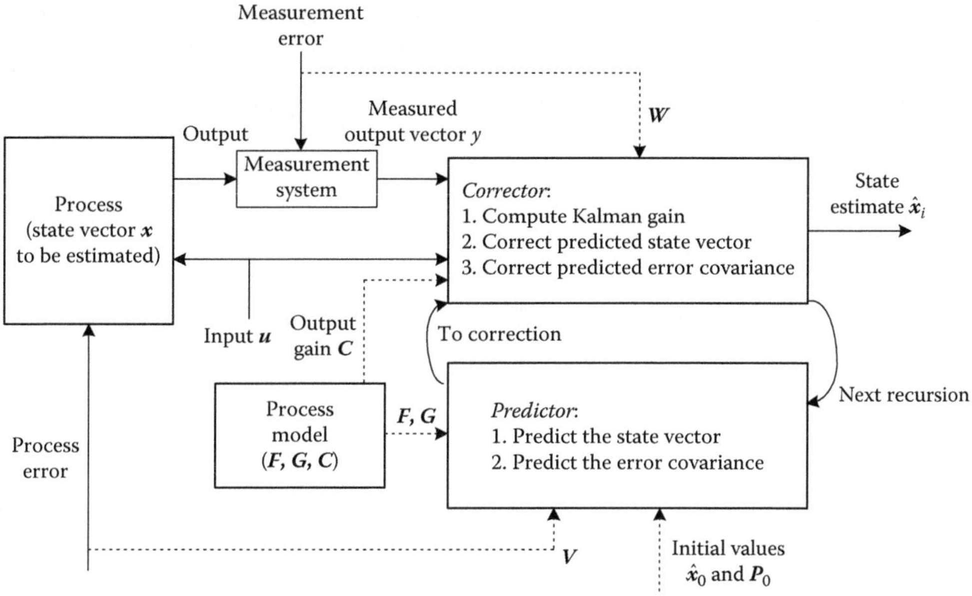

FIGURE 7.9
Computational scheme of a discrete dynamic Kalman filter.

Example 7.8

Consider the diesel engine system of Example 7.7. Suppose that the system parameters are $J = 0.5$, $K = 1.0$, $B_1 = 0.2$, and $B_2 = 0.25$. As earlier, the input is Ω, which is the diesel engine speed, and the state variables are T, which is the torque in the flexible shaft, and ω, which is the pump speed. Take the pump speed as the system output. Then, the matrices defining the system are

$$\boldsymbol{A} = \begin{bmatrix} -5.0 & -1.0 \\ 2.0 & -0.5 \end{bmatrix}; \quad \boldsymbol{B} = \begin{bmatrix} 1.0 \\ 0 \end{bmatrix}; \quad C = [0 \quad 1]; \quad \boldsymbol{D} = [0]$$

Normally, it is more difficult to measure torque in a shaft than speed, particularly when the shaft is rotating. So, in the present example, suppose that we wish to estimate the torque in the flexible shaft, when the input speed is 2.0.

Note: The system is observable (constructible) because

$$\left[C^T, A^T C^T\right] = \begin{bmatrix} 0 & 2 \\ 1 & -0.5 \end{bmatrix}$$

is nonsingular.

First, we use MATLAB to determine the eigenvalues of the system:

```
>> A=[-5.0 -1.0; 2.0 -0.5];
>> eig(A)
ans =
   -4.5000
   -1.0000
```

It is seen that the system is stable because both eigenvalues are negative. Also, the units of the eigenvalues are frequency (1/s). Hence, we select the sampling period as $T = 0.02$ s to discretize the system.

Next, we determine the discrete-time system corresponding to the given continuous-time system, using MATLAB:

```
>> B=[1.0;0];
>> C=[0 1];
>> D=[0.0];
>> [F,G]=c2d(A,B,0.02);
```

The resulting state transition matrix $\boldsymbol{F}$ and the input transition matrix $\boldsymbol{G}$ are as follows:

```
>> F
F =
     0.9045   -0.0189
     0.0379    0.9897
>> G
G =
     0.0190
     0.0004
```

Note: A zero-order hold ("zoh") has been used in the discretization.

It is given that the input speed is 2.0. The measurements of the pump speed are simulated as (approximately) its steady-state value (≈0.9) plus random noise (from a random number generator). The model (input disturbance) and measurement covariance matrices are taken as

$$\boldsymbol{V} = \begin{bmatrix} 0.05 & 0 \\ 0 & 0.02 \end{bmatrix}; \quad \boldsymbol{W} = [0.02]$$

We use the following MATLAB program to estimate the shaft torque (and also the pump speed) using the measured pump speed (with random noise).

Note: Estimated pump speed is taken as the correct value, unlike the measured value, which has noise. The error is the difference in these two quantities.

```
>> V=[0.05,0;0,0.02]; % model (input disturbance) error covariance
>> W=[0.02]; % measurement error covariance
>> xe=[0;0]; Pe=V; %initialize state estimate and estimation error
covariance
>> it=[]; meas=[]; est1=[]; est2=[]; err=[];  % declare storage vectors
for measurements, estimated states, and error between measured and
estimated outputs
>> kalmest   %call the defined function kalmest.m with the following
script:
for i=1:100
it(end+1)=i; % store the recursion number
u=2.0; % engine speed input
xe=F*xe+G*u;  % predict the state error
Pe=F*Pe*F'+V;  % predict the estimation error covariance matrix
K=Pe*C'*inv(C*Pe*C'+W);  % compute Kalman gain
y=0.9+randn/20.0; % simulate the speed measurement with noise
xe=xe+K*(y-C*xe); % correct the state error
Pe=(eye(2)-K*C)*Pe; % correct the estimation error covariance
meas(end+1)=y; % store the measured data
est1(end+1)=xe(1);  % store the first state (torque)
est2(end+1)=xe(2);  % store the second state (speed)
```

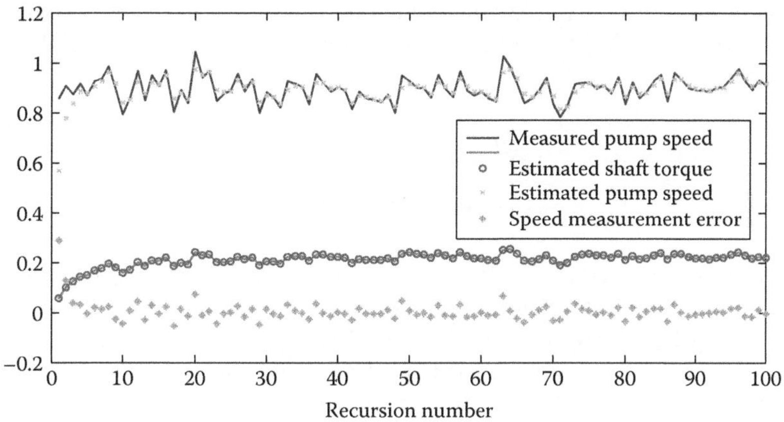

FIGURE 7.10
Results from torque estimation through Kalman filter.

```
err(end+1)=y-C*xe;    %store the speed measurement error
end

% plot the results
>> plot(it,meas,'-',it,est1,'-',it,est1,'o',it,est2,'x',it,err,'*')
>> xlabel('Recursion Number')
>> gtext('Measured speed')
>> gtext('X: Estimated pump speed')
>> gtext('Estimated shaft torque')
>> gtext('Speed measurement error')
>> Pe   %final estimation error covariance
Pe =
    0.2566    0.0049
    0.0049    0.0124
```

The results are shown in Figure 7.10. Note that in the presence of significant error in both model (i.e., input disturbance) and measurement, the Kalman filter has been effective in providing a reasonable estimate for the torque in the shaft. In particular, by comparing the measured speed (shown by the solid curve, which has significant error in view of the use of the steady-state value plus random noise to simulate it) and the estimated speed (shown by the curve with "x"), it is clear that the Kalman filter has virtually eliminated the measurement error.

7.6 Extended Kalman Filter

Often, a linear model is just an approximation to a nonlinear practical system. Also, the measured quantity may not be linearly related to the state variables. When the nonlinearities are significant, a time-invariant linear model will no longer be valid. Then, the linear Kalman filter given in Section 7.5, which assumes a linear time-invariant model, cannot be used. The "extended" Kalman filter modifies the linear Kalman filter algorithm to take into account system nonlinearities.

When a system has nonnegligible nonlinearities, the state equations (7.49) and the output (measurement) equations (7.50) need to be modified using appropriate nonlinear representations as follows:

$$x_i = f(x_{i-1}, u_{i-1}) + v_{i-1} \tag{7.61}$$

$$y_i = h(x_i) + w_i \tag{7.62}$$

where f and h are nonlinear vector functions of order n and m, respectively. *Note*: In the output equation (7.62), the input term has been omitted. As in the linear case, it can be included in a straightforward manner, if the system has "feedforward" characteristics.

As we will see, in the EKF equations, the nonlinear equations (7.61) and (7.62) can be used in that nonlinear form when dealing with the state vector and the measurement vector. However, they need to be linearized when dealing with the error covariance matrix. The reason is when a random signal is propagated through a nonlinearity, its random characteristics (specifically, the covariance matrix) will change in a complex manner. Direct incorporation of such nonlinear transformations into the Kalman filter algorithm is not a simple or straightforward task.

Specifically, for the covariance transformation, we use the following linearized state transition matrix and the output (measurement) gain matrix:

$$F_{i-1} = \frac{\partial f}{\partial x_{i-1}} \tag{7.63}$$

$$C_i = \frac{\partial h}{\partial x_i} \tag{7.64}$$

Equation 7.63 gives the Jacobian matrix (i.e., gradient) of the system and Equation 7.64 gives the Jacobian matrix of the output process. The linearized matrix terms (Equations 7.63 and 7.64) are not constant and have to be evaluated at each recursive step of the EKF, using the current values of the state vector and the input vector in that sampling period. We assume here that the functions f and h are differentiable and that they may be evaluated either analytically or numerically.

Note: $G_{i-1} = \partial f/\partial u_{i-1}$. But this result is not needed in the EKF.

7.6.1 Extended Kalman Filter Algorithm

The linear Kalman filter equations are modified, given as follows, in order to obtain the EKF equations.

Predictor step (*a priori* estimation)

$$\hat{x}_i^- = f(\hat{x}_{i-1}, u_{i-1}) \tag{7.65}$$

$$P_i^- = F_{i-1}P_{i-1}F_{i-1}^T + V \tag{7.66}$$

Corrector step (*a posteriori* estimation)

$$K_i = P_i^- C_i^T \left(C_i P_i^- C_i^T + W\right)^{-1} \tag{7.67}$$

$$\hat{x}_i = \hat{x}_i^- + K_i\left(y_i - h\left(\hat{x}_i^-\right)\right) \tag{7.68}$$

$$P_i = (I - K_i C)P_i^- \tag{7.69}$$

Next, we give an example for the application of the EKF.

Example 7.9

A passive shock-absorber unit has a piston–cylinder mechanism as schematically shown in Figure 7.11. The cylinder is fixed and rigid and is filled with an incompressible hydraulic fluid (on both sides of the piston). The piston mass is m and its area is A. It has a small opening through which the hydraulic fluid can flow from one side of the cylinder to the other side, as the piston moves. This flow generates fluid resistance, which is nonlinear (specifically, pressure drop is quadratically related to the volume flow rate through the opening). A spring of stiffness k resists the movement of the piston. Suppose that the input force applied to the shock absorber (piston) is $f(t)$. Some important variables are defined as follows:

v is the piston velocity.

Q is the fluid volume flow rate through the piston opening (taken positive from left to right).

$P = P_2 - P_1$ is the pressure difference between the two sides of the piston.

f_k is the spring force.

In a state-space model of the system, we use the following variables:

State vector: $\mathbf{x} = [v, f_k]^T$

Input vector: $\mathbf{u} = [f(t)]$

We will make the following assumptions:

1. There is no friction between the piston and the cylinder.
2. The areas of the two sides of the piston are equal (approximately) at value A. That is, neglect the area of the piston rod.

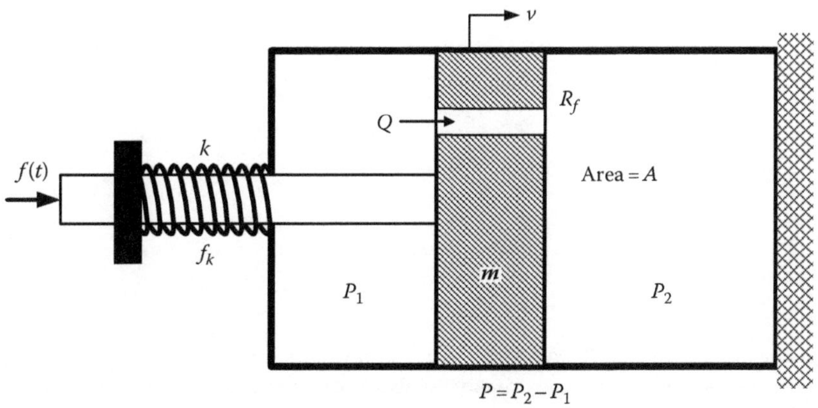

FIGURE 7.11
A passive shock-absorber unit.

We can show that the state equations of the system are

$$m\dot{v} = -cv^2 - f_k + f(t)$$
$$\dot{f}_k = kv \tag{i}$$

where c is a fluid resistance parameter. Note the nonlinearity in the first state equation. Since speed is much easier to measure than the spring force, we will measure the speed only. Hence, the output (measurement) gain matrix is $\mathbf{C} = [1\ 0]$.

Suppose that when the system is discretized, we get the following nonlinear discrete-time state-space model:

$$x_1(i) = -a_1x_1(i-1) - a_2x_1^2(i-1) - a_3x_2(i-1) + b_1u(i-1) + v_1(i-1)$$
$$x_2(i) = a_4x_1(i-1) + a_5x_2(i-1) + b_2u(i-1) + v_2(i-1) \tag{ii}$$

We have included the "additive" random model error (and/or input disturbance) terms v_1 and v_2 in the equation. We will use the following parameter values: $a_1 = 0.4$, $a_2 = 0.1$, $a_3 = 0.2$, $a_4 = 0.3$, $a_5 = 0.5$, $b_1 = 1.0$, and $b_2 = 0.2$.

In view of the model nonlinearity, in this example, we will use the EKF, not the linear Kalman filter. Furthermore, since the spring force is more difficult to measure than speed, we will estimate it through the measurement of the speed of the shock absorber.

Suppose that the input force to the shock absorber is the sinusoidal function

$$u = 2\sin 6t \tag{iii}$$

For the purpose of the present example, the measurements of the absorber speed are simulated as (approximately) values generated by the nonlinear model of the system with added random noise (from a random number generator). The model (input disturbance) and measurement covariance matrices are taken as

$$\mathbf{V} = \begin{bmatrix} 0.02 & 0 \\ 0 & 0.04 \end{bmatrix}; \quad \mathbf{W} = [0.05] \tag{iv}$$

The discrete model (ii) may be expressed as

$$\begin{bmatrix} x_1(i) \\ x_2(i) \end{bmatrix} = \begin{bmatrix} -a_1 - a_2x_1(i-1) & -a_3 \\ a_4 & a_5 \end{bmatrix} \begin{bmatrix} x_1(i-1) \\ x_2(i-1) \end{bmatrix} + \begin{bmatrix} b_1 \\ b_2 \end{bmatrix} u(i-1) + \begin{bmatrix} v_1(i-1) \\ v_2(i-1) \end{bmatrix} \tag{v}$$

On linearization (using the first term of the Taylor series expansion—first derivative), we get

$$\mathbf{F}_{i-1} = \frac{\partial f}{\partial \mathbf{x}_{i-1}} = \begin{bmatrix} -a_1 - 2a_2x_1(i-1) & -a_3 \\ a_4 & a_5 \end{bmatrix} \tag{vi}$$

Also,

$$\mathbf{G} = \begin{bmatrix} b_1 \\ b_2 \end{bmatrix} \tag{vii}$$

We use the following MATLAB program to estimate the spring force (and also the shock-absorber speed) using the measured absorber speed (with random noise).

Note: Estimated speed is taken as the correct value, while the measured speed, which has noise, is not accurate. The measurement error is the difference in these two quantities.

```
>> F=[-0.4, -0.2; 0.3, 0.5]; % define the linear F matrix (nonlinear
term will be added)
>> G=[1.0; 0.2]; % define the G matrix
>> C=[1 0]; % define the output (measurement) gain matrix
>> V=[0.02,0;0,0.04]; % model (input disturbance) error covariance
>> W=[0.05]; % measurement error covariance
>> n=2; m=1; % define system order and output order
>> extkalm
>> extkalm %call the defined function kalmest.m with the following
script:

Fsim=F; % simulation model
xe=zeros(n,1); % initialize state estimation vector
Pe=V; % initialize estimation error covariance matrix
xsim=xe; %initialize state simulation vector
it=[]; meas=[]; est1=[]; est2=[]; err=[];  % declare storage vectors
it=0; meas=0; est1=0; est2=0; err=0; %initialize plotting variables
for i=1:100
it(end+1)=i; % store the recursion number
t=i*0.02;  % time
u=2.0*sin(6*t); % absorber force input (harmonic)
x1=xe(1);
F(1,1)=-0.4-0.1*x1; % include nonlinearity
xe=F*xe+G*u;  % predict the state estimate
F(1,1)=-0.4-0.2*x1; % linearize F
Pe=F*Pe*F'+V;  % predict the estimation error covariance matrix
K=Pe*C'*inv(C*Pe*C'+W);  % compute Kalman gain
Fsim(1,1)=-0.4-0.1*xsim(1); % include nonlinearity for simulation
xsim=Fsim*xsim+G*u;
for j=1:m
yer(j,1)=rand/5.0; % simulated measurement error
end
y=C*xsim+yer; % simulate the speed measurement with noise
xe=xe+K*(y-C*xe); % corrected state estimate
Pe=(eye(2)-K*C)*Pe; % corrected estimation error covariance
x1=xe(1); % update state 1 (speed)
meas(end+1)=y(1); % store the measured data
est1(end+1)=xe(1);  % store the first state (speed)
est2(end+1)=xe(2);  % store the second state (spring force)
err(end+1)=y(1)-C*xe;  %store the speed measurement error
end
% plot the results
plot(it,meas,'-',it,est1,'-',it,est1,'o',it,est2,'x',it,err,'*')
xlabel('Recursion Number')
```

The results are shown in Figure 7.12. Note that in the presence of error in both model (input disturbance) and measurement, the EKF has been effective in providing a reasonable estimate for the spring force. In particular, by comparing the measured speed (solid-line curve), which has significant error in view of the added random noise, and the estimated speed (curve indicated by "o"), it is clear that the EKF has virtually eliminated the measurement error.

Of course, EKF is an improvement over the regular (linear) Kalman filter, when dealing with nonlinear systems. However, it has drawbacks. A major source of error is the fact that a random signal, when propagated through a nonlinear system, can have rather different statistical characteristics than what we get when the same signal is propagated through a linearized model of the nonlinear system. This problem is eliminated in UKF, which is presented next.

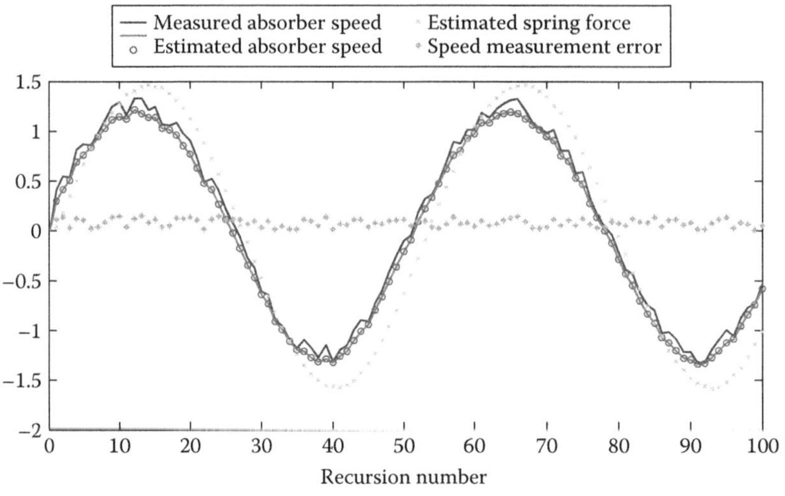

FIGURE 7.12
Results from force estimation through extended Kalman filter.

7.7 Unscented Kalman Filter

In the linear dynamic Kalman filter, we assume that the process disturbances (including model error) and the measurement noise are Gaussian (normal) white with zero mean. Also, we assume that the process model and the measurement model (output gain or measurement gain) are linear. If a process is linear, any input Gaussian random signal will remain Gaussian at the output. A Gaussian random signal only needs the mean and the variance (or std) for its complete representation. Furthermore, according to the central limit theorem, many practical (engineering) random signals (which have contributions from many independent random causes) may be approximated as Gaussian.

If the process (including the measurement process) is nonlinear, one solution is to locally linearize it about the operating point. (*Note*: Then, the operating point itself will vary with time.) The linear Kalman filter can be applied at each operating point, using the corresponding linearized model, as the operating point changes with time. This is the principle behind the EKF. An obvious drawback of EKF is the fact that Gaussian random signals, when propagated through a nonlinearity, will not remain Gaussian, and hence just mean and variance would not be adequate to represent the propagated result. More importantly, the mean and variance of the output of the linearized model will not be the same as the true mean and variance of the actual nonlinear process for the same input signal. In EKF, even if the data are propagated through the actual nonlinearity, since the covariance is propagated through the linearized model, the covariance will be corrupted. With the zero-mean assumption for noise, it is the covariance that is particularly important in Kalman filtering. Another drawback of EKF is that it needs derivatives (i.e., Jacobians) of the process nonlinearities f and the measurement nonlinearities g. In some situations (e.g., Coulomb friction), these derivatives may not exist. In summary, the following are the main shortcomings of the EKF:

1. The model has to be linearized. This is not possible if the nonlinearity is not differentiable.
2. Since the linearized model has to be recomputed at each time step, the computational effort is significantly greater than for the linear Kalman filter.
3. A Gaussian random signal does not remain Gaussian when propagated through a nonlinearity. Hence, a linearized model, which retains the Gaussian nature of a propagated signal, does not properly reflect the true random behavior of the propagated signal.

A solution to these shortcomings of the EKF has been found through the UKF. The UKF overcomes the problems of EKF by directly seeking to recover the true mean and covariance of a signal (vector) that is propagated through a nonlinearity, without actually linearizing the nonlinearity. Specifically, in the UKF, what is propagated through the nonlinearity are not the measured data but the statistical representations, called sigma points, of the data. Sigma points more authentically represent the statistical parameters (mean and covariance) of the data. In particular, it has been found that with the UKF the covariance of the results is far more accurate than what is achieved by the EKF. These two approaches of nonlinear Kalman filtering are schematically compared in Figure 7.13.

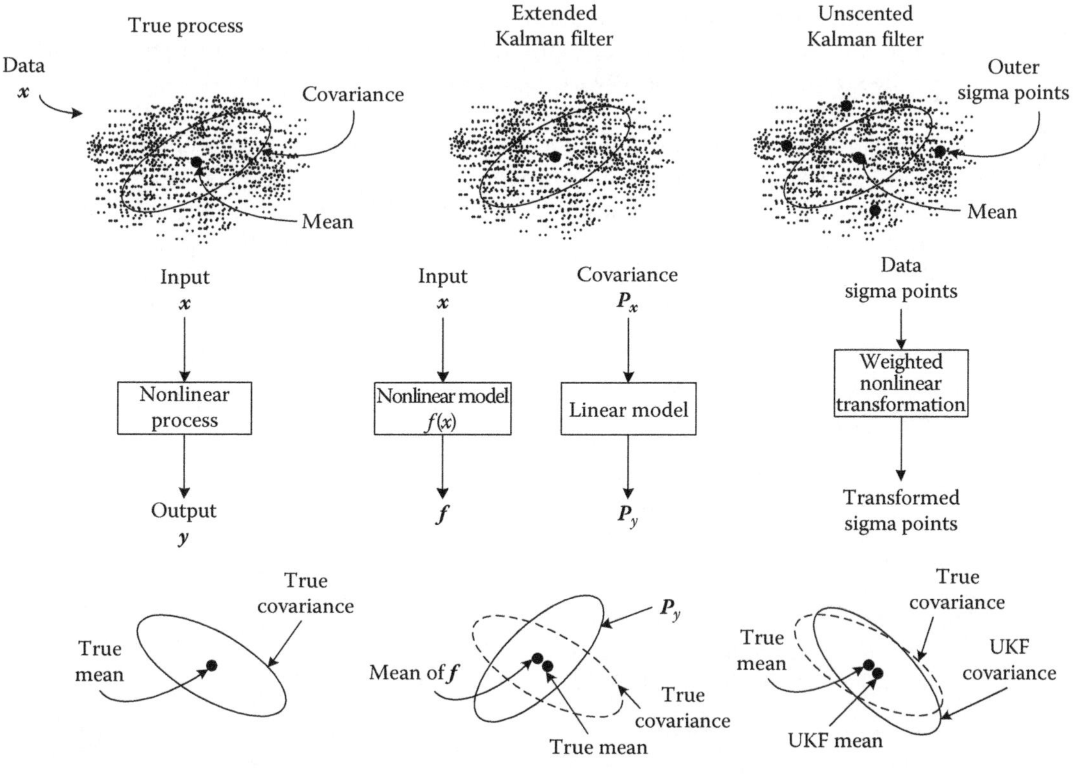

FIGURE 7.13
Schematic comparison of extended Kalman filter and unscented Kalman filter.

7.7.1 Unscented Transformation

The UKF uses the "unscented transformation" (UT), which is presented next. Suppose that a set of data vectors x (that has randomness) is propagated through nonlinearity $z = f(x)$. We use the UT to determine the statistics of the output z of the nonlinearity. The UT involves two main steps: (1) generation of the sigma-point vectors and the corresponding weights for the data set and (2) computation of the statistics (mean vector and covariance matrix) of the results (output z of the nonlinearity). These two steps are formulated in the following.

7.7.1.1 Generation of Sigma-Point Vectors and Weights

1. Formulating $2n + 1$ *sigma points,* we get

$$\begin{aligned} \chi_0 &= \hat{x} \\ \chi_j &= \hat{x} + \left[\sqrt{(n+\lambda)P_x}\right]_j \\ \chi_{j+n} &= \hat{x} - \left[\sqrt{(n+\lambda)P_x}\right]_j, \quad j = 1,2,\ldots,n \end{aligned} \tag{7.70}$$

where
$\hat{x}$ is the sample mean of the data
P_x is the sample covariance matrix of the data
$[\sqrt{A}]_j$ is the jth column of the Cholesky decomposition of the (+ve definite) matrix A
n is the dimension (order) of the data vector x
$\lambda = \alpha^2 n - n$ is the scaling parameter and α is the parameter governing the spread of the sigma points (typically of the order of 0.001)

Note: The Cholesky decomposition of a +ve definite matrix is a product of a lower triangular matrix and its conjugate transpose (an upper triangular matrix). It is somewhat similar to a matrix square root. The MATLAB result of `chol()` is an upper triangular matrix. Either the lower triangular matrix or the upper triangular matrix may be used in the present method. The product of the lower triangular matrix and the upper triangular matrix (in that order) gives the original matrix.

Example

```
>> A=[2 0.5 0.2;0.6 3 0.1;0.3 0.4 5];
>> L=chol(A)
L =
    1.4142    0.3536    0.1414
         0    1.6956    0.0295
         0         0    2.2314
>>  B=L*L'
B =
    2.1450    0.6036    0.3156
    0.6036    2.8759    0.0658
    0.3156    0.0658    4.9791
>> C=L'*L
C =
    2.0000    0.5000    0.2000
    0.5000    3.0000    0.1000
    0.2000    0.1000    5.0000
```

It is seen that the number of sigma points $(2n + 1)$ is directly related to the dimension (n) of the data vector $\boldsymbol{x}$. This is needed to give adequate coverage to the data spread. One sigma point is the mean value itself of the data. The remaining $2n$ sigma points are properly chosen around the mean to adequately cover the randomness of the data.

2. Formulating *2n + 1 weights*, we get

$$w_0^m = \frac{\lambda}{(n+\lambda)}$$

$$w_0^c = \frac{\lambda}{(n+\lambda)} + (1-\alpha^2+\beta) \tag{7.71}$$

$$w_j^m = w_j^c = \frac{1}{[2(n+\lambda)]}, \quad j = 1,2,\ldots,2n$$

where

β is the parameter representing prior knowledge of the probability distribution of $\boldsymbol{x}$ (for Gaussian distribution, $\beta = 2$)

superscript $()^m$ denotes "mean"

superscript $()^c$ denotes "covariance"

It is seen that more weight is given to the central lambda point (the mean), which is intuitively appealing. More importantly, it can be easily verified that the mean and the covariance of the weighted sum of the $2n + 1$ lambda-point vectors are equal to the mean and the covariance of the random data.

7.7.1.2 Computation of Output Statistics

1. Propagate the sigma-point vectors through the nonlinearity:

$$\xi_j = f(\chi_j), \quad j = 0,1,2,\ldots,2n \tag{7.72}$$

2. Compute the mean and the covariance matrix of the output z as the weighted sums:

$$\text{Mean } \bar{z} = \sum_{j=0}^{2n} w_j^m \xi_j$$

$$\text{Co-variance } \boldsymbol{P}_z = \sum_{j=0}^{2n} w_j^c (\xi_j - \bar{z})(\xi_j - \bar{z})^T \tag{7.73}$$

It is seen that in the UT what is propagated through the nonlinearity are the sigma-point vectors and not the actual data points. These sigma-point vectors adequately represent the random behavior of the data. UT determines the statistics of the output of the nonlinearity using the statistics of the input data. In particular, it is seen from Equation 7.70 that

1. The weighted sum of the sigma-point vectors of the data is equal to the sample mean $(\hat{x})$ of the data
2. The weighted sample covariance of the sigma-point vectors of the data is equal to the covariance $(\boldsymbol{P}_x)$ of the data

Hence, one may argue that what is propagated through the nonlinearity is an adequate characterization of the randomness of the data. The UT accomplishes this while retaining the nonlinearity of the process (i.e., without having to linearize) and also without having to propagate every data point (which results in reduced computational burden). These are important advantages of UKF over the method used in the EKF.

7.7.2 Unscented Kalman Filter Algorithm

We now use the UT, as presented earlier, to formulate the UKF. Consider the following nonlinear dynamic system, expressed in the discrete-time state-space form, as earlier:

$$\boldsymbol{x}_i = \boldsymbol{f}(\boldsymbol{x}_{i-1}, \boldsymbol{u}_{i-1}) + \boldsymbol{v}_{i-1} \tag{7.61}$$

$$\boldsymbol{y}_i = \boldsymbol{h}(\boldsymbol{x}_i) + \boldsymbol{w}_i \tag{7.62}$$

where $\boldsymbol{f}$ and $\boldsymbol{h}$ are nonlinear vector functions of order n and m, and they represent the nonlinear process dynamics and the nonlinear output/measurement relation, respectively. The vectors $\boldsymbol{v}$ and $\boldsymbol{w}$ represent the input disturbances (and/or process model error) and the output (measurement) noise, respectively, which are assumed to be additive, independent Gaussian white noise whose covariance matrices are $\boldsymbol{V}$ and $\boldsymbol{W}$.

As in other Kalman filter algorithms, the UKF also has the predictor stage (*a priori* estimation) and the corrector stage (*a posteriori* estimation), but now the UT is used in these stages. The associated computational steps are given in the following:

1. Initialize the computation (at $i = 0$) with the initial mean and covariance of the state vector: $\hat{\boldsymbol{x}}_0$ and $\boldsymbol{P}_0$. We choose $\hat{\boldsymbol{x}}_0$ as the first (central) sigma-point vector (i.e., mean) and hence $\boldsymbol{x}(0)$. We have to compute $2n$ outer sigma-point vectors for this initial time point (and recursively for the future time points i). Note that the order of the state vector is n. We select

$$\boldsymbol{P}_0 = \boldsymbol{V}$$

   This has to be a +ve definite matrix (as required by the Cholesky decomposition), which is the case for $\boldsymbol{V}$.

2. Compute the fixed parameter and the sigma-point weights:

$$\lambda = \alpha^2 n - n$$

$$\begin{aligned} w_0^m &= \frac{\lambda}{(n+\lambda)} \\ w_0^c &= w_0^m + (1 - \alpha^2 + \beta) \\ w_j^m &= w_j^c = \frac{1}{[2(n+\lambda)]}, \quad j = 1, 2, \ldots, 2n \end{aligned} \tag{7.74}$$

   An alternative choice of weights is

$$\begin{aligned} w_0^m &= w_0^c = w_0 < 1 \\ w_j^m &= w_j^c = \frac{1 - w_0}{2n} = \frac{1}{2c}, \quad j = 1, 2, \ldots, 2n \end{aligned} \tag{7.75}$$

Note: In this alternative choice, we first select w_0 as a value less than 1. Then, we select the remaining equal $2n$ weights such that the sum of all the weights is 1. In the case of this alternative choice of weighting, in Equation 7.76, use c $(=n/(1-w_0))$ in place of $n + \lambda$.

3. Recursively perform the following for all the time points $i = 1, 2, 3, \ldots$.

In the following first stage of the recursion, we compute the "predicted" estimates (i.e., *a priori* estimates) analogous to the procedure in other types of Kalman filter.

a. Compute the $2n + 1$ sigma-point vectors:

$$\begin{aligned}\chi_{0,i-1} &= \hat{x}_{i-1}\\ \chi_{j,i-1} &= \hat{x}_{i-1}+\left[\sqrt{(n+\lambda)P_{i-1}}\right]_j\\ \chi_{j+n} &= \hat{x}_{i-1}-\left[\sqrt{(n+\lambda)P_{i-1}}\right]_j, \quad j=1,2,\ldots,n\end{aligned} \tag{7.76}$$

b. Propagate the sigma-point vectors through the process nonlinearity (Equation 7.61):

$$\chi_{j,i|i-1} = f(\chi_{j,i-1}, u), \quad j = 0,1,2,\ldots,2n \tag{7.77}$$

c. Compute the "predicted" state estimate as the weighted sum:

$$\hat{x}_i^- = \sum_{j=0}^{2n} w_j^m \chi_{j,i|i-1} \tag{7.78}$$

d. Compute the "predicted" state estimation error covariance matrix as the weighted sum together with the added contribution from the input/model disturbances:

$$P_i^- = \sum_{j=0}^{2n} w_j^c \left[\chi_{j,i|i-1} - \hat{x}_i^-\right]\left[\chi_{j,i|i-1} - \hat{x}_i^-\right]^T + V \tag{7.79}$$

e. Propagate the sigma-point vectors through the measurement nonlinearity (Equation 7.62):

$$\gamma_{j,i|i-1} = h(\chi_{j,i-1}) \quad j = 0,1,2,\ldots,2n \tag{7.80}$$

f. Compute the "predicted" measurement vector as the weighted sum:

$$\hat{y}_i^- = \sum_{j=0}^{2n} w_j^m \gamma_{j,i|i-1} \tag{7.81}$$

In the following second stage of the recursion, we compute the "corrected" estimates (i.e., *a posteriori* estimates), by incorporating the actual output measurements, analogous to the procedure in other types of Kalman filter.

g. Compute the output estimation error autocovariance matrix as the weighted sum together with the added contribution from the measurement noise:

$$P_{yy,i} = \sum_{j=0}^{2n} w_j^c \left[\gamma_{j,i|i-1} - \hat{y}_i^- \right]\left[\gamma_{j,i|i-1} - \hat{y}_i^- \right]^T + W \tag{7.82}$$

h. Compute the state-output estimation error cross-covariance matrix as the weighted sum:

$$P_{xy,i} = \sum_{j=0}^{2n} w_j^c \left[\chi_{j,i|i-1} - \hat{x}_i^- \right]\left[\gamma_{j,i|i-1} - \hat{y}_i^- \right]^T \tag{7.83}$$

i. Compute the Kalman gain matrix:

$$K_i = P_{xy,i} P_{yy,i}^{-1} \tag{7.84}$$

j. Compute the "corrected" (i.e., *a posteriori*) state estimate:

$$\hat{x}_i = \hat{x}_i^- + K_i \left(y_i - \hat{y}_i^- \right) \tag{7.85}$$

where y_i is the actual output measurement at time point i.

k. Compute the "corrected" (i.e., *a posteriori*) state estimation error covariance matrix:

$$P_i = P_i^- - K_i P_{yy,i} K_i^T \tag{7.86}$$

As one might expect, economizations are possible in the computation steps of the UKF, in view of the particular structures and characteristics of the associated matrices. These considerations are beyond the scope of the present treatment.

Example 7.10

In this example, we solve the same problem presented in Example 7.9, but this time using the UKF. Specifically, we use the following MATLAB program to estimate the spring force (and also the shock-absorber speed) using measured absorber speed (with random noise).

```
>> F=[-0.4, -0.2; 0.3, 0.5]; % define the F matrix
>> G=[1.0; 0.2]; % define the G matrix
>> C=[1 0]; % define the output (measurement) gain matrix
>> V=[0.02,0;0,0.04]; % model (input disturbance) error covariance
>> W=[0.05]; % measurement error covariance
>> n=2; m=1; % system order and output order
>> ukalm   %call the defined function ukalm.m with the following script:

Fsim=F; % simulation model
xe=zeros(n,1); % initialize state estimation vector
Pe=V; % initialize estimation error covariance matrix
xsim=xe; %initialize state simulation vector
alpha=0.001; % define sigma-point spread parameter
lambda=alpha^2*n-n; % scaling parameter
beta=2; % Gaussian distribution
wm(1)=lambda/(n+lambda); % central weighting for mean
```

```
wc(1)=wm(1)+(1-alpha^2+beta); % central weighting for covariance
wm(2:2*n+1)=1/(2*(n+lambda)); % outer weighting for mean
wc(2:2*n+1)=wm(2:2*n+1); % outer weighting for covariance
it=[]; meas=[]; est1=[]; est2=[]; err=[];  % declare storage vectors
it=0; meas=0; est1=0; est2=0; err=0; %initialize plotting variables
for i=1:100
it(end+1)=i; % store the recursion number
t=i*0.02;  % time
u=2.0*sin(6*t); % input (harmonic)
xlam(:,1)=xe; % central lambda-point vector
L=chol((n+lambda)*Pe); % Cholesky cannot be used if not +ve definite
for j=2:n+1
xlam(:,j)=xe+L(:,j-1); % outer lambda-point vectors
xlam(:,n+j)=xe-L(:,j-1); % outer lambda-point vectors
end
sumx=zeros(n,1); % initialize sum
for j=1:2*n+1
x1=xlam(1,j); % first element of lambda-point vector
F(1,1)=-0.4-0.1*x1; % include nonlinearity
xlamx=xlam(:,j); % jth lambda-point vector
xlam(:,j)=F*xlam(:,j)+G*u;  % nonlinear state propagation of lambda-
point vector
sumx=sumx+xlam(:,j)*wm(j); % weighted sum
end
xe=sumx; % predicted state estimate
Pe=zeros(n,n); % initialize sum for covariance computation
for j=1:2*n+1
xlamx=xlam(:,j); % jth lambda-point vector
xlamx=xlamx-xe; %subtract mean
Pe=Pe+wc(j)*xlamx*xlamx'; % estimate error covariance matrix
end
Pe=Pe+V; % add contribution from model/input disturbances
sumy=zeros(m,1); % initialize sum
for j=1:2*n+1
xlamx=xlam(:,j); % jth lambda-point vector
xlamy(:,j)=C*xlamx;  % output (measurement) lambda-point vector
sumy=sumy+xlamy(:,j)*wm(j); % weighted sum
end
ye=sumy; % predicted output estimate
Pyy=zeros(m,m); % initialize the output auto-covariance matrix
for j=1:2*n+1
ylam=xlamy(:,j); % output lambda-point vector
ylam=ylam-ye; %subtract mean
Pyy=Pyy+wc(j)*ylam*ylam'; % estimate error covariance matrix
end
Pyy=Pyy+W; % add contribution from measurement noise
Pxy=zeros(n,m); % initialize the state-output cross-covariance matrix
for j=1:2*n+1
xlamx=xlam(:,j); % jth lambda-point vector
ylam=xlamy(:,j); % jth output lambda-point vector
xlamx=xlamx-xe; %subtract mean
ylam=ylam-ye; %subtract mean
Pxy=Pxy+wc(j)*xlamx*ylam'; % estimate error covariance matrix
end
K=Pxy*inv(Pyy);  % Kalman gain
Fsim(1,1)=-0.4-0.1*xsim(1); % include nonlinearity
xsim=Fsim*xsim+G*u;
for j=1:m
yer(j,1)=rand/5.0; % simulated measurement error
end
y=C*xsim+yer; % simulated measurement with noise
```

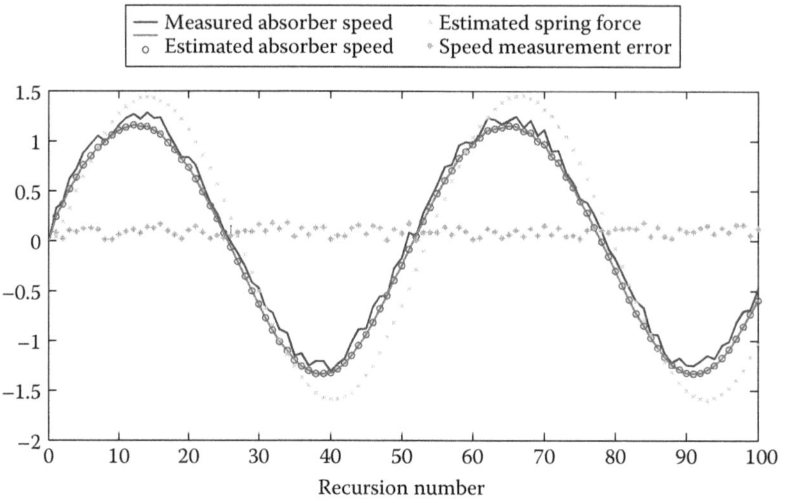

FIGURE 7.14
Results from force estimation through unscented Kalman filter.

```
xe=xe+K*(y-ye); % corrected state estimate
Pe=Pe-K*Pyy*K'; % corrected estimation error covariance
meas(end+1)=y(1); % store the measured data
est1(end+1)=xe(1);  % store the first state (speed)
est2(end+1)=xe(2);  % store the second state (spring force)
err(end+1)=y(1)-xe(1);  %store the speed measurement error
end
% plot the results
plot(it,meas,'-',it,est1,'-',it,est1,'o',it,est2,'x',it,err,'*')
xlabel('Recursion Number')
```

In this program, we have used the original scheme of weighting as given by Equation 7.74. The results are shown in Figure 7.14. It is seen that in the presence of error in both model (input disturbance) and measurement, the UKF has been effective in providing a good estimate for the spring force. In particular, by comparing the measured speed (solid-line curve), which has significant error in view of the added random noise, and the estimated speed (curve indicated by "o"), it is clear that the UKF has virtually eliminated the measurement error. Furthermore, by comparing Figure 7.14 with Figure 7.12, it is clear that speed estimate is smoother and more sinusoidal with the UKF, which is closer to the exact value (sinusoidal).

As another exercise of this example, we used the alternative weighting scheme given by Equation 7.75. The results were quite similar to what is presented in Figure 7.14.

Summary Sheet

Sources of data error: (1) Production process of measured object, (2) measurement process (mounting, error in sensor and other hardware, etc.), and (3) data processing and computation (including models used in computation).

Two main categories of error: (1) model error and (2) measurement error.

Estimation through sensed data: The estimated quantity may be (1) a parameter (e.g., mass) or variable (e.g., voltage) of a system or (2) scalar or vector. The system may be (1) static (rates of changes of variables are not important) or (2) dynamic (rates of changes of variables are nonnegligible).

Randomness in data and estimate: Assume: Data set $\{Y_1, Y_2, \ldots, Y_N\}$ is independent and identically distributed (iid). Randomness may be represented by its variance or standard deviation (std). The std of estimate ($\hat{m}$) and std data (measured quantity) are related through $\sigma_{\hat{m}} = \sigma_m/\sqrt{N}$.

Unbiased estimate: Unbiased if, mean of estimate = estimated quantity.

Sample mean ($\bar{Y}$): $\bar{Y} = 1/N\sum_{i=1}^{N} Y_i$ (unbiased estimate because $E(\bar{Y}) = E(Y_i) = \mu$).

Sample variance (S^2): $S^2 = 1/(N-1)\sum_{i=1}^{N}(Y_i - \bar{Y})^2$ (unbiased estimate because $E(S^2) = \mathrm{Var}(Y_i) = \sigma^2$).

Least-squares estimation (LSE): Minimize average squared error between data and estimate.

Least-squares point estimate: (1) Measure N data values $\{Y_1, Y_2, \ldots, Y_N\}$, and (2) minimize the sum of squared error between estimate and data: $e = \sum_{i=1}^{N}(Y_i - m)^2$. Estimate = $\hat{m} = 1/N\sum_{i=1}^{N} Y_i$ = *sample mean* of data.

Least-squares line estimate: Minimize squared error between the data set and a line (linear, quadratic, etc.).

Linear regression line: Fitted line is linear (straight line). Two parameters (slope m and intercept a) of the line are estimated. Equations:

$$m = \frac{\left(1/N\sum_{i=1}^{N} X_i Y_i - \bar{X}\bar{Y}\right)}{\left(1/N\sum_{i=1}^{N} X_i^2 - \bar{X}^2\right)}; \quad Y - \bar{Y} = m(X - \bar{X}); \quad a = \bar{Y} - m\bar{X}$$

Quality of estimate: Depends on (1) accuracy of data, (2) size of data set, (3) method of estimation, (4) model used for estimation (e.g., linear fit, quadratic fit), and (5) number of estimated parameters.

Sum of squares error (SSE): $\mathrm{SSE} = \sum_{i=1}^{N}(Y_i - Y_i)^2$; *Note*: $\hat{Y}_i$ is an estimate.

Mean square error (MSE): $\mathrm{MSE} = 1/(N-M)\sum_{i=1}^{N}(Y_i - Y_i)^2$

M = estimated number of coefficients (of fitted cure); $N - M$ = residual degrees of freedom

Root mean square error (RMSE): Square root of MSE.

***R*-Square**: $1 - \mathrm{SSE}/\left(\sum_{i=1}^{N}(Y_i - \bar{Y})^2\right)$.

Adjusted *R*-Square: 1–(MSE/VAR);
MSE = mean square error; VAR = sample variance.

Note: For a given set of data, when the number of coefficients in the fitted curve increases, the accuracy of the estimates decreases.

Maximum likelihood estimation (MLE): Maximize the likelihood of the estimated value, given the set of data that we have (i.e., we estimate the proper parameter value for the [random] process so that it would "most likely" generate the data set that we have).

Note: A Gaussian (i.e., normal) probability distribution needs only two parameters (mean and variance) for its complete representation. If the data are Gaussian, the estimates from LSE and MLE are equivalent. If not Gaussian, LSE is generally better.

MLE objective: Find the estimate $\hat{\boldsymbol{m}}$ of the parameter vector $\boldsymbol{m}$ so as to maximize the likelihood function: $L(\boldsymbol{m}|\boldsymbol{y}) = f(\boldsymbol{y}|\boldsymbol{m})$.

Note: $f(\boldsymbol{y}|\boldsymbol{m})$ = the conditional probability density function (pdf) of $\boldsymbol{y}$, given $\boldsymbol{m}$. This is a function of $\boldsymbol{m}$.

Bayes' theorem (formula): $\Pr(\boldsymbol{m}|\boldsymbol{y}) = (\Pr(\boldsymbol{y}|\boldsymbol{m})\Pr(\boldsymbol{m}))/\Pr(\boldsymbol{y})$
$\Pr(\boldsymbol{m}|\boldsymbol{y})$ = *posteriori* probability of $\boldsymbol{m}$; $\Pr(\boldsymbol{m})$ = *priori* probability of $\boldsymbol{m}$; $\Pr(\boldsymbol{y})$ = *priori* probability of $\boldsymbol{y}$:

→ $\Pr(\boldsymbol{m}|\boldsymbol{y}) \propto \Pr(\boldsymbol{m})L(\boldsymbol{m}|\boldsymbol{y})$

→ Given an estimate of parameter $\boldsymbol{m}$, the best update of the estimate (*a posteriori* estimate) is the one that maximizes the likelihood function $L(\boldsymbol{m}|\boldsymbol{y})$.

MLE with normal distribution:

$$\hat{\mu} = \frac{1}{N}\sum_{i=1}^{N} y_i; \quad \hat{\sigma}^2 = \frac{1}{N}\sum_{i=1}^{N}(y_i - \hat{\mu})^2$$

Note: This variance estimate (MLE) is not unbiased. The difference is negligible (N vs $N-1$ in the denominator) for large N.

Recursive MLE: $\hat{m}_i = \hat{m}_{i-1} - c(1/(\partial^2 L/\partial m_{i-1}^2))(\partial L/\partial m_{i-1})$
L = likelihood function, c = constant.

Recursive MLE (zero-mean Gaussian):

$$\hat{m}_i = \frac{\sigma_w^2}{\left(\sigma_{m_{i-1}}^2 + \sigma_w^2\right)}\hat{m}_{i-1} + \frac{\sigma_{m_{i-1}}^2}{\left(\sigma_{m_{i-1}}^2 + \sigma_w^2\right)} y_i; \quad \frac{1}{\sigma_{m_i}^2} = \frac{1}{\sigma_{m_{i-1}}^2} + \frac{1}{\sigma_w^2}$$

Discrete MLE: Estimate $\boldsymbol{m} = [m_1, m_2, \ldots, m_n]^T$ and measurement $\boldsymbol{y} = [y_1, y_2, \ldots, y_n]^T$

$$L(\boldsymbol{m}\,|\,\boldsymbol{y}) = P(\boldsymbol{y}\,|\,\boldsymbol{m}) = \begin{bmatrix} p_{11} & p_{12} & \cdots & p_{1n} \\ p_{21} & p_{22} & \cdots & p_{2n} \\ \cdot & \cdot & \cdot & \cdot \\ p_{n1} & p_{n2} & \cdots & p_{nn} \end{bmatrix}$$

Bayes' formula → probability vector of estimate of $\boldsymbol{m}$, given data $\boldsymbol{y}$: $P(\boldsymbol{m}|\boldsymbol{y}) = aP(\boldsymbol{y}|\boldsymbol{m})P(\boldsymbol{m})$

Note: $P(\boldsymbol{m})$ = probability vector of prior estimate; choose a such that probability elements of the vector $P(\boldsymbol{m}|\boldsymbol{y})$ add to 1.

Scalar static Kalman filter algorithm: $\hat{m}_i = \hat{m}_{i-1} + K_i(y_i - C\hat{m}_{i-1})$; $\sigma^2_{mi} = \sigma^2_{mi-1}(1 - CK_i)$, with (Kalman gain) $K_i = C\sigma^2_{mi-1}/(C^2\sigma^2_{mi-1} + \sigma^2_w)$.

Assumptions: (1) When there is no measurement error, the measured (estimated) parameter m and the measurement (data) y are linearly related through a known constant gain C (output/measurement gain). (2) Model error (e.g., manufacturing error of a product, product mounting during usage) and measurement error (e.g., sensor error, signal acquisition error) are independent and Gaussian (normal) random variables. (3) Model error has zero mean and std σ_m. Use $\sigma_{m0} = \sigma_m$ (since no measurements have been used yet). (4) Measurement error w has zero mean and std σ_w.

Note: (1) If model error is negligible (i.e., $\sigma_m = 0$) in comparison to measurement error (σ_w), estimate relies heavily on initial value (not the measurements). (2) If measurement error is negligible (i.e., $\sigma_w = 0$) in comparison to model error (σ_m), estimate relies heavily on measurements (not the initial value).

State-space model (continuous time): $\dot{\boldsymbol{x}} = \boldsymbol{Ax} + \boldsymbol{Bu}$; $\boldsymbol{y} = \boldsymbol{Cx} + \boldsymbol{Du}$ $\boldsymbol{x} = [x_1, x_2, \ldots, x_n]^T$ = nth-order state vector; $\boldsymbol{u} = [u_1, u_2, \ldots, u_r]^T$ = rth-order input vector; $\boldsymbol{y} = [y_1, y_2, \ldots, y_m]^T$ = mth-order output vector; $\boldsymbol{A}$ = system matrix; $\boldsymbol{B}$ = input distribution/gain matrix; $\boldsymbol{C}$ = output/measurement formation/gain matrix; $\boldsymbol{D}$ = input feedforward matrix.

System response: $\boldsymbol{x}(t) = e^{\boldsymbol{A}t}\boldsymbol{x}(0) + \int_0^t e^{\boldsymbol{A}(t-\tau)}\boldsymbol{B}(\tau)\boldsymbol{u}(\tau)d\tau = \boldsymbol{\Phi}(t)\boldsymbol{x}(0) + \int_0^t \Phi(t-\tau)\boldsymbol{Bu}(\tau)d\tau$

$\boldsymbol{\Phi}(t) = e^{\boldsymbol{A}t}$ = state transition matrix = $\alpha_0\boldsymbol{I} + \alpha_1\boldsymbol{A} + \cdots + \alpha_{n-1}\boldsymbol{A}^{n-1}$.

Note: α_j are functions of the eigenvalues λ_i of $\boldsymbol{A}$ and given by the solution of

$$\begin{aligned} e^{\lambda_1 t} &= \alpha_0 + \alpha_1\lambda_1 + \cdots + \alpha_{n-1}\lambda_1^{n-1} \\ &\vdots \\ e^{\lambda_n t} &= \alpha_0 + \alpha_1\lambda_n + \cdots + \alpha_{n-1}\lambda_n^{n-1} \end{aligned}$$

Controllability (reachability): State vector $\boldsymbol{x}$ can be moved to any desired value (vector) in a finite time using the inputs $\boldsymbol{u}$. Controllable iff Rank $[\boldsymbol{B}|\boldsymbol{AB}|\cdots|\boldsymbol{A}^{n-1}\mathbf{B}] = n$, which is the system order, which is the state-vector order.

Observability (constructibility): State vector $\boldsymbol{x}$ can be determined using the output vector $\boldsymbol{y}$ measured over a finite duration. Observable iff Rank $[\boldsymbol{C}^T|\boldsymbol{A}^T\boldsymbol{C}^T|\cdots|\boldsymbol{A}^{n-1}\boldsymbol{C}^T] = n$.

State-space model (discrete time): $\boldsymbol{x}_i = \boldsymbol{Fx}_{i-1} + \boldsymbol{Gu}_{i-1} + \boldsymbol{v}_{i-1}$; $\boldsymbol{y}_i = \boldsymbol{Cx}_i + \boldsymbol{Du}_i + \boldsymbol{w}_i$
$\boldsymbol{v}$ = input disturbances (model error, assumed "additive") and $\boldsymbol{w}$ = output (measurement) noise, are independent Gaussian white noise (i.e., zero-mean Gaussian random signals with a constant *power spectral density function*), and covariance matrices $\boldsymbol{V}$ and $\boldsymbol{W}$ ($\boldsymbol{V} = E[\boldsymbol{vv}^T]$ and $\boldsymbol{W} = E[\boldsymbol{ww}^T]$). $\boldsymbol{F} = \Phi(T) = e^{\boldsymbol{A}T}$; $\boldsymbol{G} = \int_0^T e^{\boldsymbol{A}\tau}d\tau\,\boldsymbol{B} = \int_0^T \Phi(\tau)d\tau\boldsymbol{B} \approx T\Phi(T)\boldsymbol{B}$.

Discrete-time controllability: Controllable (reachable) iff Rank $[\boldsymbol{G}|\boldsymbol{FG}|\cdots|\boldsymbol{F}^{n-1}\boldsymbol{G}] = n$.

Discrete-time observability: Observable (constructible) iff Rank $[\boldsymbol{F}^T|\boldsymbol{F}^T\boldsymbol{C}^T|\cdots|\boldsymbol{F}^{n-1}\boldsymbol{C}^T] = n$.

Linear Kalman filter algorithm (multivariable dynamic): Estimates ($\hat{\boldsymbol{x}}$), the state vector $\boldsymbol{x}$.

Assumptions: (1) Dynamic system is linear and time invariant, and the associated model matrices ($\boldsymbol{F}, \boldsymbol{G}, \boldsymbol{C}, \boldsymbol{D}$) are known. (2) All outputs $\boldsymbol{y}$ are measurable (measurement noise may be present). (3) Dynamic system is observable. (4) Input disturbances (or model error) and output (measurement) noise are additive, independent, Gaussian white noise, with known covariance matrices V and W. *Note*: Typically, $D = 0$ (but can be included easily).

Predictor step (*a priori* estimation): $\hat{x}_i^- = F\hat{x}_{i-1} + Gu_{i-1}$; $P_i^- = FP_{i-1}F^T + V$.

Corrector step (*a posteriori* estimation): $K_i = P_i^- C^T (CP_i^- C^T + W)^{-1}$; $\hat{x}_i = \hat{x}_i^- + K_i(y_i - C\hat{x}_i^-)$; $P_i = (I - K_i C)P_i^-$.

Initial values: $\hat{x}_0 = x(0)$; $P_0 = V$. *Note*: P_i = error covariance matrix at time point i.

Extended Kalman filter (EKF) algorithm (for nonlinear systems):
Nonlinear system: $x_i = f(x_{i-1}, u_{i-1}) + v_{i-1}$; $y_i = h(x_i) + w_i$.
Uses Jacobians (gradients) of nonlinearities: $F_{i-1} = \partial f/\partial x_{i-1}$; $C_i = \partial h/\partial x_i$.

Predictor step (*a priori* estimation): $\hat{x}_i^- = f(\hat{x}_{i-1}, u_{i-1})$; $P_i^- = F_{i-1}P_{i-1}F_{i-1}^T + V$.

Corrector step (*a posteriori* estimation): $K_i = P_i^- C_i^T (C_i P_i^- C_i^T + W)^{-1}$; $\hat{x}_i = \hat{x}_i^- + K_i(y_i - h(\hat{x}_i^-))$; $P_i = (I - K_i C)P_i^-$.

Shortcomings of EKF: (1) The model has to be linearized (not possible if nonlinearity is not differentiable). (2) Linearized model has to be computed at each time step → high computational burden. (3) A Gaussian random signal does not remain Gaussian when propagated through a nonlinearity → a linearized model (retains Gaussian nature of signal) does not properly reflect the true random behavior of the propagated signal.

Unscented Kalman filter (UKF) algorithm (for nonlinear systems):
Note: In UKF, what is propagated through the nonlinearity are not the data but the statistical representations called "sigma points" of the data. Sigma points more authentically represent the statistical parameters (mean and covariance) of the data.

Initialization: $\hat{x}_0 = x(0)$; $P_0 = V$; α = parameter governing sigma-point spread (≈0.001); $\lambda = \alpha^2 n - n$ = scaling parameter; weights: $w_0^m = \lambda/(n+\lambda)$; $w_0^c = w_0^m + (1 - \alpha^2 + \beta)$; $w_j^m = w_j^c = 1/[2(n+\lambda)]$, $j = 1, 2, \ldots, 2n$.

Note: Another choice of weights: $w_0^m = w_0^c = w_0$; $w_j^m = w_j^c = (1 - w_0)/2n = 1/(2c)$, $j = 1, 2, \ldots, 2n$.

Then, use c in place of $n + \lambda$ in the term $\sqrt{(n+\lambda)P_{i-1}}$ in the expression below.

Predicted (i.e., *a priori*) estimates:

(a) Compute the $2n + 1$ sigma-point vectors:

$$\chi_{0,i-1} = \hat{x}_{i-1}; \quad \chi_{j,i-1} = \hat{x}_{i-1} + \left[\sqrt{(n+\lambda)P_{i-1}}\right]_j;$$

$$\chi_{j+n} = \hat{x}_{i-1} - \left[\sqrt{(n+\lambda)P_{i-1}}\right]_j; \quad j = 1, 2, \ldots, n$$

(b) Propagate the sigma-point vectors through the process nonlinearity: $\chi_{j,i|i-1} = f(\chi_{j,i-1}, u)$; $j = 0, 1, 2, \ldots, 2n$.

(c) Compute the "predicted" state estimate as the weighted sum: $\hat{x}_i^- = \sum_{j=0}^{2n} w_j^m \chi_{j,i|i-1}$.

(d) Compute the "predicted" state estimation error covariance matrix as:

$$P_i^- = \sum_{j=0}^{2n} w_j^c \left[\chi_{j,i|i-1} - \hat{x}_i^-\right]\left[\chi_{j,i|i-1} - \hat{x}_i^-\right]^T + V$$

(e) Propagate the sigma-point vectors through the measurement nonlinearity: $\gamma_{j,i|i-1} = \boldsymbol{h}(\boldsymbol{\chi}_{j,i-1})$; $j = 0, 1, 2, \ldots, 2n$.

(f) Compute the "predicted" measurement vector as the weighted sum: $\hat{\boldsymbol{y}}_i^- = \sum_{j=0}^{2n} w_j^m \gamma_{j,i|i-1}$.

Corrected (i.e., *a posteriori*) estimates:

(g) Compute the output estimation error autocovariance matrix:

$$\boldsymbol{P}_{yy,i} = \sum_{j=0}^{2n} w_j^c \left[\gamma_{j,i|i-1} - \hat{\boldsymbol{y}}_i^-\right]\left[\gamma_{j,i|i-1} - \hat{\boldsymbol{y}}_i^-\right]^T + \boldsymbol{W}$$

(h) Compute the state-output estimation error cross-covariance matrix:

$$\boldsymbol{P}_{xy,i} = \sum_{j=0}^{2n} w_j^c \left[\chi_{j,i|i-1} - \hat{\boldsymbol{x}}_i^-\right]\left[\gamma_{j,i|i-1} - \hat{\boldsymbol{y}}_i^-\right]^T$$

(i) Compute the Kalman gain matrix: $\boldsymbol{K}_i = \boldsymbol{P}_{xy,i}\boldsymbol{P}_{yy,i}^{-1}$.

(j) Compute the "corrected" (i.e., *a posteriori*) state estimate: $\hat{\boldsymbol{x}}_i = \hat{\boldsymbol{x}}_i^- + \boldsymbol{K}_i(\boldsymbol{y}_i - \hat{\boldsymbol{y}}_i^-)$.
Note: $\boldsymbol{y}_i$ = actual output measurement at time point i.

(k) Compute the "corrected" (i.e., *a posteriori*) state estimation error covariance matrix: $\boldsymbol{P}_i = \boldsymbol{P}_i^- - \boldsymbol{K}_i\boldsymbol{P}_{yy,i}\boldsymbol{K}_i^T$.

Problems

7.1 Use a Gaussian random number generator (MATLAB `normrnd(μ,σ)`) with $\mu = 1.0$ and $\sigma = 0.2$ to generate a sequence of 21 random numbers Y_i. Then, recursively compute 21 values of sample mean $\bar{Y}$ and sample standard deviation S according to the following formulas:

$$\bar{Y}_1 = Y_1$$

$$\bar{Y}_{i+1} = \frac{1}{(i+1)}\left(i \times \bar{Y}_i + Y_{i+1}\right), \quad i = 1, 2, \ldots$$

$$S_1^2 = 0$$

$$S_{i+1}^2 = \frac{1}{i}\left[S_i^2 \times (i-1) + \left(Y_{i+1} - \bar{Y}_{i+1}\right)^2\right]$$

Plot the two curves of $\bar{Y}_i$ and S_i against i. Next, generate and use 51 data values to obtain the same plots. Comment on the results.

Note: In the mathematical notation, a normal distribution is denoted by $N(\mu, \sigma^2)$, where the "variance" σ^2 is used. However, the corresponding MATLAB function is `normrnd(μ,σ)` where the "standard deviation" σ is used.

7.2 The ideal calibration curve of a sensor is given by $y = ax^p$, where x is the measured quantity (measurand), y is the measurement (sensor reading), and a and p are calibration (model) parameters.

Note: In practice, x has to be determined for a measurement y, according to $(y/a)^{1/p}$.

Suppose that in a calibration process, with a set of known measurand values, the corresponding measurements are collected. Model the calibration experiment by $y = (a + v)x^p$ where v represents model error.

(a) Generate 25 points of calibration data (X_i, Y_i), $i = 1, 2, \ldots, n$ by using $a = 1.5$, $p = 2$, $v = N(0.1, 0.2^2)$ (i.e., random with Gaussian distribution of mean 0.1 and std 0.2), and $n = 25$, with $X_1 = e$ (≈2.718282) and x-increments of 0.5.
(b) Estimate the parameters a and p using linear least-squares error estimation (LSE) in log scale.
(c) Comment on the estimation results.

7.3 Consider a random signal Y whose mean is μ and the variance is σ^2. The signal is measured and N data values Y_i, $i = 1, 2, \ldots, N$ are collected, independently of one another. The sample mean and sample variance are computed using this data sample according to

$$\text{Sample mean: } \bar{Y} = \frac{1}{N}\sum_{i=1}^{N} Y_i$$

$$\text{Sample variance: } S^2 = \frac{1}{(N-1)}\sum_{i=1}^{N} (Y_i - \bar{Y})^2$$

(a) Show that these two quantities are unbiased estimates of the mean and the variance of the signal.
(b) Particularly comment on this estimate for variance.

7.4 Semiconductor strain gauges are somewhat nonlinear at high values of strain (see Chapter 9). Typically in a measurement setup, the strain gauge is part of a resistance bridge. Using the bridge circuit, the fractional change in resistance ($\delta R/R$) is measured, and using a calibration curve, the strain (ε) is computed. A typical calibration curve for a p-type semiconductor strain gauge is given by

$$\frac{\delta R}{R} = S_1\varepsilon + S_2\varepsilon^2$$

where

R is the strain-gauge resistance (Ω)
ε is the strain

In a calibration test of a p-type strain gauge, the following 15 data pairs were obtained (strain, fractional resistance increment):

```
[0,0.0095; 0.0005,0.0682; 0.0010,0.1322; 0.0015,0.1952;
0.0020,0.2541; 0.0025,0.3248; 0.0030,0.3926; 0.0035,0.4667;
0.0040,0.5411; 0.0045,0.6149; 0.0050,0.6807; 0.0055,0.7628;
0.0060,0.8355; 0.0065,0.9170; 0.0070,1.0054]
```

It is known that the true values of the parameters of the calibration curve are $S_1 = 117$ and $S_2 = 3600$. For the given data, perform

(a) A linear fit
(b) A quadratic fit

Compare and discuss the results obtained from the two least-squares fittings.

7.5 A 20-unit sample of commercial resistors of nominal resistance 100 Ω was tested, and the following resistance values were recorded:

x = [100.5377 101.8339 97.7412 100.8622 100.3188 98.6923 99.5664 100.3426 103.5784 102.7694 98.6501 103.0349 100.7254 99.9369 100.7147 99.7950 99.8759 101.4897 101.4090 101.4172].

Estimate the mean and the std using maximum likelihood estimation (MLE). Next, compute the sample mean and sample std. Compare and comment on the two results.

7.6 This problem concerns estimation of the damping parameters of a shock absorber using experimental data. In the experimental setup, one end of the shock absorber is firmly mounted on a load cell. At the other end, a velocity input is applied using a shaker (a linear actuator). The experimental setup is shown in Figure P7.6.

The velocity v that is applied by the shaker (m/s) and the resulting force f at the load cell (N) is measured and 41 pairs of data are recorded. First, obtain a simulated set of data using the following MATLAB script:

FIGURE P7.6
Experimental setup of a shock absorber.

```
% Problem 7.6
t=[]; v=[]; f=[];% declare storage vectors
dt=0.05; % time increment
v0=0.15; om= 3.0; b1=2.2; b2=0.2; % parameter values
t(1)=0.0; v(1)=0.0; f(1)=0.0; % initial values
for i=2:41
t(i)=t(i-1)+dt; % time increment
v(i)=v0*sin(om*t(i))+normrnd(0,0.01);  % velocity measurement
f(i)=b1*v(i)+b2*v(i)^2+normrnd(0.01,0.02);  % force measurement
end
t=t'; % convert to column vector
v=v'; %convert x data to a column vector
f=f'; %convert y data to a column vector
plot(t,v,'-')
plot(t,f,'-')
plot(v,f,'x')
```

(a) List possible error sources in estimating the damping parameters.
(b) Using MATLAB, curve fit the data (least-squares fit) to the linear viscous damping model $f = b_1 v + b_0$ and estimate the damping parameters b_0 and b_1. Give some statistics for estimation error and "goodness of fit."

(c) Using MATLAB, curve fit the data (least-squares fit) to the quadratic damping model $f = b_0 + b_1 v + b_2 v^2$ and estimate the damping parameters b_0, b_1, and b_2. Give some statistics for estimation error and "goodness of fit."
(d) Compare the results from the two fits. In particular, is a linear fit adequate or do you recommend quadratic (or still higher order) fit for these data?

Note: Provide plots of the data and the results of curve fitting.

7.7 A batch of bolts of nominal diameter 10 mm has been produced by a manufacturer. For the purpose of quality control, a random sample of 50 bolts was taken from the produced batch, and their diameter was accurately measured. The obtained data (mm) are given by the following MATLAB simulation:

```
% Bolt diameter data
d-[]; % dcclarc data vector
for i=1:50
d(i)=normrnd(10.0,0.1); % data
end
d=d'; % change data to column vector
```

Compute the sample mean and the sample standard deviation of the data using MATLAB (least-squares point estimate). Also, estimate the mean and the standard deviation of the data using maximum likelihood estimation in MATLAB, and compare the two sets of results. In your opinion, which result is more accurate, and why?

7.8 Give one advantage of the method of maximum likelihood estimation (MLE) over the method of least-squares estimation (LSE). Also, give one advantage of LSE over MLE.

In the process of product quality monitoring, N items are randomly chosen from a batch of products and are individually and carefully tested. During testing, each product is either accepted (denoted by "A") or rejected (denoted by "R"). Suppose that this procedure produced M rejects. It is proposed to use these data and the method of maximum likelihood estimation to estimate the most likely probability $\hat{p}$ that a randomly selected product from the batch is acceptable (A).

(a) Derive a suitable likelihood function for this estimation.
(b) Determine the maximum likelihood estimate $\hat{p}$ according to this likelihood function.
(c) Analytically verify that the result corresponds to "maximum" likelihood rather than "minimum" likelihood.

7.9 For quality monitoring of a batch of light bulbs, n bulbs are randomly selected and tested. During testing, each of the n bubs was determined to be either defective (D) or not ($\bar{\text{D}}$). Estimate the most likely probability $\hat{p}$ that a randomly selected product from the batch is acceptable (D).

7.10 Two data sets, one with 10 measurements and the other with 20 measurements, were obtained for the same quantity using a sensor under similar conditions. To simulate these measurements, the following two data sets were generated using the same Gaussian random number generator $N(1.0, 0.3^2)$:

y_1 = [0.6077, 0.8699, 1.1028, 2.0735, 1.8308, 0.5950, 1.9105, 1.2176, 0.9811, 1.2144]
y_2 = [0.9385, 0.9628, 1.4469, 1.4227, 1.4252, 1.2014, 0.6378, 1.2152, 1.4891, 1.1467, 1.3104, 1.2181, 0.9090, 1.0882, 0.7638, 1.2665, 0.6559, 0.6793, 0.7572, 0.1167]

Estimate the mean and the std using MLE, separately for the two data sets, and then for the combined data set (of 30 points), assuming a Gaussian distribution. Comment on the results and compare with results from LSE.

7.11 We are given measurements y of quantity m (which is to be estimated). The data y have both model error σ_m (also denoted by σ_v, which corresponds to the error in the model that is used to represent m and/or unknown disturbances that affect the true value of m) and measurement error σ_w. The standard deviations σ_m and σ_w are known. Obtain a recursive formula to obtain an estimate $\hat{m}$ of m. Also, determine a recursive formula for determining the variance of the estimation error.

Assume (a) Gaussian distributions and (b) zero-mean model error and zero-mean measurement error have.

7.12 A discrete sensor is used to measure the size of an object. The size m is treated as a discrete quantity, which can take one of the following three values:

m_1 = small, m_2 = medium, m_3 = large

The sensor will make one of three discrete measurements given in the vector $\mathbf{y} = [y_1\ y_2\ y_3]$ corresponding to these three object-size values.

The sensor has the following likelihood matrix:

	y_1	y_2	y_3
m_1	0.75	0.05	0.20
m_2	0.05	0.55	0.40
m_3	0.20	0.40	0.40

(a) Indicate an obvious capability of the sensor.
(b) Suppose that in the beginning of the sensing process, we have no *a priori* information about the size of an object. Then suppose that the sensor reads y_1. What is the *a posteriori* probability of the measurement? What is the measurement according to MLE?

7.13 A mobile robot has two distance sensors: (a) a laser rangefinder and (b) an ultrasonic rangefinder. While at standstill, the robot sensed the position of an obstacle using the two sensors, and 15 distance readings (m) were obtained given as follows:

(a) From the laser rangefinder: To simulate these data, generate 15 values using $N(5.0, 0.01^2)$.
(b) From the ultrasonic rangefinder: To simulate these data, generate 15 values using $N(5.0, 0.02^2)$.

Suppose that the robot position (localization) has a zero-mean random error of std σ_v = 0.04 m. Also, the laser rangefinder has a zero-mean random error of std σ_w = 0.01 m, and the ultrasonic rangefinder has a zero-mean random error of std σ_w = 0.02 m.

Using a recursive static Kalman filter, estimate and plot the distance of the obstacle from the two sets of data and the associated estimation error std. Compare the two sets of results.

7.14 A digital tachometer measures speed by counting the clock pulses per revolution (see Chapter 11). For 25 revolutions of a disk, the following numbers of clock pulses were recorded:

```
y = [803    809    789    804    802    793    798    802    818    814
793    815    804    800    804    799    799    807    807    807    803
794    804    808    802]
```

1 clock pulse = 0.5 ms.

Recursively estimate and plot the estimated speed (rev/s) and the associated estimation error std using the following:

(a) Use recursive LSE with the following algorithm:

$$\bar{Y}_1 = Y_1$$

$$\bar{Y}_{i+1} = \frac{1}{(i+1)}(i \times \bar{Y}_i + Y_{i+1}), \quad i = 1, 2, \ldots$$

$$S_1^2 = 0$$

$$S_{i+1}^2 = \frac{1}{i}[S_i^2 \times (i-1) + (Y_{i+1} - \bar{Y}_{i+1})^2]$$

(b) Use static Kalman filter with model error std = 2 pulses, and measurement error std = 3 pulses, in the neighborhood of a measurement of 800 pulses.
Compare the two results.

7.15 (a) What information is needed for estimating a variable of a dynamic system using Kalman filter?
(b) Give two general advantages of Kalman filter as a method of estimating an unknown dynamic variable using measured data.
(c) Give an advantage of the extended Kalman filter over the linear Kalman filter.
(d) Give one advantage of the unscented Kalman filter over the extended Kalman filter.

7.16 A rotary wood cutter is driven at angular speed u rad/s. The associated cutting torque T is given by $T = c|u|u$ N·m. See Figure P7.16. The angular speed is measured using a tachometer, which has noise (variance W). The cutting torque is estimated using these measurements in every sampling period ΔT s.

Consider a constant acceleration of the cutter according to $u = a_0 t$ rad/s.
Variance of speed measurement is $W = (0.2)^2$.
Parameter values are $a_0 = 2.0$ rad/s^2 and $c = 0.5$ N.m/s^2.
Sampling period is $\Delta T = 0.05$ s.
Simulate measurement noise as zero-mean Gaussian with variance W.
Use unscented transformation to estimate T.

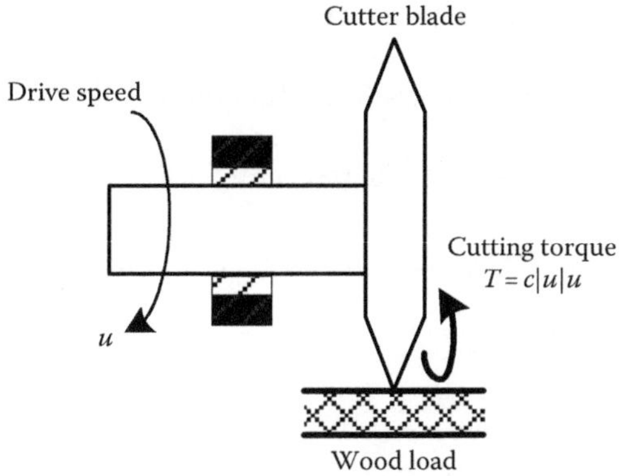

FIGURE P7.16
A rotary wood cutter.

7.17 A discrete-time model of a second-order, time-invariant, nonlinear system is given by

$$\begin{bmatrix} x_1(i) \\ x_2(i) \end{bmatrix} = \begin{bmatrix} -a_1 - a_2 x_1(i-1) & -a_3 \\ a_4 & a_5 - a_6 x_1(i-1) \end{bmatrix} \begin{bmatrix} x_1(i-1) \\ x_2(i-1) \end{bmatrix} + \begin{bmatrix} b_1 \\ b_2 \end{bmatrix} u(i-1) + \begin{bmatrix} v_1(i-1) \\ v_2(i-1) \end{bmatrix}$$

Use the parameter values: $a_1 = 0.4$, $a_2 = 0.1$, $a_3 = 0.2$, $a_4 = 0.3$, $a_5 = 0.5$, $a_6 = 0.1$, $b_1 = 1.0$, and $b_2 = 0.2$. The sampling period $T = 0.02$.

The measurement y is the first state (x_1). Use the full nonlinear model with added random noise for simulating y. The input is given by $u = 2 \sin 6t$.

Apply a linear Kalman filter with the linear system matrix

$$\boldsymbol{F} = \begin{bmatrix} -a_1 & -a_3 \\ a_4 & a_5 \end{bmatrix}$$

(i.e., $\boldsymbol{F}$ matrix corresponding to the initial state $\boldsymbol{x}(0) = 0$) to estimate the two states.

Note:

$$G = \begin{bmatrix} b_1 \\ b_2 \end{bmatrix} \quad \text{and} \quad C = [1 0]$$

Use the following covariances for the input disturbance and the measurement noise, respectively:

$$\boldsymbol{V} = \begin{bmatrix} 0.02 & 0 \\ 0 & 0.04 \end{bmatrix}; \quad \boldsymbol{W} = [0.05]$$

7.18 For the system in Problem 7.17, use an extended Kalman filter to estimate the two state variables.

7.19 For the system in Problem 7.17, use an unscented Kalman filter to estimate the two state variables.

7.20 A time-varying nonlinear system is given by

$$x_i = ae^{-pxi}x_{i-1} + d\sin\left(\frac{i}{i^2+1}x_{i-1}\right) + u_{i-1} + v_{i-1}$$

$$y_i = x_i + w_i$$

System parameters are $a = 0.8$, $p = 0.0001$, and $d = 0.5$

Variances of the input disturbance and measurement noise are $V = 0.3^2 = 0.09$ and $W = 0.5^2 = 0.25$.

Input $u = u_0 \sin 2\pi f_0 t$ with $u_0 = 2.0$ rad/s,$f_0 = 6.0$ rad/s.

Use an extended Kalman filter to estimate x.

7.21 Consider the second-order, nonlinear, time-invariant, discrete-time system:

$$x_{1,i} = a_1|x_{1,i-1}|^{1/3}\operatorname{sgn}(x_{1,i-1}) + a_2x_{2,i-1} + b_1u_{1,i-1} + v_{1,i-1}$$

$$x_{2,i} = -a_3x_{1,i-1} + a_4|x_{1,i-1}|^{1/3}\operatorname{sgn}(x_{1,i-1}) + a_5x_{2,i-1} - a_6\left|x_{2,i-1} + a_7|x_{1,i-1}|^{1/3}\operatorname{sgn}(x_{1,i-1})\right|^{1/3} + b_2u_{2,i-1} + v_{2,i-1}$$

System parameter values are $a_1 = 1.0$, $a_2 = 1.0$, $a_3 = 0.25$, $a_4 = 1.0$, $a_5 = 1.0$, $a_6 = 1.0$, $a_7 = 1.0$, $b_1 = 1.0$, and $b_2 = 1.0$.

Input $u = u_0 \sin 2\pi f_0 t$ with $u_0 = 2.0$ rad/s,$f_0 = 6.0$ rad/s.

Measurement gain matrix: $\boldsymbol{C} = [1,0]$

Calculating noise covariance matrices, we get

$$\boldsymbol{V} = \begin{bmatrix} 0.02 & 0 \\ 0 & 0.04 \end{bmatrix}; \quad \boldsymbol{W} = [0.05]$$

Measure the first state x_1 (simulate this with added Gaussian noise) and estimate the second state (x_2) using an unscented Kalman filter.

7.22 Consider the CNC (computer numerical control) milling machine system shown in Figure P7.22a. A simplified model of the milling process is shown in Figure P7.22b. *Note*: The shaft connects the motor to the milling cutter.

(a) For this model, show that a state-space representation may be given by

$$\boldsymbol{A} = \begin{bmatrix} 0 & 0 & \frac{-1}{J_m} \\ 0 & \frac{-B_c}{J_c} & \frac{B_c}{J_c} \\ K & \frac{-K}{B_c} & 0 \end{bmatrix}; \quad \boldsymbol{B} = \begin{bmatrix} \frac{1}{J_m} \\ 0 \\ 0 \end{bmatrix}; \quad \boldsymbol{C} = \begin{bmatrix} 1 & 0 & 0 \end{bmatrix}$$

with state vector $\boldsymbol{x} = [x_1, x_2, x_3]^T = [\text{Motor speed}, \text{Cutter torque}, \text{Shaft torque}]^T$, input u = voltage input to the motor armature, and measurement, motor speed $y = x_1$.

(b) Show that the system is observable (constructible).
(c) With the parameter values $J_m = 0.5$, $J_c = 1.0$, $K = 5000$, $B_c = 20.0$, show that the system is stable. From the system eigenvalues, verify that $T = 2.0 \times 10^{-3}$ s is a satisfactory sampling period (discretization period) for the system. Obtain the corresponding discrete-time state-space model.

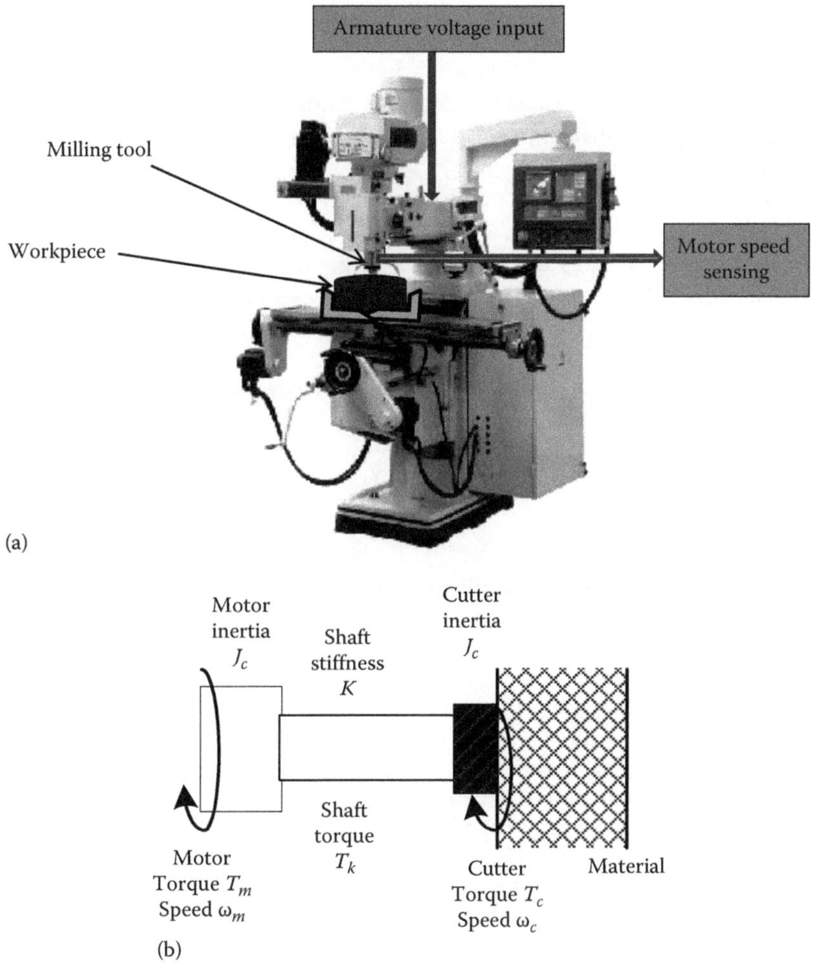

FIGURE P7.22
(a) A CNC milling machine and instrumentation and (b) a simplified model of the milling process.

7.23 Consider CNC milling machine of Problem 7.22. A voltage u is applied to the armature circuit of the drive motor of the milling machine cutter in order to accelerate (ramp up) the tool to the proper cutting speed. Since it is difficult to measure the cutting torque (which is a suitable indicator of the cutting quality and the cutter performance), the motor speed ω_m is measured instead (which is much easier to measure). The measurements of the motor speed are used in a Kalman filter to estimate the cutting torque. The following information is given:

A discrete-time, nonlinear, state-space model of the cutting system of the milling machine is given as follows:

$$x_1(i) = a_1x_1(i-1) + a_2x_2(i-1) + a_3x_2^2(i-1) - a_4x_3(i-1) + b_1u(i-1) + v_1(\text{i}-1)$$

$$x_2(i) = a_5x_1(i-1) + a_6x_1^2(i-1) + a_7x_2(i-1) + a_8x_3(i-1) + a_9x_3^2(i-1) + b_2u(i-1) + v_2(\text{i}-1)$$

$$x_3(i) = a_{10}x_1(i-1) - a_{11}x_2(i-1) - a_{12}x_2^2(i-1) + a_{13}x_3(i-1) + b_3u(i-1) + v_3(\text{i}-1)$$

where i denotes the time step.

State vector $\mathbf{x} = [x_1, x_2, x_3]^T = [\text{Motor speed, Cutter torque, Shaft torque}]^T$

Input u is the voltage input to the motor armature.

For ramping up the cutter, use $u = a(1-\exp(-bt))$ with $a = 2.0$ and $b = 30$.

Measured motor speed $y = x_1$.

The motor speed y and time t are measured with a sampling period of $T = 2.0 \times 10^{-3}$ s.

Model parameter values are the following: `a1=0.98; a2=0.001; a3=0.0002; a4=0.004; a5=0.19; a6=0.04; a7=0.95; a8=0.038; a9=0.008; a10=9.9; a11=0.48; a12=0.1; a13=0.97; b1=0.004; b2=0.0003; b3=0.02.`

Output matrix $\mathbf{C} = [1\ 0\ 0]^T$.

Input (disturbance) covariance $\mathbf{V}$ and the measurement (noise) covariance $\mathbf{W}$ are

$$\mathbf{V} = \begin{bmatrix} 0.0002 & 0 & 0 \\ 0 & 0.09 & 0 \\ 0 & 0 & 0.1 \end{bmatrix}; \quad \mathbf{W} = [0.0004]$$

Note: Both are Gaussian white, with zero mean.

As the measured data (51 points), use what is generated through simulation using the MATLAB script:

```
% Prob 7.23 Simulation of measured data
n=3; m=1; % system and output order
F=[a1, a2, -a4;a5, a7, a8;a10, -a11, a13]; % linear system matrix
G=[b1; b2; b3]; % input distribution matrix
C=[1 0 0]; % output gain matrix
Fsim=F; % simulation model
a=2.0;b=30.0; %input parameters
xe=zeros(n,1); % initialize state estimation vector
xsim=xe; %initialize state simulation vector
it=[]; meas=[]; st1=[]; st2=[]; st3=[]; err=[];  % declare storage
vectors
it=0; meas=0; st1=0; st2=0; st3=0; err=0; %initialize plotting
variables
for i=1:50
it(end+1)=i; % store the recursion number
t=i*0.002;  % time
u=a*(1-exp(-b*t)); % input
Fsim(1,2)=a2+a3*xsim(2); % include nonlinearity for simulation
Fsim(2,1)=a5+a6*xsim(1); % include nonlinearity for simulation
Fsim(2,3)=a8+a9*xsim(3); % include nonlinearity for simulation
Fsim(3,2)=-a11-a12*xsim(2); % include nonlinearity for simulation
xsim=Fsim*xsim+G*u;
```

```
for j=1:m
yer(j,1)=rand/100.0; % simulated measurement error
end
y=C*xsim+yer; % simulate the speed measurement with noise
meas(end+1)=y(1); % store the measured data
st1(end+1)=xsim(1);  % store the first state (motor speed)
st2(end+1)=xsim(2);  % store the second state (cutting torque)
st3(end+1)=xsim(3);  % store the third state (shaft torque)
err(end+1)=y(1)-C*xe;  %store the speed error
end
```

(a) Apply a linear Kalman filter to estimate the cutting torque (i.e., state x_2).
(b) Apply an extended Kalman filter to estimate the cutting torque.
(c) Apply an unscented Kalman filter to estimate the cutting torque.
(d) Compare the results from the three approaches. In particular, indicate which approach is appropriate in the present experiment and why.

Note: Provide plots of the data and the results of Kalman filtering and also the MATLAB script that you used to generate the results.

8

Analog Motion Sensors

Chapter Objectives

- Sensors and Transducers
- Terminology
- Sensor Types and Selection
- Sensors for Electromechanical Applications
- Motion Transducers
- Potentiometer
- Variable-Inductance Transducers
- Linear-Variable Differential Transformer (LVDT)
- Performance Considerations
- Resolver
- Permanent-Magnet Transducers
- DC Tachometer
- AC Tachometer
- Eddy Current Transducers
- Variable-Capacitance Transducers
- Piezoelectric Sensors
- Piezoelectric Accelerometer
- Charge Amplifier
- Gyroscopic Sensors
- Coriolis Force Devices

8.1 Sensors and Transducers

Sensors and transducers are crucial in instrumenting an engineering system. Sensors may be used in an engineering system for a variety of purposes. Essentially, sensors are needed to monitor and "learn" about the system. This knowledge will be useful in many types of applications, for example,

1. Process monitoring
2. Operating or controlling a system
3. Experimental modeling (i.e., model identification)
4. Product testing and qualification
5. Product quality assessment
6. Fault prediction, detection, and diagnosis
7. Alarm and warning generation
8. Surveillance

Specifically in a control system, sensing is used for such purposes as follows:

1. Measuring the system outputs for feedback control
2. Measuring some types of system inputs (unknown inputs, disturbances, etc.) for feedforward control
3. Measuring output signals for system monitoring, parameter adaptation, self-tuning, and supervisory control
4. Measuring input and output signal pairs for experimental modeling of the plant (i.e., for system identification)

The terms *sensor* and *transducer* are often used interchangeably to mean the same thing. However, strictly, a *sensor* senses the quantity that needs to be observed or measured (called the "measurand"), while the *transducer* converts it into a form that can be observed or used in a subsequent operation. Except when necessary, we will use the terms *sensor* and *transducer* to mean the same device. Proper selection and integration of sensors and transducers are a necessary and significant task in instrumenting an engineering system. Sometimes, we may have to design and develop new sensors or modify existing sensors, depending on the needs of the specific application. Such activities are based on a set of performance specifications for the required sensors. The characteristics of an ideal sensor and transducer are presented in Chapter 5. Even though a real sensor will not be able to achieve such ideal behavior, when designing and instrumenting an engineering system, it is desirable to use the ideal behavior of the system components as a reference with respect to which the *performance specifications* may be generated and presented. A *model* is useful in representing the behavior of a sensor. Specifically, a model may be used to analyze, simulate, design, integrate, test, and evaluate a sensor.

In this chapter, the role and significance of sensors and transducers in an engineering system are indicated; important criteria in selecting sensors and transducers for engineering applications are presented; and several representative analog sensors and transducers for motion measurement and their concepts, operating principles, models, characteristics, accessories, and applications are described. Specifically, analog motion sensors in the "macro" scale are discussed in this chapter. In particular, we study sensors for electromechanical (or mechatronic) applications. Effort sensors (for the measurement of force, torque, touch, mechanical impedance, etc.) are presented in Chapter 9. Other varieties of analog sensors including optical sensors, thermofluid sensors, and sensors that are useful in monitoring the quality of water (e.g., pH value, dissolved oxygen, and oxidation–reduction potential) are discussed in Chapter 10. Digital transducers including image sensors and Hall-effect sensors are studied in Chapter 11. Microelectromechanical system (MEMS) sensors and other practical topics such as sensor data fusion and networked sensors are studied in Chapter 12.

8.1.1 Terminology

Potentiometers, differential transformers, resolvers, tachometers, piezoelectric devices, and gyros are examples of sensors used in engineering systems.

8.1.1.1 Measurand and Measurement

The variable that is measured is termed the *measurand*. Examples are acceleration and velocity of a vehicle, torque into a robotic joint, strain in a structural member, temperature and pressure of a process plant, and current through an electric circuit. The reading (output) of the sensor unit is the *measurement*. The nature of the measurand and the nature of the sensor output are typically quite different. For example, while the measurand of an accelerometer is an acceleration, the accelerometer output may be a charge or a voltage. Similarly, the measurand of a strain-gauge bridge is a strain; the bridge output is a voltage. However, the sensor output can be calibrated in the units of the measurand (e.g., in acceleration units or strain units).

8.1.1.2 Sensor and Transducer

A measuring device passes through two main stages while measuring a signal. First, the measurand is felt or *sensed* by the sensing element. Then, the sensed signal is *transduced* (or converted) into the form of the device output. In fact the sensor, which senses the response, converts (i.e., transduces) this signal into the sensor output—the response of the sensor element. For example, a piezoelectric accelerometer senses acceleration and converts it into an electric charge; an electromagnetic tachometer senses velocity and converts it into a voltage; and a shaft encoder senses a rotation and converts it into a sequence of voltage pulses. Since sensing and transducing occur together, the terms *sensor* and *transducer* are used interchangeably to denote the entire sensor–transducer unit. The sensor and transducer stages are functional stages, and sometimes, it is not easy or even feasible to draw a line to separate them or to separately identify the physical elements associated with them. Furthermore, this separation is not very important in the use of existing devices. However, proper separation of sensor and transducer stages (physically as well as functionally) can be crucial, when designing new measuring devices.

8.1.1.3 Analog and Digital Sensor–Transducer Devices

Typically, the sensed signal is transduced (or converted) into a form that is particularly suitable for transmitting, recording, conditioning, processing, monitoring, activating a controller, or driving an actuator. For this reason, the output of a transducer is often an electrical signal. The measurand is usually an analog signal because it represents the output of a dynamic system. For example, the charge signal generated in a piezoelectric accelerometer has to be converted into a voltage signal of appropriate level using a charge amplifier. To enable its use in a digital controller, it has to be digitized using an analog-to-digital converter (ADC). Then, the analog sensor and the ADC may be taken together and treated as a digital transducer. There are other sensing devices where the output is in the pulse form, without using an ADC. In general, in a digital transducer, the output is discrete and typically a sequence of pulses. Such discrete outputs can be counted and represented in a digital form. This facilitates the direct interface of a digital transducer with a digital processor.

8.1.1.4 Sensor Signal Conditioning

A complex measuring device can have more than one sensing stage. Often, the measurand goes through several transducer stages before it is available for practical purposes. Furthermore, filtering may be needed to remove measurement noise and other types of noise and disturbances that enter into the true measurand (including process noise and external disturbance inputs). Hence, signal conditioning is usually needed between the sensor and the application. Charge amplifiers, lock-in amplifiers, power amplifiers, switching amplifiers, linear amplifiers, pulse-width modulation amplifiers, tracking filters, low-pass filters, high-pass filters, band-pass filters, and notch filters are signal-conditioning devices that are used in sensing and instrumentation applications. The subject of signal conditioning is discussed in Chapter 3, and typical signal condition devices are described there. In some literature, signal-conditioning devices such as electronic amplifiers are also classified as transducers. Since we are treating signal-conditioning and modification devices separately from measuring devices, this unified classification is avoided whenever possible, and the term *transducer* is used primarily in relation to measuring instruments. Note that it is somewhat redundant to consider electrical-to-electrical sensors–transducers as measuring devices because electrical signals need conditioning only before they are used to carry out a useful task. In this sense, electrical-to-electrical transduction should be considered as a conditioning function rather than a measuring function. Additional components, such as power supplies, isolation devices, and surge-protection units, are often needed in the instrumentation of engineering systems, but they are only indirectly related to the functions of sensing and actuation. Relays and other switching devices and modulators and demodulators (see Chapter 4) may also be included under signal conditioning (more correctly, as signal conversion). Modern sensors–transducers may have signal-conditioning circuitry integrated into them, particularly in the monolithic IC form. Then, it is rather difficult to physically separate the sensor, transducer, and signal conditioner in the overall hardware unit. A schematic representation of the process of sensing and its application is given in Figure 8.1.

8.1.1.5 Pure, Passive, and Active Devices

Pure transducers depend on nondissipative coupling in the transduction stage. Passive transducers (sometimes called "self-generating transducers") depend on their

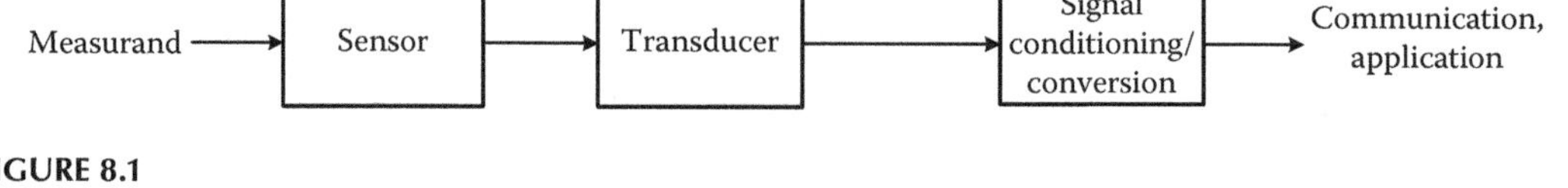

FIGURE 8.1
The stages of sensing and application.

power-transfer characteristics for operation and do not need an external power source. It follows that pure transducers are essentially passive devices. Some examples are electromagnetic, thermoelectric, radioactive, piezoelectric, and photovoltaic transducers. Active sensors/transducers need external power for their operation, and they do not depend on their own power conversion characteristics for this purpose. A good example for an active device is a resistive transducer, such as a potentiometer, which depends on its power dissipation through a resistor to generate the output signal. Specifically, an active transducer requires a separate power source (power supply) for operation, whereas a passive transducer draws its power from a measured signal (measurand). Since passive transducers derive their energy almost entirely from the measurand, they generally tend to distort (or *load*) the measured signal to a greater extent than an active transducer would. Precautions can be taken to reduce such loading effects. On the other hand, passive transducers are generally simple in design, more reliable, and less costly. In the present classification of transducers, we are dealing with power in the immediate transducer stage associated with the measurand and not the power used in subsequent signal conditioning. For example, a piezoelectric charge generation is a passive process. But a charge amplifier, which uses an auxiliary power source, would be needed by a piezoelectric device in order to condition the generated charge.

8.1.2 Sensor Types and Selection

Sensors may be categorized in various ways. One classification is based on the nature of the quantity that is measured (the measurand). Another category is based on the physical principles or technologies that are used in the sensor itself. Clearly, these two classifications are not directly related.

8.1.2.1 Sensor Classification Based on the Measurand

Paramount in the sensor selection is the nature of the quantity (variable, parameter) that needs to be measured. In a sensor classification based on this, given in the following are fields (disciplines or application areas) into which the sensors are classified and some examples of measurands in a class.

Biomedical: Motion, force, blood composition, blood pressure, temperature, flow rate, urine composition, excretion composition, ECG, breathing sound, pulse, x-ray image, ultrasonic image

Chemical: Organic compounds, inorganic compounds, concentration, heat transfer rate, temperature, pressure, flow rate, humidity

Electrical/electronic: Voltage, current, charge, passive circuit parameters, electric field, magnetic field, magnetic flux, electrical conductivity, permittivity, permeability, reluctance

Mechanical: Force (effort including torque), motion (including position and deflection), optical image, other images (x-ray, acoustic, etc.), stress, strain, material properties (density, Young's modulus, shear modulus, hardness, Poisson's ratio)

Thermofluid: Flow rate, heat transfer rate, infrared waves, pressure, temperature, humidity, liquid level, density, viscosity, Reynolds number, thermal conductivity, heat transfer coefficient, Biot number, image

8.1.2.2 Sensor Classification Based on Sensor Technology

Sensors are developed based on various physical principles and technologies and can be classified based on them. This classification is particularly useful in the design, development, and evaluation of a sensor rather than in the selection of a sensor for a particular application. Still, the concepts, principles, and technologies of a sensor are useful in modeling the sensor. The model then may be used not just to evaluate the performance of the sensor but also to study the performance of the sensor-integrated system. Some classes of sensors based on their physical principles and technologies are active, analog, digital, electric, IC, mechanical, optical, passive, piezoelectric, piezoresistive, and photoelastic.

8.1.2.3 Sensor Selection

In selecting a sensor or sensors for a particular application, we need to know the application and its purpose and what quantities (variables and parameters) need to be measured in the application. Then, by doing a thorough search, we should determine what sensors are available for carrying out the needed measurements and what quantities cannot be measured (due to inaccessibility, lack of sensors, etc.). In the latter case, the choices include the following:

1. Estimate the quantity by using other quantities that can be measured.
2. Develop a new sensor for the purpose.

As the starting step in the process of sensor selection for a particular application, we may complete a table of the form given in Table 8.1. The subsequent steps of information collection, analysis, simulation, and evaluation are aimed at matching the available sensors with the needs of the application. This is also a process of matching the specifications of the available devices with the required specifications. Here, we should go beyond simple matching of the two sets of information. Considerations such as sensitivity and bandwidth in particular (see Chapters 5 and 6 for the sensitivity-based approach and bandwidth-based approach in instrumentation and design) and a variety of performance parameters may be used for this purpose.

TABLE 8.1

Preliminary Information for Sensor Selection

Item	Information (Please Complete)
Parameters or variables that have to be measured in your application	
Nature of the information (parameters and variables) needed for the particular application (analog, digital, modulated, demodulated, power level, bandwidth, accuracy, etc.)	
Specifications for the needed measurements (measurement signal type, measurement level, range, bandwidth, accuracy, SNR, etc.)	
List of available sensors that are needed for the application and their data sheets	
Signal provided by each sensor (types—analog, digital, modulated, etc.—power level, frequency range)	
Type of signal conditioning or conversion needed for the sensors (filtering, amplification, modulation, demodulation, ADC, DAC, VFC, FVC, etc.)	
Any other comments	

If such matching is not possible, we must investigate what other hardware or modifications may be used to achieve the matching (this may include signal modification including amplification and impedance matching). If all these efforts do not lead to a proper choice of sensors, we may have to modify the specifications for the application and/or develop new sensors for the application. This process of sensor selection may be performed in several iterations before the final selection and acquisition are made.

Today, the availability of choice of sensors is rather vast and diverse. Hence, in instrumentation practice, the limitations of the system performance come not from the sensors but from other components (signal conditioners, converters, transmitters, actuators, power supplies, etc.) of the system.

8.2 Sensors for Electromechanical Applications

Now, we analyze several analog sensor–transducer devices that are commonly used in the instrumentation of engineering systems. The attempt here is not to present an exhaustive discussion of all types of sensors; rather, it is to consider a representative selection of motion sensors. Such an approach is reasonable because even though the scientific principles behind various sensors may differ, many other aspects (e.g., performance parameters and specification, selection, signal conditioning, interfacing, modeling procedures, analysis) can be common to a large extent.

The treatment of this chapter particularly concerns motion sensors for electromechanical applications or mechatronics. Specifically, we study here the main types of motion sensors (including position, proximity, rectilinear and angular velocity, and acceleration). Chapter 9 examines effort sensors (force, torque, tactile, impedance, etc.). Other types of sensors including optical sensors, ultrasonic sensors, magnetostrictive sensors, thermo-fluid sensors, and sensors that are useful in monitoring the quality of water (e.g., pH value, electrical conductivity, dissolved oxygen, and oxidation–reduction potential) are covered in Chapter 10. Digital transducers including shaft encoders and image sensors (digital camera), and some other innovative sensors including Hall-effect sensors, are studied in Chapter 11. MEMS sensors, and other advanced topics in sensing such as networked sensing, and sensor fusion are addressed in Chapter 12.

8.2.1 Motion Transducers

By motion, we particularly mean one or more of the following four kinematic variables:

1. Displacement (including position, distance, proximity, size, and gauge)
2. Velocity (rate of change of displacement)
3. Acceleration (rate of change of velocity)
4. Jerk (rate of change of acceleration)

Each type of variables in this classification is the time derivative of the preceding one.

Motion measurements are extremely useful in controlling mechanical responses and interactions in engineering systems, particularly in mechatronic systems. Numerous examples can be cited: The rotating speed of a workpiece and the feed rate of a tool are measured in controlling machining operations. Displacements and speeds (both angular and translatory) at the joints (revolute and prismatic) of a robotic manipulator or a kinematic linkage are used in controlling the manipulator trajectory. In high-speed ground transit vehicles, acceleration and jerk measurements can be used in the control of active suspensions to obtain improved ride quality. Angular speed is a crucial measurement that is used in the control of rotating machinery, such as turbines, pumps, compressors, motors, transmission units or gear boxes, machine tools, and generators in power-generating plants. Proximity sensors (to measure displacement and distance) and accelerometers (to measure acceleration) are the two most common types of measuring devices used in machine protection systems for condition monitoring, fault prediction, detection, diagnosis, and control of large and complex machinery. The accelerometer is often the only measuring device that is used in controlling dynamic test rigs (e.g., in vibration testing). Displacement measurements are used for valve control in process applications. Plate thickness (or gauge) is continuously monitored by the automatic gauge control system in steel rolling mills.

We might question the need for separate transducers to measure the four kinematic variables—displacement, velocity, acceleration, and jerk—because any one variable is related to the other through simple integration or differentiation. It should be possible, in theory, to measure only one of these four variables and use either analog processing (through analog circuit hardware) or digital processing (through a dedicated processor or microcontroller) to obtain any one of the remaining motion variables. The feasibility of this approach is highly limited, however, and it depends crucially on several factors, including the following:

1. The nature of the measured signal (e.g., steady, highly transient, periodic, narrowband, broadband)
2. The required frequency content of the processed signal (or the frequency range of interest or frequency range of operation of the sensor)
3. The signal-to-noise ratio (SNR) of the measurement
4. Available processing capabilities (e.g., analog or digital processing, limitations of the digital processor and interface, such as the speed of processing, sampling rate, and buffer size)
5. Controller requirements and the nature of the plant (e.g., time constants, delays, complexity, hardware limitations)
6. Required accuracy in the application (on which processing requirements and hardware costs depend)

For instance, differentiation of a signal (in the time domain) is often unacceptable for noisy and high-frequency narrowband signals because it will enhance the severity of the unacceptable high-frequency components. In any event, costly signal-conditioning hardware might be needed for preprocessing before differentiating a signal. As a rule of thumb, in low-frequency applications (in the order of 1 Hz), displacement measurements generally provide good accuracies. In intermediate-frequency applications (less than 1 kHz), velocity measurement is usually favored. In measuring high-frequency motions with high noise levels, acceleration measurement is preferred. Jerk is particularly useful in ground transit (ride quality), manufacturing (forging, rolling, cutting and similar impact-type operations), and shock isolation applications (for delicate and sensitive equipment), which take into account highly transient (and high frequency) signals.

8.2.1.1 Multipurpose Sensing Elements

A one-to-one relationship may not always exist between a measuring device and a measured variable. Furthermore, a particular type of sensing element may be used for multiple types of sensors. For example, although strain gauges are devices that measure strains (and, hence, stresses and forces), they can be adapted to measure displacements by using a suitable front-end auxiliary sensor element, such as a cantilever (or spring). Furthermore, a measuring device may be used to measure different variables through appropriate data interpretation techniques. For example, piezoelectric accelerometers with built-in microelectronic integrated circuitry (IC) are marketed as piezoelectric velocity transducers. Resolver signals, which provide angular displacements, are differentiated to obtain angular velocities. Pulse-generating (or digital) transducers, such as optical encoders and digital tachometers, can serve as both displacement transducers and velocity transducers, depending on whether the absolute number of pulses is counted or the pulse rate is measured. Note that pulse rate can be measured either by counting the number of pulses during a unit interval of time (i.e., pulse counting) or by gating a high-frequency clock signal through the pulse width (i.e., pulse timing). Furthermore, in principle, any force sensor can be used as an acceleration sensor, velocity sensor, or displacement sensor, depending on the specific *front-end auxiliary element* (e.g., inertia, damper, spring) that is used.

8.2.1.2 Motion Transducer Selection

In selecting a motion transducer, we need to consider several factors. Some preliminary considerations are as follows:

1. The kinetic nature of the measurand (position, proximity, displacement, speed, acceleration, etc.)
2. Rectilinear (commonly termed *linear*) or rotatory (commonly termed *rotary*) motion
3. Contact or noncontact type
4. Measurement range
5. Required accuracy
6. Required frequency range of operation (time constant, bandwidth)
7. Size
8. Cost
9. Operating environment (e.g., magnetic fields, temperature, pressure, humidity, vibration, shock)
10. Life expectancy

8.3 Potentiometer

Even though, in the early days, a potentiometer or *pot* was primarily used as a device to supply a variable voltage or a variable resistance to a circuit or some application (manually by turning a knob), we address here its use as a displacement transducer. This is an active

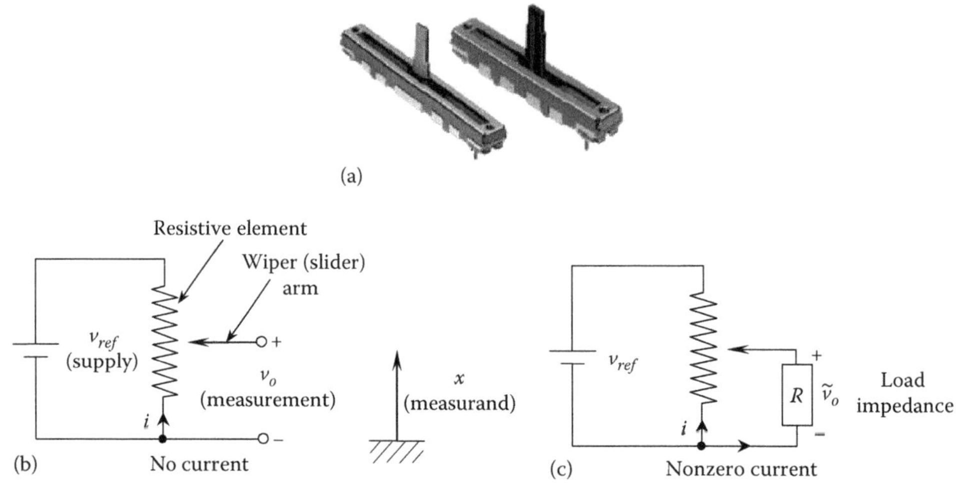

FIGURE 8.2
(a) Linear potentiometers (Courtesy of Alps Electric, Auburn Hills, MI), (b) schematic diagram of a potentiometer, and (c) potentiometer loading.

transducer that consists of a uniform coil of wire or a film of high-resistance material—such as carbon, platinum, cermet (metallic resistance element on a ceramic substrate), or conductive plastic, whose resistance is proportional to its length. This principle can be used to measure both rectilinear displacement (using a linear potentiometer) and angular displacements (using a rotary pot).

A commercial linear (or more correctly "rectilinear") potentiometer is shown in Figure 8.2a. A constant voltage v_{ref} is applied across the coil (or film) using an external direct current (DC) voltage supply. The output signal v_o of the transducer is the DC voltage between the movable contact (wiper arm or slider) sliding on the coil and the reference voltage terminal of the coil, as shown schematically in Figure 8.2b. The slider displacement x is proportional to the output voltage:

$$v_o = kx \tag{8.1}$$

This relationship is known as the "law" or the "taper" of the pot.

Loading Errors

Equation 8.1 assumes that the output terminals are in open circuit, that is, a load of infinite impedance (or resistance in the present DC case) is present at the output terminals so that the output current is zero. In actual practice, however, the electrical load (the circuitry into which the pot signal is fed—e.g., conditioning, interfacing, processing, or control circuitry) has a finite impedance. Consequently, the output current (the current through the load) is nonzero, as shown in Figure 8.2c. The output voltage thus drops to $\tilde{v}_o$, even if the reference voltage v_{ref} is assumed to remain constant under load variations (i.e., even if the output impedance of the voltage source is zero). This consequence is known as the "loading effect" of the transducer (specifically, *electrical loading*), as discussed in Chapter 2. Under these conditions, the linear relationship given by Equation 8.1 would no longer be valid, causing an error in the displacement reading.

Electrical loading can affect the transducer reading in two ways:

1. It changes the reference voltage (i.e., loads the voltage source).
2. It loads the transducer.

To reduce these effects, a voltage source that does not change its output voltage appreciably due to load variations (e.g., a regulated or stabilized power supply, which has a low output impedance) and data acquisition circuitry (including signal-conditioning circuitry) that has a high input impedance should be used.

Potentiometer, being a contact sensor, generates some *mechanical loading error* as well. Specifically, the moving part of the object whose displacement is sensed by the pot has to be directly linked to the slider of the pot, which is in contact with the resistance element. The associated sliding friction is directly exerted on the sensed object and will affect its motion. To reduce the mechanical loading effect, we must reduce the sliding friction (conducting plastics are better than carbon) and the mass of the slider.

Reduced friction (low mechanical loading), reduced wear, reduced weight, and increased resolution are advantages of using conductive plastics in potentiometers.

8.3.1 Rotatory Potentiometers

Potentiometers that measure angular (rotatory) displacements are more common and convenient, because in conventional designs of rectilinear (translatory) potentiometer, the length of the resistive element has to be increased in proportion to the measurement range or stroke. The element resistance of a pot can range from a low value on the order of 10 Ω to as high as 1 MΩ. The power rating can be 10 mW to several W. They can come in small sizes (as small as 5 mm in diameter). Figure 8.3a shows a commercial rotatory (rotary) pot. Figure 8.3b shows a circuit for a rotary pot, and Figure 8.3c indicates the external appearance including the three terminals, which correspond to the reference voltage terminal 1 (ground), power terminal 3 (hot), and the output 2 that gives the potentiometer reading (in volts). Helix-type rotatory potentiometers are available for measuring absolute angles exceeding 360°. The same function may be accomplished with a standard single-cycle rotatory pot simply by including a counter to record full 360° rotations.

Angular displacement transducers, such as rotatory potentiometers, can be used to measure large rectilinear displacements in the order of 3 m. A cable extension mechanism may be employed to accomplish this. A light cable wrapped around a spool, which moves with the rotary element of the transducer, is the cable extension mechanism. The free end of the cable is attached to the moving object, and the potentiometer housing is mounted on a stationary structure. The device is properly calibrated so that as the object moves, the rotation count and the fractional rotation reading directly provide the rectilinear displacement. A spring-loaded recoil device, such as a spring motor, winds the cable back when the object moves toward the transducer.

8.3.1.1 Loading Nonlinearity

Consider the rotatory potentiometer shown in Figure 8.3. Let us now discuss the significance of the nonlinearity error caused by the electrical loading of a purely resistive load connected to the pot. For a general position θ of the slider arm of the pot, suppose that the resistance in the output segment (pick-off terminal 2) of the coil is R_θ.

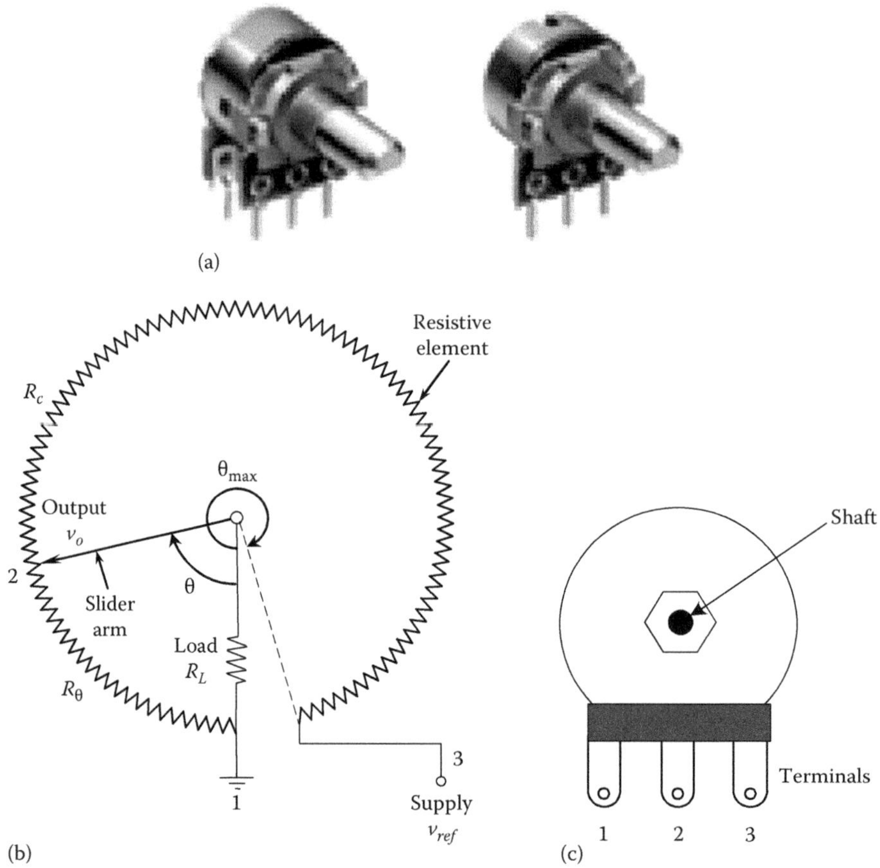

FIGURE 8.3
(a) Rotary potentiometers (Courtesy of Alps Electric, Auburn Hills, MI), (b) a rotatory potentiometer with a resistive load, and (c) external details.

Assuming a uniform coil, one has

$$R_\theta = \frac{\theta}{\theta_{max}} R_c \tag{8.2}$$

where R_c is the total resistance of the potentiometer coil. The current balance at the sliding contact point (node 2) gives

$$\frac{v_{ref} - v_o}{R_c - R_\theta} = \frac{v_o}{R_\theta} + \frac{v_o}{R_L} \tag{i}$$

where R_L is the load resistance. Multiply throughout Equation (i) by R_c and use Equation 8.2. We get $(v_{ref} - v_o)/(1-(\theta/\theta_{max})) = v_o/(\theta/\theta_{max}) + v_o/(R_L/R_c)$. By using straightforward algebra, we obtain

$$\frac{v_o}{v_{ref}} = \left[\frac{(\theta/\theta_{max})(R_L/R_c)}{(R_L/R_c + (\theta/\theta_{max}) - (\theta/\theta_{max})^2)} \right] \tag{8.3}$$

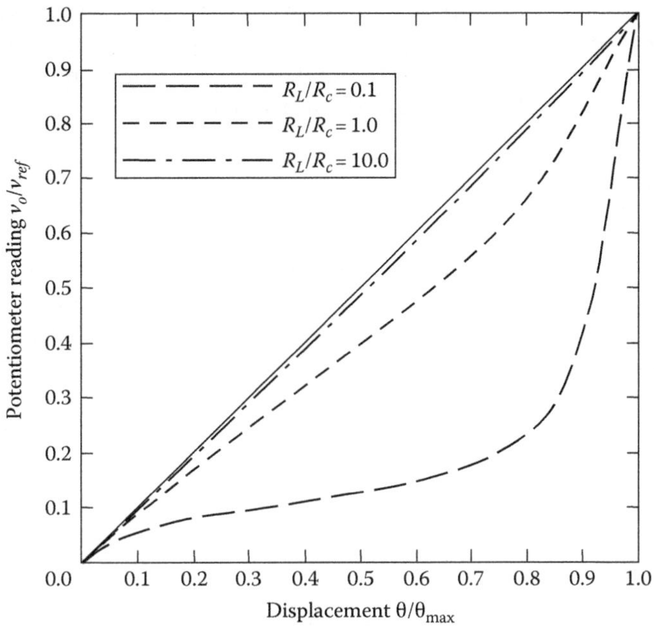

FIGURE 8.4
Electrical-loading nonlinearity in a potentiometer.

This equation is plotted in Figure 8.4. Loading error appears to be high for low values of the R_L/R_c ratio. Good accuracy is possible for $R_L/R_c > 10$, particularly for small values of θ/θ_{max}.

It should be clear that the following measures can be taken to reduce loading error in pots:

1. Increase R_L/R_c (increase load impedance, reduce coil impedance).
2. Use pots to measure small values of θ/θ_{max} (or calibrate only a small segment of the resistance element, for linear reading).

The loading nonlinearity error is defined by

$$e = \frac{(v_o/v_{ref} - \theta/\theta_{max})}{\theta/\theta_{max}} 100\% \tag{8.4}$$

The error at $\theta/\theta_{max} = 0.5$ for three values of load resistance ratio is tabulated in Table 8.2. Note that this error is always negative. Using only a segment of the resistance element as the range of the potentiometer is equivalent to adding two end resistors to the elements.

TABLE 8.2
Loading Nonlinearity Error in a Potentiometer

Load Resistance Ratio R_L/R_C	Loading Nonlinearity Error (e) at $\theta/\theta_{max} = 0.5$ (%)
0.1	−71.4
1.0	−20
10.0	−2.4

It is known that this tends to linearize the pot. If the load resistance is known to be small, a voltage follower (see Chapter 3) may be used at the potentiometer output to virtually eliminate the loading error, since this arrangement provides a high load impedance to the pot and a low impedance at the output of the amplifier.

8.3.2 Performance Considerations

A potentiometer is a resistively coupled transducer. It is an active device, which needs an external power source for its operation. The force required to move the slider arm comes from the moving object that is being measured, and the resulting energy is dissipated through friction at the slider. This energy conversion, unlike pure mechanical-to-electrical conversions, involves relatively high forces, and the energy is wasted rather than getting converted into the output signal of the transducer. Furthermore, the electrical energy from the reference source is also dissipated through the resistor element (coil or film) of the pot, resulting in an undesirable temperature rise and element degradation. These are two obvious disadvantages of a potentiometer.

8.3.2.1 Potentiometer Ratings

Stroke (for linear movement), element resistance, reference voltage, and power (at full resistance) are key rating parameters of a potentiometer. The maximum slider movement of a linear pot is called its "stroke." It can be as small as several mm to as high as 75 cm. The resistance of a pot should be chosen with care. On the one hand, an element with high resistance is preferred because this results in reduced power dissipation for a given reference voltage, which has the added benefits of reduced thermal effects and increased potentiometer life. On the other hand, increased resistance increases the output impedance of the potentiometer and results in a corresponding increase in loading nonlinearity error unless the load resistance is also increased correspondingly. Low-resistance pots have resistances less than 10 Ω. High-resistance pots can have resistances as high as 100 kΩ. Conductive plastics can provide high resistances—typically about 100 Ω/mm—and are increasingly used in potentiometers. The full resistance of the pot element is marked on the housing of the pot. Sometimes, this value is indicated using a code (e.g., 10^3 represents "10" followed by "3" zeroes or 10,000 Ω).

Another rating parameter that is important for the safety of its use is the *dielectric voltage*. This is the voltage that the insulation between the resistance element and the outside (housing and shaft) of the pot can safely withstand (say, 2.5 kV). Other precautions include using a nonmetal (say, plastic) slider arm (for linear pot) or nonmetal shaft (for rotary pot) and proper grounding.

8.3.2.2 Resolution

A coil-type pot has a finite resolution. When a coil is used as the resistance element of a pot, the slider contact jumps from one turn to the next one during movement. Accordingly, the resolution of a coil-type potentiometer is determined by the number of turns in the coil. For a coil that has N turns, the resolution r, expressed as a percentage of the output range, is given by

$$r = \frac{100}{N}\% \tag{8.5}$$

Resolutions better (smaller) than 0.1% (i.e., 1000 turns) are available with coil potentiometers. Virtually infinitesimal (sometimes incorrectly termed as *infinite*) resolutions are possible with today's high-quality resistive film potentiometers, which use conductive plastics or cermet. Then, the resolution is limited by other factors such as mechanical limitations and SNR. Nevertheless, resolutions in the order of 0.01 mm are possible with good rectilinear potentiometers.

In selecting a potentiometer for a specific application, several factors have to be considered. As noted earlier, they include element resistance, power consumption, loading, resolution, and size.

8.3.2.3 Sensitivity

The *sensitivity* of a potentiometer represents the change (Δv_o) in the output signal that results from a given small change ($\Delta\theta$) in the measurand (the object displacement). The sensitivity is usually nondimensionalized, using the actual value of the output signal (v_o) and the actual value of the displacement (θ). For a rotatory potentiometer in particular, the sensitivity S is given by

$$S = \frac{\Delta v_o}{\Delta\theta} \quad \text{or in the limit} \quad S = \frac{\partial v_o}{\partial\theta} \tag{8.6}$$

These relations may be nondimensionalized by multiplying by θ/v_o. An expression for S may be obtained by simply substituting Equation 8.3 into Equation 8.6.

Some limitations and disadvantages of potentiometers as displacement measuring devices are given in the following:

1. The force needed to move the slider (against friction and arm inertia) is provided by the displacement source (moving object that is sensed). This mechanical loading distorts the measured signal itself.
2. High-frequency (or highly transient) measurements are not feasible because of such factors as slider bounce, friction, and inertia resistance and induced voltages in the wiper arm and the primary resistor element.
3. Variations in the supply voltage cause error.
4. Electrical loading error can be significant when the load resistance is low.
5. Resolution is limited by the number of turns in the coil and by the coil uniformity (in a coil-type pot). This limits small-displacement measurements.
6. Wear and heating up (with associated oxidation) in the resistor element (coil or film) and the slider contact cause accelerated degradation.

There are several advantages associated with potentiometer devices, however, including the following:

1. They are simple in design and robust.
2. They are relatively inexpensive.
3. They provide high-voltage (low impedance) output signals, requiring no amplification in most applications.
4. Transducer impedance can be varied simply by changing the element resistance and supply voltage.

Example 8.1

A rectilinear potentiometer was tested with its slider arm moving horizontally. It was found that at a speed of 1 cm/s, a driving force of 7×10^{-4} N was necessary to maintain the speed. At 10 cm/s, a force of 3×10^{-3} N was necessary. The slider weighs 5 g, and the potentiometer stroke is ±8 cm. If this potentiometer is used to measure the damped natural frequency of a simple mechanical oscillator of mass 10 kg, stiffness 10 N/m, and damping constant 2 N/m/s, estimate the percentage error due to mechanical loading. Justify this procedure for the estimation of damping.

Solution

Denote the mass, stiffness, and damping constant of the simple oscillator by *M, K,* and *B*, respectively. The equation of free motion of the simple oscillator is given by $M\ddot{y}+B\dot{y}+Ky=0$, where y denotes the displacement of the mass from the static equilibrium position. This equation is of the form $\ddot{y}+2\zeta\omega_n\dot{y}+\omega_n^2y=0$, where ω_n is the undamped natural frequency of the oscillator and ζ is the damping ratio. By direct comparison of these two equations, it is seen that $\omega_n=\sqrt{K/M}$ and $\zeta=B/(2\sqrt{MK})$.

The damped natural frequency is $\omega_d=\sqrt{1-\zeta^2}\,\omega_n$ for $0<\zeta<1$. Hence,

$$\omega_d=\sqrt{\left(1-\frac{B^2}{4MK}\right)\frac{K}{M}}$$

Now, if the mass of the wiper arm and the damping constant of the potentiometer are denoted by m and b, respectively, the measured damped natural frequency (by using the potentiometer) is given by

$$\tilde{\omega}_d=\sqrt{\left[1-\frac{(B+b)^2}{4(M+m)K}\right]\frac{K}{(M+m)}} \tag{8.7}$$

Assuming linear viscous friction, the equivalent damping constant b of the potentiometer may be estimated as

$$b=\text{Damping force/steady-state velocity of the wiper}.$$

For this example, $b_1=7\times10^{-4}/1\times10^{-2}$ N/m/s $=7\times10^{-2}$ N/m/s at 1 cm/s and $b_2=3\times10^{-3}/10\times10^{-2}$ N/m/s $=3\times10^{-2}$ N/m/s at 10 cm/s.

We should use some form of interpolation to estimate b for the actual measuring conditions. Let us now estimate the average velocity of the wiper. The natural frequency of the oscillator is $\omega_n=\sqrt{10/10}=1\text{ rad/s}=1/(2\pi)\text{ Hz}$. Since one cycle of oscillation corresponds to a motion of 4 strokes, the wiper travels a maximum distance of $4\times8\text{ cm}=32\text{ cm}$ in one cycle. Hence, the average operating speed of the wiper may be estimated as $32/(2\pi)$ cm/s, which is approximately equal to 5 cm/s. Therefore, the operating damping constant may be estimated as the average of b_1 (at 1 cm/s) and b_2 (at 10 cm/s): $b=5\times10^{-2}$ N/m/s. With the foregoing numerical values, we get

$$\omega_d=\sqrt{\left(1-\frac{2^2}{4\times10\times10}\right)\frac{10}{10}}=0.99499\text{ rad/s}$$

$$\tilde{\omega}_d=\sqrt{\left(1-\frac{2.05^2}{4\times10.005\times10}\right)\frac{10}{10.005}}=0.99449\text{ rad/s}$$

$$\text{Percentage error}=\left[\frac{\tilde{\omega}_d-\omega_d}{\omega_d}\right]\times100\%=0.05\%$$

Although pots are primarily used as displacement transducers, they can be adapted to measure other types of signals, such as pressure and force, using appropriate auxiliary sensor (front end) elements. For instance, a bourdon tube or bellows may be used to convert pressure into displacement, and a cantilever element may be used to convert force or moment into displacement, which can then be measured using a pot.

8.3.3 Optical Potentiometer

The optical potentiometer, shown schematically in Figure 8.5a, is a displacement sensor. A layer of photoresistive material is sandwiched between a layer of ordinary resistive material and a layer of conductive material. The layer of resistive material has a total resistance of R_c, and it is uniform (i.e., it has a constant resistance per unit length). This corresponds to the element resistance of a conventional potentiometer. The photoresistive layer is practically an electrical insulator when no light is projected on it. The moving object, whose displacement is measured, causes a moving light beam to be projected on a rectangular area of the photoresistive layer. This light-activated area attains a resistance of R_p, which links the resistive layer that is above the photoresistive layer and the conductive layer that is below the photoresistive layer. The supply voltage to the potentiometer is v_{ref}, and the length of the resistive layer is L. The light spot is projected at a distance x from the reference end of the resistive element, as shown in the figure.

An equivalent circuit for the optical potentiometer is shown in Figure 8.5b. Here, it is assumed that a load of resistance R_L is present at the output of the potentiometer, with v_o

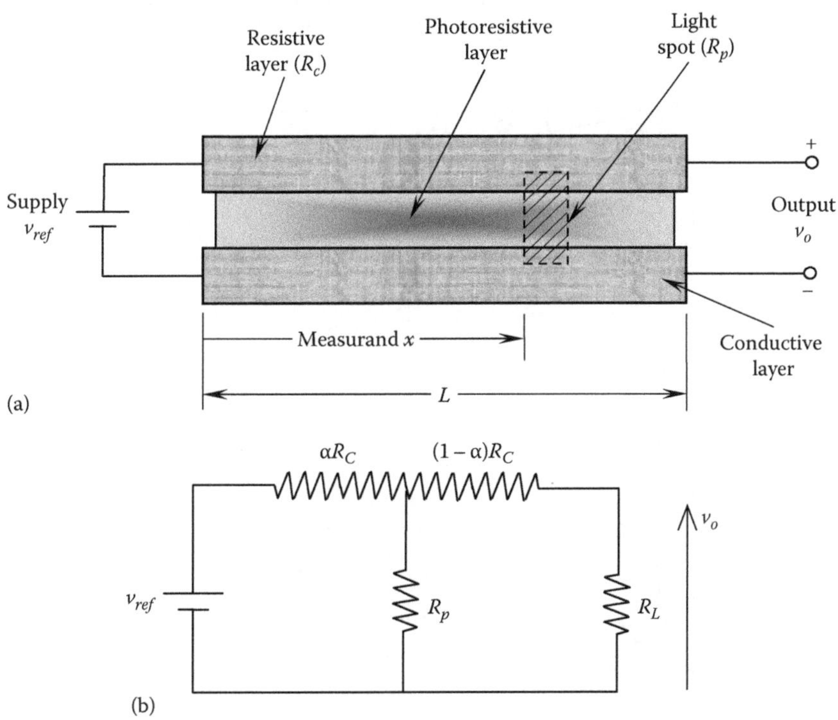

FIGURE 8.5
(a) An optical potentiometer and (b) equivalent circuit ($\alpha = x/L$).

as voltage across. The current through the load is v_o/R_L. Hence, the voltage drop across $(1-\alpha)R_c+R_L$, which is also the voltage across R_p, is given by $[(1-\alpha)R_c+R_L]\,v_o/R_L$. Note that $\alpha = x/L$ is the fractional position of the light spot. The current balance at the junction of the three resistors in Figure 8.5b gives

$$\frac{v_{ref}-[(1-\alpha)R_c+R_L]v_o/R_L}{\alpha R_c}=\frac{v_o}{R_L}+\frac{[(1-\alpha)R_c+R_L]v_o/R_L}{R_p}$$

which can be written as

$$\frac{v_o}{v_{ref}}\left\{\frac{R_c}{R_L}+1+\frac{x}{L}\frac{R_c}{R_p}\left[\left(1-\frac{x}{L}\right)\frac{R_c}{R_L}+1\right]\right\}=1 \tag{8.8a}$$

When the load resistance R_L is quite large in comparison with the element resistance R_c, we have $R_c/R_L \simeq 0$. Hence, Equation 8.8a becomes

$$\frac{v_o}{v_{ref}}=\frac{1}{\left[(x/L)(R_c/R_p)+1\right]} \tag{8.8b}$$

This relationship is still nonlinear in x/L. The nonlinearity can be decreased, however, by decreasing R_c/R_p. This is also seen in Figure 8.6 where Equation 8.8b is plotted for several values of R_c/R_p. Then, for the case of $R_c/R_p=0.1$, the original Equation 8.8a is plotted in Figure 8.7, for several values of the load resistance ratio. It is observed that, as expected,

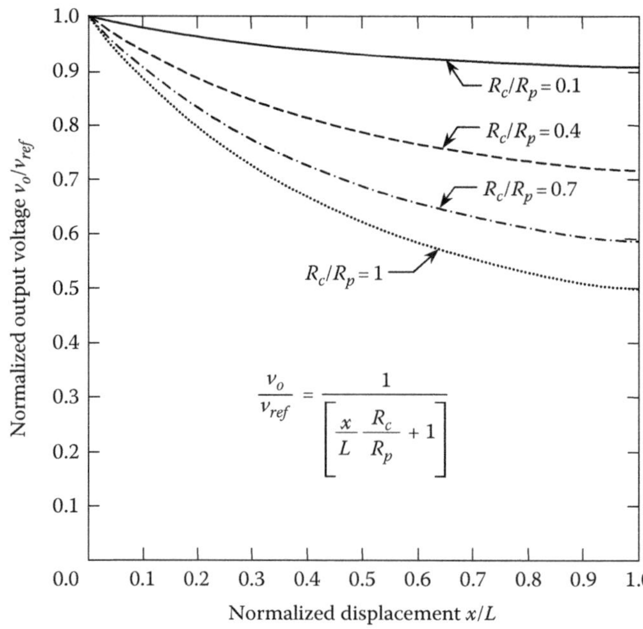

FIGURE 8.6
Behavior of the optical potentiometer at high load resistance.

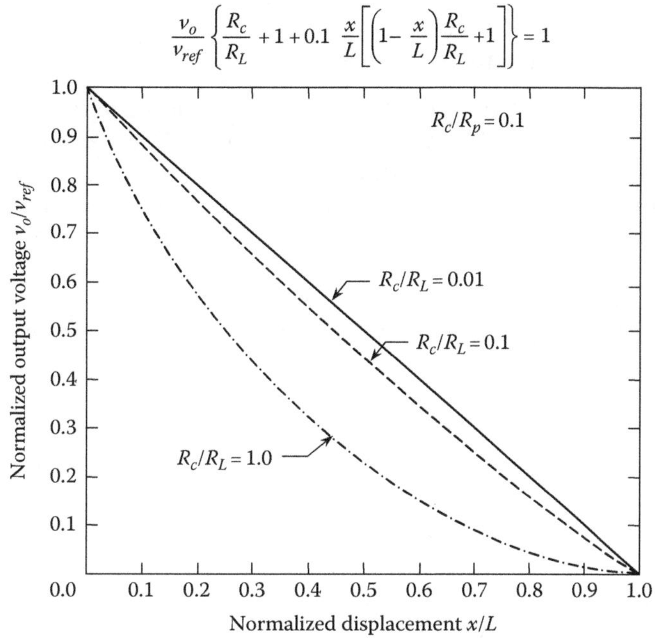

FIGURE 8.7
Behavior of the optical potentiometer for $R_c/R_p = 0.1$.

the behavior of the optical potentiometer becomes more linear for higher values of load resistance.

Note: Many other principles may be employed in potentiometers for displacement sensing. For example, an alternative possibility for optical potentiometer is to have a fixed light source and to locate a photosensor on the moving object whose displacement needs to be measured. By calibrating the device according to how the light intensity varies with the distance between the light source and the light sensor, the distance can be measured. Of course, such a device would be quite nonlinear and nonrobust (as it will be affected by environmental lighting, etc.).

8.3.3.1 Digital Potentiometer

The digital potentiometer is a device that can provide digitally incremented resistance or voltage corresponding to a digital command. The range of discrete resistance that it can provide depends on the bit size of the device (e.g., 8-bit device is able to provide 256 discrete values of resistance). The incrementing can be programmed linearly, logarithmically, etc., using a microcontroller or other digital device, depending on the application. It is clear that a digital pot is not a displacement sensor but rather a resistance splitter or voltage splitter. It is mentioned here to avoid any misconception.

The potentiometer has disadvantages such as loading problems (both mechanical and electrical), limited speed of operation, considerable time constants, wear, noise, and thermal effects. Many of these problems arise from the fact that it is a contact device where its slider has to be in intimate contact with the resistance element of the pot and also has to be an integral part of the moving object whose displacements need to be measured. Next, we consider several noncontact motion sensors that do not have these shortcomings.

8.4 Variable-Inductance Transducers

Motion transducers that employ the principle of electromagnetic induction are termed *variable-inductance transducers*. When the flux linkage (defined as magnetic flux density times the number of turns in the conductor) through an electrical conductor changes, a voltage in proportion to the rate of change of flux is induced in the conductor. This is the basis of *electromagnetic induction*. This voltage is called the "electromotive force" (emf), which in turn generates a magnetic field that opposes the original (primary) field. Hence, a mechanical force is necessary to sustain the change of flux linkage.

The rate of change in magnetic flux that "induces" the voltage in the conductor can be caused in two principal ways:

1. By changing the current that creates the magnetic field
2. By physically moving (a) the coil or the magnet that provides the magnetic field, (b) the medium (e.g., soft iron core) through which the magnetic flux links with the conductor, and (c) the conductor in which the voltage is induced, at some speed

Category 2 is particularly useful in motion sensors. In electromagnetic induction, if the change in flux linkage is brought about by a relative motion, the associated mechanical energy is directly converted (induced) into electrical energy. This is the principle of operation of electrical generators and also of *variable-inductance transducers*. Specifically, principle 2(b) in the preceding list can be utilized in a *passive* displacement sensor (e.g., *differential transformer*), and principles 2(a) and 2(b) may be utilized in a *passive* speed sensor (*tachometer*).

Note that in these transducers, the change of flux linkage is caused by a mechanical motion and mechanical-to-electrical energy transfer takes place under near-ideal conditions. The induced voltage or change in inductance is used as a measure of the motion. Hence, it is clear that they are "passive" transducers. Furthermore, it is seen that variable-inductance transducers are generally electromechanical devices coupled by a magnetic field.

Effect of environmental magnetic fields: One common property (drawback) of all variable-inductance transducers is that their reading will be affected by the magnetic fields in the environment. In view of its low field strength, the effect of the earth's magnetic field is insignificant, however, except in highly delicate instruments. When the ambient magnetic field is not negligible, protective measures have to be taken including shielding (e.g., using steel housing), noise filtering, and compensation (e.g., by sensing the ambient magnetic field).

There are many different types of *variable-inductance transducers*. Three primary types can be identified:

1. Mutual-induction transducers
2. Self-induction transducers
3. Permanent-magnet (PM) transducers

Furthermore, those variable-inductance transducers that use a nonmagnetized ferromagnetic medium to alter the *reluctance* (magnetic resistance) of the magnetic flux path are known as "variable-reluctance transducers." Some of the mutual-induction transducers and most of the self-induction transducers are of this type. Strictly speaking, PM transducers are not variable-reluctance transducers.

Mutual-Induction Transducers

The basic arrangement of a mutual-induction transducer constitutes two coils, the primary winding and the secondary winding. One of the coils (primary winding) carries an alternating-current (AC) excitation that induces a steady AC voltage in the other coil (secondary winding). The level (amplitude, root-mean-square [rms] value, etc.) of the induced voltage depends on the flux linkage between the coils. It is used as a measure of the motion, which affects the induced voltage. None of these transducers employ contact sliders or slip rings and brushes as do resistively coupled transducers (potentiometer). Consequently, they have an increased design life and low mechanical loading errors.

In mutual-induction transducers, a change in the flux linkage is effected by one of two common techniques. One technique is to move an object made of ferromagnetic material within the flux path between the primary coil and the secondary coil. This changes the reluctance of the flux path, with an associated change of the flux linkage in the secondary coil. This is, for example, the operating principle of the linear-variable differential transformer–transducer (LVDT), the rotatory-variable differential transformer–transducer (RVDT), and the mutual-induction proximity probe. All of these are displacement transducers and, in fact, are variable-reluctance transducers as well. The other common way to change the flux linkage is to move one coil with respect to the other. This is the operating principle of the resolver, the synchro transformer, and some types of AC tachometer. These are not variable-reluctance transducers, however, because a moving ferromagnetic element is not involved in their operation.

Motion can be measured by using the secondary signal (i.e., induced voltage in the secondary coil) in several ways. For example, the AC signal in the secondary coil may be demodulated (see Chapter 4) by rejecting the carrier signal (i.e., the signal component that is at the excitation frequency). The resulting signal, which represents the motion, is directly measured. This method is particularly suitable for measuring transient motions. Alternatively, the amplitude or the rms value of the secondary (induced) voltage may be measured. Another method is to measure the change of *inductance* or *reactance* in the secondary circuit directly, by using a device such as an inductance bridge circuit (see Chapter 4).

8.4.1 Inductance, Reactance, and Reluctance

The *magnetic flux linkage* ϕ is a measure of the magnetic field that is linked with a conductor (coil). It has the units of weber (Wb) and depends on the magnetic flux density, number of turns in the coil, and coil area (not the wire area). If the magnetic field is generated by a current (i.e., electromagnetism), the magnetic field depends on that current i, which has the units of amperes (A). Then, we can write

$$\phi = Li \tag{8.9}$$

where L is the *inductance* and has the units of Wb/A or henry (H).

The voltage v induced in the coil due to the rate of change of the magnetic flux is called the "electromotive force" (emf). We have

$$v = L\frac{di}{dt} \tag{8.10}$$

The generalized resistance (impedance) that results through the inductance is called "reactance" or "reactive impedance" and is denoted by *X*. This is the "imaginary part" of a complex impedance. According to Equation 8.10, in the frequency domain (where the time derivative di/dt becomes $j\omega$), the reactance is given by

$$X = Lj\omega \tag{8.11}$$

where ω is the frequency of the signal.

The ratio of the magnetic *flux density* (units in tesla or T; weber per square meter or Wb/m^2) to the magnetic *field strength* (units in ampere·turns per meter or At/m), for a magnetic circuit segment (or medium of a magnetic flux path), is called "permeability" (or magnetic permeability) and is denoted by μ. It is also the inductance per unit length for the magnetic circuit segment. Permeability has the units tesla·meter per ampere (T·m/A) or henry per meter (H/m). We have

$$\mu = \frac{B}{H} = \frac{L}{l}\ \text{T}\cdot\text{m/At or H/m} \tag{8.12}$$

where

B denotes the flux density
H denotes the field strength

The permeability of the free space is approximately $\mu_o = 4\pi \times 10^{-7} = 1.257 \times 10^{-6}$ H/m. The *relative permeability* of a magnetic path is its permeability with respect to that of the free space and is given by $\mu_r = \mu/\mu_o$, and it has no units. The relative permeability of some material is given in Table 8.3. The absolute permeability can be computed from these values since the permeability of the free space is known.

Magnetic reluctance (or simply *reluctance*) is the magnetic resistance of a magnetic circuit segment (a medium through which the magnetic field passes). It is given by

$$\mathfrak{R} = \frac{l}{\mu A} \tag{8.13}$$

where

l is the length of the magnetic circuit segment
A is the area of cross section of the magnetic circuit (e.g., coil cross section, not the wire cross section in the coil)

TABLE 8.3

Relative Permeability Values of Some Material (Approximate)

Material	Relative Permeability μ_r
Air, aluminum, concrete, copper, platinum, Teflon, water, wood	1.0
Carbon steel	100
Cobalt–iron	1.8×10^4
Iron (Fe)	2.0×10^5
Nickel	100–600
Stainless steel	40–1800

It is seen that permeability is a measure of the easiness in which the magnetic field travels in a medium and the reluctance represents the inverse of that. The inverse of reluctance is *permeance*. From Equation 8.13, it is clear that reluctance has units represented by the inverse of henry. Strictly, it also depends on the number turns in the inductor coil. Hence, reluctance has the units "turns per henry" (t/H) or "ampere·turns per weber" (At/Wb).

Note: From Equations 8.12 and 8.13, it is seen that the reluctance is inversely proportional to inductance. Hence, reluctance can be measured using an inductance bridge. Since reluctance is proportional to the length of a magnetic circuit segment, if that length changes due to some displacement in an object, we can measure the displacement by measuring the corresponding reluctance or inductance. This forms the principle of a variable-reluctance displacement sensor. Alternatively, a voltage can be induced in a conductor coil by moving it in a magnetic field. The induced voltage is proportional to the coil speed. This can form the principle of a tachometer.

8.4.2 Linear-Variable Differential Transformer–Transducer

Differential transformer is a noncontact displacement sensor, which does not possess many of the shortcomings of the potentiometer. It falls into the general category of a variable-inductance transducer and is also a variable-reluctance transducer and a mutual-induction transducer. Furthermore, unlike the potentiometer, the differential transformer is a passive device. Now, we discuss the LVDT, which is used for measuring rectilinear (or translatory) displacements. Subsequently, we describe the RVDT, which is used for measuring angular (or rotatory) displacements.

The LVDT is considered a passive transducer because the displacement, which is being measured, itself provides energy for changing the induced voltage in the secondary coil. Even though an external power supply is used to energize the primary coil, which in turn induces a steady voltage at the carrier frequency in the secondary coil, it is not relevant in the definition of a passive transducer.

In its simplest form (see Figure 8.8), the LVDT consists of an insulating, nonmagnetic form (a cylindrical structure on which a coil is wound and is integral with the housing), which has a primary coil in the midsegment and a secondary coil symmetrically wound in the two end segments, as depicted schematically in Figure 8.8b. The housing is made of magnetized stainless steel to shield the sensor from outside fields. The primary coil is energized by an AC supply of voltage v_{ref}. This generates, by mutual induction, an AC of the same frequency in the secondary coil. A core made of ferromagnetic material is inserted coaxially through the cylindrical form without actually touching it, as shown. As the core moves, the reluctance of the flux path between the primary and the secondary coils changes. The degree of flux linkage depends on the axial position of the core. Since the two secondary coils are connected in series opposition (as shown in Figure 8.9), the potentials induced in the two secondary coil segments oppose each other. Hence, the net induced voltage is zero when the core is centered between the two secondary winding segments. This is known as the "null position." When the core is displaced from this position, a nonzero-induced voltage is generated. At steady state, the amplitude v_o of this induced voltage is proportional to the core displacement x in the linear (operating) region (see Figure 8.8c). Consequently, v_o is a measure of the displacement.

Note: Because of opposed secondary windings, the LVDT provides the direction as well as the magnitude of displacement. When the output signal is demodulated, its sign gives

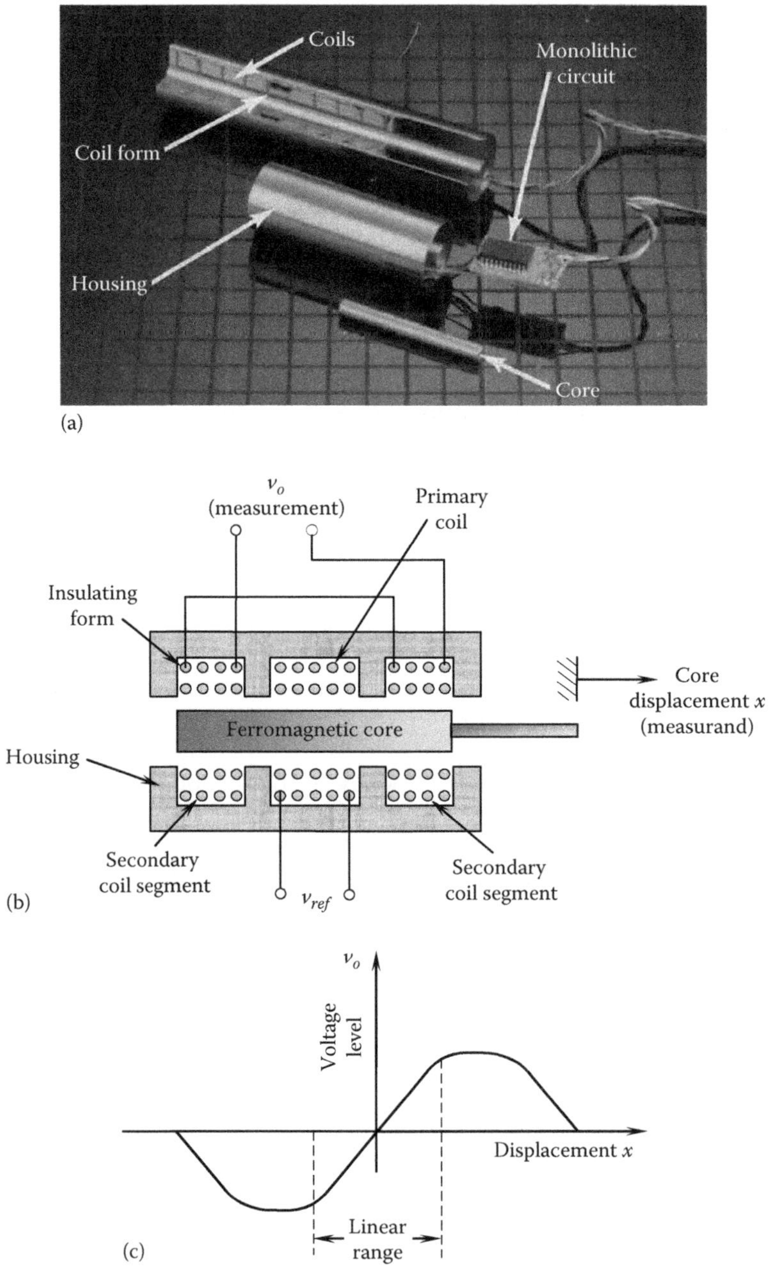

FIGURE 8.8
Linear-variable differential transformer–transducer. (a) A commercial unit (Courtesy of Scheavitz Sensors, Measurement Specialties, Inc., Hampton, VA), (b) schematic diagram, and (c) a typical operating curve.

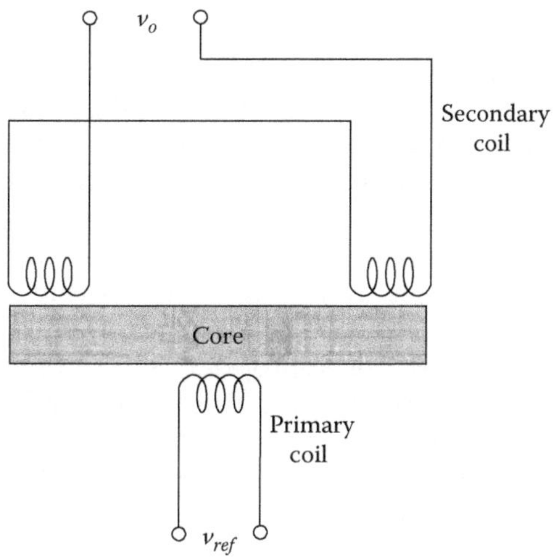

FIGURE 8.9
Series opposition connection of secondary windings of a linear-variable differential transformer–transducer.

the direction. If the output signal is not demodulated, the direction is determined by the phase angle between the primary (reference) voltage and the secondary (output) voltage, which includes the carrier signal.

For an LVDT to measure transient motions accurately, the frequency of the reference voltage (the carrier frequency) has to be at least 10 times larger than the largest significant (useful) frequency component in the measured motion and, typically, can be as high as 20 kHz. For quasi-dynamic displacements and slow transients of the order of a few hertz, a standard AC supply (at 60 Hz line frequency) is adequate. The performance (particularly sensitivity and accuracy) is known to improve with the excitation frequency, however. Since the amplitude of the output signal is proportional to the amplitude of the primary signal, the reference voltage should be regulated to get accurate results. In particular, the power source should have a low output impedance.

Commercial LVDTs normally come with accompanied signal-conditioning hardware on a single printed circuit card. It will contain such functional hardware as an oscillator, amplifier, filter, demodulator, and so on. It will have terminals for a DC power supply (e.g., 15 V). As desirable, high input impedance (e.g., 0.2 MΩ) may be provided by the signal-conditioning hardware.

8.4.2.1 Calibration and Compensation

An LVDT may be calibrated in mm/V in its linear range. In addition, a displacement offset (mm) may be provided. This typically represents the least-squares fit of a set of calibration data. Since ambient temperature and other environmental conditions will affect the LVDT output, in addition to the primary and secondary coils, a reference coil may be available for compensation of the LVDT output. Alternatively, an inductance bridge circuit, where two segments of the secondary coil form two arms of the bridge, may be employed for generating the LVDT output. Then, compensation for environmental effects (including temperature compensation) is automatically achieved as in any bridge circuit.

8.4.2.2 Phase Shift and Null Voltage

An error known as "null voltage" (or residual voltage) is present in some differential transformers. This manifests itself as a nonzero reading at the null position (i.e., at zero displacement). This is usually 90° out of phase from the main output signal and, hence, is known as the "quadrature error." Nonuniformities in the windings (unequal impedances in the two segments of the secondary winding) are a major reason for this error. The null voltage may also result from harmonic noise components in the primary signal and nonlinearities in the device. Null voltage may be ignored if it is less than 1% of the full-scale output. Typically, it is quite low (about 0.1% of full scale). This error can be eliminated from the measurements by employing appropriate signal-conditioning and calibration practices. They include removing the phase shift and the null voltage at the output by such methods as synchronized demodulation (synchronizing the output with the carrier signal) and offsetting (by measuring the null voltage and calibrating the output or by offsetting circuitry). Concepts behind them are presented now.

The output signal from a differential transformer is normally not in phase with the reference voltage. Inductance in the primary coil and the leakage inductance in the secondary coil are mainly responsible for this phase shift. Since *demodulation* involves extraction of the modulating signal by rejecting the carrier frequency component from the secondary signal, it is important to understand the size of this phase shift. This topic is addressed now. An equivalent circuit for a differential transformer is shown in Figure 8.10. The resistance in the primary coil is denoted by R_p and the corresponding inductance is denoted by L_p. The total resistance of the secondary coil is R_s. The net leakage inductance, due to magnetic flux leakage, in the two segments is denoted by L_l. The load resistance is R_L and the load inductance is L_L. First, let us derive an expression for the phase shift in the output signal.

The magnetizing voltage in the primary coil is given by $v_p = v_{ref}[(j\omega L_p)/(R_p + j\omega L_p)]$ in the frequency domain. Now, suppose that the core of length L is moved through a distance x from the null position. The induced voltage in one segment (a) of the secondary coil would be $v_a = v_p k_a(L/2 + x)$ and the induced voltage in the other segment (b) would be

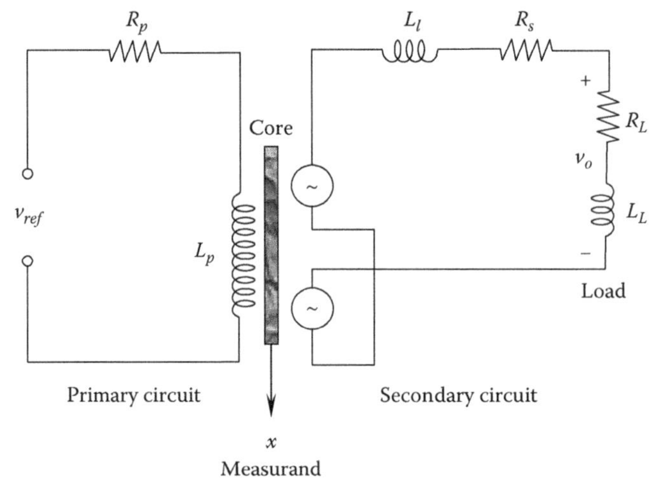

FIGURE 8.10
Equivalent circuit for a differential transformer–transducer.

$v_b = v_p k_b(L/2 - x)$. Here, k_a and k_b are nonlinear functions of the position of the core and are also complex functions of the frequency variable ω. Furthermore, each function depends on the mutual-induction properties between the primary coil and the corresponding secondary coil segment through the core element. Due to series opposition connection of the two secondary segments, the induced net secondary voltage would be

$$v_s = v_a - v_b = v_p\left[k_a\left(\frac{L}{2}+x\right) - k_b\left(\frac{L}{2}-x\right)\right] \tag{8.14}$$

In the ideal case, the two functions $k_a(\cdot)$ and $k_b(\cdot)$ would be identical. Then, at $x=0$, we have $v_s=0$. Hence, the null voltage would be zero in the ideal case. Suppose that, at $x=0$, the magnitudes of $k_a(\cdot)$ and $k_b(\cdot)$ are equal, but there is a slight phase difference. Then, the "difference vector" $k_a(L/2) - k_b(L/2)$ will have a small magnitude value, but its phase angle will be almost 90° with respect to both k_a and k_b. This is the *quadrature error.*

For small x, the Taylor series expansion in Equation 8.14 gives

$$v_s = v_p\left[k_a\left(\frac{L}{2}\right) + \frac{\partial k_a}{\partial x}\left(\frac{L}{2}\right)x - k_b\left(\frac{L}{2}\right) + \frac{\partial k_b}{\partial x}\left(\frac{L}{2}\right)x\right]$$

Then, assuming that $k_a(\cdot)=k_b(\cdot)$ and is denoted by $k_o(\cdot)$, we have $v_s = 2v_p(\partial k_o/\partial x)(L/2)x$ or $v_s = v_p kx$, where $k = 2(\partial k_o/\partial x)(L/2)$. In this case, the net induced voltage is proportional to x and is given by

$$v_s = v_{ref}\left[\frac{j\omega L_p}{R_p + j\omega L_p}\right]kx$$

It follows that the output voltage v_o at the load is given by

$$v_o = v_{ref}\left[\frac{j\omega L_p}{R_p + j\omega L_p}\right]\left[\frac{R_L + j\omega L_L}{(R_L + R_s) + j\omega(L_L + L_l)}\right]kx \tag{8.15}$$

Hence, for small displacements, the amplitude of the net output voltage of the LVDT is proportional to the displacement x. The phase lead at the output is given by

$$\phi = 90° - \tan^{-1}\frac{\omega L_p}{R_p} + \tan^{-1}\frac{\omega L_L}{R_L} - \tan^{-1}\frac{\omega(L_L + L_l)}{R_L + R_s} \tag{8.16}$$

The degree of dependence of the phase shift on the load (including the secondary circuit) can be reduced by increasing the load impedance.

8.4.2.3 Signal Conditioning

Signal conditioning associated with differential transformers includes filtering and amplification. Filtering is needed to improve the SNR of the output signal. Amplification is necessary to increase the signal strength for data acquisition, transmission, and processing. Since the reference frequency (carrier frequency) is induced into (and embedded in) the output signal, it is also necessary to interpret the output signal properly, particularly for transient motions.

The secondary (output) signal of an LVDT is an amplitude-modulated signal, where the signal component at the carrier frequency is modulated by the lower-frequency transient signal produced as a result of the core motion (x). Two methods are commonly used to interpret the crude output signal from a differential transformer: rectification and demodulation. In the first method (rectification), the AC output from the differential transformer is rectified to obtain a DC signal. This signal is amplified and then low-pass filtered to eliminate any high-frequency noise components. The amplitude of the resulting signal provides the transducer reading. In this method, the phase shift in the LVDT output has to be checked separately to determine the direction of motion. In the second method (demodulation), the carrier frequency component is rejected from the output signal by comparing it with a phase-shifted and amplitude-adjusted version of the primary (reference) signal. Phase shifting is necessary because, as discussed earlier, the output signal is not in phase with the reference signal. The result is the modulating signal (proportional to x), which is subsequently amplified and filtered.

As a result of the advances in miniature IC technology, differential transformers with built-in microelectronics for signal conditioning are commonly available today. A DC differential transformer uses a DC power supply (typically, ±15 V) to activate it. A built-in oscillator circuit generates the carrier signal. The rest of the device is identical to an AC differential transformer. The amplified full-scale output voltage can be as high as ±10 V.

The demodulation approach of signal conditioning for an LVDT is indicated in Figure 8.11a. Figure 8.11b shows a schematic diagram of a simplified signal-conditioning system for an LVDT. The system variables and parameters are as indicated in the figure. In particular, $x(t)$ is the displacement of the LVDT core (measurand, to be measured), ω_c is the frequency of the carrier voltage, and v_o is the output signal of the system (measurement). The resistances R_1, R_2, R_3, and R and the capacitance C are as marked. In addition, we may introduce a transformer parameter r for the LVDT, as required.

The primary coil of the LVDT is excited by an AC voltage of $v_p \sin \omega_c t$. The displacement of the ferromagnetic core to which the moving object is attached is $x(t)$, which is to be measured. The two secondary coils are connected in series opposition so that the LVDT output is zero at the null position and the direction of motion can be detected as well. The amplifier is a noninverting type. It amplifies the output of the LVDT, which is an AC (carrier) signal of frequency ω_c, which is modulated by the core displacement $x(t)$. The multiplier circuit generates the product of the primary (carrier) signal and the secondary (LVDT output) signal. This is an important step in demodulating the LVDT output. The product signal from the multiplier has a high-frequency ($2\omega_c$) carrier component, added to the modulating component ($x(t)$). The low-pass filter removes this unnecessary high-frequency component, to obtain the demodulated signal, which is proportional to the core displacement $x(t)$.

Amplifier equation: Potentials at the + and − terminals of the operational amplifier (op-amp) (in Figure 8.11b) are nearly equal. Also, currents through these leads are nearly zero. (These are the two common assumptions used for an op-amp.) Then, the current balance at node A gives $(v_1/R_1)+(v_2/R_2)=0$. Hence, $v_2=-kv_1$ with $k=(R_2/R_1)$ = amplifier gain.

Low-pass filter: Since the + lead of the op-amp has approximately zero potential (ground), the voltage at point B is also approximately zero. The current balance for node B gives $(v_3/R_3)+(v_o/R)+C\dot{v}_o=0$. Hence, $\tau(dv_o/dt)+v_o=-(R/R_3)v_3$, where $\tau=RC$ = filter time constant. The transfer function of the filter is $v_o/v_3=-k_o/(1+\tau s)$, with the filter gain $k_o=R/R_3$. In the frequency domain, $v_o/v_3=-k_o/(1+\tau j\omega)$.

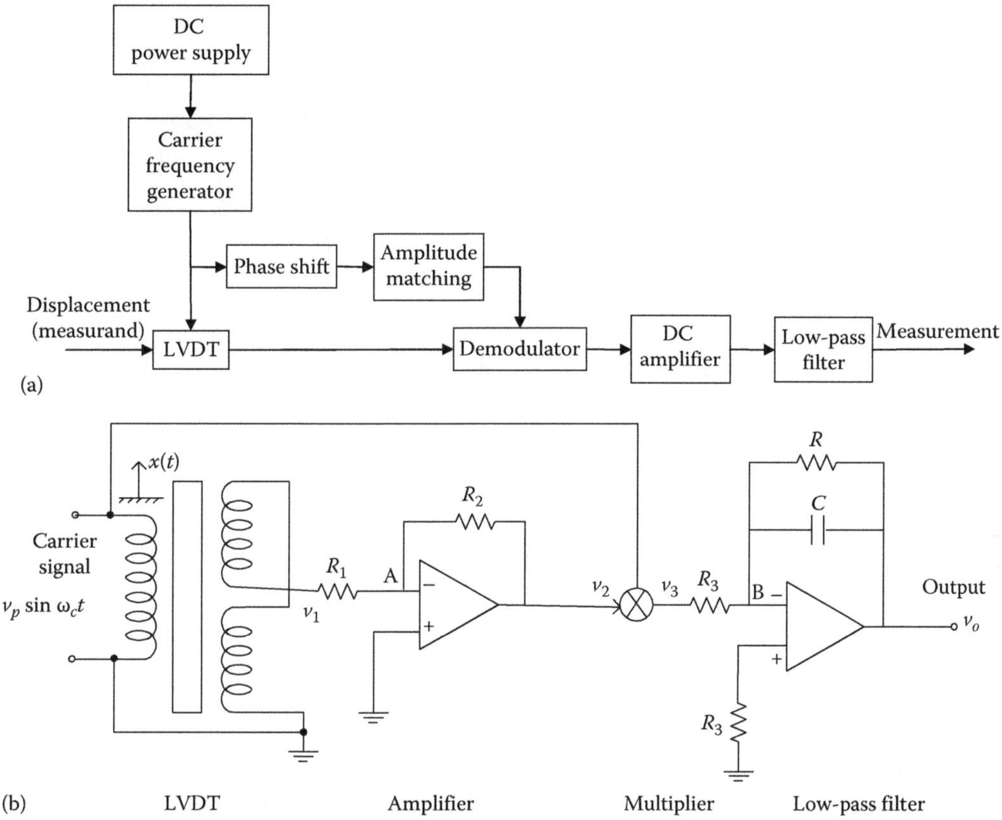

FIGURE 8.11
(a) Signal-conditioning steps for a differential transformer and (b) signal-conditioning system for a linear-variable differential transformer–transducer.

Finally, neglecting the phase shift in the LVDT, we have $v_1 = v_p r x(t) \sin \omega_c t$, $v_2 = -v_p r k x(t) \sin \omega_c t$, and $v_3 = -v_p^2 r k x(t) \sin^2 \omega_c t$ or $v_3 = -(v_p^2 r k)/2x(t)[1 - \cos 2\omega_c t]$.

The carrier signal will be filtered out by the low-pass filter with an appropriate cutoff frequency. Then, $v_o = ((v_p^2 r k k_o)/2)x(t)$.

If the displacement $x(t)$ is linearly increasing (i.e., speed is constant), the signals $u(t)$, v_1, v_2, v_3, and v_o are sketched in Figure 8.12.

Example 8.2

Suppose that in Figure 8.11b, the carrier frequency is $\omega_c = 500$ rad/s and the filter resistance $R = 100$ kΩ. If no more than 5% of the carrier component should pass through the filter, estimate the required value of the filter capacitance C. Also, what is the useful frequency range (measurement bandwidth) of the measuring device in radians per second, with these parameter values?

Solution

Filter magnitude $= k_o/\sqrt{1 + \tau^2\omega^2}$

For no more than 5% of the carrier ($2\omega_c$) component to pass through the filter, we must have $k_o/(\sqrt{1 + \tau^2(2\omega_c)^2}) \le (5/100)k_o$ or $\tau\omega_c \ge 10$ (approximately). We will pick $\tau\omega_c = 10$.

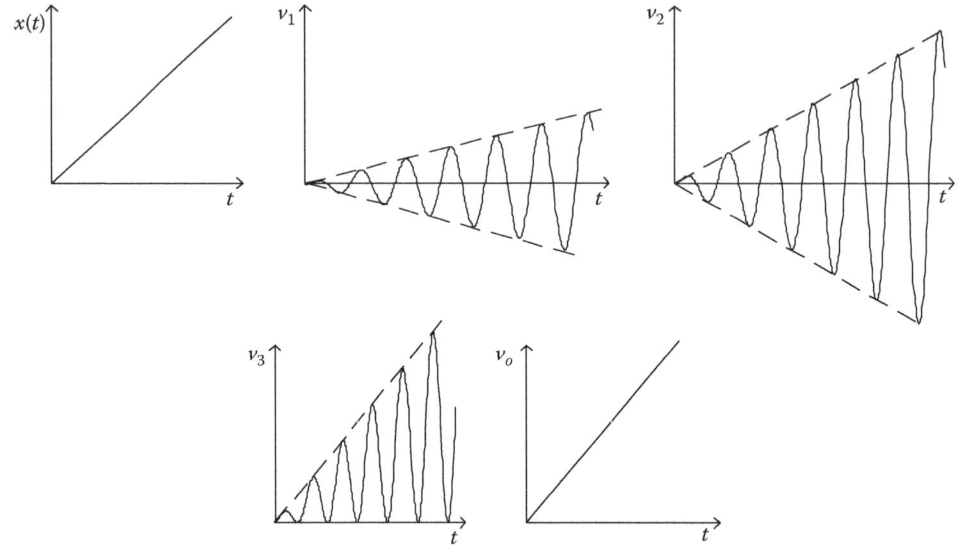

FIGURE 8.12
Nature of the signals at various locations in a linear-variable differential transformer–transducer measurement circuit.

With $R = 100$ kΩ and $\omega_c = 500$ rad/s, we have $C \times 100 \times 10^3 \times 500 = 10$.
Hence, $C = 0.2$ μF.

Based on the carrier frequency value (500 rad/s), we should be able to measure displacements $x(t)$ up to about 50 rad/s. But the flat region of the filter is about $\omega\tau = 0.1$, which, with the present value of $\tau = 0.02$ s, gives a bandwidth of only 5 rad/s for the overall measuring device (LVDT).

The advantages of using LVDT include the following:

1. It is essentially a noncontacting device with no frictional resistance. Near-ideal electromechanical energy conversion and lightweight core will result in very small resistive forces. Hysteresis (both magnetic hysteresis and mechanical backlash) is negligible.
2. It has low output impedance, typically in the order of 100 Ω. (Signal amplification is usually not needed beyond what is provided by the conditioning circuit.)
3. Directional measurements (positive/negative) are provided by it.
4. It is available in miniature sizes as well (e.g., length of 1 or 2 mm, displacement measurements of a fraction of mm, and maximum travel or "stroke" of 1 mm).
5. It has a simple and robust construction (inexpensive and durable).
6. Fine resolutions are possible (theoretically, infinitesimal resolution; practically, much better than with a coil potentiometer).

8.4.2.4 Rotatory-Variable Differential Transformer–Transducer

The RVDT operates using the same principle as the LVDT, except that in an RVDT, a rotating (rather than translating) ferromagnetic core is used as the moving member. An RVDT

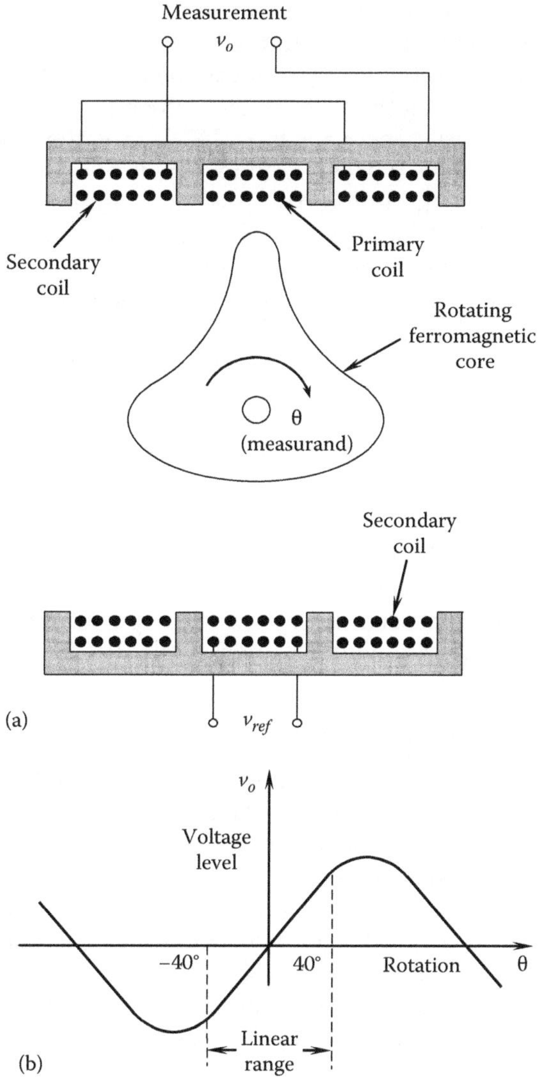

FIGURE 8.13
(a) Schematic diagram of a rotatory-variable differential transformer–transducer and (b) operating curve.

is used for measuring angular displacements. A schematic diagram of the device is shown in Figure 8.13a, and a typical operating curve is shown in Figure 8.13b. The rotating core is shaped in such way that a reasonably wide linear operating region is obtained. Advantages of the RVDT are essentially the same as those cited for the LVDT. Since the RVDT measures angular motions directly, without requiring nonlinear transformations (e.g., which is the case in resolvers), its use is convenient in angular speed applications such as position servos. The linear range is typically ±40° with a nonlinearity error less than ±0.5% of full scale.

8.4.2.4.1 Rate Error

As noted earlier, in variable-inductance devices, an induced voltage is generated through the rate of change of the magnetic flux linkage. Therefore, displacement readings are distorted by the velocity of the moving member; similarly, velocity readings are affected by the acceleration of the moving member and so on. For the same displacement value, the transducer reading depends on the velocity of the measured object at that displacement (position). This error is known as the "rate error," which increases with the ratio: (cyclic velocity of the core)/(carrier frequency), for an LVDT. Hence, the rate error can be reduced by increasing the carrier frequency. The reason for this is discussed now.

At high carrier frequencies, the induced voltage due to the transformer effect, having frequency of the primary signal, is greater than the induced voltage due to the rate (velocity) effect of the moving member. Hence, the error is small. To estimate a lower limit for the carrier frequency in order to reduce rate effects to an acceptable level, we may proceed as follows:

1. For an LVDT, let

$$\frac{\text{Maximum speed of operation}}{\text{Stroke of LVDT}} = \omega_o \tag{8.17}$$

   The excitation frequency of the primary coil (i.e., carrier frequency) should be chosen as ω_o or more.

2. For an RVDT: For the parameter ω_o in the earlier specification, use the maximum angular frequency of operation (of the rotor) of the RVDT.

8.4.3 Mutual-Induction Proximity Sensor

This displacement transducer also operates on the principle of mutual induction. A simplified schematic diagram of such a device is shown in Figure 8.14a. The insulating Ecore carries the primary winding in its middle limb. The two end limbs carry secondary windings, which are connected in series. Unlike the LVDT and the RVDT, the two voltages induced in the secondary winding segments are additive in this case. The region of the moving surface (target object) that faces the coils has to be made of ferromagnetic material so that as the object moves, the magnetic reluctance and the flux linkage between the primary and the secondary coils change. This, in turn, changes the induced voltage in the secondary coil, and this voltage change is a measure of the displacement.

Note that, unlike the LVDT, which has an axial displacement configuration, the proximity probe has a *transverse* (or lateral) displacement configuration. Hence, it is particularly suitable for measuring transverse displacements or proximities of moving objects (e.g., transverse motion of a beam or whirling shaft). We can observe from the operating curve shown in Figure 8.14b that the displacement–voltage relation of a proximity probe is rather nonlinear. Hence, these proximity sensors should be used only for measuring small displacements (e.g., in a typical linear range of 8.0 mm or 0.2 in.), unless accurate nonlinear calibration curves are available.

Since a proximity sensor is a noncontacting device, mechanical loading is small and the product life is high. Because a ferromagnetic object is used to alter the reluctance of the flux path, the mutual-induction proximity sensor is a variable-reluctance device as well. The operating frequency limit is about 1/10th the excitation frequency of the primary coil (carrier frequency). As for an LVDT, demodulation of the induced voltage (secondary voltage) is required to obtain direct (DC) output readings.

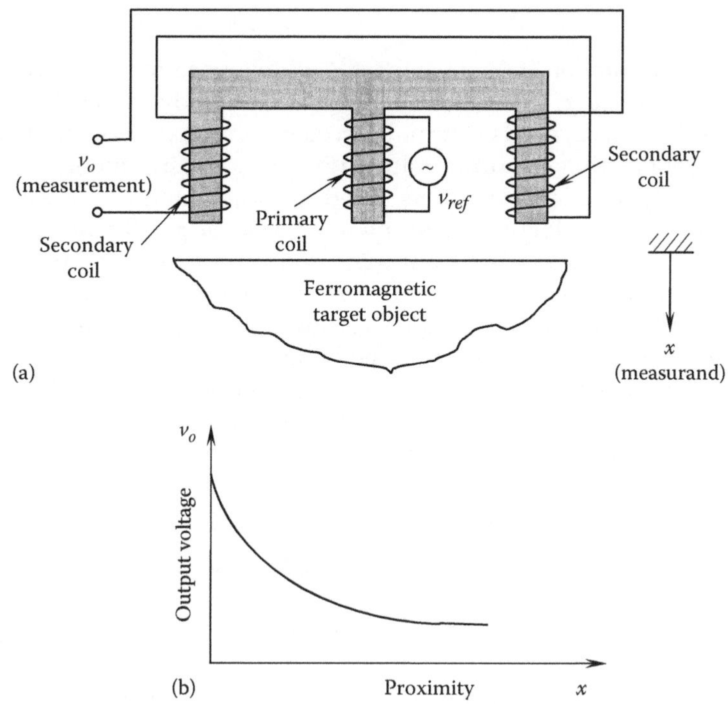

FIGURE 8.14
(a) Schematic diagram of a mutual-induction proximity sensor and (b) operating curve.

Proximity sensors are used in a wide variety of applications pertaining to noncontacting displacement sensing and dimensional gauging. Some typical applications are

1. Measurement and control of the gap between a robotic welding torch head and the work surface
2. Gauging the thickness of metal plates in manufacturing operations (e.g., rolling and forming)
3. Detecting surface irregularities in machined parts
4. Angular speed measurement, by counting the number of rotations per unit time
5. Measurement of vibration in rotating machinery and structures (for machine health monitoring and control, etc.)
6. Liquid level detection (e.g., in the filling, bottling, and chemical process industries)
7. Monitoring of bearing assembly processes

Some mutual-induction displacement transducers use the relative motion between the primary coil and the secondary coil to produce a change in flux linkage. Two such devices are the resolver and the synchro transformer. These are not variable-reluctance transducers because they do not employ a ferromagnetic moving element. The resolver is described next.

8.4.4 Resolver

Resolver is a mutual-induction transducer that is widely used for measuring angular displacements. Strictly speaking, it is a passive device as it employs magnetic induction, even though the carrier signal (AC) needs external power. It is a robust device and is used in many engineering applications that encounter tough operating environments (e.g., temperature range of −45°C–125°C), for example, robots, wind turbines, gantry mechanisms, transportation systems, and factories. High measurement accuracy (e.g., ±5 min; 1° = 60 min) is possible with the resolver even though the output is nonlinear (trigonometric).

A simplified schematic diagram of the resolver is shown in Figure 8.15. The *rotor* contains the primary coil. It consists of a single two-pole winding element, which is energized by an AC supply voltage v_{ref}. The rotor is directly attached to the object whose rotation is measured. The *stator* consists of two sets of windings placed 90° apart. If the angular position of the rotor with respect to one pair of stator windings is denoted by θ, the induced voltage in this pair of windings is given by

$$v_{o1} = av_{ref} \cos\theta \tag{8.18}$$

The induced voltage in the other pair of windings is given by

$$v_{o2} = av_{ref} \sin\theta \tag{8.19}$$

Note that these are *amplitude-modulated* signals—the carrier signal v_{ref}, which is a sinusoidal function of time, is modulated by the motion θ. The constant parameter a depends primarily on geometric and material characteristics of the device, for example, the ratio of the number of turns in the rotor and stator windings.

Either of the two output signals v_{o1} and v_{o2} may be used to determine the angular position in the first quadrant (i.e., $0 \le \theta \le 90°$). Both signals are needed, however, to determine the displacement (direction as well as magnitude) in all four quadrants (i.e., in the entire range $0 \le \theta \le 360°$) without causing ambiguity. For instance, the same sine value is obtained

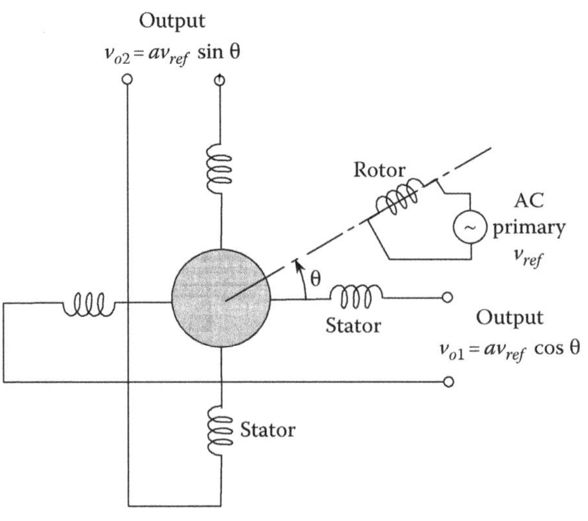

FIGURE 8.15
Schematic diagram of a resolver.

for both 90° + θ and 90° − θ (i.e., a positive rotation and a negative rotation from the 90° position), but the corresponding cosine values have opposite signs, thus providing the proper direction.

8.4.4.1 Demodulation

As for a differential transformer (LVDT and RVDT), the displacement signal (transient) from a resolver can be extracted by demodulating its (modulated) outputs. As usual, this is accomplished by filtering out the carrier signal, thereby extracting the modulating signal (which is the displacement signal). The two output signals v_{o1} and v_{o2} of a resolver are termed *quadrature signals*. Suppose that the carrier (primary) signal is

$$v_{ref} = v_a \sin \omega t \tag{8.20}$$

Then from Equations 8.18 and 8.19, the induced quadrate signals are $v_{o2} = av_a \cos\theta \sin\omega t$ and $v_{o2} = av_a \sin\theta \sin\omega t$. Multiplying these by v_{ref}, we get

$$v_{m1} = v_{o1}v_{ref} = av_a^2 \cos\theta \sin^2\omega t = \frac{1}{2}av_a^2 \cos\theta[1 - \cos 2\omega t]$$

$$v_{m2} = v_{o2}v_{ref} = av_a^2 \sin\theta \sin^2\omega t = \frac{1}{2}av_a^2 \sin\theta[1 - \cos 2\omega t]$$

Since the carrier frequency ω should be about 10 times the maximum frequency content of interest in the angular displacement θ, one can use a low-pass filter with a cutoff set at ω/10 to remove the carrier components in v_{m1} and v_{m2}. This gives the demodulated outputs:

$$v_{f1} = \frac{1}{2}av_a^2 \cos\theta \tag{8.21}$$

$$v_{f2} = \frac{1}{2}av_a^2 \sin\theta \tag{8.22}$$

Note that these equations provide both cos θ and sin θ and, hence, the magnitude and the sign of θ.

8.4.4.2 Resolver with Rotor Output

An alternative form of resolver uses two AC voltages 90° out of phase, generated from a digital signal generator, to power the two coils of the stator. The rotor contains the secondary winding in this case. The phase shift of the induced voltage determines the angular position of the rotor. An advantage of this arrangement is that it does not require slip rings and brushes to energize the windings (which are now stationary), as needed in the previous arrangement where the rotor has the primary winding. However, it will need some mechanism to pick off the output signal from the rotor. To illustrate this alternative design, suppose that the excitation signals in the two stator coils are $v_1 = v_a \sin\omega t$ and $v_2 = v_a \cos\omega t$. When the rotor coil is oriented at angular position θ with respect to the stator coil 2, it will be at an angular position π/2 − θ from the stator coil 1 (assuming that the rotor coil is in the first quadrant, $0 \le \theta \le \pi/2$). Hence, the voltage induced by stator coil 1 in the rotor coil would be $v_a \sin\omega t \sin\theta$, and the voltage induced by the stator coil 2 in the rotor coil would be $v_a \cos\,\omega t \cos\,\theta$. It follows that the total induced voltage in the rotor coil is given by $v_r = v_a \sin\,\omega t \sin\,\theta + v_a \cos\,\omega t \cos\,\theta$ or

$$v_r = v_a \cos(\omega t - \theta) \tag{8.23}$$

It is seen that the phase angle of the rotor output signal with respect to the stator excitation signals v_1 and v_2 provides both magnitude and sign of the rotor position θ.

The output signals of a resolver are nonlinear (trigonometric) functions of the angle of rotation. (Historically, resolvers were used to compute trigonometric functions or to *resolve* a vector into orthogonal components.) In robotic applications, this is sometimes viewed as an advantage. For example, in computed torque control of robotic manipulators, trigonometric functions of the joint angles are needed in order to compute the required input signals (reference joint torque values). Consequently, when resolvers are used to measure joint angles in manipulators, there is an associated reduction in the processing time of the control input signals because the trigonometric functions themselves are available as direct measurements.

The primary advantages of the resolver include

1. Fine resolution and high accuracy
2. Low output impedance (high signal levels)
3. Small size (e.g., 10 mm diameter)
4. Rugged construction (high robustness)
5. Direct availability of the sine and cosine functions of the measured angles

Its main limitations are the following:

1. Nonlinear output signals (an advantage in some applications where trigonometric functions of the rotations are needed).
2. Bandwidth is limited by the supply (carrier) frequency.
3. Slip rings and brushes would be needed if complete and multiple rotations have to be measured (which adds mechanical loading and also creates component wear, oxidation, and thermal and noise problems).

8.4.4.3 Self-Induction Transducers

These transducers are based on the principle of self-induction. Unlike mutual-induction transducers, only a single coil is employed. This coil is activated by an AC supply voltage v_{ref} of sufficiently high frequency. The current produces a magnetic flux, which is linked back with the coil itself. The level of flux linkage (or self-inductance) is varied by a moving ferromagnetic object (whose position is to be measured) within the magnetic field. This movement changes the *reluctance* of the magnetic flux linkage path and also the *inductance* in the coil. The change in the self-inductance can be measured using an inductance-measuring circuit (e.g., an inductance bridge). In this manner, the displacement of the object (measurand) can be measured. Self-induction transducers are usually variable-reluctance devices as well.

A typical self-induction transducer is a *self-induction proximity sensor*, as schematically shown in Figure 8.16. This device can be used as a displacement sensor for transverse

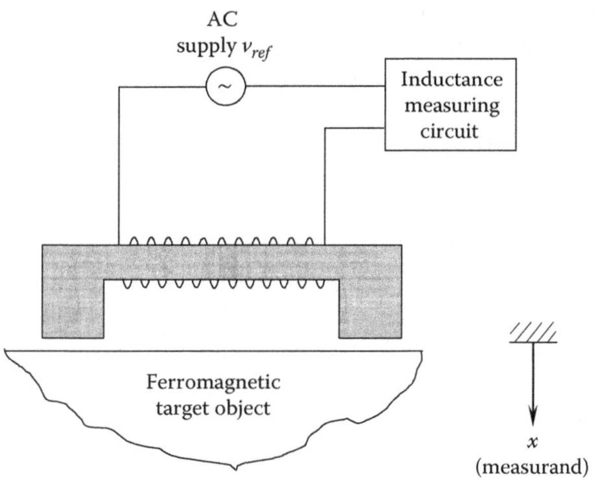

FIGURE 8.16
Schematic diagram of a self-induction proximity sensor.

displacements. For instance, the distance between the sensor tip and the ferromagnetic surface of a moving object, such as a beam or shaft, can be measured. Other applications include those mentioned for mutual-induction proximity sensors. High-speed displacement measurements can result in velocity error (rate error) when variable-inductance displacement sensors (including self-induction transducers) are used. This effect may be reduced, as in other AC-activated variable-inductance sensors, by increasing the carrier frequency.

Inductive proximity sensors come in sizes of the mm scale. The output is nonlinear, and the maximum detecting distance is normally limited to a few millimeters. The operating frequency is in the range 25–1000 Hz. If the excitation is from the line voltage (60 Hz), the bandwidth of interest of the measured signal would be limited to 5–10 Hz. The advantages and disadvantages of self-induction transducers are essentially the same as those of mutual-induction transducers.

8.5 Permanent-Magnet and Eddy Current Transducers

Now, we present transducers in the third category of variable-inductance transducer: PM transducers. In particular, we discuss several types of velocity transducers (*tachometers*). Also, we will present another class of transducers called eddy current transducers. (*Note*: Eddy current transducers are not PM transducers in general.)

PM transducers: A distinctive feature of a PM transducer is that it has a permanent magnet to generate a uniform and steady magnetic field. In a tachometer, the relative motion between the magnetic field and an electrical conductor induces a voltage, which is proportional to the speed at which the conductor crosses the magnetic field (i.e., the rate of change of flux linkage). This induced voltage is a measure of the speed. In some designs, a unidirectional magnetic field generated by a DC supply (i.e., an electromagnet) is used in place of a permanent magnet. Nevertheless, they are generally termed *permanent-magnet transducers*. PM transducers are not variable-reluctance devices in general.

Since the magnetic field of a permanent magnet is steady and independent of any other magnetic fields (such as those generated by induced voltages), PM transducers tend to be less nonlinear than other types of variable-inductance transducers. PM transducers have other advantages. For example, they can incorporate larger air gaps (greater than 1 cm) between the magnet and the moving object. In variable-reluctance transducers, false signals can be caused by ferromagnetic objects other than the target object. Such false triggering is not possible with PM transducers.

Eddy current transducers: A fluctuating magnetic field can generate currents even on a very thin and small conducting surface. If the field fluctuation is caused by a mechanical motion, the eddy currents provide a measure of that motion. This principle is used in eddy current transducers such as proximity sensors.

8.5.1 DC Tachometer

DC tachometer is a PM velocity transducer that uses the principle of electromagnetic induction where the magnetic field is generated by either a DC or a permanent magnet. A voltage is induced in a conducting coil due to the relative motion between the coil and the magnetic field (in proportion to the rate of change of the flux linkage).

Depending on the configuration, either rectilinear speeds or angular speeds can be measured. Schematic diagrams of the two configurations are shown in Figure 8.17. These are passive transducers because the energy for the output signal v_o of the transducer is derived from the motion (i.e., measured signal) itself. The entire device is usually enclosed in a steel casing to shield (to isolate) it from ambient magnetic fields.

In the rectilinear velocity transducer (Figure 8.17a), the conductor coil is wound on a core and placed centrally between two magnetic poles, which produce a cross-magnetic field.

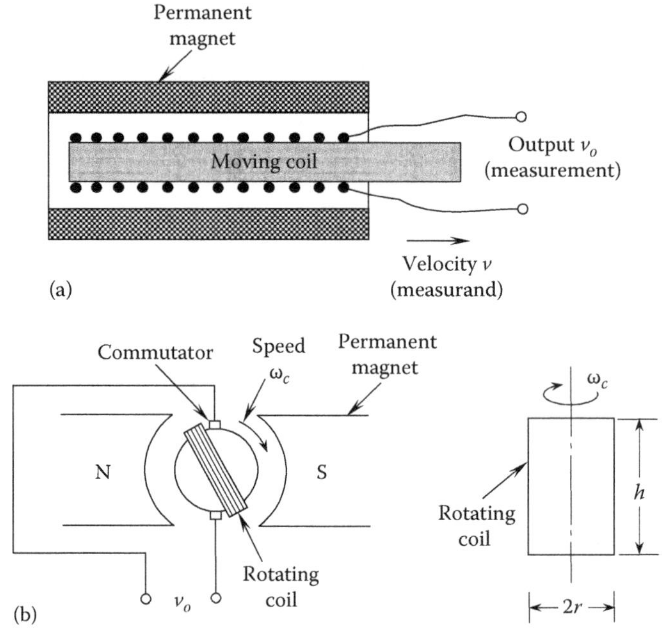

FIGURE 8.17
Permanent-magnet DC transducers. (a) Rectilinear velocity transducer and (b) DC tachometer.

The core is attached to the moving object whose velocity v is to be measured. This velocity is proportional to the induced voltage v_o. Alternatively, a moving magnet and a fixed coil may be used as the velocity transducer (rectilinear or rotatory). This latter arrangement is perhaps more desirable since it eliminates the need for any sliding contacts (slip rings and brushes) for the output leads, thereby reducing mechanical loading error, wear, and related problems.

The DC tachometer (or tachogenerator) is a common transducer for measuring angular velocities. Its principle of operation is the same as that for a DC generator (or back driving of a DC motor). This principle of operation is illustrated in Figure 8.17b. The rotor is directly connected to the rotating object. The output signal that is induced in the rotating coil is picked up as the DC voltage v_o using a suitable commutator device—typically consisting of a pair of low-resistance carbon brushes—that is stationary but makes contact with the rotating coil through split slip rings so as to maintain the direction of the induced voltage the same throughout each revolution (see commutation under DC motors). According to *Faraday's law*, the induced voltage is proportional to the rate of change of magnetic flux linkage. For a coil of height h and width $2r$ that has n turns, moving at an angular speed ω_c in a uniform magnetic field of flux density β, this is given by

$$v_o = (2nhr\beta)\omega_c = k\omega_c \tag{8.24}$$

The constant of proportionality k between v_o and ω_c is the *sensitivity* of the tachometer and is used as the scaling factor in the measurement of the speed ω_c. The proportionality constant k is also known as the "back-emf constant" or the "voltage constant" because the larger the value of k the greater the induced voltage. DC tachometers with sensitivities in the ranges 0.5–3, 1–10, 11–25, or 25–50 V per 1000 rmp are commonly available. For low-voltage tachometers, the armature resistance can be on the order of 100 Ω. For high-voltage tachometers, the armature resistance can be about 2000 Ω. The size can be about 2 cm in length. It is a rather linear device (0.1% is typical).

8.5.1.1 Electronic Commutation

Slip rings and brushes and the associated drawbacks can be eliminated in a DC tachometer by using electronic commutation. In this case, a PM rotor together with a set of stator windings is used. The output of the tachometer is drawn from the stationary (stator) coil. It has to be converted into a DC signal using an electronic switching mechanism, which has to be synchronized with the rotation of the tachometer (as in a brushless DC motors). Because of switching and associated changes in the magnetic field of the output signal, extraneous induced voltages known as "switching transients" will result. This is a drawback of electronic commutation.

8.5.1.2 Modeling of a DC Tachometer

The equivalent circuit and the mechanical free-body diagram of a DC tachometer are shown in Figure 8.18. The field windings are powered by the DC reference voltage v_f. The across variable at the input port is the measured angular speed ω_i. The corresponding torque T_i is the through variable at the input port. The output voltage v_o of the armature circuit is the across variable at the output port. The corresponding current i_o is the through variable at the output port. We now obtain a transfer-function model for this tachometer. Also, we will investigate the assumptions needed to decouple this result into a practical

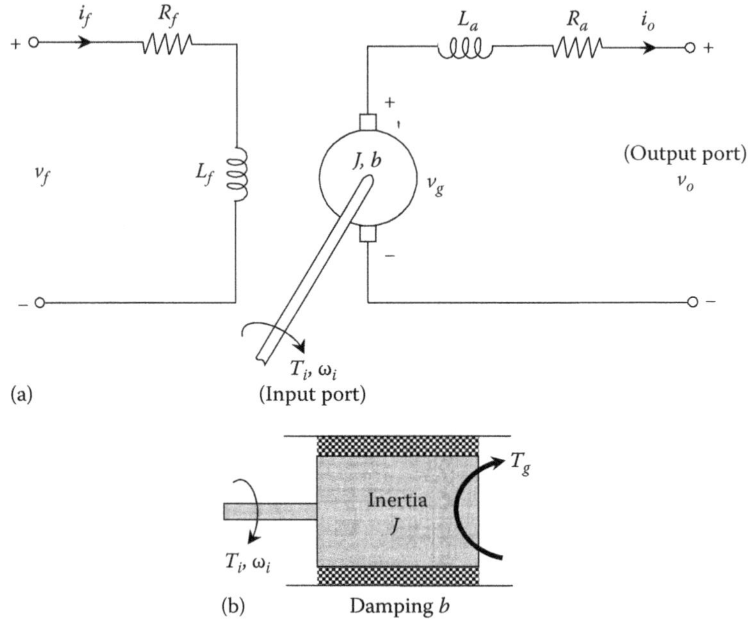

FIGURE 8.18
A DC tachometer example. (a) Equivalent circuit with an impedance load and (b) armature free-body diagram.

input–output model for a tachometer. We will discuss the corresponding design implications, particularly the significance of the mechanical time constant and the electrical time constant of the tachometer.

The generated voltage v_g at the armature (rotor) is proportional to the magnetic field strength of field windings (which, in turn, is proportional to the filed current i_f) and the speed of the armature ω_i. Hence, $v_g = K'i_f\omega_i$. Now assuming constant field current, we have $v_g = K\omega_i$. The rotor magnetic torque T_g, which resists the applied torque T_i, is proportional to the magnetic field strengths of the field windings and armature windings. Consequently, $T_g = K'i_f i_o$. Since i_f is assumed constant, we get $T_g = Ki_o$.

Note: The constant K is the *gain* or the *sensitivity* of the tachometer. It is also the *induced voltage constant* and also the *torque constant*. This is valid when the same units are used to measure mechanical power (N·m/s) and electrical power (W) and when the internal energy dissipation mechanisms are not significant in the associated internal coupling (i.e., ideal energy conversion is assumed).

The equation for the armature circuit is $v_o = v_g - R_a i_o - L_a(di_o/dt)$, where R_a is the armature resistance and L_a is the leakage inductance in the armature circuit.

With reference to Figure 8.18b, Newton's second law for a tachometer armature having inertia J and damping constant b is expressed as $J(d\omega_i/dt) = T_i - T_g - b\omega_i$. Now, we substitute into this equation the previous results in order to eliminate v_g and T_g. Next, the time derivatives are replaced by the Laplace variable s. This results in the two algebraic relations: $v_o = K\omega_i - (R_a + sL_a)i_o$ and $(b + sJ)\omega_i = T_i - Ki_o$.

Note: The variables v_i, i_o, ω_i, and T_i in these equations are in fact Laplace transforms (functions of s), not functions of t, as in the earlier time-domain equations. We keep the same symbol in both domains, for notational simplicity.

Finally, i_o in the first equation is eliminated using the second equation. This gives the matrix transfer-function relation

$$\begin{bmatrix} v_o \\ i_o \end{bmatrix} = \begin{bmatrix} K + \dfrac{(R_a + sL_a)(b+sJ)}{K} & -\dfrac{(R_a+sL_a)}{K} \\ -\dfrac{(b+sJ)}{K} & \dfrac{1}{K} \end{bmatrix} \begin{pmatrix} \omega_i \\ T_i \end{pmatrix} \tag{8.25}$$

The corresponding frequency domain relations are obtained by simply replacing s with $j\omega$, where ω represents the angular frequency (radians per second) in the frequency spectrum of a signal.

8.5.1.3 Design Considerations

Transducers are more accurately modeled as *two-port elements*, which have two variables associated with each port (see Figure 8.18). However, it is useful and often essential, in practical usage as a measuring device, to relate just one variable at the input port (*measurand*) to just one variable at the output port (*measurement*). Then, only one (scalar) transfer function (or *gain* parameter or *sensitivity*) relating these two variables need be specified. For this, some form of decoupling would be needed in the true model. If the decoupling assumptions do not hold in the range of operation of the transducer, a measurement error will result.

8.5.1.3.1 Model Decoupling

In the present tachometer example, we like to express the output voltage v_o in terms of the measured speed ω_i. For this, the off-diagonal term—$(R_a+sL_a)/K$—in Equation 8.25 has to be neglected. To do so, in the model equation (RHS of the first equation (8.25)), we have to compare the entire "coupling term" $T_i(R_a+sL_a)/K$ with the entire "direct term" $\omega_i[K+(R_a+sL_a)(b+sJ)]/K$, which include not only the parameters but also the variables T_i and ω_i. Only then we will have a valid comparison (and we would be comparing terms of the same units—voltage). It is seen that the term that should be neglected becomes smaller as we increase the *tachometer gain* (or *sensitivity*) parameter K and decrease the armature resistance R_a and the leakage inductance L_a. However, this will decrease the other term (which is retained) as well. Since the leakage inductance L_a is negligible to begin with, for a properly designed tachometer (or a motor), it is adequate to consider K and R_a only. Nevertheless, when comparing terms for decoupling the model, in addition to the values of the model parameters (K,R_a,L_a,b,J), we should consider the values (at least the extreme values) of

1. Variables at the input port (T_i and ω_i)
2. Frequency of interest of the variables (ω)

It is seen in Equation 8.25 that the dynamic terms (those containing the Laplace variable "s") will decrease as we increase K. Then, not only the entire coupling term becomes small but also the dynamic part of the direct term decreases. Both of these are desirable consequences. Note from the equations given in the model derivation that the tachometer gain K can be increased by increasing the field current i_f. This will not be feasible if the field windings are already saturated, however. Furthermore, K (or K') depends on such parameters as the number of turns, dimensions of the stator windings, and the magnetic properties of the stator core. Since there is a limitation on the physical size of the tachometer and the types of materials used in the construction, it is clear that K cannot be increased arbitrarily. The instrument designer should take such factors into consideration in developing a design that is optimal in many respects. In practical transducers, the operating range is specified in order to minimize the effect of the coupling terms (and nonlinearities, frequency dependence, etc.), and the residual errors are accounted for by using correction/calibration curves. This approach is more convenient than using the coupled model Equation 8.25, which introduces three more (scalar) transfer functions (in general) into the model.

8.5.1.3.2 Time Constants

Another desirable feature for practical transducers is to have a static (i.e., algebraic, non-dynamic) input–output relationship so that the output instantly reaches the input value (or the measured variable), and the frequency dependence of the transducer characteristic is eliminated. Then, the transducer transfer function becomes a pure gain (i.e., independent of frequency). This happens when the transducer time constants are small (i.e., the transducer *bandwidth* is high), as discussed in Chapters 5 and 6. Returning to the tachometer example, it is clear from Equation 8.25 that the transfer-function relations become static (frequency independent) when both the *electrical time constant*

$$\tau_e = \frac{L_a}{R_a} \tag{8.26}$$

and the *mechanical time constant*

$$\tau_m = \frac{J}{b} \tag{8.27}$$

are negligibly small. The electrical time constant of a motor/generator is usually an order of magnitude smaller than the mechanical time constant. Hence, one must first concentrate on the mechanical time constant. Note from Equation 8.27 that τ_m can be reduced by decreasing the rotor inertia and increasing the rotor damping. Unfortunately, rotor inertia depends on rotor dimensions, and this determines the gain parameter K, as we saw earlier. Hence, we face some design constraint in reducing K. Furthermore, when the rotor size is reduced (in order to reduce J), the number of turns in the windings should be reduced. Then, the air gap between the rotor and the stator becomes less uniform, which creates a voltage ripple in the induced voltage (tachometer output). The resulting measurement error can be significant. Next, turning to damping, it is intuitively clear that if we increase b, a larger torque T_i will be required to drive the tachometer. This will *load* the object whose speed is being measured. In other words, this will distort the measurand ω_i itself (i.e., mechanical loading). Hence, increasing b also has to be done cautiously. Now, going back to Equation 8.25, we note that the dynamic terms in the transfer function between ω_i and v_o decrease as K is increased. So we note the following benefits in increasing K:

1. Increases the sensitivity and the output signal level
2. Reduces coupling, thereby the measurement directly depends on the measurand only
3. Reduces dynamic effects (i.e., reduction of the frequency dependence of the system, thereby increasing the useful frequency range and bandwidth or speed of response)

The following are the benefits of decreasing the time constants:

1. Decreases the dynamic terms (makes the transfer function static)
2. Increases the operating bandwidth
3. Makes the sensor faster

Example 8.3

The data sheet of a commercial DC tachometer lists the following parameter values:

Armature resistance $R_a = 35\ \Omega$
Leakage inductance $L_a = 4$ mH
Rotor moment of inertia $J = 8.5 \times 10^{-7}$ kg·m²
Frictional torque $= 3.43 \times 10^{-3}$ N·m at 4000 rpm
Output voltage sensitivity $= 3.0$ V at 1000 rpm

(a) Estimate the electrical time constant, mechanical time constant, and the operating frequency range of the tachometer.
(b) Check whether the decoupling assumption is valid (i.e., the coupling input term is negligible compared to the direct input term).

Solution

Electrical time constant

$$\tau_e = \frac{L_a}{R_a} = \frac{4 \times 10^{-3}}{35} = 1.14 \times 10^{-4}\ \text{s}$$

$$4000\ \text{rmp} = \frac{4000}{60} \times 2\pi\ \text{rad/s} = 419.0\ \text{rad/s}$$

→ Estimated damping constant

$$b = \frac{3.43 \times 10^{-3}}{419}\ \text{N}\cdot\text{m/rad/s} = 8.2 \times 10^{-6}\ \text{N}\cdot\text{m/rad/s}$$

Mechanical time constant

$$\tau_m = \frac{J}{b} = \frac{8.5 \times 10^{-7}}{8.2 \times 10^{-6}} = 0.104\ \text{s}$$

Note: $\tau_e \ll \tau_m$

The operating range of the tachometer should be less than

$$\omega_o = \frac{1}{\tau_m} = \frac{1}{0.104}\ \text{s} = 9.6\ \text{rad/s}$$

$$1000\ \text{rmp} = \frac{1000}{60} \times 2\pi\ \text{rad/s} = 104.7\ \text{rad/s}$$

→ Voltage sensitivity = gain = torque constant = $K = 3.0/104.7$ V/rad/s or N·m/A = 2.9×10^{-2} V/rad/s

We take the maximum torque as 3.43×10^{-3} N·m at 419 rad/s.
Also, we take the maximum operating frequency as $\omega_o = 9.6$ rad/s.

Now, we compute the magnitudes:

Direct term

$$\omega_{\max}\left|[K+(R_a+j\omega_o L_a)(b+j\omega_o J)]/K\right|$$

$$=419\times\left|[2.9\times10^{-2}+(35+j\times9.6\times4\times10^{-3})(8.2\times10^{-6}+j\times9.6\times8.5\times10^{-7})]/2.9\times10^{-2}\right|$$

$$=419\times2.91\times10^{-2}\text{ V}=12.2\text{ V}$$

Coupling term $T_i(R_a+sL_a)/K=3.43\times10^{-3}|(35+j\times9.6\times4\times10^{-3})|/2.9\times10^{-2}=4.14\text{ V}$

It is seen that neither the dynamic terms nor the coupling term can be neglected at this maximum operating frequency (9.6 rad/s), with the given sensitivity (gain) value. Clearly, the sensor accuracy is not acceptable (errors up to 34% are possible).

Note: If we can double the tachometer sensitivity to $K=5.8\times10^{-2}$ V/rad/s, then direct term is 24.3 V and the coupling term is 2.1.

Then, the sensor accuracy will be far better (the worst error would be less than 9%).

8.5.1.4 Loading Considerations

As noted, the torque required to drive a tachometer is proportional to the current generated (in the DC output). The associated proportionality constant is called the "torque constant." With consistent units, in the case of ideal energy conversion, this constant is equal to the *voltage constant* and the *sensitivity* of the tachometer. Since the tachometer torque acts on the moving object whose speed is measured, high torque corresponds to high *mechanical loading*, which is not desirable. Hence, it is advisable to reduce the tachometer current as much as possible. This can be realized by making the *input impedance* of the signal acquisition device (i.e., voltage reading and interface hardware) for the tachometer as large as possible. Furthermore, distortion of the tachometer output signal (voltage) can occur because of the reactive (inductive and capacitive) loading of the tachometer. When a DC tachometer is used to measure transient velocities, some error will enter from the rate (acceleration) effect. This *rate error* generally increases with the maximum significant frequency that must be retained in the transient velocity signal, which in turn depends on the maximum speed that has to be measured. All these types of error can be reduced by increasing the load impedance.

For illustration, consider the equivalent circuit of a tachometer with an impedance load connected to the output port of the armature circuit shown in Figure 8.18. The induced voltage $K\omega_c$ is represented by a voltage source. The constant K depends on the coil geometry, the number of turns, and the magnetic flux density (see Equation 8.25). The coil resistance is denoted by R, and the leakage inductance is denoted by L_l. The load impedance is Z_L. From straightforward circuit analysis in the frequency domain, the output voltage at the load is given by

$$v_o=\left[\frac{Z_L}{R+j\omega L_l+Z_L}\right]k\omega_c \tag{8.28}$$

It is seen that because of the leakage inductance, the output signal attenuates more at higher frequencies ω of the velocity transient. In addition, a loading error is present. If Z_L is much larger than the coil impedance, however, the ideal proportionality, as given by $v_o=K\omega_c$, is achieved.

Note: A *digital tachometer* is a velocity transducer, which is governed by somewhat different principles. It generates voltage pulses at a frequency proportional to the angular speed. Hence, it is considered a digital transducer, as discussed in Chapter 11.

8.5.2 AC Tachometer

An AC tachometer is also a speed transducer. A typical AC tachometer has two sets of stator windings. One coil is energized by an AC carrier signal, which induces an AC signal at the same frequency in the other stator coil. The speed of rotation of the rotor will modulate the induced signal, and this can be used to measure speed. Two main types of AC tachometers are available. One uses a PM rotor and the other uses a shorted coil as the rotor.

8.5.2.1 Permanent-Magnet AC Tachometer

A PM AC tachometer has a PM rotor and two separate sets of stator windings as schematically shown in Figure 8.19a. One set of windings (primary) is energized using an AC reference (carrier) voltage. The induced voltage in the other set of windings (secondary) is the tachometer output. When the rotor is stationary or moving in a quasi-static manner, the output voltage is a constant-amplitude signal much like the reference voltage, at the same (carrier) frequency, as in an electrical transformer. As the rotor moves at some speed, an additional voltage is induced in the secondary coil, which modulates the original induced voltage of carrier frequency. This modulating signal is proportional to the rotor speed and is generated due to the rate of change of flux linkage in the secondary coil as a result of the rotating magnet.

It is seen that the overall output in the secondary coil is an *amplitude-modulated* signal (see Chapter 4). It may be *demodulated* in order to extract the transient velocity signal (i.e., the modulating signal). The direction of velocity is determined from the phase angle of the modulated signal with respect to the carrier signal. If the rotor speed is steady, the amplitude of the output signal will measure the speed (without having to demodulate the output).

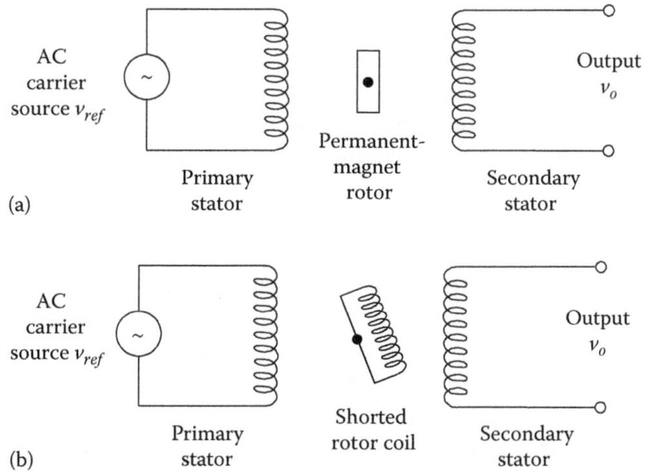

FIGURE 8.19
(a) An AC permanent-magnet tachometer and (b) an AC induction tachometer.

Note: In an LVDT, the amplitude of the AC magnetic flux (linkage) is altered by the position of the ferromagnetic core. But in an AC PM tachometer, a DC magnetic flux is generated by the magnetic rotor, and when the rotor is stationary, it does not induce a voltage in the coils. The flux linked with the stator windings changes because of the rotation of the rotor, and the rate of change of the linked flux is proportional to the speed of the rotor.

For low-frequency applications (5 Hz or less), a standard AC supply at line frequency (60 Hz) may be adequate to power an AC tachometer. For moderate-frequency applications, a 400 Hz supply may be used. For high-frequency (high bandwidth) applications, a high-frequency signal generator (*oscillator*) may be used as the primary signal. In high-bandwidth applications, carrier frequencies as high as 1.5 kHz are commonly used. Typical sensitivity of an AC PM tachometer is of the order of 50–100 mV/rad/s.

8.5.2.2 AC Induction Tachometer

AC induction tachometer is similar in construction to a two-phase induction motor. The stator arrangement is identical to that of the AC PM tachometer, as presented earlier. The rotor has windings, which are shorted and not energized by an external source, as shown in Figure 8.19b. The primary stator coil is powered by an AC supply. This induces a voltage in the secondary stator coil, as in an PM AC tachometer. As the rotor coil rotates (in the magnetic field created by the primary stator coil), a voltage is induced in it as well. This signal modulates the induced signal of carrier frequency in the secondary stator coil. Demodulation of the output signal in the secondary stator coil would be needed to extract the speed of the rotor.

8.5.2.3 Advantages and Disadvantages of AC Tachometers

The main advantage of an AC tachometer over its DC counterpart is the absence of a slip ring and brush device, since the output is obtained from the stator. The output signal from a DC tachometer usually has a voltage ripple, known as the "commutator ripple" or "brush noise," which is generated as the split ends of the slip ring pass over the brushes and as a result of contact bounce, and so forth. The frequency of the commutator ripple is proportional to the speed of operation; consequently, filtering it out using a notch filter is difficult (because a speed-tracking notch filter would be needed). Also, problems associated with frictional loading and contact bounce are present in DC tachometers, and these problems are absent in AC tachometers. Note, however, that a DC tachometer with electronic commutation does not use slip rings and brushes. But they produce switching transients that are also undesirable.

As for any sensor, in a tachometer, the noise components dominate at low levels of output signal. In particular, since the output of a tachometer is proportional to the measured speed, at low speeds, the SNR will be low. Hence, removal of noise takes an increased importance at low speeds.

AC tachometers provide drift-free measurements under steady condition. This is another advantage of it compared to a DC tachometer.

It is known that at high speeds the output from an AC tachometer is somewhat nonlinear (primarily because of the saturation effect). Furthermore, signal demodulation is necessary, particularly for measuring transient speeds. Another disadvantage of AC tachometer is that the output signal level depends on the supply voltage; hence, a regulated voltage source, which has a very small output impedance, is desirable.

Also, the measurement bandwidth (frequency limit) depends on the carrier frequency (about 1/10th of the carrier frequency).

8.5.3 Eddy Current Transducers

If a conducting (i.e., low resistivity) medium is subjected to a fluctuating magnetic field, eddy currents are generated in the medium. The strength of the eddy currents increases with the strength of the magnetic field and the frequency of the magnetic flux. This principle is used in eddy current proximity sensors. Eddy current sensors may be used as either dimensional gauging devices or displacement sensors.

A schematic diagram of an eddy current proximity sensor is shown in Figure 8.20a. Unlike variable-inductance proximity sensors, the target object of the eddy current sensor does not have to be made of a ferromagnetic material. A conducting target object is needed, but a thin film of conducting material, such as household aluminum foil glued to a nonconducting target object, would be adequate. The probe head has two identical coils that form two arms of an impedance bridge. The coil closer to the probe face is the

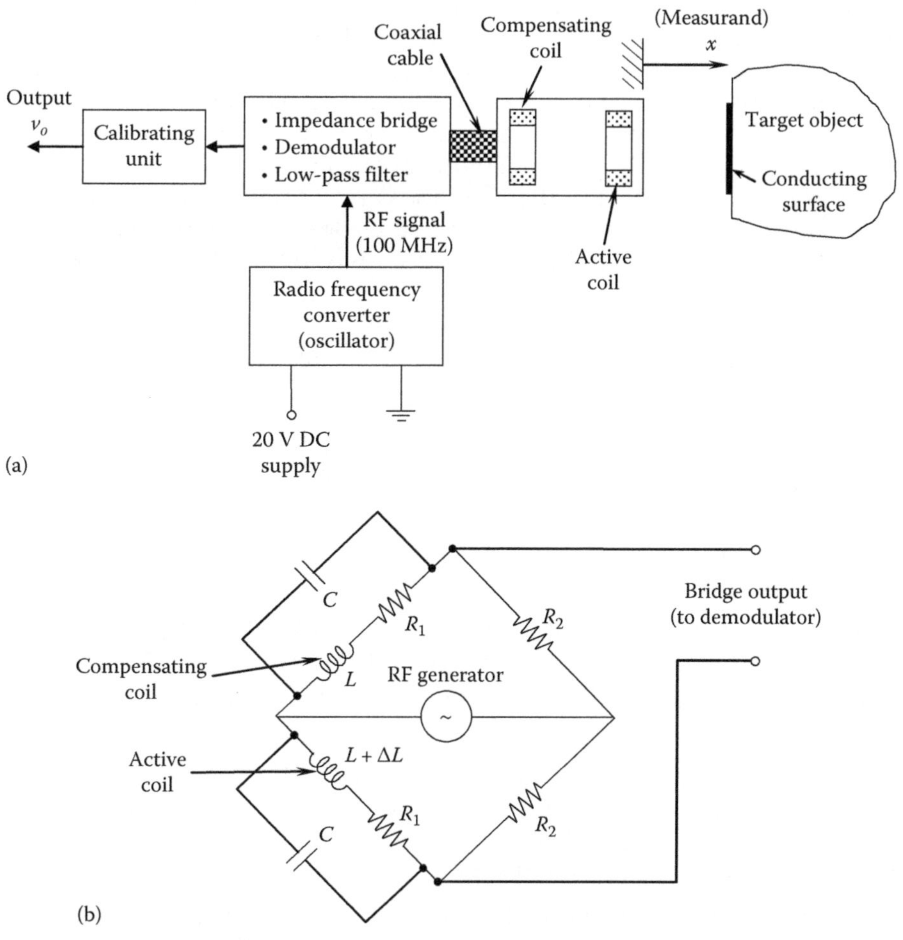

FIGURE 8.20
Eddy current proximity sensor. (a) Schematic diagram and (b) impedance bridge.

active coil. The other coil is the *compensating coil*, which compensates for ambient changes, particularly thermal effects. The remaining two arms of the bridge consist of purely resistive elements (see Figure 8.20b). The bridge is excited by a radio frequency voltage supply. The frequency may range from 1 to 100 MHz. This signal is generated from a radio frequency converter (an oscillator) that is typically powered by a 20 V DC supply. When the target object (sensed object) is absent, the output of the impedance bridge is zero, which corresponds to the balanced condition. When the target object is moved close to the sensor, eddy currents are generated in the conducting medium because of the radio frequency magnetic flux from the active coil. The magnetic field of the eddy currents opposes the primary field, which generates these currents. Hence, the inductance of the active coil increases, creating an imbalance in the bridge. The resulting output from the bridge is an amplitude-modulated signal containing the radio frequency carrier. This signal is demodulated to remove the carrier. The resulting signal (modulating signal) measures transient displacement of the target object. Low-pass filtering is used to remove high-frequency leftover noise in the output signal once the carrier is removed.

For large displacements, the output of an eddy current transducer is not linearly related to the displacement. Furthermore, the sensitivity of the transducer depends nonlinearly on the nature of the conducting medium, particularly the resistivity. For example, for low resistivities, sensitivity increases with resistivity; for high resistivities, it decreases. A calibrating unit is usually available with commercial eddy current sensors to accommodate various target objects and nonlinearities. The gauge factor is usually expressed in volts per millimeter (V/mm). Note that eddy current probes can also be used to measure resistivity and surface hardness, which affects resistivity, in metals.

In eddy current sensing, the facial area of the conducting medium of the target object has to be slightly larger than the frontal area of the eddy current probe head. If the target object has a curved surface, its radius of curvature has to be at least four times the diameter of the probe. These are not serious restrictions because the typical diameter of a probe head is about 2 mm. Eddy current sensors are medium-impedance devices; 1000 Ω output impedance is typical. Sensitivity is in the order of 5 V/mm. The advantages of eddy current sensors include the following:

1. Since the carrier frequency is very high, eddy current devices are suitable for highly transient displacement measurements—for example, bandwidths up to 100 kHz.
2. Eddy current sensor is a noncontacting device; hence, it does not exert mechanical loading on the moving (target) object.
3. Eddy current sensors are able to perform accurately even in dirty environments (as long as conductive objects do not interfere with the measurement environment).
4. It requires only a thin conducting surface that is not much wider than the probe.

8.6 Variable-Capacitance Transducers

Variable-inductance devices and variable-capacitance devices are *variable-reactance* devices. (*Note*: Reactance of an inductance L is given by $j\omega L$ and that of a capacitance C is given by $1/(j\omega C)$, since $v = L(di/dt)$ and $i = C(dv/dt)$.) For this reason, capacitive transducers fall into the general category of *reactive transducers*. They are typically high-impedance sensors, particularly at low frequencies, as clear from the impedance (reactance) expression for a capacitor. Also, capacitive sensors are noncontacting devices in the common usage. They require specific signal-conditioning hardware. In addition to analog capacitive sensors, digital (pulse generating) capacitive transducers such as digital tachometers are also available.

A capacitor is formed by two plates, which can store an *electric charge*. The stored charge generates a *potential difference* between the plates and may be maintained using an external voltage. The capacitance C of a two-plate capacitor is given by

$$C = \frac{kA}{x} \tag{8.29}$$

where

A is the common (overlapping) area of the two plates
x is the gap width between the two plates
k is the *dielectric constant* (or *permittivity* $k = \varepsilon = \varepsilon_r \varepsilon_o$; ε_r is the relative permittivity, ε_o is the permittivity of a vacuum), which depends on dielectric properties of the medium between the two plates

A change in any one of the three parameters in Equation 8.29 may be used as the principle of the sensor. To further examine this, Equation 8.29 may be written as $\ln C = -\ln x + \ln A + \ln k$. By taking the differentials of the terms in this equation, we have

$$\frac{\delta C}{C} = -\frac{\delta x}{x} + \frac{\delta A}{A} + \frac{\delta k}{k} \tag{8.30}$$

This result may be used, for example, to measure small transverse displacements, large rotations, and large fluid levels.

Note: Equation 8.30 is valid only for small increments in x and is valid even for large increments of A and k because Equation 8.29 is nonlinear in x, while linear in A and k. However, Equation 8.29 becomes linear in x as well if a log scale is used.

Schematic diagrams of capacitive sensors that use the changes in the three variable quantities in Equation 8.30 are shown in Figure 8.21. In Figure 8.21a, a transverse displacement of one of the plates results in a change in x. In Figure 8.21b, angular displacement of one of the plates causes a change in A. Finally, in Figure 8.21c, a change in k is produced as the fluid level between the capacitor plates changes. In all three cases, the associated change in the capacitance is measured directly (e.g., using a capacitance bridge or an oscillator circuit) or indirectly (e.g., output voltage from a bridge circuit or a potentiometer circuit) and is used to estimate the measurand.

8.6.1 Capacitive Sensing Circuits

In variable-capacitance transducers, a change in the capacitance is measured directly or indirectly to provide a measurement for the measurand. A change in the sensor capacitance that is not caused by a change in the measurand (e.g., due to change in humidity, temperature, and aging) causes errors in the sensor reading and should be compensated for. Common types of circuits that are used for capacitance sensing are the capacitance

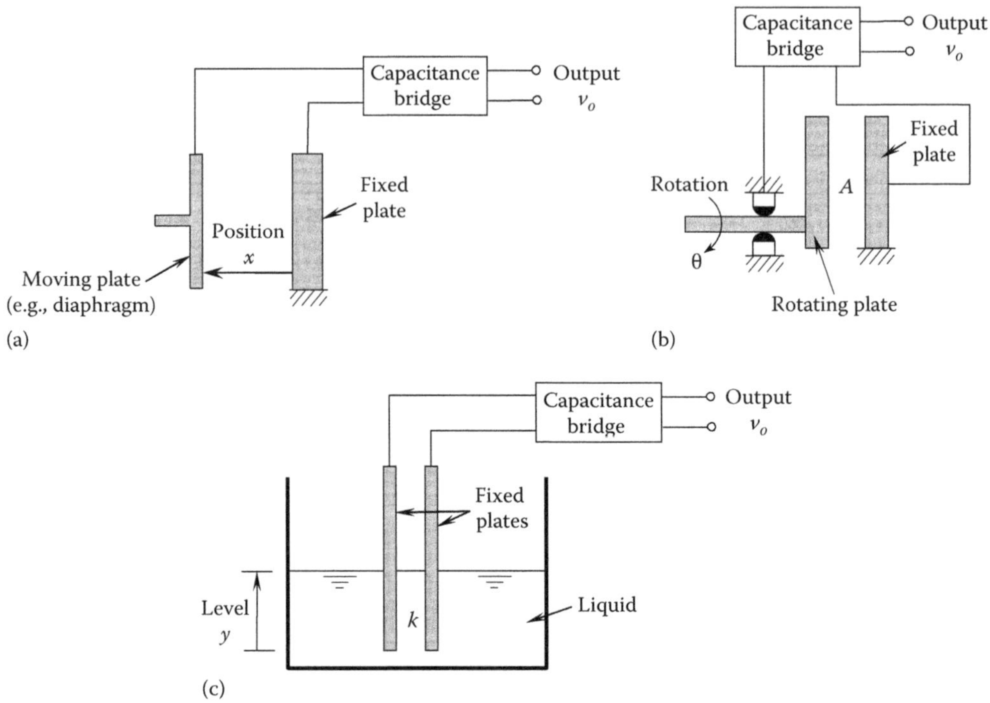

FIGURE 8.21
Schematic diagrams of capacitive sensors. (a) Capacitive displacement sensor, (b) capacitive rotation sensor, and (c) capacitive liquid level sensor.

bridge, potentiometer circuit, feedback capacitance (charge amplifier) circuit, and the inductance–capacitance (*LC*) oscillator circuit. These are outlined now.

8.6.1.1 Capacitance Bridge

A popular method for measuring a change in capacitance is to use a capacitance bridge circuit (see Chapter 4). The sensor forms one arm of the bridge, and a capacitor of similar characteristics as the sensor forms another arm. This is the compensating capacitance, which, due to ambient changes, varies in a manner similar to the sensor itself. The remaining two arms (in a full bridge) are identical impedances. The supply to the bridge is a high-frequency AC voltage. Initially, the bridge is balanced so that the output is zero. As the sensor capacitance changes in the sensing process, the bridge output is a nonzero signal, which consists of a carrier component at the same frequency as the bridge excitation (reference AC voltage) modulated by the variation of the sensor capacitance.

Consider the bridge circuit shown in Figure 8.22. In this circuit, $Z_2 = 1/(j\omega C_2)$, reactance (i.e., capacitive impedance) of the capacitive sensor (of capacitance C_2); $Z_1 = 1/(j\omega C_1)$, reactance of the compensating capacitor C_1; Z_4, Z_3, bridge completing impedances (typically, reactances); $v_{ref} = v_a \sin \omega t$. high-frequency bridge excitation AC voltage; $v_o = v_b \sin(\omega t - \phi)$, bridge output; and ϕ, phase lag of the bridge output with respect to the excitation.

Using the two assumptions for an op-amp (potentials at the negative and positive leads are equal and the current through these leads is zero; see Chapter 3), we can write the current balance equations: $(v_{ref} - v)/Z_1 + (v_o - v)/Z_2 = 0$ and $(v_{ref} - v)/Z_3 + (0 - v)/Z_4 = 0$,

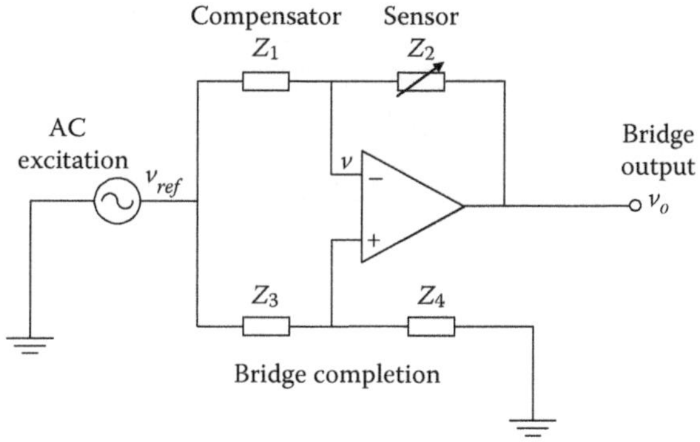

FIGURE 8.22
A bridge circuit for capacitive sensors.

where v is the common voltage at the op-amp leads. Next, eliminate v in these two equations to obtain

$$v_o = \frac{(Z_4/Z_3 - Z_2/Z_1)}{1 + Z_4/Z_3} v_{ref} \tag{8.31}$$

It is noted that the bridge output $v_o = 0$ when $Z_2/Z_1 = Z_4/Z_3$. Then, the bridge is said to be balanced. Since the sensor and the compensating capacitor are similarly affected by ambient changes, a balanced bridge will maintain that condition even under ambient changes. It follows that the ambient effects are compensated (at least up to the first order) by the bridge circuit.

From Equation 8.31, it is clear that due to a change in the measurand, when the sensor reactance Z_2 is changed by δZ, starting from a balanced state, the bridge output is given by

$$\delta v_o = -\frac{v_{ref}}{Z_1(1 + Z_4/Z_3)} \delta Z \tag{8.32}$$

The change δZ in the impedance (reactance) of the sensor capacitor modulates the carrier signal v_{ref}. For transient measurements, this "modulated" output of the bridge has to be demodulated to obtain the measurement. For steady measurements, the amplitude and the phase angle of δv_o with respect to v_{ref} are adequate to determine δZ, assuming that Z_1 and Z_4/Z_3 are known.

Note: Instead of the op-amp in Figure 8.21, an instrumentation amplifier (see Chapter 3) may be used at the bridge output, with more effective results.

8.6.1.2 Potentiometer Circuit

Instead of an impedance bridge, a simpler potentiometer circuit may be used for capacitive sensors. An example is shown in Figure 8.23.

In this circuit, the sensor Z_s is connected with a series impedance Z, which is precisely known. The circuit output is given by

$$v_o = \frac{Z_s}{Z + Z_s} v_{ref} \tag{8.33}$$

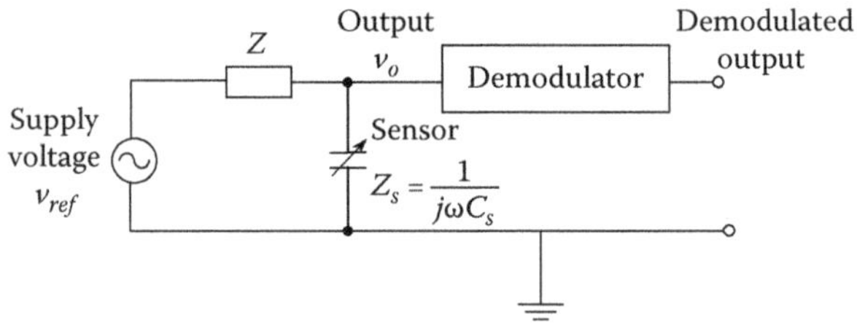

FIGURE 8.23
A potentiometer circuit for variable-capacitance transducers.

This output has the carrier signal v_{ref} modulated by the change in the sensor impedance (capacitance) Z_s. The variation in the sensor impedance (transient) may be obtained by demodulating this signal.

This is a relatively simple circuit. However, it has the disadvantages of any potentiometer circuit. For example, it is not compensated for ambient changes. Also, variations in the carrier will affect the measurement.

8.6.1.3 Charge Amplifier Circuit

An op-amp circuit with a feedback capacitor C_f, which is similar to a charge amplifier, may be used with a variable-capacitance transducer. A circuit of this type is shown in Figure 8.24. The transducer capacitance is denoted by C_s.

The charge balance at node A gives $v_{ref}C_s + v_oC_f = 0$. The circuit output is given by

$$v_o = -\frac{C_s}{C_f} v_{ref} \tag{8.34}$$

Again, this corresponds to the carrier signal modulated by the variation of the sensor capacitance. It may be demodulated to obtain the transducer measurement, under transient conditions.

8.6.1.4 Inductance–Capacitance Oscillator Circuit

An alternative method of sensing the sensor capacitance, which is variable, is to make the sensor an integral part of an *LC* oscillator circuit with a precisely known inductance *L*.

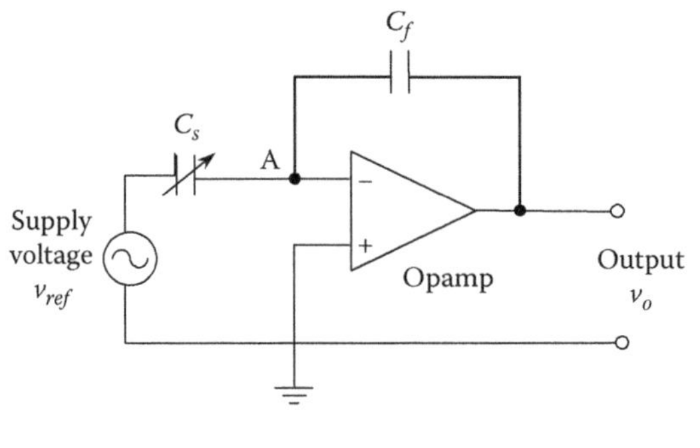

FIGURE 8.24
A feedback capacitor circuit for a variable-capacitance transducer.

The resonant frequency of this oscillator circuit is $1/\sqrt{LC}$. Hence, the sensor capacitance can be measured by measuring the resonant frequency of the circuit. *Note*: This method may be used to measure inductance as well.

8.6.2 Capacitive Displacement Sensor

The arrangement shown in Figure 8.21a provides a sensor for measuring transverse displacements and proximities (x). One of the capacitor plates is attached to the moving object (*Note*: The moving object itself can form the moving capacitor plate) and the other plate is kept stationary. The sensor relationship, as given by Equation 8.29, is nonlinear in this case. If we include the entire nonlinearity (not small increments as in Equation 8.30), the change in sensor capacitance due to the change in displacement is given by

$$\Delta C = kA\left[\frac{1}{x+\Delta x} - \frac{1}{x}\right] \quad \text{or} \quad \frac{\Delta C}{C} = \left[\frac{1}{1+\Delta x/x} - 1\right] \tag{8.35}$$

Note: For small increments of x, this nonlinear relationship may be approximated by the linear relationship $\delta C/C = -\delta x/x$, as given by Equation 8.30.

A simple way to linearize the transverse displacement sensor so that the result is valid for a displacement change of any size, without losing accuracy, is to use an inverting amplifier, as shown in Figure 8.25. Note that C_{ref} is a fixed reference capacitance, whose value is accurately known. Since the gain of the operational amplifier is very high, the voltage at the negative lead (node point A) is zero for most practical purposes (because the positive lead is grounded). Furthermore, since the input impedance of the op-amp is also very high, the current through the input leads is negligible. These are the two common assumptions used in op-amp analysis (see Chapter 3). Accordingly, the charge balance equation for node point A is $v_{ref}C_{ref} + v_oC = 0$. Now, substituting Equation 8.29, we get the following linear relationship for the output voltage v_o in terms of the displacement x:

$$v_o = -\frac{v_{ref}C_{ref}}{K}x \tag{8.36}$$

where $K = kA$. Hence, the circuit output of v_o may be linearly calibrated to give the displacement. The sensitivity of the device can be increased by increasing v_{ref} and C_{ref}. The frequency of the reference excitation (carrier) voltage may be as high as 25 kHz (for high-bandwidth measurements). The output voltage, as given by Equation 8.36, is a

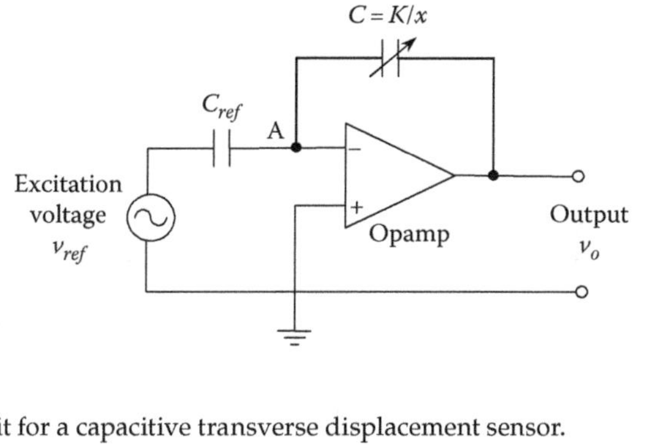

FIGURE 8.25
Linearizing amplifier circuit for a capacitive transverse displacement sensor.

modulated signal, which has to be demodulated to measure transient displacements, as discussed before.

8.6.3 Capacitive Sensors for Rotation and Angular Velocity

In the arrangement shown in Figure 8.21b, one plate of the capacitor is (or attached to) the sensed object (shaft), which rotates. The other plate is kept stationary. Since the common area A is proportional to the angle of rotation θ, from Equation 8.29, the sensor equation may be written as

$$C = K\theta \tag{8.37}$$

where K is the sensor gain.

This is a linear relationship between C and θ. The angle of rotation θ may be measured by measuring the capacitance using any convenient method, as discussed before. Then, the sensor may be linearly calibrated to give the angle of rotation.

A schematic diagram for an angular velocity sensor that uses a rotating-plate capacitor is shown in Figure 8.26. It has a DC supply voltage v_{ref} and a current sensor. Since the current sensor must have a negligible resistance, the voltage across the capacitor is almost equal to v_{ref}, which is constant. It follows that the current in the circuit is given by $i = (d/dt)(Cv_{ref}) = v_{ref}(dC/dt)$, which, in view of Equation 8.37, may be expressed as

$$\frac{d\theta}{dt} = \frac{i}{Kv_{ref}} \tag{8.38}$$

This is a linear relationship for angular velocity in terms of the measured current i. Care must be exercised, however, to ensure that the current-measuring device does not interfere with (e.g., does not load) the basic circuit.

8.6.4 Capacitive Liquid Level Sensor

The arrangement shown in Figure 8.21c can be used for measuring liquid level (y). This is based on the variation of the dielectric constant (k) of the capacitor, while A and x are kept constant in Equation 8.29. Since the voltage in the air segment (a) and the liquid segment (l) of the capacitor is the same while the charges are additive, the capacitances are additive:

$$C = C_a + C_l = \frac{1}{x}(k_a A_a + k_l A_l) = \frac{b}{x}[k_a \times (h-y) + k_l y]$$

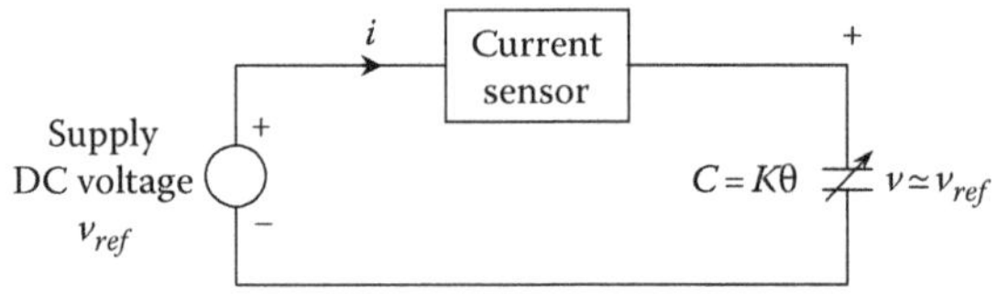

FIGURE 8.26
Rotating-plate capacitive angular velocity sensor.

Hence, we can express the liquid level as

$$y = \frac{xC}{b(k_l - k_a)} - \frac{h}{(k_l/k_a - 1)} \tag{8.39}$$

where
x is the plate gap
h is the plate height
b is the plate width
k_a is the permittivity of air
k_l is the permittivity of the liquid

The liquid level y can be determined by measuring the capacitance C using any method as discussed before.

The arrangement shown in Figure 8.21c can be used as well for displacement sensing. In this case a solid dielectric element, which is free to move in the longitudinal direction of the capacitor plates, is attached to the moving object whose displacement is to be measured. The dielectric constant of the capacitor will change when the common area between the dielectric element and the capacitor plates varies due to the motion. Hence, Equation 8.39 can be used to determine the displacement. In this case, "l" denotes the solid dielectric medium that moves with the moving object whose displacement needs to be measured.

8.6.4.1 Permittivity of Dielectric Medium

Apart from level and displacement sensors, many other types of sensors are based on the permittivity of the dielectric medium of a capacitor. Essentially, the measurand (e.g., humidity) changes the permittivity of the dielectric medium, which is measured by measuring the resulting change in the capacitance. The *relative permittivity* of some material is listed in Table 8.4. These values are expressed with respect to the permittivity of vacuum, $\varepsilon_o = 8.8542 \times 10^{-12}$ F/m, which is almost equal to that of air. Hence, the relative permittivity of vacuum = 1 ≈ relative permittivity of air.

TABLE 8.4

Relative Permittivity Values of Some Material (Approximate)

Material	Relative Permittivity ε_r
Barium titanate	1,250–10,000
Concrete	4.5
Ethylene glycol	37
Glycerol	43
Graphite	10–15
Lead zirconium titanate	500–6,000
Paper, silicon dioxide	4
Polystyrene, nylon, Teflon	2.3
Pyrex (glass)	4–10
Rubber	7
Salt	3–15
Silicon	12
Titanium dioxide	85–170
Water	80

8.6.5 Applications of Capacitive Sensors

In view of their advantages, capacitive sensors are applied directly or indirectly in many practical situations. They have several disadvantages as well.

8.6.5.1 Advantages and Disadvantages

There are many advantages of capacitive sensors that include the following:

1. They are noncontacting devices (the moving plate is integral with the target object; mechanical loading effects are negligible).
2. Very fine measurement can be made at high resolution (e.g., subnanometer, capacitance resolution of 10^{-5} picofarads, 1 pF = 10^{-12} F).
3. The measurement is not sensitive to the material of the target object (capacitor plate).
4. The compensation of extraneous capacitances (e.g., cable capacitance) is straightforward (using a charge amplifier, bridge circuit, etc.).
5. Relatively less costly and small.
6. Linear and high-bandwidth measurements are possible (e.g., 10 kHz bandwidth; 0–10 V bridge output; displacement measurement range, 10–500 μm; probe diameter 8 mm, mass 8 gm, linearity 0.25%).

The main disadvantages include the following:

1. Operation needs a clean environment (a capacitor is affected by moisture, temperature, pressure, dirt, dust, aging, etc.).
2. Large plate gaps result in high error.
3. Low sensitivity (for a transverse displacement transducer, the sensitivity <1 pF = 10^{-12} F). *Note*: High supply voltage and amplifier circuitry can be used to increase the sensor sensitivity.

Since a capacitor relies on its charge and the resulting electric field, any situation that affects this field will cause an error. A capacitive sensor typically has a guard to create an additional field around it. This field is created by the same voltage that is present in the sensor capacitor. The guard essentially acts as a compensating capacitor, which compensates for any extraneous capacitances that affect the sensor.

8.6.5.2 Applications of Capacitive Sensors

Capacitive sensors may be used directly to measure a quantity that affects its capacitance (i.e., a measurand that changes x, A, or k in Equation 8.29) or a quantity that is related (say, through an auxiliary element) to such a measurand (e.g., using deflection sensing for force measurement using a load cell, pressure sensing, and acceleration sensing). Applications of capacitive transducers include the following:

1. Motion sensing (e.g., wafer positioning in semiconductor industry, disk drive control, machine tool control)
2. Gauging and metrology (e.g., thickness of plates, gauging manufactured parts for quality control)

3. Object detection (e.g., counting parts in production lines, detecting cap placement in bottling plants, touch button switches of elevators)
4. Liquid level sensing (e.g., in process plants)
5. Material testing (e.g., surface properties of objects, detecting water in fuels)
6. Environmental sensing (e.g., moisture, soil)

Example 8.4

A capacitive relative-humidity sensor has an average sensitivity of 2.0%RH/μF and an offset of −5.0%RH. What is the percentage relative humidity (%RH) corresponding to a capacitance reading of 50 μF?

Solution

Assume a linear sensor whose calibration curve is given by $RH = RH_0 + m \times C$.
Note: This assumption is confirmed to be satisfactory according to experimental data.
Given: $RH_0 = -5.0\%$ and $m = 2.0\%\text{RH}/\mu\text{F}$
For $C = 50\ \mu\text{F}$, we have $RH = -5 + 2.0 \times 50 = 95\%\text{RH}$.

8.7 Piezoelectric Sensors

Some substances such as barium titanate, single-crystal quartz, lead zirconate–titanate (PZT), lanthanum-modified PZT, lithium niobate, and piezoelectric polymeric polyvinylidene fluoride generate an electrical charge and an associated potential difference when they are subjected to mechanical stress or strain. This piezoelectric effect is the governing principle in *piezoelectric transducers*. These are *passive* sensors because energy conversion (*electromechanical coupling*) through the piezoelectric effect is used in sensing the measurand.

The direct application of the piezoelectric effect is found in pressure- and strain-measuring devices, touch screens of computer monitors, sophisticated microphones, knock sensors in automotive engines, temperature sensors (crystal resonant frequency, which changes nonlinearly with temperature, may be used for temperature sensing; for example, the resonant frequency increases from approximately −20°C to +20°C and decreases from +20°C to +50°C), and a variety of microsensors. Many indirect applications also exist. They include piezoelectric accelerometers and velocity sensors, piezoelectric torque sensors, and force sensors. Of course, along with a "passive" piezoelectric sensor, signal conditioning (e.g., a *charge amplifier*, which needs external power) has to be used.

It is also interesting to note that a piezoelectric material deforms when subjected to a potential difference (or charge or electric field) and can serve as an *actuator*. This is the "reverse" *piezoelectric effect*. Some delicate test equipment (e.g., in nondestructive, dynamic testing) use such piezoelectric actuating elements (which undergo reverse piezoelectric action) to create fine motions. Also, piezoelectric valves (e.g., flapper valves and fuel injectors), with direct actuation using voltage signals, are used in pneumatic and hydraulic control applications, in ink-jet printers, and in automotive engines. Piezoelectric actuators are used to generate acoustic waves in a variety of applications including medical imaging and sophisticated speakers. Miniature stepper motors based on the reverse piezoelectric action are available as well. Microactuators that use the piezoelectric effect are found in a number of applications including hard disk drives (HDDs). This multifunctional character (sensing and actuation) of a piezoelectric material makes it a "smart material," which is used in sophisticated engineering applications and MEMS—see Chapter 12.

Piezoelectric Effect

The piezoelectric effect is caused by the charge polarization in an anisotropic material (having nonsymmetric molecular structure), as a result of an applied strain. Specifically, a charge (or electric field) is released when the material is strained. This is a reversible effect. In particular, when an electric field is applied to the material, it changes the ionic polarization, and the material can deform or shed an existing strain (i.e., the strain is removed and the material regains its original shape). Natural piezoelectric materials are by and large crystalline, whereas synthetic piezoelectric materials tend to be ceramics. When the direction of the electric field and the direction of the strain (or stress) are the same, we have direct sensitivity. Cross sensitivities can be defined as well, in a 6×6 matrix with reference to three orthogonal direct axes and three rotations about these axes.

Consider a piezoelectric crystal in the form of a disk with two electrodes plated on the two opposite faces. Since the crystal is a dielectric medium, this device is essentially a *capacitor*, which may be modeled by a capacitance. Accordingly, a piezoelectric sensor may be represented as a charge source q_s with a capacitance C_s in parallel, as shown in Figure 8.27. This is the *Norton equivalent circuit*. Of course, there is an internal resistance as well in the piezoelectric element (between the electrodes), which can be represented in series with the charge source. But, it will have no effect on the charge source (as it is in series) and is omitted (or considered as internal to the charge source) in the present circuit. The other effects that are ignored in the circuit in Figure 8.27 are insulating resistance of the piezoelectric element (which is very high) in parallel with the charge source (*Note*: Any charge leakage through the insulation, which can be neglected, would be through this resistance) and the cable capacitance in parallel with the charge source (*Note*: The effects of the cable can be included separately or compensated for).

Another equivalent circuit (Thevenin equivalent representation) can be given as well, where the capacitor is in series with an equivalent voltage source. This is completely equivalent to the Norton circuit given in Figure 8.27.

The impedance from the capacitor is given by

$$Z_s = \frac{1}{j\omega C_s} \tag{8.40}$$

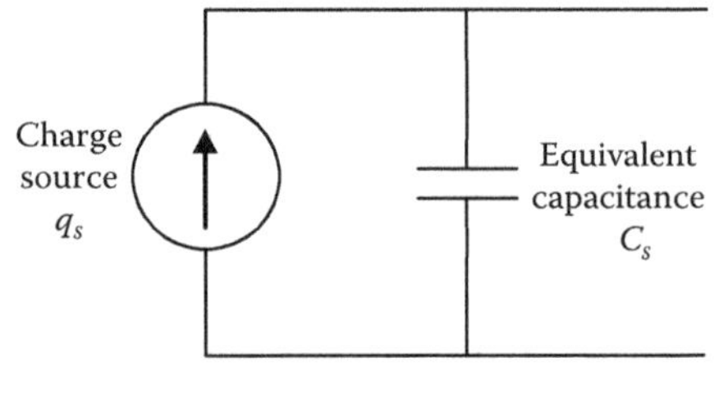

FIGURE 8.27
Equivalent circuit (Norton) representation of a piezoelectric sensor.

It is clear from Equation 8.40 that the *output impedance* of a piezoelectric sensor can be very high, particularly at low frequencies. For example, a quartz crystal may present an impedance of many megohms at 100 Hz, increasing hyperbolically with decreasing frequencies. This is one reason why piezoelectric sensors have a limitation on the useful lower frequency, where the charge leakage cannot be neglected. As we will see, a charge amplifier can resolve this problem.

Note: Even though the resistance (DC) of a piezoelectric crystal is also extremely high, it will have no effect on the charge source (as it is in series). The insulating resistance R_l provides a leakage path in parallel with the piezoelectric element (i.e., capacitance C_s). However, the output impedance is governed by C_s because its impedance (reactance) is considerably smaller than the insulator resistance. Nevertheless, it will be seen that later all these extraneous impedances are compensated for by means of a charge amplifier.

8.7.1 Charge Sensitivity and Voltage Sensitivity

The sensitivity of a piezoelectric crystal may be represented either by its charge sensitivity or by its voltage sensitivity. The *charge sensitivity* S_q is defined as

$$S_q = \frac{\partial q}{\partial F} \tag{8.41a}$$

where

q denotes the generated charge

F denotes the applied force

For a crystal with surface area A, Equation 8.41a may be expressed as

$$S_q = \frac{1}{A}\frac{\partial q}{\partial p} \tag{8.41b}$$

where p is the stress (normal or shear) or pressure applied to the crystal surface. The *voltage sensitivity* S_v is given by the change in voltage due to a unit increment in pressure (or stress) per unit thickness of the crystal. In the limit, we have

$$S_v = \frac{1}{d}\frac{\partial v}{\partial p} \tag{8.42}$$

where d is the crystal thickness. Now, since the capacitance equation for a piezoelectric element is given by $\delta q = C\delta v$, by using $C = (kA)/d$ (see Equation 8.29), the following relationship between charge sensitivity and voltage sensitivity is obtained:

$$S_q = kS_v \tag{8.43}$$

where k is the dielectric constant (permittivity) of the crystal capacitor. The overall sensitivity of a piezoelectric device can be increased through the use of properly designed multielement structures (i.e., *bimorphs*).

TABLE 8.5

Direct Sensitivities of Several Piezoelectric Materials

Material	Charge Sensitivity S_q (pC/N)	Voltage Sensitivity S_v (mV·m/N)
Lead zirconate titanate	110	10
Barium titanate	140	6
Quartz	2.5	50
Rochelle salt	275	90

Example 8.5

A barium titanate crystal has a charge sensitivity of 150.0 picocoulombs per newton (pC/N). (*Note*: 1 pC = 1×10^{-12} coulombs (C); coulombs = farads × volts). The dielectric constant for the crystal is 1.25×10^{-8} farads per meter (F/m). From Equation 8.43, the voltage sensitivity of the crystal may be computed as

$$S_v = \frac{150.0\ \text{pC/N}}{1.25 \times 10^{-8}\ \text{F/m}} = \frac{150.0 \times 10^{-12}\ \text{C/N}}{1.25 \times 10^{-8}\ \text{F/m}} = 12.0 \times 10^{-3}\ \text{V} \cdot \text{m/N} = 12.0\ \text{mV} \cdot \text{m/N}$$

The sensitivity of a piezoelectric element is dependent on the direction of loading. This is because the sensitivity depends on the molecular structure (e.g., crystal axis). Direct sensitivities of several piezoelectric materials along their most sensitive crystal axis are listed in Table 8.5.

Electromechanical coupling: The piezoelectric effect is a result of electromechanical coupling. Specifically, when an element is strained, an external device does "mechanical work" on the element. This work deforms the element, and the energy that goes into the element is stored as "strain energy." It can be modeled as a spring, which stores "elastic potential energy." An electric charge is released in the process. This is the *direct piezoelectric effect*. The device acts as a mechanical displacement *sensor*. An electric field has to be applied by an external means to "undeform" the element and release the stored strain energy. Alternatively, an initially unstrained element can be deformed as well by means of an external electric field (using an external electrical energy source). This is the *reverse piezoelectric effect*, where the element acts like an *actuator*. However, the circuit shown in Figure 8.27 includes only the electrical dynamics. For a complete representation of a piezoelectric element, the mechanical dynamics has to be included as well. In the circuit shown in Figure 8.27, this is hidden in the charge source q_s. A simplified mechanical model may include just the stiffness and inertia of the element. Then, the charge q_s will be a result of strain, stress, or pressure in the element and can be related through the charge sensitivity (see Equation 8.41). The corresponding stress, pressure, or force will cause dynamics in the inertia (mass) of the sensor element, as governed by Newton's second law. In essence, this coupled device can be represented by a *two-port element* where one port has mechanical energy flow and the other port has electrostatic energy flow.

8.7.2 Charge Amplifier

Piezoelectric signals cannot be acquired using low-impedance devices. The two primary reasons for this are as follows:

1. High output impedance in the sensor will result in small output signal levels and large loading errors.
2. The charge can quickly leak out through the load.

To overcome these problems to a great extent, a charge amplifier is commonly used as the primary signal-conditioning device for a piezoelectric sensor. The input impedance of a charge amplifier is quite high. Hence, electrical loading on the piezoelectric sensor is reduced. Furthermore, because of impedance transformation, the output impedance of a charge amplifier is rather small (very much smaller than the output impedance of a piezoelectric sensor). This virtually eliminates the loading error and provides a low-impedance output for such purposes as monitoring, signal acquisition, communication, recording, processing, and control. Also, by using a charge amplifier circuit with a relatively large time constant, the speed of charge leakage can be reduced.

A charge amplifier is simply an op-amp with a capacitive feedback (C_f). Typically, we include a resistive feedback (R_f) as well. As an example, consider the circuit of a piezoelectric sensor and charge amplifier combination, as shown in Figure 8.28. We will examine how the rate of the charge leakage is reduced and extraneous capacitances (and other impedances/resistances) are compensated for by using this arrangement. The sensor capacitance, feedback capacitance of the charge amplifier, and feedback resistance of the charge amplifier are denoted by C_s, C_f, and R_f, respectively. The capacitance of the cable, which connects the sensor to the charge amplifier, is denoted by C_c. The insulator resistance of the piezoelectric sensor (which provides a path for charge leakage) is denoted by R_l. As usual, the internal resistance of the piezoelectric element is not included in the circuit because it is in series with the charge source and hence it has no effect on the circuit equations.

Since the potentials at the two input leads of the op-amp are almost equal, they are at the ground potential (zero). Hence, the current/charge leakage through the parallel paths of the sensor is negligible (*Note*: This is further supported by the fact that the corresponding impedances are very high as well). The current balance at node A gives

$$\dot{q} + C_f \dot{v}_o + \frac{v_o}{R_f} = 0 \tag{8.44}$$

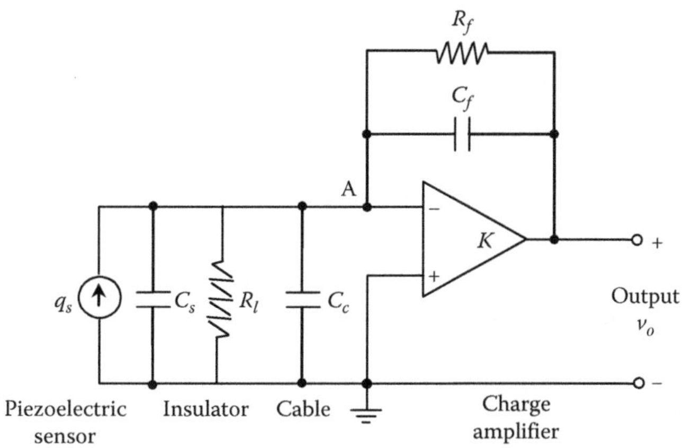

FIGURE 8.28
Piezoelectric sensor and charge amplifier combination.

The corresponding transfer function is

$$\frac{v_o(s)}{q(s)} = -\frac{R_f s}{[R_f C_f s + 1]} \tag{8.45a}$$

where s is the Laplace variable. Now, in the frequency domain ($s = j\omega$), we have

$$\frac{v_o(j\omega)}{q(j\omega)} = -\frac{R_f j\omega}{[R_f C_f j\omega + 1]} \tag{8.45b}$$

By properly calibrating the charge amplifier (with respect to [wrt] the factor $-1/C_f$), the frequency transfer function of the overall system can be written as

$$G(j\omega) = \frac{j\tau_s\omega}{[j\tau_s\omega + 1]} \tag{8.46}$$

Hence, the sensor–amplifier unit is represented by a first-order system with its time constant τ_s given by

$$\tau_s = R_f C_f \tag{8.47}$$

This is completely governed by the feedback elements of the charge amplifier, which can be precisely and appropriately chosen.

The output of the device is zero at zero frequency ($\omega = 0$). Hence, a piezoelectric sensor cannot be used for measuring constant (DC) signals. On the other hand, at very high frequencies, the transfer-function magnitude $M = (\tau_s\omega)/\sqrt{\tau_s^2\omega^2 + 1}$ approaches unity. Hence, at infinite frequency there is no sensor error. Measurement accuracy depends on the closeness of M to 1.

Example 8.6

For a piezoelectric accelerometer with a charge amplifier, an accuracy level better than 99% is obtained if $(\tau_s\omega)/\sqrt{\tau_s^2\omega^2 + 1} > 0.99 \rightarrow \tau_s\omega > 7.0$.

The minimum frequency of a transient sensor signal that can tolerate this level of accuracy is $\omega_{min} = 7.0/\tau_s$.

It follows that, for a specified level of accuracy, a specified lower limit on frequency of operation may be achieved by increasing the time constant (i.e., by increasing R_f, C_f, or both). The feasible lower limit on the frequency of operation (ω_{min}) can be set by adjusting the time constant.

Advantages of piezoelectric sensors include the following:

1. They provide high speed of response and high operating bandwidth (very small time constant).
2. They can be produced in small size (as microminiature devices).
3. They are passive and hence robust and relatively simple in operation.

4. The sensitivity can be increased by using proper piezoelectric material.
5. Extraneous effects can be easily compensated for by means of straightforward signal conditioning (e.g., charge amplifier).
6. They are multifunctional (can serve as a sensor or an actuator in the same system).

Shortcomings of piezoelectric sensors include the following:

1. High output impedance
2. High temperature sensitivity
3. In view of the multifunctional capability (an advantage), one function can be affected by the other function (e.g., stray electric field can affect the sensing accuracy).
4. Not suitable for low-frequency or DC sensing.

8.7.3 Piezoelectric Accelerometer

Accelerometers: It is known from Newton's second law that a force (f) is necessary to accelerate a mass (or inertia element), and its magnitude is given by the product of mass (m) and acceleration (a). This product (ma) is commonly termed the "inertia force." The rationale for this terminology is that if a force of magnitude ma was applied to the accelerating mass in the direction opposing the acceleration, then the system could be analyzed using static equilibrium considerations. This is known as "d'Alembert's principle" (Figure 8.29). The force that causes acceleration is itself a measure of the acceleration (*Note*: Mass is kept constant). Accordingly, a mass can serve as a front-end auxiliary element to convert acceleration into force. This is the principle of operation of common accelerometers. There are many different types of accelerometers, ranging from strain-gauge devices to those that use electromagnetic induction. For example, the force that causes the acceleration may be converted into a proportional displacement using a spring element, and this displacement may be measured using a convenient displacement sensor. Examples of this type are *differential transformer accelerometers, potentiometer accelerometers,* and *variable-capacitance accelerometers.* Alternatively, the strain at a suitable location of a member that was deflected due to the inertia force may be determined using a strain gauge. This method is used in *strain-gauge accelerometers.* Vibrating-wire accelerometers use the accelerating force to tension a wire. The force is measured by detecting the natural frequency of vibration of the

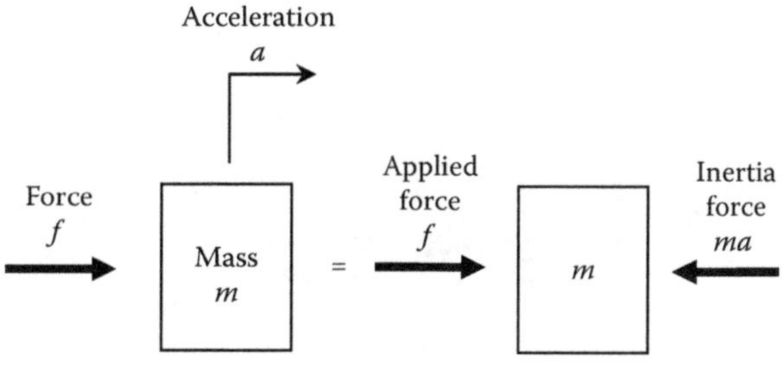

FIGURE 8.29
Illustration of d'Alembert's principle.

wire, which is proportional to the square root of tension. In servo force-balance (or null balance) accelerometers, the inertia element is restrained from accelerating by detecting its motion and feeding back a force (or torque) to exactly cancel out the accelerating force (torque). This feedback force is determined, for instance, by knowing the motor current, and it is a measure of the acceleration.

8.7.3.1 Piezoelectric Accelerometer

The piezoelectric accelerometer (or *crystal accelerometer*) is an acceleration sensor, which uses a piezoelectric element to measure the inertia force caused by acceleration. A *piezoelectric velocity transducer* is simply a piezoelectric accelerometer with a built-in integrating amplifier in the form of a miniature IC.

The advantages of piezoelectric accelerometers over other types of accelerometers are their light weight and high-frequency response (up to about 1 MHz). However, piezoelectric transducers are inherently high output impedance devices, which generate small voltages (in the order of 1 mV). For this reason, special impedance-transforming amplifiers (e.g., charge amplifiers) have to be employed to condition the output signal and to reduce loading error.

A schematic diagram for a *compression-type piezoelectric accelerometer* is shown in Figure 8.30. The crystal and the inertia element (mass) are restrained by a spring of very high stiffness. Consequently, the fundamental natural frequency or *resonant frequency* of the device becomes high (typically 20 kHz). This gives a rather wide useful frequency range or operating range (typically up to 5 kHz) and high speed of operation. The lower limit of the useful frequency range (typically 1 Hz) is set by factors such as the limitations of the signal-conditioning system, the mounting method, charge leakage in the piezoelectric element, time constant of the charge-generating dynamics, and the SNR. A typical frequency response curve of a piezoelectric accelerometer is shown in Figure 8.31.

For an accelerometer, acceleration is the signal that is measured (the measurand). Hence, accelerometer sensitivity is commonly expressed in terms of electrical charge per unit acceleration or voltage per unit acceleration (compare this with Equations 8.60 and 8.61). Acceleration is measured in units of acceleration due to gravity (g) and charge is measured in picocoulombs (pC), which are units of 10^{-12} coulombs (C). Typical accelerometer sensitivities are 10 pC/g and 5 mV/g. Sensitivity depends on the piezoelectric properties (see Table 8.5), the way in which the inertia force is applied to the piezoelectric element

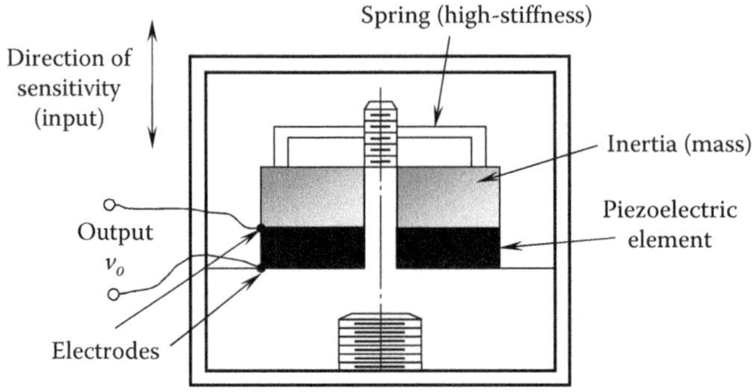

FIGURE 8.30
A compression-type piezoelectric accelerometer.

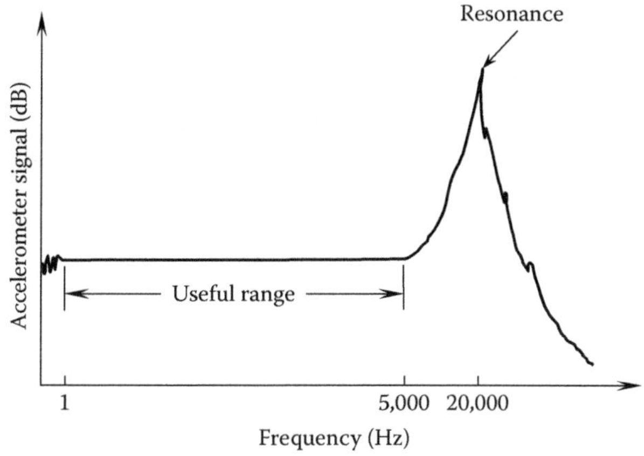

FIGURE 8.31
A typical frequency response curve for a piezoelectric accelerometer.

(e.g., compressive, tensile, shear), and the mass of the inertia element. If a large mass is used, the reaction inertia force on the crystal becomes large for a given acceleration, thus generating a relatively large output signal. Large accelerometer mass results in several disadvantages, however. In particular,

1. The accelerometer mass distorts the measured motion variable (mechanical loading effect).
2. A heavy accelerometer has a lower resonant frequency and hence a lower useful frequency range (Figure 8.31).

In a compression-type crystal accelerometer, the inertia force is sensed as a compressive normal stress in the piezoelectric element. There are also piezoelectric accelerometers where the inertia force is applied to the piezoelectric element as a shear strain or as a tensile strain.

Accelerometer configurations: For a given accelerometer size, better sensitivity can be obtained by using the shear–strain configuration rather than normal strain. In the shear configuration, several shear layers can be used (e.g., in a delta arrangement) within the accelerometer housing, thereby increasing the effective shear area and hence the sensitivity, which is proportional to the shear area. Another factor that should be considered in selecting an accelerometer is its *cross sensitivity* or *transverse sensitivity*. Cross sensitivity is present because a piezoelectric element can generate a charge in response to forces and moments (or torques) in orthogonal directions as well. The problem can be aggravated due to manufacturing irregularities of the piezoelectric element, including material unevenness and incorrect orientation of the sensing element, and due to poor design. Cross sensitivity should be less than the maximum error (percentage) that is allowed for the device (typically 1%).

Mounting methods: The technique employed to mount an accelerometer on an object can significantly affect the useful frequency range of the accelerometer. Some common mounting techniques are

1. Screw-in base
2. Glue, cement, or wax

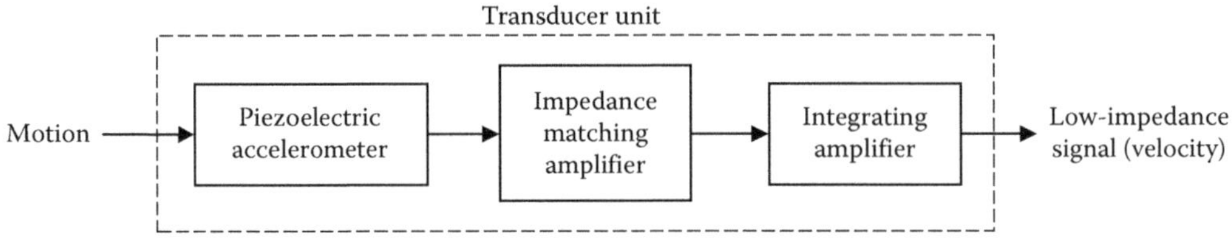

FIGURE 8.32
Schematic diagram of a piezoelectric velocity transducer.

3. Magnetic base
4. Spring base
5. Handheld probe

Drilling holes in the object can be avoided by using the second through fifth methods, but the useful frequency range can decrease significantly when spring-base mounts or handheld probes are used (typical upper limit of 500 Hz). The first two methods usually maintain the full useful range (e.g., 5 kHz), whereas the magnetic attachment method reduces the upper frequency limit to some extent (typically 3 kHz).

In theory, it is possible to measure velocity by first converting velocity into a force using a viscous damping element and measuring the resulting force using a piezoelectric sensor. This principle may be used to develop a piezoelectric velocity transducer. However, the practical implementation of an ideal velocity–force transducer is quite difficult, primarily due to nonlinearities in damping materials. Hence, commercial piezoelectric velocity transducers use a piezoelectric accelerometer and a built-in (miniature) integrating amplifier. A schematic diagram of the arrangement of such a piezoelectric velocity transducer is shown in Figure 8.32. The overall size of the unit can be as small as 1 cm. With double integration hardware, a piezoelectric displacement transducer is obtained using the same principle. A homing method is needed to identify the reference position (initial condition) when a position is measured using integration. Furthermore, numerical integration slows down the sensing process (operating frequency limit). Alternatively, an ideal spring element (or cantilever), which converts displacement into a force (or bending moment or strain), may be employed to stress the piezoelectric element, resulting in a displacement transducer. Such devices are usually not practical for low-frequency (few hertz) applications because of the poor low-frequency characteristics of piezoelectric elements.

8.8 Gyroscopic Sensors

Gyroscopic sensors are used for measuring angular orientations and angular speeds in a variety of applications including aircraft, ships, vehicles, robots, missiles, radar systems, machinery, camera stabilization, and other mechanical devices. These sensors are commonly used in control systems for stabilizing vehicle systems. Since a spinning body (a gyroscope) requires an external torque to turn (precess) its axis of spin, if this gyro is mounted (in a frictionless manner) on a rigid vehicle so that there are a sufficient number of frictionless degrees of freedom (at most three) between the gyro and the vehicle, the spin axis will remain unchanged in space, regardless of the motion of the vehicle. Hence, the axis of spin of the gyro provides a reference with respect to which the vehicle orientation (e.g., azimuth or yaw, pitch, and roll angles) and angular speed can be measured. The orientation can be measured by using angular sensors at the pivots of the structure that mounts the gyro on the vehicle. The angular speed about an orthogonal axis can be determined, for example, by measuring the precession torque (which is proportional to the angular speed) using a strain-gauge sensor or by measuring, using a position sensor such as a resolver, the deflection of a torsional spring that restrains the precession. In the latter case, the angular deflection is proportional to the precession torque and hence the angular speed.

8.8.1 Rate Gyro

A rate gyro is used to measure angular speeds. The arrangement shown in Figure 8.33a may be used to explain its principle of operation.

A rigid disk (gyroscopic disk) of polar moment of inertia J is spun at angular speed ω about frictionless bearings using a constant-speed motor, which is spinning about an axis. The angular momentum H about this axis is given by

$$H = J\omega \tag{8.48}$$

This vector is shown by the solid line in Figure 8.33b. Due to the angular speed (rate) Ω, which is the quantity to be measured (measurand or sensor input), the vector $\boldsymbol{H}$ will turn through angle $\Omega \cdot \Delta t$ in an infinitesimal time Δt, as shown. The magnitude of the resulting change in angular momentum is $\Delta H = J\omega \cdot \Omega \cdot \Delta t$ or the rate of change of angular momentum is $dH/dt = J\omega \cdot \Omega$. To perform this rotation (precession), a torque has to be applied in the orthogonal direction (the direction of ΔH in Figure 8.33b), which is the same as the direction of rotation θ in Figure 8.33a. If this direction is restrained by a torsional spring of stiffness K and a damper with rotational damping constant B, the corresponding resistive torque is $K\theta + B\dot{\theta}$. Newton's second law (torque = rate of change of angular momentum) gives $J\omega\Omega = K\theta + B\dot{\theta}$ or

$$\Omega = \frac{K\theta + B\dot{\theta}}{J\omega} \tag{8.49}$$

From this result, it is seen that, when B is very small, the angular rotation θ at the gimbal bearings (measured, e.g., by a resolver) will be proportional to the angular speed to be measured (Ω). This reading can be properly calibrated to read the angular speed.

A main source of error in a gyroscope is the drift. Recalibration has to be done routinely to eliminate this error. This is done by zeroing the reading when the measurand is zero.

8.8.2 Coriolis Force Devices

Consider a mass m moving at velocity v relative to a rigid frame. If the frame itself rotates at an angular velocity $\boldsymbol{\omega}$, it is known that the absolute acceleration of m (which is used in the equation of Newton's second law) has a term given by $2\boldsymbol{\omega} \times v$. This is known as the "Coriolis acceleration." The associated force $2m\boldsymbol{\omega} \times v$ is the *Coriolis force*. This force can be sensed either by directly using a force sensor or by measuring a resulting deflection in a flexible element and may be used to determine the variables ($\boldsymbol{\omega}$ or v) in the Coriolis force.

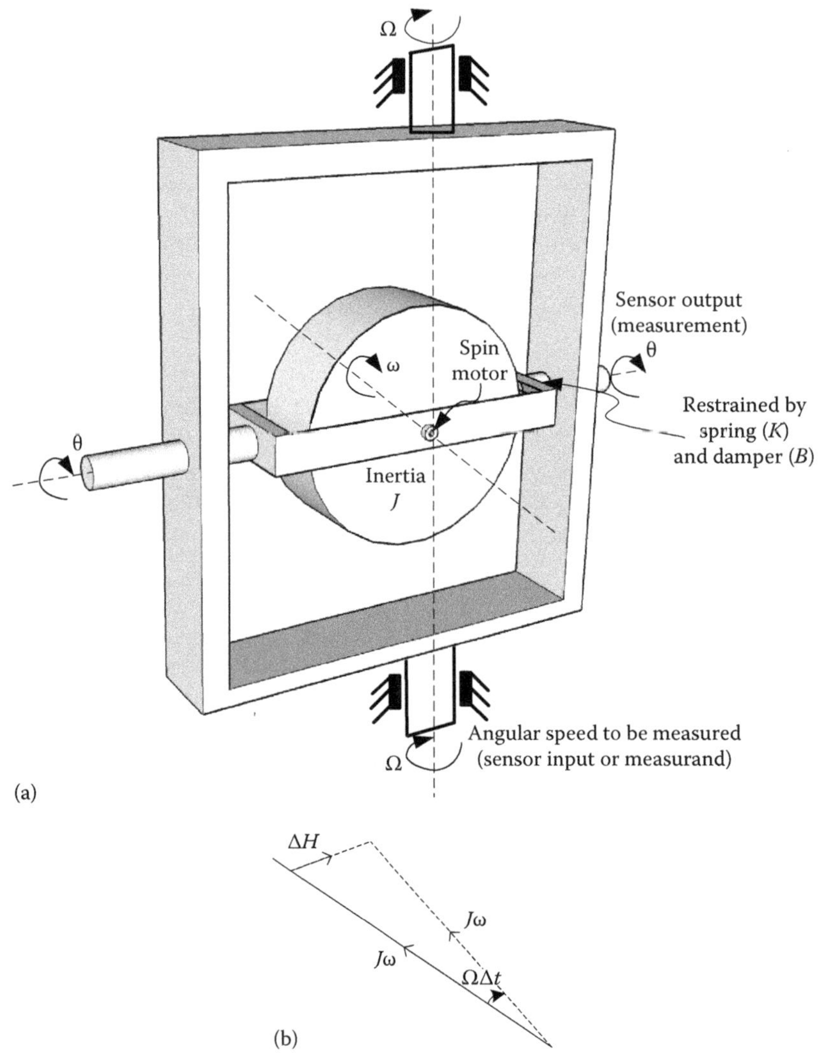

FIGURE 8.33
(a) A rate gyro and (b) gyroscopic torque needed to change the direction of an angular momentum vector.

Note that Coriolis force is somewhat similar to gyroscopic force even though the concepts are different. For this reason, devices based on the Coriolis effect are also commonly termed gyroscopes. Coriolis concepts are gaining popularity in MEMS-based sensors, which use MEMS technologies (see Chapter 12).

Summary Sheet

Measurand: Variable that is being measured.

Measurement: Output or reading of the measuring device.

Two stages in a measuring device: (1) Measurand is "felt" or *sensed* and (2) measured signal is *transduced* (or converted) into the form of device output.

Sensor applications: Process monitoring; testing and qualification; system identification (experimental modeling); product quality assessment; fault prediction, detection, and diagnosis; warning generation; surveillance; controlling a system.

Use in control: Measure: outputs for feedback control; some types of inputs (unknown inputs, disturbances, etc.) for feedforward control; outputs for system monitoring, parameter adaptation, self-tuning, and supervisory control; input–output signal pairs for experimental modeling (i.e., system identification).

Sensor system: May mean, (1) multiple sensors, sensor/data fusion, or (2) sensor and its accessories.

Human sensory system: Five senses: sight (visual), hearing (auditory), touch (tactile), smell (olfactory), and taste (flavory). Process: Receptor/neural dendrites (sensors and transducers) → neural axons (communication) → central nervous system (signal processing).

Other sensory features of human: Sense of balance, pressure, temperature, pain, and motion.

Sensor classification:

(a) Based on technology: *Active*: Power for sensing does not come from sensed object (comes from external source); *Analog*: Output is analog; *Digital*: Output is digital, pulses, counts, etc.; electric, integrated circuit, mechanical, optical, etc; *Passive*: Power for sensing comes from sensed object; *Piezoelectric*: Pressure on sensor element generates a charge or voltage; *Piezoresistive*: Pressure/stress/strain on sensor element changes its electrical resistance; *Photoelastic*: Stress/strain on the sensor element changes its optical properties.

(b) Based on measurand: *Biomedical*: Motion, force, blood composition, blood pressure, temperature, flow rate, urine composition, excretion composition, ECG (EKG), breathing sound, pulse, x-ray image, ultrasonic image; *Chemical*: Organic compounds, inorganic compounds, concentration, heat transfer rate, temperature, pressure, flow rate, humidity; *Electrical/electronic*: Voltage, current, charge, passive circuit parameters, electric field, magnetic field, magnetic flux, electrical conductivity, permittivity, permeability, reluctance; *Mechanical*: Force (effort including torque), motion (including position and deflection), optical image, other images (x-ray, acoustic, etc.), stress, strain, and material properties (density, Young's modulus, shear modulus, hardness, Poisson's ratio); *Thermofluid*: Flow rate, heat transfer rate, infrared waves, pressure, temperature, humidity, liquid level, density, viscosity, Reynolds number, thermal conductivity, heat transfer coefficient, Biot number, and image.

Sensor selection: "Match" sensor (ratings) with the application (requirements/specifications). (a) Study the application, its purpose, and what quantities (variables and parameters) need to be measured and (b) determine what sensors are available and what quantities cannot be measured (due to inaccessibility, lack of sensors, etc.), and; if it cannot be measured, estimate using other quantities that can be measured or develop a new sensor for the purpose.

Sensor selection process: (1) What parameters or variables have to be measured in the application; (2) nature of the information (parameters and variables) needed for the application (analog, digital, modulated, demodulated, power level, bandwidth, accuracy, etc.); (3) specifications for the needed measurements (measurement signal type, measurement level, range, bandwidth, accuracy, signal-to-noise, etc.); (4) available sensors for the application and their data sheets; (5) signal provided by each sensor (types—analog, digital, modulated, etc.; power level; frequency range); (6) type of signal conditioning or conversion needed for the sensors (filtering, amplification, modulation, demodulation, ADC, DAC, voltage–frequency conversion, frequency–voltage conversion, etc.).

Pure transducers: Depend on nondissipative coupling in the transduction stage (no wastage of signal power).

Perfect measurement device: (1) Measuring device output instantly reaches the measured value (*fast response*); (2) transducer output is sufficiently large (*high gain, low output impedance, high sensitivity*); (3) device output remains at the measured value (without drifting or being affected by environmental effects and other undesirable disturbances and noise) unless the measurand (what is measured) itself changes (*stability and robustness*); and (4) the output signal level of the transducer varies in proportion to the signal level of the measurand (*static linearity*); the measuring device does not distort the measurand itself (*loading effects are absent and impedances are matched*); and power consumption is small (*high-input impedance*).

Motion transducers: Displacement (position, distance, proximity, size, gauge, etc.), velocity, acceleration, and jerk. *Note*: Each variable = time derivative of preceding one.

Front-end auxiliary element: *Inertia element*: converts acceleration into force; *damping element*: converts velocity into force; *spring element*: converts displacement into force.

Limitations of conversion among motion variables: (1) Nature of measured signal (steady, highly transient, periodic, narrow/broadband, etc.); (2) required frequency content of processed signal (frequency range of interest); (3) Signal-to-noise ratio (SNR) of the measurement; (4) available processing capabilities (e.g., analog or digital processing, limitations of digital processor and interface, speed of processing, sampling rate, buffer size, etc.); (5) controller requirements and nature of plant (e.g., control bandwidth, operating bandwidth, time constants, delays, complexity, hardware limitations); (6) required task accuracy (processing requirements, hardware costs depend on this).

Example: Differentiation is unacceptable for noisy and high-frequency narrowband signals. Costly signal-conditioning hardware may be needed for pre-preprocessing.

Rule of thumb: *Low-frequency applications* (*~1 Hz*): use displacement measurement; *intermediate-frequency applications* (*<1 kHz*): use velocity measurement; *high-frequency motions with high noise levels*: use acceleration measurement.

Motion transducers: Potentiometers (resistively coupled devices); variable-inductance transducers (electromagnetically coupled devices); eddy current transducers; variable-capacitance transducers; piezoelectric transducers.

Considerations in motion transducer selection: Kinetic nature of measurand (position, proximity, displacement, speed, acceleration, etc.), rectilinear (commonly termed linear) or rotatory (commonly termed rotary) motion, contact or noncontact type, measurement range, required accuracy, required frequency range of operation (time constant, bandwidth), size, cost, operating environment (e.g., magnetic fields, temperature, pressure, humidity, vibration, shock), and life expectancy.

Potentiometer characteristics: Output voltage drops when a load with finite impedance is connected → loading effect → linear relationship is not valid (also affects supply [reference] voltage).

Reduced loading effects use: Use regulated/stabilized power supply with low output impedance and signal-conditioning circuitry with high input impedance.

Note: High element resistance → reduced power dissipation; less thermal effects (shortcoming: increases output impedance → increased loading nonlinearity error).

Pot resistances: 10–1 MΩ. *Conductive plastics*: High resistances (100 Ω/mm), low friction (low mechanical loading), reduced wear, reduced weight, and increased resolution. *Rotary pot loading*, $v_o/v_{ref} = [((\theta/\theta_{max})(R_L/R_C))/(R_L/R_c + (\theta/\theta_{max}) - (\theta/\theta_{max})^2)]$ (nonlinear) → Error $e = ((v_o/v_{ref} - \theta/\theta_{max})/(\theta/\theta_{max}))100\%$.

Performance limitations: Force needed to move the slider (against friction and arm inertia) ⇒ mechanical loading ⇒ distorts measured signal; high-frequency (highly transient) measurements difficult due to slider bounce, friction, inertia resistance, induced voltages in wiper arm and primary coil; variations in the supply voltage; high electrical loading error at low load resistance; wear and heating up (oxidation).

Advantages: Robust, simple, and relatively inexpensive; and provides high-voltage (low impedance) output signals; impedance can be varied by changing the coil resistance.

Optical potentiometer:

$$\frac{v_o}{v_{ref}}\left\{\frac{R_c}{R_L}+1+\frac{x}{L}\frac{R_c}{R_p}\left[\left(1-\frac{x}{L}\right)\frac{R_c}{R_L}+1\right]\right\}=1$$

$$\rightarrow \frac{v_o}{v_{ref}}=\frac{1}{[(x/L)(R_c/R_p)+1]} \quad \text{(for high load resistance)}$$

Variable-inductance transducers: (1) Mutual-induction transducers, (2) self-induction transducers, and (3) permanent magnet transducers.

Variable-reluctance transducers: Variable-inductance transducers that use a nonmagnetized ferromagnetic medium to alter reluctance (magnetic resistance) of flux path (magnetic circuit).

Note: Some mutual-induction transducers and most self-induction transducers are of this type.

Note: Permanent magnet transducers are not considered variable-reluctance transducers.

Definitions: *Magnetic flux linkage* $\phi = Li$, Units: webers (Wb); Depends on magnetic flux density, number of turns in coil, and coil area (not wire area), i = current generating magnetic field; Units: amperes (A), L = inductance; Units: Wb/A or henry (H).

Induced voltage: Electromotive force (emf) $v = L(di/dt)$.

Reactance = reactive impedance of inductance = $X = Lj\omega$; Units: ohms (Ω).

Permeability $\mu = B/H$ = [magnetic flux density; Units: tesla or T; weber per square meter or Wb/m^2]/[magnetic field strength; Units: ampere.turns per meter or At/m] = L/l = inductance per unit length; Units: tesla.meter per ampere (T.m/A) or henry per meter (H/m) → measure of easiness magnetic field passage.

Relative permeability = permeability wrt free space = $\mu_r = \mu/\mu_o$; $\mu_o = 4\pi \times 10^{-7} = 1.257 \times 10^{-6}$ H/m

Reluctance = magnetic resistance of a magnetic circuit segment $\mathfrak{R} = l/(\mu A)$, l = length, A = area of X-section of magnetic circuit; Units: "turns per henry" (t/H) or "ampere. turns per weber" (At/Wb); inversely proportional to inductance → measured using an inductance bridge

Permeance = inverse of reluctance.

Mutual-induction transducers: (1) Moving a ferromagnetic material in flux path (e.g., LVDT, RVDT, mutual-induction proximity probe) and (2) moving one coil with respect to the other (e.g., resolver, synchro transformer, AC induction tachometer).

Differential transformer: (1) Linear-variable (LVDT), and (2) rotary-variable (RVDT). Primary coil with AC, two secondary coil segments in series opposition, a movable ferromagnetic core. It is a: variable-inductance transducer, variable-reluctance transducer, mutual-induction transducer, and passive transducer. Induced voltage is amplitude modulated. *Demodulation*: Multiply by carrier and low-pass filter.

Advantages of DT (LVDT): Noncontacting, low output impedance ~100 Ω, directional measurements (positive/negative), available in small sizes (e.g., 2 mm long with stroke of 1 mm), simple and robust construction (inexpensive and durable), fine resolutions.

Rate error: Displacement measurements are distorted by velocity; velocity measurements are distorted by acceleration; can be reduced by increasing carrier frequency (Suitable ratio >5).

Mutual-induction proximity sensor: Primary coil with AC, secondary coil, transverse ferromagnetic object. This is used for: measuring transverse displacements and also small displacements (nonlinear) and detecting the presence or absence of a pulse (e.g., limit switch).

Resolver: Mutual-induction transducer for angular displacements; rotor has primary winding and is energized by supply AC voltage; stator has two sets of windings placed 90° apart. Secondary voltages (sine $v_{f2} = (1/2)av_a^2 \sin\theta$ and cosine $v_{f1} = (1/2)av_a^2 \cos\theta$) are demodulated → gives direct and magnitude of rotation; nonlinear (geometric), an advantage in robotic applications.

Alternative design: Excite the two stator windings at 90° phase shift. *Output*: Rotor coil voltage $v_r = v_a \cos(\omega t - \theta)$.

Self-induction transducers: Single coil (self-induction); coil is activated by AC supply voltage → The current produces a magnetic flux that links with same coil. These transducers are based on the principle of self-induction: the level of flux linkage (self-inductance) is varied by a moving ferromagnetic object (target object) within the magnetic field.

Permanent magnet (PM) transducer: Uses a permanent magnet to generate magnetic field, which is used in sensing; e.g., DC tachometer; relative speed between magnetic field and electrical conductor → induced voltage; induced voltage ∝ speed of conductor crossing the magnetic field. Depending on the configuration, (1) rectilinear speeds and (2) angular speeds can be measured.

Characteristics: *Sensitivity ranges*: 0.5–3, 1–10, 11–25, or 25–50 V per 1000 rmp; *Armature resistance*: 100 Ω for low-voltage tachos: 2000 Ω for high-voltage tachos; *Size*: 2 cm in length; good linearity (0.1% is typical).

Model: Two-port; *inputs*: angular speed ω, torque T_i; *outputs*: armature voltage v_o, armature circuit current i_o; field windings are powered by DC voltage v_f (constant)

$$\begin{bmatrix} v_o \\ i_o \end{bmatrix} = \begin{bmatrix} K + \dfrac{(R_a + sL_a)(b + sJ)}{K} & -\dfrac{(R_a + sL_a)}{K} \\ -\dfrac{(b + sJ)}{K} & \dfrac{1}{K} \end{bmatrix} \begin{pmatrix} \omega_i \\ T_i \end{pmatrix}$$

Electrical time constant: $\tau_e = L_a/R_a$; mechanical time constant $\tau_m = J/b$.

Benefits of increasing *K*: Increases sensitivity and output signal level; reduces coupling, thereby the measurement directly depends on the measurand only; reduces dynamic effects (i.e., reduction of frequency dependence of system → increases useful frequency range and bandwidth or speed of response);

Benefits of decreasing time constants: Decreases dynamic terms (makes the transfer function static); increases operating bandwidth; makes the sensor faster.

PM AC tachometer: Rotor is a permanent magnet; two stator coils are used, and one of them is energized by an AC. When the rotor is stationary or moving quasi-statically the induced (output) voltage of the second stator coil will be constant. As the rotor moves → an additional voltage is induced in the secondary coil, which modulates the original induced voltage of carrier frequency. This modulating signal is proportional (∝) to the rotor speed; Induced output is equal to the amplitude-modulated signal and is proportional (∝) to rotor speed. It may be demodulated to measure transient speeds. The direction is obtained from phase angle of output. For low-frequency applications (~5 Hz), a supply with 60 Hz is adequate; The typical *sensitivity*: of a PM AC tachometer is 50–100 mV/rad/s.

AC induction tachometer: The stator windings are same as in PM AC tachometer; similar to induction motor; rotor windings are not powered (i.e., shorted); demodulate as in PM AC tachometer.

Advantages of AC tachometer (over DC tachometers): No slip rings or brushes; no output drift (at steady state). *Disadvantages*: demodulation is needed for transient measurements; output is nonlinear.

Eddy current transducer: Has an active coil and compensating coil; a high-frequency (radio frequency) voltage applied to active coil; two coils form two arms of inductance bridge. *Principle*: Conducting materials when subjected to a fluctuating magnetic field produce eddy currents; when a target object is moved closer to the sensor, the inductance of the active coil changes. *Bridge output*: AM signal. Demodulate → displacement signal.

Characteristics, advantages: *Target object*: small, thin layer of conducting material (e.g., glued aluminum foil); typical diameter ~2 mm (large 75 mm); target object, slightly > probe frontal area; output impedance ~1 kΩ (medium impedance); sensitivity ~5 V/mm; measurement range: 0.25–30 mm; suitable for transient measurements (up to ~100 kHz). *Applications*: displacement/proximity sensing, machine health monitoring, fault detection, metal detection, and braking.

Variable-capacitance transducers: $C = (kA)/x$, A = common (overlapping) area of the two plates, x = gap width between the two plates, and k = *dielectric constant* of medium (*permittivity*, $k = \varepsilon = \varepsilon_r \varepsilon_o$; ε_r = relative permittivity and ε_o = permittivity of a vacuum); *Sensing*: A change in any one of these parameter. *Examples*: transverse displacement (x), rotation (A), fluid level, or moisture content $k \to \delta C/C = -(\delta x/x) + (\delta A/A) + (\delta k/k)$; *Signal acquisition methods*: Capacitance bridge, charge amplifier, and inductance capacitance oscillator circuit.

Capacitive displacement sensor (with linearizing op-amp): $v_o = -((v_{ref}C_{ref})/K)x$.

Capacitive rotation sensor: $C = K\theta$.

Capacitive angular speed sensor: $d\theta/dt = i/Kv_{ref}$ (i = capacitor current).

Level measurement: Liquid level $y = (xC)/(b(k_l - k_a)) - h/(k_l/k_a - 1)$; x = plate gap, h = plate height, b = plate width, k_a = permittivity of air, and k_l = permittivity of the liquid.

Displacement sensor through k: Attach moving object (sensed object) to a solid dielectric element placed between the plates. k measures the displacement.

Piezoelectric sensors: $BaTiO_3$ (barium titanate), SiO_2 (quartz in crystalline form), lead zirconate titanate (PZT), etc. generate an electric charge when subjected to stress (strain).

Applications: Pressure and strain measuring devices, touch screens, accelerometers, and torque/force sensors.

Reverse effect: Piezoelectric materials deform when a voltage is applied.

Applications: Piezoelectric valves, microactuators, and MEMS.

Properties: High output impedance (varies with frequency; ~MΩ at 100 Hz).

Charge sensitivity: $S_q = \partial q/\partial F = (1/A)(\partial q/\partial p)$; *voltage sensitivity*: $S_v = (1/d)(\partial v/\partial p)$ (d = thickness)

$\delta q = C\delta v$, $\delta q = C\delta v \rightarrow S_q = kS_v$, and k = dielectric constant of crystal capacitor.

Piezoelectric accelerometer: Light weight, high frequency response (1 MHz), high output impedance → small voltages ~1 mV, sensitivity 10 pC/g (picocoulomb per gravity) or 5 mV/g (depends on the piezoelectric properties and how inertia force is applied).

Charge amplifier: Op-amp with R–C feedback → low output impedance, reduces charge leakage of piezoelectric sensor.

Gyroscopic sensors: Sensed speed from rate gyro $\Omega = (K\theta + B\dot{\theta})/J\omega$; θ = measured angular rotation at gimbal bearings, J = gyro disk polar moment of inertia, ω = spinning angular speed, K = torsional spring stiffness at gimbal, and B = rotational damping constant at gimbal.

Coriolis force devices: They use $2m\omega \times v$, similar (but not identical) to gyros.

Problems

8.1 In each of the following examples, indicate at least one (unknown) input, which should be measured and used for feedforward control to improve the accuracy of the control system.

(a) A servo system for positioning a mechanical load. The servomotor is a field-controlled DC motor, with position feedback using a potentiometer and velocity feedback using a tachometer.

(b) An electric heating system for a pipeline carrying a liquid. The exit temperature of the liquid is measured using a thermocouple and is used to adjust the power of the heater.

(c) A room heating system. The room temperature is measured and compared with the set point. If it is low, a valve of a steam radiator is opened; if it is high, the valve is shut.
(d) An assembly robot, which grips a delicate part to pick it up without damaging the part.
(e) A welding robot, which tracks the seam of a part to be welded.

8.2 A typical input variable is identified for each of the following examples of dynamic systems. Give at least one output variable for each system.

(a) Human body: neuroelectric pulses
(b) Company: information
(c) Power plant: fuel rate
(d) Automobile: steering wheel movement
(e) Robot: voltage to joint motor

8.3 Measuring devices (sensors–transducers) are useful in measuring the outputs of a process for feedback control.

(a) Give other situations in which signal measurement would be important.
(b) List at least five different sensors that are used in an automobile engine.

8.4 Give one situation where output measurement is needed and give another where input measurement is needed for proper control of the chosen system. In each case justify the need.

8.5 Giving examples, discuss situations in which the measurement of more than one type of kinematic variables using the same measuring device is (a) an advantage or (b) a disadvantage.

8.6 Indicate the main steps or guidelines for selecting sensors for a specific application.

8.7 Derive the expression for electrical loading nonlinearity error (percentage) in a rotatory potentiometer in terms of the angular displacement, maximum displacement (stroke), potentiometer element resistance, and load resistance. Plot the percentage error as a function of the fractional displacement, for the three cases $R_L/R_c = 0.1$, 1.0, and 10.0.

8.8 Determine the angular displacement (nondimensional) of a rotatory potentiometer at which the loading nonlinearity error is the largest. Use the result of Problem 8.7.

8.9 It is said that end resistors can help linearize a potentiometer output. The present problem examines this possibility. A potentiometer circuit with an element of resistance R_c, equal end resistors R_e, and load resistor R_L is shown in Figure P8.9.

(a) Derive the corresponding displacement–output voltage relation. Normalize the relationship with respect to the maximum displacement (stroke) and the maximum output voltage. Comment on the effect of the end resistors on the sensor output (or sensitivity) and variations in the supply voltage.
(b) Consider the case where the load resistance R_L is equal to the element resistance R_c. Determine the required end resistance value for the error in the potentiometer output at midstroke to become zero.
(c) Plot the normalized output of the potentiometer against the normalized displacement for this "best" value of the end resistor and compare it with the four cases: $R_e/R_c = 0$, 0.1, 1.0, and 10.0.

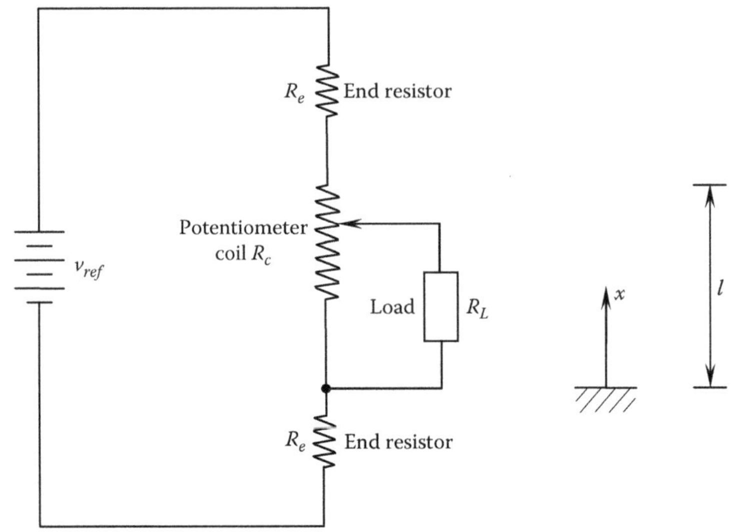

FIGURE P8.9
A potentiometer circuit with end resistors.

8.10 Derive an expression for the sensitivity (normalized) of a rotatory potentiometer as a function of displacement (normalized). Plot the corresponding curve in the nondimensional form for the three load values given by $R_L/R_c = 0.1$, 1.0, and 10.0. Where does the maximum sensitivity occur? Verify your observation using the analytical expression.

8.11 The range of a coil-type potentiometer is 10 cm. If the wire diameter is 0.1 mm, determine the resolution of the device.

8.12 A high-precision mobile robot uses a coil-type potentiometer attached to the drive wheel to record its travel during autonomous navigation. The required resolution for robot motion is 1 mm, and the diameter of the drive wheel of the robot is 20 cm. Examine the design considerations for a standard (single coil) rotatory potentiometer to be used in this application.

8.13 The data acquisition system connected at the output of a differential transformer (say, an LVDT) has a very high resistive load. Obtain an expression for the phase lead of the output signal (at the load) of the differential transformer, with reference to the supply to the primary windings of the transformer, in terms of the impedance of the primary windings only.

8.14 At the null position, the impedances of the two secondary winding segments of an LVDT were found to be equal in magnitude but slightly unequal in phase. Show that the quadrature error (null voltage) is approximately 90° out of phase with reference to the predominant component of the output signal under open-circuit conditions.

Hint: This may be proved either analytically or graphically by considering the difference between two rotating directed lines (phasors) that are separated by a very small angle.

8.15 A vibrating system has an effective mass M, an effective stiffness K, and an effective damping constant B in its primary mode of vibration at point A with respect to the coordinate y.

(a) Write expressions for the undamped natural frequency, the damped natural frequency, and the damping ratio for this first mode of vibration of the system.
(b) A displacement transducer is used to measure the fundamental undamped natural frequency and the damping ratio of the system by subjecting the system to an initial excitation and recording the displacement trace at a suitable location (point A along y in Figure P8.15) in the system. This trace provides the period of damped oscillations and the logarithmic decrement of the exponential decay. From these values the required parameters can be computed using well-known relations. However, it was found that the mass m of the moving part of the displacement sensor and the associated equivalent viscous damping constant b are not negligible. Using the model shown in Figure P8.15, derive expressions for the measured undamped natural frequency and damping ratio.
(c) Suppose that $M=10$ kg, $K=10$ N/m, and $B=2$ N/m/s. Consider an LVDT whose core weighs 5 g and has negligible damping and a potentiometer whose slider arm weighs 5 g and has an equivalent viscous damping constant of 0.05 N/m/s. Estimate the percentage error of the results for the undamped natural frequency and damping ratio, as measured using each of these two displacement sensors.

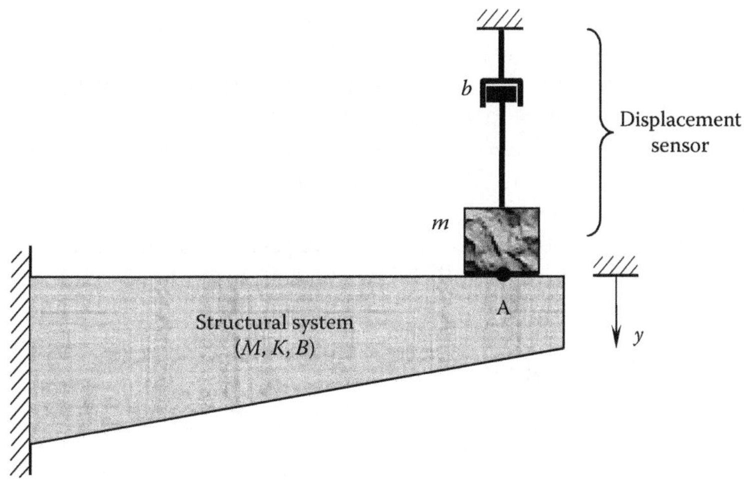

FIGURE P8.15
The use of a displacement sensor to measure the natural frequency and the damping ratio of a structure.

8.16 In many applications, rectilinear motion is produced from a rotary motion (say, of a motor) through a suitable transmission device, such as rack and pinion or lead screw and nut. In these cases, rectilinear motion can be determined by measuring the associated rotary motion, assuming that errors due to backlash, flexibility, and so forth, in the transmission device can be neglected. For the direct measurement of rectilinear motions, standard rectilinear displacement sensors such as the LVDT and the potentiometer may be employed. Displacements up to 25 cm may be measured by this approach. Within this range, accuracies as high as ±0.2% can be obtained. For measuring large displacements in the order of 3 m, cable extension displacement sensors, which have an angular displacement sensor as the basic sensing unit, may be used. In this method, an angular motion sensor with a spool rigidly coupled to the rotating part of the sensor (e.g., the encoder disk; see Chapter 11) and a cable that wraps around the spool is used. The other end of the cable is attached to the object whose rectilinear motion is to be sensed. The housing of the rotary sensor is firmly mounted on a stationary platform, such as the support structure of the system that is monitored, so that the cable can extend in the direction of motion. When the object moves, the cable extends, causing the spool to rotate. This angular motion is measured by the rotary sensor. With proper calibration, this device can give rectilinear measurements directly. One such displacement sensor uses a rotatory potentiometer and a light cable, which wraps around a spool that rotates with the wiper arm of the pot. A spring motor winds the cable back as the cable retracts. Using a sketch, describe the operation of this displacement sensor. Discuss the shortcomings of this device.

8.17 The factors that should be considered in selecting an LVDT for a particular application include linearity, sensitivity, response time, size and weight of core, size of the housing, primary excitation frequency, output impedance, phase change between the primary and secondary voltages, null voltage, stroke, and environmental effects (temperature compensation, magnetic shielding, etc.). Explain why and how each of these factors is an important consideration.

8.18 The signal-conditioning system for an LVDT has the following components: power supply, oscillator, synchronous demodulator, filter, and voltage amplifier. Using a schematic block diagram, show how these components are connected to the LVDT. Describe the purpose of each component. A high-performance LVDT has a linearity rating of 0.01% within its output range of 0.1–1.0 V AC. The response time of the LVDT is known to be 10 ms. What should be a suitable frequency for the primary excitation (carrier AC)?

8.19 List merits and shortcomings of a potentiometer (pot) as a displacement sensing device, in comparison with an LVDT. Give several ways to improve the measurement linearity of a potentiometer.

Suppose that a resistance R_l is added to the conventional potentiometer circuit as shown in Figure P8.19. With $R_l = R_L$ show that

$$\frac{v_o}{v_{ref}} = \frac{(R_L/R_c + 1 - x/x_{\max})x/x_{\max}}{[R_L/R_c + 2x/x_{\max} - 2(x/x_{\max})^2]}$$

where

R_c is the potentiometer coil resistance (total)
R_L is the load resistance
v_{ref} is the supply voltage to the coil
v_o is the output voltage
x is the slider displacement
$x_{\max}$ is the slider stroke (maximum displacement)

Explain why R_l produces a linearizing effect.

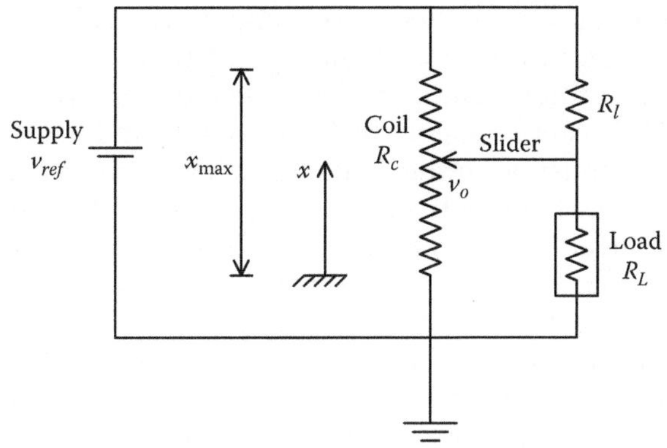

FIGURE P8.19
Potentiometer circuit with a linearizing resistor.

8.20 Consider again the potentiometer circuit shown in Figure P8.19. Show that the output equation is

$$\frac{v_o}{v_{ref}} = \frac{[1/(1-\alpha)+(1/\beta)]}{[(1/\alpha)+(1/(1-\alpha))+(1+\beta)+(1+\gamma)]}$$

where

$$\frac{x}{x_{max}} = \alpha, \quad \frac{R_l}{R_c} = \beta, \quad \frac{R_L}{R_c} = \gamma$$

Determine the sensor output for the midstroke ($x/x_{max} = \alpha = 0.5$) when $(R_L/R_c) = \gamma = 1$. Show that $R_l = R_c$ gives zero error. Compare the performance with no R_l, small R_l, and large R_l (with respect to R_c).

8.21 Suppose that a sinusoidal carrier frequency is applied to the primary coil of an LVDT. Sketch the shape of the output voltage of the LVDT when the core is stationary at (a) null position, (b) left of the null position, and (c) right of the null position.

8.22 For directional sensing using an LVDT, it is necessary to determine the phase angle of the induced signal. In other words, *phase-sensitive demodulation* would be needed.

(a) First, consider a linear core displacement starting from a positive value, moving to zero, and then returning to the same position in an equal time period. Sketch the output of the LVDT for this triangular core displacement.

(b) Next, sketch the output if the core continued to move to the negative side at the same speed.

By comparing the two outputs, show that phase-sensitive demodulation would be needed to distinguish between the two cases of displacement.

8.23 The "synchro" is somewhat similar in operation to a resolver. The main differences are that the synchro employs two identical rotor–stator pairs and each stator has three sets of windings, which are placed 120° apart around the rotor shaft. A schematic diagram for this arrangement is shown in Figure P8.23. Both rotors have single-phase windings. One of the rotors is energized with an AC supply voltage v_{ref}. This induces voltages in the three winding segments of the corresponding stator. These voltages have different amplitudes, which depend on the angular position of the rotor. (*Note*: The resultant magnetic field from the induced currents in these three stator winding sets must be in the same direction as the rotor magnetic field.) This drive rotor–stator pair is known as the "transmitter." The other rotor–stator pair is known as the "receiver" or the "control transformer." The windings of the transmitter stator are connected correspondingly to the windings of the receiving stator, as shown in Figure P8.23. Accordingly, the resultant magnetic field of the receiver stator must be in the same direction as the resultant magnetic field of the transmitter stator (and of course the transmitter rotor). This resultant magnetic field in the receiver stator induces a voltage v_o in the rotor of the receiver. Suppose that the angle between the transmitter rotor and one set of windings in its stator (the same reference winding set as what is used to measure the angle of the transmitter rotor) is denoted by θ_t. The receiver rotor angle is denoted by θ_r.

(a) Write equations to describe the operation of the synchro transformer as a position servo system. Indicate the necessary signal-conditioning procedures.
(b) List some applications of the device. Also, indicate some advantages and disadvantages of a synchro transformer.

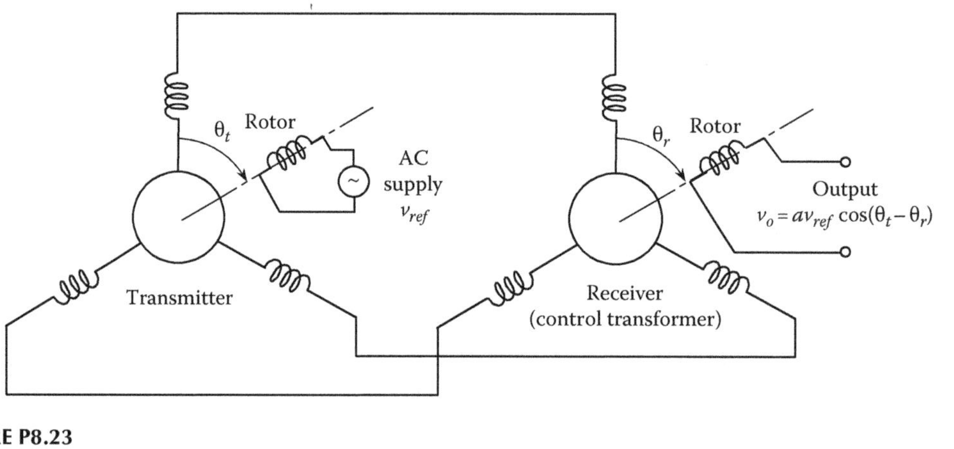

FIGURE P8.23
Schematic diagram of a synchro transformer.

8.24 Joint angles and angular speeds are the two basic measurements used in the direct (low level) control of robotic manipulators. One type of robot arm uses resolvers to measure angles and differentiates these signals (digitally) to obtain angular speeds. A gear system is used to step up the measurement (typical gear ratio, 1:8). Since the gear wheels are ferromagnetic, an alternative measuring device would be a self-induction or mutual-induction proximity sensor located at a gear wheel. This arrangement, known as a pulse tachometer, generates a pulse (or near sine) signal, which can be used to determine both angular displacement and angular speed. Discuss the advantages and disadvantages of these two arrangements (resolver and pulse tachometer) in this particular application.

8.25 Why is motion sensing important in trajectory, following control of robotic manipulators? Identify five types of motion sensors that may be used in robotic manipulators.

8.26 Compare and contrast the principles of operation of DC tachometer and AC tachometer (both PM and induction types). What are the advantages and disadvantages of these two types of tachometers?

8.27 Describe three different types of proximity sensors. In some applications, it may be required to sense just two states (e.g., presence or absence, go or no-go). Proximity sensors can be used in such applications, and in that context, they are termed proximity switches (or limit switches). For example, consider a parts handling application in automated manufacturing in which a robot end effector grips a part and picks it up to move it from a conveyor to a machine tool. We can identify four separate steps in the gripping process:

(a) Make sure that the part is at the expected location on the conveyor.
(b) Make sure that the gripper is open.
(c) Make sure that the end effector has moved to the correct location so that the part is in between the gripper fingers.
(d) Make sure that the part did not slip when the gripper was closed.

Explain how proximity switches may be used for sensing in each of these four tasks.

Note: A similar use of limit switches is found in lumber mills, where tree logs are cut (bucked) into smaller logs, bark removed (debarked), cut into square or rectangular logs using a "chip-n-saw" operation, and sawed into smaller dimensions (e.g., two by four cross sections) for marketing.

8.28 Discuss the relationships among displacement sensing, distance sensing, position sensing, and proximity sensing. Explain why the following characteristics are important in using some types of motion sensors:

(a) Material of the moving (or target) object
(b) Shape of the moving object
(c) Size (including mass) of the moving object
(d) Distance (large or small) of the target object
(e) Nature of motion (transient or not, what speed, etc.) of the moving object
(f) Environmental conditions (humidity, temperature, magnetic fields, dirt, lighting conditions, shock, vibration, etc.)

8.29 In some industrial processes, it is necessary to sense the condition of a system at one location and, depending on that condition, activate an operation at a location far from that location. For example, in a manufacturing environment, when the count of the finished parts exceeds some value, as sensed in the storage area, a milling machine may have to be shut down or started up. A proximity switch may be used for sensing and a networked (e.g., Ethernet based) control system for process control. Since activation of the remote process usually requires a current that is larger than the rated load of a proximity switch, it may be necessary to use a relay circuit, which is operated by the proximity switch. One such arrangement is shown in Figure P8.29. The relay circuit can be used to operate a device such as a valve, a motor, a pump, or a heavy-duty switch. Discuss an application of the arrangement shown in Figure P8.29 in the food-packaging industry. A mutual-induction proximity sensor with the following ratings is used in this application:

Sensor diameter = 1 cm

Sensing distance (proximity) = 1 mm

Supply to primary windings = 110 AC at 60 Hz

Load current rating (in secondary) = 200 mA

Discuss the limitations of this proximity sensor.

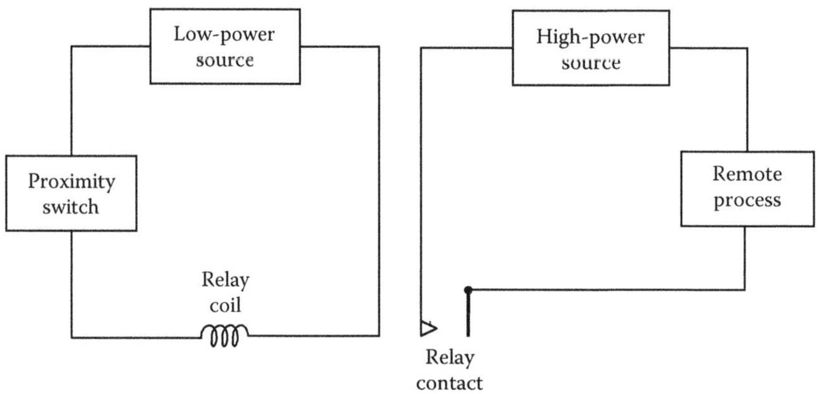

FIGURE P8.29
Proximity switch-operated relay circuit.

8.30 Compression molding is used in making parts of complex shapes and varying sizes. Typically, the mold consists of two platens, the bottom platen fixed to the press table and the top platen operated by a hydraulic press. Metal or plastic sheets—for example, for the automotive industry—can be compression molded in this manner. The main requirement in controlling the press is to position the top platen accurately with respect to the bottom platen (say, with a 0.001 in. or 0.025 mm tolerance), and it has to be done quickly (say, in a few seconds). How many degrees of freedom have to be sensed (how many position sensors are needed) in controlling the mold? Suggest typical displacement measurements that would be made in this application and the types of sensors that could be employed. Indicate sources of error that cannot be perfectly compensated for in this application.

8.31 Seam tracking in robotic arc welding needs precise position control under dynamic conditions. The welding seam has to be accurately followed (tracked) by the welding torch. Typically, the position error should not exceed 0.2 mm. A proximity sensor could be used for sensing the gap between the welding torch and the welded part. The sensor has to be mounted on the robot end effector in such a way that it tracks the seam at some distance (typically 1 in. or 2.5 cm) ahead of the welding torch. Explain why this is important. If the speed of welding is not constant and the distance between the torch and the proximity sensor is fixed, what kind of compensation would be necessary in controlling the end effector position? Sensor sensitivity of several volts per millimeter is required in this application of position control. What type of proximity sensor would you recommend?

8.32 An angular motion sensor, which operates somewhat like a conventional resolver, has been developed at the Wright State University. The rotor of this resolver is a permanent magnet. A 2 cm diameter Alnico-2 disk magnet, diametrically magnetized as a two-pole rotor, has been used. Instead of the two sets of stationary windings placed at 90° in a conventional resolver, two Hall-effect sensors (see Chapter 11) placed at 90° around the PM rotor are used for detecting quadrature signals. *Note*: Hall-effect sensors can detect moving magnetic sources. Describe the operation of this modified resolver and explain how this device could be used to measure angular motions continuously. Compare this device with a conventional resolver, giving advantages and disadvantages.

8.33 Obtain expressions for the sensitivity of a variable-capacitance lateral displacement sensor and a rotary angle sensor. Discuss the implications of these results.

8.34 Consider the capacitor shown in Figure P8.34 where the two end plates are fixed and the middle plate is attached to a moving object whose displacement (δx) needs to be measured. Suppose that the capacitor plates are connected to the bridge circuit of Figure 4.18a, forming the reactances Z_3 and Z_4. Initially, the middle plate is placed at an equal separation of x from either end plate. Obtain a relationship for the bridge output v_o and the plate movement δx. This relationship is linear.

Note: This arrangement is a differential (push–pull) displacement sensor.

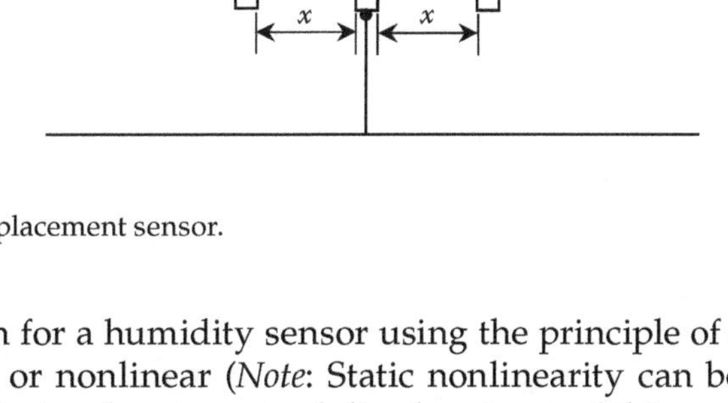

FIGURE P8.34
A capacitive push–pull displacement sensor.

8.35 Propose a design for a humidity sensor using the principle of capacitance. Will this device be linear or nonlinear (*Note*: Static nonlinearity can be removed by proper calibration)? Indicate advantages and disadvantages of this sensor.

8.36 Discuss factors that decide the lower frequency and upper frequency limits of the output in the following sensors:

(a) Potentiometer
(b) LVDT
(c) Resolver
(d) Eddy current proximity sensor
(e) DC tachometer
(f) Piezoelectric transducer

8.37 An active suspension system is proposed for a high-speed ground transit vehicle in order to achieve significant improvements in ride quality. The system senses jerk (rate of change of acceleration) that results from the road disturbances and adjusts system parameters accordingly.

(a) Draw a suitable schematic diagram for the proposed control system and describe appropriate measuring devices.
(b) Suggest a way to specify the desired ride quality for a given type of vehicle. (Would you specify one value of jerk, a jerk range, or a jerk curve with respect to time or frequency?)
(c) Discuss the drawbacks and limitations of the proposed control system with respect to such factors as reliability, cost, feasibility, and accuracy.

8.38 A design objective in many control system applications is to achieve small time constants. An exception is the time constant requirements for a piezoelectric sensor. Explain why a large time constant, in the order of 1.0 s, is desirable for a piezoelectric sensor in combination with its signal-conditioning system.

An equivalent circuit for a piezoelectric accelerometer, which uses a quartz crystal as the sensing element, is shown in Figure P8.38. The generated charge is denoted by q, and the output voltage at the end of the accelerometer cable is v_o. The piezoelectric sensor capacitance is modeled by C_p, and the overall capacitance experienced at the sensor output, whose primary contribution is due to cable capacitance, is denoted by C_c. The resistance of the electric insulation in the accelerometer is denoted by R. Write a differential equation relating v_o to q. What is the corresponding transfer function? Using this result, show that the accuracy of the accelerometer improves when the sensor time constant is large and when the frequency of the measured acceleration is high. For a quartz crystal sensor with $R = 1 \times 10^{11}\ \Omega$ and $C_p = 300$ pF and a circuit with $C_c = 700$ pF, compute the time constant.

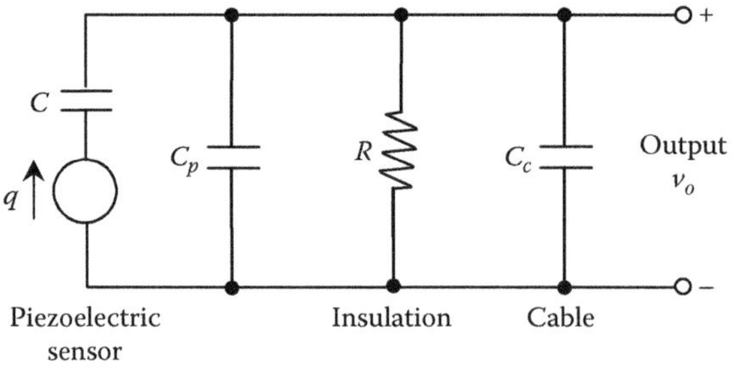

FIGURE P8.38
Equivalent circuit for a quartz crystal (piezoelectric) accelerometer.

8.39 Applications of accelerometers are found in the following areas:

(a) Transit vehicles (automobiles—microsensors for airbag sensing, in particular—aircraft, ships, etc.)
(b) Power cable monitoring
(c) Control of machine tools

(d) Monitoring of buildings and other civil engineering structures
(e) Shock and vibration testing
(f) Position and velocity sensing

Describe one direct use of acceleration measurement in each application area.

8.40 (a) A standard accelerometer whose mass is 100 g is mounted on a test object that has an equivalent mass of 3 kg. Estimate the accuracy in the first natural frequency of the object when measured using this arrangement, considering mechanical loading due to the accelerometer mass alone. If a miniature accelerometer that weighs 0.5 g is used instead, what is the resulting accuracy?

(b) A strain-gauge accelerometer uses a semiconductor strain gauge mounted at the root of a cantilever element, with the seismic mass mounted at the free end of the cantilever. Suppose that the cantilever element has a square cross section with dimensions 1.5×1.5 mm². The equivalent length of the cantilever element is 25 mm, and the equivalent seismic mass is 0.2 g. If the cantilever is made of an aluminum alloy with Young's modulus $E = 69 \times 10^9$ N/m², estimate the useful frequency range of the accelerometer in Hertz.

Hint: When force F is applied to the free end of a cantilever, the deflection y at that location may be approximated by the formula $y = Fl^3/3EI$, where l is the cantilever length, I is the second moment area of the cantilever cross section about the bending axis $= bh^3/12$, b is the cross-section width, and h is the cross-section height.

8.41 Applications of piezoelectric sensors are numerous: push-button devices and switches, airbag MEMS sensors in vehicles, pressure and force sensing, robotic tactile sensing, accelerometers, glide testing of computer HDD heads, excitation sensing in dynamic testing, respiration sensing in medical diagnostics, wearable ambulatory monitoring units that include an accelerometer and a gyroscope for human mobility sensing, and graphics input devices for computers.

Discuss the advantages and disadvantages of piezoelectric sensors. What is cross sensitivity of a sensor? Indicate how the anisotropy of piezoelectric crystals (i.e., charge sensitivity being quite large along one particular crystal axis) is useful in reducing cross-sensitivity problems in a piezoelectric sensor.

8.42 As a result of advances in microelectronics, piezoelectric sensors (such as accelerometers and impedance heads) are now available in miniature form with built-in charge amplifiers in a single integral package. When such units are employed, additional signal conditioning is usually not necessary. An external power supply unit is needed, however, to provide power for the amplifier circuitry. Discuss the advantages and disadvantages of a piezoelectric sensor with built-in microelectronics for signal conditioning.

A piezoelectric accelerometer is connected to a charge amplifier. An equivalent circuit for this arrangement is shown in Figure P8.42.

(a) Obtain a differential equation for the output v_o of the charge amplifier, with acceleration a as the input, in terms of the following parameters: S_a is the charge sensitivity of the accelerometer (charge/acceleration), R_f is the feedback resistance of the charge amplifier, and τ_c is the time constant of the system (charge amplifier).

(b) If an acceleration pulse of magnitude a_o and duration T is applied to the accelerometer, sketch the time response of the amplifier output v_o. Show how this response varies with τ_c. Using this result, show that the larger the τ_c the more accurate the measurement.

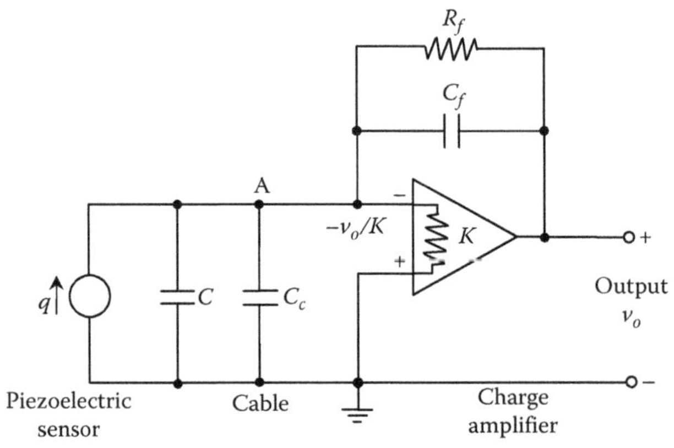

FIGURE P8.42
Equivalent circuit of piezoelectric sensor and charge amplifier.

8.43 Give typical values for the output impedance and the time constant of the following measuring devices: (a) potentiometer, (b) differential transformer, (c) resolver, and (d) piezoelectric accelerometer.

A resistance temperature detector (see Chapter 10) has an output impedance of 500 Ω. If the loading error has to be maintained near 5%, estimate a suitable value for the load impedance.

8.44 A signature verification pen has been developed by the IBM Corporation. The purpose of the pen is to authenticate the signature, by detecting whether the user is trying to forge someone else's signature. The instrumented pen has analog sensors. Sensor signals are conditioned using microcircuitry built into the pen and sampled into a microcontroller through a wireless communication link, at the rate of 80 samples/s. Typically, about 1000 data samples are collected per signature. Before the pen's use, authentic signatures are collected off-line and stored in a reference database. When a signature and the corresponding identification code are supplied to the computer for verification, a program in the processor retrieves the authentic signature from the database, by referring to the identification code, and then compares the two sets of data for authenticity. This process takes about 3 s. Discuss the types of sensors that could be used in the pen. Estimate the total time required for signal verification. What are the advantages and disadvantages of this method in comparison with a procedure where the user keys in an identification code alone or provides the signature without an identification code?

8.45 A simple rate gyro, which may be used to measure angular speeds, is shown in Figure P8.45. The angular speed of spin is ω and is kept constant at a known value. The angle of rotation of the gyro about the gimbal axis (or the angle of twist of the torsional spring) is θ and is measured using a displacement sensor. The angular

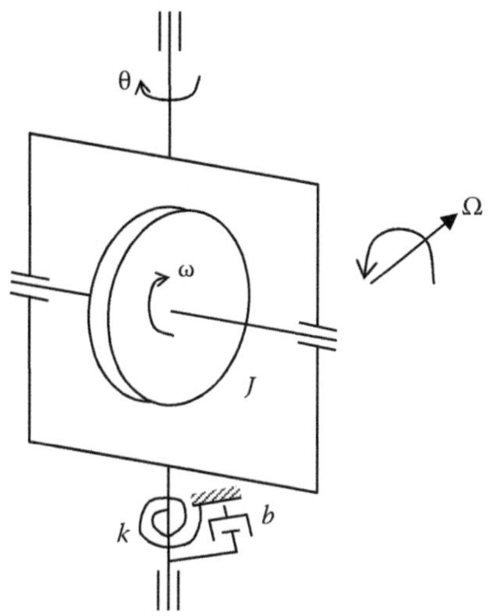

FIGURE P8.45
A rate gyro speed sensor.

speed of the gyro about the axis that is orthogonal to both gimbal axis and spin axis is Ω. This is the angular speed of the supporting structure (vehicle), which needs to be measured. Obtain a relationship between Ω and θ in terms of such parameters as the following: *J* is the moment of inertia of the spinning wheel, *k* is the torsional stiffness of the spring restraint at the gimbal bearings, and *b* is the damping constant of rotational motion about the gimbal axis and the spinning speed. How would you improve the sensitivity of this device?

9

Effort Sensors

Chapter Objectives

- Introduction to Effort Sensors
- Force Sensors for Motion Measurement
- Force Sensor Location
- Strain Gauges
- Automatic (Self-) Compensation for Temperature
- Torque Sensors
- Design Considerations
- Deflection Torque Sensors
- Reaction Torque Sensors
- Motor Current Torque Sensors
- Tactile Sensing
- Impedance Sensing

9.1 Sensors

The previous chapter studied analog motion sensors. This chapter covers a different class of sensors, called effort sensors, which include force, torque, tactile, and impedance. Other types of sensors including optical sensors, ultrasonic sensors, magnetostrictive sensors, thermofluid sensors, and sensors that are useful in monitoring the quality of water (e.g., pH value, dissolved oxygen, and oxidation–reduction potential) are discussed in Chapter 10. Digital transducers including shaft encoders, image sensors (digital camera), binary transducers, and Hall-effect sensors are studied in Chapter 11. Microelectromechanical system (MEMS) sensors, and other advanced topics in sensing such as networked sensing, and sensor fusion are addressed in Chapter 12.

9.1.1 Effort Sensors

A mechanical system *responds* (generates an *output*) to an *excitation* (*input*) made through an "effort" such as a *force* or a *torque* applied to it. In addition to point forces and torques, an effort may be applied as a distributed force or torque of "touch" such as *tactile force*. In this sense, the effort is what *drives* the system and is an important consideration in application that involves a mechanical dynamic system. Furthermore, many applications exist whose *performance specifications* are made in terms of forces and torques. Examples include machine-tool operations, such as grinding, cutting, forging, extrusion, and rolling; manipulator tasks, such as parts handling, assembly, engraving, and robotic fine manipulation; devices of haptic (force feedback) teleoperation; and actuation tasks, such as locomotion. Also, "mechanical impedance" is related to effort since it is defined as the ratio (generalized force)/(generalized motion) in the frequency domain (where the independent variable is frequency). Hence, mechanical impedance sensing may also be treated under effort sensing.

The forces and torques that are present in a dynamic system are generally functions of time. Performance monitoring and evaluation, failure detection and diagnosis, testing, experimental modeling (model identification), and control of mechanical dynamic systems can depend heavily on the accurate measurement of associated forces and torques. One example where the sensing of forces (and torques) can be very useful is that of a drilling robot. The drill bit is held at the end effector by the gripper of the robot, and the workpiece is rigidly fixed to a support structure by clamps. Although a displacement sensor such as a potentiometer or a differential transformer can be used to measure the drill motion in the axial direction, this alone does not determine the drill performance. Depending on the material properties of the workpiece (e.g., hardness) and the nature of the drill bit (e.g., degree of wear), a small misalignment or slight deviation in the feed (axial movement) or speed (rotational speed of the drill) can create large normal (axial) and lateral forces and resistance torques. This can create problems such as excessive vibrations and chattering, uneven drilling, excessive tool wear, and the resulting poor product quality. Eventually, this may lead to a major mechanical failure or hazard. Sensing the axial force or motor torque, for example, and using that information to adjust process variables (speed, feed rate, etc.), or even to provide warning signals and eventually stop the process, can significantly improve the system performance. Another example in which force sensing is useful is in nonlinear feedback control (or feedback linearization technique) of mechanical systems such as robotic manipulators. There, the system nonlinearities, as represented through measured forces, are fed back into the system to remove the nonlinear behavior of the system.

Since both force and torque are effort variables, the term *force* (more properly, generalized force) may be used to represent both these variables. This generalization is adopted here except when discrimination between the two types of effort is necessary—for example, when specifically discussing torque sensors and specific applications of them.

9.1.2 Force Sensors for Motion Measurement

At least in principle, any force sensor can be used as an acceleration sensor, velocity sensor, or displacement sensor, depending on the specific *front-end auxiliary element* that is used. Specifically, we can use

1. An inertia element (to convert acceleration into force, in proportion)
2. A damping element (to convert velocity into force, in proportion)
3. A spring element (to convert displacement into force, in proportion)

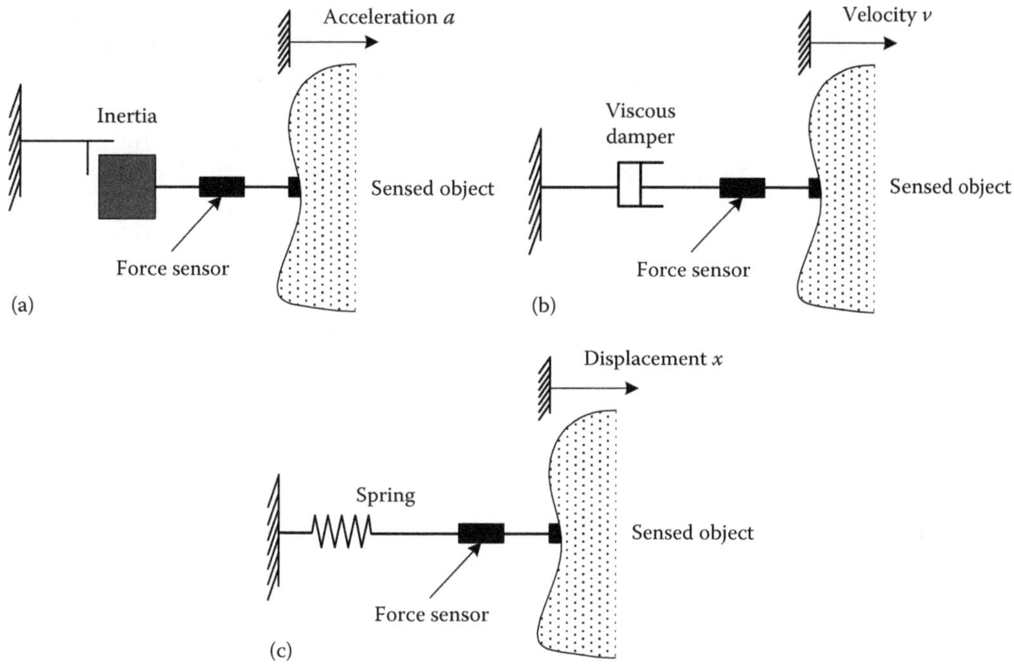

FIGURE 9.1
The use of an auxiliary front-end element and force sensing to measure: (a) acceleration, (b) velocity, and (c) displacement.

Then, as schematically shown in Figure 9.1, we are able to use force sensing to measure acceleration, velocity, or displacement.

Note: The practical implementation of an ideal velocity-to-force transducer is quite difficult, primarily due to nonlinearities in damping elements (the assumption of a linear viscous damper is not very realistic in many practical implementations).

9.1.3 Force Sensor Location

From the point of view of accuracy, the force sensor has to be located exactly at the place where the force information is needed. Sometimes, however, it may be difficult (or even impossible) to place the sensor at the required location (due to inaccessibility, motion of the monitored location, hazard, etc.). Then, one may place the sensor at a different location and then "estimate" the force at the required location using the measured data.

There can be other issues related to sensor location. From the point of view of stability of a feedback control system, for example, it is best to locate the sensors at the drive location even when the load whose motion needs to be measured (e.g., for feedback control) is farther away from the driving point (e.g., motor location). Specifically in force feedback control, the location of the force sensor with respect to the location of actuation can have a crucial effect on the system performance, particularly stability. For example, in robotic manipulator applications, it has been experienced that with some locations and configurations of a force-sensing wrist at the robot end effector, dynamic instabilities were present in the manipulator response for some (large) values of control gains in the force feedback loop. These instabilities were found to be limit-cycle-type motions in most cases.

Generally, it is known that when the force sensors are more remotely located with respect to the drive actuators of a mechanical system, the system is more likely to exhibit instabilities under force feedback control. Hence, it is desirable to make force measurements very close to the actuator locations when force feedback is used.

Consider a mechanical processing task. The tool actuator generates the processing force, which is applied to the workpiece. The force transmitted to the workpiece by the tool is measured by a force sensor and is used by a feedback controller to generate the correct actuator force. The machine tool is a dynamic system, which consists of a tool subsystem (dynamic) and a tool actuator (dynamic). The workpiece is also a dynamic system.

Relative location of the tool actuator with respect to the force sensor (at the tool–workpiece interface) can affect the stability of the feedback control system. In general, the closer the actuator to the sensor, the more stable the feedback control system. Two scenarios are shown in Figure 9.2, which can be used to study the stability of the overall system. In both cases, the processing force at the interface between the tool and the workpiece is measured using a force sensor and is used by the feedback controller to generate the actuator drive signal. In Figure 9.2a, the tool actuator, which generates the drive signal of the actuator, is located next to the force sensor. In Figure 9.2b, the tool actuator is separated from the force sensor by a dynamic system of the processing machine. It is known that the arrangement (b) is less stable than the arrangement (a). The reason is simple. Arrangement (b) introduces more dynamic delay into the feedback control loop. It is well known that time delay has a destabilizing effect on a feedback control system, particularly at high control gains.

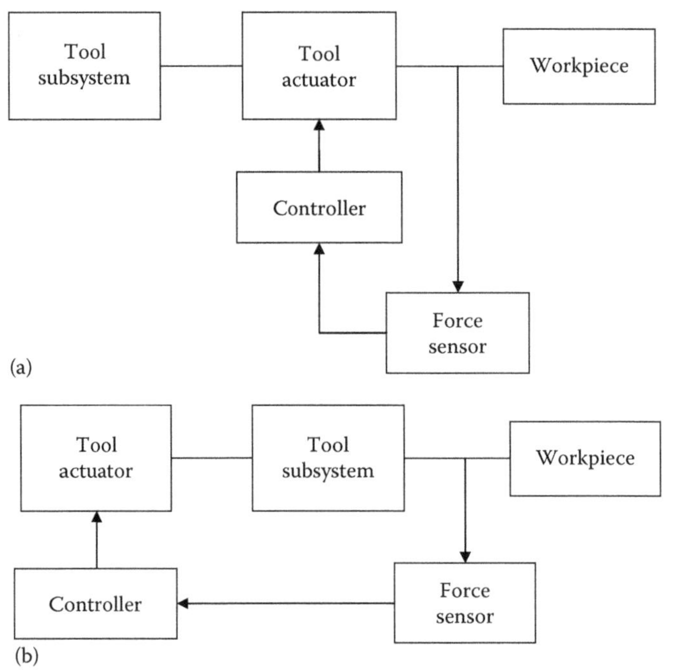

FIGURE 9.2
(a) Force sensor located next to the actuator and (b) force sensor separated from the actuator by a dynamic subsystem.

9.2 Strain Gauges

Many types of force and torque sensors employ strain-gauge measurements. Although strain gauges measure strain, the measurements can be directly related to stress and force. Hence, it is appropriate to discuss strain gauges under force and torque sensors.

Note: Strain gauges may be used in a somewhat indirect manner (using auxiliary front-end elements) to measure other types of variables, including displacement, acceleration, pressure, and temperature. In those situations, the front-end element physically coverts the quantity that needs to be measured (i.e., the *measurand*) into a strain, which is then measured by the strain gauge.

Two common types of resistance strain gauges are discussed next. Specific types of force and torque sensors are dealt in the subsequent sections.

9.2.1 Equations for Strain-Gauge Measurement

The change in electrical resistance of a material when mechanically deformed is the property that is used in resistance-type strain gauges. The resistance R of a conductor of length ℓ and area of cross section A is given by

$$R = \rho \frac{\ell}{A} \tag{9.1}$$

where ρ is the resistivity of the material. Taking the logarithm of Equation 9.1, we have $\log R = \log \rho + \log(\ell/A)$. Now, taking the differential of each term, we obtain

$$\frac{\delta R}{R} = \frac{\delta \rho}{\rho} + \frac{\delta(\ell/A)}{\ell/A} \tag{9.2}$$

The first term on the RHS of this equation is the fractional change in resistivity, and the second term represents fractional deformation. It follows that the change in resistance in the material comes from the change in shape as well as from the change in resistivity (a material property) of the material. For linear deformations, the two terms on the RHS of Equation 9.2 are linear functions of strain ε; the proportionality constant of the second term, in particular, depends on Poisson's ratio of the material. Hence, the following relationship can be written for a strain-gauge element:

$$\frac{\delta R}{R} = S_s \varepsilon \tag{9.3}$$

The constant S_s is known as the "gauge factor" or "sensitivity" of the strain-gauge element. The numerical value of this parameter ranges from 2 to 6 for most metallic strain-gauge elements and from 40 to 200 for semiconductor (SC) strain gauges. These two types of strain gauges are discussed later. The change in resistance of a strain-gauge element, which determines the associated strain (Equation 9.3), is measured using a suitable electrical circuit (typically, a bridge circuit).

Indirect strain-gauge sensors: Many variables—including displacement, acceleration, pressure, temperature, liquid level, stress, force, and torque—can be determined using strain measurements. Some variables (e.g., stress, force, and torque) can be determined by measuring the strain of the dynamic object itself at suitable locations. In other situations,

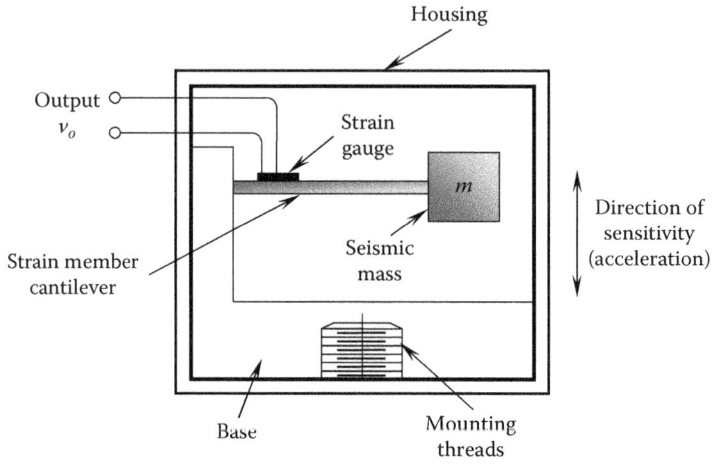

FIGURE 9.3
A strain-gauge accelerometer.

an auxiliary front-end device may be required to convert the measurand into a proportional strain. For instance, pressure or displacement may be measured by converting them to a measurable strain using a diaphragm, bellows, or bending element. Acceleration may be measured by first converting it into an inertia force of a suitable mass (seismic mass) element, then subjecting a cantilever (strain member) to that inertia force and, finally, measuring the strain at a high-sensitivity location of the cantilever element (see Figure 9.3). Temperature may be measured by measuring the thermal expansion or deformation in a bimetallic element.

Thermistors are temperature sensors made of SC material whose resistance changes with temperature. Resistance temperature detectors operate by the same principle, except that they are made of metals, not of SC material. These temperature sensors (see Chapter 10), and the piezoelectric sensors discussed in Chapter 8, should not be confused with strain gauges. Resistance strain gauges are based on resistance change as a result of strain or the *piezoresistive* property of materials.

Early strain gauges were fine metal filaments. Modern strain gauges are manufactured primarily as metallic foil (e.g., using the copper–nickel alloy known as constantan) or SC elements (e.g., silicon with trace impurity boron). They are manufactured by first forming a thin film (foil) of metal or a single crystal of SC material and then cutting it into a suitable grid pattern, either mechanically or by using photoetching (optochemical) techniques. This process is much more economical and is more precise than making strain gauges with metal filaments. The strain-gauge element is formed on a backing film of electrically insulated material (e.g., polyamide plastic). This element is cemented or bonded using epoxy onto the member whose strain is to be measured. Alternatively, a thin film of insulating ceramic substrate is melted onto the measurement surface, on which the strain gauge is mounted directly. The direction of sensitivity is the major direction of elongation of the strain-gauge element (Figure 9.4a). To measure strains in more than one direction, multiple strain gauges (e.g., various rosette configurations) are available as single units. These units have more than one direction of sensitivity. The principal strains in a given plane (the surface of the object on which the strain gauge is mounted) can be determined by using these multiple strain-gauge units.

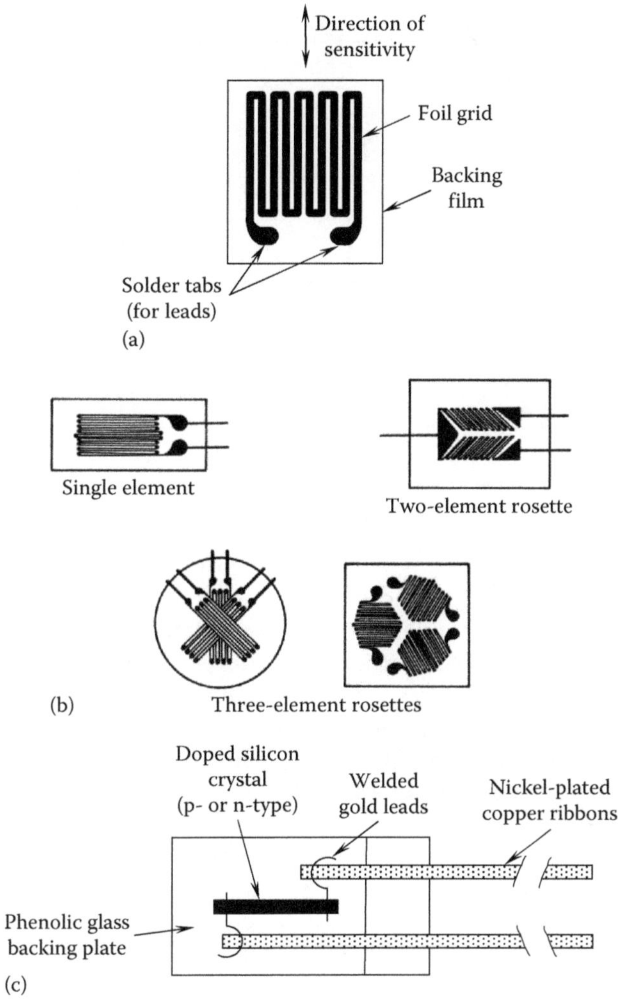

FIGURE 9.4
(a) Strain-gauge nomenclature, (b) typical foil-type strain gauges, and (c) a semiconductor strain gauge.

Typical foil-type gauges are shown in Figure 9.4b, and an SC strain gauge is shown in Figure 9.4c.

A direct way to obtain strain-gauge measurement is to apply a constant DC voltage across a series-connected strain-gauge element (of resistance R) and a suitable (complementary) resistor R_c and to measure the output voltage v_o across the strain gauge under open-circuit conditions (i.e., using a device of high input impedance). This arrangement is known as a "potentiometer circuit" or "ballast circuit" and has several weaknesses. Any ambient temperature variation directly introduces some error because of associated change in the strain-gauge resistance and the resistance of the connecting circuitry. Also, measurement accuracy will be affected by possible variations in the supply voltage v_{ref}. Furthermore, the electrical loading error will be significant unless the load impedance is very high. Perhaps the most serious disadvantage of this circuit is that the change in signal due to strain is usually a small fraction of the total signal level in the circuit output. This problem can be reduced to some extent by decreasing v_o, which may be accomplished by increasing

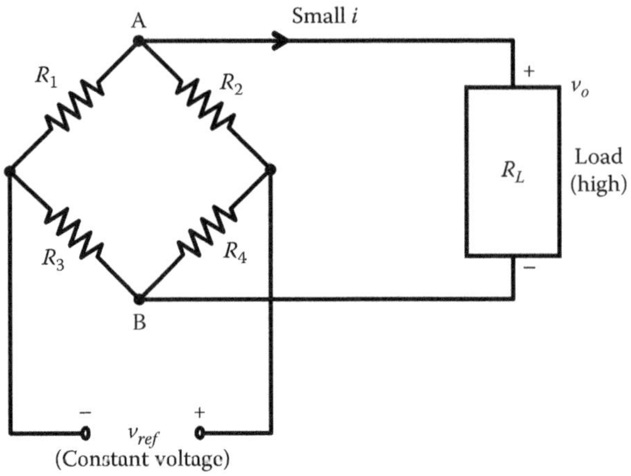

FIGURE 9.5
Wheatstone bridge circuit.

the resistance R_c. This, however, reduces the sensitivity of the circuit. Any changes in the strain-gauge resistance due to ambient changes will directly affect the strain-gauge reading unless R and R_c have identical coefficients with respect to ambient changes.

A more favorable circuit for use in strain-gauge measurements is the Wheatstone bridge, as discussed in Chapter 4. One or more of the four resistors R_1, R_2, R_3, and R_4 in the bridge (Figure 9.5) may represent strain gauges. The output relationship for the Wheatstone bridge circuit is given by (see Chapter 4)

$$v_o = \frac{R_1 v_{ref}}{(R_1 + R_2)} - \frac{R_3 v_{ref}}{(R_3 + R_4)} = \frac{(R_1 R_4 - R_2 R_3)}{(R_1 + R_2)(R_3 + R_4)} v_{ref} \tag{9.4}$$

When this output voltage is zero, the bridge is balanced. It follows from Equation 9.4 that for a balanced bridge,

$$\frac{R_1}{R_2} = \frac{R_3}{R_4} \tag{9.5}$$

This equation is valid for any value of the load resistance R_L (the resistance of the device that is connected to the bridge output) not just for large R_L, because when the bridge is balanced, current i through the load becomes zero, even for small R_L.

9.2.1.1 Bridge Sensitivity

Strain-gauge measurements are calibrated with respect to a balanced bridge. When a strain gauge in the bridge deforms, the balance is upset. If one of the arms of the bridge has a variable resistor, it can be adjusted to restore the balance. The amount of this adjustment measures the amount by which the resistance of the strain gauge has changed, thereby measuring the applied strain. This is known as the "null-balance method" of strain measurement. This method is inherently slow because of the time required to balance the bridge each time a reading is taken. A more common method, which is particularly suitable for making dynamic readings from a strain-gauge bridge, is to measure the output voltage resulting from the imbalance caused by the deformation of an active strain gauge in the bridge. To determine the calibration constant of a strain-gauge bridge, the sensitivity of the bridge output to changes in the four resistors in the bridge should be known. For small changes in resistance, using straightforward calculus (by taking the differential of Equation 9.4 and rearranging the result), the calibration constant may be determined as

$$\frac{\delta v_0}{v_{ref}} = \frac{(R_2\delta R_1 - R_1\delta R_2)}{(R_1 + R_2)^2} - \frac{(R_4\delta R_3 - R_3\delta R_4)}{(R_3 + R_4)^2} \tag{9.6}$$

This result is subject to Equation 9.5 because changes are measured from the balanced condition. Note from Equation 9.6 that if all four resistors are identical (in value and material), the changes in resistance due to ambient effects cancel out among the first-order terms (δR_1, δR_2, δR_3, δR_4), producing no net effect on the output voltage from the bridge. Closer examination of Equation 9.6 reveals that only the adjacent pairs of resistors (e.g., R_1 with R_2 and R_3 with R_4) have to be identical in order to achieve this environmental compensation. Even this requirement can be relaxed. In fact, compensation is achieved if R_1 and R_2 have the same temperature coefficient and if R_3 and R_4 have the same temperature coefficient (which can be different from the coefficient of R_1 and R_2).

Example 9.1

Suppose that in Figure 9.5 R_1 represents the only active strain gauge and R_2 represents an identical dummy gauge. The other two elements of the bridge are bridge-completion resistors, which do not have to be identical to the strain gauges. For a balanced bridge, we must have $R_3 = R_4$, but they need not be equal to the resistance of the strain gauge. Let us determine the output of the bridge.

In this example, only R_1 changes. Hence, from Equation 9.6, we have

$$\frac{\delta v_o}{v_{ref}} = \frac{\delta R}{4R} \tag{9.7*}$$

where R denotes the strain-gauge resistance.

9.2.1.2 Bridge Constant

Equation 9.7* assumes that only one resistance (strain gauge) in the Wheatstone bridge (Figure 9.5) is active. Numerous other activating combinations are possible, however, for example, tension in R_1 and compression in R_2, as in the case of two strain gauges mounted symmetrically at 45° about the axis of a shaft in torsion. In this manner, the overall sensitivity of a strain-gauge bridge can be increased. It is clear from Equation 9.6 that if all four resistors in the bridge are active, the best sensitivity is obtained if, for example, R_1 and R_4 are in tension and R_2 and R_3 are in compression so that all four differential terms have the same sign. If more than one strain gauge is active, the bridge output may be expressed as

$$\frac{\delta v_o}{v_{ref}} = k\frac{\delta R}{4R} \tag{9.7}$$

where k = (bridge output in the general case)/(bridge output if only one strain gauge is active).

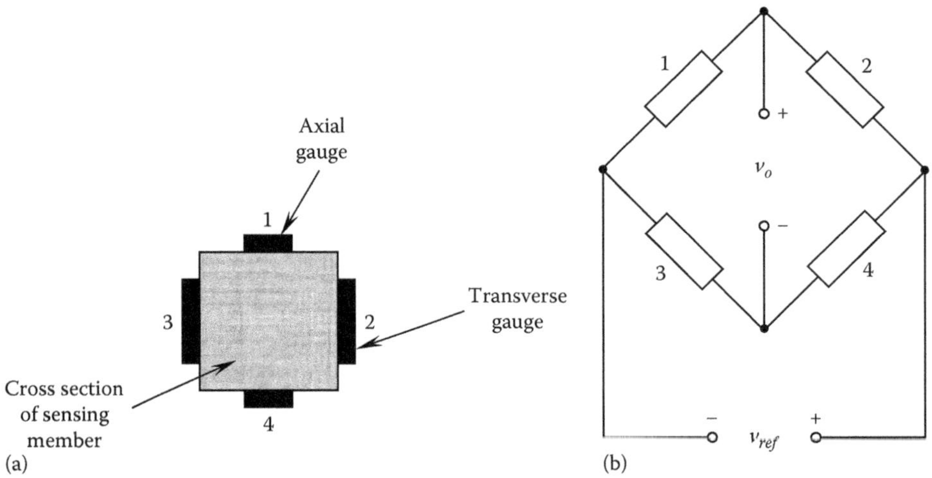

FIGURE 9.6
An example of four active strain gauges. (a) Mounting configuration on the load cell and (b) bridge circuit.

This constant is known as the "bridge constant." The larger the bridge constant, the better the sensitivity of the bridge.

Example 9.2

A strain-gauge load cell (force sensor) consists of four identical strain gauges, forming a Wheatstone bridge, which are mounted on a rod that has a square cross section. One opposite pair of strain gauges is mounted axially and the other pair is mounted in the transverse direction, as shown in Figure 9.6a. To maximize the bridge sensitivity, the strain gauges are connected to the bridge as shown in Figure 9.6b. Determine the bridge constant k in terms of Poisson's ratio ν of the rod material.

Solution

Suppose that $\delta R_1 = \delta R$. Then, for the given configuration, we have

$$\delta R_2 = -\nu\delta R, \quad \delta R_3 = -\nu\delta R, \quad \delta R_4 = \delta R$$

Note that from the definition of Poisson's ratio,

$$\text{Transverse strain} = (-\nu) \times \text{longitudinal strain}$$

Now, it follows from Equation 9.6 that $\delta v_o/v_{ref} = 2(1 + \nu)(\delta R/4R)$. According to this, the bridge constant is given by $k = 2(1 + \nu)$.

9.2.1.3 Calibration Constant

The calibration constant C of a strain-gauge bridge relates the strain that is measured to the output of the bridge. Specifically,

$$\frac{\delta v_o}{v_{ref}} = C\varepsilon \qquad (9.8)$$

Now, in view of Equations 9.3 and 9.7, the calibration constant may be expressed as

$$C = \frac{k}{4} S_s \tag{9.9}$$

where

k is the bridge constant

S_s is the sensitivity (or *gauge factor*) of the strain gauge

Ideally, the calibration constant should remain constant over the measurement range of the bridge (i.e., should be independent of strain ε and time t) and should be stable (drift-free) with respect to ambient conditions. In particular, there should not be any creep, nonlinearities such as hysteresis or thermal effects.

Example 9.3

A schematic diagram of a strain-gauge accelerometer is shown in Figure 9.7a. A lumped mass of weight W is used as the acceleration sensing element. A light cantilever with rectangular cross section, mounted inside the accelerometer casing, converts the inertia force of the mass into a strain (*Note*: This is the front-end auxiliary element). The maximum bending strain at the root of the cantilever is measured using four identical active SC strain gauges. Two of the strain gauges (A and B) are mounted axially on the top surface of the cantilever, and the remaining two (C and D) are mounted on the bottom surface, as shown in Figure 9.7b. In order to maximize the sensitivity of the accelerometer, indicate the manner in which the four strain gauges—A, B, C, and D—should be connected to a Wheatstone bridge circuit. What is the bridge constant of the resulting circuit?

Obtain an expression relating the applied acceleration a (in units of g, which denotes acceleration due to gravity) to the bridge output δv_o (measured using a bridge that is balanced at zero acceleration) in terms of the following parameters:

$W = Mg$ is the weight of the seismic mass at the free end of the cantilever element, E is Young's modulus of the cantilever, ℓ is the length of the cantilever, b is the cross-sectional

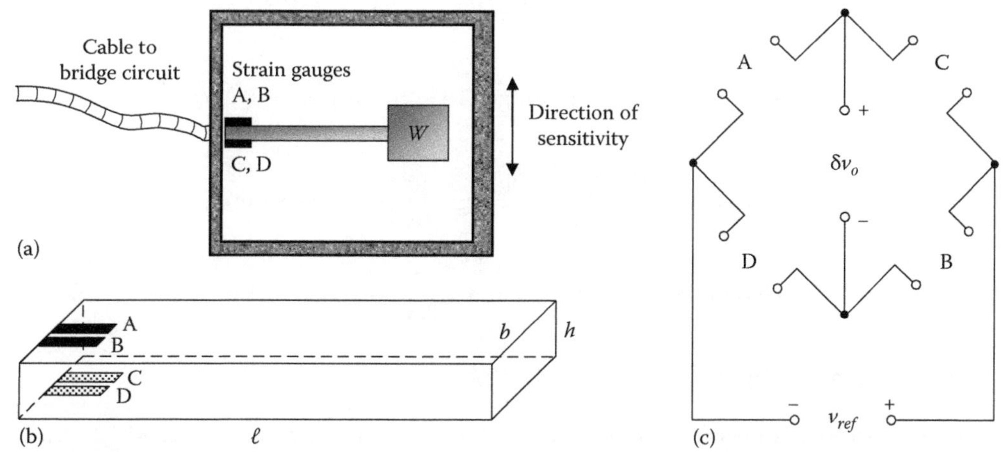

FIGURE 9.7
A miniature accelerometer using strain gauges: (a) schematic diagram, (b) mounting configuration of the strain gauges, and (c) bridge connection.

width of the cantilever, h is the cross-sectional height of the cantilever, S_s is the gauge factor (sensitivity) of each strain gauge, and v_{ref} is the supply voltage to the bridge.

If $M = 5$ g, $E = 5 \times 10^{10}$ N/m², $\ell = 1$ cm, $b = 1$ mm, $h = 0.5$ mm, $S_s = 200$, and $v_{ref} = 20$ V, determine the sensitivity of the accelerometer in microvolts per gram.

If the yield strength (yield stress) of the cantilever element is 5×10^7 N/m² (or 5×10^7 Pa), what is the maximum acceleration that could be measured using the accelerometer? If the analog-to-digital converter (ADC) that reads the strain signal into a microcontroller has the range 0–10 V, how much amplification (bridge amplifier gain) would be needed at the bridge output so that this maximum acceleration corresponds to the upper limit of the ADC (10 V)?

Is the cross sensitivity (i.e., the sensitivity to tension and other direction of bending) small with your arrangement of the strain-gauge bridge? Explain.

Hint: For a cantilever subjected to force F at the free end, the maximum stress at the root is given by $\sigma = (6F\ell)/(bh^2)$ with the present notation.

Note: MEMS accelerometers where the cantilever member, inertia element, and the strain gauge are all integrated into a single SC (silicon) unit are available in commercial applications such as air bag activation sensors for automobiles (see Chapter 12).

Solution

Clearly, the bridge sensitivity is maximized by connecting the strain gauges A, B, C, and D to the bridge as shown in Figure 9.7c. This follows from Equation 9.6, noting that the contributions from all four strain gauges are positive when δR_1 and δR_4 are positive and δR_2 and δR_3 are negative. The bridge constant for the resulting arrangement is $k = 4$. Hence, from Equation 9.7, $\delta v_o/v_{ref} = \delta R/R$ or from Equations 9.8 and 9.9, $\delta v_o/v_{ref} = S_s\varepsilon$. Also, $\varepsilon = \sigma/E = (6F\ell)/Ebh^2$ where F denotes the inertia force: $F = (W/g)\ddot{x} = Wa$.

Note: $\ddot{x}$ is the acceleration in the direction of sensitivity and $\ddot{x}/g = a$ is the acceleration in units of g.

Thus,

$$\varepsilon = \frac{6W\ell}{Ebh^2}a \quad \text{or} \quad \delta v_o = \frac{6W\ell}{Ebh^2}S_s v_{ref}\, a$$

Substitute values:

$$\frac{\delta v_o}{a} = \frac{6\times 5\times 10^{-3}\times 9.81\times 1\times 10^{-2}\times 200\times 20}{5\times 10^{10}\times 1\times 10^{-3}\times (0.5\times 10^{-3})^2}\,\text{V/g} = 0.94\ \text{V/g}$$

$$\frac{\varepsilon}{a} = \frac{1}{S_s v_{ref}}\frac{\delta v_o}{a} = \frac{0.94}{200\times 20}\,\text{strain/g} = 2.35\times 10^{-4}\ \varepsilon/\text{g} = 235.0\ \mu\varepsilon/\text{g}$$

$$\text{Yield strain} = \frac{\text{Yield strength}}{E} = \frac{5\times 10^7}{5\times 10^{10}} = 1\times 10^{-3}\ \text{strain}$$

$$\rightarrow \text{Number of } g\text{'s to yield point} = \frac{1\times 10^{-3}}{2.35\times 10^{-4}}\,\text{g} = 4.26\ \text{g}$$

Corresponding voltage = 0.94 × 4.26 V = 4.0 V → amplifier gain = 10.0/4.0 = 2.25.

Cross sensitivity comes from accelerations in the two directions y and z, which are orthogonal to the direction of sensitivity (x). In the lateral (y) direction, the inertia force causes lateral bending. This produces equal tensile (or compressive) strains in B and D and equal compressive (or tensile) strains in A and C. According to the bridge circuit, we see that these contributions cancel each other. In the axial (z) direction, the inertia force causes equal tensile (or compressive) stresses in all four strain gauges. These also cancel out, as is clear from the relationship in Equation 9.6 for the bridge, which gives

$$\frac{\delta v_o}{v_{ref}} = \frac{(\delta R_A - \delta R_C - \delta R_D + \delta R_B)}{4R}$$

It follows that this arrangement compensates for cross-sensitivity problems.

9.2.1.4 Data Acquisition

For measuring dynamic strains, either the servo null-balance method or the imbalance output method should be employed (see Chapter 2). A schematic diagram for the imbalance output method is shown in Figure 9.8. In this method, the output from the active bridge is directly measured as a voltage signal and calibrated to provide the measured strain. Figure 9.8 corresponds to the use of an AC bridge. In this case, the bridge is powered by an AC voltage. The supply frequency should be about 10 times the maximum frequency of interest in the dynamic strain signal (bandwidth). A supply frequency in the order of 1 kHz is typical. This signal is generated by an oscillator and is fed into the bridge. The transient component of the output from the bridge is very small (typically less than 1 mV and possibly a few microvolts). This signal has to be amplified, demodulated (especially if the signals are transient), and filtered to provide the strain reading. The calibration constant of the bridge should be known in order to convert the output voltage to strain.

Strain-gauge bridges powered by DC voltages are common. They have the advantages of simplicity with regard to the necessary circuitry and portability. The advantages of AC bridges include improved stability (reduced drift) and accuracy and reduced power consumption.

9.2.1.5 Accuracy Considerations

Foil gauges are available with resistances as low as 50 Ω and as high as several kilohms. The power consumption of a bridge circuit decreases with increased resistance. This has the added advantage of decreased heat generation. Bridges with a high range of measurement (e.g., a maximum strain of 0.04 m/m) are available. The accuracy depends on the linearity of the bridge, environmental effects (particularly temperature), and mounting techniques. For example, zero shifts, due to strains produced when the cement or epoxy that is used to mount the strain gauge dries, result in calibration error. Creep introduces errors during static and low-frequency measurements. Flexibility and hysteresis of the bonding cement

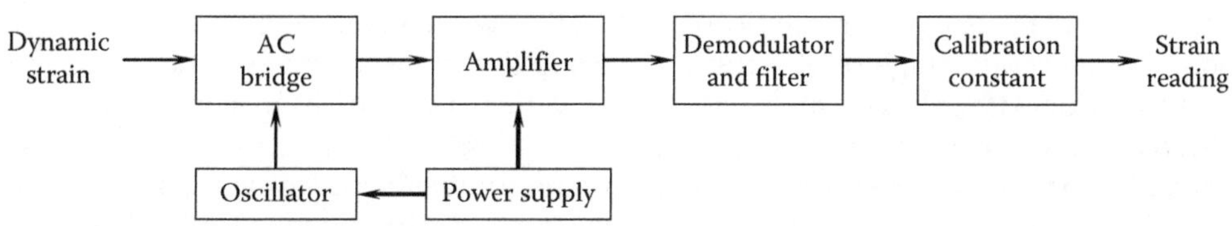

FIGURE 9.8
Measurement of dynamic strains using an AC bridge.

(or epoxy) bring about errors during high-frequency strain measurements. Resolutions in the order of 1 μm/m (i.e., one *microstrain*) are common.

As noted earlier, the cross sensitivity of a strain gauge is the sensitivity to strains that are orthogonal to the measured strain. This cross sensitivity should be small (say, less than 1% of the direct sensitivity). Manufacturers usually provide cross-sensitivity factors for their strain gauges. This factor, when multiplied by the cross strain that is present in a given application, gives the error in the strain reading due to cross sensitivity.

Sensing of moving members: Often, strains in moving members are sensed in engineering applications. Examples include real-time monitoring and failure detection in machine tools, measurement of power, measurement of force and torque for feedforward and feedback control in dynamic systems, instrumentation of biomechanical devices, and tactile sensing using instrumented hands in industrial robots. A strain gauge mounted on a moving member needs power for the connected circuitry (typically, from a stationary source) and some means of acquiring the sensed signal (strain, change in resistance or bridge output) by a stationary device (e.g., computer). If the motion is small or the moving member has a limited stroke, strain gauges mounted on the moving member can be directly connected to the power source, signal-conditioning circuitry, and data acquisition system by using coiled flexible cables. For large motions, particularly in rotating shafts, some form of *commutation* arrangement has to be used. Slip rings and brushes may be used for this purpose. When AC bridges are used, a mutual-induction device (*rotary transformer*) may be used, with one coil located on the moving member and the other coil kept stationary. To accommodate and compensate for errors (e.g., losses and glitches in the output signal) that are caused by commutation, it is desirable to place all four arms of the bridge, rather than just the active arms, on the moving member. A more modern approach is to use telemetry or wireless communication (at radio frequency) from the moving member to a stationary local device of data acquisition. The signal-conditioning electronics may be located as well on the moving member since monolithic microminiature hardware is available for this purpose. The sensor and the local hardware on the moving element may be powered through energy harvesting (e.g., magnetic induction, photoelectricity) as well.

9.2.2 Semiconductor Strain Gauges

In some low-strain applications (e.g., dynamic torque measurement), the sensitivity of foil gauges is not adequate to produce an acceptable strain-gauge signal. SC strain gauges are particularly useful in such situations. The strain element of an SC strain gauge is made of a single crystal of piezoresistive material such as silicon, doped with a trace impurity such as boron. A typical construction is shown in Figure 9.9. The gauge factor (sensitivity) of an SC strain gauge (typically, 40–200) is about two orders of magnitude higher than that of a metallic foil gauge, as seen for silicon, from the data given in Table 9.1.

The resistivity is also higher, providing reduced power consumption and lower heat generation. Another advantage of SC strain gauges is that they deform elastically to fracture (a common property of brittle material). In particular, mechanical hysteresis is negligible. Furthermore, they are smaller and lighter, providing less cross sensitivity, reduced distribution error (i.e., improved spatial resolution), and negligible error from mechanical loading. The maximum strain that is measurable using an SC strain gauge is typically 0.003 m/m (i.e., 3000 με). The strain-gauge resistance can be an order of magnitude greater for an SC strain gauge, for example, several hundred ohms for a metal foil strain gauge (typically,

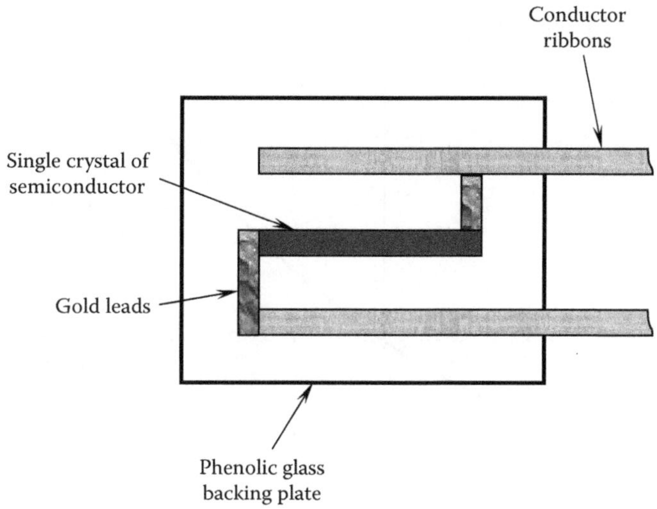

FIGURE 9.9
Component details of a semiconductor strain gauge.

TABLE 9.1
Properties of Common Strain-Gauge Material

Material	Composition	Gauge Factor (Sensitivity)	Temperature Coefficient of Resistance (10^{-6}/°C)
Constantan	45% Ni, 55% Cu	2.0	15
Isoelastic	36% Ni, 52% Fe, 8% Cr, 4% (Mn, Si, Mo)	3.5	200
Karma	74% Ni, 20% Cr, 3% Fe, 3% Al	2.3	20
Monel	67% Ni, 33% Cu	1.9	2000
Silicon	p-type	100–170	70–700
Silicon	n-type	−140 to −100	70–700

120 or 350 Ω), while several thousand ohms (5000 Ω) for an SC strain gauge. There are several disadvantages associated with SC strain gauges, however, which can be interpreted as advantages of foil gauges. Undesirable characteristics of SC gauges include the following:

1. The strain–resistance relationship is more nonlinear.
2. They are brittle and difficult to mount on curved surfaces.
3. The maximum strain that can be measured is one to two orders of magnitude smaller (typically, less than 0.001 m/m).
4. They are more costly.
5. They have much larger temperature sensitivity.

The first disadvantage is illustrated in Figure 9.10. There are two types of SC strain gauges: the p-type, which are made of an SC (e.g., silicon) doped with an acceptor impurity (e.g., boron), and the n-type, which are made of an SC doped with a donor impurity (e.g., arsenic). In p-type strain gauges, the direction of sensitivity is along the (1, 1, 1)

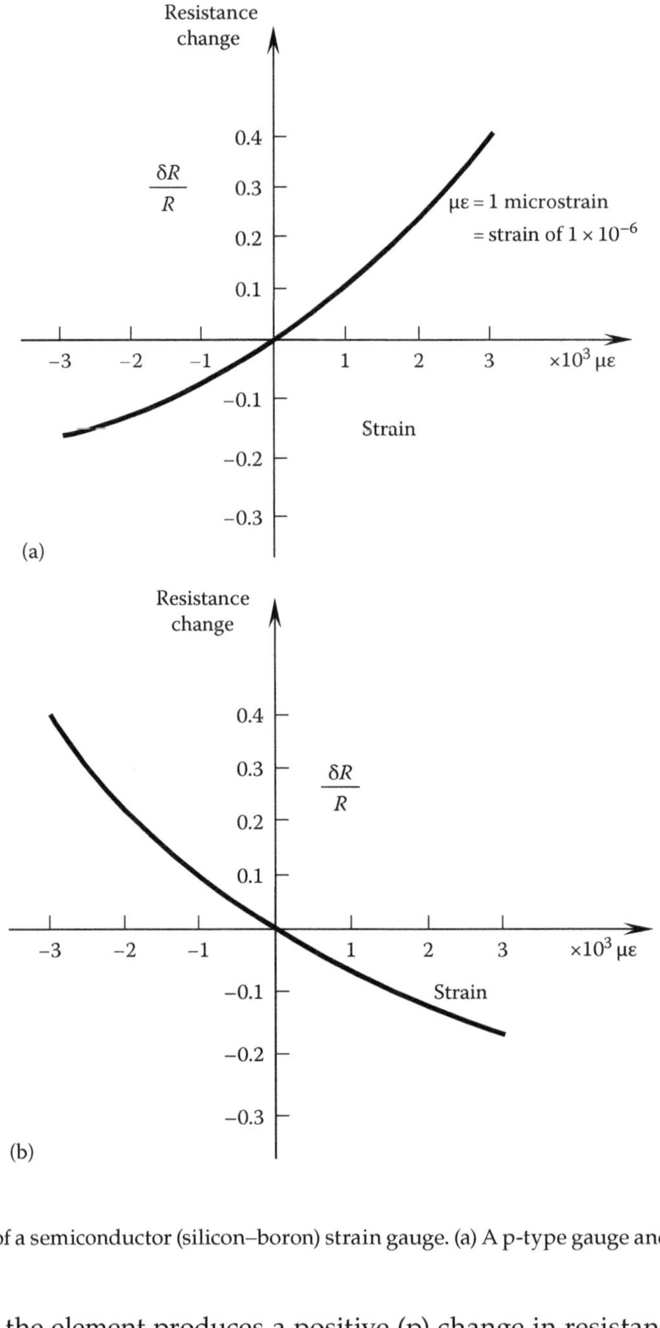

FIGURE 9.10
Nonlinear behavior of a semiconductor (silicon–boron) strain gauge. (a) A p-type gauge and (b) an n-type gauge.

crystal axis, and the element produces a positive (p) change in resistance in response to a positive strain. In n-type strain gauges, the direction of sensitivity is along the (1, 0, 0) crystal axis, and the element responds with a negative (n) change in resistance to a positive strain. In both types, the response is nonlinear and can be approximated by the quadratic relationship

$$\frac{\delta R}{R} = S_1\varepsilon + S_2\varepsilon^2 \tag{9.10}$$

The parameter S_1 represents the linear gauge factor (linear sensitivity), which is positive for p-type gauges and negative for n-type gauges. Its magnitude is usually somewhat larger for p-type gauges, corresponding to better sensitivity. The parameter S_2 represents the degree of nonlinearity, which is usually positive for both types of gauges. Its magnitude, however, is typically somewhat smaller for p-type gauges. It follows that p-type gauges are less nonlinear and have higher strain sensitivities. The nonlinear relationship given by Equation 9.10 or the nonlinear characteristic curve (Figure 9.10) should be used when measuring moderate to large strains with SC strain gauges. Otherwise, the nonlinearity error would be excessive.

Example 9.4

For an SC strain gauge characterized by the quadratic strain–resistance relationship, Equation 9.10, obtain an expression for the equivalent gauge factor (sensitivity) S_s, using the linear least-squares error approximation (see Chapter 7) and assuming that strains in the range $\pm\varepsilon_{max}$ have to be measured. Derive an expression for the percentage nonlinearity.

Taking $S_1 = 117$, $S_2 = 3600$, and $\varepsilon_{max} = 1 \times 10^{-2}$, calculate S_s and the percentage nonlinearity.

Solution

The linear approximation of Equation 9.10 may be expressed as

$$\left[\frac{\delta R}{R}\right]_L = S_s\varepsilon$$

The error is given by

$$e = \frac{\delta R}{R} - \left[\frac{\delta R}{R}\right]_L = S_1\varepsilon + S_2\varepsilon^2 - S_s\varepsilon = (S_1 - S_s)\varepsilon + S_2\varepsilon^2 \tag{i}$$

The quadratic integral error is

$$J = \int_{-\varepsilon_{max}}^{\varepsilon_{max}} e^2 d\varepsilon = \int_{-\varepsilon_{max}}^{\varepsilon_{max}} \left[(S_1 - S_s)\varepsilon + S_2\varepsilon^2\right]^2 d\varepsilon \tag{ii}$$

We have to determine S_s that results in a minimum J, and we apply $\partial J/\partial S_s = 0$. Hence, from Equation ii,

$$\int_{-\varepsilon_{max}}^{\varepsilon_{max}} (-2\varepsilon)\left[(S_1 - S_s)\varepsilon + S_2\varepsilon^2\right]^2 d\varepsilon = 0$$

On performing the integration and solving the equation, we get

$$S_s = S_1 \tag{9.11}$$

The quadratic curve and the linear approximation are shown in Figure 9.11. The maximum error occurs at $\varepsilon = \pm\varepsilon_{max}$. The maximum error value is obtained from Equation i, with $S_s = S_1$ and $\varepsilon = \pm\varepsilon_{max}$, as $e_{max} = S_2\varepsilon_{max}^2$.

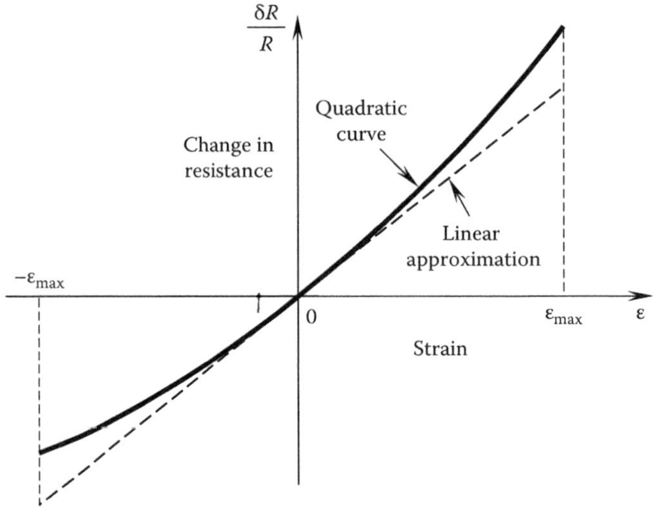

FIGURE 9.11
Least-squares linear approximation for a semiconductor strain gauge.

The true change in resistance (nondimensional) from $-\varepsilon_{max}$ to $+\varepsilon_{max}$ is obtained using Equation 9.10 as

$$\frac{\Delta R}{R} = \left(S_1\varepsilon_{max} + S_2\varepsilon_{max}^2\right) - \left(-S_1\varepsilon_{max} + S_2\varepsilon_{max}^2\right) = 2S_1\varepsilon_{max}$$

Hence, the percentage nonlinearity is given by

$$N_p = \frac{\text{Max error}}{\text{Range}} \times 100\% = \frac{S_2\varepsilon_{max}^2}{2S_1\varepsilon_{max}} \times 100\%$$

or

$$N_p = \frac{50S_2\varepsilon_{max}}{S_1}\% \tag{9.12}$$

Substitute the given numerical values. We have

$$S_s = 117 \quad \text{and} \quad N_p = \frac{50 \times 3600 \times 1 \times 10^{-2}}{117}\% = 15.4\%$$

We obtained this high value for nonlinearity because the given strain limits were high. Usually, the linear approximation is adequate for strains up to $\pm 1 \times 10^{-3}$.

The higher-temperature sensitivity, which is listed as a disadvantage of SC strain gauges over metal ones, may be considered an advantage in some situations. For instance, it is this property of high-temperature sensitivity that is used in piezoresistive temperature sensors. Furthermore, using the fact that the temperature sensitivity of an SC strain gauge can be determined very accurately, precise methods can be employed for temperature compensation in strain-gauge circuitry, and temperature calibration can also be done accurately. In particular, a passive SC strain gauge may be used as an accurate temperature sensor for compensation purposes.

9.2.3 Automatic (Self-) Compensation for Temperature (for Semiconductor Strain Gauges)

In foil gauges, the change in resistance due to typical temperature variations is small. Then, the linear (first order) approximation is adequate to represent the contribution from each arm of the bridge to the output signal, as given by Equation 9.6. Furthermore, these contributions cancel out if we pick strain-gauge elements and bridge-completion resistors properly—for example, R_1 identical to R_2 and R_3 identical to R_4. If this is the case, the only remaining effect of temperature change on the bridge output signal comes from the variations in the parameter values k and S_s (see Equations 9.8 and 9.9). For foil gauges, such changes are also typically negligible. Hence, for small to moderate temperature changes, as experienced in normal operating conditions, additional compensation is not required when foil gauges are employed in bridge circuits.

In SC gauges, as the temperature varies (and as the strain varies), not only the change in resistance but also the change in S_s is larger when compared with the corresponding changes in foil gauges. Hence, the linear approximation given by Equation 9.6 might not be adequately accurate for SC gauges under conditions of variable temperature. Furthermore, the bridge sensitivity may change significantly with temperature. Under such conditions, compensation for temperature becomes necessary.

A straightforward way to account for temperature changes is by directly measuring the temperature and correcting the strain-gauge readings by using the data for thermal calibration. Another method of temperature compensation is described now. This method assumes that the linear approximation given by Equation 9.6 is valid, and hence, Equation 9.8 is applicable.

The resistance R and strain sensitivity (or gauge factor) S_s of an SC strain gauge are highly dependent on the concentration of the trace impurity, in a nonlinear manner. The typical behavior of the temperature coefficients of these two parameters for a p-type SC strain gauge is shown in Figure 9.12. The *temperature coefficient of resistance* α and the *temperature coefficient of sensitivity* β are defined by

$$R = R_o(1 + \alpha \cdot \Delta T) \tag{9.13}$$

$$S_s = S_{so}(1 + \beta \cdot \Delta T) \tag{9.14}$$

where ΔT is the temperature increase. Note from Figure 9.12 that β is a negative quantity and that for some dope concentrations, its magnitude is less than the value of the temperature coefficient of resistance (α). This property can be used in self-compensation with regard to temperature for a p-type SC (silicon) strain gauge.

Consider a constant-voltage bridge circuit with a compensating resistor R_c connected to the supply lead, as shown in Figure 9.13a. It can be shown that self-compensation can result if R_c is set to a value that is predetermined on the basis of the temperature coefficients of the strain gauges. Consider the case where the load impedance is very high and the bridge

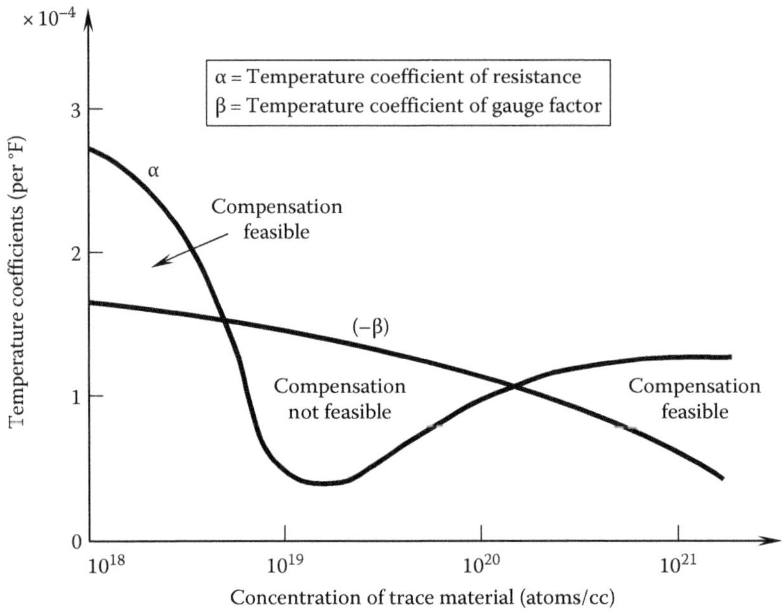

FIGURE 9.12
Temperature coefficients of resistance and gauge factor of p-type semiconductor strain gauge.

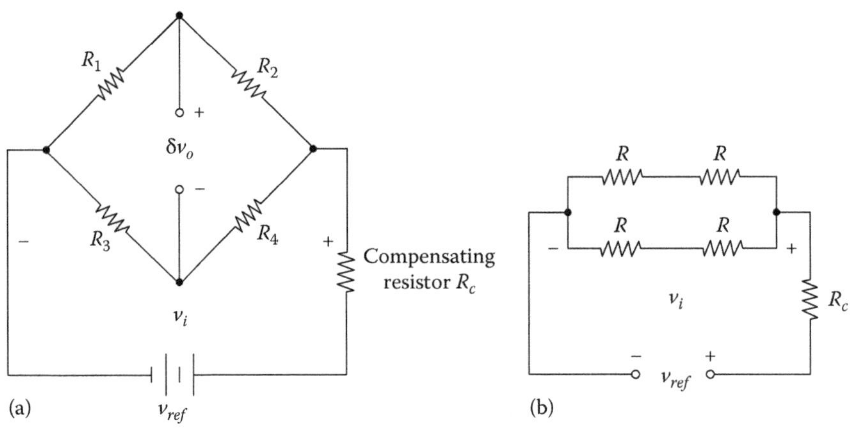

FIGURE 9.13
A strain-gauge bridge with a compensating resistor. (a) Constant-voltage DC bridge and (b) equivalent circuit with high load impedance.

has four identical SC strain gauges, each of which has resistance R. In this case, the bridge can be represented by the circuit shown in Figure 9.13b.

Since series impedances and parallel admittances (inverse of impedance) are additive, the equivalent resistance of the bridge is R. Hence, the voltage supplied to the bridge, allowing for the voltage drop across R_c, is not v_{ref} but v_i, as given by

$$v_i = \frac{R}{(R + R_c)} v_{ref} \tag{9.15}$$

Now, from Equations 9.8 and 9.9, we have

$$\frac{\delta v_o}{v_{ref}} = \frac{R}{(R+R_c)}\frac{kS_s}{4}\varepsilon \tag{9.16}$$

Note: Here, we assume that the bridge constant k does not change with temperature. Otherwise, the following procedure still holds, provided that the calibration constant C is used in place of the gauge factor S_s (see Equation 9.9).

For self-compensation, we must have the same output after the temperature has changed through ΔT. Hence, from Equation 9.16, we must have

$$\frac{R_o}{(R_o+R_c)}S_{so} = \frac{R_o(1+\alpha\cdot\Delta T)}{[R_o(1+\alpha\cdot\Delta T)+R_c]}S_{so}(1+\beta\cdot\Delta T)$$

where the subscript o denotes values before the temperature change. Cancelation of the common terms and cross multiplication gives $R_o\beta + R_c(\alpha + \beta) = (R_o + R_c)\alpha\beta\Delta T$. Now, since both $\alpha\cdot\Delta T$ and $\beta\cdot\Delta T$ are usually much smaller than unity, we may neglect the second-order term (on the RHS) in the preceding result. This gives the following expression for the compensating resistance:

$$R_c = -\left[\frac{\beta}{\alpha+\beta}\right]R_o \tag{9.17}$$

Note: Compensation is possible because the temperature coefficient of the strain-gauge sensitivity (β) is negative.

The feasible ranges of operation, which correspond to positive R_c, are indicated in Figure 9.12. This method requires the R_c to be maintained constant at the chosen value under temperature variations. One way of accomplishing this is by selecting a material with negligible temperature coefficient of resistance for R_c. Another way is to locate R_c in a separate, temperature-regulated environment (e.g., an ice bath).

9.3 Torque Sensors

The sensing of torque and force is useful in many applications, including the following:

1. In robotic tactile (distributed touch) and manufacturing applications such as grasping, handling, fine manipulating, surface gauging, and material forming, exerting an adequate load on an object is a primary purpose of the task.
2. In the control of fine motions (e.g., fine manipulation and micromanipulation) and in assembly tasks, a small error in motion can cause large and damaging forces or performance degradation.
3. In control systems that are not fast enough when motion feedback alone is employed, force feedback and feedforward force control can be used to improve accuracy and bandwidth (speed).

4. In process testing, monitoring, and diagnostic applications, torque sensing can detect, predict, and identify abnormal operation, malfunction, component failure, or excessive wear (e.g., in monitoring of machine tools such as milling machines and drills).
5. In the measurement of power transmitted through a rotating device, power is given by the product of torque and angular velocity in the same direction.
6. In controlling complex, nonlinear mechanical systems, measurement of force and acceleration can be used to estimate the unknown nonlinear terms. Nonlinear feedback of the estimated terms will linearize or simplify the system (this method is called nonlinear feedback control or linearizing feedback technique or LFT).
7. In experimental modeling (i.e., in model identification where a model is determined by analyzing input–output data), the system input is a torque.

In most applications, sensing is done by detecting either an effect of torque or the cause of torque. Also, there are methods for measuring torque directly. Common methods of torque sensing include the following:

1. Measuring the *strain* in a sensing member between the drive element (or actuator) and the driven element (or load), using a strain-gauge bridge
2. Measuring the *displacement* in a sensing member (as in the first method)—either directly, using a displacement sensor, or indirectly, by measuring a variable such as magnetic inductance or capacitance that varies with displacement
3. Measuring the *reaction* in support structure or housing (e.g., by measuring the force and the associated lever arm length that is required to hold it down)
4. In electric motors, measuring the *field current* or *armature current*, which produces motor torque, and in hydraulic or pneumatic actuators, measuring the *actuator pressure*
5. Measuring the torque *directly*, using piezoelectric sensors, for example
6. Employing a *servo method*—balancing the unknown torque with a feedback torque generated by an active device (say, a servomotor) whose torque characteristics are precisely known
7. Measuring the *angular acceleration* caused by the unknown torque in a known inertia element

The remainder of this section is devoted to a study of the torque measurement using several of these methods. Force sensing is analogous to torque sensing and may be accomplished by using essentially the same techniques. For the sake of brevity, however, we limit our treatment primarily to torque sensing, which may be interpreted as sensing of a "generalized force." The extension of torque-sensing techniques to force sensing is somewhat challenging, however.

9.3.1 Strain-Gauge Torque Sensors

The most straightforward method of torque sensing is to connect a torsion member between the drive unit (e.g., actuator) and the driven device (load) in series, as shown in Figure 9.14, and to measure the torque in the torsion member.

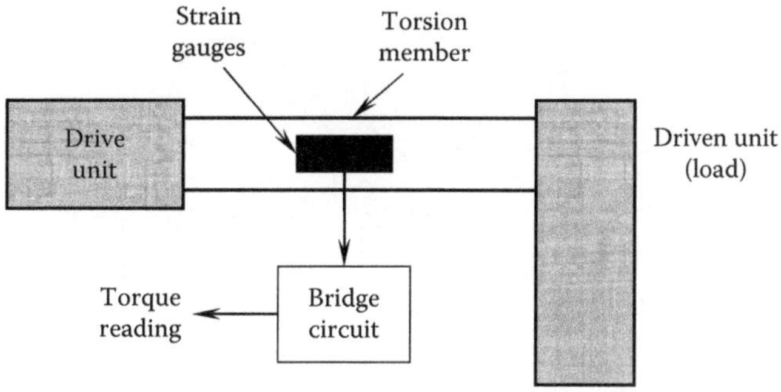

FIGURE 9.14
Torque sensing using a torsion member.

If a circular shaft (solid or hollow) is used as the torsion member, the torque–strain relationship becomes relatively simple and is given by

$$\varepsilon = \frac{r}{2GJ} T \tag{9.18}$$

where
T is the torque transmitted through the member
ε is the principal strain (which is at 45° to shaft axis) at radius r within the member
J is the polar moment of area of cross section of the member
G is the shear modulus of the material

Moreover, the shear stress τ at a radius r of the shaft is given by

$$\tau = \frac{Tr}{J} \tag{9.19}$$

It follows from Equation 9.18 that torque T can be determined by measuring the direct strain ε on the shaft surface along a principal stress direction (i.e., at 45° to the shaft axis). This is the basis of torque sensing using strain measurements. Using the general bridge Equation 9.8 along with Equation 9.9 in Equation 9.18, we obtain torque T from the bridge output $\delta\nu_o$:

$$T = \frac{8GJ}{kS_s r} \frac{\delta v_o}{v_{ref}} \tag{9.20}$$

where S_s is the gauge factor (or sensitivity) of the strain gauges. The bridge constant k depends on the number of active strain gauges used. Strain gauges are assumed to be mounted along a principal direction. Three possible configurations are shown in Figure 9.15. In configurations (a) and (b), only two strain gauges are used and the bridge constant $k = 2$. *Note*: Both axial and bending loads are compensated for with the given configurations because the resistance in both gauges changes by the same amount (the same sign and the same magnitude), which cancel out up to the first order, for the bridge circuit connection shown in Figure 9.15.

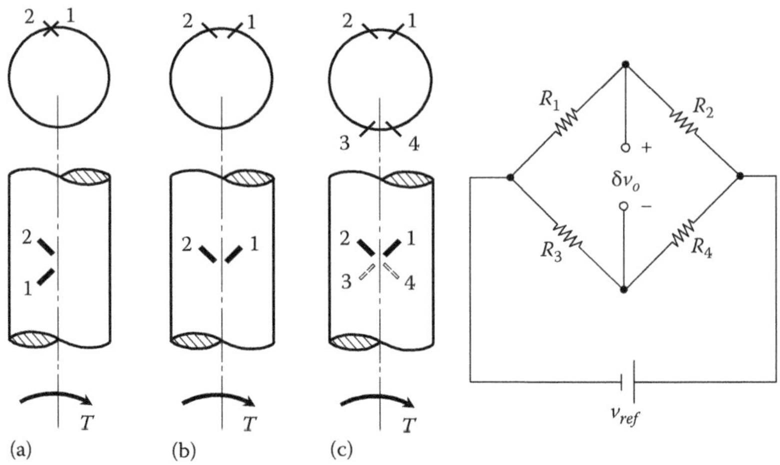

Bridge constant (k)	2		2		4
Axial loads compensated	Yes		Yes		Yes
Bending loads compensated	Yes		Yes		Yes
Configuration	(a)		(b)		(c)

Strain-gauge bridge

FIGURE 9.15
Strain-gauge configurations for a circular shaft torque sensor.

Configuration (c) has two pairs of gauges mounted on the two opposite surfaces of the shaft. The bridge constant is doubled in this configuration, and here again, the sensor self-compensates for axial and bending loads up to the first order [$O(\delta R)$].

9.3.2 Design Considerations

Two conflicting requirements in the design of a torsion element for torque sensing are sensitivity and bandwidth. The element has to be sufficiently flexible in order to get an acceptable level of sensor sensitivity (i.e., a sufficiently large output signal). According to Equation 9.18, this requires a small torsional rigidity GJ to produce a large strain for a given torque. Unfortunately, since the torsion-sensing element is connected in series between a drive element and a driven element, an increase in flexibility of the torsion element results in reduction of the overall stiffness of the system. Specifically, with reference to Figure 9.16, the overall stiffness K_{old} before connecting the torsion element is given by

$$\frac{1}{K_{old}} = \frac{1}{K_m} + \frac{1}{K_L} \tag{9.21}$$

and the stiffness K_{new} after connecting the torsion member is given by

$$\frac{1}{K_{new}} = \frac{1}{K_m} + \frac{1}{K_L} + \frac{1}{K_s} \tag{9.22}$$

where

K_m is the equivalent stiffness of the drive unit (motor)
K_L is the equivalent stiffness of the load
K_s is the stiffness of the torque-sensing element

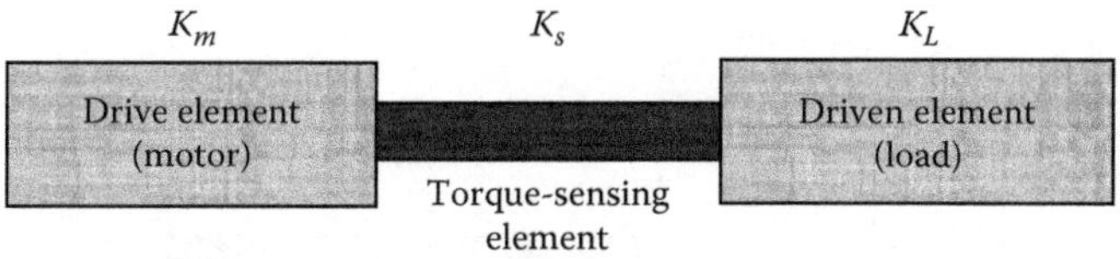

FIGURE 9.16
Stiffness degradation due to flexibility of the torque-sensing element.

It is clear from Equations 9.21 and 9.22 that $1/K_{new} > 1/K_{old}$. Hence, $K_{new} < K_{old}$. This reduction in stiffness is associated with a reduction in natural frequency and bandwidth, resulting in slower response to control commands in the overall system. Furthermore, a reduction in stiffness causes a reduction in the loop gain. As a result, the steady-state error in some motion variables can increase, which calls for more effort from the controller to achieve a required level of accuracy. One aspect in the design of the torsion element is to guarantee that the element stiffness is small enough to provide adequate sensitivity but large enough to maintain adequate bandwidth and system gain. In situations where K_s cannot be increased adequately without seriously jeopardizing the sensor sensitivity, the system bandwidth can be improved by decreasing either the load inertia or the drive unit (motor) inertia.

Example 9.5

Consider a rigid load, which has a polar moment of inertia J_L and driven by a motor with a rigid rotor, which has inertia J_m. A torsional member of stiffness K_s is connected between the rotor and the load, as shown in Figure 9.17a, to measure the torque transmitted to the load.

(a) Determine the transfer function between the motor torque T_m and the angle of twist θ of the torsion member. What is the torsional natural frequency ω_n of the system? Discuss why the system bandwidth depends on ω_n. Show that the bandwidth can be improved by increasing K_s, by decreasing J_m, or by decreasing J_L. Give some advantages and disadvantages of introducing a gearbox at the motor output.
(b) If a torsion member of stiffness $0.5K_s$ is mounted at the load end of the shaft (in series), by what percentage the original torsional bandwidth of the system (representative of the allowable operating frequency range for the torque sensor) is reduced?

Solution

From the free-body diagram shown in Figure 9.17b, the following motion equations are written:

$$\text{For motor:} \quad J_m\ddot{\theta}_m = T_m - K_s(\theta_m - \theta_L) \tag{i}$$

$$\text{For load:} \quad J_L\ddot{\theta}_L = K_s(\theta_m - \theta_L) \tag{ii}$$

where
θ_m is the motor rotation
θ_L is the load rotation

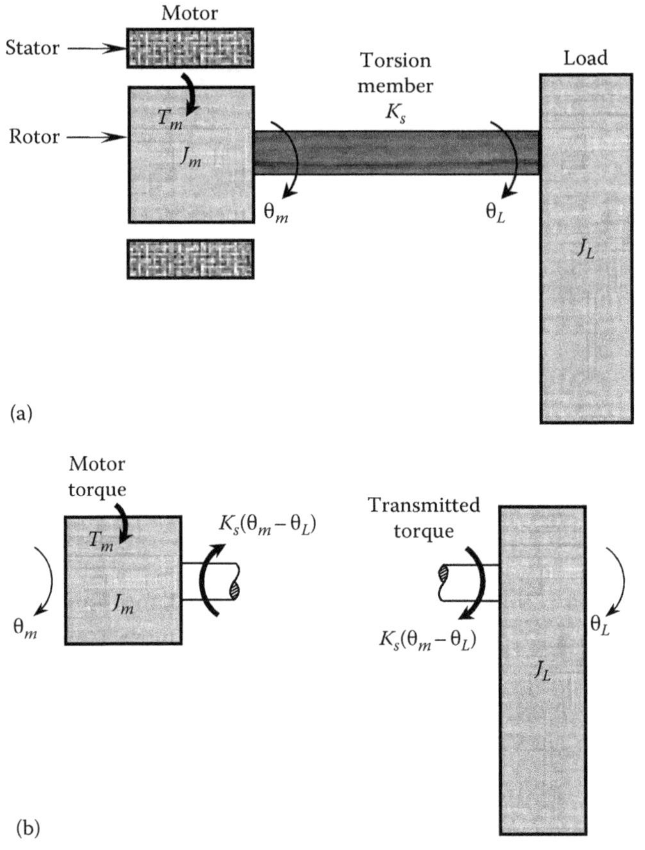

FIGURE 9.17
An example of bandwidth analysis of a system with a torque sensor. (a) System model and (b) free-body diagram.

Divide Equation i by J_m, divide Equation ii by J_L, and subtract the second equation from the first:

$$\ddot{\theta}_m - \ddot{\theta}_L = \frac{T_m}{J_m} - \frac{K_s}{J_m}(\theta_m - \theta_L) - \frac{K_s}{J_L}(\theta_m - \theta_L)$$

This equation can be expressed in terms of the angle of twist:

$$\theta = \theta_m - \theta_L \quad \text{(iii)}$$

$$\ddot{\theta} + K_s\left(\frac{1}{J_m} + \frac{1}{J_L}\right)\theta = \frac{T_m}{J_m} \quad \text{(iv)}$$

This equation corresponds to the twisting dynamic mode (*torsional mode*) of the system. The transfer function $G(s)$ between input T_m and output θ is obtained by introducing the Laplace variable s in place of the time derivative d/dt. Specifically, we have

$$G(s) = \frac{1/J_m}{s^2 + K_s(1/J_m + 1/J_L)} \quad \text{(v)}$$

The characteristic equation of the twisting system is

$$s^2 + K_s\left(\frac{1}{J_m} + \frac{1}{J_L}\right) = 0 \tag{vi}$$

It follows that the torsional (twisting) natural frequency ω_n is given by

$$\omega_n = \sqrt{K_s\left(\frac{1}{J_m} + \frac{1}{J_L}\right)} \tag{vii}$$

In addition to this natural frequency, there is a zero natural frequency in the overall system, which corresponds to rotation of the entire system as a rigid body without any twisting (deformation) in the torsional member. This is the *rigid-body mode*. Both natural frequencies are obtained if the output is taken as either θ_m or θ_L rather than the angle of twist θ. When the output is taken as θ, the response is measured relative to the rigid-body mode; hence, the zero-frequency term disappears from the characteristic equation, and only the torsional vibration mode (*twisting mode*) remains in the dynamic equation.

The transfer function given by Equation v may be written as

$$G(s) = \frac{1/J_m}{s^2 + \omega_n^2} \tag{viii}$$

In the frequency domain $s = j\omega$, and the resulting frequency transfer function is

$$G(j\omega) = \frac{1/J_m}{\omega_n^2 - \omega^2} \tag{ix}$$

It follows that if ω is small in comparison to ω_n, the transfer function can be approximated by

$$G(j\omega) = \frac{1/J_m}{\omega_n^2} \tag{x}$$

This is a static relationship, implying an instantaneous response without any dynamic delay. Since system bandwidth represents the excitation frequency range ω within which the system responds sufficiently fast (which corresponds to the sufficiently flat region of the transfer function magnitude), it follows that the system bandwidth improves when ω_n is increased. Hence, ω_n is a measure of the system bandwidth.

Now, observe from Equation vii that ω_n (and the system bandwidth) increases when K_s is increased, when J_m is decreased, or when J_L is decreased. If a gearbox is added to the system, the equivalent inertia increases and the equivalent stiffness decreases. This reduces the system bandwidth, resulting in a slower response. Another disadvantage of a gearbox is the backlash and friction, which are nonlinearities that enter the system. The main advantage, however, is that the torque transmitted to the load is amplified through the speed reduction between the motor and the load. However, high torques and low speeds can be achieved by using torque motors without employing any speed reducers or by using backlash-free transmissions such as harmonic drives and traction (friction) drives.

For series-connected two torsional segments of stiffness K_s and $0.5K_s$, the equivalent stiffness K_e is given by

$$\frac{1}{K_e} = \frac{1}{K_s} + \frac{1}{0.5K_s} = \frac{3}{K_s} \rightarrow K_e = \frac{K_s}{3}$$

For a given moment of inertia, the natural frequency is proportional to the square root of the stiffness.

→Bandwidth is reduced by a factor of $1/\sqrt{3} \approx 0.58$.

→BW is reduced by approximately 42%.

The design of a torsion element for torque sensing can be viewed as the selection of the polar moment of area J of the element to meet the following four requirements:

1. The strain capacity limit specified by the strain-gauge manufacturer is not exceeded.
2. A specified upper limit on nonlinearity for the strain gauge is not exceeded, for linear operation.
3. Sensor sensitivity is acceptable in terms of the output signal level of the differential amplifier (see Chapter 2) in the bridge circuit.
4. The overall stiffness (bandwidth, steady-state error, and so on) of the system is acceptable.

In this situation, the torque sensor not only performs the sensing function but also becomes an integral part of the structure of the original system. In particular, the strength, dynamics, and bandwidth of the overall system are affected by the torque sensor. Hence, the sensor design takes a special meaning here, and specific considerations of system dynamics have to be taken into account. Now, we develop design criteria for the four requirements listed earlier for a torque sensor.

9.3.2.1 Strain Capacity of the Gauge

The maximum strain handled by a strain-gauge element is limited by factors such as strength, creep problems associated with the bonding material (epoxy), and hysteresis. This limit ε_{max} is specified by the strain-gauge manufacturer. For a typical SC gauge, the maximum strain limit is in the order of 3000 με. If the maximum torque that the sensor should handle is T_{max}, we have, from Equation 9.18:

$$\frac{r}{2GJ}T_{max} \le \varepsilon_{max} \rightarrow$$

$$J \ge \frac{r}{2G}\frac{T_{max}}{\varepsilon_{max}} \tag{9.23}$$

where ε_{max} and T_{max} are specified.

9.3.2.2 Strain-Gauge Nonlinearity Limit

For large strains, the characteristic equation of a strain gauge becomes increasingly nonlinear. This is particularly true for SC gauges. If we assume the quadratic equation (Equation 9.10), the percentage nonlinearity N_p is given by Equation 9.12. For a specified nonlinearity, an upper limit for strain can be determined using this result:

$$\frac{r}{2GJ}T_{max} = \varepsilon_{max} \le \frac{N_p S_1}{50 S_2} \tag{9.24}$$

The corresponding J is given by

$$J \geq \frac{25S_2}{GS_1}\frac{T_{max}}{N_p} \tag{9.25}$$

where N_p and T_{max} are specified.

9.3.2.3 Sensitivity Requirement

The output signal from the strain-gauge bridge is provided by a differential amplifier (see Chapter 3), which detects the voltages at the two output nodes of the bridge (A and B in Figure 9.5), takes the difference, and amplifies it by a gain K_a. This output signal is supplied to an ADC (see Chapter 4), which provides a digital signal to the computer for performing further processing and control. The signal level of the amplifier output has to be sufficiently high so that the signal-to-noise ratio is adequate. Otherwise, serious noise problems can result. Typically, a maximum voltage in the order of ±10 V is desired.

Amplifier output v is given by

$$v = K_a \delta v_o \tag{9.26}$$

where δv_o is the bridge output before amplification. It follows that the desired signal level can be obtained by simply increasing the amplifier gain. There are limits to this approach, however. In particular, a large gain increases the susceptibility of the amplifier to saturation and instability problems, such as drift, and errors as a result of parameter changes. Hence, sensitivity has to be improved as much as possible through mechanical considerations.

By substituting Equation 9.20 into Equation 9.26, we get the signal level requirement as

$$v_o \leq \frac{K_a k S_s r v_{ref}}{8GJ} T_{max}$$

where v_o is the specified lower limit on the output signal from the bridge amplifier and T_{max} is also specified. Then, the limiting design value for J is given by

$$J \leq \frac{K_a k S_s r v_{ref}}{8G}\frac{T_{max}}{v_o} \tag{9.27}$$

where v_o and T_{max} are specified.

9.3.2.4 Stiffness Requirement

The lower limit of the overall stiffness of the system is constrained by such factors as speed of response (represented by the system bandwidth) and steady-state error (represented by the system gain). The polar moment of area J should be chosen such that the stiffness of the torsional element does not fall below a specified limit K. First, we have to obtain an expression for the torsional stiffness of a circular shaft. For a shaft of length L and radius r, an angle of twist θ corresponds to a shear strain of

$$\gamma = \frac{r\theta}{L} \tag{9.28}$$

on the outer surface. Accordingly, shear stress is given by

$$\tau = \frac{Gr\theta}{L} \tag{9.29}$$

Now in view of Equation 9.19, the torsional stiffness of the shaft is given by

$$K_s = \frac{T}{\theta} = \frac{GJ}{L} \tag{9.30}$$

Note that the stiffness can be increased by increasing GJ. However, this decreases the sensor sensitivity because, in view of Equation 9.18, the measured direct strain ε decreases for a given torque, when GJ is increased. There are two other parameters—outer radius r and length L of the torsion element—which we can manipulate. Although for a solid shaft J increases (to the fourth power) with r, for a hollow shaft, it is possible to manipulate J and r independently, with practical limitations. For this reason, hollow members are commonly used as torque-sensing elements. With these design freedoms, for a given value of GJ, we can increase r to increase the sensitivity of the strain-gauge bridge without changing the system stiffness, and we can decrease L to increase the system stiffness without affecting the bridge sensitivity.

Assuming that the shortest possible length L is used in the sensor, for a specified stiffness limit K, we should have $GJ/L \geq K$. Then, the limiting design value for J is given by

$$J \geq \frac{L}{G}K \tag{9.31}$$

where K is specified.

Overall design problem: The design problem of a circular torsional member for a torque sensor may be carried out using the formulas (inequalities) derived earlier. The governing formulas for the polar moment of area J of a torque sensor, based on the four criteria discussed earlier, are summarized in Table 9.2. In particular, note the direction of each inequality. It is chosen so that any value within the inequality would satisfy the particular specification (albeit conservatively) and the best value is the one corresponding to the equality. Also, it is clear that out of the three "≥" values for J, we must pick the largest one.

TABLE 9.2

Design Criteria for a Strain-Gauge Torque-Sensing Element

Criterion	Specification	Governing Formula for Polar Moment of Area (J)
Strain capacity of strain-gauge element	ε_{max} and T_{max}	$\geq \frac{r}{2G} \cdot \frac{T_{max}}{\varepsilon_{max}}$
Strain-gauge nonlinearity	N_p and T_{max}	$\geq \frac{25rS_2}{GS_1} \cdot \frac{T_{max}}{N_p}$
Sensor sensitivity	v_o and T_{max}	$\leq \frac{K_a k S_s r v_{ref}}{8G} \cdot \frac{T_{max}}{v_o}$
Sensor stiffness (system bandwidth and gain)	K	$\geq \frac{L}{G} \cdot K$

Then, the other two specifications would be satisfied conservatively. If this largest lower-limit value for J is less than the upper-limit value for J as determined for the sensor sensitivity specification (third inequality of Table 9.2), then the largest lower limit is the best choice for J. If the latter is greater than the former, then there is no proper design choice, and we have to change some specifications and repeat the design calculations.

Note: Even after satisfying all four requirements given in Table 9.2, there may be other requirements that need to be addressed in the sensor design. For example, the wall thickness of the torsion member that is optimal under the four criteria in Table 9.2 might be too small and this could introduce the danger of structural instability (e.g., buckling). When such considerations are taken into account, the final design of the sensing member may not be "optimal" with regard to the four criteria in Table 9.2.

Example 9.6

A joint of a direct-drive robotic arm is sketched in Figure 9.18. The rotor of the drive motor is an integral part of the driven link, and there are no gears or any other speed reducers. Also, the motor stator is an integral part of the drive link. A tachometer measures the joint speed (relative), and a resolver measures the joint rotation (relative).

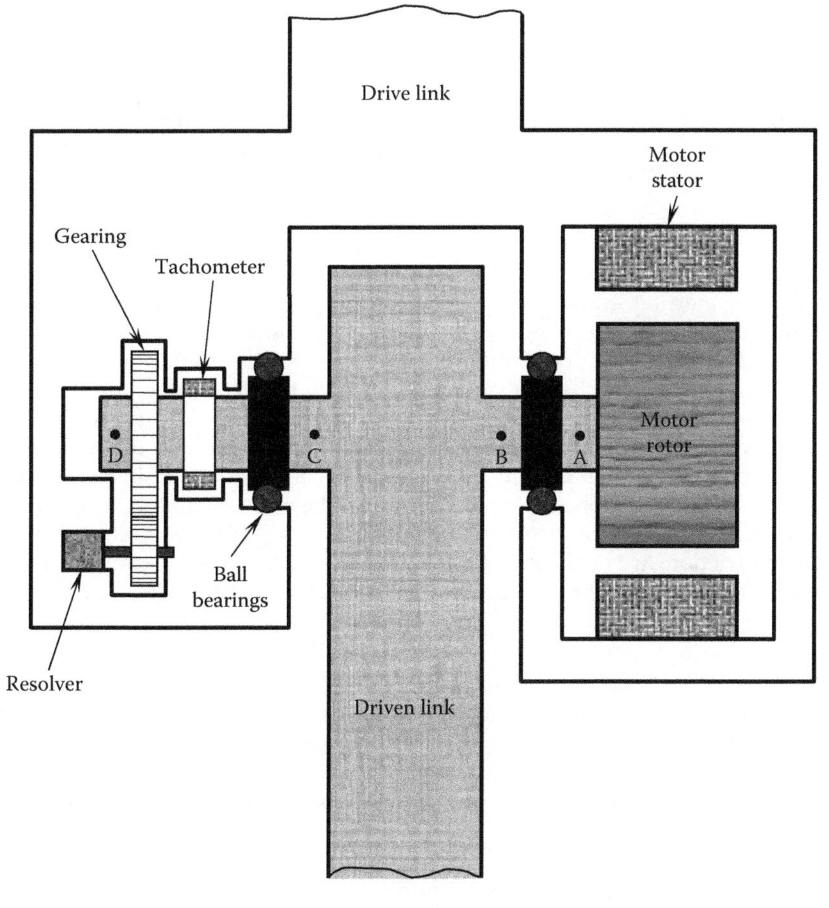

FIGURE 9.18
A joint of a direct-drive robotic arm.

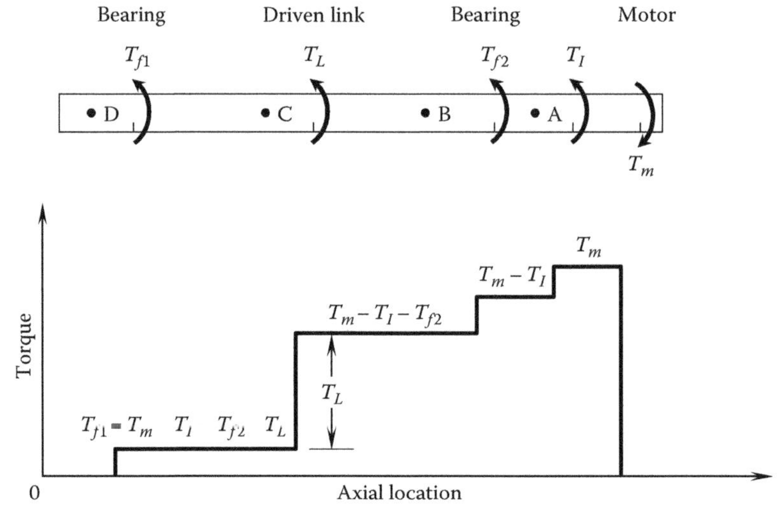

FIGURE 9.19
Torque distribution along the axis of a direct-drive manipulator joint.

Gearing is used to improve the performance of the resolver, and it does not affect the load transfer characteristics of the joint. Neglecting the mechanical loading from the sensors and the gearing, but including the bearing friction, sketch the torque distribution along the joint axis. Suggest a location (or locations) for measuring using a strain-gauge torque sensor the net torque transmitted to the driven link.

Solution

For simplicity, assume point torques. Denoting the motor (magnetic) torque by T_m, the total rotor inertia torque and the frictional torque in the motor by T_I, and the frictional torques at the two bearings by T_{f1} and T_{f2}, the torque distribution is sketched in Figure 9.19. The net torque transmitted to the driven link is T_L. The locations available to install strain gauges include A, B, C, and D. Note that T_L is given by the difference between the torques at B and C. Hence, strain-gauge torque sensors should be mounted at B and C and the difference of the readings should be taken for accurate measurement of T_L. Since bearing friction is small for most practical purposes, a single torque sensor located at B provides reasonably accurate results. The motor torque T_m is approximately equal to the transmitted torque when the effects of bearing friction and motor loading (inertia and friction) are negligible. This is the reason behind using motor current (field or armature) to measure joint torque in some applications (e.g., in robots).

Example 9.7

Consider the design of a tubular torsional element. Using the notation in Table 9.2, the following design specifications are given: $\varepsilon_{max} = 3000\ \mu\varepsilon$, $N_p = 5\%$, $v_o = 10$ V, and $K = 2.5 \times 10^3$ N·m/rad to achieve a system bandwidth of 50 Hz. A bridge with four active strain gauges is used to measure torque in the torsional element. The following parameter values are provided:

1. For the strain gauges: $S_s = S_1 = 115$ and $S_2 = 3500$.
2. For the torsion element: outer radius $r = 2$ cm, shear modulus $G = 3 \times 10^{10}$ N/m^2, and length $L = 2$ cm.

3. For the bridge circuitry: $v_{ref} = 20$ V and $K_a = 100$.
4. The maximum torque that is expected is $T_{\max} = 10$ N·m.

Using these values, design a torsional element for the sensor. Compute the operating parameter limits for the designed sensor.

Solution

Let us assume a factor of safety of 1 (i.e., use the actual limiting values of the design formulas). We can compute the polar moment of area J using each of the four criteria given in Table 9.2:

1. For $\varepsilon_{\max} = 3000$ με: $J = \dfrac{0.02\times 10}{2\times 3\times 10^{10}\times 3\times 10^{-3}}\,\text{m}^4 = 1.11\times 10^{-9}\ \text{m}^4$.
2. For $N_p = 5$: $J = \dfrac{25\times 0.02\times 3500\times 10}{3\times 10^{10}\times 115\times 5}\,\text{m}^4 = 1.01\times 10^{-9}\ \text{m}^4$.
3. For $v_o = 10$ V: $J = \dfrac{100\times 4\times 115\times 0.02\times 20\times 10}{8\times 3\times 10^{10}\times 10}\,\text{m}^4 = 7.67\times 10^{-8}\ \text{m}^4$.
4. For $K = 2.5\times 10^3$ N·m/rad: $J = \dfrac{0.02\times 2.5\times 10^3}{3\times 10^{10}}\,\text{m}^4 = 1.67\times 10^{-9}\ \text{m}^4$.

It follows that for an acceptable sensor, we should satisfy

$$J \ge (1.11\times 10^{-9}) \text{ and } (1.01\times 10^{-9}) \text{ and } (1.67\times 10^{-9}) \text{ and } J \le 7.67\times 10^{-8}\ \text{m}^4$$

$$\rightarrow 1.67\times 10^{-9} \le J \le 7.67\times 10^{-8}\ \text{m}^4$$

We pick $J = 7.67\times 10^{-8}$ m^4, which is the largest J that satisfies all the design specifications. This is not the optimal choice if only the four design specifications given in Table 9.2 are considered. However, this choice is made so that the tube thickness is sufficiently large to transmit the load without buckling or yielding. To illustrate this point, let us compare this "nonoptimal" design choice with the optimal value.

For a tubular shaft, $J = (\pi/2)(r_o^4 - r_i^4)$ where r_o is the outer radius and r_i is the inner radius $\rightarrow 7.67\times 10^{-8} = (\pi/2)(0.02^4 - r_i^4) \rightarrow r_i = 1.8$ cm.

Now, with the chosen value for J:

$$\varepsilon_{\max} = \frac{7.67\times 10^{-8}}{1.11\times 10^{-9}}\times 3000\ \mu\varepsilon = 2.07\times 10^5\ \mu\varepsilon$$

$$N_p = \frac{1.01\times 10^{-9}}{7.67\times 10^{-8}}\times 5\% = 0.07\%$$

$$v_o = 10\ \text{V}$$

$$K = \frac{7.67\times 10^{-8}}{1.67\times 10^{-9}}\times 2.5\times 10^3 = 1.15\times 10^5\ \text{N}\cdot\text{m/rad}$$

Since natural frequency is proportional to the square root of stiffness, for a given inertia, we note that a bandwidth of $50\sqrt{(1.15\times 10^5)/(2.5\times 10^3)} = 339$ Hz is possible with this design.

Note: In this situation of torque sensing, the bandwidth of the sensor (i.e., operating frequency range of torque sensing) and the mechanical bandwidth of the overall dynamic system (governed by the torsional natural frequency of the two inertias connected by a flexible shaft) are intimately related. Hence, even though we may be specifying the sensor bandwidth for the measurement process, we are indirectly constraining the mechanical bandwidth of the overall system. Such intimate coupling of sensor bandwidth and system bandwidth may not be present in some other sensing situations.

Now, consider the optimal value for J, which is $J = 1.67 \times 10^{-9}$ m^4.

We have $1.67 \times 10^{-9} = (\pi/2)(0.02^4 - r_i^4) \rightarrow r_i = 1.997$ cm.

Clearly, this is not a good choice for J because the corresponding wall thickness is very small and can easily cause buckling and other structural problems. Also, this choice will lead to very high sensitivity for the bridge output (much larger than 10 V). So, it is desirable to stay with the "nonoptimal" choice given earlier.

Note: Typically, hollow sensor elements are used for sensing torque up to about 50 N·m and solid sensor elements are used for higher torques.

The manner in which the strain gauges are configured on a torque sensor can be exploited to compensate for cross-sensitivity effects arising from factors such as tensile and bending loads, which cause error in a torque measurement. However, it is advisable to use a torque-sensing element that inherently possesses low sensitivity to these factors. The tubular torsion element discussed in this section is convenient for analytical purposes because of the simplicity of the associated expressions for design parameters. Its mechanical design and integration into a practical system are convenient as well. Unfortunately, this member is not optimal with respect to rigidity (stiffness) for the transmission of both bending and tensile loads. Alternative shapes and structural arrangements have to be considered when inherent rigidity (insensitivity) to cross-loads is needed. Furthermore, a tubular element has the same principal strain at all locations on the element surface. This does not give us a choice with respect to the mounting locations of the strain gauges in order to maximize the torque sensor sensitivity. Another disadvantage of the basic tubular torsional member is that, due to its curved surface, much care is needed in mounting fragile SC gauges, which could be easily damaged even with slight bending. Hence, a sensor element that has flat surfaces to mount the strain gauges would be desirable.

A torque-sensing element that has the foregoing desirable characteristics (i.e., inherent insensitivity to cross-loading, nonuniform strain distribution on the surface, and

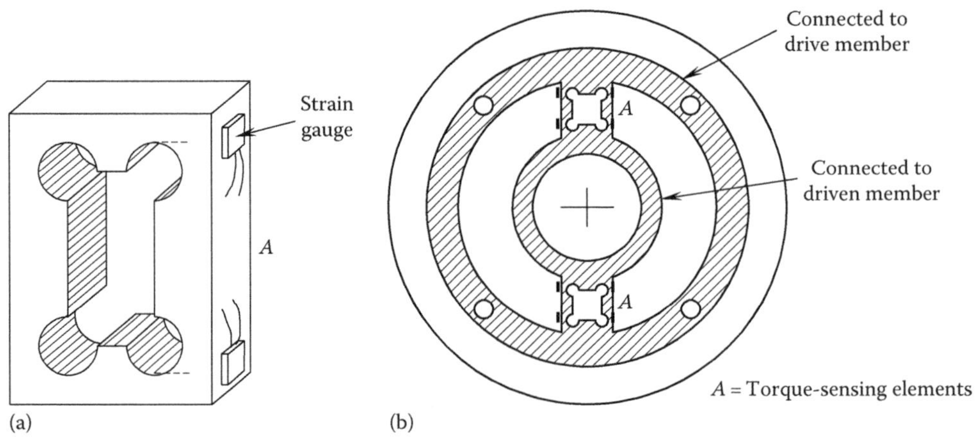

FIGURE 9.20
A bending element for torque sensing. (a) Shape of the sensing element. (b) Element locations (two radially placed elements).

availability of flat surfaces to mount strain gauges) is shown in Figure 9.20. Note that two sensing elements are connected radially between the drive unit and the driven member. In this design, the sensing elements undergo bending to transmit a torque between the driver and the driven member. Bending strains are measured at locations of high sensitivity and are taken to be proportional to the transmitted torque. Analytical determination of the calibration constant is not easy for such complex sensing elements, but experimental determination is straightforward. Finite element analysis may be used as well for this purpose.

Note: Strain-gauge torque sensors measure the direction as well as the magnitude of the torque transmitted through it.

Surface acoustic wave (SAW) torque sensors: SAW sensor is a microminiature acoustic resonator made of a piezoelectric material whose resonant frequency (in the MHz range) varies with the surface strain at the sensor location. Hence, it can be considered as a strain sensor and can be used to measure torque. The frequency variation is sensed by a stationary detector. The advantages of a SAW torque sensor include wireless operation (useful for sensing of moving parts) and high measurement bandwidth (in the kHz range).

9.3.3 Deflection Torque Sensors

Instead of the strain in the sensor element, the actual deflection or deformation (twisting or bending) may be measured and used to determine the torque, through a suitable calibration constant. For a circular-shaft (solid or hollow) torsional element, the governing relationship for the angle of twist (θ) corresponding to an applied torque (T) is given by Equation 9.30, which may be written in the form

$$T = \frac{GJ}{L}\theta \tag{9.32}$$

The calibration constant is GJ/L, and it has to be small in order to achieve high sensitivity. This means that the element stiffness should be low. This limits the bandwidth, which measures the speed of response, and the gain, which determines the steady-state error, of the overall system. For a system with high bandwidth, the angle of twist θ should be very small (e.g., a fraction of a degree). Hence, very accurate measurement of θ is required in this type of torque sensors. Three types of displacement- or deformation-based torque sensors are described next. One sensor directly measures the angle of twist. The second sensor uses the change in magnetic induction associated with sensor deformation. The third sensor uses reverse magnetostriction.

9.3.3.1 Direct-Deflection Torque Sensor

Direct measurement of the angle of twist between two axial locations in a torsional member, using an angular displacement sensor, may be used to determine torque. The difficulty in this method is that, under dynamic conditions, relative deflection has to be measured while the torsion element is rotating. One type of displacement sensor that could be used here is a synchro transformer. Suppose that the two rotors of the synchro are mounted at the two ends of the torsion member. The synchro output gives the relative angle of rotation of the two rotors.

Another type of displacement sensor that may be used for the same objective is shown in Figure 9.21a. Two ferromagnetic gear wheels are splined at two axial locations of

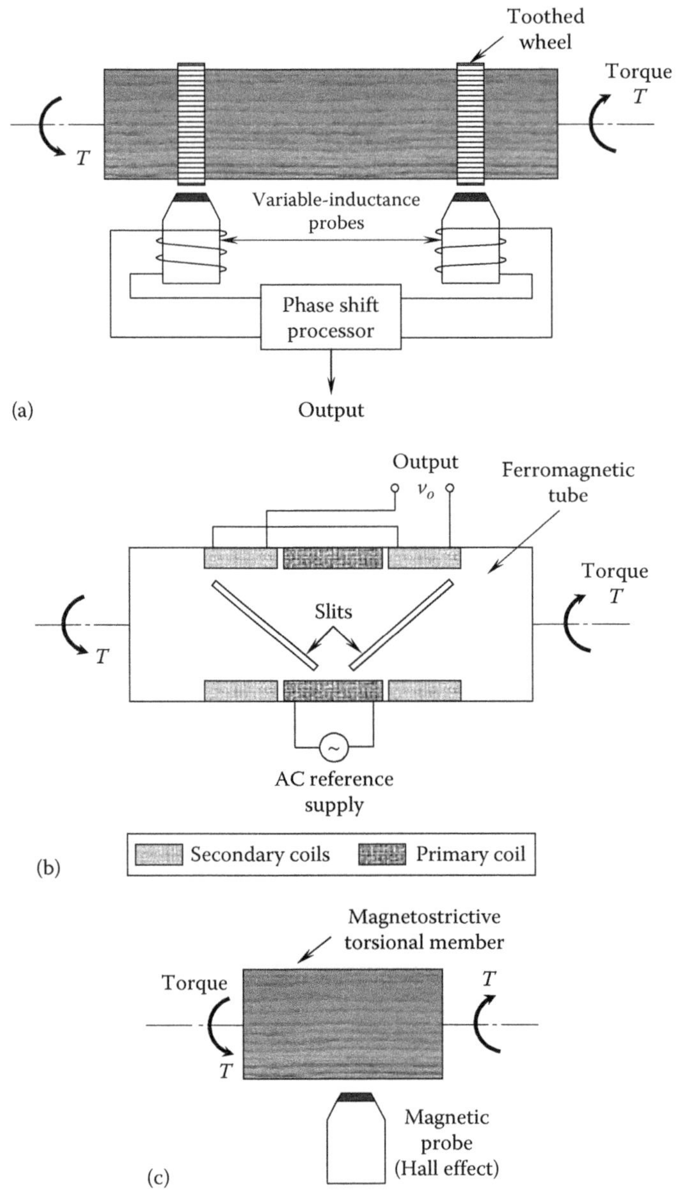

FIGURE 9.21
Deflection torque sensors. (a) A direct-deflection torque sensor, (b) a variable-reluctance torque sensor, and (c) a magnetostrictive torque sensor.

the torsional element. Two stationary proximity probes of the magnetic induction type (self-induction or mutual induction) are placed radially, facing the gear teeth, at the two locations. As the shaft rotates, the gear teeth cause a change in flux linkage with the proximity sensor coils. The resulting output signals of the two probes are pulse sequences, shaped somewhat like sine waves. The phase shift of one signal with respect to the other determines the relative angular deflection of one gear wheel with respect to the other, assuming that the two probes are synchronized under no-torque conditions. Both the magnitude and the direction of the transmitted torque are determined using this method. A 360° phase shift corresponds to a relative deflection by an integer multiple of the gear pitch. It follows that in this arrangement, deflections less than half the gear-tooth pitch can be measured without ambiguity. Assuming that the output signals of the two probes are sine waves (narrowband filtering can be used to achieve this), the phase shift ϕ is proportional to the angular twist θ. If the gear wheel has n teeth, a primary phase shift of 2π corresponds to an angle of twist of $2\pi/n$ radians. Hence, $\theta = \phi/n$ and from Equation 9.32, we get

$$T = \frac{GJ\phi}{Ln} \tag{9.33}$$

where

- G is the shear modulus of the torsion element
- J is the polar moment of area of the torsion element
- ϕ is the phase shift between the two proximity probe signals
- L is the axial separation of the proximity probes
- n is the number of teeth in each gear wheel

Note: Proximity probes are noncontact devices. Eddy current proximity probes (Chapter 8) and Hall-effect proximity probes may be used instead of magnetic-induction probes in this method of torque sensing.

9.3.3.2 Variable-Reluctance Torque Sensor

A torque sensor that is based on the deformation of the sensor element and that does not require a contacting commutator is shown in Figure 9.21b. This is a variable-reluctance device, which operates like a differential transformer (RVDT or LVDT) as studied in Chapter 8. The torque-sensing element is a ferromagnetic tube, which has two sets of slits, typically oriented along the two principal stress directions of the tube (i.e., at 45° to the axial direction) under torsion. When a torque is applied to the torsion member, one set of gaps closes and the other set opens as a result of the principal stresses normal to the slit axes. Primary and secondary coils are placed around the slit tube, and they remain stationary. One segment of the secondary coil is placed around one set of slits, and the secondary segment is placed around the other (perpendicular) set. The primary coil is excited by an AC supply, and the induced voltage v_o in the secondary coil is measured. As the tube deforms, it changes the magnetic reluctance in the flux linkage path, thus changing the induced voltage. To obtain the best sensitivity, the two segments of the secondary coil, as shown in Figure 9.21b, should be connected so that the induced voltages are absolutely additive (algebraically subtractive) because one voltage increases and the other decreases. The output signal should be demodulated (by removing the carrier frequency component) to effectively measure transient torques.

Note: The direction of torque is given by the sign of the demodulated signal.

9.3.3.3 Magnetostriction Torque Sensor

Magnetostriction torque sensor uses the principle of *reverse magnetostriction*. In the direct magnetostriction, a magnetostrictive material deforms when subjected to a magnetic field.

TABLE 9.3

Some Magnetostrictive Materials

Material	Stress Sensitivity $\partial B/\partial\sigma$ (T·m²/N)
82% Ni–18% Co alloy	12.7×10^{-9}
Alfer (86% Fe–14% Al)	6.5×10^{-9}
Nickel (Ni)	6.1×10^{-9}

Note: Stress = Young's modulus × strain. Strain sensitivities can be determined through this relation.

In the reverse magnetostriction (or *Villari effect*), the deformation of a magnetostrictive material changes its magnetization. *Note*: The energy conversion is between elastic potential (mechanical) energy and magnetic energy. The change in magnetization in a magnetostrictive torsional member may be sensed by a stationary probe (e.g., a Hall-effect sensor) and from it the deformation of the member (and hence the torque carried by it) can be sensed.

Common magnetostrictive materials are nickel and its alloys, some ferrites, some rare earths, and alfer (86% iron and 14% aluminum alloy). An important property in their use in mechanical sensing (e.g., torque sensing) is the stress sensitivity $\partial B/\partial\sigma$, which is the change in magnetic flux density (unit: weber/m² = tesla [T]) for a unit change in stress (unit: N/m²).

Some magnetostrictive materials and their stress sensitivities are given in Table 9.3. A schematic representation of a magnetostrictive torque sensor is given in Figure 9.21c.

9.3.4 Reaction Torque Sensors

The methods of torque sensing that were described thus far use a sensing element that is connected between the drive member (actuator) and the driven member (load). There are two major drawbacks in this arrangement of torque sensing:

1. The sensing element modifies the original system in an undesirable manner, particularly by decreasing the system stiffness and adding inertia. As a result, not only does the overall bandwidth of the system decrease, but the original torque is also changed (due to mechanical loading) because of the inclusion of an auxiliary sensing element.
2. Under dynamic conditions, the sensing element is in motion, thereby making torque measurement more difficult. Then, some form of commutation (e.g., slip ring and brush), rotary transformer, or wireless telemetry would be needed in reading the sensor signal.

The reaction method of torque sensing eliminates these problems to a large extent. In particular, this method can be conveniently used to measure torque in a rotating machine.

In this method, the supporting structure (or housing) of the rotating machine (e.g., motor, pump, compressor, turbine, generator) is cradled by releasing the fixtures, and the effort that is necessary to keep the structure from moving (i.e., to hold down) is measured. A schematic representation of the method is shown in Figure 9.22a. Typically, a lever arm is mounted on the cradled housing, and the force that is required to maintain the housing

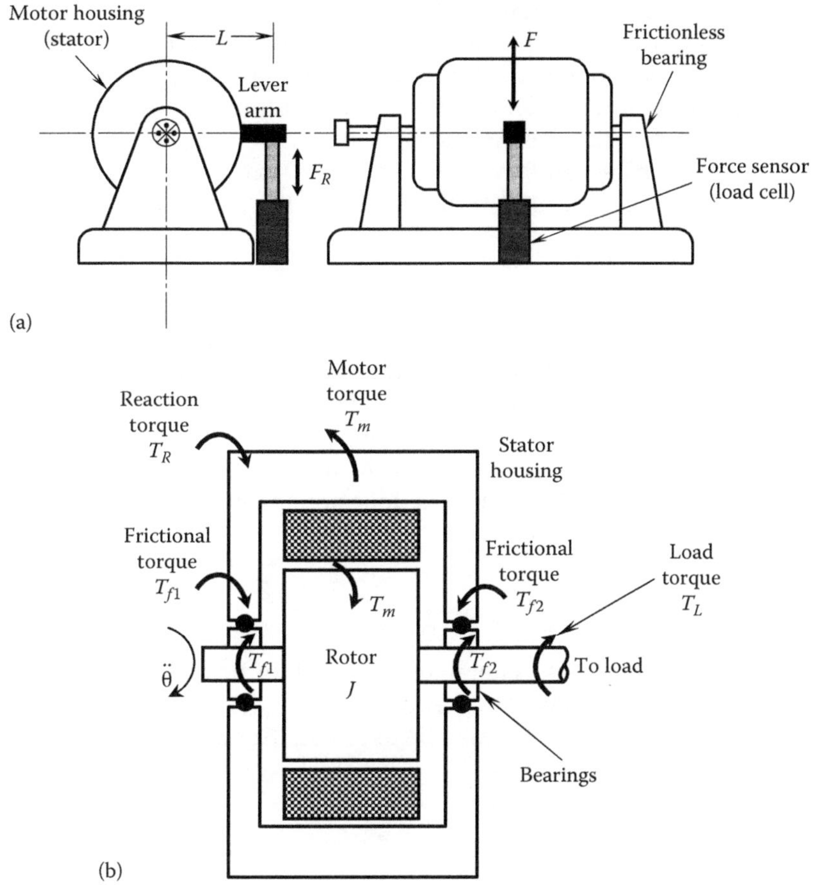

FIGURE 9.22
(a) Schematic representation of a reaction torque sensor setup (reaction dynamometer) and (b) the relationship between reaction torque and load torque.

stationary is measured using a force sensor (load cell). The reaction torque on the housing is given by

$$T_R = F_R \cdot L \tag{9.34}$$

where
F_R is the reaction force that is measured using the load cell
L is the lever arm length

Alternatively, strain gauges or other types of force sensors may be mounted directly at the fixture locations (e.g., on the mounting bolts) of the housing, to measure the reaction forces without actually having to cradle the housing. Then, the reaction torque is determined with the knowledge of the distance of the fixture locations from the shaft axis.

The reaction-torque method of torque sensing is widely used in dynamometers (reaction dynamometers), which determine the transmitted power in rotating machinery through the measurement of torque and shaft speed. A drawback of the reaction-type torque sensors can be explained using Figure 9.22b. A motor of rotor inertia J, which rotates at angular acceleration $\ddot{\theta}$, is shown. By Newton's third law (action = reaction), the electromagnetic torque T_m generated at the rotor of the motor is reacted back onto the stator and housing. In the figure, T_{f1} and T_{f2} denote the frictional torques at the two bearings and T_L is the torque transmitted to the driven load.

When applying Newton's second law to the entire system, note that the frictional torques and the motor (magnetic) torque all cancel out, giving $J\ddot{\theta} = T_R - T_L$ or

$$T_L = T_R - J\ddot{\theta} \tag{9.35}$$

Note: T_L is what must be measured.

Under accelerating or decelerating conditions, the reaction torque T_R, which is measured, is not equal to the actual torque T_L that is transmitted. A method of compensating for this error is to measure the shaft acceleration, compute the inertia torque, and adjust the measured reaction torque using this inertia torque.

Note: The frictional torque in the bearings does not enter into the final equation, which is another advantage of this method.

9.3.5 Motor Current Torque Sensors

The torque in an electric motor is generated as a result of the electromagnetic interaction between the rotor magnetic field and the stator magnetic field of the motor. Hence, the current that generates the magnetic field may be used to estimate the motor torque. Here, we will consider both DC and AC motors.

9.3.5.1 DC Motors

In a DC motor, the rotor may have armature windings and the stator may have field windings. Consider a DC motor where both rotor and stator have electromagnets (windings carrying current). The resulting (magnetic) torque T_m is given by

$$T_m = k i_f i_a \tag{9.36}$$

where

i_f is the field current
i_a is the armature current
k is the torque constant

It is seen from Equation 9.36 that the motor torque can be determined by measuring either i_a or i_f while the other is kept constant at a known value (or the corresponding magnetic field is provided by a permanent magnet). In particular, i_f is assumed constant in armature control and i_a is assumed constant in field control.

As noted earlier (e.g., see Figure 9.22b), the magnetic torque of a motor is not quite equal to the transmitted torque, the latter being what needs to be sensed in most applications. It follows that the motor current provides only an approximation for the needed torque. The actual torque that is transmitted through the motor shaft (the load torque) is different from the motor (magnetic) torque generated at the stator–rotor interface of the motor. This difference is necessary for overcoming the inertia torque of the moving parts of the

FIGURE 9.23
A current sensor. (Courtesy of Alps Electric, Auburn Hills, MI.)

motor unit (particularly the rotor inertia) and the frictional torque (particularly bearing friction). Methods are available to adjust (compensate for) the magnetic toque so as to estimate the transmitted torque at sufficient accuracy. One approach is to incorporate a suitable dynamic model for the electromechanical system of the motor and the load in a Kalman filter (see Chapter 7) whose input is the measured current and the estimated output is the transmitted load. A detailed presentation of this approach is beyond the present scope.

The motor current can be measured by several ways, for example, by sensing the voltage across a known resistor (of low resistance) placed in series with the current circuit or by sensing the magnetic field generated by the current (e.g., by using a Hall-effect sensor). Currents as high as 100 A may be sensed at fast response times (1 μs) using miniature conductors (60 μΩ). A commercial current sensor (size: 1 cm) that can be used in such applications as motor drivers; power-conditioning systems; building heating, ventilation, and air-conditioning systems; and industrial machinery is shown in Figure 9.23.

9.3.5.2 AC Motors

In the past, DC motors were predominantly used in complex control applications. Although AC synchronous motors were limited mainly to constant-speed applications in the past, they are finding numerous uses in variable-speed applications (e.g., robotic manipulators) and servo systems because of rapid advances in solid-state drives. Today, AC motor drive systems incorporate both frequency control and voltage control using advanced SC technologies.

The torque in an AC motor may also be determined by sensing the motor current. For example, consider the three-phase synchronous motor shown schematically in Figure 9.24.

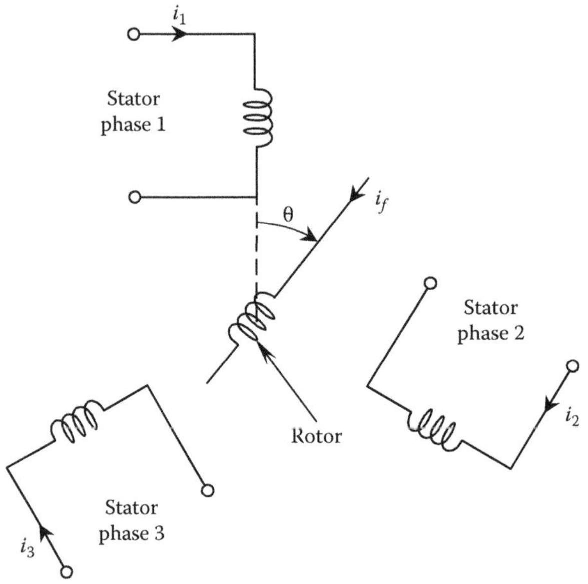

FIGURE 9.24
Schematic representation of a three-phase synchronous motor.

The armature windings of a conventional synchronous motor are carried by the stator (in contrast to the case of a DC motor). Suppose that the currents in the three phases (armature currents) are denoted by i_1, i_2, and i_3. The DC field current in the rotor windings is denoted by i_f. Then, the motor torque T_m can be expressed as

$$T_m = ki_f\left[i_1 \sin\theta + i_2 \sin\left(\theta - \frac{2\pi}{3}\right) + i_3 \sin\left(\theta - \frac{4\pi}{3}\right)\right] \tag{9.37}$$

where
θ is the angular rotation of the rotor
k is the torque constant of the synchronous motor

Since i_f is assumed fixed, the motor torque can be determined by measuring the phase currents. For the special case of a balanced three-phase supply, we have $i_1 = i_a \sin \omega t$, $i_2 = i_a \sin(\omega t - (2\pi/3))$, and $i_3 = i_a \sin(\omega t - (4\pi/3))$, where ω is the line frequency (frequency of the current in each supply phase) and i_a is the amplitude of the phase current. Substituting these equations into Equation 9.37 and simplifying by using well-known trigonometric identities, we get $T_m = 1.5ki_f i_a \cos(\theta - \omega t)$. The angular speed of a three-phase synchronous motor with one pole pair per phase is equal to the line frequency ω. Accordingly, $\theta = \theta_0 + \omega t$, where θ_0 is the angular position of the rotor at $t = 0$. It follows that with a balanced three-phase supply, the torque of a synchronous motor is given by

$$T_m = 1.5ki_f i_a \cos\theta_0 \tag{9.38}$$

This expression is quite similar to the one for a DC motor, as given by Equation 9.36.

9.3.6 Force Sensors

Force sensors are useful in numerous applications. For example, cutting forces generated by a machine tool may be monitored to detect tool wear and an impending *failure* and to *diagnose* the causes of it; to control the machine tool, through feedback; to determine an experimental model; or to evaluate *product quality*. In vehicle testing, force sensors are used to monitor impact forces on the vehicles and crash-test dummies. Robotic handling and assembly tasks are controlled by measuring the forces generated at the end effector. Haptic teleoperation using a master manipulator and a slave manipulator may use force sensing at the robot end effector when interacting with the work environment. This information can then be used to generate a realistic "feel" to the operator of the master manipulator (which remotely guides the task execution) of the operating conditions at the slave manipulator (where the task is executed). Measurement of excitation forces and corresponding responses is employed in *experimental modeling* (model identification) of mechanical systems. Direct measurement of forces is useful in *nonlinear feedback control* of mechanical systems, where the measured force is used to determine the nonlinear dynamics, which can be subtracted from the overall dynamics of the system, using a suitable controller. This will make the system behave in a linear manner.

Force sensors that employ strain-gauge elements or piezoelectric (quartz) crystals with built-in microelectronics are common. For example, thin-film and foil sensors that employ the strain-gauge principle for measuring forces and pressures are commercially available. A sketch of an industrial load cell, which uses the strain-gauge method, is shown in Figure 9.25. Both impulsive forces and slowly varying forces can be monitored using this sensor. Some types of force sensors are based on measuring a deflection caused by the force. Relatively high deflections (fraction of an mm) would be necessary for this technique to be feasible. Commercially available sensors range from sensitive devices, which can detect forces in the order of thousandth of a newton to heavy-duty load cells, which can handle very large forces (e.g., 10,000 N). The techniques of torque sensing that have been discussed (e.g., magnetostrictive, RAW) can be extended in a straightforward manner to force sensing. Hence, further discussion of the topic is not undertaken here. Typical rating parameters for several types of sensors are given in Table 9.4.

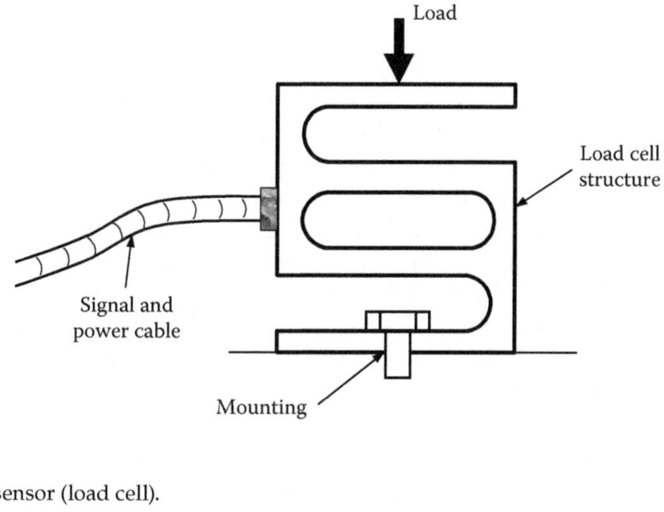

FIGURE 9.25
An industrial force sensor (load cell).

TABLE 9.4

Rating Parameters of Several Sensors and Transducers

Transducer	Measurand	Measurand Frequency Max/Min	Output Impedance	Typical Resolution	Accuracy	Sensitivity
Potentiometer	Displacement	10 Hz/DC	Low	≤0.1 mm	0.1%	200 mV/mm
LVDT	Displacement	2500 Hz/DC (max, limited by carrier frequency)	Moderate	≤0.001 mm	0.1%	50 mV/mm
Resolver	Angular displacement	500 Hz/DC (max, limited by carrier frequency)	Low	2 min	0.2%	10 mV/°
DC tachometer	Velocity	700 Hz/DC	Moderate (50 Ω)	0.2 mm/s	0.5%	5 mV/mm/s 75 mV/rad/s
Eddy current proximity sensor	Displacement	100 kHz/DC	Moderate	0.001 mm, 0.05% full scale	0.5%	5 V/mm
Piezoelectric accelerometer	Acceleration (and velocity, etc.)	25 kHz/1 Hz	High	1 mm/s²	0.1%	0.5 mV/m/s²
Semiconductor strain gauge	Strain (displacement, acceleration, etc.)	1 kHz/DC (limited by fatigue)	200 Ω	1–10 με (1 με = 10^{-6} strain)	0.1%	1 V/ε, 2000 με·max
Load cell	Force (1–1000 N)	500 Hz/DC	Moderate	0.01 N	0.05%	1 mV/N
Laser	Displacement/ shape	1 kHz/DC	100 Ω	1.0 μm	0.5%	1 V/mm
Optical encoder	Motion	100 kHz/DC	500 Ω	10 bit	±½ bit	10^4 pulses/ rev

9.4 Tactile Sensing

Tactile sensing is usually interpreted as *touch sensing*, but tactile sensing is different from a situation of simple "clamping" where a very few discrete force measurements may be made. In tactile sensing, a force "distribution" is measured, using a closely spaced array of force sensors and usually exploiting the skin-like properties of the sensor array.

Tactile sensing is particularly important in two types of operations: (1) grasping, handling, and fine manipulation and (2) object identification. In grasping, handling, and fine manipulation, the object has to be held in a stable manner without being allowed to slip and without being damaged. Object identification includes recognizing or determining the shape, location, and orientation of an object as well as detecting or identifying surface properties (e.g., density, hardness, texture, flexibility) and defects. Ideally, these tasks would require two types of sensing:

1. Continuous spatial sensing of time-variable contact forces
2. Sensing of surface deformation profiles (time variable)

These two types of data are generally related through the constitutive relations (physical characteristics such as stress–strain relations) of the touch surface of the tactile sensor or of the object that is being grasped. As a result, either the almost-continuous spatial sensing of tactile forces or the sensing of a tactile deflection profile, separately, is often termed *tactile sensing*. Note that "learning" also can be an important part of tactile sensing. For example, picking up a fragile object such as an egg and picking up an object that has the same shape but is made of a flexible material are not identical processes; they require some learning through touch, particularly when vision capability is not available. Also, learning of "impending" slip may be of importance in some applications of object handling.

9.4.1 Tactile Sensor Requirements

Significant advances in tactile sensing have taken place in the robotics area. Applications, which are very general and numerous, include automated inspection of surface profiles and joints (e.g., welded or glued parts) for defects, material handling or parts transfer (e.g., pick and place), detection of material properties using touch (e.g., the texture of a piece of herring roe), parts assembly (e.g., parts mating), parts identification and gauging in manufacturing applications (e.g., determining the size and shape of a turbine blade picked from a bin), haptic teleoperation of a slave robot at a distant location using a master manipulator, and fine-manipulation tasks (e.g., production of arts and craft, robotic engraving, and robotic microsurgery). *Note*: Some of these applications might need only simple touch (force–torque) sensing if the parts being grasped are properly oriented and if adequate information about the process and the objects is already available.

Naturally, the frequently expressed design objective for a tactile sensing device has been to mimic the capabilities of human fingers. Specifically, a tactile sensor should have a compliant covering with skin-like properties, along with adequate degrees of freedom for flexibility and dexterity, adequate sensitivity and resolution for information acquisition, adequate robustness and stability to accomplish various tasks, and some local intelligence for identification and learning purposes. Although the spatial resolution of a human fingertip is about 2 mm, still finer spatial resolutions (less than 1 mm) can be realized if information through other senses (e.g., vision), prior experience, and intelligence are used simultaneously during the touch. The force resolution (or tactile feel) of a human fingertip is on the order of 1 gm. Also, human fingers can predict "impending slip" during grasping so that corrective actions can be taken before the object actually slips. At an elementary level, this requires the knowledge of the shear stress distribution and friction properties at the common surface between the object and the hand. Additional information and an "intelligent" processing capability are also needed to accurately predict and to take corrective actions to prevent slipping. These are, of course, somewhat ideal goals for a tactile sensor, but they are not entirely unrealistic.

Sensor density or resolution, dynamic range, response time or bandwidth, strength and physical robustness, size, stability (dynamic robustness), linearity, flexibility, and localized intelligence (including data processing, learning, and reorganization) are important factors. They have to be incorporated in the analysis, design, or selection of a tactile sensor. Because of the necessity of a large number of sensor elements, the signal conditioning and processing for tactile sensors present challenges. Typical specifications for an industrial tactile sensor are as follows:

1. Spatial resolution of about 1 mm (about 100 sensor elements)
2. Force resolution of about 2 g

3. Dynamic range of 60 dB
4. Force capacity (maximum touch force) of about 1 kg
5. Response time of 5 ms or less (a bandwidth of over 200 Hz)
6. Low hysteresis (low energy dissipation)
7. Durability (and strength) under harsh and strenuous working conditions
8. Robustness and insensitivity to change in environmental conditions (temperature, dust, humidity, vibration, etc.)
9. Capability to detect and even predict slip

The chosen specifications depend on the particular application and associated constraints (size, shape, strength, cost, etc.).

Although the technologies of tactile sensing have not peaked yet and the widespread use of tactile sensors in industrial applications is still to come, tactile sensors that meet and even exceed the foregoing specifications are commercially available. In the design and development of tactile sensors, two separate groups of issues are commonly addressed:

1. Improvement of the mechanical characteristics and design of a tactile sensor so that accurate data with high resolution can be acquired quickly using the sensor
2. Improvement of the capabilities of signal acquisition, analysis, and processing so that useful information can be extracted accurately and quickly from the data acquired through tactile sensing

Under the second category, we also consider techniques of using tactile information in the feedback control of dynamic processes. In this context, the development of control algorithms, rules, and inference techniques for intelligent controllers that use tactile information is relevant.

9.4.1.1 Dexterity

Dexterity is an important consideration in sophisticated manipulators and robotic hands that employ tactile sensing. The dexterity of a device is conventionally defined as the ratio

$$\frac{\text{[Number of degrees of freedom in the device]}}{\text{[Motion resolution of the device]}}$$

We will call this "motion dexterity."

We can define another type of dexterity called "force dexterity," as follows:

$$\text{Force dexterity} = \frac{\text{Number of degrees of freedom}}{\text{Force resolution}} \tag{9.39}$$

Both types of dexterity are useful in mechanical manipulation where tactile sensing is used.

9.4.2 Construction and Operation of Tactile Sensors

The touch surface of a tactile sensor is usually made of an elastomeric pad or flexible membrane. Starting from this common basis, the principle of operation of a tactile sensor can differ primarily depending on whether the distributed force is sensed or the deflection of the tactile surface is sensed. The common methods of tactile sensing include the following:

1. Use a closely spaced set of strain gauges or other types of force sensors to sense the distributed force.
2. Use a conductive elastomer as the tactile surface. The change in its resistance, as it deforms, will determine the distributed force.
3. Use a closely spaced array of deflection sensors or proximity sensors (e.g., optical sensors) to determine the deflection profile of the tactile surface.

Since force and deflection are related through a constitutive law for the tactile sensor (touch pad), only one type of measurement, not both force and deflection, is needed in tactile sensing. A force distribution profile or a deflection profile that is obtained in this manner may be treated as a 2D array or an "image" and may be processed (filtered, function fitted, etc.) and displayed as a "tactile image" or used in applications (object identification, manipulation control, etc.).

The contact force distribution in a tactile sensor is commonly measured using an array of force sensors located under the flexible membrane. Arrays of piezoelectric sensors (see Chapter 8) and metallic or SC strain gauges (piezoresistive sensors) in sufficient density (number of elements per unit area) may be used for the measurement of the tactile force distribution. SC elements are poor in mechanical strength but have good sensitivity. Alternatively, the skin-like membrane itself can be made from a conductive elastomer (e.g., graphite-leaded neoprene rubber) whose change in resistance can be sensed and used in determining the distribution of force and deflection. In particular, as the tactile pressure increases, the resistance of the particular elastomer segment decreases and the current conducted through it (due to an applied constant voltage) will increase. Conductors can be etched underneath the elastomeric pad to detect the current distribution in the pad through proper signal acquisition circuitry. Common problems with conductive elastomers are electrical noise, nonlinearity, hysteresis, low sensitivity, drift, low bandwidth, and poor material strength.

The deflection profile of a tactile surface may be determined using a matrix of proximity sensors or deflection sensors. Electromagnetic and capacitive sensors may be used in obtaining this information. The principles of operation of these types of sensors have been discussed in Chapter 8. Optical tactile sensors use light-sensitive elements (photosensors) to sense the intensity of light (or laser beams) reflected from the tactile surface (see Chapter 10). Optical methods have the advantages of being free from electromagnetic noise and safe in explosive environments, but they can have errors due to stray light reaching the sensor, variation in intensity of the light source, and changes in environmental conditions (e.g., dirt, humidity, and smoke).

Example 9.8

A tactile sensor pad consists of a matrix of conductive-elastomer elements. The resistance R_t in each tactile element is given by $R_t = a/F_t$, where F_t is the tactile force applied to the element and a is a constant. The circuit shown in Figure 9.26 is used to acquire the tactile sensor signal v_o, which measures the local tactile force F_t. The entire matrix of tactile elements may be scanned by addressing the corresponding elements through an appropriate switching arrangement.

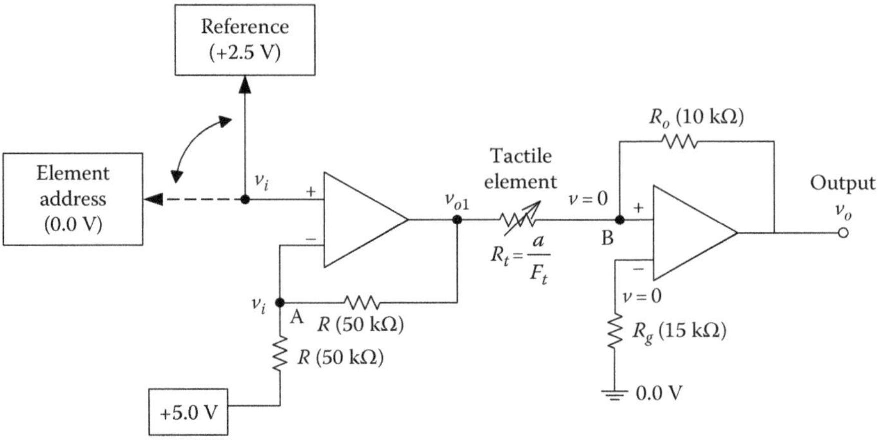

FIGURE 9.26
A signal acquisition circuit for a conductive-elastomer tactile sensor.

For the signal acquisition circuit shown in Figure 9.26, obtain a relationship for the output voltage v_o in terms of the parameters a, R_o and others if necessary and the variable F_t. Show that $v_o = 0$ when the tactile element is not addressed (i.e., when the circuit is switched to the reference voltage 2.5 V).

Solution

Define: v_i is the input to the circuit (2.5 or 0.0 V) and v_{o1} is the output of the first op-amp. We use the following properties of an op-amp (see Chapter 3):

1. Voltages at the two input leads are equal.
2. Currents through the two input leads are zero.

Hence, note the same v_i at both input leads of the first op-amp (and at node A) and the same zero voltage at both input leads of the second op-amp (and at node B) because one of the leads is grounded.

Current balance at A:

$$\frac{5.0 - v_i}{R} = \frac{v_i - v_{o1}}{R} \Rightarrow v_{o1} = 2v_i - 5.0 \tag{i}$$

Current balance at B:

$$\frac{v_{o1} - 0}{R_t} = \frac{0 - v_o}{R_o} \Rightarrow v_o = -v_{o1}\frac{R_o}{R_t} \tag{ii}$$

Substitute Equation i into Equation ii and also substitute the given expression for R_t. We get $v_o = (R_o/a)F_t(5.0 - 2v_i)$. Substitute the two switching values for v_i. We have

$$\begin{aligned} v_o &= \frac{5R_o}{a}F_t \quad \text{when addressed} \\ &= 0 \quad \text{when referenced} \end{aligned}$$

9.4.3 Optical Tactile Sensors

A schematic representation of an optical tactile sensor (built at the Man–Machine Systems Laboratory in Massachusetts Institute of Technology) is shown in Figure 9.27, which uses the principle of optical proximity sensor (see Chapter 10). In the system, the flexible tactile element consists of a thin, light-reflecting surface embedded within an outer layer (touch pad) of high-strength rubber and an inner layer of transparent rubber. Optical fibers are uniformly and rigidly mounted across this inner layer of rubber so that light can be projected directly onto the reflecting surface.

The light source, the beam splitter, and the solid-state digital camera form an integral unit, which can be moved laterally in known steps in order to scan the entire array of optical fiber if a single image frame of the camera does not cover the entire array. The splitter plate reflects part of the light from the light source onto a bundle of optical fiber. This light is reflected by the reflecting surface and is received by the camera. Since the intensity of the light received by the camera depends on the proximity of the reflecting surface, the grayscale intensity image detected by the camera will determine the deflection profile of the tactile surface. Using appropriate constitutive relations for the tactile sensor pad, the tactile force distribution can be determined as well. The image processor will condition (filter, segment, etc.) the successive image frames received by the frame grabber and will compute the deflection profile and the associated tactile force distribution in this manner. The image resolution will depend on the pixel size of each image frame (e.g., 512 × 512 pixels, 1024 × 1024 pixels, etc.) as well as the spacing of the fiber

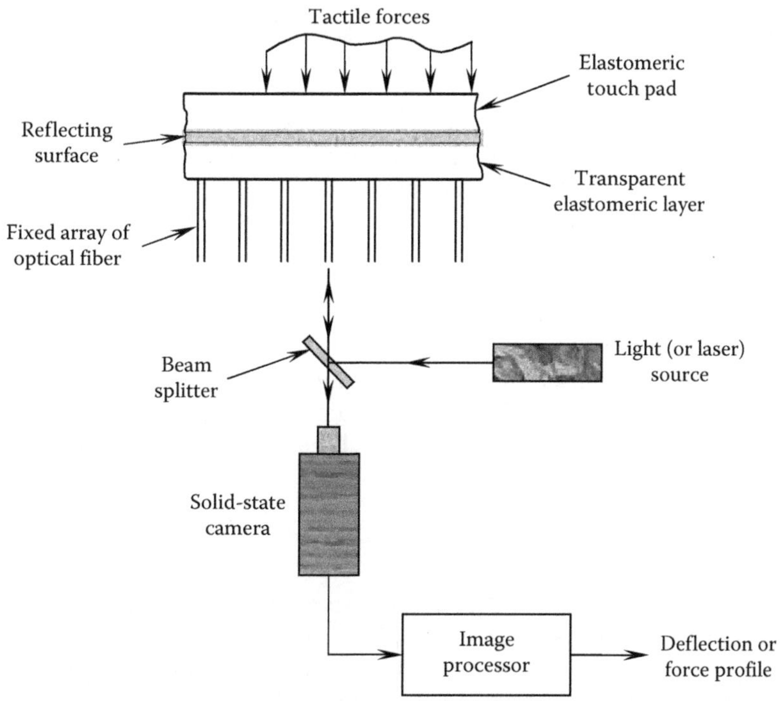

FIGURE 9.27
Schematic representation of a fiber-optic tactile sensor.

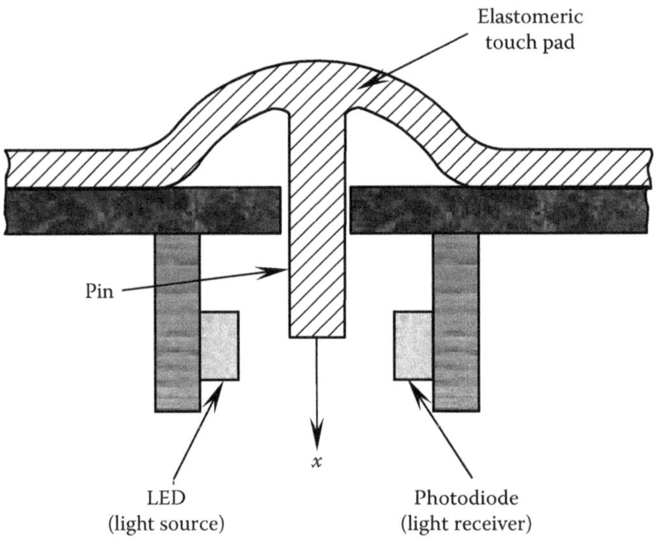

FIGURE 9.28
An optical tactile sensor with localized light sources and photosensors.

optic matrix. The force resolution (or tactile feel) of the tactile sensor can be improved at the expense of the thickness of the elastomeric layer, which determines the robustness of the sensor.

In the described fiber-optic tactile sensor (Figure 9.27), the optical fibers serve as the medium through which the light or laser rays are transmitted to the tactile site. This is an "extrinsic" use of fiber optics for sensing. Alternatively, an "intrinsic" application can be developed where an optical fiber serves as the sensing element itself. Specifically, the tactile pressure is directly applied to a mesh of optical fibers. Since the amount of light that is transmitted through a fiber will decrease due to deformation caused by the tactile pressure, the light intensity at a receiver can be used to determine the tactile pressure distribution.

Yet another alternative of an optical tactile sensor is available. In this design, the light source and the receiver are located at the tactile site itself; optical fibers are not used. The principle of operation of this type of tactile sensor is shown in Figure 9.28. When the elastomeric touch pad is pressed at a particular location, a pin attached to the pad at that point moves (in the x-direction), thereby obstructing the light received by the photodiode from the light-emitting diode. The output signal of the photodiode measures the pin movement.

9.4.4 Strain-Gauge Tactile Sensor

A strain-gauge tactile sensor has been developed by the Eaton Corporation in Troy, Michigan. The concept behind it can be employed to determine the size and location of a point-contact force, which is useful, for example, in parts-mating applications. A square plate of length a is simply supported by frictionless hinges at its four corners on strain-gauge load cells, as shown in Figure 9.29a. The magnitude, direction, and location of a point force P, applied normally to the plate, can be determined using the readings of the four (strain gauge) load cells.

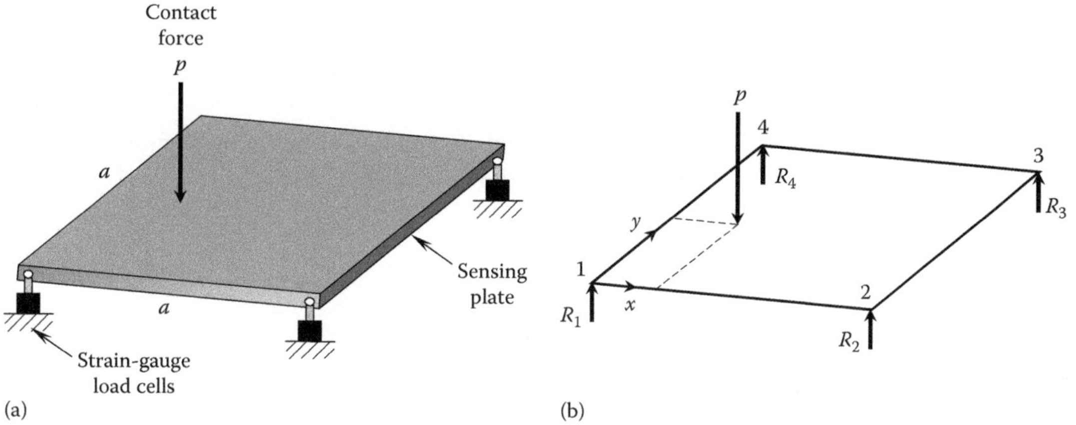

FIGURE 9.29
(a) Schematic representation of a strain-gauge point-contact sensor and (b) free-body diagram.

To illustrate this principle, consider the free-body diagram shown in Figure 9.29b. The location of force P is given by the coordinates (x, y) in the Cartesian coordinate system (x, y, z), with the origin located at 1, as shown. The load cell reading at location i is denoted by R_i. Equilibrium in the z-direction gives the force balance:

$$P = R_1 + R_2 + R_3 + R_4 \tag{9.40}$$

Equilibrium about the y-axis gives the moment balance:

$$Px = R_2 a + R_3 a \rightarrow$$

$$x = \frac{a}{P}(R_2 + R_3) \tag{9.41}$$

Similarly, equilibrium about the x-axis gives

$$y = \frac{a}{P}(R_3 + R_4) \tag{9.42}$$

It follows from Equations 9.40 through 9.42 that the force P (direction as well as magnitude) and its location (x, y) are completely determined by the load cell readings. Typical values for the plate length a and the maximum force P are 5 cm and 10 kg, respectively.

Example 9.9

In a particular parts-mating process using the principle of strain-gauge tactile sensor, suppose that the tolerance on the measurement error of the force location is limited to δr. Determine the tolerance δF on the load cell error.

Solution

Take the differentials of Equations 9.40 and 9.41:

$$\delta P = \delta R_1 + \delta R_2 + \delta R_3 + \delta R_4 \quad \text{and} \quad P\delta x + x\delta P = a\delta R_2 + a\delta R_3$$

Direct substitution gives

$$\delta x = \frac{a}{P}(\delta R_2 + \delta R_3) - \frac{x}{P}(\delta R_1 + \delta R_2 + \delta R_3 + \delta R_4)$$

Note that x lies between 0 and a, and each δR_i can vary up to $\pm\delta F$. Hence, the largest error in x is given by $(2a/P)\delta F$. This is limited to δr. Hence, we have

$$\delta r = \frac{2a}{P}\delta F$$

or

$$\delta F = \frac{P}{2a}\delta r$$

This gives the tolerance on the force error. The same result is obtained by considering y instead of x.

9.4.5 Other Types of Tactile Sensors

Another type of tactile sensor (piezoresistive) uses an array of SC strain gauges mounted under the touch pad on a rigid base. In this manner, the force distribution on the touch pad is measured directly.

Ultrasonic tactile sensors are based, for example, on pulse-echo ranging. In this method, the tactile surface consists of two membranes separated by an air gap. The time taken for an ultrasonic pulse to travel through the gap and be reflected back onto a receiver depends, in particular, on the thickness of the air gap. Since this time interval changes with deformation of the tactile surface, it can serve as a measure of the deformation of the tactile surface at a given location.

Other possibilities for tactile sensors include the use of chemical effects that might be present when an object is touched and the influence of grasping on the natural frequencies of an array of sensing elements.

Example 9.10

When is tactile sensing preferred over sensing of a few point forces? A piezoelectric tactile sensor has 25 force-sensing elements per square centimeter. Each sensor element in the sensor can withstand a maximum load of 40 N and can detect load changes on the order of 0.01 N. What is the force resolution of the tactile sensor? What is the spatial resolution of the sensor? What is the dynamic range of the sensor in decibels?

Solution

Tactile sensing is preferred when it is not a simple-touch application. Shape, surface characteristics, and flexibility characteristics of a manipulated (handled or grasped) object can be determined using tactile sensing.

$$\text{Force resolution} = 0.01\ \text{N}$$

$$\text{Spatial resolution} = \frac{\sqrt{1}}{\sqrt{25}}\ \text{cm} = 2\ \text{mm}$$

$$\text{Dynamic range} = 20\log_{10}\left(\frac{40}{0.01}\right) = 72\ \text{dB}$$

9.5 Impedance Sensing and Control

Consider a mechanical operation where we push against a spring that has constant stiffness. Here, the value of the force completely determines the displacement, and similarly, the value of the displacement completely determines the force. It follows that, in this example, we are unable to control force and displacement independently at the same time. Also, it is not possible, in this example, to apply a command force that has an arbitrarily specified relationship with displacement. In other words, stiffness control is not possible. Now suppose that we push against a complex dynamic system, not a simple spring element. In this case, we should be able to command a pushing force in response to the displacement of the dynamic system so that the ratio of force to displacement varies in a specified manner. This is a stiffness control (or compliance control) action.

Dynamic stiffness is defined as the ratio: (Output force)/(Input displacement), expressed in the frequency domain. Dynamic flexibility or compliance or receptance is the inverse of dynamic stiffness. Mechanical impedance is defined as the ratio: (Output force)/(Input velocity), in the frequency domain. Mobility is the inverse of mechanical impedance. Note that stiffness and impedance both relate force and motion variables in a mechanical system. The objective of impedance control is to make the impedance function equal to some specified function (without separately controlling or independently constraining the associated force and velocity variables).

Force control and motion control can be considered limiting cases of impedance control (and stiffness control). Since the objective of force control is to keep the force variable from deviating from a desired level, in the presence of independent variations of the associated motion variable (an input), force is the output variable, whose deviation (increment) from the desired value must be made zero, under control. Hence, force control can be interpreted as zero-impedance control, when velocity is chosen as the motion variable (or zero-stiffness control, when displacement is chosen as the motion variable). Conversely, displacement control can be considered as infinite-stiffness control and velocity control can be considered as infinite-impedance control.

Impedance control has to be accomplished through active means, generally by generating forces as specified functions of associated displacements. Impedance control is particularly useful in mechanical manipulation against physical constraints that are not "hard," which is the case in compliant assembly and machining tasks. In particular, very high impedance is naturally present in the direction of a motion constraint and very low impedance in the direction of a free motion. Problems that arise using motion control in applications where small motion errors would create large forces can be avoided to some extent if stiffness control or impedance control is applied. Furthermore, the stability of the overall system can be guaranteed and the robustness of the system improved by properly bounding the values of impedance parameters.

Impedance control can be particularly useful in tasks of fine and flexible manipulation. Application examples include processing of flexible and inhomogeneous natural material such as meat, robotic surgery, and routing machine that cut flat objects (e.g., leather, wood, and plastic) into different shapes. In these tasks, the mechanical impedance of the task interface (i.e., in the region where the mechanical processor or the cutting tool interacts with the processed object) provides valuable characteristics of the process, which can be used in fine control of the processing task. Since impedance relates the input velocity to the output force, it is a transfer function. The concepts of impedance control can be applied as well to situations where the input is not a velocity and the output is not a force.

Still, the term *impedance control* is used in the literature, even though the corresponding transfer function is, strictly speaking, not an impedance.

Example 9.11

The control of processes such as machine tools and robotic manipulators may be addressed from the viewpoint of impedance control. For example, consider a milling machine that performs a straight cut on a workpiece, as shown in Figure 9.30a. The tool position is stationary, and the machine table, which holds the workpiece, moves along a horizontal axis at speed v—the feed rate. The cutting force in the direction of feed is f. Suppose that the machine table is driven using the speed error, according to the law:

$$F = Z_d(V_{ref} - V) \tag{9.43}$$

where
Z_d is the drive impedance of the table
V_{ref} is the reference (command) feed rate

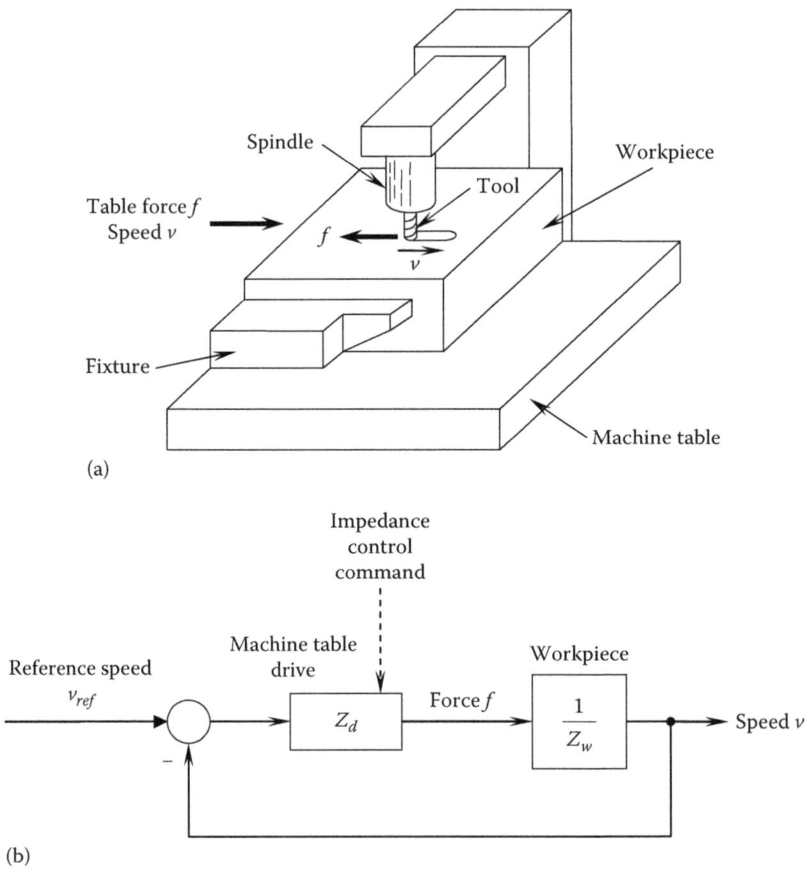

FIGURE 9.30
(a) A straight-cut milling operation. (b) Impedance block diagram representation.

(The uppercase letters are used to represent frequency-domain variables of the system.) Cutting impedance Z_w of the workpiece satisfies the relation:

$$F = Z_w V \tag{9.44}$$

Note that Z_w depends on system properties, and we usually do not have a direct control over it. The overall system is represented by the block diagram in Figure 9.30b. An impedance control problem would be to adjust (or adapt) the drive impedance Z_d so as to maintain the feed rate near V_{ref} and the cutting force near F_{ref}. We will determine an adaptive control law for Z_d.

Solution

The control objective is satisfied by minimizing the objective function:

$$J = \frac{1}{2}\left[\frac{F - F_{ref}}{f_o}\right]^2 + \frac{1}{2}\left[\frac{V - V_{ref}}{v_o}\right]^2 \tag{9.45}$$

where
 f_o is the force tolerance
 v_o is the speed tolerance

For example, if we desire stringent control of the feed rate, we need to choose a small value for v_o, which corresponds to a heavy weighting on the feed rate term in J. Hence, these two tolerance parameters are weighting parameters as well, in the cost function.

The optimal solution is given by

$$\frac{\partial J}{\partial Z_d} = 0 = \frac{(F - F_{ref})}{f_0^2}\frac{\partial F}{\partial Z_d} + \frac{(V - V_{ref})}{v_0^2}\frac{\partial V}{\partial Z_d} \tag{9.46}$$

Now, from Equations 9.43 and 9.44, we obtain

$$V = \left[\frac{Z_d}{Z_d + Z_w}\right]V_{ref} \tag{9.47}$$

$$F = \left[\frac{Z_d Z_w}{Z_d + Z_w}\right]V_{ref} \tag{9.48}$$

On differentiating these equations, we get

$$\frac{\partial V}{\partial Z_d} = \frac{Z_w}{(Z_d + Z_w)^2}V_{ref} \tag{9.49}$$

and

$$\frac{\partial F}{\partial Z_d} = \frac{Z_w^2}{(Z_d + Z_w)^2}V_{ref} \tag{9.50}$$

Next, we substitute Equations 9.49 and 9.50 into Equation 9.46 and divide by the common term to get

$$\frac{(F - F_{ref})}{f_0^2}Z_w + \frac{(V - V_{ref})}{v_0^2} = 0 \tag{9.51}$$

This equation is expanded after substituting Equations 9.47 and 9.48 in order to get the required expression for Z_d:

$$Z_d = \left[\frac{Z_0^2 + Z_w Z_{ref}}{Z_w - Z_{ref}} \right] \tag{9.52}$$

where

$$Z_o = \frac{f_o}{v_o} \quad \text{and} \quad Z_{ref} = \frac{F_{ref}}{V_{ref}}$$

This equation is the impedance control law for the table drive. Specifically, since Z_w—which depends on the workpiece characteristics, the tool bit characteristics, and the rotating speed of the tool bit—is known through a suitable model or might be experimentally determined (identified) by monitoring v and f and since Z_d and Z_{ref} are specified, we are able to determine the necessary drive impedance Z_d using Equation 9.52. Parameters of the table drive controller—particularly gain—can be adjusted to match this optimal impedance. Unfortunately, exact matching is virtually impossible because Z_d is generally a function of frequency. If the component bandwidths are high, we may assume that the impedance functions are independent of frequency, and this somewhat simplifies the impedance control task.

Note from Equation 9.44 that for the ideal case of $V = V_{ref}$ and $F = F_{ref}$, we have $Z_w = Z_{ref}$. Then, from Equation 9.52, it follows that a drive impedance of infinite magnitude is needed for exact control. This is impossible to achieve in practice, however. Of course, an upper limit for the drive impedance should be set in any practical scheme of impedance control.

Summary Sheet

Strain Gauges

Metallic foil (copper–nickel alloy—constantan): $\delta R/R = S_s\varepsilon$; gauge factor (sensitivity) $S_s = 2$–4; more linear and smaller temperature coefficient of resistance α.

Semiconductor (silicon with impurity): $\delta R/R = S_1\varepsilon + S_2\varepsilon^2$; gauge factor $S_s = 40$–200; resistivity is higher (5 kΩ) → reduced power consumption; smaller and lighter. More nonlinear → nonlinearity $N_p = 50S_2\varepsilon_{max}/S_1\%$.

Sensing: Change in resistance → bridge circuit:

$$\frac{\delta v_0}{v_{ref}} = \frac{(R_2\delta R_1 - R_1\delta R_2)}{(R_1 + R_2)^2} - \frac{(R_4\delta R_3 - R_3\delta R_4)}{(R_3 + R_4)^2}$$

Measurements: Displacement, acceleration, pressure, temperature, liquid level, stress, force torque, etc. *Note*: Indirect measurements: convert measurand into stress (strain) using a front-end auxiliary device.

Bridge output: $\delta v_o/v_{ref} = k(\delta R/4R)$.

Bridge constant: k = (Bridge output in the general case)/(Bridge output if only one strain gauge is active).

Calibration constant: $C = (k/4)S_s$ with $\delta v_o/v_{ref} = C\varepsilon$.

If all four resistors are active → largest k → best sensitivity.

Self-compensation for temperature (for p-type SC strain gauge): Uses series resistor R_c with power supply $R_c = -[\beta/(\alpha + \beta)]R_o$; β = temperature coefficient of sensitivity.

Strain-Gauge Torque Sensor

Equations: Principal strain (45° to axis) at radius r of sensor, $\varepsilon = (r/2GJ)T$; torque transmitted through member, $T = (8GJ/kS_s r)(\delta v_o/v_{ref})$; G = shear modulus of material; J = polar moment of area of cross section.

Design considerations:

$$\text{Stiffness reduction} \leftarrow \frac{1}{K_{new}} = \frac{1}{K_m} + \frac{1}{K_L} + \frac{1}{K_s} = \frac{1}{K_{old}} + \frac{1}{K_s}$$

Torsional (twisting) natural frequency $\omega_n = \sqrt{K_s((1/J_m)+(1/J_L))}$; sensor stiffness $K_s = T/\theta = (GJ/L)$; strain capacity of strain-gauge element $J \geq (r/2G)\cdot(T_{\max}/\varepsilon_{\max})$; strain-gauge nonlinearity $J \geq (25rS_2/GS_1)\cdot(T_{\max}/N_p)$; sensor sensitivity $J \leq (K_a kS_s rv_{ref}/8G)\cdot(T_{\max}/v_o)$; and sensor stiffness (system bandwidth and gain) $J \geq (L/G)\cdot K$.

Surface Acoustic Wave (SAW) sensor: Microminiature acoustic resonator made of piezoelectric material; its resonant frequency (MHz range) varies with strain → strain sensor; advantages: wireless operation (useful for sensing of moving parts) and high measurement bandwidth (kHz).

Direct-deflection torque sensor: Measures twist angle (e.g., proximity probes on toothed wheels → phase shift → twist → torque (both magnitude and direction)).

Variable-reluctance torque sensor: Ferromagnetic tube with two slits in principal stress directions: applied torque → one slit opens and other closes → reluctance changes → induced voltage → torque.

Magnetostriction torque sensor: Uses "reverse" magnetostriction (*Villari effect*) → deformation of magnetostrictive material changes its magnetization → sense by stationary magnetic field probe (e.g., Hall-effect sensor); materials: nickel and its alloys, some ferrites, some rare earths, and alfer (86% iron and 14% aluminum alloy).

Reaction torque sensors: Cradle the housing of rotating machine → measure cradling force.

Advantages: Torque-sensing element reduces the system stiffness and bandwidth and adds extra loading to the system ← reaction torque sensors eliminate these problems; frictional torques do not affect the measurement. *Note*: Instead of cradling, reactions at support locations can be sensed to get reaction torque.

Motor current torque sensors: DC motor: $T_m = ki_f i_a$; three-phase synchronous motor: $T_m = 1.5ki_f i_a \cos\theta_0$.

Problems

9.1 Giving examples for suitable auxiliary front-end elements, discuss the use of a force sensor to measure (a) displacement, (b) velocity, and (c) acceleration.

9.2 Under what conditions can displacement control effectively replace force control? Describe a situation in which this is not feasible.

9.3 Consider the joint of a robotic manipulator, shown schematically in Figure P9.3. Torque sensors are mounted at locations 1, 2, and 3. Denoting the magnetic torque generated at the motor rotor by T_m, write equations for the torque transmitted to link 2, the frictional torque at bearing A, the frictional torque at bearing B, and the reaction torque on link 1, in terms of the measured torques, the inertia torque of the rotor, and T_m.

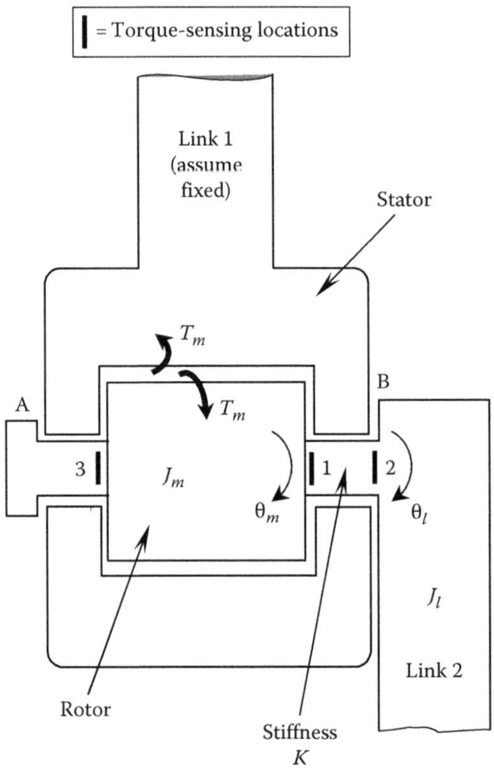

FIGURE P9.3
Torque-sensing locations for a manipulator joint.

9.4 The connecting shaft between the motor rotor and the link of a robotic joint is solid circular with radius r and shear modulus G. It was too hard and unsuitable for mounting strain gauges. In order to make strain measurements for determining the joint torque, a softer sleeve was placed in tight fit, along the shaft (see Figure P9.4). The outer radius of the sleeve is R and the shear modulus is G_s. Determine an expression for the transmitted torque T in terms of the measured principal strain ε on the outer surface of the sleeve (at 45° from axial direction).

Hint: From *Mechanics of Materials*, the equivalent value of GJ for a composite circular shaft is given by $GJ + G_sJ_s$.*

* For example, see De Silva, C.W., *Mechanics of Materials*, CRC Press, Taylor & Francis Group, Boca Raton, FL, 2014.

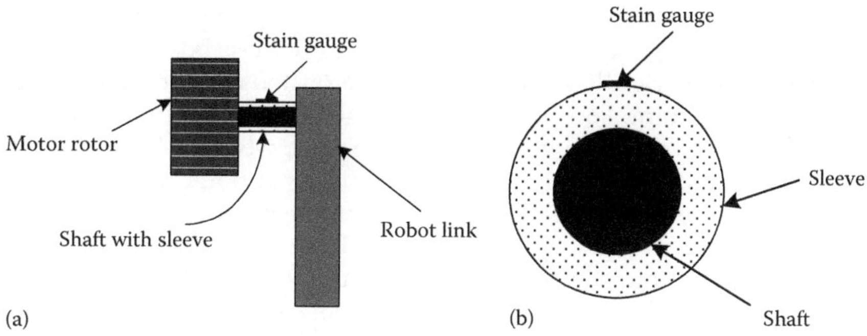

FIGURE P9.4
(a) A robotic joint shaft with a torque-sensing sleeve and (b) cross section of the shaft with sleeve.

9.5 A model for a machining operation is shown in Figure P9.5. The cutting force is denoted by f, and the cutting tool with its fixtures is modeled by a spring (stiffness k) together with a viscous damper (damping constant b) and a mass m. The actuator (hydraulic) with its controller is represented by an active stiffness g. Assuming linear g, obtain a transfer relation between the actuator input u and the cutting force f. Now determine an approximate expression for the gradient $\partial g/\partial u$. Discuss a control strategy for counteracting the effects from random variations in the cutting force. *Note*: This is important for controlling the product quality.

Hint: You may use a reference-adaptive feedforward control strategy where the reference values of g and u are the inputs to the machine tool. The reference g is adapted using the gradient $\partial g/\partial u$, as u changes by Δu.

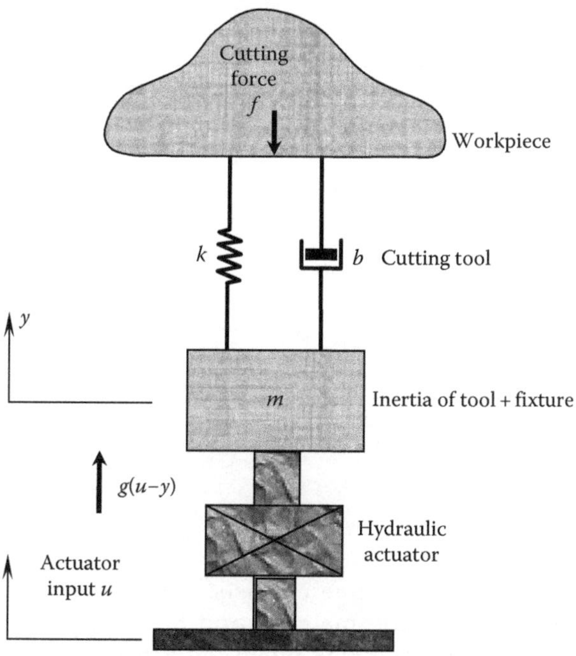

FIGURE P9.5
A model for a machining operation.

9.6 A strain-gauge sensor to measure the torque T_m generated by a motor is schematically shown in Figure P9.6. The motor is floated on frictionless bearings. A uniform rectangular lever arm is rigidly attached to the motor housing, and its projected end is restrained by a pin joint. Four identical strain gauges are mounted on the lever arm, as shown. Three of the strain gauges are at point A, which is located at a distance a from the motor shaft, and the fourth strain gauge is at point B, which is located at a distance $3a$ from the motor shaft. The pin joint is at a distance l from the motor shaft. The strain gauges 2, 3, and 4 are on the top surface of the lever arm, and the strain gauge 1 is on the bottom surface. Obtain an expression for T_m in terms of the bridge output δv_o and the following additional parameters: S_s is the gauge factor (strain-gauge sensitivity), v_{ref} is the supply voltage to the bridge, b is the width of the lever arm cross section, h is the height of the lever arm cross section, and E is Young's modulus of the lever arm.

Verify that the bridge sensitivity does not depend on a and l. Describe means to improve the bridge sensitivity. Explain why the sensor reading is only an approximation to the torque transmitted to the load. Give a relation to determine the net normal reaction force at the bearings, using the bridge output.

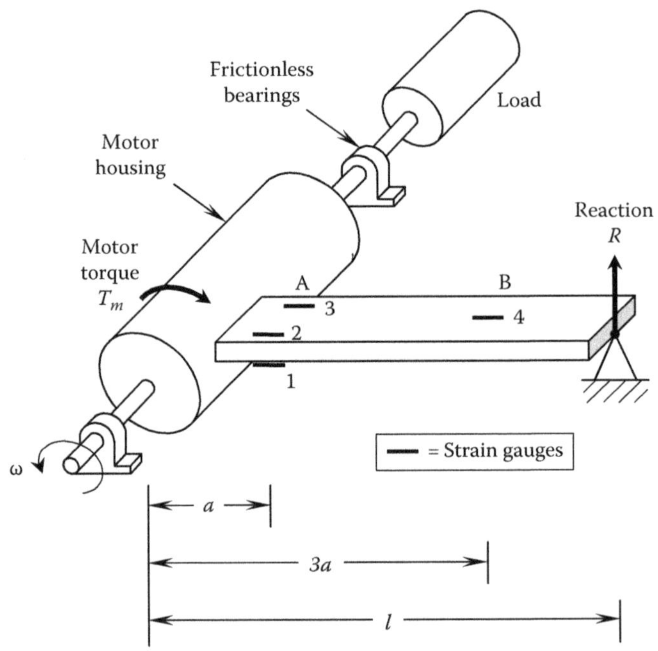

FIGURE P9.6
A strain-gauge sensor for measuring motor torque.

9.7 The sensitivity S_s of a strain gauge consists of two parts: the contribution from the change in resistivity of the material and the direct contribution due to the change in shape of the strain gauge when deformed. Show that the second part may be approximated by $(1 + 2\nu)$, where ν denotes Poisson's ratio of the strain-gauge material.

9.8 Discuss the advantages and disadvantages of the following techniques in the context of measuring transient signals:

(a) DC bridge circuits versus AC bridge circuits
(b) Slip ring and brush commutators versus AC transformer commutators
(c) Strain-gauge torque sensors versus variable-inductance torque sensors
(d) Piezoelectric accelerometers versus strain-gauge accelerometers
(e) Tachometer velocity transducers versus piezoelectric velocity transducers
(f) Wireless telemetry commutation versus transformer commutation

9.9 For a semiconductor strain gauge that is characterized by the quadratic strain–resistance relationship $\delta R/R = S_1\varepsilon + S_2\varepsilon^2$, obtain an expression for the equivalent gauge factor (sensitivity) S_s using the least-squares error linear approximation. Assume that only positive strains up to ε_{max} are measured with the gauge. Derive an expression for the percentage nonlinearity. Taking $S_1 = 117$, $S_2 = 3600$, and $\varepsilon_{max} = 0.01$ strain, compute S_s and the percentage nonlinearity.

9.10 Briefly describe how strain gauges may be used to measure the following:

(a) Force
(b) Displacement
(c) Acceleration
(d) Pressure
(e) Temperature

Show that if a compensating resistance R_c is connected in series with the supply voltage v_{ref} to a strain-gauge bridge that has four identical members, each with resistance R, the output equation is given by $\delta v_o/v_{ref} = (R/(R + R_c))(kS_s/4)\varepsilon$, in the usual rotation.

A foil-gauge load cell uses a simple (1D) tensile member to measure force. Suppose that k and S_s are insensitive to temperature change. If the temperature coefficient of R is α_1, that of the series compensating resistance R_c is α_2, and that of Young's modulus of the tensile member is $(-\beta)$, determine an expression for R_c that would result in automatic (self-) compensation for temperature effects. Under what conditions is this arrangement realizable?

9.11 Draw a block diagram for a single joint of a robot, identifying the inputs and outputs. Using the diagram, explain the advantages of torque sensing in comparison to displacement and velocity sensing at the joint. What are the disadvantages of torque sensing?

9.12 Figure P9.12 shows a schematic diagram of a measuring device.

(a) Identify the various components in this device.
(b) Describe the operation of the device, explaining the function of each component and identifying the nature of the measurand and the output of the device.
(c) List the advantages and disadvantages of the device.
(d) Describe a possible application of this device.

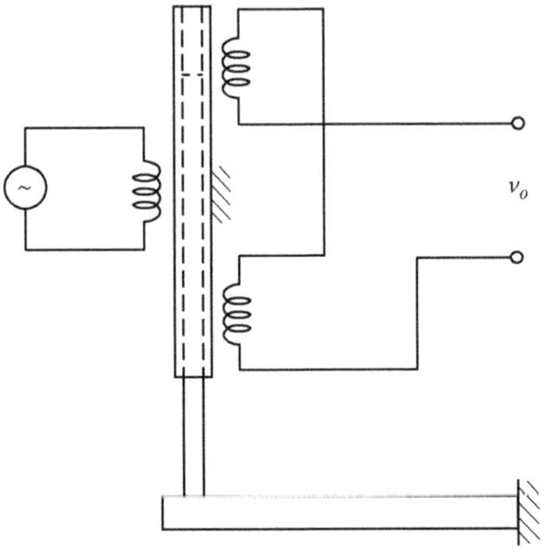

FIGURE P9.12
An analog sensor.

9.13 Discuss the advantages and disadvantages of torque sensing by the motor current method. Show that for a synchronous motor with a balanced three-phase supply, the electromagnetic torque generated at the rotor–stator interface is given by

$$T_m = k i_f i_a \cos(\theta - \omega t)$$

where
- i_f is the DC in the rotor (field) winding
- i_a is the amplitude of the supply current to each phase in the stator (armature)
- θ is the angle of rotation
- ω is the frequency (angular) of the AC supply
- t is the time
- k is the motor torque constant

9.14 Discuss factors that limit the lower frequency and upper frequency limits of measurements obtained from the following devices:

(a) Strain gauge
(b) Rotating shaft torque sensor
(c) Reaction torque sensor

9.15 Briefly describe a situation in which tension in a moving belt or cable has to be measured under transient conditions. What are some of the difficulties associated with measuring tension in a moving member? A strain-gauge tension sensor for a belt-drive system is shown in Figure P9.15. Two identical active strain gauges, G_1 and G_2, are mounted at the root of a cantilever element with rectangular cross section, as shown. A light, frictionless pulley is mounted at the free end of the cantilever element. The belt makes a 90° turn when passing over this idler pulley.

(a) Using a circuit diagram, show the Wheatstone bridge connections necessary for the strain gauges G_1 and G_2 so that the strains that result from the axial forces in the cantilever member have no effect on the bridge output (i.e., the effects of axial loads are compensated for) and the sensitivity to bending loads is maximized.

(b) Obtain an equation relating the belt tension T and the bridge output δv_o in terms of the following additional parameters:

S_s is the gauge factor (sensitivity) of each strain gauge
E is Young's modulus of the cantilever element
L is the length of the cantilever element
b is the width of the cantilever cross section
h is the height of the cantilever cross section
In particular, show that the radius of the pulley does not enter into this equation. Give the main assumptions made in your derivation.

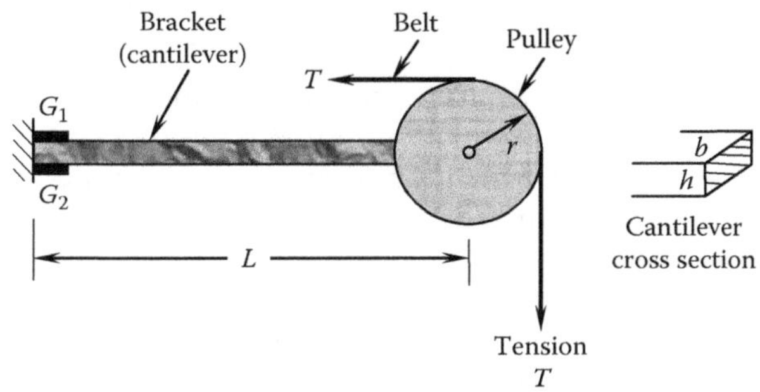

FIGURE P9.15
A strain-gauge tension sensor for a moving belt.

9.16 The read–write head in a hard disk drive of a digital computer should float at a constant but small height (say, fraction of a μm) above the disk surface. Because of the aerodynamics resulting from the surface roughness and the surface deformations of the disk, the head can be excited into vibrations that can cause head–disk contacts. These contacts, which are called head–disk interferences (HDIs), are clearly undesirable. They can occur at very high frequencies (say, 1 MHz). The purpose of a glide test is to detect HDIs and to determine the nature of these interferences. Glide testing can be used to determine the effect of parameters such as the flying height of the head and the speed of the disk and to qualify (certify the quality of) disk drive units for specific types of operating conditions. Indicate the basic instrumentation that is needed in glide testing. In particular, suggest the types of sensors that could be used and their advantages and disadvantages.

9.17 Torque, force, and tactile sensing can be very useful in many applications, particularly in the manufacturing industry. For each of the following applications, indicate the types of sensors that would be useful for properly performing the task:

(a) Controlling the operation of inserting printed circuit boards into card cages using a robotic end effector
(b) Controlling a robotic end effector that screws a threaded part into a hole
(c) Failure prediction and diagnosis of a drilling operation
(d) Grasping a fragile, delicate, and somewhat flexible object by a robotic hand without damaging the object
(e) Grasping a metal part using a two-fingered gripper
(f) Quickly identifying and picking a complex part from a bin that contains many different parts

9.18 A weight sensor is used in a robotic wrist. What would be the purpose of this sensor? How can the information obtained from the weight sensor be used in controlling the robotic manipulator?

Describe four advantages and four disadvantages of a weight sensor that uses semiconductor strain gauges.

9.19 (a) List three advantages and three disadvantages of a semiconductor strain gauge when compared with a foil strain gauge.

(b) A flywheel device is schematically shown in Figure P9.19. The wheel consists of four spokes that carry lumped masses at one end and are clamped to a rotating hub at the other end, as shown. Suppose that the inertia of the spokes can be neglected in comparison with that of the lumped masses.

Four active strain gauges are used in a bridge circuit for measuring speed.

(i) If the bridge can be calibrated to measure the tensile force F in each spoke, express the dynamic equation, which may be used to measure the rotating speed (ω). The following parameters may be used: m is the mass of the lumped element at the end of a spoke and r is the radius of rotation of the center of mass of the lumped element.

(ii) For good results with regard to high sensitivity of the bridge and also to compensate for secondary effects such as out-of-plane bending, indicate where the four strain gauges (1, 2, 3, 4) should be located on the spokes and in what configuration they should be connected in a DC bridge.

(iii) Compare this method of speed sensing to that using a tachometer and a potentiometer by giving three advantages and two disadvantages of the strain-gauge method.

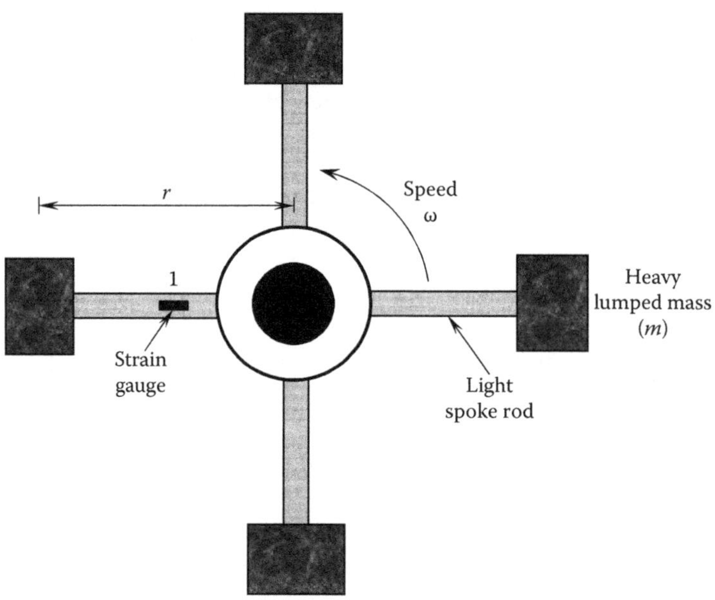

FIGURE P9.19
Strain-gauge speed sensor for a flywheel.

9.20 (a) Consider a simple mechanical manipulator. Explain why in some types of manipulation tasks, motion sensing alone might not be adequate for accurate control, and torque or force sensing might be needed as well.

(b) Discuss what factors should be considered when installing a torque sensor to measure the torque that is transmitted from an actuator to a rotating load.

(c) A harmonic drive consists of the following three main components:

(i) Input shaft with the elliptical wave generator (cam)

(ii) Circular flexispline with external teeth

(iii) Rigid circular spline with internal teeth

Consider the free-body diagrams shown in Figure P9.20. The following variables are defined: ω_i is the speed of the input shaft (wave generator), ω_o is the speed of the output shaft (rigid spline), T_o is the torque transmitted to the driven load by the output shaft (rigid spline), T_i is the torque applied on the harmonic drive by the input shaft, T_f is the torque transmitted by the flexispline to the rigid spline, T_r is the reaction torque on the flexispline at the fixture, and T_w is the torque transmitted by the wave generator.

If strain gauges are to be used to measure the output torque T_o, suggest suitable locations for mounting them and discuss how the torque measurement can be obtained in this manner. Using a block diagram for the system, indicate whether you consider T_o to be an input to or an output of the harmonic drive. What are the implications of this consideration?

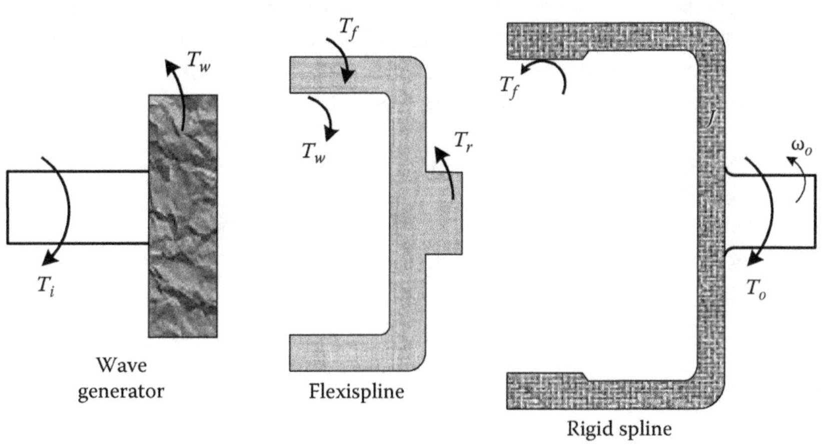

FIGURE P9.20
Free-body diagram of a harmonic drive.

9.21 (a) Describe three different principles of torque sensing. Discuss relative advantages and disadvantages of the three approaches.

(b) A torque sensor is needed for measuring the drive torque that is transmitted to a link of a robot (i.e., joint torque). What characteristics and specifications of the sensor and the requirement of the system should be considered in selecting a suitable torque sensor for this application?

9.22 What are the typical requirements for an industrial tactile sensor? Explain how a tactile sensor differs from a simple touch sensor. Define spatial resolution and force resolution (or sensitivity) of a tactile sensor.

The spatial resolution of your fingertip can be determined by a simple experiment using two pins and a helper. Close your eyes. Instruct the helper to apply one pin or both pins randomly to your fingertip so that you feel the pressure of the tips of the pins. You should respond by telling the helper whether you feel both pins or just one pin. If you feel both pins, the helper should decrease the spacing of the two pins in the next round of testing. The test should be repeated in this manner by successively decreasing the spacing between the pins until you feel only one pin when actually both pins are applied. Then, measure the distance between the two pins in millimeters. The largest spacing between the two pins that will result in this incorrect sensation corresponds to the spatial resolution of your fingertip. Repeat this experiment for all your fingers, repeating the test several times on each finger. Compute the average and the standard deviation. Then, perform the test on other subjects. Discuss your results. Do you notice large variations in the results?

9.23 Discuss whether there is any relationship between the dexterity and the stiffness of a manipulator hand. The stiffness of a robotic hand can be improved during grasping operations by temporarily decreasing the number of degrees of freedom of the hand using suitable fixtures. What purpose does this serve?

9.24 The *motion dexterity* of a device is defined as the ratio of [number of degrees of freedom in the device]/[motion resolution of the device]. The *force dexterity* may be defined as the ratio of [number of degrees of freedom in the device]/[force resolution of the device]. Give a situation where both types of dexterity mean the same thing and a situation where the two terms mean different things. Outline how the force dexterity of a device (say, a robotic end effector) can be improved by using tactile sensors. Provide the dexterity requirements for the following tasks by indicating whether motion dexterity or force dexterity is preferred in each case:

(a) Grasping a hammer and driving a nail with it
(b) Threading a needle
(c) Seam tracking of a complex part in robotic arc welding
(d) Finishing the surface of a complex metal part using robotic grinding

10

Miscellaneous Sensors

Chapter Objectives

- Lasers
- Laser Interferometer
- Laser Doppler Interferometer
- Fiber-Optic Position Sensor
- Fiber-Optic Gyroscope
- Light Sensors
- Ultrasonic Sensors
- Magnetostrictive Displacement Sensor
- Acoustic Emission Sensor
- Pressure Sensors
- Flow Sensors
- Temperature Sensors
- pH Value Sensor
- Dissolved Oxygen Sensor
- Oxidation–Reduction Potential Sensor

10.1 Introduction

The previous chapters studied analog motion sensors and effort sensors, which included force, torque, tactile, and impedance. Several other types of sensors including optical sensors, ultrasonic sensors, magnetostrictive sensors, acoustic emission (AE) sensors, thermo-fluid sensors, and sensors that are useful in monitoring the quality of water (e.g., pH value, dissolved oxygen [DO], and oxidation–reduction potential [ORP]) are discussed in this chapter. Digital transducers including shaft encoders, image sensors (digital camera), binary transducers, and Hall-effect sensors are studied in Chapter 11. Microelectromechanical system (MEMS) sensors, other advanced topics in sensing such as networked sensing, and sensor fusion are addressed in Chapter 12.

The so-called "optical" sensors such as lasers may operate outside the visible spectrum, for example, in the infrared and ultraviolet ranges. Regardless, they have specific advantages and applications over other types of sensors. Furthermore, they can operate in the "intrinsic" mode where it serves as the sensor itself or in the "extrinsic" mode where it serves as the medium to transmit the sensor signal. Similarly, ultrasonic sensors use frequencies greater than those of audible sound. They too can operate in the intrinsic and extrinsic modes. Thermofluid sensors include those for temperature, pressure, and flow rate.

10.2 Optical Sensors and Lasers

There are many sensors that use light or laser as the basis of measurement. Also, camera images are widely used for sensing purposes. This section addresses optical sensors. The use of digital camera as a sensor is specifically discussed in Chapter 11.

10.2.1 Lasers

Laser (*l*ight *a*mplification by *s*timulated *e*mission of *r*adiation) is a light source that emits a concentrated beam of electromagnetic radiation in the ultraviolet, visible, or infrared bands of the spectrum, which will propagate typically at one or two frequencies (wavelengths) and in phase. Usually, the frequency band is extremely narrow (i.e., *monochromatic*), and the waves in each frequency are in phase (i.e., *coherent*). Furthermore, the energy of a laser is highly concentrated (power densities of the order of one billion W/cm^2). Consequently, a laser beam can travel in a straight line over a long distance (e.g., 30 m) with very little dispersion (i.e., with negligible beam spread). The operation of a laser makes use of the oscillations of atoms or molecules of various elements.

The laser may be considered as a "structured" light source that is monochromatic and coherent. Hence, it can produce a very narrow beam, which can travel a long distance virtually unaltered and can be focused at a point on an object. It is useful in fiber optics. But it can also be used directly in sensing and gauging applications. For example, lasers are useful in imaging, gauging, and aligning applications. They can be used in a wide variety of sensors (e.g., motion sensors, tactile sensors, laser Doppler velocity sensors, 3D imaging sensor) that employ photosensing and fiber optics. Also, lasers are used in medical applications, microsurgery in particular. Lasers have been used in manufacturing and material removal applications such as precision welding, cutting, and drilling of different types of materials, including metals, glass, plastics, ceramics, rubber, leather, and cloth. Lasers are used in inspection (detection of faults and irregularities) and gauging (measurement of dimensions) of parts. Other applications of lasers include heat treatment of alloys, holographic methods of nondestructive testing, communication, information processing, and high-quality printing.

Lasers may be classified as *solid*, *liquid*, *gas*, and *semiconductor*. In a solid laser (e.g., *ruby* laser, *glass* laser), a solid rod with reflecting ends is used as the laser medium. The laser medium of a liquid laser (e.g., *dye* laser, *salt solution* laser) is a liquid such as an organic solvent with a dye or an inorganic solvent with dissolved salt compound. Very high peak power levels are possible with liquid lasers. Gas lasers (e.g., *helium–neon* [He–Ne] laser, *helium–cadmium* [He–Cd] laser, *carbon dioxide* [CO_2] laser) use gas as the laser medium.

Semiconductor lasers (e.g., *gallium arsenide* laser) use a semiconductor diode similar to an edge-emitting light-emitting diode (LED). Some lasers have their main radiation components outside the visible spectrum of light. Lasers are also characterized according to their wavelength in a vacuum. For example, CO_2 laser (wavelength of about 110,000 Å) primarily emits infrared radiation. The helium–neon (He–Ne) laser and the semiconductor laser are commonly used in optical sensor applications. In theory, a laser has a single wavelength. However, in practice, a laser may contain several slightly different wavelengths. However, their method of generation is stimulated emission.

In a conventional laser unit, the laser beam is generated by first originating an excitation to create a light flash. This will initiate a process of emitting photons from molecules within the laser medium. This light is then reflected back and forth between two reflecting surfaces before the light beam is finally emitted as a laser. These waves will be limited to a very narrow frequency band (*monochromatic*) and will be in phase (*coherent*). For example, consider the He–Ne laser unit schematically shown in Figure 10.1. The helium and neon gas mixture in the *cavity resonator* is heated by a filament lamp and ionized using a high DC voltage (2000 V). Electrons released in the process are accelerated by the high voltage and collide with the atoms, thereby releasing *photons* (light). These photons collide with other molecules, releasing more photons. This process is known as "lasing." The light generated in this manner is reflected back and forth by the silvered surface and the partially reflective lens (beam splitter) in the cavity resonator, thereby stimulating it. This is somewhat similar to a resonant action. The stimulated light is concentrated into a narrow beam by a glass tube and emitted as a laser beam, through the partially silvered lens.

A *semiconductor laser* is somewhat similar to an LED. The laser element is typically made of a *pn* junction (diode) of semiconductor material such as gallium arsenide (GaAs) or indium gallium arsenide phosphide (InGaAsP). The edges of the junction are reflective (naturally or by depositing a film of silver). As voltage is applied to the semiconductor laser, the ionic injection and spontaneous recombination that take place near the *pn* junction emit

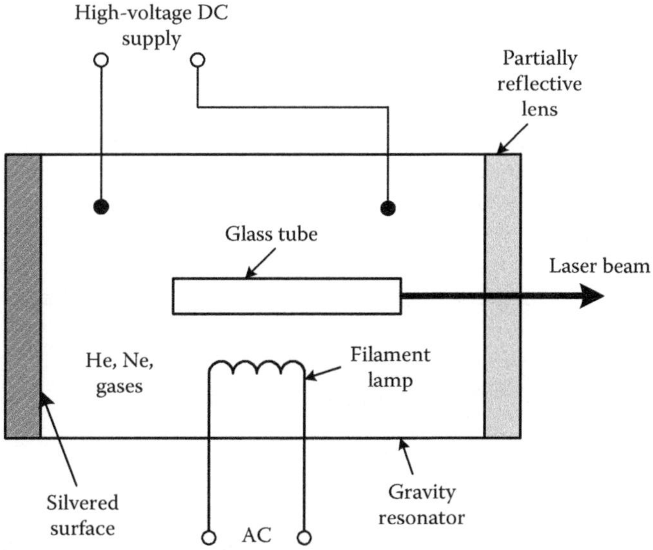

FIGURE 10.1
Helium–neon laser.

TABLE 10.1

Properties of Several Types of Lasers

Laser Type	Wavelength (Å)	Output Power (W)
Solid		
Ruby	7,000	0.1–100
Glass	1,000	0.1–500
Liquid		
Dye	4,000–10,000	0.001–1
Gas		
Helium–neon	6,330	0.001–2
Helium–cadmium	4,000	0.001–1
Carbon dioxide	110,000	$1–1 \times 10^4$
Semiconductor		
GaAs	9,000	0.002–0.01
InGaAsP	13,000	0.001–0.005

Note: $1\text{ Å} = 1 \times 10^{-10}$ m.

light as in an LED. This light is reflected back and forth between the reflective surfaces, passing along the depletion region many times and creating more photons. The stimulated light (laser) beam is emitted through an edge of the *pn* junction. Semiconductor lasers are often maintained at very low temperatures in order to obtain a reasonable component life. Semiconductor lasers can be manufactured in very small sizes. They are lower in cost and require less power in comparison to the conventional lasers. Wavelength and power output characteristics of several types of lasers are given in Table 10.1.

Laser sensor applications: Laser sensors are used in distance measurement, object localization, dimensional gauging, object quality monitoring and system inspection (e.g., manufactured parts and joints), profiling, and machinery (e.g., crane) positioning.

Example 10.1

A time-of-flight (TOF) sensor is used to measure the liquid level in a tank. It can measure levels of up to 3 m with accuracy levels better than 2 mm. In a measurement, the TOF of a laser pulse was found to be 15 ns. Estimate the depth of the liquid surface from the sensor head. Indicate some advantages and disadvantages of this method (say, compared to an ultrasonic sensor).

Solution

Taking the speed of light as 300×10^6 m/s, the measured depth d (half the distance of flight) is given by

$$d = \frac{1}{2} \times 300 \times 10^{-6}(\text{m/s}) \times 15 \times 10^{-9}(\text{s}) = 2.25\,\text{m}$$

10.2.1.1 Advantages

1. The reading is virtually unaffected by the ambient temperature, pressure, and dielectric properties.
2. It can sense from very large heights.

3. It can make very fast measurements.
4. It can provide very fine resolutions.
5. It is a noncontact sensor.

10.2.1.2 Disadvantages

1. It can be affected by dirt and dust in the environment.
2. The complete system tends to be relatively expensive.
3. It may provide some hazard (e.g., to the eyes).
4. It may be affected by the nature of the liquid and the liquid surface.

10.2.2 Fiber-Optic Sensors

The characteristic component in a fiber-optic sensor is a bundle of glass fibers (typically a few hundred) that can carry light. Each optical fiber may have a diameter on the order of a few μm to about 0.01 mm. There are two basic types of fiber-optic sensors. In one type—the "indirect" or the *extrinsic* type—the optical fiber acts just as the medium in which the sensor light is transmitted. In this type, the sensing element itself does not consist of optical fibers. In the second type—the "direct" or the *intrinsic* type—the optical fiber itself acts as the sensing element. When a condition of the sensed medium changes, the light propagation properties of the optical fiber change as well (e.g., due to microbending of a straight fiber as a result of an applied force), providing a measurement of the change in the condition. Examples of the first (extrinsic) type of sensor include fiber-optic position sensors, proximity sensors, and tactile sensors. The second (intrinsic) type of sensor is found, for example, in fiber-optic gyroscopes, fiber-optic hydrophones, and some types of microscale displacement or force sensors. As an *intrinsic* application of fiber optics in sensing, consider a straight optical fiber element that is supported at the two ends. In this configuration, almost 100% of the light at the source end will transmit through the optical fiber and reach the detector (receiver) end. Now, suppose that a slight load is applied to the optical fiber segment at its mid span. It will deflect slightly due to the load, and as a result, the amount of light received at the detector can drop significantly. For example, a microdeflection of just 50 μm can result in a drop in intensity at the detector by a factor of 25. Such an arrangement may be used in deflection, force, and tactile sensing. Another intrinsic application is the fiber-optic gyroscope, as described next.

10.2.2.1 Advantages

The advantages of fiber optics include insensitivity to electrical and magnetic noise (due to optical coupling); safe operation in explosive, high-temperature, corrosive, and hazardous environments; and high sensitivity. Furthermore, mechanical loading and wear problems do not exist because fiber-optic position sensors are noncontact devices with no moving parts.

10.2.2.2 Disadvantages

The disadvantages of fiber optics include direct sensitivity to variations in the intensity of the light source and dependence on ambient conditions (temperature, dirt, moisture, smoke, etc.). Compensation can be made, however, with respect to temperature.

10.2.3 Examples of Optical Sensors

Several types of sensors that employ lasers and fiber optics are described next.

10.2.3.1 Laser Interferometer

A laser interferometer is useful in the accurate measurement of small displacements. This is an *extrinsic* application of fiber optics where optical fiber is used for light transmission rather than for light sensing. In this fiber-optic position sensor, the same bundle of fibers is used for sending and receiving a monochromatic beam of light (a laser). Alternatively, monomode fibers, which transmit only monochromatic light (of a specific wavelength), may be used for this purpose. In either case, as shown in Figure 10.2, a beam splitter (A) is used so that part of the light is directly reflected back to the bundle tip and the other part reaches the target object (as in Figure 10.2) and is reflected back from it (using a reflector mounted on the object) onto the bundle tip. In this manner, part of the light returning through the bundle had not traveled beyond the beam splitter while the other part had traveled between the beam splitter (A) and the object (through an extra distance equal to twice the separation between the beam splitter and the object). As a result, the two components of light will have a phase difference ϕ, which is given by

$$\phi = \frac{2x}{\lambda} \times 2\pi \tag{10.1}$$

where
  x is the distance of the target object from the beam splitter
  λ is the wavelength of monochromatic light

The returning light is directed to a light sensor using a beam splitter (B). The sensed signal is processed using principles of interferometry to determine ϕ and, from Equation 10.1, the distance x. Very fine resolutions better than a fraction of a micrometer (μm) can be obtained using this type of fiber-optic position sensors.

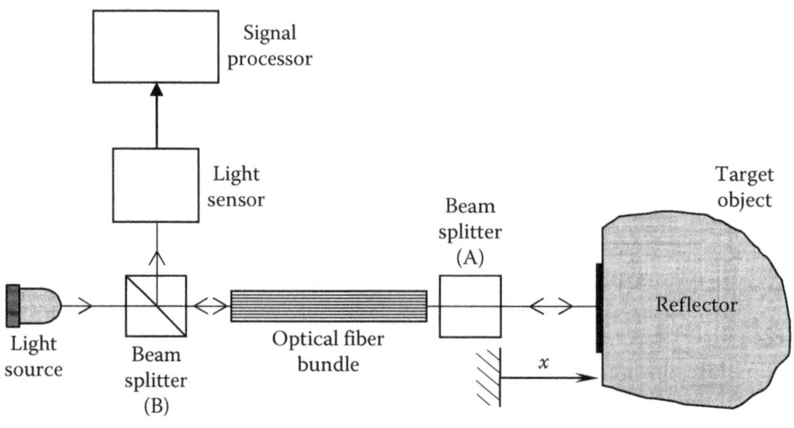

FIGURE 10.2
Laser interferometer position sensor.

10.2.3.2 Laser Doppler Interferometer

The laser Doppler interferometer is used for accurate measurement of speed. It is based on two principles: the Doppler effect and the light wave interference. The latter phenomenon is used in the laser interferometer position sensor, which was discussed earlier. To explain the former phenomenon, consider a wave source (e.g., a light source or sound source) that is moving with respect to a receiver (observer). If the source moves toward the receiver, the frequency of the received wave appears to have increased, and if the source moves away from the receiver, the frequency of the received wave appears to have decreased. The change in frequency is proportional to the velocity of the source relative to the receiver. This phenomenon is known as the "Doppler effect."

Now consider a monochromatic (single-frequency) light wave of frequency f (say, 5×10^{14} Hz) emitted by a laser source. If this ray is reflected by a target object and received by a light detector, the frequency of the received wave would be $f_2 = f + \Delta f$. The frequency increase Δf will be proportional to the velocity v of the target object, which is assumed positive when moving toward the light source. Specifically,

$$\Delta f = \frac{2f}{c} v = kv \tag{10.2}$$

where c is the speed of light in the particular medium (typically, air). Now by comparing the frequency f_2 of the reflected wave with the frequency $f_1 = f$ of the original wave, we can determine Δf and hence the velocity v of the target object.

The change in frequency Δf due to the Doppler effect can be determined by observing the fringe pattern due to light wave interference. To understand this, consider the two waves $v_1 = a \sin 2\pi f_1 t$ and $v_2 = a \sin 2\pi f_2 t$. If we add these two waves, the resulting wave would be $v = v_1 + v_2 = a(\sin 2\pi f_1 t + \sin 2\pi f_2 t)$, which can be expressed as

$$v = 2a \sin \pi(f_2 + f_1)t \cos \pi(f_2 - f_1)t \tag{10.3}$$

It follows that the combined signal beats at the beat frequency $\Delta f/2$. Since f_2 is very close to f_1 (because Δf is small compared to f), these beats will appear as dark and light lines (fringes) in the resulting light wave. This is known as "wave interference." The frequency change Δf can be determined by two methods:

1. By measuring the spacing of the fringes
2. By counting the beats in a given time interval or by timing successive beats using a high-frequency clock signal

The velocity of the target object is determined in this manner. Displacement can be obtained simply by digital integration (or by accumulating the count).

A schematic diagram for the laser Doppler interferometer is shown in Figure 10.3. Industrial interferometers usually employ a helium–neon laser, which has waves of two frequencies close together. In that case, the arrangement shown in Figure 10.3 has to be modified to take into account the two frequency components.

Note: The laser interferometer discussed earlier (Figure 10.2) directly measures displacement rather than speed. It is based on measuring the phase difference between the direct and returning laser beams, not the Doppler effect (frequency difference).

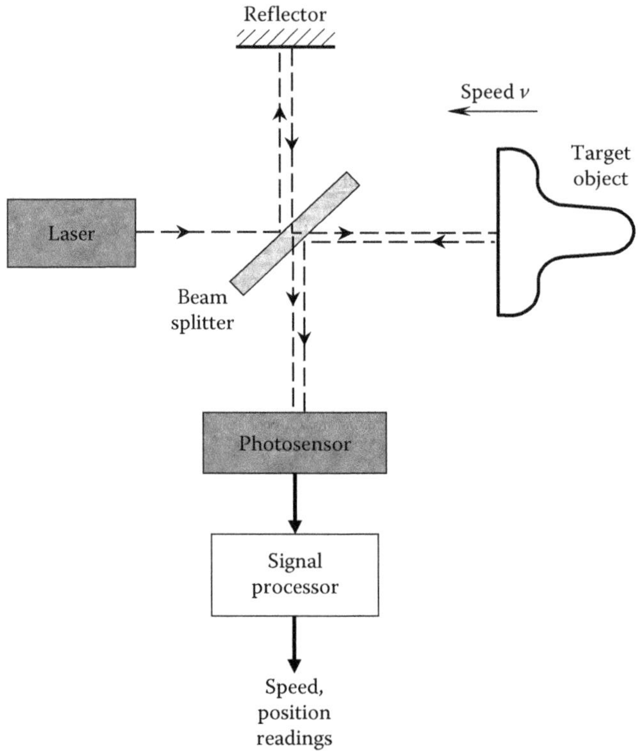

FIGURE 10.3
A laser Doppler interferometer for measuring velocity and displacement.

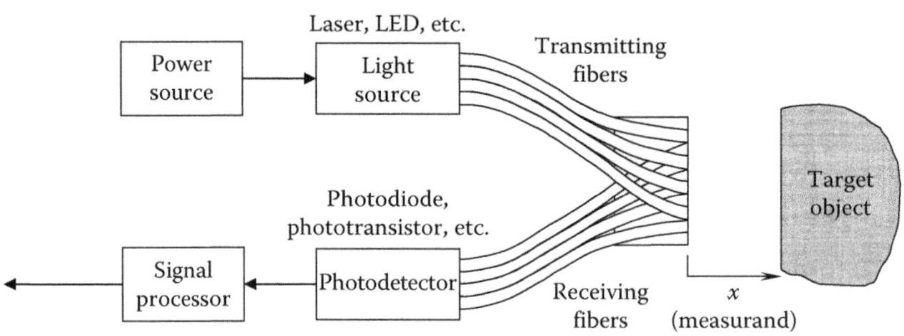

FIGURE 10.4
A fiber-optic position sensor.

10.2.3.3 Fiber-Optic Position Sensor

A schematic representation of a fiber-optic position sensor (or proximity sensor or displacement sensor) is shown in Figure 10.4.

The optical fiber bundle is divided into two groups: transmitting fibers and receiving fibers. Light from the light source is transmitted along the first bundle of fibers to the target object whose position is being measured. Light reflected (or diffused) onto the receiving fibers by the surface of the target object is carried to a photodetector.

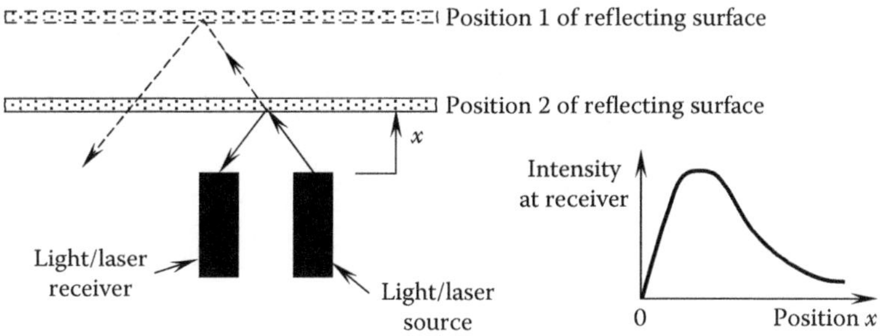

FIGURE 10.5
The principle of a fiber-optic proximity sensor.

The intensity of the light received by the photodetector will depend on position x of the target object. In particular, if $x = 0$, the transmitting bundle will be completely blocked off and the light intensity at the receiver will be zero. As x is increased, the intensity of the received light will increase, because more and more light will be reflected onto the tip of the receiving bundle. This will reach a peak at some value of x. When x is increased beyond that value, more and more light will be reflected outside the receiving bundle; hence, the intensity of the received light will drop. In general, then the proximity–intensity curve for an optical proximity sensor will be nonlinear and will have the shape shown in Figure 10.5. Using this (calibration) curve, we can determine the position (x) once the intensity of the light received at the photosensor is known. The light source could be a laser (structured light), infrared light source, or some other type, such as an LED. A device such as a photodiode or a photo field-effect transistor (photo-FET) may be used as the light sensor (photodetector). This type of fiber-optic sensors can be used, with a suitable front-end device (such as bellows and springs), to measure pressure, force, touch (tactile sensor; see Chapter 9), and so on.

10.2.3.4 Fiber-Optic Gyroscope

This is an angular speed sensor that uses fiber optics. Contrary to its name, however, it is not a gyroscope in the conventional sense (see Chapter 8). Two loops of optical fiber wrapped around a cylinder are used in this sensor, and they rotate with the cylinder, at the same angular speed, which needs to be sensed. One loop carries a monochromatic light (laser) beam in the clockwise direction, and the other loop carries a beam from the same light source in the counterclockwise direction (see Figure 10.6). Since the laser beam traveling in the direction of rotation of the cylinder attains a higher frequency than that of the other beam, the difference in frequencies (known as the "Sagnac effect") of the two laser beams received at a common location will measure the angular speed of the cylinder. This may be accomplished through interferometry, because the combined signal is a sine beat. As a result, light and dark patterns (fringes) will be present in the detected light, and they will measure the frequency difference and hence the rotating speed of the optical fibers.

In a laser (ring) gyroscope, it is not necessary to have a circular path for the laser. Triangular and square paths are used as well. In general, the beat frequency $\Delta\omega$ of the combined light from two laser beams traveling in opposite directions is given by

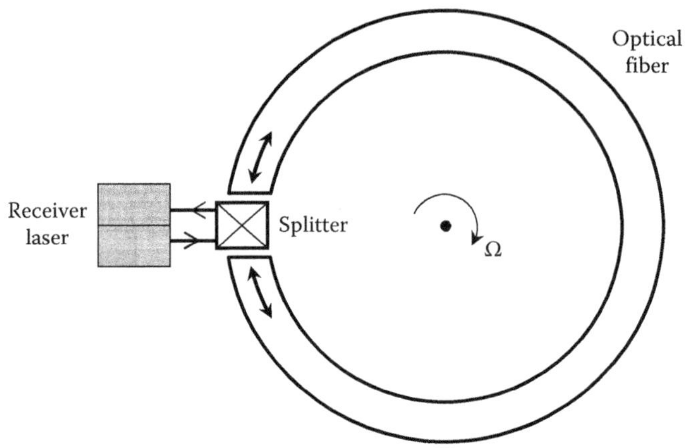

FIGURE 10.6
A fiber-optic, laser gyroscope.

$$\Delta\omega = \frac{4A}{p\lambda}\Omega \tag{10.4}$$

where
- A is the area enclosed by the travel path (πr^2 for a cylinder of radius r)
- p is the length (perimeter) of the traveled path ($2\pi r$ for a cylinder)
- λ is the wavelength of the laser
- Ω is the angular speed of the object (or optical fiber)

The length of the optical fiber wound around the rotating object can exceed 100 m and can reach even 1 km. Angular displacements can be measured with a laser gyro simply by counting the number of cycles and clocking fractions of cycles. Acceleration can be determined by digitally determining the rate of change of speed. In a laser gyro, there is an alternative to the use of two separate loops of optical fiber, wound in opposite direction. The same loop can be used to transmit light from the same laser from the opposite ends of the fiber. A beam splitter has to be used in this case, as shown in Figure 10.6.

10.2.4 Light Sensors

Semiconductor-based light sensors as well as light sources are needed in optoelectronics. A light sensor (also known as a "photodetector" or "photosensor") is a device that is sensitive to light. Usually, it is a part of an electrical circuit with associated signal conditioning (amplification, filtering, etc.) so that an electrical signal that is representative of the intensity of light falling on the photosensor is obtained. Some photosensors can serve as energy sources (*cells*) as well. A photosensor may be an integral component of an optoisolator or other optically coupled system. In particular, a commercial optical coupler typically has an LED light source and a photosensor in the same package, with leads for connecting it to other circuits, together with power leads.

By definition, the purpose of a photodetector or photosensor is to sense visible light. But there are many applications where sensing of adjoining bands of the electromagnetic spectrum, namely, *infrared* radiation and *ultraviolet* radiation, would be useful. In particular, these nonvisible frequency bands are not corrupted by ambient light (visible). Since objects emit reasonable levels of infrared radiation even at low temperatures, infrared sensing can be used in applications where imaging of an object in the dark is needed. Applications include infrared photography, security systems, and missile guidance. Also, since infrared radiation is essentially *thermal energy*, infrared sensing can be effectively used in thermal sensing and thermal control systems. Ultraviolet sensing is not as widely applied as infrared sensing.

Typically, a photosensor is a resistor, diode, or transistor element that brings about a change (e.g., generation of a potential or a change in resistance) in an electrical circuit, in response to light that is falling on the sensor element. The power of the output signal may be derived primarily from the power source that energizes the electrical circuit. Hence, they are active sensors. Alternatively, a photocell can be used as a photosensor. In this latter case, the energy of the light falling on the cell is converted into electrical energy of the output signal. Hence, photocells are passive sensors (or energy sources). Typically, a photosensor is available as a tiny cylindrical element with a sensor head consisting of a circular window (lens). Several types of photosensors are described next.

10.2.4.1 Photoresistor

A photoresistor (or *photoconductor*) has the property of decreasing its electrical resistance (increasing the conductivity) as the intensity of light falling on it increases. Typically, the resistance of a photoresistor could change from very high values (megohms) in the dark to reasonably low values (less than 100 Ω) in bright light. As a result, very high sensitivity to light is possible. Some photocells can function as photoresistors because their impedance decreases (output increases) as the light intensity increases. Photocells used in this manner are termed *photoconductive cells*. The circuit symbol of a photoresistor is given in Figure 10.7a. A photoresistor may be formed by sandwiching a photoconductive crystalline material such as *cadmium sulfide* (CdS) or *cadmium selenide* (CdSe) between two electrodes. Lead sulfide (PbS) or lead selenide (PbSe) may be used in infrared photoresistors.

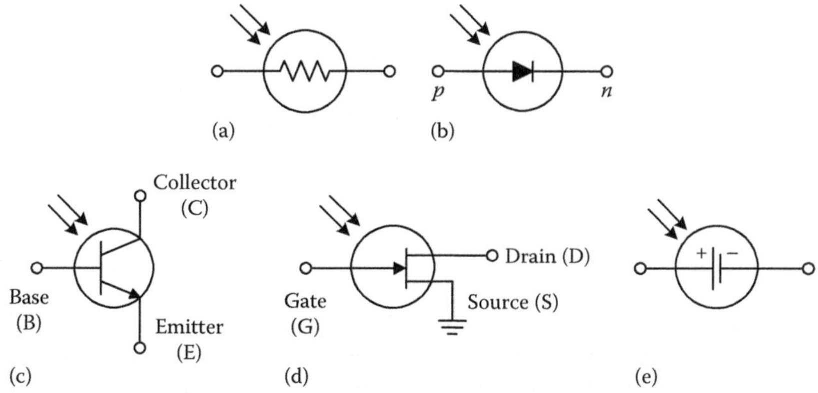

FIGURE 10.7
Circuit symbols of some photosensors: (a) photoresistor, (b) photodiode, (c) phototransistor (*npn*), (d) photo **field-effect transistor** (*n*-channel), and (e) photocell.

10.2.4.2 Photodiode

A photodiode is a *pn* junction of semiconductor material that produces electron–hole pairs in response to light. The symbol for a photodiode is shown in Figure 10.7b. Two types of photodiodes are available. A *photovoltaic* diode generates a sufficient potential at its junction in response to light (photons) falling on it. Hence, an external bias source is not necessary for a photovoltaic diode. A *photoconductive* diode undergoes a resistance change at its junction in response to photons. This type of photodiode is operated in reverse-biased form; the *p*-lead of the diode is connected to the negative lead of the circuit and the *n*-lead is connected to the positive lead of the circuit. The breakdown condition may occur at about 10 V, and the corresponding current will be nearly proportional to the intensity of the light falling on the photodiode. Hence, this current can be used as a measure of the light intensity. The sensitivity of a photodiode is rather low particularly due to the reverse-biased operation. Since the output current level is usually low (a fraction of a milliampere), amplification might be necessary before using it in a subsequent application (e.g., signal transmission, actuation, control, display). Semiconductor materials such as silicon, germanium, cadmium sulfide, and cadmium selenide are commonly used in photodiodes. The response speed of a photodiode is high. A diode with an intrinsic layer (a *pin diode*) can still provide faster response than with a regular *pn* diode.

10.2.4.3 Phototransistor

Any semiconductor photosensor with amplification circuitry built into the same package (chip) is popularly called a phototransistor. Hence, a photodiode with an amplifier circuit in a single unit might be called a phototransistor. Strictly, a phototransistor is manufactured in the form of a conventional *bipolar junction transistor* with *base* (B), *collector* (C), and *emitter* (E) leads.

Symbolic representation of a phototransistor is shown in Figure 10.7c. This is an *npn* transistor. The base is the central (*p*) region of the transistor element. The collector and the emitter are the two end regions (*n*) of the element. Under operating conditions of the phototransistor, the collector–base junction is *reverse biased* (i.e., a positive lead of the circuit is connected to the collector and a negative lead of the circuit is connected to the base of an *npn* transistor). Alternatively, a phototransistor may be connected as a two-terminal device with its base terminal floated and the collector terminal properly biased (positive for an *npn* transistor). For a given level of source voltage (usually applied between the emitter lead of the transistor and load, the negative potential being at the emitter lead), the collector current (current through the collector lead) i_c is nearly proportional to the intensity of the light falling on the collector–base junction of the transistor. Hence, i_c can be used as a measure of the light intensity. Germanium or silicon is the semiconductor material that is commonly used in phototransistors.

10.2.4.4 Photo-FET

A photo-FET is similar to a conventional FET. The symbol shown in Figure 10.7d is for an *n*-channel photo-FET. This consists of an *n*-type semiconductor element (e.g., *silicon* doped with *boron*), called "channel." A much smaller element of *p*-type material is attached to the *n*-type element. The lead on the *p*-type element forms the *gate* (G). The *drain* (D) and the *source* (S) are the two leads on the channel. The operation of an FET depends on the electrostatic fields created by the potentials applied to the leads of the FET.

Under operating conditions of a photo-FET, the gate is reverse biased (i.e., a negative potential is applied to the gate of an *n*-channel photo-FET). When light is projected at the gate, the drain current i_d will increase. Hence, the drain current (current at the D lead) can be used as a measure of light intensity.

10.2.4.5 Photocell

Photocells are similar to photosensors except that a photocell is used as an electricity source rather than a sensor of radiation. Solar cells, which are more effective in sunlight, are commonly available. A typical photocell is a semiconductor junction element made of a material such as single-crystal silicon, polycrystalline silicon, and cadmium sulfide. Cell arrays are used in moderate-power applications. Typical power output is 10 mW per square centimeter of surface area, with a potential of about 1.0 V. The circuit symbol of a photocell is given in Figure 10.7e.

10.2.4.6 Charge-Coupled Device

A charge-coupled device (CCD) is an integrated circuit element (a *monolithic device*) of semiconductor material. A CCD made from silicon is schematically represented in Figure 10.8. A silicon wafer (*p*-type or *n*-type) is oxidized to generate a layer of SiO_2 on its surface. A matrix of metal electrodes is deposited on the oxide layer and is linked to the CCD output leads. When light falls onto the CCD element (from an object), *charge packets* are generated within the substrate *silicon wafer.* Now, if an external potential is applied to a particular electrode of the CCD, a *potential well* is formed under the electrode and a charge packet is deposited there. This charge packet can be moved across the CCD to an output circuit by sequentially energizing the electrodes using pulses of external voltage. Such a charge packet corresponds to a *pixel* (a picture element) of an image of the object. The circuit output is the video signal of the image (see Chapter 11). The pulsing rate can be higher than 10 MHz. CCDs are commonly used in imaging applications, particularly in cameras. A typical CCD element with a facial area of a few square centimeters may detect 576 × 485 pixels, but larger elements (e.g., 4096 × 4096 pixels) are available as well. A *charge injection device* (CID) is similar to a CCD. In a CID, however, there is a matrix of

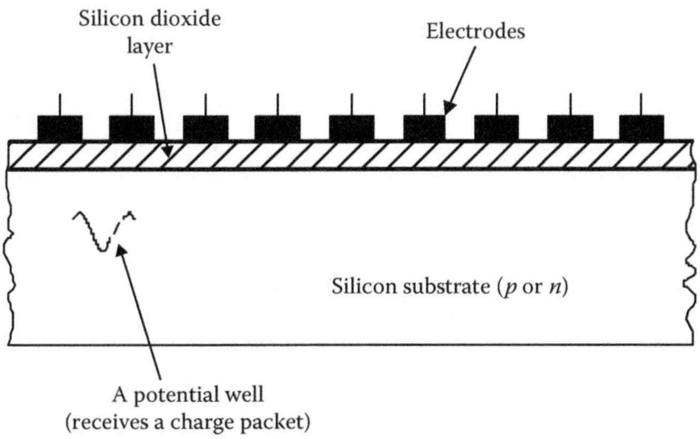

FIGURE 10.8
A charge-coupled device.

semiconductor capacitor pairs. Each capacitor pair can be directly addressed through voltage pulses. When a particular element is addressed, the potential well there will shrink, thereby injecting minority carriers into the substrate. The corresponding signal, tapped from the substrate, forms the video signal. The signal level of a CID is substantially smaller than that of a CCD, as a result of higher capacitance.

10.2.5 Applications of Optically Coupled Devices

One direct application of optical coupling is in the isolation of electric circuitry. When two circuits are directly connected through hard electrical connections (cables, wires, etc.), a two-way path for the electrical signals is created at the interface. In other words, signals in circuit A will affect circuit B and signals in circuit B will affect circuit A. This interaction means that "noise" in one circuit will directly affect the other. Furthermore, there will be loading problems; the source will be affected by the load. Both these situations are undesirable. If the two circuits are optically coupled, however, there is only a one-way interaction between the two circuits (see Figure 10.9). Variations in the *output circuit* (*load circuit*) will not affect the *input circuit*. Hence, the input circuit is *isolated* from the output circuit. The connecting cables in an electrical circuit can introduce noise components such as *electromagnetic interference*, *line noise*, and *ground-loop noise* (see Chapter 3). The likelihood of these noise components affecting the overall system is also reduced by using optical coupling. In summary, isolation between two circuits and isolation of a circuit from noise can be achieved by optical coupling. For these reasons, optical coupling is widely used in communication networks (telephones, computers, etc.) and in circuitry for high-precision signal conditioning (e.g., for sophisticated sensors and control systems).

The medium through which the light passes from a light source to a photosensor can create noise problems, however. If the medium is open (see Figure 10.9), then ambient lighting conditions will affect the output circuit, resulting in an error. Also environmental impurities (dust, dirt, smoke, moisture, etc.) will affect the light received by the photosensor. Hence, a more controlled medium of transmission would be desirable. Linking the light source and the photosensor using *optical fibers* is a good way to reduce problems due to ambient conditions in an optically coupled system.

Optical coupling may be used in *relay circuits* where a low-power circuit is used to operate a high-power circuit. If the relay that operates the high-power circuit is activated using an optical coupler, reaction effects (noise and loading) on the low-power circuit can be eliminated. Optical coupling is used in *power electronics* and *control systems* in this manner.

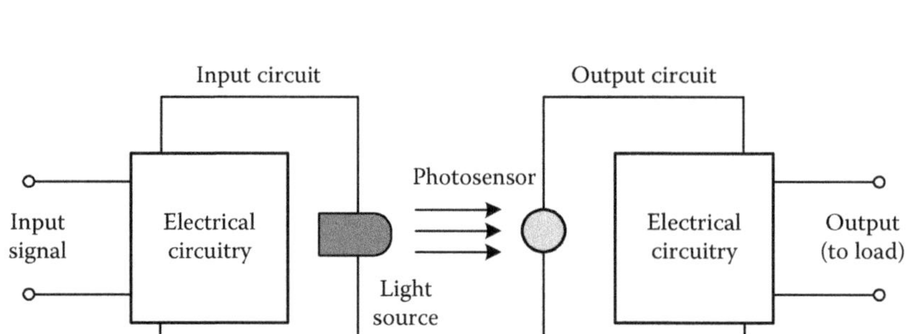

FIGURE 10.9
An optically coupled device.

Many types of sensors and transducers that are based on optical methods do indeed employ optical coupling (e.g., optical encoders, fiber-optic tactile sensors). Optical sensors are widely used in industry for parts counting, parts detection, and level detection. In these sensors, a light beam is projected from a source to a photodetector, both units being stationary. An interruption of the beam through the passage of a part will generate a pulse at the detector, and this pulse is read by a counter or a parts detector (see Chapter 11). Furthermore, if the light beam is located horizontally at a required height, its interruption when the material filled into a container reaches that level may be used for filling control in the packaging industry. Note that the light source and the sensor can be located within a single package if a mirror is used to reflect light from the source back onto the detector. Further applications include computer disk drive systems, for example, to detect the write protect notch as well as the position of the recording head.

10.3 Miscellaneous Sensor Technologies

Three other sensors that are used in engineering applications are presented next. They are ultrasonic, magnetostrictive, and AE sensors.

10.3.1 Ultrasonic Sensors

Audible sound waves have frequencies in the range of 20 Hz to 20 kHz. Ultrasound waves are pressure waves, just like sound waves, but their frequencies are higher ("ultra") than the audible frequencies. Ultrasonic sensors are used in many applications, including medical imaging, ranging systems for cameras with autofocusing capability, level sensing, and speed sensing. In medical applications, ultrasound probes of frequencies 40 kHz, 75 kHz, 7.5 MHz, and 10 MHz are commonly used.

Ultrasound can be generated according to several principles. For example, high-frequency (gigahertz) oscillations in a piezoelectric crystal subjected to an electrical potential are able to generate very high-frequency ultrasound. Another method is to use the *magnetostrictive* property of a material, which deforms when subjected to a magnetic field. Respondent oscillations generated by this principle can produce ultrasonic waves. Another method of generating ultrasound is to apply a high-frequency voltage to a metal-film capacitor. A microphone can serve as an ultrasound detector (receiver).

Intrinsic and extrinsic sensors: From analogous to fiber-optic sensing, there are two common ways of employing ultrasound in a sensor. In one approach—the *intrinsic* method—the ultrasound signal undergoes changes as it passes through an object, due to acoustic impedance and absorption characteristics of the object. The resulting signal (image) may be interpreted to determine the properties of the object, such as texture, firmness, and deformation. This approach has been utilized, for example, in an innovative firmness sensor for herring roe. It is also the principle used in medical ultrasonic imaging. In the other approach—the *extrinsic* method—the TOF of an ultrasound burst from its source to an object and then back to the receiver is measured. This approach is used in distance and position measurement, in level sensing, and in dimensional gauging. For example, an ultrasound sensor of this category has been used in the thickness measurement of fish. This is also the method used in camera autofocusing.

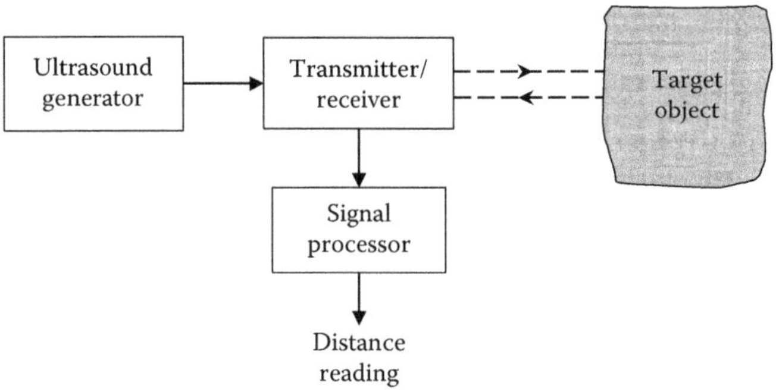

FIGURE 10.10
An ultrasonic **time-of-flight** position sensor.

In distance (range, proximity, displacement) measurement using ultrasound, a burst of ultrasound is projected at the target object, and the time taken for the echo to be received is clocked. A signal processor computes the position of the target object, possibly compensating for environmental conditions. This configuration is shown in Figure 10.10. The applicable relation is

$$x = \frac{ct}{2} \tag{10.5}$$

where
t is the TOF of the ultrasound pulse (from generator to receiver)
x is the distance between the ultrasound generator/receiver and the target object
c is the speed of sound in the medium (typically, air)

Distances as small as a few cm to several meters may be accurately measured by this approach, with fine resolution (e.g., a millimeter or less). Since the speed of ultrasonic wave propagation depends on the medium and the temperature of the medium (typically air), errors will enter into the ultrasonic readings unless the sensor is compensated for the variations in the medium, particularly for temperature.

Alternatively, the velocity of the target object can be measured, using the Doppler effect, by measuring (clocking) the change in frequency between the transmitted wave and the received wave. The "beat" phenomenon is employed here. The applicable relation is Equation 10.2; now, f is the frequency of the ultrasound signal and c is the speed of sound.

10.3.2 Magnetostrictive Displacement Sensor

The magnetostrictive property and how it may be used in the sensing of strain or stress have been discussed in Chapter 9. Alternatively, the ultrasound-based TOF method may be used in a magnetostrictive displacement sensor (e.g., the sensor manufactured by Temposonics). The principle behind this method is illustrated in Figure 10.11. The sensor head generates an interrogation current pulse, which travels along the magnetostrictive wire or rod (called the "waveguide") that is enclosed in a protective cover. A timer is started as the interrogation pulse is sent. This pulse, which carries a magnetic field, interacts with the magnetic field of the permanent magnet and generates an ultrasound

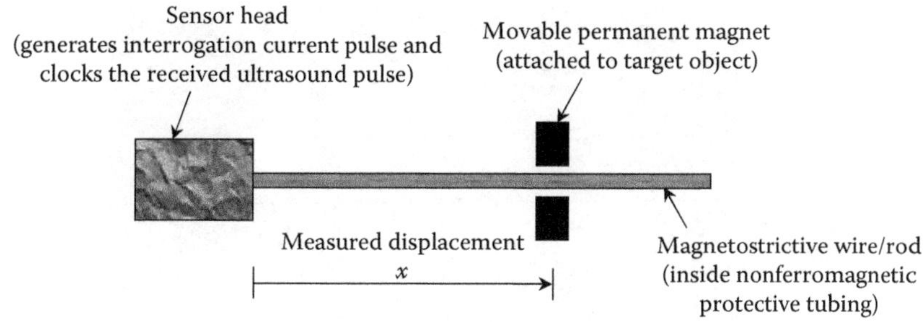

FIGURE 10.11
A magnetostrictive ultrasound displacement sensor.

(strain) pulse (by the magnetostrictive action in the waveguide). This pulse is received at the sensor head and timed. The TOF is proportional to the distance of the magnet from the sensor head. The target object is attached to the magnet of the sensor, and its position (x) is determined using the TOF as usual.

Strokes (maximum displacements) ranging from a few cm to 1 or 2 m at resolutions better than 50 μm are possible with these sensors. With a 15 V DC power supply, the sensor can provide a DC output in the range ±5 V. Since the sensor uses a magnetostrictive medium with protective nonferromagnetic tubing, some of the common sources of error in ultrasonic sensors that use air as the medium of propagation can be avoided.

10.3.3 Acoustic Emission Sensors

When an object undergoes rapid and irreversible change in its internal microscopic structure, it releases strain energy in the form of acoustic waves. Such AE can be sensed and used to localize (i.e., to determine the location) the object that emits it and also to determine the nature of the change (crack, fracture, wear, etc.) that generated it. In this manner, AE may be applied in predicting, detecting, and diagnosing failure and malfunction, machine health monitoring, planning structural maintenance, nondestructive testing of objects, quality control, process control, development of new material, and so on. Specific applications of AE sensing include assessing the quality of welding; monitoring cracks and other flaws in pressure vessels, airplanes, and ground vehicles; detecting corrosion in concrete structures; and monitoring pipelines that transport natural gas and oil.

An AE sensor is any sensor that can detect an acoustic wave. Typically, a piezoelectric element is used for this purpose (e.g., PZT; see Chapter 8). Hence, an AE sensor may look like a piezoelectric accelerometer. The sensor signal needs some preprocessing, preamplification in particular. Some AE sensors may contain a built-in preamplifier. A cable is used to connect the sensor to a signal processor or a data acquisition board of a computer. AE sensors come in very small sizes (e.g., 2 mm and a few mg). An AE sensor may be mounted on the monitored object in the same manner as an accelerometer. The common methods of mounting include screw-in base, magnetic base, and glue.

AE waves may be present in frequencies in the ultrasonic range of tens of kHz up to as high as several MHz. The sensor typically uses the condition of resonance to detect an AE signal, thereby improving the sensor sensitivity. Wideband models, which do not specifically use the method of resonance, are available as well. In particular, in a wideband sensor, a damper is bonded to the piezoelectric element, thereby flattening the resonance of the element. AE sensors may also be categorized into low-frequency (e.g., 20–100 kHz),

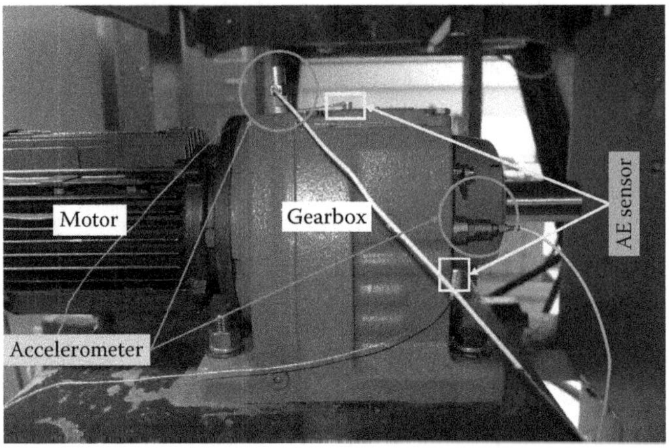

(a)

FIGURE 10.12
Use of acoustic emission (AE) sensors to diagnose a faulty gear transmission. (a) The experimental setup. (b) AE signals under normal and faulty conditions. (c) Corresponding frequency spectra.

medium-frequency (e.g., 100–400 kHz), and high-frequency (e.g., >400 kHz) types. The choice of an AE sensor for an application may depend on such considerations as nature, size, and material of the monitored object, background noise (sound and other disturbances), and the frequency range of interest.

As an application of AE sensors, consider the setup shown in Figure 10.12a. Its purpose is to monitor the performance of the gear transmission of an industrial machine. Both AE sensors and accelerometers are used in the monitoring system. Figure 10.12b presents a pair of AE signals obtained under normal conditions and faulty conditions. The corresponding frequency spectrum magnitudes are shown in Figure 10.12c. In this situation, the faulty condition corresponds to a damaged bearing of the gear transmission. In the time signals, we notice a clear difference in the two cases, and the damaged gear transmission has resulted in increased intensity in the AE signal. The frequency spectra display similar characteristic and also indicate the critical frequency bands, which can be particularly useful in diagnosing the faulty condition.

10.4 Thermofluid Sensors

Common thermofluid sensors include those measuring pressure, fluid flow rate, temperature, and heat transfer rate. Such sensors are useful in a variety of engineering applications. Several common types of sensors in this category are presented next.

10.4.1 Pressure Sensors

The common methods of pressure sensing are the following:

1. Balance the pressure with an opposing force (or head) and measure this force. Examples are liquid manometers and pistons.

2. Subject the pressure to a flexible front-end (auxiliary) member and measure the resulting deflection. Examples are Bourdon tube, bellows, and helical tube.
3. Subject the pressure to a front-end auxiliary member and measure the resulting strain (or stress). Examples are diaphragms and capsules.

Some of these devices are illustrated in Figure 10.13.

Manometer: In the manometer shown in Figure 10.13a, the liquid column of height h and density ρ provides a counterbalancing pressure head to support the measured pressure p with respect to the reference (ambient) pressure p_{ref}. Accordingly, this device measures the gauge pressure as given by

$$p - p_{ref} = \rho g h \tag{10.6}$$

where g is the acceleration due to gravity.

Counterbalance piston: In the pressure sensor shown in Figure 10.13b, a frictionless piston of area A supports the pressure load by means of an external force F. The governing equation is

$$p = \frac{F}{A}. \tag{10.7}$$

The pressure is determined by measuring F using a force sensor.

Bourdon tube: The Bourdon tube shown in Figure 10.13c deflects with a straightening motion as a result of the internal pressure. This deflection can be measured using a displacement sensor (typically, a rotatory sensor) or indicated by a moving pointer.

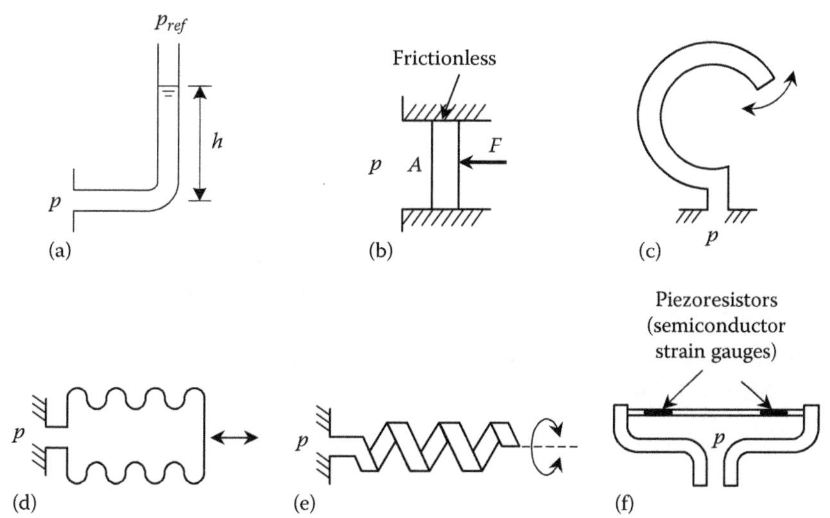

FIGURE 10.13
Typical pressure sensors: (a) manometer, (b) counterbalance piston, (c) Bourdon tube, (d) bellows, (e) helical tube, and (f) diaphragm.

Bellows: The bellows unit deflects as a result of the internal pressure, causing a linear motion, as shown in Figure 10.13d. The deflection can be measured using a sensor such as LVDT or a capacitive sensor (see Chapter 8) and can be calibrated to indicate pressure.

Helical tube: The helical tube shown in Figure 10.13e undergoes a twisting (rotational) motion when deflected by the internal pressure. This deflection can be measured by an angular displacement sensor (RVDT, resolver, potentiometer, etc.; see Chapter 8), to provide a pressure reading through proper calibration.

Diaphragm pressure sensors: Figure 10.13f illustrates the use of a diaphragm to measure pressure. The membrane (typically metal) is strained due to pressure. The pressure can be measured by means of strain gauges (i.e., piezoresistive sensors; see Chapter 9) mounted on the diaphragm. MEMS pressure sensors that use this principle are available (see Chapter 12). In one such device, the diaphragm has a silicon wafer substrate integral with it. Through proper doping (using boron, phosphorus, etc.), a microminiature semiconductor strain gauge can be formed. In fact, more than one piezoresistive sensor can be etched on the diaphragm, and used in a bridge circuit to provide the pressure reading, through proper calibration. The most sensitive locations for the piezoresistive sensors are closer to the edge of the diaphragm, where the strains reach the maximum. Magnetostrictive strain gauges can be used as well in pressure sensors, by using a magnetostrictive material in the diaphragm.

10.4.2 Flow Sensors

The volume flow rate Q of a fluid is related to the mass flow rate Q_m through $Q_m = \rho Q$, where ρ is the mass density of the fluid. In addition, for a flow across an area A at average velocity v, we have

$$Q = Av \tag{10.8}$$

When the flow is not uniform, a suitable correction factor has to be included depending on what velocity is used in this equation. Next, according to Bernoulli's equation for incompressible, ideal flow (no energy dissipation), we have

$$p + \frac{1}{2}\rho v^2 = \text{constant} \tag{10.9}$$

This theorem may be interpreted as conservation of energy. Moreover, note that the pressure p due to a fluid head of height h is given by (gravitational potential energy) ρgh. Using Equations 10.8 and 10.9 and allowing for dissipation (friction), the flow across a constriction (i.e., a fluid resistance element such as an orifice, nozzle, valve, and so on) of area A can be shown to obey the relation

$$Q = c_d A\sqrt{\frac{2\Delta p}{\rho}} \tag{10.10}$$

where

Δp is the pressure drop across the constriction
c_d is the discharge coefficient for the constriction

The common methods of measuring fluid flow may be classified as follows:

1. Measure pressure across a known constriction or opening. Examples include nozzles, Venturi meters, and orifice plates.
2. Measure the pressure head, which brings the flow to static conditions. The examples include pitot tube and liquid level sensing using floats.
3. Measure the flow rate (volume or mass) directly. Turbine flowmeter and angular momentum flowmeter are examples.
4. Measure the flow velocity. Coriolis meter, laser Doppler velocimeter, and ultrasonic flowmeter are examples.
5. Measure an effect of the flow and estimate the flow rate using that information. Hot-wire (or hot-film) anemometer and magnetic induction flowmeter are examples.

Several examples of flowmeters are shown in Figure 10.14.

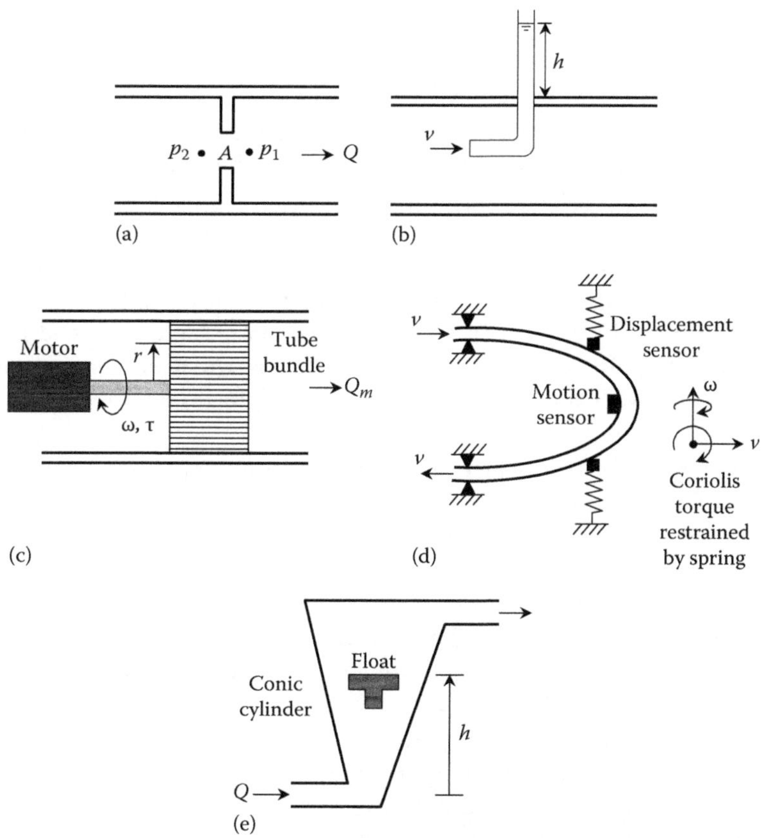

FIGURE 10.14
Several flowmeters: (a) orifice flowmeter, (b) pitot tube, (c) angular momentum flowmeter, (d) Coriolis velocity meter, and (e) rotameter.

Orifice flowmeter: For the orifice meter shown in Figure 10.14a, Equation 10.10 is applied to measure the volume flow rate. The pressure drop is measured using the techniques outlined earlier.

Pitot tube: For the pitot tube shown in Figure 10.14b, Bernoulli's Equation 10.9 is applicable, noting that the fluid velocity at the free surface of the tube is zero. This gives the flow velocity

$$v = \sqrt{2gh} \tag{10.11}$$

Note: A correction factor is needed when determining the flow rate because the velocity is not uniform across the flow section.

Angular momentum flowmeter: In the angular momentum method shown in Figure 10.14c, the tube bundle through which the fluid flows is rotated by a motor. The motor torque τ and the angular speed ω are measured. As the fluid mass passes through the tube bundle, it imparts an angular momentum at a rate governed by the mass flow rate Q_m of the fluid. The motor torque provides the torque needed for this rate of change of angular momentum. Neglecting losses, the governing equation is

$$\tau = \omega r^2 Q_m \tag{10.12}$$

where r is the radius of the centroid of the rotating fluid mass.

Turbine flowmeter: In a turbine flowmeter, the rotation of the turbine wheel located in the flowing fluid can be calibrated to directly give the flow rate.

Coriolis velocity meter: In the Coriolis method shown in Figure 10.14d, the fluid is made to flow through a "U" segment, which is hinged to oscillate out of plane (at angular velocity ω) and restrained by springs (of known stiffness) in the lateral direction. If the fluid velocity is v, the resulting Coriolis force (due to Coriolis acceleration $2\boldsymbol{\omega} \times \boldsymbol{v}$) is supported by the springs. The out-of-plane angular speed is measured by a motion sensor. In addition, the spring force is measured using a suitable sensor (e.g., displacement sensor). This information determines the Coriolis acceleration of the fluid particles and hence their velocity.

Laser Doppler velocimeter: In the laser Doppler velocimeter, a laser beam is projected on the fluid flow (through a window), and its frequency shift due to the Doppler effect is measured (see the section "Optical Sensors"). This is a measure of the speed of the fluid particles.

Ultrasonic flow sensor: In this method of sensing the velocity of a fluid, an ultrasonic burst is sent in the direction of the flow and the TOF is measured. The increase in the speed of propagation is due to the fluid velocity and may be determined as usual (see the section "Ultrasonic Sensors").

Hot-wire anemometer: In the hot-wire anemometer, a conductor carrying current (i) is placed in the fluid flow. The temperatures of the wire (T) and of the surrounding fluid (T_f) are measured along with the current. The coefficient of heat transfer (forced convection) at the boundary of the wire and the moving fluid is known to vary with $\sqrt{v}$, where v is the fluid velocity. Under steady conditions, the heat loss from the wire into the fluid is exactly balanced by the heat generated by the wire due to its resistance (R). The heat balance equation gives

$$i^2R = c(a+\sqrt{v})(T-T_f) \tag{10.13}$$

This relation can be used to determine v. Instead of a wire, a metal film (e.g., platinum-plated glass tube) may be used.

Rotameter: A rotameter (see Figure 10.14e) is another device for measuring fluid flow. This device consists of a conic tube with uniformly increasing cross-sectional area, which is vertically oriented. A cylindrical object is floated in the conic tube, through which the fluid flows. The weight of the floating object is balanced by the pressure differential on the object. When the flow speed increases, the object rises within the conic tube, thereby allowing more clearance between the object and the tube, for the fluid to pass. The pressure differential, however, still balances the weight of the object and is constant. Equation 10.10 is used to measure the fluid flow rate, since A increases quadratically with the height of the object. Consequently, the level of the object can be calibrated to give the flow rate.

There are other indirect methods of measuring fluid flow rate. In one method, the drag force on an object that is suspended in the flow using a cantilever arm is measured (using a strain-gauge sensor at the clamped end of the cantilever). This force is known to vary quadratically with the fluid speed.

10.4.3 Temperature Sensors

In most (if not all) temperature measuring devices, the temperature is sensed through heat transfer from the source to the measuring device. The physical (or chemical) change in the device that is caused by this heat transfer is the transducer stage of the sensing device. Several temperature sensors are outlined next.

10.4.3.1 Thermocouple

When the temperature changes at the junction formed by joining two dissimilar conductors, its electron configuration changes due to the resulting heat transfer. This electron reconfiguration produces a voltage (electromotive force [emf]) and is known as the "Seebeck effect" or "thermoelectric effect." Two junctions (or more) of a thermocouple are made with two dissimilar conductors such as iron and constantan, copper and constantan, chrome and alumel, and so on. One junction is placed in a reference source (cold junction) with temperature T_0 and the other in the temperature source (hot junction) of temperature T, as shown in Figure 10.15. The voltage V across the two junctions is measured to give the temperature of the hot junction with respect to the cold junction. The associated relationship (approximately) is

$$V = \alpha(T-T_0)+\gamma\left(T^2-T_0^2\right) \tag{10.14}$$

The presence of any other junctions, such as the ones formed by the wiring to the voltage sensor, does not affect the reading as long as these leads are maintained at the same temperature. Very low temperatures (e.g., –250°C) as well as very high temperatures (e.g., 3000°C) can be measured using a thermocouple. Since the temperature–voltage relationship is nonlinear, a correction has to be made when measuring changes in temperature;

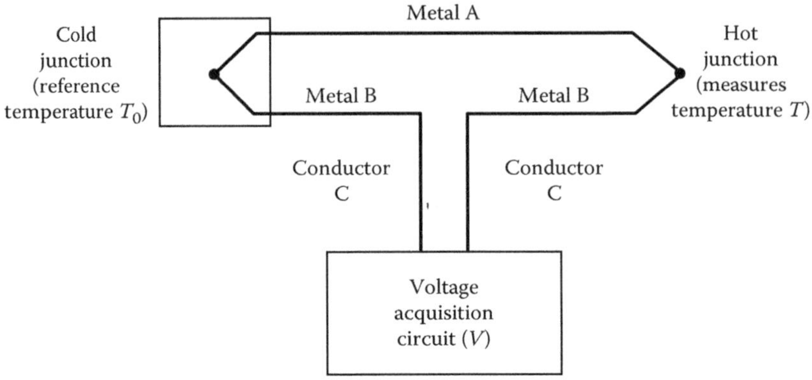

FIGURE 10.15
A thermocouple.

usually by using polynomial relations. The thermocouple sensitivity is about 0.1 mV/°C and depends on the metal pair. Typically, signal conditioning will be needed before using the sensor signal. The thermocouple "type" is based on the metal pair that is used, for example, Type E (chromel–constantan), Type J (iron–constantan), Type K (chromel–alumel), Type N (nicrosil–nisil), and Type T (copper–constantan). Of these, Type E has the highest sensitivity (70 μV/°C). Fast measurements are possible with miniature thermocouples having low time constants (e.g., 1 ms). Important considerations in selecting a thermocouple (or any temperature sensor) include (1) temperature range, (2) sensitivity, (3) speed (time constant), (4) robustness (to vibration, and to environment including chemicals), and (5) ease of use (installation, etc.).

10.4.3.2 Resistance Temperature Detector

A resistance temperature detector (RTD) is a *thermoresistive* temperature sensor. It consists of a metal element (in a ceramic tube) whose resistance typically increases with temperature, according to a known function. A linear approximation is given by

$$R = R_0(1 + \alpha T) \tag{10.15}$$

where α is the temperature coefficient of resistance. An RTD measures temperature through its change in resistance (which is measured, say, using a bridge circuit; see Chapter 4). Equation 10.15 is adequate when the temperature change is not too large. Metals used in RTDs include platinum, nickel, copper, and various alloys. The temperature coefficient of resistance (α) of several metals, which can be used in RTDs, is given in Table 10.2.

TABLE 10.2
Temperature Coefficients of Resistance of Several RTD Metals

Metal	Temperature Coefficient of Resistance α (/K)
Copper	0.0043
Nickel	0.0068
Platinum	0.0039

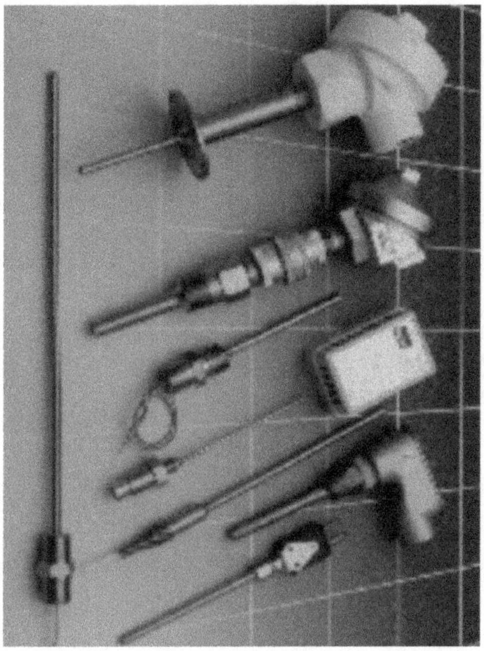

FIGURE 10.16
A commercial resistance temperature detector unit. (From RdF Corporation, Hudson, NH. With permission.)

The useful temperature range of an RTD is about −200°C to +800°C. At high temperatures, these devices may tend to be less accurate than thermocouples. The speed of response can be lower as well (e.g., fraction of a second). A commercial RTD unit is shown in Figure 10.16.

10.4.3.3 Thermistor

Unlike an RTD, a thermistor is made of a semiconductor material (e.g., metal oxides such as those of chromium, cobalt, copper, iron, manganese, and nickel), which usually has a negative change in resistance with temperature (i.e., negative α). The resistance change is detected through a bridge circuit or a voltage divider circuit. Even though the accuracy provided by a thermistor is usually better than that of an RTD, the temperature–resistance relation is far more nonlinear, as given by

$$R = R_0 \exp\left[\beta\left(\frac{1}{T} - \frac{1}{T_0}\right)\right] \tag{10.16}$$

Here, temperature T is in K. Typically, $R_0 = 5000\,\Omega$ at $T_0 = 298\,\text{K}$ (i.e., 25°C). The characteristic temperature β (about 4200 K) itself is temperature dependent, thereby adding to the overall nonlinearity of the device. Hence, proper calibration is essential when operating in a wide temperature range (say, greater than 50°C). Thermistors are quite robust, and they provide a fast response and high sensitivity (compared with RTDs) particularly because of their high resistance (several kΩ) and hence high change in resistance.

Example 10.2

The Steinhart–Hart model for a thermistor (a semiconductor) is given by

$$\frac{1}{T} = A + B\ln(R) + C[\ln(R)]^3 \tag{10.17}$$

where
  T is the temperature (in K)
  R is the resistance at temperature T (in Ω)
  A, B, and C are the Steinhart–Hart coefficients

These model coefficients depend on the characteristics of the thermistor and the temperature range of operation.

Consider the data (temperature and the corresponding resistance) given in Table 10.3. These data agree with the thermistor model given by Equation 10.16. Using the first three entries of these data (i.e., the first three rows of data in the table), determine the model coefficients of Equation 10.17. Then, using the Steinhart–Hart model, determine the resistance values corresponding to the temperatures 310 and 350 K. Compare these results with the given data and comment on the accuracy of the model.

Solution

Use the following MATLAB.m file:

```
%%% Example 10.2
clc;
clear;
T = [273.0, 298.0, 320.0];
R = [18176.0, 5000.0, 1897.4];
syms A B C
eq1=1/T(1)-A-B*log(R(1))-C*((log(R(1)))^3);
eq2=1/T(2)-A-B*log(R(2))-C*((log(R(2)))^3);
eq3=1/T(3)-A-B*log(R(3))-C*((log(R(3)))^3);
[A,B,C] = solve(eq1,eq2,eq3,'A','B','C');
A
B
C

%%%% T=310, calculate R
T=310.0;
syms R
eq = 1/T-A-B*log(R)-C*((log(R))^3);
result = real(double(solve(eq,'R')));
vpa(result,7)
%%%% T=350, calculate R
```

TABLE 10.3

Data from a Thermistor

T (K)	Data for R (Ω)	R Values Computed from SH Model
273.0	18,176.0	17,176.0
298.0	5,000.0	5,000.0
320.0	1,897.4	1,897.4
310.0	2,897.5	2,897.556
350.0	616.0	616.0075

```
T=350.0;
syms R
eq = 1/T-A-B*log(R)-C*((log(R))^3);
result = real(double(solve(eq,'R')));
vpa(result,7)
```

The computed values of the model coefficients are as follows: A = 78213551462176, B = 134868437908, and C = −7867863224518618.

The corresponding values of the sensor resistance are given in the third column of Table 10.3. It is seen that results from the model (10.17) are almost identical to the given data and hence the model (10.16). It can be concluded that two thermistor models (10.16) and (10.17) are almost identical and are accurate. In fact, it can be easily seen that model (10.16) is a special case of model (10.17), corresponding to when the coefficient $C = 0$.

10.4.3.4 Bimetal Strip Thermometer

Unequal thermal expansion of dissimilar materials is used in the temperature measurement by a bimetal strip thermometer. If strips of the two materials (typically metals) are firmly bonded, thermal expansion causes this element to bend toward the material with the lower expansion. This motion can be measured using a displacement sensor or indicated using a needle and scale. Household thermostats commonly use this principle for temperature sensing and control (on–off).

10.4.3.5 Resonant Temperature Sensors

Resonant temperature sensors use the temperature dependency of the resonant frequency of single-crystal silicon dioxide (SiO_2). The associated relation is quite accurate and precise, and the sensitivity is relatively high. Consequently, these temperature sensors are very precise, and they are particularly suitable for measuring very small temperature changes.

10.5 Sensors for Water Quality Monitoring

Monitoring the quality of water is important for many reasons including human survival and protection of the natural environment. Many parameters affect the quality of water. They include temperature, pH value, DO, ORP, electrical conductivity, flow rate, turbidity, nitrogen (including nitrates, nitrites, and compounds of ammonia), phosphate content, organic carbon, and various organic matter and bacteria. At least some of these parameters have to be monitored in order to determine the quality of water and the associated trends. Based on that information, decisions can be made in order to make predictions, provide warnings and advisories, and take corrective actions. Sensing of flow rate and temperature has been addressed in the previous section. Several other types of sensors that are useful in determining the quality of water are described next.

10.5.1 pH Value Sensor

The term pH stands for "power of hydrogen" and is a measure of the acidity or alkalinity of a water solution. In particular, it indicates the ability of the water solution to receive or donate hydrogen ions. It is measured by the current that is generated from the activity of the hydrogen ion (H^+). Specifically, the pH value is given by

$$\mathrm{pH} = -\log_{10}(a_{\mathrm{H}^+}) = \log_{10}\left(\frac{1}{a_{\mathrm{H}^+}}\right) \tag{10.18}$$

where a_{H^+} denotes the hydrogen ion activity in the water solution. Pure water is neutral, being neither acidic nor alkaline (basic), and has a pH value of 7. Solutions with a pH value less than 7 are acidic and solutions with a pH value greater than 7 are basic.

In a pH sensor, it is the potential E of the electrode of the sensor (ion-selective electrode) that is read, which gives the pH value. This is given by the Nernst equation

$$E = E^0 + \frac{RT}{F}\ln(a_{\mathrm{H}^+}) = E^0 - \frac{2.303RT}{F}\mathrm{pH} \tag{10.19}$$

where

E^0 is the standard electrode potential
R is the gas constant
T is the temperature in Kelvin
F is the Faraday constant

It is seen that the sensor reading depends on the temperature. This means calibration (compensation) for temperature is a requirement in pH sensing. This calibration may be done either by using the sensor model (Equation 10.19) itself or through reference data from the sensor at different temperatures.

10.5.2 Dissolved Oxygen Sensor

DO in water is essential for aquatic life. In particular, fish rely on DO for their respiration. The DO in water may come from several sources including the atmospheric oxygen and living organisms in the water.

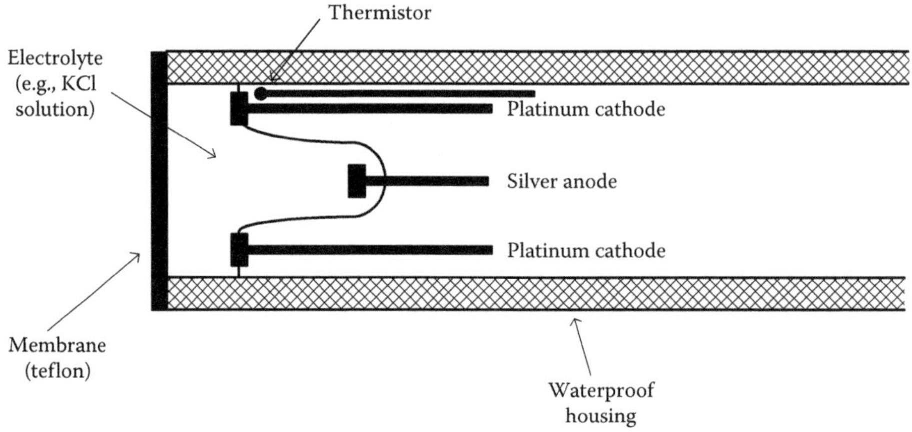

FIGURE 10.17
A dissolved oxygen sensor.

The operating principle of a DO sensor is shown in Figure 10.17. The sensor consists of two electrodes—the anode (e.g., silver) and the cathode (e.g., platinum) in contact with an electrolyte (e.g., potassium chloride [KCl] solution in water). There is a Teflon membrane at the end of the probe, which traps the electrolyte inside the probe. The entire unit is placed within a tubular, waterproof cover. There is a built-in thermistor for temperature compensation.

When the probe is placed in water, the DO in the water will diffuse through the membrane into the electrolyte of the probe. An oxidation reaction will take place at the anode and electrons will be released (i.e., a current will be generated) in proportion to the rate of diffusion. The voltage corresponding to this current can be read through the anode, which forms the sensor reading. The reading can be calibrated in mg/L (milligram per liter). In addition to temperature, compensation for salinity and pressure may be needed as well.

10.5.3 Oxidation–Reduction Potential Sensor

ORP measures the ability of a water solution to release or accept electrons. This is similar to the ability to receive or donate hydrogen ions, as represented by the pH value and is a measure of the quality of water. By accepting (i.e., removing) electrons from a substance that enters the water, the affinity of the substance to react with oxygen (i.e., to oxidize) will increase. This is an oxidizing system and an "oxidation potential" is present in the water solution then. On the other hand, when electrons are released (given) to a substance that enters the water, the ability of the substance to release oxygen (or to "reduce") will increase. This is a reducing system, and then a "reduction potential" is present in the water.

The ORP value of water will change depending on the nature and the concentration of the material (oxidizing or reducing) that enters the water. This is quite useful in the context of water quality. For example, the survival duration of bacteria in water will depend on the ORP of the water. ORP is measured using an electrode placed in the water, with respect to a reference electrode. A sensor node, which has been developed at our laboratory (Industrial Automation Lab, The University of British Columbia, Vancouver, British Columbia, Canada) for monitoring the quality of water, is shown in Figure 10.18. It has five probes for sensing: ORP, DO, pH value, temperature, conductivity, total dissolved solids, and salinity. Also, it has a microcontroller unit and data communication hardware. The following information are relevant here:

Global System for Mobile communication (*GSM*): This is a European standard and is applicable for cellular networks (for cell phones). For its incorporation (i.e., for network access), the sensor node needs a subscriber identity module card.

2G (second generation) GSM networks operate in 900 or 1800 MHz bands. Rarely, 400 and 450 MHz frequency bands are used (because *1G* system used those frequencies previously). Most *3G* networks in Europe operate in the 2100 MHz frequency band.

SD (*Secure Digital*) *card* is a nonvolatile memory card.

SD card shield: This is the shield that carries the SD card.

The antenna is required for signal communication. *Wi-Fi* may be used for the Internet access. For communication between several devices over short distances, *Bluetooth* (a wireless technology standard) may be incorporated. It uses short-wavelength radio waves in the frequency range of 2.4–2.485 GHz.

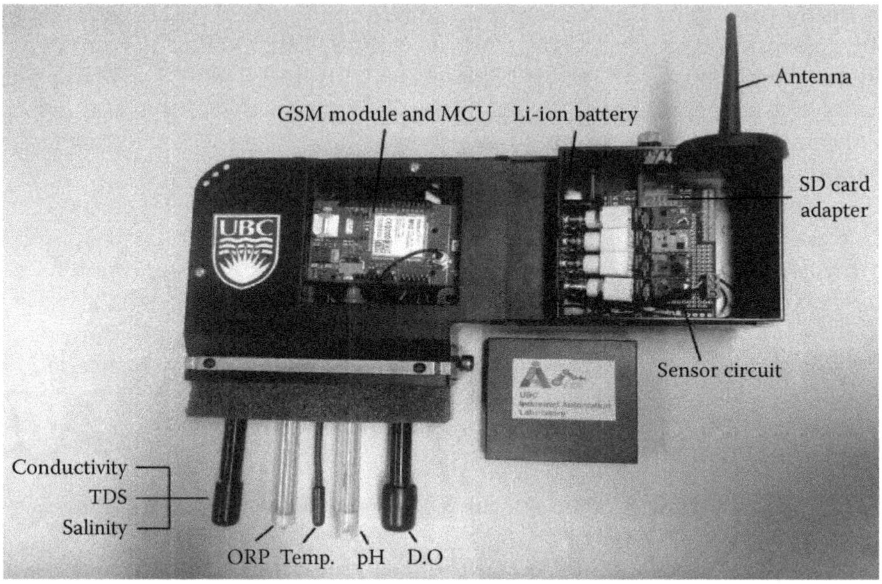

FIGURE 10.18
A sensor node for water quality monitoring.

Summary Sheet

Laser: *L*ight *a*mplification by *s*timulated *e*mission of *r*adiation. Produces electromagnetic radiation in ultraviolet, visible, or infrared bands. Can provide single-frequency (*monochromatic*) source. *Coherent* (all waves have constant phase angles). Uses oscillations of atoms or molecules. Useful in fiber optics. Can also be used directly in sensing and gauging applications. Helium–neon (He–Ne) and semiconductor lasers are common.

Fiber-optic sensor: Bundle of glass fibers (few hundred, diameter few mm to 0.01 mm) that can carry light. *Indirect* (*extrinsic*) *type*: Optical fiber acts only as medium of light transmission. Examples: position sensor, proximity sensor, and tactile sensor. *Direct* (*intrinsic*) *type*: optical fiber itself acts sensing element → measurand changes light-propagation properties of optical fiber. Examples: fiber-optic gyroscopes, fiber-optic hydrophones, and some displacement or force sensors.

Laser interferometer: *Extrinsic*: Accurate measurement of small displacements. Same bundle of fibers send and receive a monochromatic beam of light (laser). Interferometry → phase difference → displacement.

Laser Doppler interferometer: *Extrinsic*: Accurate measurement of speed. Frequency change of reflected light → speed of reflecting object. Measure by: 1. Spacing of the fringes and 2. Beats in a time period.

Fiber-optic gyroscope: Two loops of optical fiber wrapped in opposite directions around a cylinder whose angular speed is sensed are used in this sensor. The difference in frequencies (*Sagnac effect*) of the two laser beams received at a common location will measure the angular speed of the cylinder (e.g., by interferometry).

Light sensors: Photoresistor, photodiode, phototransistor, photo-FET, photocell, and charge-coupled device (CCD).

Optically coupled devices: Devices are connected through fiber optics. Noise/disturbance of one device will not affect the other devices.

Ultrasound: Audible sound waves have frequencies from 20 Hz to 20 kHz. Ultrasound → >20 kHz. For example, medical ultrasound probes of frequencies 40 kHz, 75 kHz, 7.5 MHz, and 10 MHz. Generated by high-frequency (gigahertz) oscillations in a piezoelectric crystal subjected to electrical potential, *magnetostrictive* property of material (deforms when subjected to magnetic fields), or applying a high-frequency voltage to a metal-film capacitor. Ultrasound detector (receiver) is a microphone.

Ultrasonic sensors: *Intrinsic* method: Ultrasound signal undergoes changes as it passes through an object, due to acoustic impedance and absorption characteristics of object. *Extrinsic* method: time of flight of ultrasound burst from source to sensed object and return to receiver. Doppler effect: measure change in frequency of reflected a wave of the object, whose speed is measured.

Magnetostrictive displacement sensor: Sensor head generates interrogation current pulse that travels along magnetostrictive wire/rod (*waveguide*) enclosed in protective cover. Magnetic field of the pulse interacts with magnetic field of permanent magnet connected to sensed object → ultrasound (strain) pulse (by magnetostrictive action in waveguide). Time this ultrasonic pulse → distance of magnet/object from the sensor head.

Acoustic emission (AE) sensors: Rapid and irreversible change in the internal microscopic structure releases strain energy (acoustic waves). This is AE. Frequencies of tens of kHz to several MHz. Typically a piezoelectric element is used as the sensor.

Pressure sensors: Principles:

1. Balance the pressure with an opposing force (or head) and measure it (e.g., liquid manometers and pistons).
2. Subject the pressure to a flexible front-end (auxiliary) member and measure the resulting deflection (e.g., Bourdon tube, bellows, helical tube).
3. Subject the pressure to a front-end auxiliary member and measure the resulting strain or stress (e.g., diaphragms and capsules).

Flow sensors: Principles:

1. Measure pressure across known constriction (e.g., nozzles, venturi meters, and orifice plates; volume flow rate $Q = c_d A\sqrt{2\Delta p/\rho}$, Δp = pressure drop across the constriction, c_d = discharge coefficient of constriction, and ρ = mass density of the fluid).
2. Measure pressure head that brings flow to static conditions (e.g., pitot tube $v = \sqrt{2gh}$; liquid level sensing using floats).

3. Measure flow rate (volume or mass) directly (e.g., turbine flowmeter, angular momentum flowmeter $\tau = \omega r^2 Q_m$: mass flow rate Q_m, motor torque τ, and angular speed ω are measured).
4. Measure flow velocity (e.g., Coriolis meter, laser Doppler velocimeter, ultrasonic flowmeter).
5. Measure an effect of flow and estimate flow rate using that (e.g., hot-wire or hot-film anemometer $i^2R = c(a+\sqrt{v})(T-T_f)$: conductor carrying current i is placed in fluid flow, temperatures of wire T and surrounding fluid T_f; magnetic induction flowmeter).

Temperature sensors: Heat transfer from source to measuring device → measure resulting physical (or chemical) change in the device (transducer stage). *Selection considerations*: (1) Temperature range, (2) sensitivity, (3) speed (time constant), (4) robustness (to vibration, and to environment including chemicals, etc.), and (5) ease of use (installation, etc.).

Thermocouple: Temperature change at the junction formed by joining two unlike conductors → voltage (emf or electromotive force), known as the "Seebeck effect" or "thermoelectric effect." Nonlinear: $V = \alpha(T-T_0)+\gamma(T^2-T_0^2)$; reference source (cold junction) temperature = T_0 and source (hot junction) temperature = T. Can measure very low temperatures (e.g., –250°C) and very high temperatures (e.g., 3000°C). Type E (chromel–constantan), Type J (iron–constantan), Type K (chromel–alumel), Type N (nicrosil–nisil), Type T (copper–constantan). Of these, Type E has the highest sensitivity (70 μV/°C). Fast measurements are possible (e.g., 1 ms).

Resistance temperature detector (RTD): Thermoresistive temperature sensor. Metal element (platinum, nickel, copper, and various alloys in a ceramic tube) → resistance increases with temperature: $R = R_0(1 + \alpha T)$; α = temperature coefficient of resistance. Linear; useful temperature range: –200°C to + 800°C; less accurate than thermocouples at high temperatures. Lower speed of response (e.g., fraction of a second).

Thermistor: Made of a semiconductor material (e.g., oxides of chromium, cobalt, copper, iron, manganese, nickel). Sense resistance change: $R = R_0 \exp [\beta((1/T)\beta(1/T_0))]$. Typically, $R_0 = 5000\,\Omega$ at $T_0 = 298$ K (i.e., 25°C); characteristic temperature β (about 4200 K). Robust, fast response, and high sensitivity (compared with RTDs) due to high resistance (several kΩ) → high change in resistance. Nonlinear.

Bimetal strip thermometer: Two metal strips firmly bonded → thermal expansion causes bending → measure displacement (e.g., household thermostat).

Resonant temperature sensor: Uses temperature dependency of the resonant frequency of single-crystal silicon dioxide (SiO_2). Accurate and precise, sensitivity is high, suitable for measuring very small temperature changes.

pH sensor: "Power of hydrogen." Measures acidity or alkalinity of a water solution → ability of the water solution to receive or donate hydrogen ions. Measured by current generated from hydrogen ion (H^+) activity a_{H^+}; $pH = -\log_{10}(a_{H^+}) = \log_{10}(1/(a_{H^+}))$; potential E of sensor electrode (ion-selective electrode) is given by the Nernst equation:

$$E = E^0 + \frac{RT}{F}\ln(a_{H^+}) = E^0 - \frac{2.303RT}{F}\text{pH}$$

E^0 is the standard electrode potential; R is the gas constant; T is the temperature in Kelvin; F is the Faraday constant. Depends on temperature → calibration (compensation) is required.

Dissolved oxygen (DO) sensor: Dissolved oxygen in water will diffuse through sensor membrane into the electrolyte pool of sensor → oxidation reaction at anode → electrons released → current proportional to diffusion rate. Calibrated in mg/L (milligram per liter)

Oxidation-reduction potential (ORP) sensor: Measures ability of a water solution to release or accept electrons. Opposite of pH sensor. Accepting (removing) electrons from a substance → affinity of substance to react with oxygen (oxidize) will increase → oxidation potential; Releasing (giving) electrons to a substance → ability of substance to release oxygen (reduce) will increase → reduction potential. Measured using an electrode with respect to a reference electrode.

Problems

10.1 Level sensors are used in a wide variety of applications, including soft drink bottling, food packaging, monitoring of storage vessels, mixing tanks, and pipelines. Consider the following types of level sensors, and briefly explain the principle of operation of each type in level sensing. Additionally, what are the limitations of each type?

(a) Capacitive sensors
(b) Inductive sensors
(c) Ultrasonic sensors
(d) Vibration sensors

10.2 Consider the following types of position sensors: inductive, capacitive, eddy current, fiber optic, and ultrasonic. For the following conditions, indicate which of these types are not suitable and explain why:

(a) Environment with variable humidity
(b) Target object made of aluminum
(c) Target object made of steel
(d) Target object made of plastic
(e) Target object several feet away from the sensor location
(f) Environment with significant temperature fluctuations
(g) Smoke-filled environment

10.3 The manufacturer of an ultrasonic gauge states that the device has applications in measuring cold roll steel thickness, determining parts positions in robotic assembly, lumber sorting, measurement of particle board and plywood thickness, ceramic tile dimensional inspection, sensing the fill level of food in a jar, pipe diameter gauging, rubber tire positioning during fabrication, gauging of fabricated automotive components, edge detection, location of flaws in products, and parts identification. Discuss whether the following types of sensors are also equally suitable for some or all of the foregoing applications: (a) fiber-optic position sensors, (b) self-induction proximity sensors, (c) eddy current proximity sensors, (d) capacitive gauges, (e) potentiometers, and (f) differential transformers.

In each situation where a particular sensor is not suitable for a given application, give reasons to support that claim.

10.4 (a) Consider the motion control system that is shown by the block diagram in Figure P10.4.

(i) Giving examples of typical situations, explain the meaning of the block represented as the "load" in this system.

(ii) Indicate the advantages and shortcomings of moving the motion sensor location from the motor shaft to the load response point, as indicated by the broken lines in the figure.

(b) Indicate, giving reasons, what type of sensors will be suitable for the following applications:

(i) In a soft drink bottling line, for online detection of improperly fitted metal caps on glass bottles

(ii) In a paper processing plant, to simultaneously measure both the diameter and eccentricity of rolls of newsprint

(iii) To measure the dynamic force transmitted from a robot to its support structure, during operation

(iv) In a plywood manufacturing machine, for online measurement of the thickness of plywood

(v) In a food canning plant, to detect defective cans (e.g., damage to flange and side seam, bulging of the lid)

(vi) To read codes on food packages

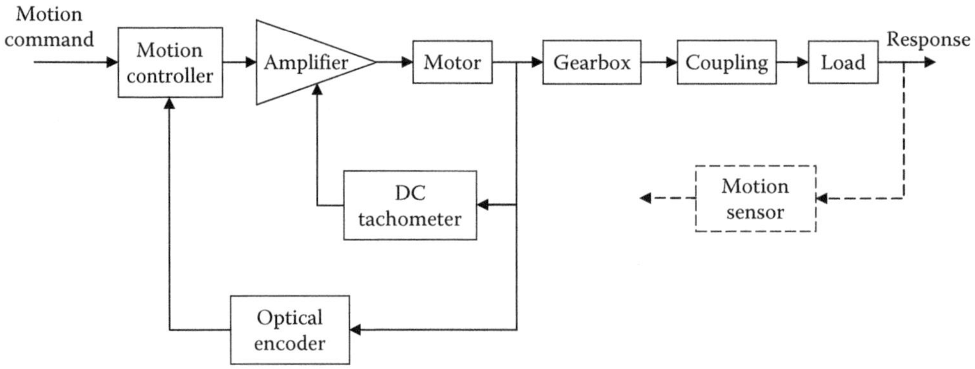

FIGURE P10.4
Block diagram of a motion control system.

10.5 In an application of machine health monitoring, both AE sensors and accelerometers are mounted at several different locations of a gear transmission of the machine.

(a) In what locations should these two types of sensors be placed and why?
(b) What are the reasons for using both types of sensors and at several locations?
(c) What are advantages and disadvantages of one type of sensor over the other?

10.6 Some sensors such as AE sensors and accelerometers have built-in signal-conditioning (e.g., preamplifier) circuitry. By integrating a built-in preamplifier, a passive sensor is converted into an active sensor. What are the advantages and disadvantages of integrating the signal-conditioning hardware into a sensor?

10.7 Compare and contrast the following temperature sensors with regard to their advantages: thermocouple, RTD, and thermistor.

10.8 (a) Compare the two models of a thermistor:

$$\frac{1}{T} = A + B\ln(R) + C[\ln(R)]^3 \quad \text{and} \quad R = R_o \exp\left[\beta\left(\frac{1}{T} - \frac{1}{T_o}\right)\right]$$

where R is the resistance of the thermistor at temperature T. The rest are model parameters. Which model is more accurate?

(b) Consider the thermistor model $R = R_o \exp\left[\beta\left(\frac{1}{T} - \frac{1}{T_o}\right)\right]$.

(i) What do the model parameters T_o and R_o represent? Explain.

(ii) By determining the derivative of the expression on the right-hand side of the model equation, determine a linearized model about the operating point (T_0, R_0). Use the incremental variables $\hat{T}$ of T and $\hat{R}$ of R to express your linear model.

(iii) Typically, the linearized model is valid only over a small range about the operating point. To illustrate the degree of nonlinearity, sketch the curves of T/T_0 versus R/R_0 for both linear model and nonlinear model for values of T/T_0 in the interval 0.95 – 1.05, with $\beta = 4200\,\text{K}$ and $T_0 = 298\,\text{K}$.

(iv) Suggest another method to linearize the given thermistor model so that it will be accurate for any value of T and R.

10.9 The following table gives some specifications of five commercial sensors.

Sensor Specifications	Temperature	pH Value	Dissolved Oxygen	Conductivity	ORP
Time constant	750 ms	378 ms	650 ms	1 s	1 s
Impedance (Ω)	2000	1650	702	165	227.6
Resolution	0.0625°C	0.01	0.01 mg/L	0.1 µs	0.1 mV
Accurate reading range	−55–125°C	0.01–14.00	0–20 mL	5–200,000+ µs	−1019.9–1019.9 mV

Give an application where all these five sensors are useful. What type of compensation may be needed for obtaining accurate data using these sensors? Are the given sensor specifications adequate for the particular application? Why?

10.10 It has been noted that the pH value variation (with time) in a water solution is the opposite of the variation of the ORP. According to this, an ORP sensor may be substituted by a pH sensor and vice versa. However, it can be argued that both types of sensors are important for accurate sensing of the quality of water. Explain.

10.11 Discuss the advantages and disadvantages of fiber-optic sensors. Consider the fiber-optic position sensor. In the curve of intensity of received light versus x, in which region would you prefer to operate the sensor and what are the corresponding limitations?

11

Digital Transducers

Chapter Objectives

- Advantages of Digital Transducers
- Incremental Optical Encoder and Hardware Features
- Direction, Position, and Speed Sensing
- Resolution and Error Considerations
- Absolute Optical Encoder
- Linear Encoder
- Digital Binary Sensors
- Digital Resolver, Tachometer
- Moiré Fringe Displacement Sensors
- Hall-Effect Sensors
- Digital Camera and Image Acquisition

11.1 Innovative Sensor Technologies

Sensors and transducers are useful in a variety of engineering applications. Numerous examples are found in transit systems, computing systems, process monitoring and control, energy systems, material processing, manufacturing, mining, food processing, service sector, forestry, civil engineering structures and systems, and so on. The previous chapters studied analog motion sensors; effort sensors, which included force, torque, tactile, and impedance; and several other types of sensors including optical sensors, ultrasonic sensors, magnetostrictive sensors, acoustic emission sensors, thermofluid sensors, and sensors that are useful in monitoring the quality of water (e.g., pH value, dissolved oxygen, and oxidation–reduction potential).

In this chapter, we study digital transducers and some other innovative sensor technologies. Our primary focus here is on transducers in mechatronic systems including motion sensors. As noted in Chapter 9 in particular, by using a suitable auxiliary front-end sensor, other measurands, such as force, torque, temperature, and pressure, may be converted into a motion and subsequently measured using a motion transducer. For example, altitude (or pressure) measurements in aircraft and aerospace applications are made using a pressure-sensing front end, such as a bellows or diaphragm device, in conjunction with an optical encoder (which is a digital transducer) to measure the resulting displacement. Similarly, a bimetallic element may be used to convert temperature into a displacement, which may be measured using a displacement sensor. Other important sensor technologies such as microelectromechanical system sensors, multisensor data fusion, and wireless sensor networks are studied in Chapter 12.

It is acceptable to call an analog sensor as an analog transducer because both the sensor stage and the transducer stage of it are analog. Typically, the sensor stage of a digital transducer is analog as well. For example, motion, as manifested in a physical system, is continuous in time. Therefore, we cannot generally speak of digital motion sensors. In a digital measuring device, it is the transducer stage that generates a discrete output signal (e.g., pulse train, count, frequency, encoded data). Hence, digital sensing devices may be termed *digital transducers* rather than *digital sensors*.

11.1.1 Analog versus Digital Sensing

Any measuring device that presents information as discrete samples and does not introduce a *quantization error* when the reading is represented in the digital form may be classified as a digital transducer. According to this definition, for example, an analog sensor such as a thermocouple that is integrated with an analog-to-digital converter (ADC) is not a digital transducer. This is so because a quantization error is introduced by the ADC process (see Chapter 4). In particular, a measuring device that falls into one of the following types may be classified as a digital transducer:

1. A measuring device that produces a discrete or digital output without using an ADC
2. A transducer whose output is a pulse signal or a count
3. A transducer whose output is a frequency (which can be precisely converted into a count or a rate)

Note: When the output is a pulse signal, a counter is used to count the pulses or to count the number of clock cycles over a pulse duration, both of which are discrete readings.

Comparison example: To compare the basic characteristics of a digital transducer and an analog sensor, consider the sensing arrangement shown in Figure 11.1. The system

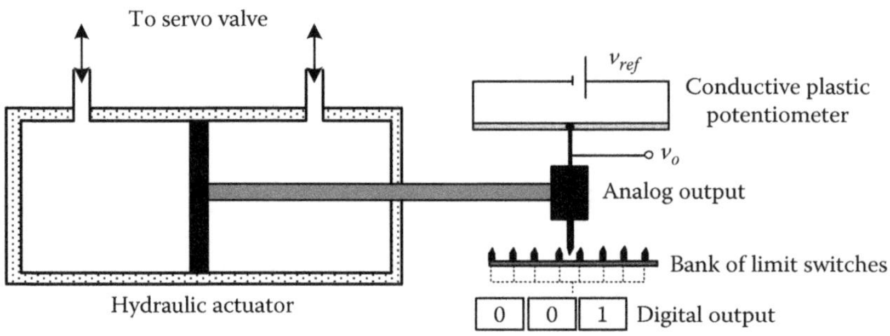

FIGURE 11.1
Analog and digital methods for displacement sensing.

has a hydraulic actuator, which moves a load along a straight line (i.e., a linear actuator). On one side of the load, there is a potentiometer whose resistive element is made of conductive plastic (see Chapter 8) that generates a continuous output voltage v_o, which is proportional to the displacement of the load. On the other side of the load, there is a pointer that is able to trigger a limit switch as the load moves past it. There are eight such limit switches in the system. Clearly, a 3-bit register, which can represent eight discrete values, can sense the absolute location of the load as it touches the corresponding limit switch.

Let us compare the two approaches of sensing in this example, with regard to accuracy, complexity, cost, usefulness, robustness, and so on. For a fair comparison, let us assume that the potentiometer output v_o is sampled and digitized by a 3-bit ADC. Then, both analog and digital devices present the data in a 3-bit register.

1. Analog sensing method: use of a potentiometer with 3-bit analog-to-digital converter
   a. An ADC is required to acquire the data by a computer.
   b. Data accuracy is lost in sampling (i.e., aliasing error) and cannot be recovered; signal/sensor noise directly enters into the reading.
   c. Can sense continuous signals with fine resolution. The resolution of the digitized signal can be improved by using an ADC of larger bit size (say, 4 bits).
   d. Less robust due to reasons b, f, and g.
   e. Sensing is direct and simple, but data acquisition (DAQ) into a computer is more complex and costly (e.g., filter and amplifier, sample and hold, ADC).
   f. The device entirely fails if the sensor (potentiometer) fails.
   g. Quantization error is introduced when a sampled data value is digitized (represented in 3-bit form).
   h. Relatively slow (sensor time constant, signal conditioning, sampling, digitizing, and registering).
2. Digital sensing method: use of eight limit switches
   a. This method makes data acquisition into a computer easier (e.g., the 1-bit output of a limit switch is typically transistor-to-transistor logic [TTL] compatible and can be directly acquired by a microcontroller).
   b. The 3-bit accuracy is precisely retained even if the limit switch signal has high noise (because only a 1-bit information [i.e., triggered or not triggered] is needed from a limit switch).
   c. The resolution is fixed by the number of limit switches.
   d. More robust due to reasons b and f.
   e. More components (potentially less reliable) but operates even if a limit switch fails and provides perfect accuracy with respect to the remaining limit switches.
   f. There is no issue of quantization error. The actual positions of the limit switches are determined precisely.
   g. Relatively fast (a limit switch is binary; no further signal processing, sampling, and digitizing are needed).

It is seen that the digital approach to sensing has clear advantages while the analog approach has advantages as well.

11.1.2 Advantages of Digital Transducers

As noted earlier, there are benefits to using digital devices over analog devices for sensing. Digital sensing devices (or digital transducers, as they are commonly known) generate discrete output signals, such as pulse trains or encoded data, which present further advantages in their subsequent use. In particular, the output of a digital transducer can be directly read by a digital processor, without needing the stages of sampling and digitization (or ADC). A digital processor plays a key role in the utilization of the sensed data, by facilitating complex processing of measured signals and other known quantities. For example, it can serve as the controller in a digital control system, which generates control signals for the plant (i.e., the system that is being controlled). On the other hand, if the measured signals are available in the analog form, sampling and digitization stages are necessary before processing by means of a digital processor.

Nevertheless, the sensor stage itself of a digital sensing device is usually quite similar to that of an analog counterpart. There are digital measuring devices that incorporate microprocessors to locally perform numerical manipulations and conditioning and provide output signals in either digital form or analog form. These measuring systems are particularly useful when the required variable is not directly measurable but could be computed using one or more measured outputs (e.g., power = force × speed). Although a microprocessor is an integral part of the measuring device in this case, it performs a conditioning task rather than a measuring task. In this context, we consider the two tasks separately.

When the output of a digital transducer is a pulse signal, a common method of reading the signal is by using a counter, either to count the pulses (for high-frequency pulses) or to count the number of clock cycles over one pulse duration (for low-frequency pulses). The count is placed as a digital word in a buffer/register, which can be accessed by the computer, typically at a constant frequency (called the sampling rate). If the output of a digital transducer is available in a coded form (e.g., natural binary code or gray code), it can be directly read by a computer. Then, the coded signal is normally generated by a parallel set of pulse signals; each pulse transition generates one bit of the digital word, and the numerical value of the word is determined by the pattern of the generated pulses. This is the case, for example, with *absolute encoders*, as discussed later in this chapter. DAQ from (i.e., computer interfacing) a digital transducer is commonly carried out by using a general-purpose input–output (I/O) or DAQ card (see Chapter 4), for example, a motion control (servo) card, which may be able to accommodate multiple transducers (e.g., eight channels of encoder inputs with 24-bit counters), or by using a DAQ card specific to the particular transducer.

Digital transducers (or digital representation of information) have several advantages in comparison with analog methods. The following are notable:

1. They do not introduce quantization error.
2. Digital signals are less susceptible to noise, disturbances, or parameter variation in instruments because data can be generated, represented, transmitted, and processed as binary words consisting of bits, which possess two identifiable states (the noise threshold is half a bit).
3. Complex signal processing with very high accuracy and speed is possible through digital means (hardware implementation is faster than software implementation).
4. High reliability in a system can be achieved by minimizing the analog hardware components.

5. Large amounts of data can be stored using compact, high-density data storage methods.
6. Data can be stored or maintained for very long periods of time without any drift or disruption by adverse environmental conditions.
7. Fast data transmission is possible through existing communication means over long distances with no attenuation and with less dynamic delays, compared to analog signals.
8. Digital signals use low voltages (e.g., 0–12 V DC) and low power.
9. Digital devices typically have low overall cost.

These advantages help build a strong case in favor of digital measuring and signal transmission devices for engineering systems.

11.2 Shaft Encoders

Any transducer that generates a coded (digital) reading of a measurement can be termed as an encoder. Shaft encoders are digital transducers that measure angular displacements and angular velocities. Applications of these devices include motion measurement in performance monitoring and control of robotic manipulators, machine tools, industrial processes (e.g., food processing and packaging, pulp and paper), digital data storage devices, positioning tables, satellite mirror positioning systems, vehicles, construction machinery, planetary exploration devices, battlefield equipment, and rotating machinery such as motors, pumps, compressors, turbines, and generators. High resolution (which depends on the word size of the encoder output and the number of pulses generated per revolution of the encoder), high accuracy (particularly due to noise immunity and reliability of digital signals and superior construction), and relative ease of adoption in digital systems (because transducer output can be read as a digital word), with the associated reduction in system cost and improvement of system reliability, are some of the relative advantages of digital transducers in general and shaft encoders in particular, in comparison with their analog counterparts.

11.2.1 Encoder Types

Shaft encoders can be classified into two categories depending on the nature and the method of interpretation of the transducer output: (1) incremental encoders and (2) absolute encoders.

Incremental encoder: The output of an incremental encoder is a pulse signal, which is generated when the transducer disk rotates due to the motion that is measured. By counting the pulses or by timing the pulse width using a clock signal, both angular displacement and angular velocity can be determined. With an incremental encoder, displacement is measured with respect to some reference point. The reference point can be the home position of the moving component (say, determined by a limit switch) or a reference point on the encoder disk, as indicated by a reference pulse (index pulse) generated at that location on the disk. Furthermore, the index pulse count determines the number of full revolutions.

Absolute encoder: An absolute encoder (or whole-word encoder) has many pulse tracks on its transducer disk. When the disk of an absolute encoder rotates, several pulse trains—equal in number to the tracks on the disk—are generated simultaneously. At a given instant, the magnitude of each pulse signal will have one of two signal levels (i.e., a binary state), as determined by a level detector (or edge detector). This signal level corresponds to a binary digit (0 or 1). Hence, the set of pulse trains gives an encoded binary number at any instant. The windows in a track are not equally spaced but are arranged in a specific pattern to obtain coded output data from the transducer. The pulse windows on the tracks can be organized into some pattern (code) so that the generated binary number at a particular instant corresponds to the specific angular position of the encoder disk at that time. The pulse voltage can be made compatible with some digital interface logic (e.g., TTL). Consequently, the direct digital readout of an angular position is possible with an absolute encoder, thereby expediting digital DAQ and processing, and also eliminating error retention if, for example, a pulse is missed (unlike with an incremental encoder). Absolute encoders are commonly used to measure fractions of a revolution. However, complete revolutions can be measured using an additional track, which generates an I pulse, as in the case of an incremental encoder.

The same signal generation (and pickoff) mechanism may be used in both types (incremental and absolute) of transducers.

11.2.1.1 Encoder Technologies

Four techniques of transducer signal generation may be identified for shaft encoders:

1. Optical (photosensor) method
2. Sliding contact (electrical conducting) method
3. Magnetic saturation (reluctance) method
4. Proximity sensor method

By far, the optical encoder is most common and cost-effective. The other three approaches may be used in special circumstances, where the optical method may not be suitable (e.g., under extreme temperatures or in the presence of dust, smoke) or may be redundant (e.g., where a code disk such as a toothed wheel is already available as an integral part of the moving member). For a given type of encoder (incremental or absolute), the method of signal interpretation is identical for all four types of signal generation listed previously. Now, we briefly describe the principle of signal generation for all four techniques and consider only the optical encoder in the context of signal interpretation and processing.

11.2.1.1.1 Optical Encoder

The optical encoder uses an opaque disk (code disk) that has one or more circular tracks, with some pattern (sequence) of identical transparent windows (slits) in each track. A parallel beam of light (e.g., from a set of light-emitting diodes [LEDs]) is projected to all tracks from one side of the disk. The transmitted light is picked off using a bank of photosensors on the other side of the disk, which typically has one photosensor for each track. This arrangement is shown in Figure 11.2a, which indicates just one track and one pickoff sensor. The light sensor could be a silicon photodiode or a phototransistor (see Chapter 10). Since the light from the source is interrupted by the opaque regions of the track, the output signal from the photosensor is a series of voltage pulses. This signal can be interpreted

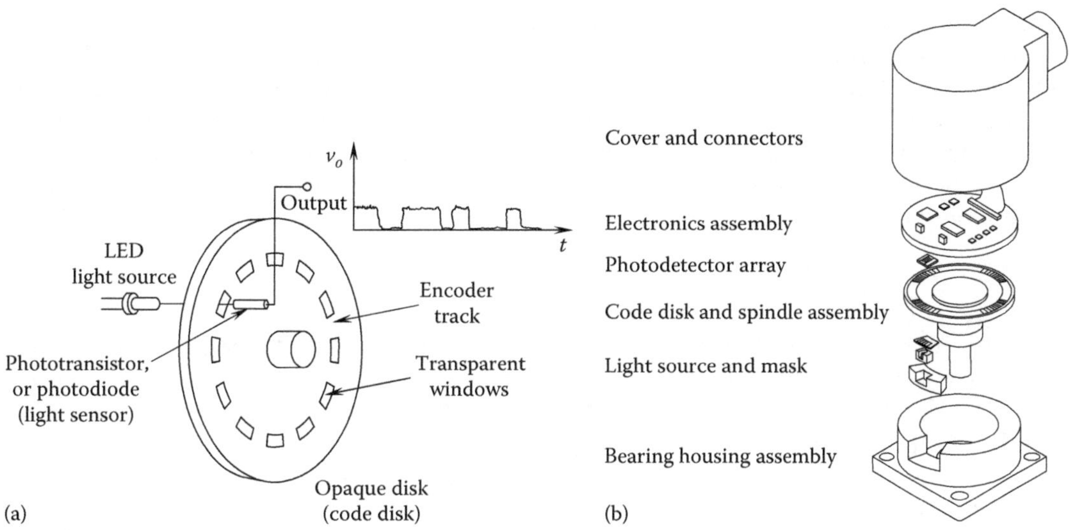

FIGURE 11.2
(a) Schematic representation of an (incremental) optical encoder and (b) components of a commercial incremental encoder. (Courtesy of BEI Electronics, Inc., Franksville, WI.)

(e.g., through edge detection or level detection) to obtain the increments in the angular position and also the angular velocity of the disk. In standard terminology, the sensor element of such a measuring device is the encoder disk, which is connected to the rotating object (directly or through a gear mechanism). The transducer stage corresponds to the conversion of the disk motion (analog) into the pulse signals, which can be coded into a digital word.

The opaque background of the transparent windows (the window pattern) on an encoder disk may be produced by contact printing techniques. The precision of this production procedure is a major factor that determines the accuracy of an optical encoder. If the direction of rotation is fixed (or not important), an incremental encoder disk requires only one primary track that has equally spaced and identical window (pickoff) regions. A reference track that has just one window may be used to generate the I pulse, to initiate pulse counting for angular position measurement, and to detect complete revolutions.

Note: A transparent disk with a track of opaque spots will work equally well as the encoder disk of an optical encoder. In either form, the track has a 50% duty cycle (i.e., the length of the transparent region is equal to the length of the opaque region). Components of a commercially available optical encoder are shown in Figure 11.2b.

11.2.1.1.2 Sliding Contact Encoder

In a sliding contact encoder, the transducer disk is made of an electrically insulating material. Circular tracks on the disk are formed by implanting a pattern of conducting areas. These conducting regions correspond to the transparent windows of an optical encoder disk. All conducting areas are connected to a common slip ring on the encoder shaft. A constant voltage v_{ref} is applied to the slip ring using a brush mechanism. A sliding contact such as a brush touches each track, and as the disk rotates, a voltage pulse signal is picked off by it. The pulse pattern depends on the conducting–nonconducting pattern on each track as well as the nature of rotation of the disk. The signal interpretation is done as it is for optical encoders. The advantages of sliding contact encoders include high sensitivity (depending on the supply voltage) and simplicity of construction (low cost). The disadvantages include the familiar drawbacks of contacting and commutating devices (e.g., friction, wear, brush bounce due to vibration, and signal glitches and metal oxidation due to electrical arcing). A transducer's accuracy is very much dependent on the precision of the conducting patterns of the encoder disk. One method of generating the conducting pattern on the disk is electroplating.

11.2.1.1.3 Magnetic Encoder

A magnetic encoder has high-strength magnetic regions imprinted on the encoder disk using techniques such as etching, stamping, or recording (similar to magnetic data recording). These magnetic regions correspond to the transparent windows on an optical encoder disk. The signal pickoff device is a microtransformer, which has primary and secondary windings on a circular ferromagnetic core. This pickoff sensor resembles a core storage element in a historical mainframe computer. A high-frequency (typically 100 kHz) primary voltage induces a voltage in the secondary windings of the sensing element at the same frequency, operating as a transformer. A magnetic field of sufficient strength can saturate the core, however, thereby significantly increasing the reluctance and dropping the induced voltage. By demodulating the induced voltage, a pulse signal is obtained. This signal can be interpreted in the usual manner. A pulse peak corresponds to a nonmagnetic area, and a pulse valley corresponds to a magnetic area on each track. Magnetic encoders have noncontacting pickoff sensors, which is an advantage. They are more costly than the contacting devices, however, primarily because of the cost of the transformer elements and the demodulating circuitry for generating the output signal.

11.2.1.1.4 Proximity Sensor Encoder

A proximity sensor encoder uses a proximity sensor as the signal pickoff element. Any type of proximity sensor may be used, for example, a magnetic-induction probe or an eddy current probe, as discussed in Chapter 8. In the magnetic-induction probe, for example, the disk is made of ferromagnetic material. The encoder tracks have raised spots of the same material, serving a purpose analogous to that of the windows on an optical encoder disk. As a raised spot approaches the probe, the flux linkage increases due to the associated decrease in reluctance. This raises the induced voltage level. The output voltage is a pulse-modulated signal at the frequency of the supply (primary) voltage of the proximity sensor. This is then demodulated, and the resulting pulse signal is interpreted. Instead of a disk with a track of raised regions, a ferromagnetic toothed wheel may be used along with a proximity sensor placed in a radial orientation. In principle, this device operates like a conventional digital tachometer. If an eddy current probe is used, the pulse areas in the track have to be plated with a conducting material.

Direction sensing: As will be explained in detail later, an incremental encoder needs a second probe placed at quarter-pitch separation from the first probe (pitch is the center-to-center distance between adjacent windows) to generate a *quadrature signal*, which will identify the direction of rotation. Some designs of incremental encoders have two identical tracks, one with a quarter-pitch offset from the other, and the two pickoff sensors placed radially without offset. The two (quadrature) signals obtained with this arrangement will be similar to those with the previous arrangement. Including the track that generates the reference pulse, an incremental encoder may have three tracks on its code disk.

In many applications, encoders are built into the monitored device itself rather than being externally fitted onto a rotating shaft. For instance, in a robot arm, the encoder may be an integral part of the joint motor and may be located within its housing. This reduces coupling errors (e.g., errors due to backlash, shaft flexibility, and resonances added by the transducer and fixtures), installation errors (e.g., misalignment and eccentricity), and overall cost. Encoders are available in sizes as small as 2 cm and as large as 15 cm in diameter.

Since the techniques of signal interpretation are quite similar for the various types of encoders with different principles of signal generation, we limit our further discussion to optical encoders only. Signal interpretation differs depending on whether the particular optical encoder is an incremental device or an absolute device.

11.3 Incremental Optical Encoder

There are two possible designs of track and probe configuration for an incremental encoder disk with the direction sensing capability:

1. Offset probe configuration (has two probes and one track)
2. Offset track configuration (has two probes and two tracks)

The first configuration is schematically shown in Figure 11.3. The disk has a single circular track with identical and equally spaced transparent windows. The area of the opaque region between adjacent windows is equal to the window area. *Note*: An output pulse is on for half the period and off for the other half, giving a 50% duty cycle. Two photodiode sensors (probes 1 and 2 in Figure 11.3) are positioned facing the track at a quarter-pitch (half the window length) apart. The forms of their output signals (v_1 and v_2), after passing them through pulse-shaping circuitry (idealized), are shown in Figure 11.4a and b for the two directions of rotation.

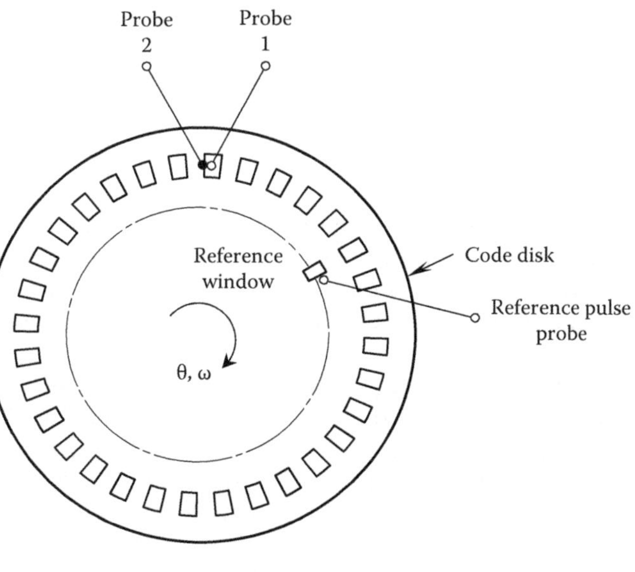

FIGURE 11.3
An incremental encoder disk (offset probe configuration).

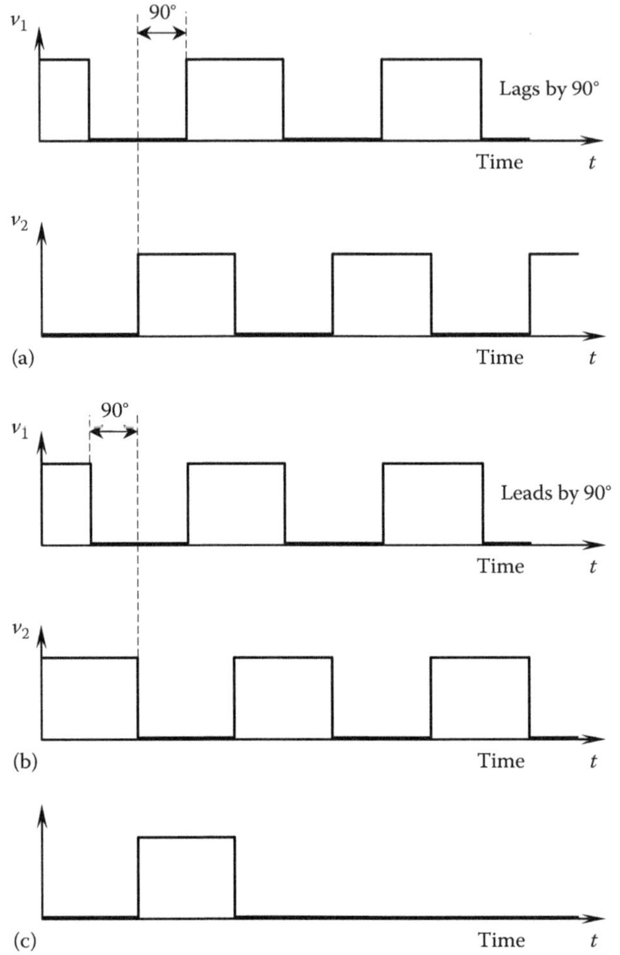

FIGURE 11.4
Shaped pulse signals from an incremental encoder. (a) Clockwise rotation, (b) counterclockwise rotation, and (c) reference pulse signal.

Note: The circumferential offset between the two probes can be increased by an integer number of periods, giving more room to place the probes. *Note*: A period is the angular separation between two successive similar windows. The delay between the two signals will change by an integer multiple of 360° (assume constant speed over the delay), that is, no change.

In the second configuration of an incremental encoder, two identical tracks are used, one at an offset from the other by a quarter-pitch. Each track has its own probe (light sensor), which is oriented facing the corresponding track. The two probes are positioned along a radial line of the disk, without any circumferential offset unlike the previous configuration. The output signals from the two sensors are the same as earlier, however (Figure 11.4).

In both configurations, an additional track with a lone window and associated probe is also usually present. This track generates a reference pulse (I pulse) per revolution of the disk (see Figure 11.4c). This pulse is used to initiate the counting operation and also to count complete revolutions, which is required in measuring absolute angular rotations.

Note: When the disk rotates at a constant angular speed, the pulse width and pulse-to-pulse period (encoder cycle) are constant (with respect to time) in each sensor output. When the disk accelerates, the pulse width decreases continuously; when the disk decelerates, the pulse width increases.

11.3.1 Direction of Rotation

An incremental encoder typically has the following five pins:

1. Ground
2. I channel
3. A channel
4. +5 V DC power
5. B channel

Pins for channel A and channel B give the quadrature signals shown in Figure 11.4a and b, and the I pin gives the reference pulse signal shown in Figure 11.4c.

The quarter-pitch offset in the probe location (or in track placement) is used to determine the direction of rotation of the disk. For example, Figure 11.4a shows the shaped (idealized) sensor outputs (v_1 and v_2) when the disk rotates in the clockwise (cw) direction, and Figure 11.4b shows the outputs when the disk rotates in the counterclockwise (ccw) direction. Several methods can be used to determine the direction of rotation using these two "quadrature signals." For example,

1. By the phase angle between the two signals
2. By the clock counts between two adjacent rising edges of the two signals
3. By checking for rising or falling edge of one signal when the other is at "high"
4. For a high-to-low transition of one signal check the next transition of the other signal

Method 1: It is clear from Figure 11.4a and b that in the cw rotation, v_1 lags v_2 by a quarter of a cycle (i.e., a phase lag of 90°), and in the ccw rotation, v_1 leads v_2 by a quarter of a cycle. Hence, the direction of rotation may be obtained by determining the phase difference of the two output signals, using phase-detecting circuitry.

Method 2: A rising edge of a pulse can be determined by comparing successive signal levels at fixed time periods (this can be done in both hardware and software). Time to a rising edge can be measured using pulse counts of a high-frequency clock. Suppose that the counting (timing) begins when the v_1 signal begins to rise (i.e., when a rising edge is detected). Let n_1 be the number of clock cycles (time) up to the time when v_2 begins to rise and n_2 be the number of clock cycles up to the time when v_1 begins to rise again. Then, the following logic applies:

$$\text{If } n_1 > n_2 - n_1 \Rightarrow \text{cw rotation}$$

$$\text{If } n_1 < n_2 - n_1 \Rightarrow \text{ccw rotation}$$

This logic for direction detection should be clear from Figure 11.4a and b.

Method 3: In this case, we first detect a high level (logic high or binary 1) in signal v_2 and then check whether the edge in signal v_1 rises or falls during this "high" period of v_2. It is clear from Figure 11.4a and b that the following logic applies:

If edge is rising in v_1 when v_2 is at logic high $\Rightarrow$ cw rotation

If edge is falling in v_1 when v_2 is at logic high $\Rightarrow$ ccw rotation

Method 4: Detect a high-to-low transition in signal v_1.

If the next transition in signal v_2 is from low to high $\rightarrow$ cw rotation
If the next transition in signal v_2 is from high to low $\rightarrow$ ccw rotation

11.3.2 Encoder Hardware Features

The actual hardware of a commercial encoder is not as simple as that suggested by Figures 11.2a and 11.3. The main components of a commercial encoder are identified in Figure 11.2b. A more detailed schematic diagram of the signal generation mechanism of an optical incremental encoder is shown in Figure 11.5a. The light generated by the LED is collimated (forming parallel rays) using a lens. This pencil of parallel light passes through a window of the rotating code disk. The masking (grating) disk is stationary and has a track

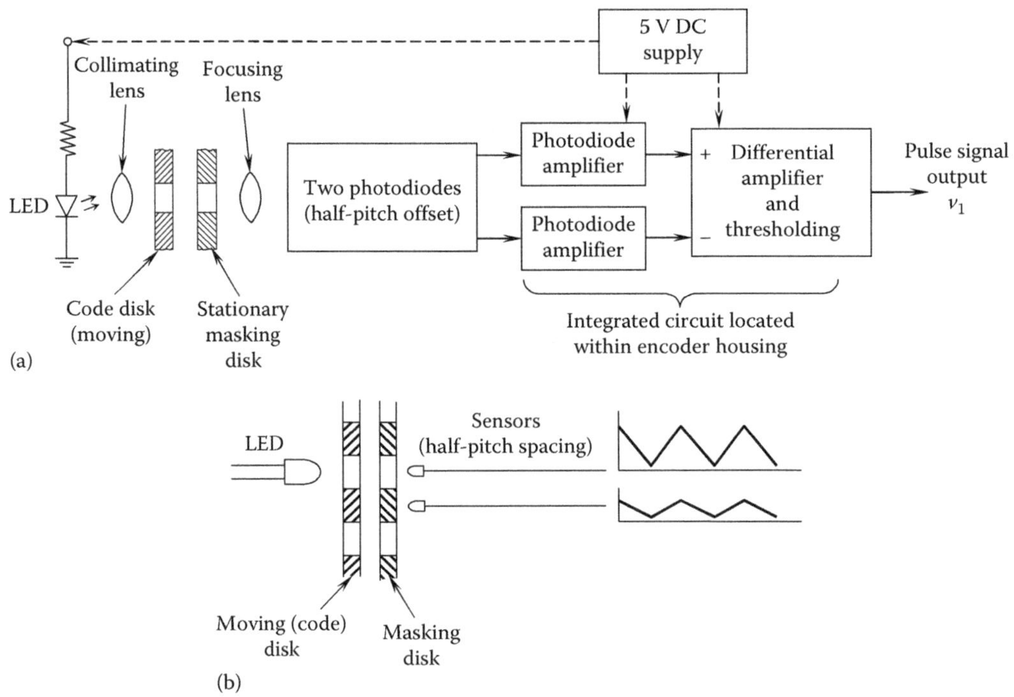

FIGURE 11.5
(a) Internal hardware of an optical incremental encoder and (b) use of two sensors at 180° spacing to generate an enhanced pulse.

of windows identical to that in the code disk. Because of the presence of the masking disk, the light from the LED will pass through more than one window of the code disk, thereby improving the intensity of the light received by the photosensor while not introducing any error due to the diameter of the pencil of light by being larger than the window length. When the windows of the code disk face the opaque areas of the masking disk, virtually no light is received by the photosensor. When the windows of the code disk face the transparent areas of the masking disk, the maximum amount of light reaches the photosensor. Hence, as the code disk moves, a sequence of triangular (and positive) pulses of light is received by the photosensor.

Note: The width of a resulting triangular pulse is a full cycle (i.e., it corresponds to the window pitch and not a half cycle of a 50% duty rectangular pulse). A rectangular pulse sequence can be obtained by thresholding the triangular pulse sequence.

11.3.2.1 Signal Conditioning

Fluctuation in the supply voltage to the encoder light source directly influences the light level received by the photosensor. If the sensitivity of the photosensor is not high enough, a low light level might be interpreted as no light, which would result in measurement error. Such errors due to instabilities and changes in the supply voltage can be eliminated by using two photosensors, one placed half a pitch away from the other along the window track, as shown in Figure 11.5b. This arrangement is used for contrast detection, and it should not be confused with the quarter-of-a-pitch offset arrangement that is required for direction detection. The sensor that is facing the opaque region of the masking disk will always read a low signal. The other sensor will read a triangular signal whose peak occurs when a moving window completely overlaps with a window of the masking disk and whose valley occurs when a moving window faces an opaque region of the masking disk. The two signals from these two sensors are amplified separately and fed into a differential amplifier (see Chapter 3). The result is a high-intensity triangular pulse signal. A shaped (or binary) pulse signal can be generated by subtracting a threshold value from this signal and identifying the resulting positive (or binary 1) and negative (or binary 0) regions. This procedure will produce a more distinct (or binary) pulse signal that is immune to noise.

Signal amplifiers are monolithic integrated circuit (IC) devices and are housed within the encoder itself. Additional pulse-shaping circuitry may also be present. The power is provided through an encoder from an external power supply. The voltage level and the pulse width of the output pulse signal are logic compatible (e.g., TTL) so that they may be read directly using a digital DAQ card. Note that if the output level v_1 is positive high, we have a logic high (or binary 1) state. Otherwise, we have a logic low (or binary 0) state. In this manner, a stable and accurate digital output can be obtained even under conditions of unstable voltage supply. The schematic diagram in Figure 11.5 shows the generation of only one (v_1) of the two quadrature pulse signals. The other pulse signal (v_2) is generated using identical hardware but at a quarter-of-a-pitch offset. The I pulse (reference pulse) signal is also generated in a similar manner. The cable of the encoder (usually, a ribbon cable) has a multipin connector (for the five pins mentioned earlier).

Note: The only moving part in the system shown in Figure 11.5 is the code disk.

11.3.3 Linear Encoders

In a rectilinear encoder (popularly called linear encoder, where "linear" does not imply linearity but refers to rectilinear or straight-line motion), a rectangular flat plate that moves rectilinearly, instead of a rotating disk, is used with the same type of signal generation and interpretation mechanism as for a shaft (rotatory) encoder. A transparent plate with a series of opaque lines arranged in parallel in the transverse direction forms the stationary plate (grating plate or phase plate) of the transducer. This is called the mask plate. A second transparent plate, with an identical set of ruled lines, forms the moving plate (or the code plate). The lines on both plates are evenly spaced, and the line width is equal to the spacing between the adjacent lines. A light source is placed on the side of the moving plate, and the light that is transmitted through the common area of the two plates is detected on the other side using one or more photosensors. When the lines on the two plates coincide, the maximum amount of light will pass through the common area of the two plates. When the lines on one plate fall on the transparent spaces of the other plate, virtually no light will pass through the plates. Accordingly, as one plate moves relative to the other, a pulse train is generated by the photosensor, and it determines the rectilinear displacement and velocity, as in the case of an incremental encoder.

A suitable arrangement is shown in Figure 11.6. The code plate is attached to the moving object whose rectilinear motion is to be measured. An LED light source and a phototransistor light sensor are used to detect the motion pulses, which can be interpreted just like the way it is done for a rotatory encoder. The phase plate is used, as with a shaft encoder, to enhance the intensity and the discrimination of the detected signal. Two tracks of windows in quadrature (i.e., at quarter-pitch offset) would be needed to determine the direction of motion, as shown in Figure 11.6. Another track of windows at half-pitch offset from the main track (not shown in Figure 11.6) may be used as well on the phase plate, to further enhance the discrimination of the detected pulses. Specifically, when the sensor at the main track reads a high intensity (i.e., when the windows on the code plate and the phase plate are aligned), the sensor at the track that is half-pitch away will read a low intensity (because the corresponding windows of the phase plate are blocked by the solid regions of the code plate).

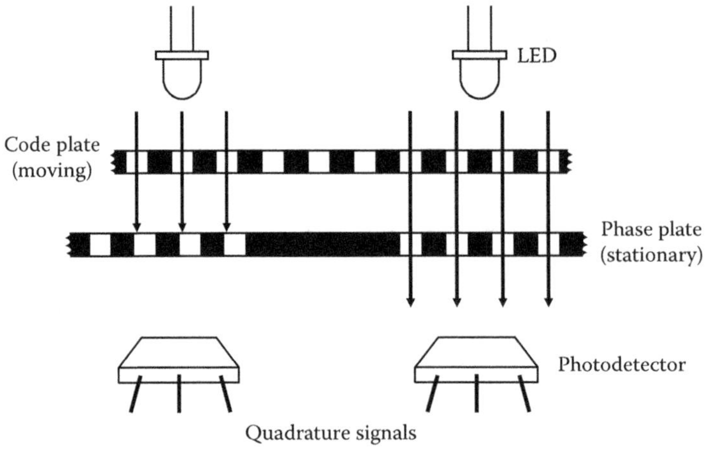

FIGURE 11.6
A rectilinear optical encoder.

11.4 Motion Sensing by Encoder

An optical encoder can measure both displacement and velocity. Also, depending on the encoder design (linearly moving code plate or rotating code disk), rectilinear motions or angular motions can be measured. An incremental encoder measures displacement as a pulse count, and it measures velocity as a pulse frequency. A digital processor is able to express these readings in engineering units (radians, degrees, rad/s, rev/s, rpm, etc.) using pertinent conversion factors, which depend on the parameter values of the physical system.

We will consider angular motions only because the same concepts can be directly extended to linear (i.e., rectilinear) motions. We will present formulas for computing displacement and velocity using encoder outputs. Also, we will discuss the important concept of encoder resolution with regard to both displacement and velocity.

11.4.1 Displacement Measurement

Suppose that the maximum count possible from an incremental encoder is M pulses and the range of the encoder is $\pm\theta_{max}$. The angular position θ corresponding to a count of n pulses is computed as

$$\theta = \frac{n}{M}\theta_{max} \tag{11.1}$$

11.4.1.1 Digital Resolution

The resolution of an encoder represents the smallest change in measurement that can be measured realistically. Since an encoder can be used to measure both displacement and velocity, we can identify a resolution for each case. Here, we consider the displacement resolution, which is governed by the number of windows N in the code disk and the digital size (number of bits) of the buffer or register where the counter output is stored. Now, we discuss digital resolution.

The displacement resolution of an incremental encoder is given by the change in displacement corresponding to a unit change in the count (n). It follows from Equation 11.1 that the displacement resolution is given by

$$\Delta\theta = \frac{\theta_{max}}{M} \tag{11.2}$$

The digital resolution corresponds to a unit change in the bit value. Suppose that the encoder count is stored as digital data of r bits. Allowing for a sign bit, we have $M = 2^{r-1}$. By substituting this into Equation 11.2, we have the digital resolution

$$\Delta\theta_d = \frac{\theta_{max}}{2^{r-1}} \tag{11.3*}$$

Typically, $\theta_{max} = \pm 180°$ or 360°. Then,

$$\Delta\theta_d = \frac{180°}{2^{r-1}} = \frac{360°}{2^r} \tag{11.3}$$

Note: The minimum count corresponds to the case where all the bits are zero and the maximum count corresponds to the case where all the bits are unity. Suppose that these two readings represent the angular displacements θ_{min} and θ_{max}. We have

$$\theta_{max} = \theta_{min} + (M-1)\Delta\theta \tag{11.4}$$

or substituting $M = 2^{r-1}$ we have $\theta_{max} = \theta_{min} + (2^{r-1}-1)\Delta\theta_d$. This gives the conventional definition for digital resolution:

$$\Delta\theta_d = \frac{(\theta_{max} - \theta_{min})}{(2^{r-1}-1)} \tag{11.5}$$

This result is exactly the same as that given by Equation 11.3.

If θ_{max} is 2π and θ_{min} is 0, then θ_{max} and θ_{min} will correspond to the same position of the code disk. To avoid this ambiguity, we use

$$\theta_{min} = \frac{\theta_{max}}{2^{r-1}} \tag{11.6}$$

Note that if we substitute Equation 11.6 into Equation 11.5 we get Equation 11.3* as required. Then, the digital resolution is given by $(360° - 360°/2^r)/(2^r-1)$, which is identical to Equation 11.3.

11.4.1.2 Physical Resolution

The physical resolution of an encoder is governed by the number of windows N in the code disk. If only one pulse signal is used (i.e., no direction sensing) and if only the rising edges of the pulses are detected (i.e., full cycles of the encoder signal are counted), the physical resolution is given by the pitch angle of the track (i.e., angular separation between adjacent windows), which is $(360/N)°$. However, when quadrature signals (i.e., two pulse signals, one out of phase with the other by 90° or quarter-of-a-pitch angle) are available and the capability to detect both rising and falling edges of a pulse is also present, four counts can be made per encoder cycle, thereby improving the resolution by a factor of 4. Under these conditions, the physical resolution of an encoder is given by

$$\Delta\theta_p = \frac{360°}{4N} \tag{11.7}$$

To understand this, note in Figure 11.4a (or Figure 11.4b) that when the two signals v_1 and v_2 are added, the resulting signal has a transition at every quarter of the encoder cycle. This is illustrated in Figure 11.7. By detecting each transition (through edge detection or level detection), four pulses can be counted within every main cycle. It should be mentioned that each signal (v_1 or v_2) has a resolution of half a pitch separately, provided that all transitions (rising edges and falling edges) are detected and counted instead of counting pulses

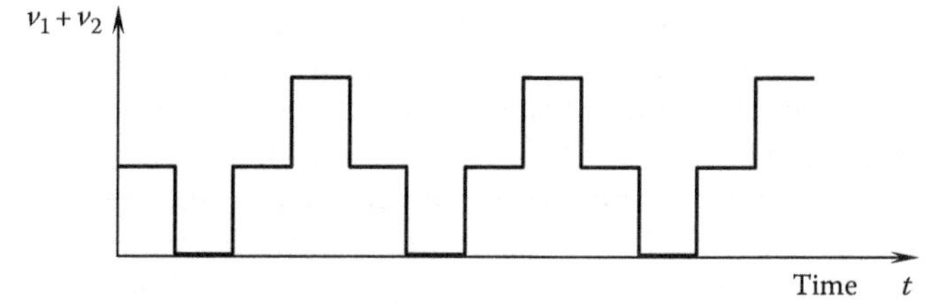

FIGURE 11.7
Quadrature signal addition to improve physical resolution.

(or high signal levels). Accordingly, a disk with 10,000 windows has a resolution of 0.018° if only one pulse signal is used (and both transitions, rise and fall, are detected). When two signals (with a phase shift of a quarter of a cycle) are used, the resolution improves to 0.009°. This resolution is achieved directly from the mechanics of the transducer; no interpolation is involved. It assumes, however, that the pulses are nearly ideal and, in particular, that the transitions are perfect. In practice, this cannot be realized if the pulse signals are noisy. Then, pulse shaping will be necessary as mentioned earlier. The larger value of the two resolutions given by the digital resolution in Equation 11.3 and the physical resolution in Equation 11.7 governs the displacement resolution of an encoder.

Example 11.1

For an ideal design of an incremental encoder, obtain an equation relating the parameters: d is the diameter of encoder disk, w is the number of windows per unit diameter of disk, and r is the word size (bits) of the angle measurement. Assume that quadrature signals are available.

If $r = 12$ and $w = 500$/cm, determine a suitable disk diameter.

Solution

In this problem, we take the ideal design as the case where the physical resolution is equal to the digital resolution. The position resolution due to physical constraints (assuming that quadrature signals are available) is given by Equation 11.7. Hence, $\Delta\theta_p = (1/4)(360/wd)°$. The resolution limited by the digital word size of the buffer is given by Equation 11.3: $\Delta\theta_d = (360/2^r)°$. For an ideal design, we need $\Delta\theta_p = \Delta\theta_d$, which gives $(1/4)(360°/wd) = (360°/2^r)$. Simplifying, we have $wd = 2^{r-2}$.

Substitute $r = 12$ and $w = 500$/cm to obtain $d = (2^{12-2}/500)$ cm $= 2.05$ cm.

11.4.1.3 Step-Up Gearing

The physical resolution of an encoder can be improved by using step-up gearing so that one rotation of the moving object that is monitored corresponds to several rotations of the code disk of the encoder. This improvement is directly proportional to the step-up gear ratio (p). Specifically, from Equation 11.7, we have

$$\Delta\theta_p = \frac{360°}{4pN} \tag{11.8}$$

Backlash in the gearing introduces a new error, however. For best results, this backlash error should be several times smaller than the resolution with no backlash.

The digital resolution will not improve by gearing if the size of the buffer/register where the encoder count is stored corresponds to the maximum angle of rotation of the moving object (say, 360°). Then, the change in the least significant bit (LSB) of the buffer corresponds to the same change in the angle of rotation of the moving object. In fact, the overall displacement resolution can be harmed in this case if excessive backlash is present. However, if the buffer or register size corresponds to a full rotation of the code disk (i.e., a rotation of 360°/p in the object) and if the output register (or buffer) is cleared at the end of each such rotation and a separate count of full rotations of the code disk is kept, then the digital resolution will also improve by a factor of p. Specifically, from Equation 11.3, we get the digital resolution

$$\Delta\theta_d = \frac{180°}{p2^{r-1}} = \frac{360°}{p2^r} \tag{11.9}$$

Example 11.2

By using high-precision techniques to imprint window tracks on the code disk, it is possible to attain a window density of 500 windows/cm of diameter. Consider a 3000 window disk. Suppose that step-up gearing is used to improve resolution and the gear ratio is 10. If the word size of the output register is 16 bits, examine the displacement resolution of this device for the two cases where the register size corresponds to (1) a full rotation of the object and (2) a full rotation of the code disk.

Solution

First, consider the case in which gearing is not present. With quadrature signals, the physical resolution is $\Delta\theta_p = (360°/(4 \times 3000)) = 0.03°$.

For a range of measurement given by ±180°, a 16-bit output provides a digital resolution of $\Delta\theta_d = (180°/2^{15}) = 0.005°$.

Hence, in the absence of gearing, the overall displacement resolution is 0.03°.

Next, consider a geared encoder with gear ratio of 10 and neglect gear backlash. The physical resolution improves to 0.003°. However, in Case 1, the digital resolution remains unchanged at best. Hence, the overall displacement resolution improves to 0.005° as a result of gearing. In Case 2, the digital resolution improves to 0.0005°. Hence, the overall displacement resolution becomes 0.003°.

In summary, the displacement resolution of an incremental encoder depends on the following factors:

1. Number of windows on the code track (or disk diameter)
2. Gear ratio
3. Word size of the measurement register

Example 11.3

A positioning table uses a backlash-free high-precision lead screw of lead 2 cm/rev, which is driven by a servo motor with a built-in optical encoder for feedback control. If the required positioning accuracy is ±10 μm, determine the number of windows required in the encoder track. In addition, what is the minimum bit size required for the digital data register/buffer of the encoder count?

Solution

The required accuracy is ±10 μm. To achieve this accuracy, the required resolution for a linear displacement sensor is ±5 μm. The lead of the lead screw is 2 cm/rev. To achieve the required resolution, the number of pulses per encoder revolution is

$$\frac{2\times 10^{-2}\ \text{m}}{5\times 10^{-6}\ \text{m}} = 4000\ \text{pulses/rev}$$

Assuming that quadrature signals are available (with a resolution improvement of 4), the required number of windows in the encoder track is 1000. The percentage value of physical resolution = (1/4000) × 100% = 0.025%. Consider a buffer size of r bits, including a sign bit. Then, we need $2^{r-1} = 4000$ or $r = 13$ bits.

11.4.1.4 Interpolation

The output resolution of an encoder can be further enhanced by interpolation. This is accomplished by adding equally spaced pulses in between every pair of pulses generated by the encoder circuit. These auxiliary pulses are not true measurements, and they can be interpreted as a linear interpolation scheme between true pulses. One method of accomplishing this interpolation is by using the two probe signals that are generated by the encoder (quadrature signals). These signals are nearly sinusoidal (or triangular) before shaping (say, by level detection). They can be filtered to obtain two sine signals that are 90° out of phase (i.e., a sine signal and a cosine signal). By weighted combination of these two signals, a series of sine signals can be generated such that each signal lags the preceding signal by any integer fraction of 360°. By level detection or edge detection (of rising and falling edges), these sine signals can be converted into square wave signals. Then, by logical combination of the square waves, an integer number of pulses can be generated within each encoder cycle. These are the interpolation pulses that are added to improve the encoder resolution. In practice, about 20 interpolation pulses can be added between 2 adjacent main pulses.

11.4.2 Velocity Measurement

Two methods are available for determining velocities using an incremental encoder:

1. Pulse-counting method
2. Pulse-timing method

In the first method, the pulse count over a fixed time period (the successive time period at which the data register is read) is used to calculate the angular velocity. For a given period of data reading, there is a lower speed limit below which this method is not very accurate. To compute the angular velocity ω using this method, suppose that the count during a time period T is n pulses. Hence, the average time for one pulse cycle (i.e., window-to-window pitch angle) is T/n. If there are N windows on the disk, assuming that quadrature signals are not used, the angle moved during one pulse period is $2\pi/N$ radians. Hence,

$$\text{For pulse-counting method:}\quad \text{Speed } \omega = \frac{2\pi/N}{T/n} = \frac{2\pi n}{NT} \tag{11.10}$$

If quadrature signals are used, replace N by $4N$ in Equation 11.10.

In the second method, the time for one encoder pulse cycle (i.e., window-to-window pitch angle) is measured using a high-frequency clock signal. This method is particularly suitable for accurately measuring low speeds. In this method, suppose that the clock frequency is f Hz. If m cycles of the clock signal are counted during an encoder pulse period (i.e., window pitch, which is the interval between two adjacent windows, assuming that quadrature signals are not used), the time for that encoder cycle (i.e., the time to rotate through one encoder pitch) is given by m/f. With a total of N windows on the track, the angle of rotation during this period is $2\pi/N$ radians as earlier. Hence,

$$\text{For pulse-timing method:} \quad \text{Speed } \omega = \frac{2\pi/N}{m/f} = \frac{2\pi f}{Nm} \tag{11.11}$$

If quadrature signals are used, replace N by $4N$ in Equation 11.11.

We note that a single incremental encoder can serve as both a position sensor and a speed sensor. Hence, for example, in a control system a position loop and a speed loop can be closed using a single encoder, without having to use a conventional (analog) speed sensor such as a tachometer (see Chapter 8). The speed resolution of the encoder (which depends on the method of speed computation—pulse counting or pulse timing) can be chosen to meet the accuracy requirements for the speed control loop. A further advantage of using an encoder rather than a conventional (analog) motion sensor is that an ADC would be unnecessary. For example, the pulses generated by the encoder may be directly read into a microcontroller. Alternatively, the pulses may be used as interrupts for a computer. These interrupts are then directly counted (by an up/down counter or indexer) or timed (by a clock in the DAQ system) within the computer, thereby providing position and velocity readings.

11.4.2.1 Velocity Resolution

The velocity resolution of an incremental encoder depends on the method that is employed to determine the velocity. As both the pulse-counting method and the pulse-timing method are based on counting, the velocity resolution is given by the change in angular velocity that corresponds to a change (increment or decrement) in the count by one.

For the pulse-counting method, it is clear from Equation 11.10 that a unity change in the count n corresponds to a speed change of

$$\Delta\omega_c = \frac{2\pi}{NT} \tag{11.12}$$

where

N is the number of windows in the code track

T is the time period over which a pulse count is read

This gives the velocity resolution by this method. Note that the engineering value (in rad/s) of this resolution is independent of the angular velocity itself, but when expressed as percentage of the speed, the resolution becomes better (smaller) at higher speeds. Note further from Equation 11.12 that the resolution improves with the number of windows and the count reading (sampling) period. Under transient conditions, the accuracy of a velocity reading decreases with increasing T (because according to Shannon's sampling theorem—see Chapter 6—the sampling frequency has to be at least double the highest frequency of interest in the velocity signal). Hence, the sampling period should not be increased indiscriminately. As usual, if quadrature signals are used, N in Equation 11.12 has to be replaced by $4N$ (i.e., the resolution improves by ×4).

In the pulse-timing method, the velocity resolution is given by (see Equation 11.11)

$$\Delta\omega_t = \frac{2\pi f}{Nm} - \frac{2\pi f}{N(m+1)} = \frac{2\pi f}{Nm(m+1)} \tag{11.13*}$$

where f is the clock frequency. For large m, $(m + 1)$ can be approximated by m. Then, by substituting Equation 11.11 into Equation 11.13*, we get

$$\Delta\omega_t \approx \frac{2\pi f}{Nm^2} = \frac{N\omega^2}{2\pi f} \tag{11.13}$$

Note that in this case the resolution degrades quadratically with speed. Also, the resolution degrades with the speed even when it is considered as a fraction of the measured speed:

$$\frac{\Delta\omega_t}{\omega} = \frac{N\omega}{2\pi f} \tag{11.14}$$

This observation confirms the previous suggestion that the pulse-timing method is appropriate for low speeds. For a given speed and clock frequency, the resolution further degrades with increasing N. This is true because when N is increased, the pulse period shortens, and hence, the number of clock cycles per pulse period also decreases. The resolution can be improved, however, by increasing the clock frequency.

Example 11.4

An incremental encoder with 500 windows in its track is used for speed measurement. Suppose the following:

(a) In the pulse-counting method, the count (in the buffer) is read at the rate of 10 Hz.
(b) In the pulse-timing method, a clock of frequency 10 MHz is used.

Determine the percentage resolution for each of these two methods when measuring a speed of (1) 1 rev/s and (2) 100 rev/s.

Solution

Assume that quadrature signals are not used.

Case 1: Speed = 1 rev/s

With 500 windows, we have 500 pulses/s.

(a) Pulse-counting method

$$\text{Counting period} = \frac{1}{10\ \text{Hz}} = 0.1\ \text{s}$$

TABLE 11.1

Comparison of Speed Resolution from an Incremental Encoder

Speed (rev/s)	Percentage Resolution Pulse-Counting Method (%)	Pulse-Timing Method (%)
1.0	2	0.005
100.0	0.02	0.5

$$\text{Pulse count (in 0.1 s)} = 500 \times 0.1 = 50$$

$$\text{Percentage resolution} = \frac{1}{50} \times 100\% = 2\%$$

(b) Pulse-timing method

$$\text{At 500 pulses/s, pulse period} = \frac{1}{500}\ \text{s} = 20 \times 10^{-3}\ \text{s}$$

$$\text{With a 10 MHz clock, clock count} = 10 \times 10^6 \times 2 \times 10^{-3} = 20 \times 10^3$$

$$\text{Percentage resolution} = \frac{1}{20 \times 10^3} \times 100\% = 0.005\%$$

Case 2: Speed = 100 rev/s

With 500 windows, we have 50,000 pulses/s.

(a) Pulse-counting method

$$\text{Pulse count (in 0.1 s)} = 50{,}000 \times 0.1 = 5{,}000$$

$$\text{Percentage resolution} = \frac{1}{5000} \times 100\% = 0.02\%$$

(b) Pulse-timing method

$$\text{At 50,000 pulses/s, pulse period} = \frac{1}{50{,}000}\ \text{s} = 20 \times 10^{-6}\ \text{s}$$

$$\text{With a 10 MHz clock, clock count} = 10 \times 10^6 \times 20 \times 10^{-6} = 200$$

$$\text{Percentage resolution} = \frac{1}{200} \times 100\% = 0.5\%$$

The results are summarized in Table 11.1.

Results given in Table 11.1 confirm that in the pulse-counting method the resolution improves with speed, and hence, it is more suitable for measuring high speeds. Furthermore, in the pulse-timing method the resolution degrades with speed, and hence, it is more suitable for measuring low speeds.

11.4.2.2 Velocity with Step-Up Gearing

Consider an incremental encoder with a track having N windows per track and connected to a rotating shaft through a gear unit with step-up gear ratio p. Formulas for computing angular velocity of the shaft by (1) pulse-counting method and (2) pulse-timing method can be easily determined from Equations 11.10 and 11.11, for this case of step-up gearing.

Specifically, the angle of rotation of the shaft corresponding to one window spacing (pitch) of the encoder disk now is $2\pi/(pN)$. Hence, the corresponding formulas for speed can be obtained by replacing N by pN in Equations 11.10 and 11.11. We have

$$\text{For pulse-counting method: } \omega = \frac{2\pi n}{pNT} \tag{11.15}$$

$$\text{For pulse-timing method: } \omega = \frac{2\pi f}{pNm} \tag{11.16}$$

Note: These relations may be obtained in a more straightforward manner by simply dividing the encoder disk speed by the gear ratio p, which gives the object speed.

11.4.2.3 Velocity Resolution with Step-Up Gearing

As before, the speed resolution is given by the change in speed corresponding to a unity change in the count. Hence,

$$\text{For the pulse-counting method: } \Delta\omega_c = \frac{2\pi(n+1)}{pNT} - \frac{2\pi n}{pNT} = \frac{2\pi}{pNT} \tag{11.17}$$

It follows that in the pulse-counting method, step-up gearing causes an improvement in the resolution.

$$\text{For the pulse-timing method: } \Delta\omega_t = \frac{2\pi f}{pNm} - \frac{2\pi f}{pN(m+1)} = \frac{2\pi f}{pNm(m+1)} \cong \frac{pN}{2\pi f}\omega^2 \tag{11.18}$$

Note: In the pulse-timing approach, for a given speed, the resolution degrades with increasing p.

In summary, the speed resolution of an incremental encoder depends on the following factors:

1. Number of windows N
2. Count reading (sampling) period T
3. Clock frequency f
4. Speed ω
5. Gear ratio p

In particular, gearing up has a detrimental effect on the speed resolution in the pulse-timing method, but it has a favorable effect in the pulse-counting method.

11.5 Encoder Data Acquisition and Processing

An incremental encoder typically has five pins corresponding to the (1) ground, (2) index (I) channel, (3) A channel, (4) +5 V DC power, and (5) B channel. The channels A and B

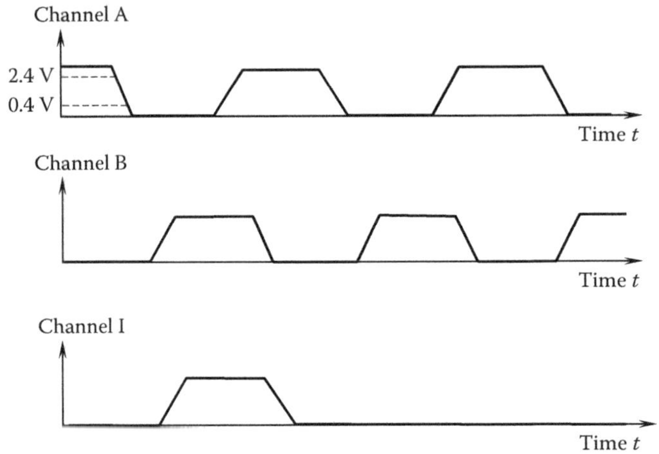

FIGURE 11.8
Outputs from an incremental encoder.

provide the quadrature signals (the position increment signals that are 90° out of phase) and I channel gives the index (full rotation) pulses. The photosensor signals that generate the signals for these three channels are conditioned by the integrated circuitry within the encoder. What comes out are digital signals (TTL compatible) that can be directly read by a microcontroller or any DAQ computer. The nature of the TTL compatible output signals from an incremental encoder is shown in Figure 11.8.

11.5.1 Data Acquisition by Microcontroller

The output pins of an encoder may be directly connected to the pins of a microcontroller. In order to avoid distortion (loading) of the encoder output due to the TTL load (i.e., microcontroller) that reads the encoder data, the output pins of the encoder may have to be connected to pull-up resistors of high resistance (e.g., 3 kΩ). Often, the load (microcontroller) itself may provide the needed pull-up resistance (i.e., internal pull-up resistance of the microcontroller). A TTL output is considered "low" (or binary 0) when the voltage is between 0 and 0.4 V and is considered "high" (or binary 1) when the voltage is between 2.6 and 5 V (see Figure 11.8). *Note*: Accommodating a voltage range of this type is needed for noise immunity (up to 0.4 V in this case). An example application that uses an incremental encoder and a microcontroller to monitor and control the motion of a DC motor is shown in Figure 11.9.

An encoder pulse can be counted and the direction of rotation can be determined (the pulse count in the microcontroller register is changed based on them) by detecting the levels (high or low) and transitions (from high to low or from low to high) in the output signals of the encoder, as discussed earlier. For example,

If channel A goes from high to low and channel B is low → increment count

If channel A goes from high to low and channel B is high → decrement count

This logic should be clear from Figure 11.4. Such operations are done with reference to the internal clock of the microcontroller. The required frequency of the operation depends

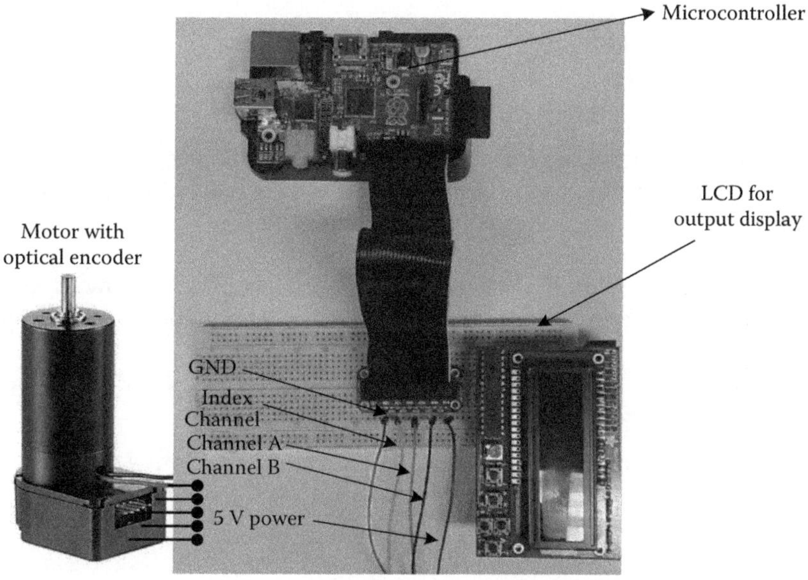

FIGURE 11.9
Use of an incremental encoder and a microcontroller for motion sensing of a motor.

on the nature of the encoder and the requirements of the particular sensing application. Specifically, first a suitable encoder is chosen for the application. Next, the maximum number of pulses per second from the encoder is estimated. Usually, twice this maximum pulse rate is adequate for the frequency of the counting operation.

The driver software for the microcontroller should be installed, the programming library should be imported, and the microcontroller should be programmed (say, using a desktop computer with the microcontroller connected using a USB cable) to read the encoder and compute the displacement and velocity. A pseudocode (a high-level description of the computer program) for this purpose is given in the following:

```
IMPORT GPIO
IMPORT time
SET a GPIO port for A channel
SET a GPIO port for B channel
COMPUTE angle of each window in the disk
WHILE until the pulse of B channel is high
      OBTAIN GPIO input value of B channel
ENDWHILE
WHILE UNTIL the detection of high to low transition in B channel
  OBTAIN GPIO input value of B channel
  IF GPIO input value of B channel is low THEN
    OBTAIN GPIO input value of A channel
    IF GPIO input value of A channel is low
      DISPLAY "Froward rotation"
    ELSE
      DISPLAY "Backward rotation"
    ENDIF
  ENDIF
```

```
ENDWHILE
COMPUTE the displacement
COMPUTE the time
COMPUTE the speed
```

This program imports libraries, set the port for DAQ (a GPIO port, see Chapter 4), and carries out the DAQ and motion computations in the "while loop." Computation of the angle of rotation and the speed may be done as follows:

1. Count clock pulses from B channel transition (from high to low or from low to high) to the very next A channel transition (from high to low or from low to high). Call it n.
2. Compute $A = 2\pi/(4N)$ where N is the number of windows in a track of the encoder disk (given). *Note*: This can be computed off-line.
3. Update displacement and time, $D = D + A$, and time, $T = T + n \times \Delta T$, where ΔT is the clock-pulse period (in seconds).
4. Compute speed: $W = A/(n \times \Delta T)$.

11.5.2 Data Acquisition by Desktop Computer

Interfacing an incremental encoder to a desktop computer may be done through a standard DAQ card that is placed in an expansion slot of the computer (see Chapter 4). However, this is a far more expensive solution when compared to using a microcontroller. Since the encoder outputs are TTL compatible pulses, three digital channels are the maximum requirement for the DAQ. A modern DAQ can easily satisfy this requirement. The main operations of DAQ are schematically shown in Figure 11.10.

The pulse signals from the encoder are fed into the digital channels of the DAQ. An up/down counter will detect the pulses (e.g., by rising-edge detection, by falling-edge detection, or by level detection) and will determine the direction of motion. A pulse in one direction (say, clockwise) will increase the count by one (an upcount), and a pulse in the

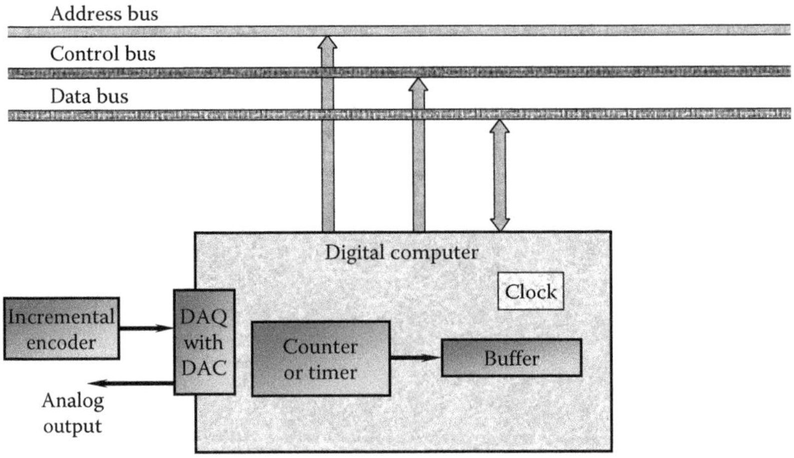

FIGURE 11.10
Computer interfacing of an incremental encoder.

opposite direction will decrease the count by one (a downcount). The count is transferred to a latch buffer so that the measurement is read from the buffer rather than from the counter itself. This arrangement provides an efficient means of DAQ because the counting process can continue without interruption while the computer reads the count from the latch buffer.

The computer identifies various components in the measurement system using addresses, and this information is communicated to the individual components through the address bus. The start, end, and nature of an action (e.g., data read, clear the counter, clear the buffer) are communicated to various devices by the computer through its control bus. The computer can command an action to a component in one direction of the bus, and the component can respond with a message (e.g., job completed) in the opposite direction. The data (e.g., the count) are transmitted through the data bus. While the computer reads (samples) data from the buffer, the control signals guarantee that no data are transferred to that buffer from the counter. It is clear that DAQ consists of "handshake operations" between the main processor of the computer and auxiliary components. More than one encoder may be addressed, controlled, and read by the same three buses of the computer. *Note*: As mentioned in Chapter 4, the buses are conductors, for example, multicore cables carrying signals in parallel logic. The internal electronics of the encoder may be powered by 5 V DC supply from the computer.

While measuring the displacement (position) of an object using an incremental encoder, the pulse count is read by the computer only at finite time intervals (say, 5 ms). The net count gives the displacement. Since a cumulative count is required in displacement measurement, the buffer is not cleared once the count is read by the computer.

In velocity measurement by the pulse-counting method, the buffer is read at fixed time intervals of T, which is also the counting-cycle time. The counter is cleared every time a count is transferred to the buffer so that a new count can begin. With this method, a new reading is available at every sampling instant.

In the pulse-timing method of velocity computation, the counter is actually a timer. The encoder cycle is timed using a clock (internal or external), and the count is passed on to the buffer. The counter is then cleared, and the next timing cycle is started. The buffer is periodically read by the computer. With this method, a new reading is available at every encoder cycle. Note that under transient velocities, the encoder-cycle time is variable and is not directly related to the data sampling period. In the pulse-timing method, it is desirable to make the sampling period slightly smaller than the encoder-cycle time so that no count is missed by the processor.

More efficient use of the digital processor may be achieved by using an interrupt routine. With this method, the counter (or buffer) sends an interrupt request to the processor when a new count is ready. The processor then temporarily suspends the current operation and reads in the new data. In this case, the processor does not continuously wait for a reading.

11.6 Absolute Optical Encoders

An absolute encoder directly generates a coded digital word to represent each discrete angular position (sector) of its code disk. This is accomplished by producing a set of pulse signals (data channels) equal in number to the word size (number of bits) of the reading.

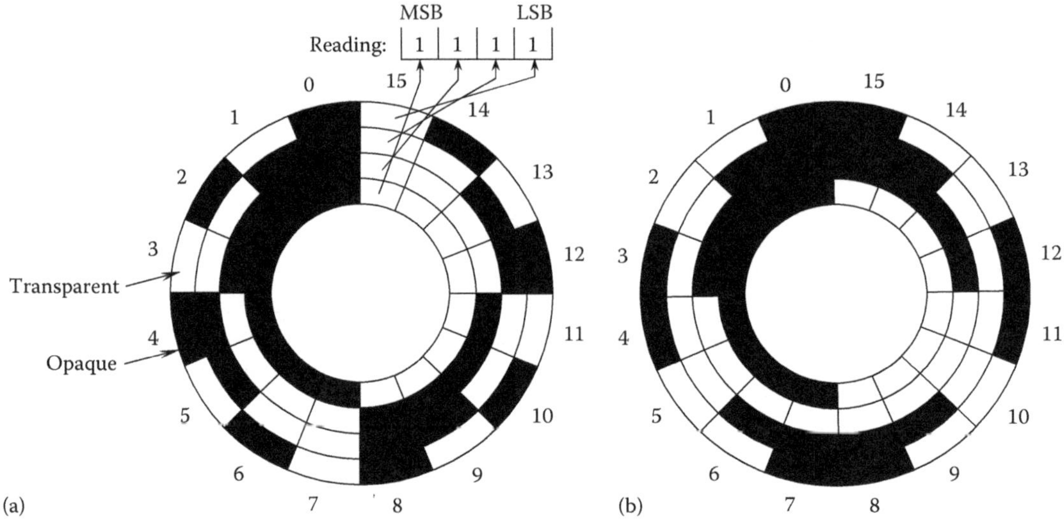

FIGURE 11.11
Illustration of the code pattern of an absolute encoder disk: (a) a binary code and (b) a gray code.

Unlike with an incremental encoder, no pulse counting is involved. An absolute encoder may use various techniques (e.g., optical, sliding contact, magnetic saturation, and proximity sensor) to generate the sensor signal, as described earlier for an incremental encoder. The optical method, which uses a code disk with transparent and opaque regions and pairs of light sources and photosensors, is the most common technique of transducer signal generation, however.

A simplified code pattern on the disk of an absolute encoder, which uses the direct binary code, is shown in Figure 11.11a. The number of tracks (n) in this case is 4, but in practice n is in the order of 14, and may even be as high as 22. The disk is divided into 2^n sectors. Each partitioned area of the matrix thus formed corresponds to a bit of data. For example, a transparent area will correspond to binary 1 and an opaque area to binary 0. Each track has a probe similar to what is used in an incremental encoder. The set of n probes is arranged along a radial line and facing the tracks on one side of the disk. A light source (e.g., LED) illuminates the other side of the disk. As the disk rotates, the bank of probes generates pulse signals, which are sent to n parallel data channels (or pins). At a given instant, the particular combination of signal levels in the data channels will provide a coded data word that uniquely determines the position of the disk at that time.

11.6.1 Gray Coding

In an absolute encoder, there is a data interpretation problem associated with the straight binary code. Note in Table 11.2 that with the straight binary code, the transition from one sector to an adjacent sector may require more than one switching of bits in the binary data. For example, the transition from 0011 to 0100 or from 1011 to 1100 requires three-bit switchings, and the transition from 0111 to 1000 or from 1111 to 0000 requires four-bit switchings. If the light probes are not properly aligned along a radius of the encoder disk, or if the manufacturing error tolerances for imprinting the code pattern on the disk were high,

TABLE 11.2

Sector Coding for a 4-Bit Absolute Encoder

Sector Number	Straight Binary Code (MSB → LSB)	Gray Code (MSB → LSB)
0	0000	0000
1	0001	0001
2	0010	0011
3	0011	0010
4	0100	0110
5	0101	0111
6	0110	0101
7	0111	0100
8	1000	1100
9	1001	1101
10	1010	1111
11	1011	1110
12	1100	1010
13	1101	1011
14	1110	1001
15	1111	1000

or if environmental effects have resulted in large irregularities in the sector matrix, or if there is uneven delay in the electronics, then the bit switching from one reading to the next may not take place simultaneously. This will result in ambiguous readings during the transition period. For example, in changing from 0011 to 0100, if the LSB switches first, the reading becomes 0010. In the decimal form, this incorrectly indicates that the rotation was from angle 3 to angle 2, whereas it was actually a rotation from angle 3 to angle 4. Such ambiguities in data interpretation can be avoided by using a gray code, as shown in Figure 11.11b for this example. The coded representation of the sectors is given in Table 11.2. *Note*: In the case of gray code, each adjacent transition involves only one-bit switching.

For an absolute encoder, a gray code is not essential for removing the ambiguity in bit switchings. For example, for a given absolute reading, the two adjacent absolute readings are automatically known. A reading can be checked against these two valid possibilities (or a single possibility if the direction of rotation is known) to see whether the reading is correct. Another approach is to introduce a delay (e.g., Schmitt trigger) to the reading of the output. In this manner, a reading will be taken only after all the bit switchings have taken place, thereby eliminating the possibility of an intermediate ambiguous reading.

11.6.1.1 Code Conversion Logic

A disadvantage of using a gray code is that it requires additional logic to convert the gray-coded number to the corresponding binary number. This logic may be provided in hardware or software. In particular, an "exclusive-OR" gate can implement the necessary logic, as given by

$$\begin{aligned} B_{n-1} &= G_{n-1} \\ B_{k-1} &= B_k \oplus G_{k-1}, \quad k = n-1,\ldots,1 \end{aligned} \tag{11.19}$$

This logic converts an n-bit gray-coded word $[G_{n-1}, G_{n-2}, \ldots, G_0]$ into an n-bit binary-coded word $[B_{n-1}, B_{n-2}, \ldots, B_0]$, where the subscript $n-1$ denotes the most significant bit (MSB) and 0 denotes the LSB. For a small word size, the code may be given as a lookup table (see Table 11.2). The gray code itself is not unique. Other gray codes that provide single-bit switching between adjacent numbers are available.

11.6.2 Resolution

The resolution of an absolute encoder is limited by the word size of the output data. Specifically, the displacement (position) resolution is given by the sector angle, which is also the angular separation between adjacent transparent and opaque regions on the outermost track of the code disk:

$$\Delta\theta = \frac{360^\circ}{2^n} \tag{11.20}$$

where n is the number of tracks on the disk (which is equal to the number of bits in the digital reading). In Figure 11.11a, the word size of the data is 4 bits. This can represent decimal numbers from 0 to 15, as given by the 16 sectors of the disk. In each sector, the outermost element is the LSB and the innermost element is the MSB. The direct binary representation of the disk sectors (angular positions) is given in Table 11.2. The angular resolution for this simplified example is $(360/2^4)^\circ$ or 22.5°. If $n=14$, the angular resolution improves to $(360/2^{14})^\circ$ or 0.022°. If $n=22$, the resolution further improves to 0.000086°.

For an absolute encoder, a step-up gear mechanism may be employed to improve the encoder resolution. However, this has the same disadvantages as mentioned under incremental encoders (e.g., backlash, added weight and loading, and increased cost). Furthermore, when a gear is included, the absolute nature of a reading will be limited to a fraction of rotation of the main shaft, specifically, 360°/gear ratio. We can overcome this limitation by counting the total rotations of the code disk as well.

An ingenious method of improving the resolution of an absolute encoder is available through the generation of auxiliary pulses in between the bit switchings of the coded word. This requires an auxiliary track (usually placed as the outermost track) with a sufficiently finer pitch than the LSB track and some means of direction sensing (e.g., two light probes placed at quarter-pitch apart, to generate quadrature signals). This is equivalent to having an incremental encoder of finer resolution along with an absolute encoder, in a single integral unit. Knowing the reading of the absolute encoder (from its coded output, as usual) and the direction of motion (from the quadrature signals), it is possible to determine the angle corresponding to the successive incremental pulses (from the finer track) until the next reading of the absolute word. Of course, if a data failure occurs in between the absolute readings, the additional accuracy (and resolution) that is provided by the incremental pulses will be lost.

11.6.3 Velocity Measurement

An absolute encoder can be used for angular velocity measurement as well. For this purpose, either the pulse-timing method or the angle-measurement method may be used. With the first method, the interval between two consecutive readings is strobed (i.e., timed) using a high-frequency strobe (clock) signal, as in the case of an incremental encoder. Typical strobing frequency is 1 MHz. The start and stop of strobing are triggered by the coded data from the encoder. The clock cycles are counted by a counter, as in the case of an incremental encoder, and the count is reset (cleared) after each counting cycle. The angular speed can be computed using these data, as discussed earlier for an incremental encoder. With the second method, the change in angle is measured from one absolute angle reading to the next, and the angular speed is computed as [angle change]/[sampling period].

11.6.4 Advantages and Drawbacks

The main advantage of an absolute encoder is its ability to provide absolute angle readings (for a full 360° rotation). Hence, if a reading is missed, it will not affect the next reading. Specifically, the digital output uniquely corresponds to a physical rotation of the code disk, and hence, a particular reading is not dependent on the accuracy of a previous reading. This provides immunity to data failure. A missed pulse (or a data failure of some sort) in an incremental encoder would carry an error into the subsequent readings until the counter is cleared.

An incremental encoder has to be powered throughout the operation of the device. Thus, a power failure can introduce an error unless the reading is reinitialized (or calibrated). An absolute encoder has the advantage that it needs to be powered and monitored only when a reading is taken.

Because the code matrix on the disk is more complex in an absolute encoder and because more light sensors are required, an absolute encoder can be nearly twice as expensive as an incremental encoder. Also, since the resolution depends on the number of tracks present, it is more costly to obtain finer resolutions. An absolute encoder does not require digital counters and buffers, unless resolution enhancement is effected using an auxiliary track or pulse timing is used for velocity calculation.

11.7 Encoder Error

Errors in shaft encoder readings can come from several factors. The primary sources of these errors are as follows:

1. Quantization error (due to digital word size limitations)
2. Assembly error (eccentricity of rotation, etc.)
3. Coupling error (gear backlash, belt slippage, loose fit, shaft flexibility, etc.)
4. Structural limitations (disk deformation and shaft deformation due to loading)
5. Manufacturing tolerances (errors from inaccurately imprinted code patterns, inexact positioning of the pickoff sensors, limitations and irregularities in the hardware of signal generation and sensing, etc.)
6. Ambient effects (vibration, temperature, light noise, humidity, dirt, smoke, etc.)

These factors can result in inexact readings of displacement and velocity and erroneous detection of the direction of motion.

One form of error in an encoder reading is hysteresis. For a given position of the moving object, if the encoder reading depends on the direction of motion, the measurement has a hysteresis error. In that case, if the object rotates from position A to position B and back to position A, for example, the initial and the final readings of the encoder will not match. The causes of hysteresis include backlash in gear couplings, loose fits, mechanical deformation in the code disk and shaft, delays in electronic circuitry and components (electrical time constants, nonlinearities, etc.), and noisy pulse signals that make the detection of pulses (say, by level detection or edge detection) less accurate.

The raw pulse signal from an optical encoder is somewhat irregular and does not constitute perfect rectangular pulses. This is primarily due to the variation (more or less triangular) of the intensity of light received by the optical sensor, as the code disk moves through a window and because of noise in the signal generation circuitry including the noise created by imperfect light sources and photosensors. Noisy pulses have imperfect edges. As a result, pulse detection through edge detection can result in errors such as multiple triggering for the same edge of a pulse. This can be avoided by including a Schmitt trigger (a logic circuit with electronic hysteresis) in the edge detection circuit so that slight irregularities in the pulse edges will not cause erroneous triggering, provided that the noise level is within the hysteresis band of the trigger. A disadvantage of this method, however, is that hysteresis will be present even when the encoder itself is perfect. Virtually noise-free pulses can be generated if two photosensors are used to simultaneously detect adjacent transparent and opaque areas on a track, and a separate circuit (a comparator) is used to create a pulse that depends on the sign of the voltage difference of the two sensor signals. This method of pulse shaping has been described earlier, with reference to Figure 11.5.

11.7.1 Eccentricity Error

Eccentricity (denoted by e) of an encoder is defined as the distance between the center of rotation C of the code disk and the geometric center G of the circular code track. Nonzero eccentricity causes a measurement error known as the eccentricity error. The primary contributions to eccentricity are as follows:

1. Shaft eccentricity (e_s)
2. Assembly eccentricity (e_a)
3. Track eccentricity (e_t)
4. Radial play (e_p)

Shaft eccentricity results if the rotating shaft on which the code disk is mounted is imperfect so that its axis of rotation does not coincide with its geometric axis. Assembly eccentricity is caused if the code disk is improperly mounted on the shaft such that the center of the code disk does not fall on the shaft axis. Track eccentricity comes from irregularities in the imprinting process of the code track so that the center of the track circle does not coincide with the nominal geometric center of the disk. Radial play is caused by any looseness in the assembly in the radial direction. All four of these parameters are random variables. Let their mean values be μ_s, μ_a, μ_t, and μ_p and the standard deviations be σ_s, σ_a, σ_t, and σ_p, respectively. A very conservative upper bound for the mean value of the overall eccentricity is given by the sum of the individual absolute (i.e., considered positive) mean values. A more reasonable estimate is provided by the root-mean-square (rms) value, as given by

$$\mu = \sqrt{\mu_s^2 + \mu_a^2 + \mu_t^2 + \mu_p^2} \tag{11.21}$$

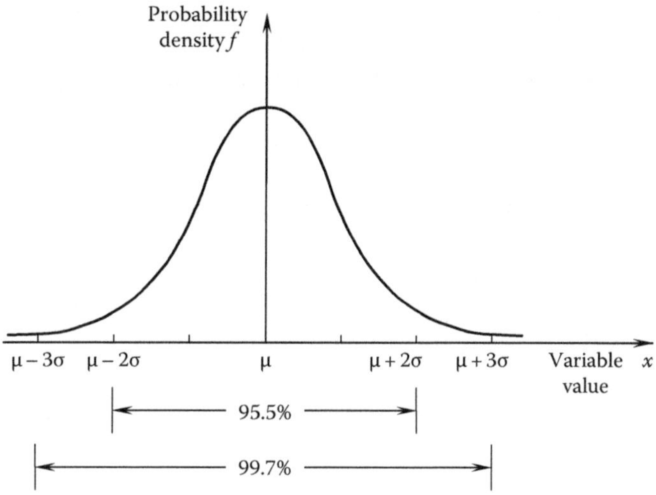

FIGURE 11.12
Gaussian (normal) probability density function.

Furthermore, assuming that the individual eccentricities are independent random variables, the standard deviation of the overall eccentricity is given by

$$\sigma = \sqrt{\sigma_s^2 + \sigma_a^2 + \sigma_t^2 + \sigma_p^2} \tag{11.22}$$

By knowing the mean value μ and the standard deviation σ of the overall eccentricity, it is possible to obtain a reasonable estimate for the maximum eccentricity that can occur. It is reasonable to assume that the eccentricity has a Gaussian (or normal) distribution, as shown in Figure 11.12. The probability that the eccentricity lies between two given values is obtained by the area under the probability density curve within these two values (points) on the x-axis (see Appendix C). In particular, for normal distribution, the probability that the eccentricity lies within $\mu - 2\sigma$ and $\mu + 2\sigma$ is 95.5%, and the probability that the eccentricity lies within $\mu - 3\sigma$ and $\mu + 3\sigma$ is 99.7%. We can state, for example, that at a confidence level of 99.7%, the net eccentricity will not exceed $\mu + 3\sigma$.

Example 11.5

The mean values and the standard deviations of the four primary contributions to eccentricity in a shaft encoder (in millimeters) are given as follows: shaft eccentricity (0.1, 0.01), assembly eccentricity (0.2, 0.05), track eccentricity (0.05, 0.001), and radial play (0.1, 0.02).

Estimate the overall eccentricity at a confidence level of 96%.

Solution

Using Equation 11.21, the mean value of the overall eccentricity is estimated as the rms value of the individual means: $\mu = \sqrt{0.1^2 + 0.2^2 + 0.05^2 + 0.1^2} = 0.25$ mm.

Using Equation 11.22, the standard deviation of the overall eccentricity is estimated as $\sigma = \sqrt{0.01^2 + 0.05^2 + 0.001^2 + 0.02^2} = 0.055$ mm.

Now, assuming a Gaussian distribution, an estimate for the overall eccentricity at a confidence level of 96% is given by

$$\hat{e} = 0.25 + 2 \times 0.055 = 0.36 \text{ mm}$$

Once the overall eccentricity is estimated in the foregoing manner, the corresponding measurement error can be determined. Suppose that the true angle of rotation is θ and the corresponding measurement is θ_m. The eccentricity error is given by

$$\Delta\theta = \theta_m - \theta \tag{11.23}$$

Figure 11.13 presents the maximum error, which can be shown to exist when the line of eccentricity (CG) is symmetrically located within the angle of rotation. For this configuration, the sine rule for triangles gives $(\sin(\Delta\theta/2)/e) = (\sin(\theta/2)/r)$, where r is the code track radius, which for most practical purposes can be taken as the disk radius. Hence, the eccentricity error is given by

$$\Delta\theta = 2\sin^{-1}\left(\frac{e}{r}\sin\frac{\theta}{2}\right) \tag{11.24}$$

It is intuitively clear that the eccentricity error should not enter measurements of complete revolutions, and this can be verified by substituting $\theta = 2\pi$ into Equation 11.24, when we have $\Delta\theta = 0$. For multiple revolutions, the eccentricity error is periodic with period 2π.

The inverse sine of a small quantity is approximately equal to the quantity itself, in radians. Hence, for small e, the eccentricity error in Equation 11.24 may be expressed as

$$\Delta\theta = \frac{2e}{r}\sin\frac{\theta}{2} \tag{11.25}$$

Furthermore, for small angles of rotation, the fractional eccentricity error is given by

$$\frac{\Delta\theta}{\theta} = \frac{e}{r} \tag{11.26}$$

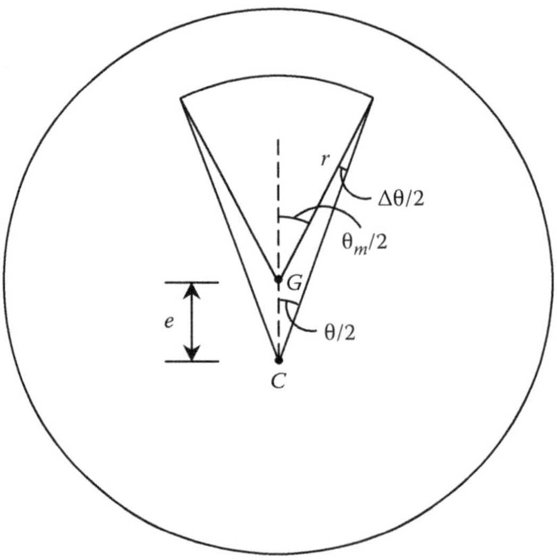

FIGURE 11.13
Nomenclature for eccentricity error (C is the center of rotation and G is the geometric center of the code track).

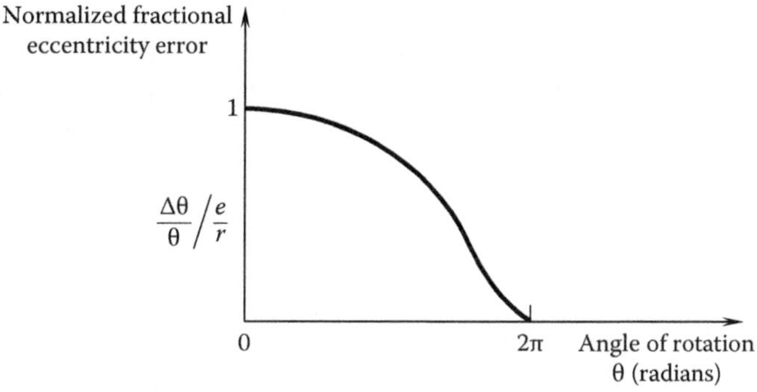

FIGURE 11.14
Fractional eccentricity error variation of an encoder with respect to the angle of rotation.

which is in fact the worst-case fractional error. As the angle of rotation increases, the fractional error decreases (as shown in Figure 11.14), reaching the zero value for a full revolution. From the point of view of gross error, the worst value occurs when $\theta = \pi$, which corresponds to half a revolution. From Equation 11.24, it is clear that the maximum gross error due to eccentricity is given by

$$\Delta\theta_{max} = 2\sin^{-1}\frac{e}{r} \tag{11.27}$$

If this value is less than half the resolution of the encoder, the eccentricity error becomes inconsequential. For all practical purposes, since e is much less than r, we may use the following expression for the maximum eccentricity error:

$$\Delta\theta_{max} = \frac{2e}{r} \tag{11.28}$$

Example 11.6

Suppose that in Example 11.5 the radius of the code disk is 5 cm. Estimate the maximum error due to eccentricity. If each track has 1000 windows, determine whether the eccentricity error is significant.

Solution

With the given level of confidence, we have calculated the overall eccentricity to be 0.36 mm. Now, from Equation 11.27 or 11.28, the maximum angular error is given by

$$\Delta\theta_{max} = \frac{2\times 0.36}{50} = 0.014 \text{ rad} = 0.83°$$

Assuming that quadrature signals are used to improve the encoder resolution, we have

$$\text{Resolution} = \frac{360°}{4\times 1000} = 0.09°$$

Note that the maximum error due to eccentricity is more than 10 times the encoder resolution. Hence, eccentricity will significantly affect the accuracy of the encoder.

The eccentricity of an incremental encoder also affects the phase angle between the quadrature signals if a single track and two probes (with circumferential offset) are used. This error can be reduced by using the two-track arrangement, with the two probes positioned along a radial line, so that the eccentricity would equally affect the two outputs.

11.8 Miscellaneous Digital Transducers

Now, several other types of digital transducers that are useful in engineering applications are described. Typical applications include conveyor systems of industrial processes, x–y positioning tables, machine tools, valve actuators, read–write heads in hard disk drive and other data storage systems, and robotic manipulators (e.g., at prismatic joints) and robot hands. For these devices, the techniques of signal acquisition, interpretation, conditioning, and so on are more or less the same as those described so far.

11.8.1 Binary Transducers

Digital binary transducers are two-state sensors. The information provided by such a device takes only two states (on/off, present/absent, go/no-go, high/low, etc.) that can be represented by 1 bit. For example, a limit switch is a sensor that is used in detecting whether an object has reached a particular position (or limit) and is useful in sensing presence/absence and in object counting. In this sense, a limit switch is considered a digital transducer. Additional logic is needed if the direction of contact is also needed. Limit switches are available for both rectilinear and angular motions. A commercial limit switch is shown in Figure 11.15. This can detect an object reaching from either direction (i.e., it is bidirectional).

A limit of a movement can be detected by mechanical means using a simple contact mechanism to close a circuit or trigger a pulse. Although a purely mechanical device consisting of linkages, gears, ratchet wheels and pawls, and so forth can serve as a limit switch, electronic and solid-state switches are usually preferred for such reasons as speed, accuracy, durability, low activating force (practically zero) requirement, low cost, and

FIGURE 11.15
A bidirectional limit switch. (Courtesy of Alps Electric, Auburn Hills, MI.)

small size. Any proximity sensor may serve as the sensing element of a limit switch, to detect the presence of an object. The proximity sensor signal is then used in a desired manner—for example, to activate a counter, a mechanical switch, or a relay circuit—or simply as an input to a computer or a digital controller to indicate the position (presence) of the object in order to take further action. A microswitch is a solid-state switch that can be used as a limit switch. Microswitches are commonly used in counting operations—for example, to keep a count of completed products in a factory warehouse.

There are many types of binary transducers that are applicable in detection and counting of objects. They include

1. Electromechanical switches
2. Photoelectric devices
3. Magnetic (Hall effect, eddy current) devices
4. Capacitive devices
5. Ultrasonic devices

An electromechanical switch is a mechanically activated and spring-loaded electric switch. The contact with an arriving object turns on the switch, thereby completing a circuit and providing an electrical signal. This signal provides the "presence" state of the object. When the object is removed, the contact is lost and the switch is turned off by the retracting spring. This corresponds to the "absence" state.

In the other four types of binary transducers listed earlier, a signal (light beam, magnetic field, electric field, or ultrasonic wave) is generated by a source (emitter) and received by a receiver. A passing object interrupts the signal. This event can be detected by usual means, using the signal received at the receiver. In particular, the signal level, a rising edge, or a falling edge may be used to detect the event. The following three arrangements of the emitter–receiver pair are common:

1. Through (opposed) configuration
2. Reflective (reflex) configuration
3. Diffuse (proximity, interceptive) configuration

In the through configuration (Figure 11.16a), the receiver is placed directly facing the emitter. In the reflective configuration, the emitter–source pair is located in a single package. The emitted signal is reflected by a reflector, which is placed facing the emitter–receiver package (Figure 11.16b). In the diffuse configuration as well, the emitter–reflector pair is in a single package. In this case, a conventional proximity sensor can serve the purpose of detecting the presence of an object (Figure 11.16c) by using the signal diffused from the intercepting object. When the photoelectric method is used, an LED may serve as the emitter and a phototransistor may serve as the receiver (see Chapter 10). Infrared LEDs are preferred emitters for phototransistors because their peak spectral responses match and also because they are not affected by ambient light.

Many factors govern the performance of a digital transducer for object detection. They include

1. Sensing range (operating distance between the sensor and the object)
2. Response time

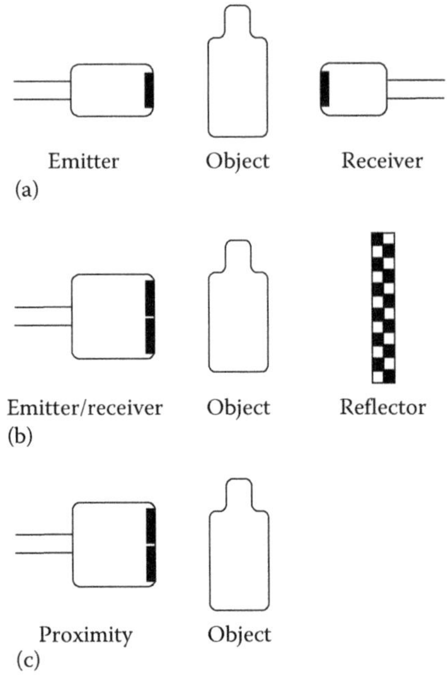

FIGURE 11.16
Two-state transducer configurations. (a) Through (opposed), (b) reflective (reflex), and (c) interceptive (proximity).

3. Sensitivity
4. Linearity
5. Size and shape of the object
6. Material of the object (e.g., color, reflectance, permeability, permittivity)
7. Orientation and alignment (optical axis, reflector, object)
8. Ambient conditions (light, dust, moisture, magnetic field, etc.)
9. Signal-conditioning considerations (modulation, demodulation, shaping, etc.)
10. Reliability, robustness, and design life

Example 11.7

The response time of a binary transducer for object counting is the fastest (shortest) time that the transducer needs to detect an absent-to-present condition or a present-to-absent condition and generate the counting signal (say, a pulse). Consider the counting process of packages on a conveyor as shown in Figure 11.17. Suppose that, typically, packages of length 20 cm are placed along the conveyor at 15 cm spacing. A transducer of response time 10 ms is used for counting the packages. Estimate the allowable maximum operating speed of the conveyor.

Solution

Let the conveyor speed be v cm/ms. Then,

$$\text{Package-present time} = \frac{20.0}{v}\,\text{ms}$$

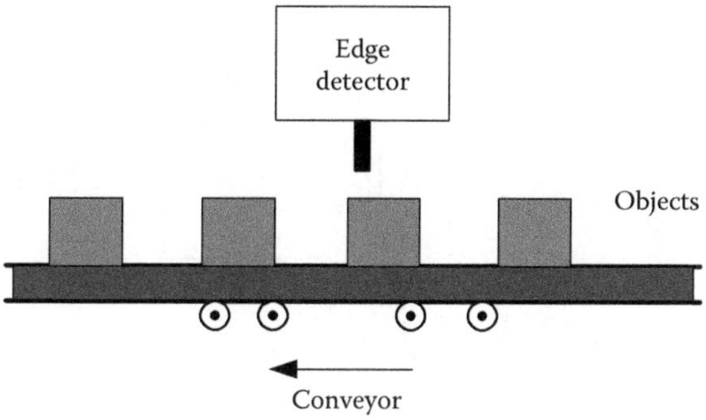

FIGURE 11.17
Object counting on a conveyor.

$$\text{Package-absent time} = \frac{15.0}{v}\ \text{ms}$$

In the sensor selection, we must use the shorter of these two times. Hence, the transducer response time must be at most $(15.0/v)$ ms$\rightarrow 10.0 \le (15.0/v)$ or $v \le 1.5$ cm/ms.

The maximum allowable operating speed is 1.5 cm/ms or 15.0 m/s. This corresponds to a counting rate of 1.5/(20.0 + 15.0) packages/ms or about 43 packages/s.

11.8.2 Digital Resolvers

Digital resolvers, or *mutual-induction encoders,* operate somewhat like analog resolvers, using the principle of mutual induction. They are commercially known as "Inductosyns." A digital resolver has two disks facing each other (but not in contact), one (the stator) stationary and the other (the rotor) coupled to the rotating object whose motion is measured. The rotor has a fine electric conductor foil imprinted on it, as schematically shown in Figure 11.18. The printed pattern is pulse shaped, closely spaced, and connected to a high-frequency AC supply (carrier) of voltage v_{ref}. The stator disk has two separate printed patterns that are identical to the rotor pattern, but one pattern on the stator is shifted by a quarter-pitch from the other pattern (*Note*: Pitch is the spacing between two successive crests of the foil). The primary voltage in the rotor circuit induces voltages in the two secondary (stator) foils at the same frequency; that is, the rotor and the stator are *inductively coupled*. These induced voltages are quadrature signals (i.e., 90° out of phase). As the rotor turns, the level of the induced voltage changes, depending on the relative position of the foil patterns on the two disks. When the foil pulse patterns coincide, the induced voltage is a maximum (positive or negative), and when the rotor foil pattern has a half-pitch offset from the stator foil pattern, the induced voltage in the adjacent segments cancels each other, producing a zero output. The output (induced) voltages v_1 and v_2 in the two foils of the stator have a carrier component at the supply frequency and a modulating component corresponding to the rotation of the disk. The latter (modulating component) can be extracted through demodulation (see Chapters 4 and 8) and converted into a proper pulse signal using pulse-shaping circuitry, as for an incremental encoder. When the rotating speed is constant, the two modulating components are periodic and nearly sinusoidal, with a phase shift of 90° (i.e., in quadrature). When the speed is not constant, the pulse width will vary with time.

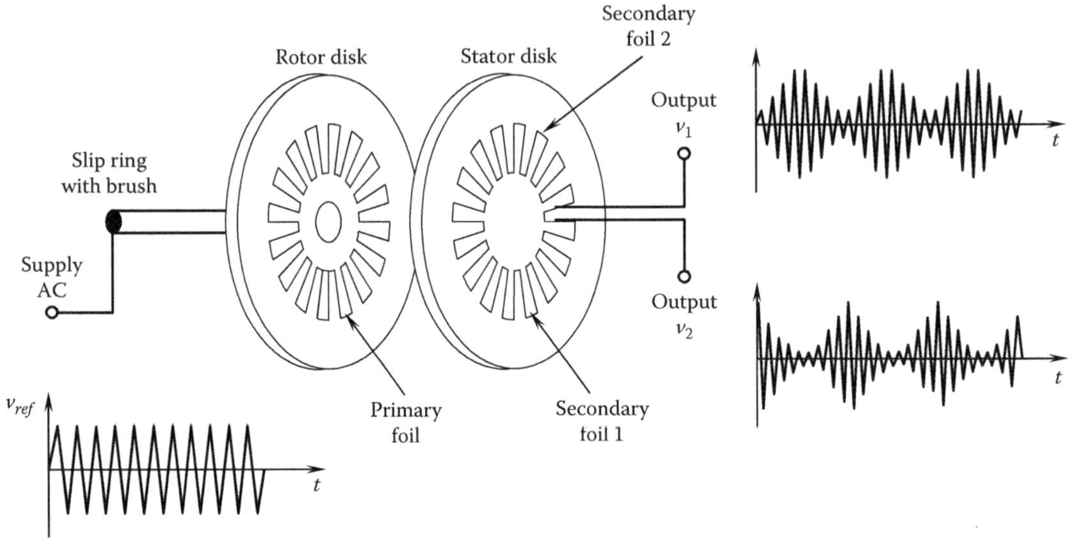

FIGURE 11.18
Schematic diagram of a digital resolver.

As in the case of an incremental encoder, angular displacement is determined by counting the pulses, and angular velocity is determined either by counting the pulses over a fixed time period (counter sampling period) or by timing a pulse. The direction of rotation is determined by the phase difference in the two modulating (output) signals. (In one direction, the phase shift is 90°; in the other direction, it is −90°.) Very fine resolutions (e.g., 0.0005°) may be obtained from a digital resolver, and it is usually not necessary to use step-up gearing or other techniques to improve the resolution. These transducers are usually more expensive than optical encoders. The use of a slip ring and brush to supply the carrier signal may be viewed as a disadvantage.

Consider the conventional resolver discussed in Chapter 8. Its outputs may be converted into digital form using appropriate hardware. Strictly speaking, such a device cannot be classified as a digital resolver.

11.8.3 Digital Tachometers

A pulse-generating transducer whose pulse train is synchronized with a mechanical motion may be treated as a digital transducer for motion measurement. In particular, pulse counting may be used for displacement measurement, and the pulse rate (or pulse timing) may be used for velocity measurement. As studied in Chapter 8, tachometers are devices for measuring angular velocities. According to this terminology, a shaft encoder (particularly, an optical incremental encoder) can be considered as a digital tachometer. According to popular terminology, however, a digital tachometer is a device that employs a toothed wheel to measure angular velocities.

Magnetic-induction digital tachometer: A schematic diagram of a digital tachometer is shown in Figure 11.19. This is a magnetic-induction, pulse tachometer of the variable-reluctance type. The teeth on the wheel are made of a ferromagnetic material. The two magnetic-induction (and variable reluctance) proximity probes are placed radially facing the teeth, at quarter-pitch apart (pitch is the tooth-to-tooth spacing). When the toothed

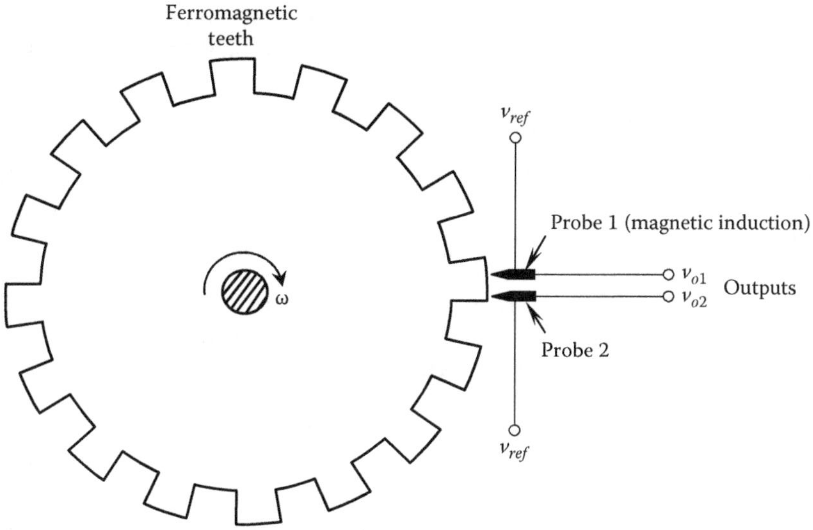

FIGURE 11.19
Schematic diagram of a pulse tachometer.

wheel rotates, the two probes generate output signals that are 90° out of phase (i.e., quadrature signals). One signal leads the other in one direction of rotation and lags the other in the opposite direction. In this manner, a directional reading (i.e., velocity rather than speed) is obtained. The speed is computed either by counting the pulses over a sampling period or by timing the pulse width, as in the case of an incremental encoder.

Eddy current digital tachometer: Alternative types of digital tachometers use eddy current proximity probes or capacitive proximity probes (see Chapter 8). In the case of an eddy current tachometer, the teeth of the pulsing wheel are made of (or plated with) electricity-conducting material. The probe consists of an active coil connected to an AC bridge circuit excited by a radio frequency (i.e., in the range 1–100 MHz) signal. The resulting magnetic field at radio frequency is modulated by the tooth-passing action. The bridge output may be demodulated and shaped to generate the pulse signal. In the case of a capacitive tachometer, the toothed wheel forms one plate of the capacitor; the other plate is the probe and is kept stationary. As the wheel turns, the gap width of the capacitor fluctuates. If the capacitor is excited by an AC voltage of high frequency (typically 1 MHz), a nearly pulse-modulated signal at that carrier frequency is obtained. This can be detected through a bridge circuit as for the eddy current probe, but using a capacitance bridge rather than an inductance bridge. In particular, by demodulating the output signal, the modulating signal can be extracted, which can be shaped to generate the pulse signal. The pulse signal that is generated in this manner is used in the angular velocity computation.

Advantages: The advantages of digital (pulse) tachometers over optical encoders include simplicity, robustness, immunity to environmental effects and other common fouling mechanisms (except magnetic effects), and low cost. Both are noncontacting devices.

Disadvantages: The disadvantages of a pulse tachometer include poor resolution—determined by the number of teeth and size (bigger and heavier than optical encoders)—mechanical errors due to loading, hysteresis (i.e., output is not symmetric and depends on the direction of motion), and manufacturing irregularities. *Note*: Mechanical loading will not be a factor if the toothed wheel already exists as an integral part of the original system that is sensed. The digital resolution depends on the word size used for DAQ.

11.8.4 Moiré Fringe Displacement Sensors

Suppose that a piece of transparent fabric is placed on another similar fabric. If one piece is moved or deformed with respect to the other, we will notice various designs of light and dark patterns (lines) in motion. Dark lines of this type are called moiré fringes. In fact, the French term *moiré* refers to a silklike fabric, which produces moiré fringe patterns. An example of a moiré fringe pattern is shown in Figure 11.20. Consider the rectilinear encoder, which was described earlier. When the window slits of one plate overlap with the window slits of the other plate, we get an alternating light and dark pattern. This is a special case of moiré fringes. A moiré device of this type may be used to measure rigid-body movements of one plate of the sensor with respect to the other.

Application of the moiré fringe technique is not limited to sensing rectilinear motions. This technology can be used to sense angular motions (rotations) and, more generally, distributed deformations (e.g., elastic deformations) of one plate with respect to the other. Consider two plates with gratings (optical lines) of identical pitch (spacing) p. Suppose that initially the gratings of the two plates perfectly coincide. Now, if one plate is deformed in the direction of the grating lines, the transmission of light through the two plates will not be altered. However, if a plate is deformed in the perpendicular direction to the grating lines, then the window width of that plate will be deformed accordingly. In this case, depending on the nature of the plate deformation, some transparent lines of one plate will be completely covered by the opaque lines of the other plate, and some other transparent

FIGURE 11.20
A moiré fringe pattern.

lines of the first plate will have coinciding transparent lines on the second plate. Thus, the observed image will have dark lines (moiré fringes) corresponding to the regions with clear–opaque overlaps of the two plates and bright lines corresponding to the regions with clear–clear overlaps of the two plates. The resulting moiré fringe pattern will provide the deformation pattern of one plate with respect to the other. Such 2D fringe patterns can be detected and observed by arrays of optical sensors using charge-coupled device (CCD) elements and by photographic means. In particular, since the presence of a fringe is a binary piece of information, binary optical sensing techniques (as for optical encoders) and digital imaging techniques may be used with these transducers. Accordingly, these devices may be classified as digital transducers. With the moiré fringe technique, very small resolutions (e.g., 0.002 mm) can be realized because finer line spacing (in conjunction with wider light sensors) can be used.

To further understand and analyze the fundamentals of moiré fringe technology, consider two grating plates with identical line pitch (spacing between the windows) p. Let us keep one plate stationary. This is the plate of *master gratings* (or reference gratings or main gratings). The other plate, which is the plate containing *I gratings* or model gratings, is placed over the fixed plate and rotated so that the I gratings form an angle α with the master gratings, as shown in Figure 11.21. The shown lines are in fact the opaque regions, which are identical in size and spacing to the windows in between the opaque regions. A uniform light source is placed on one side of the overlapping pair of plates and the light

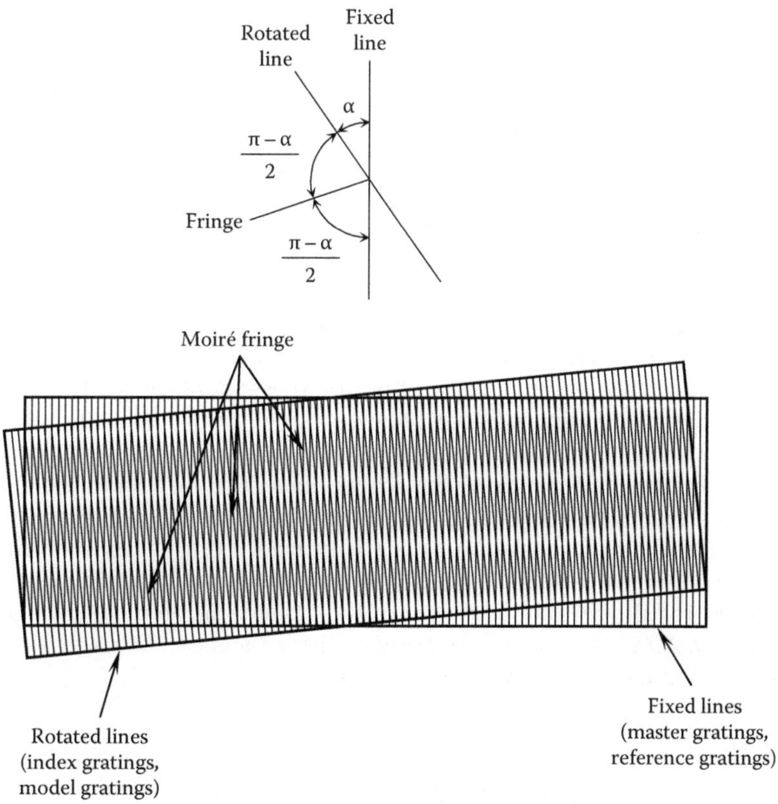

FIGURE 11.21
Formation of moiré fringes.

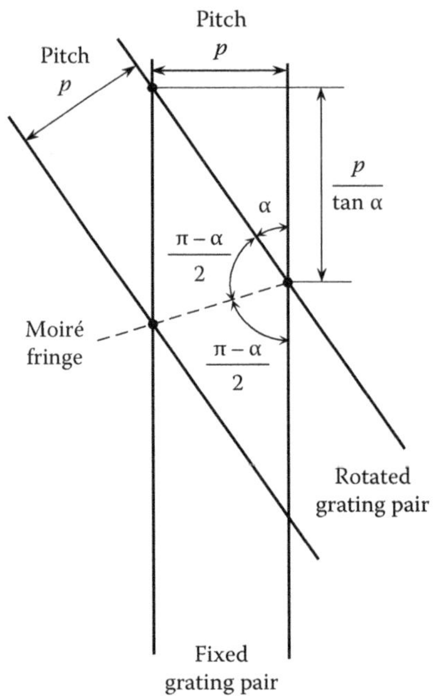

FIGURE 11.22
Orientation of moiré fringes.

transmitted through them is observed on the other side. Dark bands called moiré fringes are seen as a result, as in Figure 11.21.

A moiré fringe corresponds to the line joining a series of points of intersection of the opaque lines of the two plates because no light can pass through such points. This is further shown in Figure 11.22. Note that in the present arrangement the line pitch of the two plates is identical and equal to p. A fringe line that is formed is shown as the broken line in Figure 11.22. Since the line pattern in the two plates is identical, by symmetry of the arrangement, the fringe line should bisect the obtuse angle $(\pi-\alpha)$ formed by the intersecting opaque lines. In other words, a fringe line makes an angle of $(\pi-\alpha)/2$ with the fixed gratings. Furthermore, the vertical separation (or the separation in the direction of the fixed gratings) of the moiré fringes is seen to be $p/\tan\alpha$.

In summary, then the rotation of the I plate with respect to the reference plate can be measured by sensing the orientation of the fringe lines with respect to the fixed (master or reference) gratings. Furthermore, the period of the fringe lines in the direction of the reference gratings is $p/\tan\alpha$, and when the I plate is moved rectilinearly by a distance of one grating pitch, the fringes also shift vertically by its period of $p/\tan\alpha$ (see Figure 11.22). It is clear then that the rectilinear displacement of the I plate can be measured by sensing the fringe spacing. In a 2D pattern of moiré fringes, these facts can be used as local information to sense full-field motions and deformations.

Example 11.8

Suppose that each plate of a moiré fringe deformation sensor has a line pitch of 0.01 mm. A tensile load is applied to one plate in the direction perpendicular to the lines. Five moiré fringes are observed over a length of 10 cm of the moiré image under tension. What is the tensile strain in the plate?

Solution

There is one moiré fringe in every 10/5 = 2 cm of the plate. Hence, extension of a 2 cm portion of the plate is 0.01 mm, and

$$\text{Tensile strain} = \frac{0.01 \text{ mm}}{2 \times 10 \text{ mm}} = 0.0005\ \varepsilon = 500\ \mu\varepsilon$$

In this example, we have assumed that the strain distribution (or deformation) of the plate is uniform. Under nonuniform strain distributions, the observed moiré fringe pattern generally will not be parallel straight lines but rather complex shapes.

11.9 Hall-Effect Sensors

Consider a semiconductor element subject to a DC voltage v_{ref}. If a magnetic field is applied perpendicular to the direction of this voltage, a voltage v_o will be generated in the third orthogonal direction, within the semiconductor element. This is known as the Hall effect (observed by E.H. Hall in 1879). A schematic representation of a Hall-effect sensor is shown in Figure 11.23.

11.9.1 Hall-Effect Motion Sensors

The Hall effect may be used for motion sensing in many ways, for example, as an analog proximity sensor, a limit switch (digital), or a shaft encoder. Because the output voltage v_o

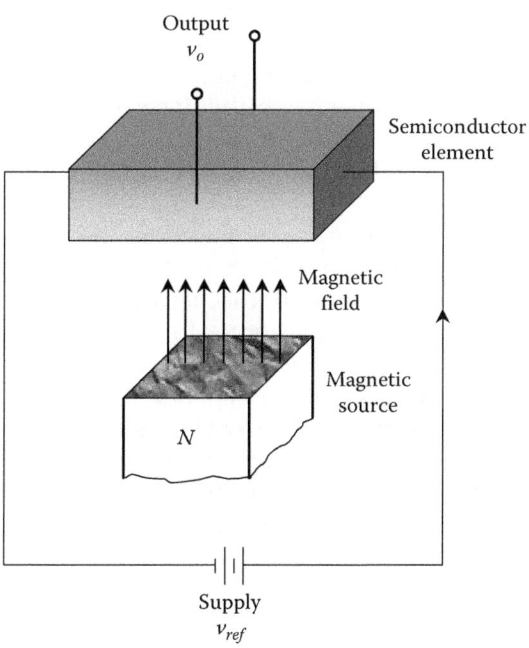

FIGURE 11.23
Schematic representation of a Hall-effect sensor.

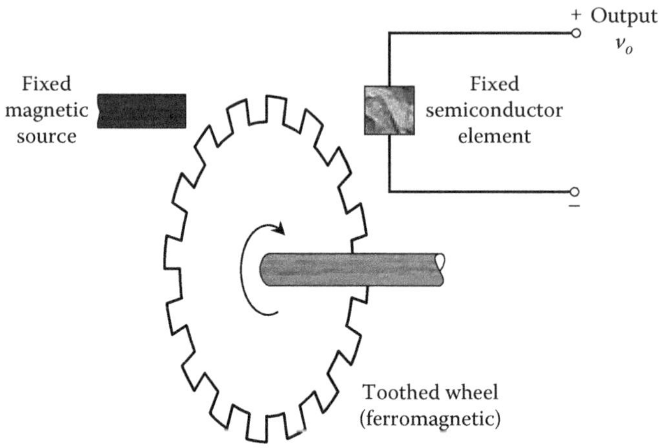

FIGURE 11.24
A Hall-effect shaft encoder or digital tachometer.

in Figure 11.23 increases as the distance from the magnetic source to the semiconductor element decreases, the output signal v_o can be used as a measure of proximity. This is the principle behind an analog proximity sensor. Alternatively, a certain threshold level of the output voltage v_o can be used to generate a binary output, which represents the presence/absence of an object. This principle is used in a digital limit switch. Alternatively, a shaft encoder based on Hall effect can be constructed by using a toothed ferromagnetic wheel (as in a digital tachometer) to alter the magnetic flux.

The longitudinal arrangement of a proximity sensor, in which the moving element approaches the sensor head-on, is not suitable when there is the danger of overshooting the target, since it will damage the sensor. A more desirable configuration is the lateral arrangement, in which the moving member slides by the sensing face of the sensor. The sensitivity will be lower, however, with this lateral arrangement. The relationship between the output voltage v_o and the distance x of a Hall-effect sensor, measured from the moving member, is nonlinear. Linear Hall-effect sensors use calibration to linearize their output.

A practical arrangement for a motion sensor based on the Hall effect would be to have the semiconductor element and the magnetic source fixed relative to one another in a single package. As a ferromagnetic member approaches the sensor package, the flux linkage will vary. The output voltage v_o will change accordingly. This arrangement is suitable for both an analog proximity sensor and a limit switch. By using a toothed ferromagnetic wheel as in Figure 11.24 to change v_o and then by shaping the resulting signal, it is possible to generate a pulse train in proportion to the wheel rotation. This provides a shaft encoder or a digital tachometer. Apart from the familiar applications of motion sensing, Hall-effect sensors are used for *electronic commutation* of brushless DC motors where the field circuit of the motor is appropriately switched depending on the angular position of the rotor with respect to the stator.

11.9.2 Properties

The sensitivity of a practical Hall-effect sensor element is of the order of 10 V/T (*Note*: T denotes "tesla," which is the unit of magnetic flux density; 1 T = 1 Wb/m^2). For a Hall-effect device, the temperature coefficient of resistance is positive and the temperature coefficient of sensitivity is negative. In view of these properties, self-compensation for temperature may be achieved (as for a semiconductor strain gauge—see Chapter 9).

Hall-effect motion transducers are rugged devices and have many advantages. They are not affected by rate effects (specifically, the generated voltage is not affected by the rate of change of the magnetic field). In addition, their performance is not severely affected by common environmental factors, except magnetic fields. They are noncontacting sensors with associated advantages as mentioned earlier. Some hysteresis will be present, but it is not a serious drawback in digital transducers. Another possible drawback is the contamination of the sensor output by ambient magnetic fields. Miniature Hall-effect devices (mm scale) are available.

11.10 Image Sensors

An image of an object is indeed a valuable source of information about that object. In this context, an imaging device is a sensor, and an image is the sensed data. Depending on the imaging device, an image can be generated in many forms such as optical, thermal or infrared, x-ray, ultraviolet, acoustic, ultrasound, and so on. Since the image-processing methods are rather similar among these imaging devices, we will consider here only the *digital camera* as a sensor. This is a very popular optical imaging device and is used in a variety of engineering applications such as industrial process monitoring and vision-guided robotics.

11.10.1 Image Processing and Computer Vision

An image may be analyzed to obtain a more refined image from which useful information such as edges, contours, areas, and other geometrical information can be determined. This is called image processing. Computer vision goes beyond image processing and performs such operations as object recognition, pattern recognition and classification, abstraction, and knowledge-based decision-making using information extracted through image processing. It follows that computer vision involves higher-level operations than image processing and is akin to what humans infer based on what they see.

11.10.2 Image-Based Sensory System

A complete image-based sensory system consists of a camera (e.g., CCD camera or complementary metal–oxide–semiconductor [CMOS] camera), DAQ system (e.g., frame grabber), computer, and associated software. Such a system is schematically shown in Figure 11.25. Not included in the figure are other useful components such as a structured lighting source, which may be needed to capture clear images of good quality, without shadows and so on.

11.10.3 Camera

A digital camera has an array or matrix of semiconductor elements that are sensitive to the brightness of light coming from an object (through the camera lens). The image sensor of

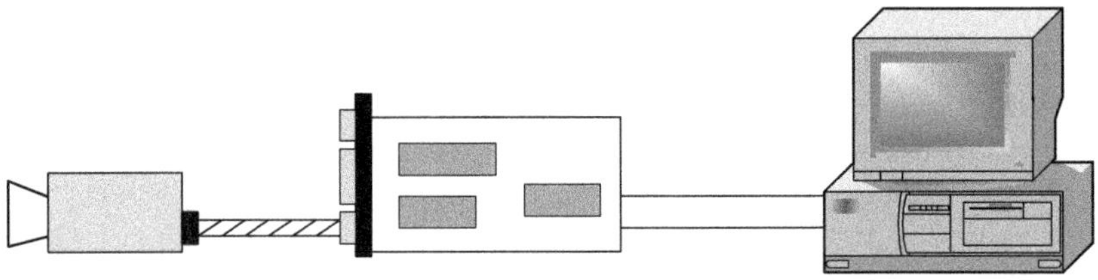

FIGURE 11.25
A camera-based sensory system.

today's digital cameras commonly uses CCD technology or CMOS technology. Less common technologies such as charge injection device are found as well. CMOS image sensors are less expensive because they employ the same processes that are used to mass produce IC chips and they use less power. Also, a CMOS image sensor may be considered as a matrix of digital elements that correspond to the picture elements (pixels), which can be directly accessed (in parallel) for image data retrieval. On the other hand, the generated charges in the cells of a CCD image sensor are retrieved in a somewhat sequential fashion and then digitized for image generation. The CCD technology is more mature and generates better quality images than those from the CMOS technology. However, once a digital camera generates an image, it can be acquired and processed by a computer in the same manner (e.g., using a frame grabber board or a USB link) regardless of what technology is used in its image sensor. For this reason, we will only consider CCD image sensors in the subsequent discussion.

11.10.3.1 Charge-Coupled Device Image Sensor

Suppose that a 2D beam of light coming from the sensed object is directed by the camera lens onto the CCD matrix (e.g., 4000×4000) located on the focal plane of the lens in the back of the camera. Each cell of the CCD sensor generates a charge that is proportional to the brightness of the light. An integrated circuit device in the camera reads these charge levels of the cells row by row (from the bottom to the top row, sequentially) through a row-shifting operation controlled and synchronized by a clock and other hardware. The analog signal from each CCD cell is digitized and represented as a "picture element" or *pixel*. The number of bits in a pixel is representative of the number of *gray levels* it can store. For example, an 8-bit pixel can represent $2^8 = 256$ gray levels from 0 to 255 (black to white). This procedure of generating the pixels of a 2D image is represented in Figure 11.26.

11.10.4 Image Frame Acquisition

Image pixels from the buffer of a CCD camera are arranged into an image frame of digital data and provided to the image-processing computer. This DAQ device (often called a "frame grabber board") may be placed in the card cage of the image-processing computer. Once the associated driver software is located in the computer, images can be acquired at high speed (e.g., 200 MB/s). With a USB camera, the USB image stream acquisition process

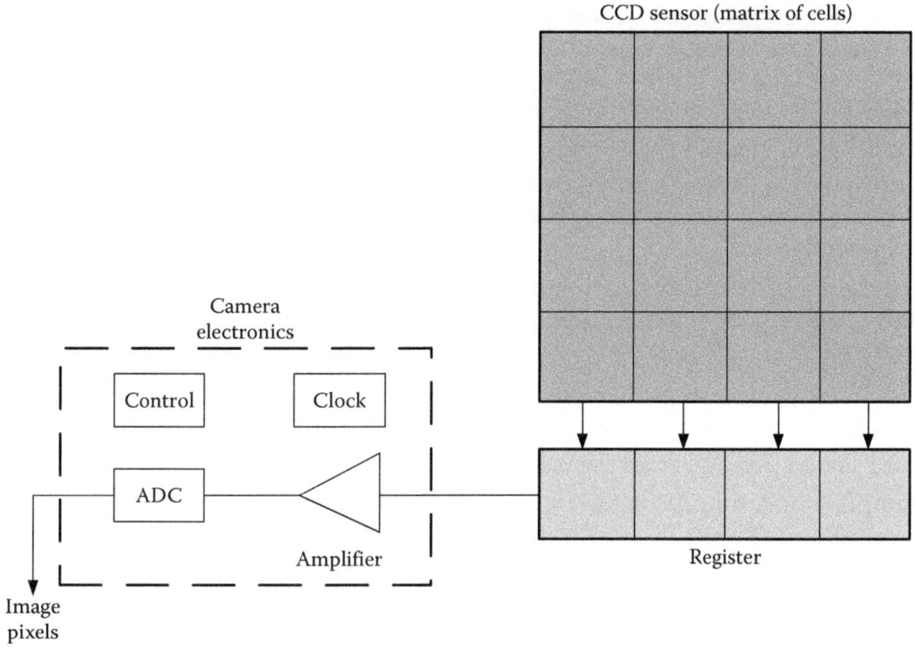

FIGURE 11.26
Digital image generation by a charge-coupled device camera.

may be carried out in a convenient manner. The process is somewhat slow where images are copied from the camera storage into the computer storage as a process of file transfer under the control of the computer.

11.10.5 Color Images

A grayscale image can be represented by a single image frame. On the other hand, at least three image frames are needed to represent a color image. For example, in the RGB model, a red (R) image, a green (G) image, and a blue (B) image are formed using red, green, and blue filters. The resulting three separate image frames can be combined to form the original color image. Even though human eye is quite sensitive to the colors R, G, and B, humans cannot typically perceive/describe visual images in terms of their RGB components. In terms of human perception/description of a visual image, a more appropriate model is the HIS model. In this model, hue (H) is a representative of the dominant color in the image, saturation (S) represents the degree of white light mixed with a dominant color in the image, and intensity (I) represents the level of brightness of the image. There are analytical relationships that convert an RGB model into an HIS model.

11.10.6 Image Processing

Some degree of "analog" image processing is also carried out by the electronics in a digital camera (e.g., analog filtering and amplification). However, the focus here is the "digital" image processing done by a computer. The objective of digital image processing is to remove unwanted elements and noise in the image, enhance the important features, and extract the needed geometric information from the processed data. Several useful operations of image processing are listed in the following:

1. Filtering (to remove noise and enhance the image) including directional filtering (to enhance edges, for edge detection)
2. Thresholding (to generate a two-level black-and-white image where the gray levels above a specified threshold are assigned white and those below the threshold are assigned black)
3. Segmentation (to subdivide an enhanced image, identify geometric shapes/objects, and capture properties such as area and dimensions of the identified geometric entities)
4. Morphological processing (sequential shrinking, filtering, stretching, etc., to prune out unwanted image components and extract those that are important)
5. Subtraction (e.g., subtract the background from the image)
6. Template matching (to match a processed image to a template—useful in object recognition)
7. Compression (to reduce the quantity of data that is needed to represent the useful information of an image)

11.10.7 Some Applications

The applications of image-based sensors are numerous. Several are listed as follows:

1. Measurement of a location of an object for grasping, manipulating, transporting, assembling, mechanical processing (machining, cutting, grinding), etc.
2. Measurement/estimation of size, shape, weight, color, texture, firmness, etc., for quality assessment or grading of a product.
3. Visual servoing. Here, the actual position of an object is measured (using camera images) and compared with the position of a robotic end effector (gripper, hand, tool, etc.). The difference (error) is used to generate a motion command for the robot so that the end effector would reach the object. The same approach can be used in the navigation of mobile robots, automated vehicles, unmanned aerial vehicles, etc.
4. Object recognition in various applications of security, safety, machine health monitoring, and automated processing.
5. In telemedicine, to examine a patient from a distant location.

Summary Sheet

Digital transducer: A measuring device that produces a discrete or digital output without using an ADC; or a transducer whose output is a pulse signal or a count; or a transducer whose output is a frequency (which can be precisely converted into a count or a rate).

Advantages of digital transducers: No quantization error; less susceptible to noise, disturbances, or parameter variation (bits → two states → noise threshold = half a bit); complex signal processing with very high accuracy and speed (hardware implementation is faster than software implementation); high reliability (fewer analog hardware components); large amounts of data can be stored, maintained for very long periods of time accurately; fast and accurate data transmission through existing communication means over long distances; use low voltages (e.g., 0–12 V DC) and low power; typically low overall cost.

Incremental encoder: Output is a pulse signal in proportion to displacement.

Absolute encoder: Output is a digital word representing the absolute displacement.

Techniques of encoder signal generation: Optical (photosensor); sliding contact (electrical conducting); magnetic saturation (reluctance); proximity sensor.

Direction sensing: Use quadrature signals (90° out of phase). Methods: (1) phase angle between the two signals, (2) clock counts to two adjacent rising edges of the two signals (If $n_1 > n_2 - n_1 \Rightarrow$ cw rotation, if $n_1 < n_2 - n_1 \Rightarrow$ ccw rotation), (3) rising or falling edge of one signal when the other is at "high," and (4) for a high-to-low transition of one signal check, the next transition of the other signal.

Displacement measurement: Angular position corresponding to a count of n pulses $\theta = (n/M)\theta_{max}$, range of the encoder $= \pm\theta_{max}$.

Encoder displacement resolution: Corresponds to a unit change in pulse count $\rightarrow$ $\Delta\theta = (\theta_{max}/M)$, $\Delta\theta_d = (\theta_{max} - \theta_{min})/(2^{r-1} - 1)$.

Digital resolution: Corresponds to a unit change in the bit value. $M = 2^{r-1} \rightarrow \Delta\theta_d = (\theta_{max}/2^{r-1})$. Typically, $\theta_{max} = \pm 180°$ or $360° \rightarrow \Delta\theta_d = (180°/2^{r-1}) = (360°/2^r)$. $\Delta\theta_d = (\theta_{max} - \theta_{min})/(2^{r-1} - 1)$ with $\theta_{min} = (\theta_{max}/2^{r-1})$.

Physical resolution: Governed by number of windows N in code disk $\rightarrow \Delta\theta_p = (360°/4N)$ with quadrature signals.

With step-up gearing: $\Delta\theta_p = (360°/4pN)$ and $\Delta\theta_d = (180°/p2^{r-1}) = (360°/p2^r)$, p = gear ratio.
Note: Max count $\leftarrow$ encoder disk full rotation.

Velocity measurement: For pulse-counting method,

$$\text{Speed } \omega = \frac{2\pi/N}{T/n} = \frac{2\pi n}{NT}$$

resolution

$$\Delta\omega_c = \frac{2\pi}{NT}$$

For pulse-timing method,

$$\text{Speed } \omega = \frac{2\pi/N}{m/f} = \frac{2\pi f}{Nm}$$

resolution

$$\Delta\omega_t = \frac{2\pi f}{Nm} - \frac{2\pi f}{N(m+1)} = \frac{2\pi f}{Nm(m+1)} \rightarrow \Delta\omega_t \approx \frac{2\pi f}{Nm^2} = \frac{N\omega^2}{2\pi f}$$

With quadrature signals, replace N by $4N$.

With step-up gearing:

$$\text{Pulse-counting method,} \quad \omega = \frac{2\pi n}{pNT}$$

$$\text{Pulse-timing method,} \quad \omega = \frac{2\pi f}{pNm}$$

Velocity resolutions:

$$\text{For pulse-counting method,} \quad \Delta\omega_c = \frac{2\pi(n+1)}{pNT} - \frac{2\pi n}{pNT} = \frac{2\pi}{pNT}$$

$$\text{For pulse-timing method,} \quad \Delta\omega_t = \frac{2\pi f}{pNm} - \frac{2\pi f}{pN(m+1)} = \frac{2\pi f}{pNm(m+1)} \cong \frac{pN}{2\pi f}\omega^2$$

Digital binary transducers: Two-state sensors. Electromechanical switches; photoelectric devices; magnetic (Hall-effect, eddy current) devices; capacitive devices; ultrasonic devices.

Configurations: Through (opposed); reflective (reflex); diffuse (proximity, interceptive).

Factors governing performance: Sensing range (operating distance between sensor and object); response time; sensitivity; linearity; size and shape of object; material of object (e.g., color, reflectance, permeability, permittivity); orientation and alignment (optical axis, reflector, object); ambient conditions (light, dust, moisture, magnetic field, etc.); signal conditioning considerations (modulation, demodulation, shaping, etc.); reliability, robustness, and design life.

Digital resolver: Mutual-induction encoder, Inductosyn. Has a stationary disk (stator), and a rotating disk (rotor) coupled to sensed object. Rotor has fine imprinted electric conductor foil (pulse shaped, closely spaced, connected to high-frequency AC carrier). Stator has two printed patterns identical to rotor pattern, but with quarter-pitch shift.

Digital tachometer: Pulse tachometer. Magnetic-induction type uses ferromagnetic teeth on wheel. Probe is a magnetic-induction proximity sensor. Alternatively, eddy current proximity probe (with conducting teeth) or capacitive proximity probe with dielectric teeth may be used.

Moiré fringe sensors: Window slits of one plate overlap with window slits of other plate → alternating light and dark pattern (special case of moiré fringes). Used in sensing rectilinear motions, angular motions (rotations), and distributed deformations (e.g., elastic deformations) of one plate with respect to the other. 2D fringe patterns can be detected and observed by arrays of optical sensors (digital camera).

Hall effect: If a semiconductor element is subject to a DC voltage in one direction and the magnetic field is applied in perpendicular direction of this voltage, a voltage will be generated in the third orthogonal direction.

Hall-effect sensor: Analog proximity sensor, a limit switch (digital), shaft encoder, magnetic field sensor (e.g., for dc motor commutation).

Image sensor: Optical, thermal or infrared, X-ray, ultraviolet, acoustic, ultrasound, etc. Image processing methods are similar.

Digital camera: Charge-coupled device (CCD) technology and complementary metal oxide semiconductor (CMOS) technology are common.

CMOS image sensor: Employs same processes as IC chips → less expensive. Less power. Matrix of digital elements corresponding to picture elements (pixels), directly accessed (in parallel) for image data retrieval.

CCD image sensor: Generated charges in sensor cells are retrieved sequentially and digitized. More mature technology. Generates better quality images than those from the CMOS technology.

Image frame acquisition: Once digital camera generates an image, it is acquired and processed by computer in similar manner (e.g., using a frame grabber board or a USB link).

Image processing: Filtering (remove noise and enhance image) including directional filtering (enhance edges); thresholding (generate black-and-white image ← grey levels > threshold are white, and < threshold are black; segmentation (subdivide enhanced image, identify geometric shapes/objects, and capture properties—area, dimensions of identified entities); morphological processing (sequential shrinking, filtering, stretching, etc. to prune out unwanted image components and extract important ones); subtraction (e.g., subtract background from image); template Matching (useful in object recognition); compression (reduce the needed data to represent the useful information).

Problems

11.1 Identify active transducers among the following types of shaft encoders and justify your claims. Also, discuss the relative merits and drawbacks of the following four types of encoders:

(a) Optical encoder
(b) Sliding contact encoder
(c) Magnetic encoder
(d) Proximity sensor encoder

11.2 Consider the two quadrature pulse signals (say, *A* and *B*) from an incremental encoder. Using sketches of these signals, show that in one direction of rotation, signal *B* is at a high level during the up transition of signal *A* and, in the opposite direction of rotation, signal *B* is at a low level during the up transition of signal *A*. The direction of motion can be determined in this manner, by using level detection of one signal during the up transition of the other signal.

11.3 Explain why the speed resolution of a shaft encoder depends on the speed itself. What are some of the other factors that affect speed resolution? The speed of a DC motor was increased from 50 to 500 rpm. How would the speed resolution change if the speed were measured using an incremental encoder,

(a) By the pulse-counting method?
(b) By the pulse-timing method?

11.4 Describe methods of improving the displacement resolution and the velocity resolution in an encoder. An incremental encoder disk has 5000 windows. The word size of the output data is 12 bits. What is the angular displacement resolution of the device? Assume that quadrature signals are available but that no interpolation is used.

11.5 An incremental optical encoder that has N windows per track is connected to a shaft through a gear system with gear ratio p. Derive formulas for calculating angular velocity of the shaft by the

(a) Pulse-counting method
(b) Pulse-timing method

What is the speed resolution in each case? What effect does step-up gearing have on the speed resolution?

11.6 What is hysteresis in an optical encoder? List several causes of hysteresis and discuss ways to minimize hysteresis.

11.7 An optical encoder has n windows per centimeter of diameter (in each track). What is the eccentricity tolerance e below which the readings are not affected by eccentricity error?

11.8 Show that in the single-track, two-sensor design of an incremental encoder, the phase angle error (in quadrature signals) due to eccentricity is inversely proportional to the second power of the radius of the code disk for a given window density. Suggest a way to reduce this error.

11.9 Suppose that an encoder with 1000 windows in its track is capable of providing quadrature signals. What is the displacement resolution $\Delta\theta_r$ in radians? Obtain a value for the nondimensional eccentricity e/r below which the eccentricity error has no effect on the sensor reading. For this limiting value, what is $\Delta\theta_r/(e/r)$? Typically, the values for this parameter, as given by the encoder manufacturer, range from 3 to 6. *Note*: e is the track eccentricity and r is the track radius.

11.10 What is the main advantage of using a gray code instead of the straight binary code in an encoder? Give a table corresponding to a gray code that is different from what is given in Table 11.2 for a 4-bit absolute encoder. What is the code pattern on the encoder disk in this case?

11.11 Discuss construction features and operation of an optical encoder for measuring rectilinear (straight line) displacements and velocities.

11.12 A particular type of multiplexer can handle 96 sensors. Each sensor generates a pulse signal with variable pulse width. The multiplexer scans the incoming pulse sequences, one at a time, and passes the information onto a control computer.

(a) What is the main objective of using a multiplexer?
(b) What type of sensors may be used with this multiplexer?

11.13 A centrifuge is a device that is used to separate components in a mixture. In an industrial centrifugation process, the mixture to be separated is placed in the centrifuge and rotated at high speed. The centrifugal force on a particle depends on the mass, radial location, and angular speed of the particle. This force is responsible for separating the particles in the mixture.

Angular motion and the temperature of the container are the two key variables that have to be controlled in a centrifuge. In particular, a specific centrifugation curve is used, which consists of an acceleration segment, a constant-speed segment, and a braking (deceleration) segment, and this corresponds to a trapezoidal speed profile. An optical encoder may be used as the sensor for microcontroller-based speed control in the centrifuge. Discuss whether an absolute encoder is preferred for this purpose. Give the advantages and possible drawbacks of using an optical encoder in this application.

11.14 Suppose that a feedback control system (see Figure P11.14) is expected to provide an accuracy within $\pm\Delta y$ for a response variable y. Explain why the sensor that measures y should have a resolution of $\pm(\Delta y/2)$ or better for this accuracy to be possible. An x–y table has a travel of 2 m. The feedback control system is expected to provide an accuracy of ±1 mm. An optical encoder is used to measure the position for feedback in each direction (x and y). What is the minimum bit size that is required for each encoder output buffer? If the motion sensor that is used is an absolute encoder, how many tracks and how many sectors should be present on the encoder disk?

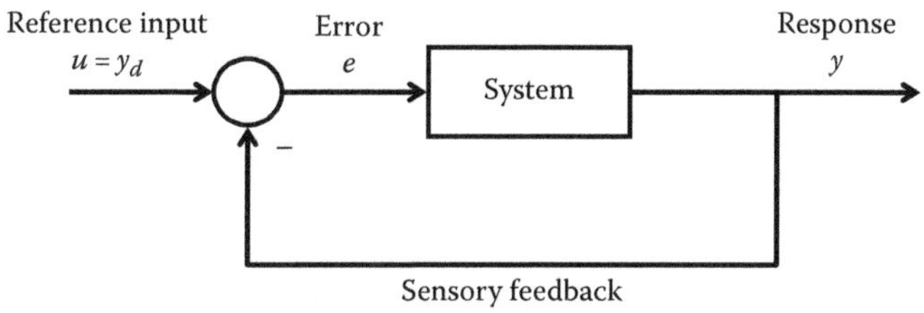

FIGURE P11.14
A feedback control loop.

11.15 Encoders that can provide 50,000 counts/turn with ±1 count accuracy are commercially available. What is the resolution of such an encoder? Describe the physical construction of an encoder that has this resolution.

11.16 The pulses generated by the coding disk of an incremental optical encoder are approximately triangular (actually, upward shifted sinusoidal) in shape. Explain the reason for this. Describe a method for converting these triangular (or shifted sinusoidal) pulses into sharp rectangular pulses.

11.17 Explain how the resolution of a shaft encoder can be improved by pulse interpolation. Specifically, consider the arrangement shown in Figure P11.17. When the masking windows are completely covered by the opaque regions of the moving disk, no light is received by the photosensor. The peak level of light is received when the windows of the moving disk coincide with the windows of the masking disk. The variation in the light intensity from the minimum level to the peak level is approximately linear (generating a triangular pulse) but more accurately sinusoidal and may be given by

$$v = v_o\left(1 - \cos\frac{2\pi\theta}{\Delta\theta}\right)$$

where

- θ is the angular position of the encoder window with respect to the masking window, as shown
- $\Delta\theta$ is the window pitch angle

In the context of rectangular pulses, the pulse corresponds to the motion in the interval $\Delta\theta/4 \le \theta \le 3\Delta\theta/4$. By using this sinusoidal approximation for a pulse, as given previously, show that one can improve the resolution of an encoder indefinitely simply by measuring the shape of each pulse at clock cycle intervals using a high-frequency clock signal.

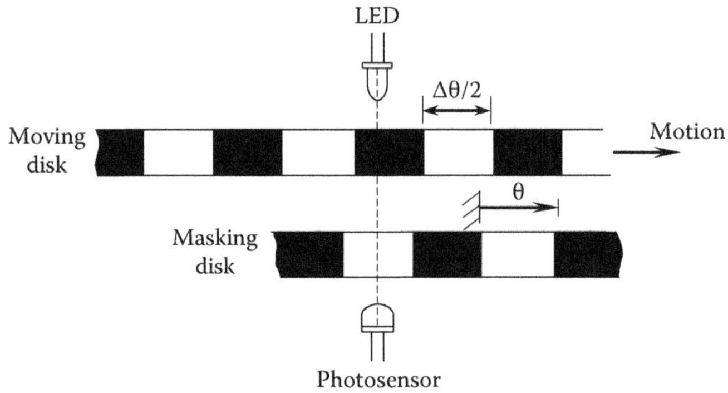

FIGURE P11.17
An encoder with a masking disk.

11.18 A Schmitt trigger is a semiconductor device that can function as a level detector or a switching element, with hysteresis. The presence of hysteresis can be used, for example, to eliminate chattering during switching as caused by noise in the switching signal. In an optical encoder, a noisy signal that is detected by the photosensor may be converted into a clean signal of rectangular pulses by this means. The I/O characteristic of a Schmitt trigger is shown in Figure P11.18a. If the input signal is as shown in Figure P11.18b, determine the output signal.

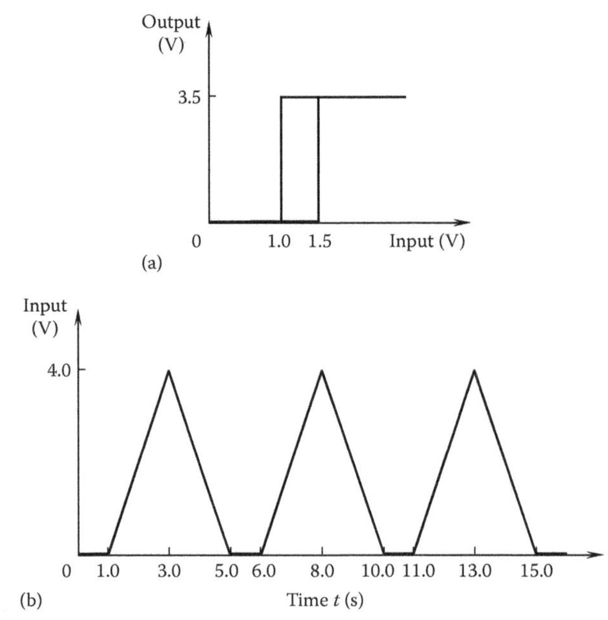

FIGURE P11.18
(a) The input–output characteristic of a Schmitt trigger and (b) a triangular input signal.

11.19 Displacement sensing and speed sensing are essential in a position servo. If a digital controller is employed to generate the servo signal, one option would be to use an analog displacement sensor and an analog speed sensor, along with ADCs to produce the necessary digital feedback signals. Alternatively, an incremental encoder may be used to provide both displacement and speed feedbacks. In this latter case, ADCs are not needed. Encoder pulses will provide interrupts to the digital controller. Displacement is obtained by counting the interrupts. The speed is obtained by timing the interrupts. In some applications, analog speed signals are needed. Explain how an incremental encoder and a frequency-to-voltage converter may be used to generate an analog speed signal.

11.20 Compare and contrast an optical incremental encoder against a potentiometer, by giving advantages and disadvantages, for an application involving the sensing of a rotatory motion.

A schematic diagram for the servo control loop of one joint of a robotic manipulator is given in Figure P11.20. The motion command for each joint of the robot is generated by the robot controller, in accordance with the required trajectory. An optical incremental encoder is used for both position and velocity feedback in each servo loop. For a six-degree-of-freedom robot, there will be six such servo loops. Describe the function of each hardware component shown in the figure and explain the operation of the servo loop.

After several months of operation, the motor of one joint of the robot was found to be faulty. An enthusiastic engineer quickly replaced the motor with an identical one without realizing that the encoder of the new motor was different. In particular, the original encoder generated 200 pulses/rev, whereas the new encoder generated 720 pulses/rev. When the robot was operated, the engineer noticed an erratic and unstable behavior at the repaired joint. Discuss reasons for this malfunction and suggest a way to correct the situation.

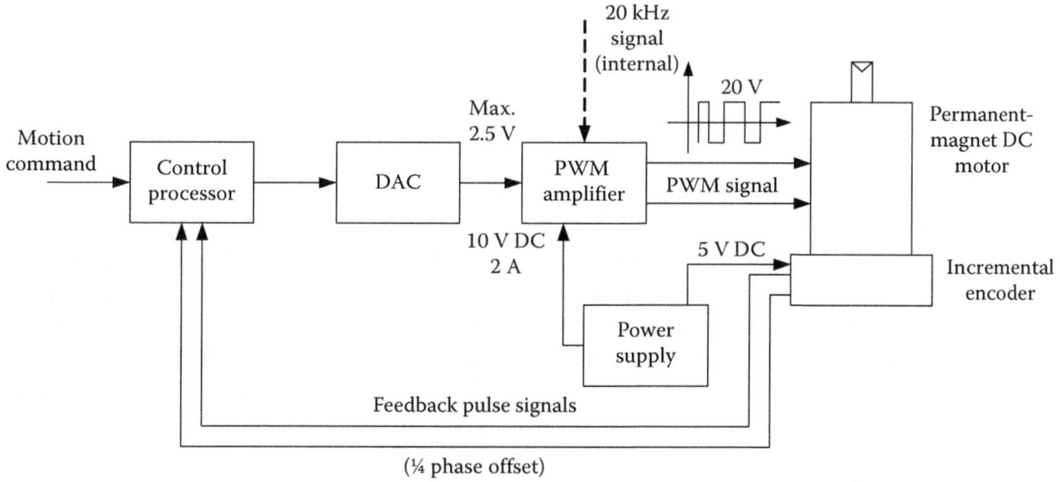

FIGURE P11.20
A servo loop of a robot.

11.21 (a) A position sensor is used in a microcontroller-based feedback control system for accurately moving the cutter blades of an automated meat-cutting machine. The machine is an integral part of the production line of a meat-processing plant. What are the primary considerations in selecting the position sensor for this application? Discuss the advantages and disadvantages of using an optical encoder in comparison to a linear variable differential transformer (see Chapter 8) in this context.

(b) Figure P11.21 illustrates one arrangement of the optical components in a linear incremental encoder.

The moving code plate has uniformly spaced windows as usual, and the fixed masking plate has two groups of identical windows, one above each of the two photodetectors. These two groups of fixed windows are positioned in half-pitch out of phase so that when one detector receives light from its source directly through the aligned windows of the two plates, the other detector has the light from its source virtually blocked by the masking plate.

Explain the purpose of the two sets of photodiode-detector units, giving a schematic diagram of the necessary electronics. Can the direction of motion be determined with the arrangement shown in Figure P11.21? If so, explain how this could be done. If not, describe a suitable arrangement for detecting the direction of motion.

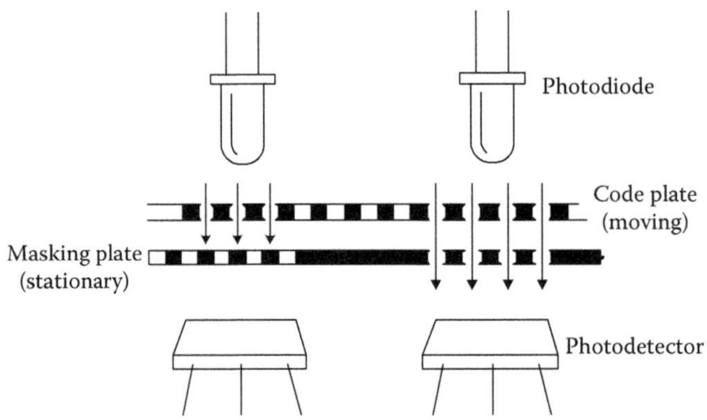

FIGURE P11.21
Photodiode-detector arrangement of a linear optical encoder.

11.22 (a) What features and advantages of a digital transducer will distinguish it from a purely analog sensor?

(b) Consider a linear incremental encoder that is used to measure rectilinear (straight line) positions and speeds. The moving element is a nonmagnetic plate containing a series of identically magnetized areas uniformly distributed along its length. The pickoff transponder is a mutual-induction-type proximity sensor (i.e., a transformer) consisting of a toroidal core with a primary winding and a secondary winding. A schematic diagram of the encoder is shown in Figure P11.22. The primary excitation v_{ref} is a high-frequency sine wave.

Explain the operation of this position encoder, clearly indicating what types of signal conditioning would be needed to obtain a pure pulse signal. Also, sketch the output v_o of the proximity sensor as the code plate moves very slowly. Which position of the code plate does a high value of the pulse signal represent and which position does a low value represent?

Suppose that the pulse period timing method is used to measure speed (v) using this encoder. The pitch distance of the magnetic spots on the plate is p, as shown in Figure P11.22. If the clock frequency of the pulse period timer is f, give an expression for the speed v in terms of the clock cycle count m.

Show that the speed resolution Δv for this method may be approximated by $\Delta v = (v^2/pf)$.

It follows that the dynamic range $v/\Delta v = pf/v$.

If the clock frequency is 20 MHz, the code pitch is 0.1 mm and the required dynamic range is 100 (i.e., 40 dB), what is the maximum speed in m/s that can be measured by this method?

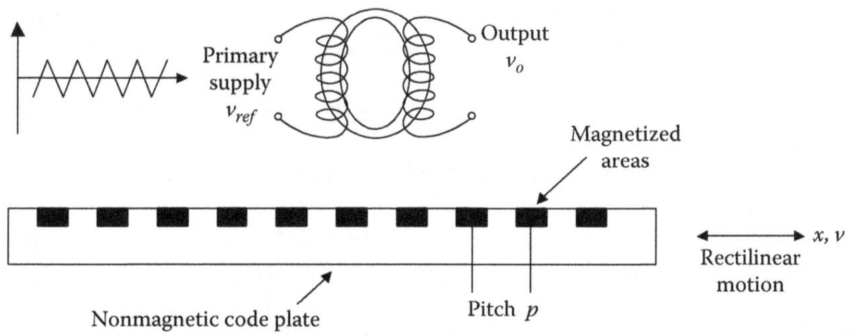

FIGURE P11.22
A linear incremental encoder of the magnetic-induction type.

11.23 Consider an absolute encoder of 4-bit output and a very accurate rotary potentiometer with a 4-bit ADC, for use in measuring angular motions of an object up to one full rotation. It is stated that even though the pot is very accurate, the absolute encoder may provide more accurate results in this application because it precisely gives the absolute position of the object. Examine this claim.

11.24 (a) Define the term *resolution* of a sensor or transducer.
(b) List four advantages of digital transducers over analog sensors.
(c) An incremental encoder provides quadrature signals v_1 and v_2. Counting of clock pulses begins at a rising edge of a pulse of v_1. The clock-pulse count up to the next rising edge of v_2 is n_1, and the total clock-pulse count up to the next rising edge of v_1 is n_2. Consider the two cases: (i) $n_1 = 60$ and $n_2 = 100$ and (ii) $n_1 = 80$ and $n_2 = 100$. Is the disk rotating in the same direction or opposite direction in the two cases? Give sketches (idealized) of the encoder signals to justify your answer.

11.25 The code disk of an incremental encoder (optical) has 1500 windows. The word size of the output register of the encoder is 12 bits, and it corresponds to a full rotation (signed) of the code disk. *Note*: Assume throughout that quadrature signals are available and used in the resolution determination.

(a) Giving the key steps, determine the overall resolution of the encoder in the measurement of the angle of rotation of an object.
(b) If backlash-free step-up gearing with gear ratio of 5 is used from the measured object to the encoder disk, what is the overall displacement resolution then? Give the main steps of your derivation.

11.26 You are given an incremental encoder to measure the angle of rotation and angular speed (in rad and rad/s) including direction (of a shaft). The pins of the encoder are (1) ground, (2) index, (3) A channel, (4) +5 V DC power, and (5) B channel.

The A channel and the B channel give the quadrature pulse signals (i.e., 90° out of phase). The direction of rotation may be determined as follows: When B channel output is "low," if the A channel transition is from low to high → cw rotation or if the A channel transition is from high to low → ccw rotation.

In your application, the following have to be performed: (1) Acquire both pulse sequences (A and B) from the encoder (into a microcontroller, which you need to buy). (2) Determine the direction of rotation. (3) Compute the angle of rotation (in radians, from a reference position). (4) Compute the speed (in rad/s) at any time.

(a) Do a search and select other main hardware (in addition to the incremental optical encoder) that you will need.
(b) Give a sketch to indicate how these hardware components are connected in the final system.
(c) Give a pseudocode (or a C++ code or a MATLAB® code if you wish) to perform the four operations listed earlier, in the application.

11.27 Object counting on the moving conveyor of a production process is done as follows: The objects are placed single file with some spacing. The conveyor moves at the required speed of the process. A fixed sensor senses an object and the product count is incremented by 1. Suppose that the object size (in the direction of the conveyor motion) ranges from 2 to 4 cm and the product spacing on the conveyor can range from 1 to 2 cm. The conveyor speed ranges from 0.5 to 1.0 m/s.

(a) If a binary (two-state) transducer (e.g., limit switch) is used for object counting, what is the largest allowable response time for the transducer (give the key steps of your computation)?
(b) Instead, suppose that an analog sensor (e.g., proximity sensor) is used in object counting. Specifically, the signal from the sensor is sampled and processed to determine the presence of an object. What is the smallest allowable sampling speed for the sensor (give the key steps of your computation)?
(c) Which one of the two methods do you recommend for this application? Why?

11.28 What is a Hall-effect tachometer? Discuss the advantages and disadvantages of a Hall-effect motion sensor in comparison with an optical motion sensor (e.g., an optical encoder).

11.29 Discuss the advantages of solid-state limit switches over mechanical limit switches. Solid-state limit switches are used in many applications, particularly in the aircraft and aerospace industries. One such application is in landing gear control, to detect up, down, and locked conditions of the landing gear. High reliability is of utmost importance in such applications. A mean time between failures (MTBF) of over

100,000 h is possible with solid-state limit switches. Using your engineering judgment, give an MTBF value for a mechanical limit switch.

11.30 Mechanical force switches are used in applications where only a force limit, rather than a continuous force signal, has to be detected. Examples include detecting closure force (torque) in valve closing, detecting fit in parts assembly, automated clamping devices, robotic grippers and hands, overload protection devices in process/machine monitoring, and product filling in containers by weight. Expensive and sophisticated force sensors may not be needed in such applications because a continuous history of a force signal is not needed. Moreover, force limit switches are generally robust and reliable and can safely operate in hazardous environments. Using a sketch, describe the construction of a simple spring-loaded force switch.

11.31 Consider the following three types of photoelectric object counters (or object detectors or limit switches):

(a) Through (opposed) type
(b) Reflective (reflex) type
(c) Diffuse (proximity, interceptive) type

Classify these devices into long-range (up to several meters), intermediate-range (up to 1 m), and short-range (up to fraction of a meter) detection.

11.32 A brand of autofocusing camera uses a feedback control system consisting of a CCD imaging system, a microcontroller, a drive motor, and an optical encoder. The purpose of the control system is to focus the camera automatically based on the image of the subject as sensed by a matrix of CCD cells. *Note*: As an alternative to a CCD image sensor, a metal–oxide–semiconductor (CMOS) image sensor may be used as well. The light rays from the subject that pass through the lens will fall onto the CCD matrix. This will generate a matrix of charge signals that are shifted one at a time, row by row, digitized, and placed in a data buffer of the microcontroller. The image data are analyzed by the microcontroller to determine whether the camera is in focus. If not, the lens is moved by the motor so as to achieve focusing. Draw a schematic diagram for the autofocusing control system and explain the function of each component in the control system, including the encoder.

11.33 Measuring devices with frequency outputs may be considered as digital transducers. Justify this statement.

11.34 A smart seat belt that can alert the drivers when they fall asleep at the wheel is being developed. This is based on sensing the heart rate and breathing of the drivers and then alerting them.

(a) Apart from the heart rate and breathing, what other aspects of the driver may be sensed for this purpose?
(b) What type of sensors may be used in this application?
(c) What type of alerting mechanism may be appropriate?

12

Microelectromechanical Systems and Multisensor Systems

Chapter Objectives

- Microelectromechanical System (MEMS) Sensors
- MEMS Characteristics and Modeling
- MEMS Materials and Fabrication
- Wireless Sensor Network (WSN) Architecture
- Advantages and Applications of WSN
- Energy Management in WSN
- Communication Issues and Standards of WSN
- Localization Using WSN
- Nature and Types of Multisensor Data Fusion
- Bayesian Approach to Sensor Fusion
- Kalman Filter Approach to Sensor Fusion
- Dempster–Shafer Evidence Theory in Sensor Fusion
- Fuzzy-Neural Network Approach to Sensor Fusion

12.1 Advanced and Multisensor Technologies

In the previous chapters, we studied many issues of sensors and sensor systems covering their physics, theory, and practice. They included component interconnection, signal acquisition, conditioning and signal conversion (filtering amplification, digital-to-analog conversion, analog-to-digital conversion, bridge circuits, etc.), performance specification, data sampling and bandwidth considerations, sensing accuracy, error propagation, and estimation through measured data. Also, we studied different types of sensors and transducers such as analog motion sensors; effort sensors, which included force, torque, tactile, and impedance; several other types of sensors including optical sensors, ultrasonic sensors, magnetostrictive sensors, acoustic emission sensors, thermofluid sensors, and sensors that are useful in monitoring the quality of water (e.g., pH value, dissolved oxygen [DO], and oxidation–reduction potential [ORP]); digital transducers; and some other innovative sensor technologies. In this concluding chapter, we will study further advanced considerations of sensing and sensor systems. In particular, we will study sensors that are based on microelectromechanical system (MEMS) technologies. Also, we will study multisensor systems that operate in a wireless sensor network (WSN). Another important issue of sensor systems is the multisensor data fusion. We will study that subject as well while presenting several popular methods of sensor fusion.

12.2 MEMS Sensors

MEMSs are microminiature devices consisting of microelectronic components such as sensors, actuators, and signal processing integrated and embedded into a single chip while exploiting both their electrical/electronic and their mechanical features. The device size can be in the submillimeter scale (0.01–1.0 mm), and the component size can be as small as a micrometer (micron), in the range 0.001–0.1 mm. Since MEMSs exploit the integrated circuit (IC) technologies in their fabrication, many components can be integrated into a single device (e.g., a few to a million).

12.2.1 Advantages of MEMS

The advantages of MEMS are primarily the advantages of IC devices. These advantages include

- Microminiature size and weight
- Large surface-area-to-volume ratio (when compared in the same measurement units)
- Large-scale integration (LSI) of components/circuits
- High performance
- High speed (20 ns switching speeds)
- Low power consumption
- Easy mass production
- Low cost (in mass production)

In particular, the microminiature size also means negligible mechanical *loading*, fast response, and negligible power consumption (and the related electrical loading is low).

12.2.1.1 Special Considerations

A typical MEMS device has integrated functions, primarily

- Sensing
- Actuation
- Signal processing

within a common electromechanical *structure*. Different types of MEMS sensors are available. They include accelerometers (piezoelectric, capacitive, etc.), flow sensors (based on differential pressure sensing, temperature sensing of a heated element in the flow medium, etc.), gyroscopes (Coriolis, etc.), humidity sensors (capacitive, etc.), light sensors (semiconductor photodetector, etc.), magnetometers (to measure magnetic field, magnetoresistivity, etc.), microfluidic sensors (involve multiple sensing [temperature, pressure, flow, current, etc.] and microactuation), microphones (piezoelectric, etc.), pressure sensors (piezoresistive diaphragm, piezoelectric, etc.), proximity sensors (capacitive, etc.), and temperature sensors (Zener breakdown voltage, etc.).

12.2.1.2 Rating Parameters

Notwithstanding the distinct advantages, MEMS sensors use many of the same rating parameters as those for macro sensors to represent their performance. They include sensitivity, bandwidth, linearity, dynamic range, resolution, stability (drift-free performance), and robustness (signal-to-noise ratio [SNR], ability to withstand shock and other disturbances, compensation for environmental factors including temperature). These parameters are discussed in Chapter 5. Furthermore, typically, MEMS devices are implemented in macro-level applications (e.g., automobiles, factories, consumer electronics, medical diagnosis and treatment systems, mechanical system monitoring).

12.2.2 Modeling of MEMS Sensors

Typically, MEMS sensors use some of the same technologies as their macro (or miso or meso) counterparts (e.g., piezoelectric, capacitive, electromagnetic, piezoresistive). However, due to their microminiature size, some of the conventional physical equations of sensor modeling may not be exactly valid in MEMS sensors. Furthermore, even when the analytical models of MEMS devices are similar to those of familiar electromechanical models of macro devices, the actual physical principles may deviate at the microscale. Hence, in general, all the physical concepts at the macroscale cannot be directly extended to the modeling and analysis of MEMS devices.

Furthermore, normally, an MEMS device may contain many components of multiple functions in a circuit-like structure. There exist combined MEMS sensors that have multiple sensing capabilities and even multisensor fusion embedded into them. For example, a six-axis *inertial measurement unit* (IMU) consists of a three-axis accelerometer and a three-axis gyroscope. An integrated MEMS package may contain the functions of an accelerometer, displacement sensor, gyroscope, magnetometer, and pressure sensor. Multifunctional motion sensing capabilities of this type are predominant in consumer electronics such as smartphones and tablets.

12.2.2.1 Energy Conversion Mechanism

The associated mechanism of energy conversion is a primary consideration in understanding and modeling the physics of an MEMS sensor. In particular, piezoelectric, electrostatic, and electrostatic energy conversions are relevant. These are addressed in Figure 12.1.

Piezoelectric: See Figure 12.1a. Mechanical strain in a piezoelectric material causes a charge separation across the material producing a voltage. Strain energy produced by the *mechanical work* that is needed to deform the material is converted into electrostatic energy. This is a passive device.

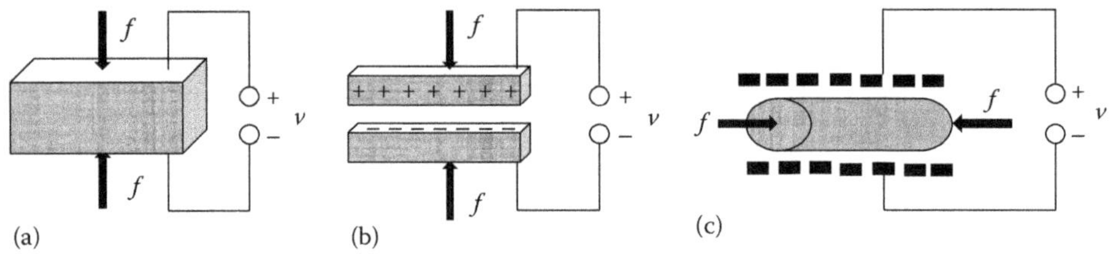

FIGURE 12.1
Energy conversion in a microelectromechanical system device: (a) piezoelectric, (b) electrostatic, and (c) electrostatic.

Electrostatic: See Figure 12.1b. A voltage causes positive (+) and negative (−) charges to separate into the capacitor plates. The attraction force between the plates is supported by an external "mechanical" force. If the plates are moved apart, some mechanical work is done, the capacitance is reduced, and the voltage is increased. Hence, mechanical energy is converted into electrical energy. This is a passive device.

Electromagnetic: See Figure 12.1c. As a coil moves in a magnetic field, a current is *induced* in the coil. In this process, mechanical energy is converted into electrical energy. This is a passive device.

12.2.3 Applications of MEMS

Applications of MEMS sensors and related devices are numerous, including transportation, structural monitoring, smartphones, energy exploration, human health monitoring, and medical treatment. There is a variety of MEMS sensors, particularly in the categories of biomedical, mechanical (including thermofluid and material engineering), chemical, industrial, defense, energy, service, and telecommunication. In particular, MEMS technologies in biological and medical applications are gaining such prominence that the term *BioMEMS* is used to refer to them. The market of MEMS devices and technologies is reaching US $25 billion and is growing at the rate close to 10% per year. The main obstacle to this growth is not the technological capabilities of MEMS but rather implementation and operation in harsh practical environments. In this context, packaging and robustness of MEMS devices are becoming key practical considerations. MEMS sensors and actuators are found in a variety of applications. They include

- Automotive (e.g., accelerometers and gyroscopes or IMUs for airbag deployment, handling control, safety and collision avoidance, ride quality, and dynamic stability, brakes, car tire pressure sensors)
- Biomedical applications (BioMEMS and microfluidics including lab-on-chip that uses bodily fluids for diagnosis, HIV/AIDS testing, and pregnancy testing; micro total analysis [biosensor and chemosensor]; implants including stents; microsurgical tools including microrobots for angioplasty, catheterization, endoscopy, laparoscopy, and neurosurgery; tissue engineering including applications of cell biology, proteomics, and genomics; disposable blood pressure sensors, sensors for intraocular pressure in eyes, intracranial pressure inside skull, and intrauterine pressure; IMUs in defibrillators and pacemakers; microphone and hearing aids;

microneedles, patches, etc., for controlled drug delivery/release, biosignal recording electrodes, bodily fluid extraction and sampling, cancer therapy, and microdialysis; and prosthetics, orthotics, wheelchairs, etc.)

- Computers, consumer electronics, and home appliances (touch screen controllers; inkjet printer nozzles and cartridges; IMUs and microphones for cell phones, laptops, tablets, game controllers, personal media players, digital cameras, and headsets; hard disk drives, computer peripherals, wireless devices, etc., in computers; and interferometric modulator display applications such as flat-panel displays)
- Heavy machinery, transportation, and civil engineering structures (vehicles, airplanes [wing surface sensing and control], etc.; construction machinery, aerospace industry, sports, and recreation machinery; sensors for stress and strain in buildings, bridges, guideways, etc.; wireless transmission and control)
- Optical MEMS (micromirrors, scanners, picoprojectors, fog-free lenses, light sensors for IR imaging, high-speed optical switching devices—20 ns)
- Energy sector (sensor-driven heating and cooling of buildings, oil and gas exploration, energy harvesting, microcooling)
- Global positioning system (GPS) sensors (for vehicles, smartphones, and courier package tracking and handling).

12.2.4 MEMS Materials and Fabrication

The key features in the fabrication process of an MEMS device are as follows: the device has to be microminiature, the device will have many integrated components having various functional structures, and many devices have to be manufactured in a batch. The functional structure (e.g., that of a sensor or an actuator) is similar to the circuit structure in an IC chip. Furthermore, and fortunately, the quite mature processes of IC (semiconductor) fabrication can be used in the fabrication of MEMS devices as well. Just like an IC chip, an MEMS device is formed by forming the required functional structure on a substrate. The substrate can be silicon (as for IC chips), polymer (cheaper and easier to fabricate), metal (e.g., gold, nickel, aluminum, copper, chromium, titanium, tungsten, platinum, silver), or ceramic (e.g., a nitride of silicon, aluminum or titanium, silicon carbide; they provide desirable material characteristics for sensors, actuators, etc.).

12.2.4.1 IC Fabrication Process

Since the fabrication process of an MEMS device is similar to that of an IC chip, we will first summarize the fabrication process of an IC chip. The main steps in the fabrication process of an IC package are substrate preparation, film growth, doping, lithography, etching, photoresist removal, dicing, and packaging. These are outlined next.

Substrate preparation: The process of IC fabrication starts with a thin slice of substrate on which the circuit (consisting of the equivalent of many millions of interconnected transistors and other components) is formed. Typically, this is a thin slice of polished silicon (Si) wafer.

Film growth: On the substrate, a thin film (of silicon, silicon dioxide, silicon nitride, polycrystalline silicon, or metal) is deposited. It is on this film that the components and interconnections of the IC circuit are fabricated. The film should have a well-defined crystalline orientation with respect to the substrate crystal structure.

Doping: A controlled trace amount of doping material (atomic impurity) is injected into the film (e.g., by thermal diffusion or ion implantation). This low concentration of doping material (e.g., boron, phosphorus, arsenic, antimony) will make the subsequent formation of electronic circuit structures feasible.

Photolithography: A thin uniform layer of photosensitive material (photoresist) is formed on the substrate through spin coating and prebaking). A pattern, which corresponds to the circuit structure, is transferred to the photoresist by applying intense light through a "mask" (a glass plate coated with a circuit pattern of chromium film).

Etching: A chemical agent (wet or dry) is used to remove the regions of the film or substrate that are not protected by the photoresist pattern.

Photoresist removal: The photoresist, which helped form the circuit structure on the substrate, is removed. This process is called ashing.

Dicing: The wafer containing the IC structure is cut into square shape.

Packaging: The diced wafer is packaged in protective casing. The casing also forms the electrical contacts that connect the IC chip to a circuit board.

The final step is to test the IC chip.

12.2.4.2 MEMS Fabrication Processes

The basic processes of MEMS fabrication are essentially the same as those for an IC chip. In particular, they are deposition (deposition of a film on the substrate; e.g., physical or chemical deposition), patterning (transfer of the pattern or MEMS structure onto the film; typically, lithography is used for this purpose), etching (removal of unwanted parts of film or substrate, outside of the MMS structure; wet etching where material is dissolved when immersed in a chemical solution or dry etching where material is sputtered or dissolved using reactive ions or a vapor phase etchant may be used), die preparation (removal of individual dies that formed MEMS structures on the wafer), and dicing (cutting or grinding the wafer to proper shape, say, a thin square).

Complex functional structures (sensing, actuation, signal processing, etc.) of MEMS devices are fabricated in several ways. The primary are

1. Bulk micromachining
2. Surface micromachining
3. Micromolding

Bulk micromachining: The required MEMS structures are constructed (etched) on the substrate in three dimensions. A wafer may be bonded with other wafers as well to form special functional structures (e.g., piezoelectric, piezoresistive, and capacitive sensors and bridge circuits).

Surface micromachining: The required MEMS structures are formed layer by layer on the substrate with multiple deposition and etching (micromachining) processes of film material. Some layers may form the necessary gaps between structural layers (e.g., plate gap of capacitors).

Micromolding: In micromolding, the required MEMS structures are fabricated using molds to deposit the structural layers. Hence, etching is not required (unlike bulk micromachining and surface micromachining). After application, the mold is dissolved using a chemical that does not affect the deposited MEMS structural material.

12.2.5 Examples of MEMS Sensors

MEMS sensors that use piezoelectric, capacitive, piezoresistive (strain gauge), and electromagnetic principles are common. Since these principles have been addressed earlier, they are not repeated here. Instead, we will present examples of MEMS sensors in several application categories to illustrate the range, diversity, and utility of MEMS sensors.

MEMS accelerometer: The approach used in an MEMS accelerometer is to convert acceleration into the inertia force of a proof mass, which bends a microminiature cantilever. One arrangement is to have a cantilever element with a point mass (proof mass) at its free end. As the cantilever deforms (bends) due to the inertia force of the point mass, the associated displacement may be sensed by the capacitive, piezoresistive, or piezoelectric methods. Another structure that uses the capacitance method has two "combs," one fixed and the other supported on a cantilever (spring) and carrying a proof mass at the other end (see Figure 12.2). The teeth of the combs form the capacitor plates. The fixed plates are located in between the movable plates. As the comb moves due to the inertia force of the proof mass, the capacitance changes. This change in capacitance is measured, which gives the acceleration. Some rating parameters of an MEMS cantilever/capacitor accelerometer are range = ±70 g, sensitivity = 16 mV/g, bandwidth (3 dB) = 22 kHz, supply voltage = 3–6 V, and supply current = 5 mA.

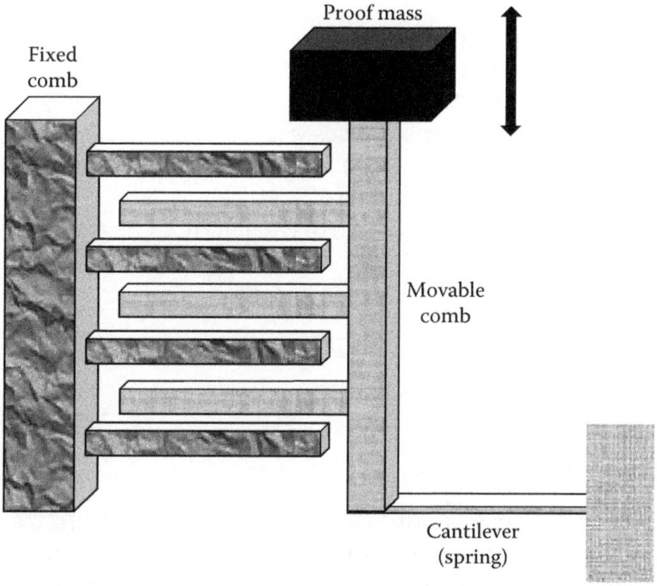

FIGURE 12.2
Microelectromechanical systems capacitive accelerometer.

MEMS thermal accelerometer: In this sensor, a heated air bubble takes the place of the proof mass. Two thermistors are used for temperature sensing. Its principle is as follows: Due to acceleration, the air bubble moves in that direction between the two thermistor elements. As a result, the temperature of one thermistor increases and the other decreases. The temperature difference gives the acceleration.

MEMS gyroscope: A gyro measures angle (orientation) and a rate gyro measures angular speed. Both types of sensing use the Coriolis force or gyroscopic torque that is generated as a velocity or angular momentum vector changes orientation. The force or torque may be sensed (e.g., through the capacitance change of a cantilever comb, as in an MEMS accelerometer) to produce the gyro reading. Three-axis MEMS gyros that depend on the Coriolis force are available. Rating parameters of an MEMS Coriolis device are range = ±250°/s, sensitivity = 7 mV/°/s, bandwidth (3 dB) = 2.5 kHz, supply voltage = 4–6 V, and supply current = 3.5 mA.

MEMS blood cell counter: The device has two electrodes by which current pulses can be applied to the path between the electrodes. A blood sample is injected across the electrode path. The resulting change in electrical resistance (or resistance pulses) is measured, which is an indicator of the quantity of blood cells. *Note*: Both pulse count and pulse height are sensed. Pulse count gives the number of blood cells; pulse height can be used to differentiate between red and white blood cells.

MEMS pressure sensor: This sensor uses a suspended membrane (Si substrate) between two electrodes to measure pressure. The capacitance between the electrodes changes as the membrane (with capacitor plate attached) moves due to the pressure. The measurement of capacitance gives the pressure reading. Another type of MEMS pressure sensor uses a piezoresistive (strain gauge) cantilever. As it deforms due to the pressure difference on the two sides, the resulting change in resistance is measured using a bridge circuit, which gives the differential pressure. The rating parameters include measurement range of 260–1260 mbar, supply voltage of 1.7–3.6 V, and sensor weight of 10 mg. This sensor is used in biomedical applications such as diagnosis and treatment of neuromuscular diseases.

MEMS magnetometer: This sensor uses the magnetoresistive property of an MEMS element to measure magnetic field. It is used in applications of electronic compass, GPS navigation, and detection of magnetic field.

MEMS temperature sensor: This sensor uses a Zener diode whose breakdown voltage is proportional to the absolute temperature. The rating parameters include sensitivity of 10 mV/K and current range of 450 μA to 5 mA.

MEMS humidity sensor: It uses the principle of capacitance change in a polymer dielectric planar capacitor due to humidity. Typically, MEMS sensors are available, which measure both humidity and temperature using the same device.

12.3 Wireless Sensor Networks

A WSN consists of several sensor nodes that are in wireless (radio) communication with each other and with a base station (gateway). Each sensor node contains one or more sensors, a microcontroller, data acquisition system, and a radio transceiver. Many practical applications require multiple sensors that are geographically distributed throughout the system. Cabling to connect sensors may be difficult, costly, or even infeasible in many situations. When many sensors are needed, the cost per sensor is also a consideration. Hence, a WSN may be the best sensory solution for many applications. For these reasons, when addressing networks of sensors, typically, we focus only on wireless networks. Embedded system technologies and the integration of sensors, radio communication, and digital electronics into a single IC package are key enablers of WSNs. Technologies of swam intelligence and multirobot cooperation (localization, optimal navigation, energy optimization, networked communication) can help the advancement of WSN technologies. Furthermore, WSNs are an integral part of the Internet of Things (IoT).

WSNs have significantly advanced the scale and resolution of data collection, analysis, distribution, and decision-making, in many applications. Even though the concepts and technologies of WSN originated over a decade ago, their full potential and advantages have not been realized yet. This may be due to high development costs when compared with cabled systems. Some obstacles to wider deployment of WSN are that the application scale is limited by communication bandwidth, power source constraints and power management, required robustness of software and hardware, limited simulation capability, high cost of field testing, system security issues, and nonuniform, complex, and evolving standards.

12.3.1 WSN Architecture

A WSN, typically, consists of the following components:

- A group of *sensor nodes* with wireless communication.
- Nodes are arranged in a specific *architecture*.
- Nodes *communicate* information (possibly preprocessed, compressed, and aggregated) to a *base station* (*gateway*) using a *radio transceiver*.
- Base station forwards information (possibly after further processing) to the *application server/user*.

This structure of a WSN is shown in Figure 12.3.

A WSN may contain just a few nodes or thousands of nodes, depending on the application. Scalability to the scale of the application is an important consideration of a WSN.

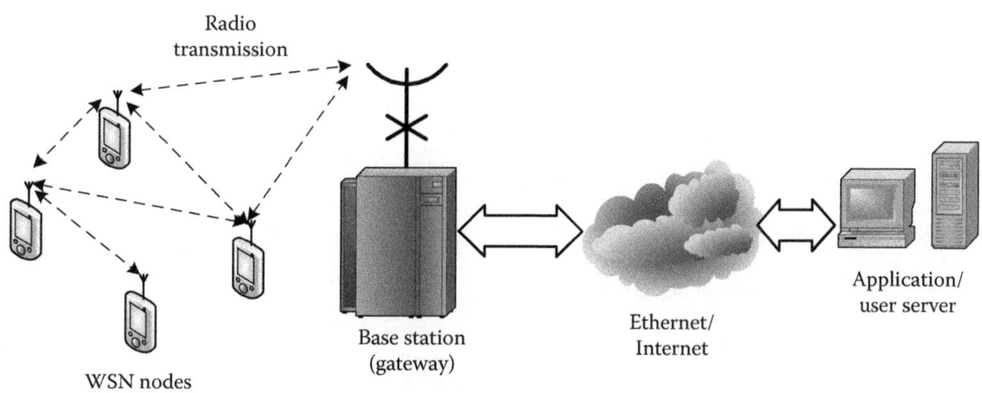

FIGURE 12.3
Typical structure of a wireless sensor network.

Even though a sensor node may have just one sensor, it is quite possible to incorporate several sensors in a given sensor node, where measurements are typically acquired from one sensor at a time. The base station (gateway) collects data from the sensor nodes through wireless radio frequency (RF) transmission. Since it is not economical or feasible to transmit all the data from all the sensors to the base station, due to such limitations as data capacity, range of transmission, power usage, and accuracy requirements, sensory data are preprocessed and condensed at the sensor node before transmitting to the base station. The base station will further process the information that is collected from the sensor nodes and transmit the processed information to the server at the application (user) site for use in the application.

Note: In some WSN architectures, a *leader* (*master or mother*) *node* may receive and process data from other nodes in its cluster and transmit the processed data to the base station.

12.3.1.1 Sensor Node

A sensor node contains one or more sensors, processing capability (microcontroller with an operating system [OS], CPU, memory, and I/O), data acquisition hardware and software, a power source, and an RF transceiver with antenna, which is omnidirectional (transmitting uniformly in all directions in two dimensions). A sensor node is software programmable. An actuator may be integrated at a node, depending on the application, but this is not a required function of a sensor node. The actuator may control the sensory activities or other functions at the node, using external commands (from the base station, user, etc.) or a control program in the microcontroller. The size of a sensor node can range from 1 to 10 cm or more. The components of a typical sensor node are shown in Figure 12.4.

Hardware of a sensor node is characterized by simplicity, low cost, limited functionality, and low power usage (and high power efficiency). In particular, the microcontroller of a sensor node need not be a complex and general-purpose platform of extensive capabilities. Its *OS* should be simple, should support a convenient high-level programming language (e.g., C/C++), and should be specific to WSN applications (e.g., TinyOS, which supports event-based programming rather than multithreading). Intel Galileo, Arduino Uno, and Raspberry Pi are all possible microcontrollers for a sensor node. However, some of these microcontrollers may be more powerful than what is needed for a particular application.

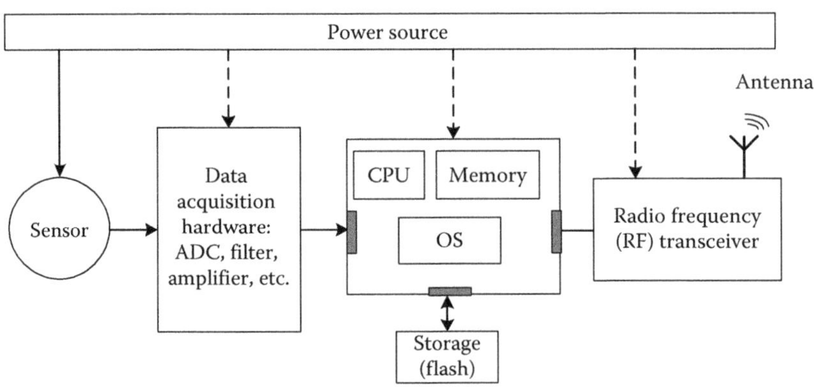

FIGURE 12.4
Components of a sensor node.

Motes: Tiny and low-cost sensor nodes (*motes*) of the millimeter scale, with limited sensing, processing, and transmission capabilities may be used in special applications (e.g., defense, environmental monitoring). These may be deployed over large areas using mobile deployers (e.g., sowed aerially by a helicopter or a drone). Motes use low-cost and simple power sources (e.g., self-generation, photoelectric).

12.3.1.2 WSN Topologies

The nodes in a WSN may be interconnected according to different topologies. They include star, ring, bus, tree, mesh, and fully connected topologies. Some examples are given in Figure 12.5. The decision on the appropriate topology is mainly dependent on the application. Resource limitations, communication bandwidth, and cost are relevant issue.

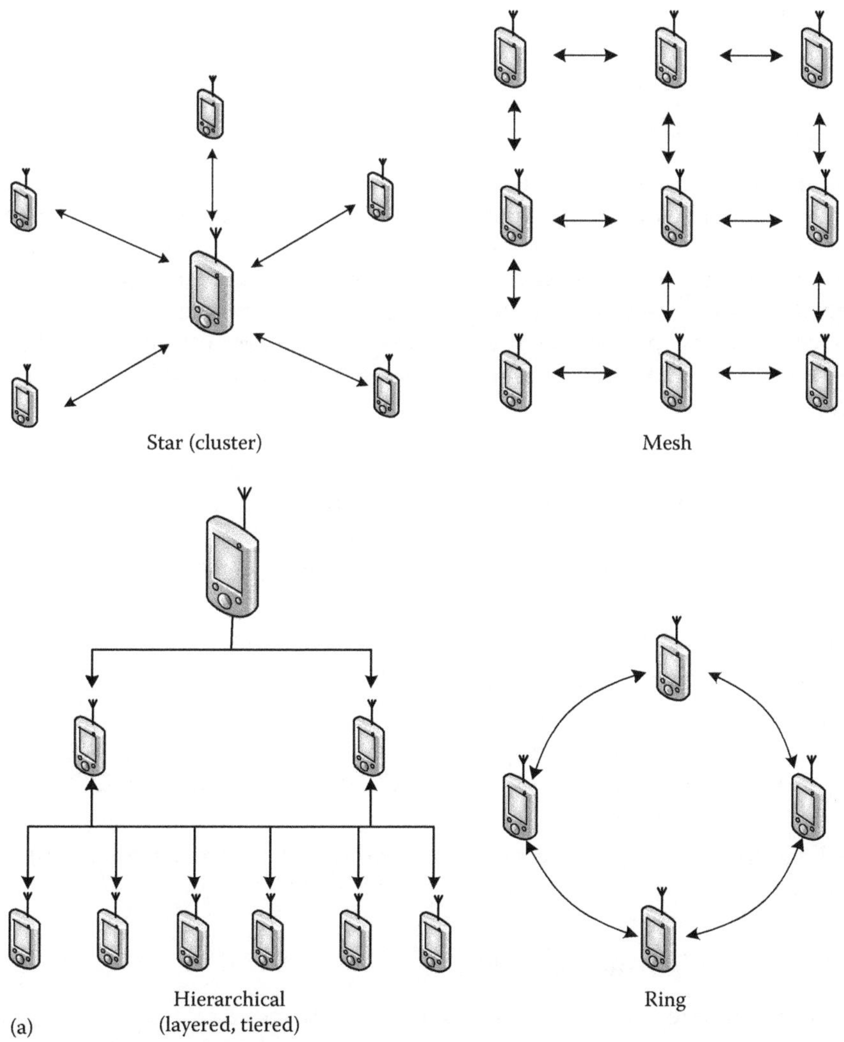

FIGURE 12.5
Network topologies. *(Continued)*

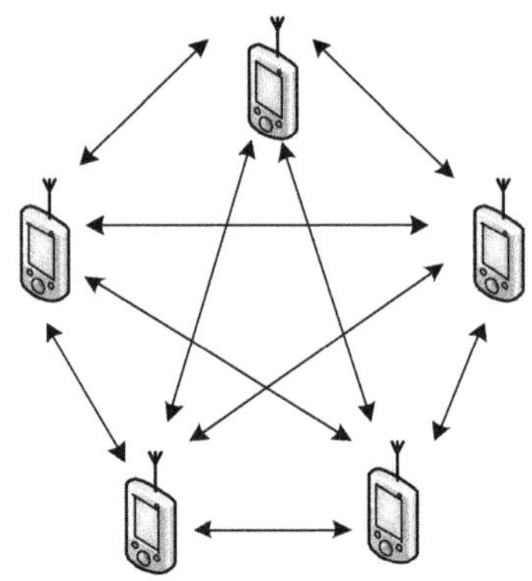

(b) Fully connected

FIGURE 12.5 (*Continued*)
Network topologies.

12.3.1.3 Operating System of WSN

The OS operates the microcontroller at a sensor node. It provides interfaces between applications and hardware at the node and schedules tasks at the node. Open-source OSs for WSNs include TinyOS, Contiki, and OpenWSN. An OS for a WSN node should be far less complex than a general-purpose OS of a computer. In particular, an embedded system OS would be adequate (e.g., eCos or uC/OS, which is application specific and of low cost and low power). Virtual memory is unnecessary and real-time operation may also be an added luxury for a sensor node. Some examples of OSs for WSNs are given here:

TinyOS is specifically designed for WSN. It uses event-driven programming model (not multithreading). It is simple and power efficient. It is composed of *event handlers* and *tasks* with run-to-completion semantics—when an external event occurs (e.g., incoming data packet; sensor reading), it signals the appropriate event handler.

LiteOS provides UNIX-like abstraction and supports C.

Contiki uses a simpler programming style of C.

RIOT implements a microkernel architecture, provides multithreading with standard application programming interface, and supports C/C++. RIOT also supports common IoT protocols like 6LoWPAN, IPv6, RPL, TCP, and UDP. *Note*: RIOT may be excessive for WSN.

ERIKA Enterprise is an open-source kernel (multicore, memory protection) and supports C.

12.3.2 Advantages and Issues of WSNs

In fact, many advantages of WSNs emanate from the disadvantages of cabling. In particular, cabling is difficult in some terrains (e.g., underwater, urban landscape, extensive and deep remote areas, complex terrain such as rock formations and hills). Hence, it introduces high cost of installation and maintenance. The cost increases with the number of sensors and the location and area of operation. Corresponding to this, the power usage also increases and the efficiency decreases. The wasteful nature of cabling may primarily come from power dissipation, which increases with the extent of deployment of the sensor network. They are messy (cables tangle) and less flexible (cannot conveniently adapt to mobile sensors, site relocation, or expansion). Furthermore, cables (bundles of wire including optical fiber) are less reliable. In particular, cables can encounter breakage and connector failure (due to aging, wear and tear, accidents, malicious action, etc.). The main advantages of WSNs are listed as follows:

- Use wireless connection
- Can have many nodes and cover very large areas
- Are easily scalable to the application scale (e.g., thousands of nodes)
- Require low cost of installation and operation
- Have reliable connectivity, robustness, and security
- Have flexible structure of nodes (restructuring, mobility, rescaling, etc.)
- Are autonomous and self-organizing
- Can operate in harsh environments
- Comprise distributed architecture with distributed processing and decision-making (smart)
- Function accurately, efficiently, and fast

12.3.2.1 Key Issues of WSN

Issues common to sensors and sensing are an important subset of the issues of WSNs. In addition, there are specific issues, which include multiplicity and distribution of nodes, wireless transmission, communication network, power constraints, available technologies, applications, power management, network topology, autonomous operation, self-organization, reliability, communication technologies and protocols, node localization, data rate, congestion of communication network, network/node mobility, voids (for some nodes there may not be receiver nodes in the transmission range, e.g., if the nodes are moving), synchronization (of node activities), and standards (evolving).

Note: Cost of problem resolution can exceed possible savings (at least initially).

12.3.2.2 Engineering Challenges

The main engineering challenges of WSNs include the following:

Component life and network life: A network may function even if some components fail; an alkaline battery life is 2–5 W·h, and two batteries will last approximately a month at a power of 9 mW. Harvesting energy and/or being energy efficient is important.

Response speed: Response to events, user query, etc.; Reduces lifetime.

Robustness: In harsh deployment environment; low cost is equal to low quality, low robustness, and low reliability.

Scalability: This is scaling to the application scale (thousands of nodes may be required); distributed, hierarchical architecture (not centralized) needs local processing.

Autonomous (unattended) operation: This includes self-localization, self-calibration, synchronization, and self-organization.

Unfamiliar and dynamic environments: This includes adapting activities and protocol to maximize performance and the use of "learning" and data models to minimize transmission.

Resource constraints: E.g., power; Limits on features/robustness/security; but application-specific implementation will help.

12.3.2.3 Power Issues

Power is critical in WSN applications. In fact, energy efficiency is more important than data processing and transmission efficiency. For example, sending 1 bit of data consumes three orders of magnitude more energy than processing one instruction. Furthermore, computing technologies are more advanced than flexible power technologies (e.g., battery, energy harvesting) and evolve much faster. Energy is not easily accessible in many WSN applications, and the lifetime of an energy source is not easily predictable as many unknown and random factors can affect it. Batteries are not the best solution for WSNs as they are bulky and not renewable/rechargeable at remote sites and their lifetime may not be adequate.

12.3.2.4 Power Management

Design for power efficiency (i.e., design the system components and integrate them so as to optimize the power usage and power efficiency, which is a subject of mechatronic design) is an important consideration in WSNs. Technologies of power generation (development and use of appropriate, efficient, and low-cost power technologies) are key aspects in this regard. Power conservation (taking measures to reduce usage and wastage of power during system operation) and power control (e.g., using the microcontroller control the usage of power during system operation) should be used in power management in WSNs. Some important considerations and approaches to power management in WSNs are indicated here:

Energy management paradigms: This includes multihop communication, routing control (optimize transmission routes by selecting power-efficient nodes with respect to a performance index), duty cycling, data preprocessing (process/compress data locally before transmission), passive participation (when two nodes have the same information, transmit only one), adaptive sampling, adaptive sensing and transmission (stop sensing and transmission when the sensed quantity does not change), and use of efficient power technologies (e.g., efficient and low-cost batteries, harvesting energy from the environment such as solar, vibration, wind, waves, and geothermal; some nodes may use line—AC power).

Multihop communication: Radio transmission uses the most power in WSN. Transmission power (consumption/dissipation) increases exponentially with transmission range (and signal reliability and strength decrease). Transmission-power/

sensor-hardware-power ratio increases with frequency (increases by an order of magnitude or more as the frequency doubles). Multihop communication may be used to reach a targeted destination (with intermediate nodes, to reduce transmission range). This improves power efficiency, accuracy, and robustness. Furthermore, the hopping strategy can be optimized (i.e., by optimal choice of intermediate nodes).

Duty cycling: This is an important approach to power management and is controlled by the microcontroller of the node. The methods include the following. Fixed duty cycling uses sleep–awake duty cycles for all components (sensor, data acquisition hardware, data processor). Adaptive duty cycling powers a component only when it is needed. Sleep at other times uses sentries (i.e., nodes that are always awake). A minimum set of sentries is needed to maintain the WSN coverage. Other nodes may sleep when not needed. Messaging/communication optimization minimizes messaging. Communication that is driven by "sensing event" may be used (i.e., activate transmission only when sensing is performed—this minimizes both sensor power and communication power).

Data preprocessing: Data should be processed/compressed locally in the node microcontroller before transmission (so that only compact and smaller amounts of information are transmitted). This is important in power management because sending 1 bit of data consumes three orders of magnitude more energy than processing one instruction. Preprocessing may include data compression, aggregation, and modeling, which are protocol dependent. For example, we may use such operations as min, max, and mean to compress data. Also, a discrete data set $\{x_i\}$ may be represented by a model such as (linear) $ax + b$.

Adaptive sampling: Faster sampling generates more data. Some sensors may need high power and may waste energy. Fast sensing is not needed if the measurand does not change rapidly. In such situations, adaptive sampling may be used. The approach is to change the sampling rate and the period of data sampling depending on the sensor and application; specifically, increase the sampling rate when measurand changes rapidly and decrease otherwise.

12.3.3 Communication Issues

Data communication is a key function of WSNs. The related issues include network topology, communication protocols, communication standards, multihop communication, network traffic (e.g., low data rate, bursty traffic, monitoring-type applications. *Note*: Burst mode provides a way to dedicate the entire channel for the transmission of data from one source), and data centricity (i.e., a programming paradigm that is centered on processing and relaying data). Some of these issues have been addressed already. Other main issues are discussed next.

12.3.3.1 Communication Protocol of WSN

A communication protocol defines the format and the order of message exchanges and what subsequent actions would be taken among entities of a communication network. A protocol model is arranged into layers. In a WSN, many nodes and the base station would want to send or receive data at a given time. Hence, a WSN needs a communication protocol. WSN protocol requirements are constrained by resource limitations (small memory and code size, variable conditions).

Medium access control (*MAC*) *protocol layer*: This is an important sublayer of a WSN protocol model. Its functions include coordinating transmission among neighboring nodes and in a shared channel (to optimize and avoid packet collision); coordinating the actions for a shared channel (here are the steps: test if busy; if not busy, transmit; if busy, wait and try again); communication interactions (request to send [RTS] and clear to send [CTS]); sleep mode (for nodes that are not active), which saves energy and simplifies communication for other nodes; low-power "listening mode" to decide sleep or awake; and packet backoff (wait if sending is not urgent). It is consistent with the IEEE open standard 802.15.4 and is widely used, but it is somewhat complex.

Note: Another version of MAC protocol layer is B-MAC.

IEEE standard 802.15.4 characteristics (*specific for WSN*): Transmission frequencies are 868 MHz/902–928 MHz/2.48–2.5 GHz; data rates are 20 kbps (868 MHz band), 40 kbps (902 MHz band), and 250 kbps (2.4 GHz band); it supports star and peer-to-peer (mesh) network connections; it does encryption of transmitted data, for security; it determines link quality (useful for multihop mesh networking algorithms); it provides robust data communication.

12.3.3.2 Routing of Communication in WSN

Data have to be properly routed in a WSN. For this, a routing algorithm is used. WSN does not need complex routing methods as in the Internet. Typical methods of routing for WSN are as follows: (1) Discover neighboring nodes (ID and location). *Note*: Node knows its own characteristic (location, capability, remaining power, etc.). (2) Pick the best node. (3) Send message to that destination node.

Note 1: Once the node location is known, messages are sent to the location coordinates, not to the node ID (called geographic forwarding [GF]).

Note 2: WSN should not send messages to a sleeping node unless it is awakened first.

Issues of routing in WSN: It includes time delay, reliability, remaining energy, and data aggregation to reduce transmission cost. Multihopping may be used. Unicast semantics is message sent to a specific node; multicast semantics is message sent to several nodes simultaneously; anycast semantics is message sent without specifying any nodes (diffusing or flooding).

Routing protocol: The routing protocol functions as follows: Given a destination address, properly route data to that destination. It should be robust to node failures and unintentional disconnection. Also, it should be power efficient and not too complex. TCP/IP is a general-purpose protocol. It is too complex for WSN and not energy efficient. A routing algorithm is used in a routing protocol. Multihopping may be used to optimize power, reliability, etc. Internet Engineering Task Force (IETF) is standardizing Routing Over Low Power and Lossy Networks (ROLL). This is relevant for WSN.

Routing protocols for WSN: Some routing protocols for WSNs are given as follows. In low-energy adaptive cluster hierarchy (LEACH), operation is divided into rounds. Each round uses a different cluster of nodes with cluster heads (CHs). A node selects the closest CH and joins that cluster to transmit data. In power-efficient gathering in sensor information systems (PEGASIS), there is no cluster formation. Each node communicates only with the closest neighbor (by adjusting its power signal to be only heard by the closest neighbor. Signal strength is used to measure the distance of travel). After chain formation, a leader is chosen from the chain (which has the most residual energy). Virtual grid architecture (VGA) utilizes data aggregation and in-network processing to maximize the network lifetime. It is energy efficient.

12.3.3.3 WSN Standards

Standards are needed to achieve component compatibility (interoperability of devices from different manufacturers), proper communication, and so on.

Examples of WSN Communication Standards

- *Wi-Fi*: It is the most common brand name of wireless local-area network (WLAN) and uses 2.4 GHz UHF and 5 GHz SHF radio signals, which is based on IEEE standard 802.11. Its application is for networking of devices in a local environment.
- *Bluetooth*: A wireless technology standard. It uses short-wavelength radio waves in the frequency range of 2.4–2.485 GHz. This technology is IEEE standard 802.15 based that uses wireless personal area network (WPAN). It is a short-range RF technology for communication among electronic devices and the Internet. It also has user-transparent data synchronization.
- *ZigBee*: It builds on the physical and MAC layer based on IEEE standard 802.15.4. It is more secure. It supports hybrid star–mesh network topology. It is cost-effective, consumes low power, and uses wireless connection.
- *6LoWPAN*:
  - It is addressable as an IPv6 device, for example, by your PC.
  - Standards in progress are as follows.
    - RFC4919: 6LoWPAN overview
    - RFC6775: Neighbor discovery
    - RFC6282: Compression format for IPv6 datagrams
    - RFC6606, 6568: Routing requirements and design space
- WirelessHART/IEC 62591
  - It is an alternative to ZigBee for industrial applications but is more costly.
  - It consumes lower power and is more robust to interference than ZigBee, but IEEE standard 802.15.4e will be competitive.

Global System for Mobile Communication (*GSM*): This is a European standard and is applicable for cellular networks (for cell phones). For its incorporation (i.e., for network access), the sensor node needs a subscriber identity module (*SIM*) card.

2G (second generation) GSM networks operate in 900 or 1800 MHz bands. Rarely, 400 and 450 MHz frequency bands are used (because *1G* system used those frequencies previously). Most 3G networks in Europe operate in the 2100 MHz frequency band.

Secure Digital (SD) card is a nonvolatile memory card.

SD card shield: This is the shield that carries the SD card.

12.3.3.4 Other Software of WSN

Time synchronization: This is important because most data are only meaningful with a time reference (time series).

Reprogramming: This is needed for updating firmware of all the nodes in the network (feature addition, bug/security fix) over the air and over multiple hops. Security measures are essential to prevent hackers from installing their firmware.

12.3.4 Localization

Localization involves determination of the geographic location of nodes of a WSN. It is needed for locating and tracking the nodes. The uses of localization include monitoring the spatial evolution of a WSN, which is needed, for example, in spatial data mining and determining spatial statistics; determining the quality of node coverage; achieving load balancing of nodes; facilitating routing (e.g., optimal, multihop routing); turning on and off or moving (changing the location of) a node, to optimize the WSN operation, as needed; and optimizing communication.

In a WSN, data need both time reference and location reference (e.g., target tracking, intrusion detection, control of dynamic/mobile nodes). For localization, the coordinate system and the algorithm can be application specific.

The steps of localization are

1. Establish the location of a selected (reference) set of nodes (anchors/beacons/landmarks).
2. Measure the distances to them from the node to be localized.

The issues of localization include accuracy, speed, communication range, and energy requirement and whether indoor or outdoor, 2D or 3D, hostile or environment friendly, which nodes to localize, how often to localize, where the computation of localization is performed, and how to localize (i.e., localization method).

12.3.4.1 Methods of Localization

The primary methods of distance measurement in localization include time of flight of signal and radio signal strength at reception. These two approaches are as follows:

1. First, determine the time of flight of the RF signal from node to node. Then use geometry to compute the coordinates of the required node.
2. Use the energy of the received signal (i.e., energy loss during transmission) to determine the distance of signal transmission to the node.

Note: Both methods need "beacon nodes (landmarks or reference nodes)" whose locations are known (and a node that can send/receive signals to/from them).

Indirect method: Count the number of hops between the nodes. Then use the "average" distance per hop to estimate distance between the two nodes.

For determining the absolute position, GPS or GPS with a mobile deployer may be used. This method cannot be used indoors.

12.3.4.2 Localization by Multilateration

Multilateration concerns estimation of the position of a node (i.e., localize it) using the distances from it to three or more landmarks (with known locations). The needed formula for localization is derived now. In Figure 12.6, the dotted circles represent the transmission ranges of the landmark nodes. Consider a general landmark node i, whose location (coordinates with respect to a planar Cartesian coordinate frame) is known. The distance from this node to the node to be localized is measured by some means (e.g., using time of flight or energy loss of the transmission signal).

Expressed in the reference Cartesian coordinate frame, the known quantities are (x_i, y_i), which are the coordinates of the ith landmark, $i = 1, 2, \ldots, n$ for n landmark nodes, and d_i, which is the distance of the ith landmark from the node to be localized.

Let δ_i be the error in the measurement of distance d_i.

What we need to determine is (x, y), which are the coordinates of the node to be localized.

Applying the Pythagoras theorem $(d_i + \delta_i')^2 = d_i^2 + 2d_i\delta_i' + \delta_i'^2 \approx d_i^2 + 2d_i\delta_i' = d_i^2 + \delta_i$ because $\delta_i' \ll d_i$

$$\rightarrow (x - x_i)^2 + (x - x_i)^2 = d_i^2 + \delta_i, i = 1, 2, \ldots, n$$

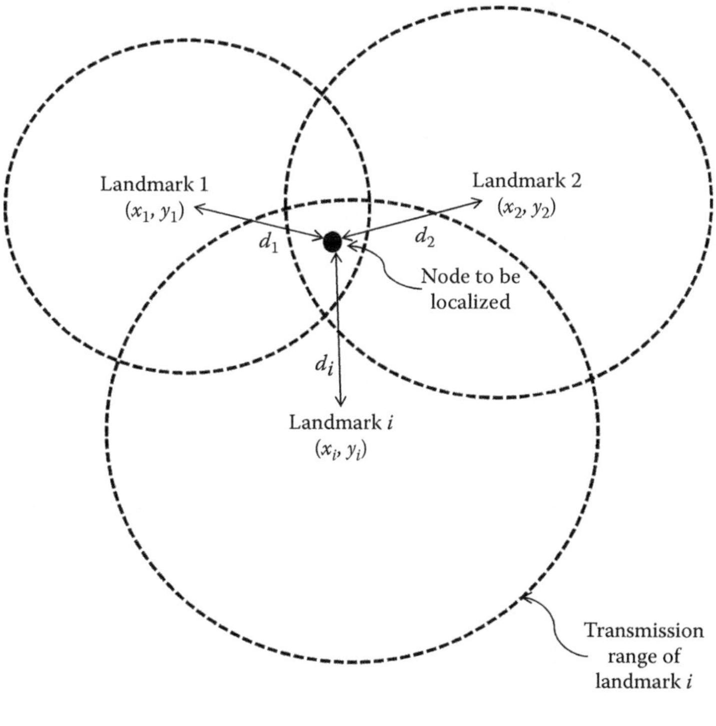

FIGURE 12.6
Node localization by multilateration.

Subtract the last (nth) equation from the first ($n - 1$) equations. We get the $n - 1$ equations:

$$y = X\beta + \varepsilon \quad \text{or} \quad \varepsilon = y - X\beta \tag{12.1}$$

where

$$X = \begin{bmatrix} 2(x_1 - x_n) & 2(y_1 - y_n) \\ \vdots & \vdots \\ 2(x_{n-1} - x_n) & 2(y_{n-1} - y_n) \end{bmatrix}$$

$$y = \begin{bmatrix} x_1^2 - x_n^2 + y_1^2 - y_n^2 + d_n^2 - d_1^2 \\ \vdots \\ x_{n-1}^2 - x_n^2 + y_{n-1}^2 - y_n^2 + d_n^2 - d_{n-1}^2 \end{bmatrix}$$

$$\beta = \begin{bmatrix} x \\ y \end{bmatrix}$$

$$\varepsilon = \begin{bmatrix} \delta_1 - \delta_n \\ \vdots \\ \delta_{n-1} - \delta_n \end{bmatrix}$$

Note: ε denotes a vector of measurement error.

$$\text{Squared error}: \quad E = \varepsilon^T \varepsilon = (y - X\beta)^T (y - X\beta)$$

To determine the least-squares error estimate of $\boldsymbol{\beta}$, we proceed as follows:

$$\text{Minimize } E: \quad \frac{\partial E}{\partial \beta} = 0 \rightarrow -2X^T(y - X\beta) = 0 \rightarrow X^T y - X^T X\beta = 0$$

We get

$$\beta = \left[X^T X\right]^{-1} X^T y \tag{12.2}$$

Example 12.1

In a node localization exercise with three landmark nodes, the following three data vectors were obtained:

$$\begin{bmatrix} x_1 \\ y_1 \\ d_1 \end{bmatrix} = \begin{bmatrix} 1 \\ 1 \\ 1 \end{bmatrix}; \quad \begin{bmatrix} x_2 \\ y_2 \\ d_2 \end{bmatrix} = \begin{bmatrix} 2 \\ -1 \\ 1 \end{bmatrix}; \quad \begin{bmatrix} x_3 \\ y_3 \\ d_3 \end{bmatrix} = \begin{bmatrix} -1 \\ 2 \\ 2 \end{bmatrix}$$

Determine the location (coordinates) of the node that is localized.

Solution

$$X = \begin{bmatrix} 2\times 2 & -1\times 2 \\ 3\times 2 & -3\times 2 \end{bmatrix} = \begin{bmatrix} 4 & -2 \\ 6 & -6 \end{bmatrix}; \quad y = \begin{bmatrix} 1^2-1^2+1^2-2^2+2^2-1^2 \\ 2^2-1^2+1^2-2^2+2^2-1^2 \end{bmatrix} = \begin{bmatrix} 0 \\ 3 \end{bmatrix}$$

$$\beta = \left[\begin{bmatrix} 4 & 6 \\ -2 & -6 \end{bmatrix}\begin{bmatrix} 4 & -2 \\ 6 & -6 \end{bmatrix}\right]^{-1}\begin{bmatrix} 4 & 6 \\ -2 & -6 \end{bmatrix}\begin{bmatrix} 0 \\ 3 \end{bmatrix} = \frac{1}{2}\left[\begin{bmatrix} 2 & 3 \\ -1 & -3 \end{bmatrix}\begin{bmatrix} 2 & -1 \\ 3 & -3 \end{bmatrix}\right]^{-1}\begin{bmatrix} 2 & 3 \\ -1 & -3 \end{bmatrix}\begin{bmatrix} 0 \\ 3 \end{bmatrix}$$

$$\rightarrow = \frac{1}{2}\begin{bmatrix} 13 & -11 \\ -11 & 10 \end{bmatrix}^{-1}\begin{bmatrix} 2 & 3 \\ -1 & -3 \end{bmatrix}\begin{bmatrix} 0 \\ 3 \end{bmatrix} = \frac{1}{2}\begin{bmatrix} 13 & -11 \\ -11 & 10 \end{bmatrix}^{-1}\begin{bmatrix} 9 \\ -9 \end{bmatrix}$$

$$\rightarrow \beta = \frac{1}{2\times(13\times 10 - 11\times 11)}\begin{bmatrix} 10 & 11 \\ 11 & 13 \end{bmatrix}\begin{bmatrix} 9 \\ -9 \end{bmatrix} = \frac{1}{2}\begin{bmatrix} -1 \\ -2 \end{bmatrix} = \begin{bmatrix} -0.5 \\ -1.0 \end{bmatrix}$$

12.3.4.3 Distance Measurement Using Radio Signal Strength

In wireless RF transmission from node to node, a signal is sent by one node and is received by the second node.

12.3.4.3.1 Advantage of RF Signals (Electromagnetic Spectrum)

The advantages of using wireless RF transmission in communication (particularly in WSNs) are as follows: no cabling is needed (wireless); it can penetrate objects such as walls; it can transmit long distances; and it can accommodate mobile nodes. WLAN technologies use local-area radio channels (for distances ranging from 10 m to several hundred m). Cellular technologies use wide-area radio channels for larger distances (tens of km).

12.3.4.3.2 Signal Distortion during Transmission

The transmission signals are electromagnetic—travel at the speed of light. There will be signal degradation during transmission. In this context, we define the SNR as follows:

SNR is a measure (dB) of the strength of the received signal with respect to the signal degradation due to transmission.

Note: Larger SNR means easier (and more faithful) recovery of the original signal from the received signal (by eliminating the background noise).

Issues of signal degradation include the following:

1. Signal strength decreases (signal will disperse) even in free space. This is called the path loss.
2. Interference with other signals (particularly transmitted signals in the same frequency band, environmental electromagnetic noise from other devices, etc.) will decrease the SNR.
3. Objects that obstruct in the transmission path cause signal degradation (reflection, absorption, shadowing, etc.). Moving objects cause more serious problems.
4. Bit error rate (BER) is the probability that the transmitted bit is received in error. It decreases with SNR and increases with the transmission rate (megabit per second [Mbps]).

12.3.4.3.3 Method

Signal degradation during transmission can be used for distance estimation. An immediate advantage of the method is that we make use of the existing communication hardware and resources of the WSN and there is no need for extra hardware for localization. In this method, the power of the received signal is determined (from the received signal strength indicator [RSSI]) many times, and the sample mean $\bar{P}_{i,j}$ is computed. Next, using a known reference distance d_0 and reference power P_0, the following "shadowing model" of signal strength (path loss) is used to estimate the distance:

$$\hat{d}_{i,j} = d_0 \left(\frac{\bar{P}_{i,j}}{P_0} \right)^{-(1/\eta)} \tag{12.3}$$

where η is the path loss exponent ~2.

12.3.4.3.4 Use of RSSI (According to IEEE Standard 802.11-1999)

The RSSI has a value ranging from 0 up to RSSI max. It is provided by a sublayer of the radio transmission protocol (8-bit RSSI) as

Power (dBm) = RSSI_VAL + RSSI_OFFSET

Typical accuracy = ±6 dB

Note: dBm denotes decibel-milliwatts. It is an abbreviation for power ratio in decibels (dB) of measured power in milliwatts referenced to one 1 mW. Since the signals are "power," we use $10\log_{10}$(power ratio), not $20\log_{10}()$, to convert into dB.

12.3.5 WSN Applications

The applications of WSNs are essentially those of multiple sensors. These may include both distributed sensing (sensing geographically extensive systems) and sensor fusion (improving the accuracy and reliability of a specific sensory decision/objective by combining and aggregating information from multiple sensors to determine a particular measurand). However, distributed sensing (geographically) is the most natural application category of WSNs. Some key applications of WSNs are associated with the following:

- Defense, surveillance, and security: For example, VigilNet with hierarchical architecture consisting of (1) application components, (2) middleware components, and (3) TinyOS system components
- Environmental monitoring: Pollution, water quality, forest fires, natural disasters, nuclear accidents and contamination, etc.; spatial distribution from 1 cm to 100 m; temporal sampling from 1 ms to several days; sensor size from 1 to 10 cm
- Transportation (ground, air, water, and underwater)
- Monitoring of machinery and civil engineering structures (e.g., for condition-based maintenance, detecting onset of seismic activity, at low sampling rates; once an activity is detected, sampling is done at much higher rate.)
- Industrial automation

- Robotics (e.g., multirobot cooperation in rescue, defense, home care, and future cities)
- Entertainment
- Intelligent workspaces
- Medical and assisted living
- Energy (exploration, production, transmission, management)

Medical and assisted living: A specific implementation of WSN is the architecture of AlarmNet for telemedicine, telehealth, home care, etc. Its features include

- *Body networks and front ends*: Patient sensors
- *Emplaced sensor network*: Patient's living space–environment network
- *Backbone*: Connects interfacing devices like laptop, cell phone, and iPad to the network
- *In-network databases*: Used for real-time processing, temporary storage, etc.
- *Back-end databases*: Used for long-term archiving, data mining, etc., at a central server
- *Human interface*: Patient and caregiver interface using PDAs, cell phones, iPads, etc.; used with wearable sensor nodes and environmental sensor nodes

Note: Some are mobile nodes.

Tasks: Localization, patient identification, monitoring, data collection, preprocessing and aggregation, storing, transmission, and action.

Structural health monitoring: Strain gauges, accelerometers, cameras, etc., may be used for monitoring bridges, buildings, and other civil engineering structures through a WSN. Flexible, low-cost, high-resolution monitoring is possible in this manner. Through continuous monitoring, large amount of data are generated. Record keeping, data analysis, and intensive diagnosis and prediction of impending problems are done with the monitored data. Condition-based maintenance can be carried out using the generated information. Power for the sensor nodes is a challenge. Energy harvesting (through vibration, solar, etc.) would be desirable for this purpose.

Other applications include telemedicine and water quality monitoring, which make use of WSN and are developed in our laboratory (Industrial Automation Laboratory, the University of British Columbia). These projects are outlined in Chapter 1.

12.3.6 Example of WSN Implementation

The present application concerns optimally self-adaptive, intelligent, and dynamic agent network with a common system architecture, for multiple engineering applications. This is an innovative paradigm of networked and automated or autonomous operation of multiple applications. It uses a common, self-adaptive, and intelligent system architecture and a common application platform, with resource sharing and optimized operation.

Networked operation of multiple agents (containing sensors, actuators, controllers, and other devices) is found in existing engineering applications. Furthermore, system optimization, intelligent systems, and adaptive control have been extensively investigated and applied. In this backdrop, the following innovations may be incorporated for networked engineering systems:

1. In the targeted multiple engineering applications (e.g., spatiotemporal monitoring of the water quality in natural sources, inspection of pipelines that distribute oil and gas, telemedicine and automated home care, multirobot cooperation in emergency response and human rescue), at least some of the networked agents (e.g., robots, unmanned aerial vehicles [UAVs] or drones, and sensor nodes, containing sensors, actuators, effectors, controllers, etc.) will be dynamic/mobile.
2. The system operating environment will be dynamic, unstructured, and unknown (or partially known).
3. The system will be self-adaptive to optimize its performance, particularly by utilizing the dynamic components in addition to parameter adjustment/tuning and structural reorganization.
4. The system operation will be further optimized by sharing resources among the applications.
5. Dynamic/mobile sensors will receive "feedback" from themselves, to improve their sensing effectiveness (e.g., data/information quality, relevance of their data, speed, confidence).
6. The networked agents will possess some degree of "intelligence" to facilitate autonomous operation and to achieve the desired performance.
7. The system will be able to predict, detect, and diagnose malfunctions and faults in it and accommodate or self-repair them.

The architecture of such a networked system is shown in Figure 12.7. Its development will involve technologies and methodologies that pertain to sensor/data fusion and adaptive sensing; multiagent cooperation; multiobjective and parameter/structure optimization; fault prediction, detection, diagnosis, and resolution; self-organization/adaptation; and distributed/networked intelligent control. In this activity, it is needed to determine/quantify the design constraints, performance limits, trade-offs, and development/operation guidelines and benchmarks for this type of networked applications. Such a system should lead to significant improvements in performance, developmental/operational costs, productivity, resource requirements, energy efficiency, safety, fault tolerance, reliability, autonomy, and sustainability of networked engineering systems.

Importance: Product/service quality, productivity and performance, quality and safety of the work environment, reliability, autonomy, energy efficiency, sustainability, and the costs of development, operation, and maintenance are all important for engineering systems and industrial processes. Optimally sharing the network resources (sensors, effectors, and other devices) and self-reorganization/tuning of the system to achieve optimal performance, which includes power management and energy efficiency, will lead to realizing these goals. There are reliability and safety advantages as well, which improve the work environment. Through methodologies of fault tolerance and self-repair, the system downtime and other adverse effects of faults can be reduced. Operational efficiency and reduction of energy utilization have direct environmental benefits. Through networked communication, system problems may be resolved even from remote locations, and hazardous production facilities can be isolated from densely populated areas. Furthermore, engineers, managers, and operators will be able to handle multiple facilities in a coordinated,

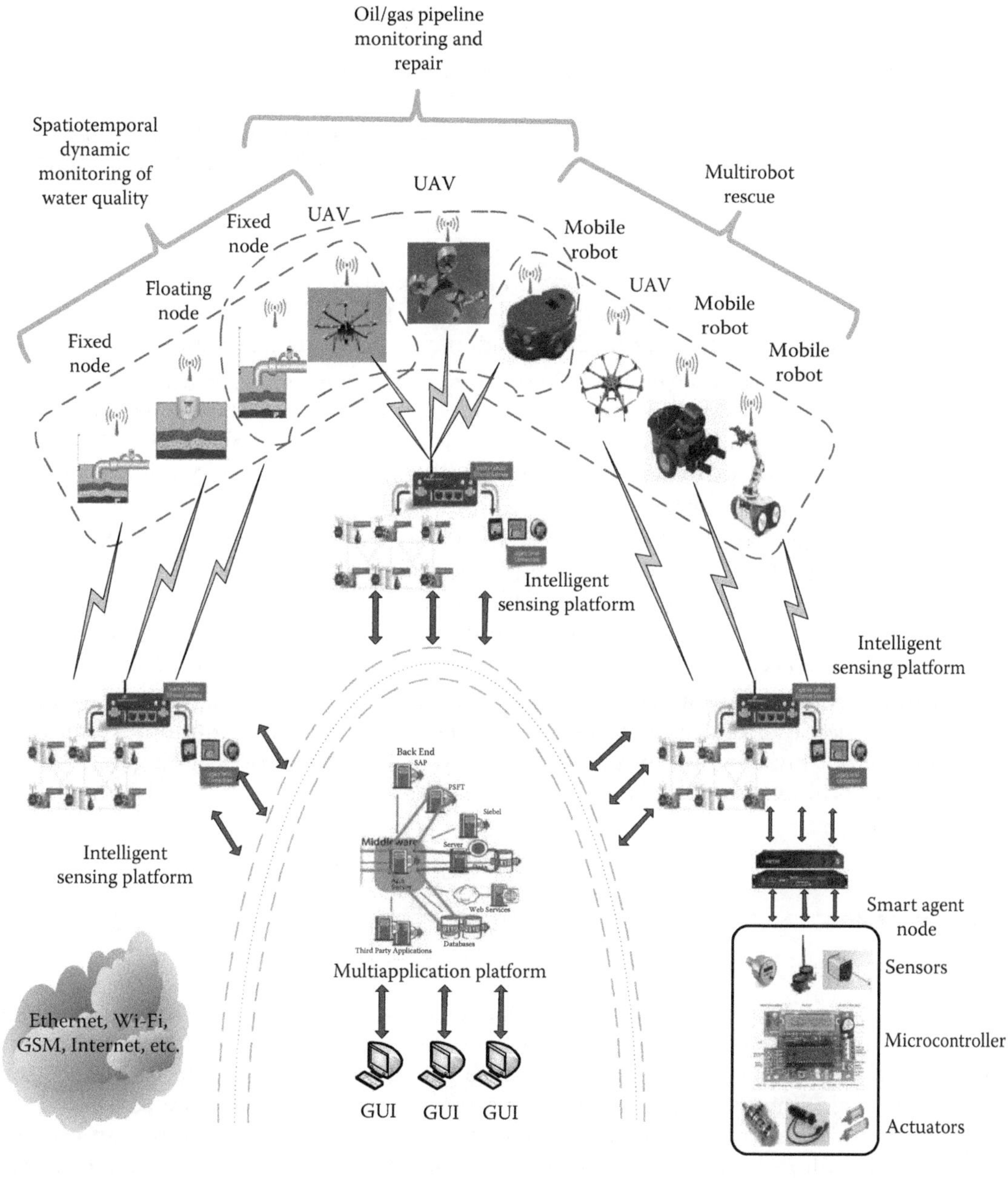

FIGURE 12.7
A common dynamic wireless sensor network for multiple engineering applications.

interactive and efficient manner, without being present at the same location. The resulting improvements in the quality of life are clear.

Originality: The key innovations in the proposed approach to WSN application are as follows: (1) the use of processed sensory data to improve the quality (e.g., accuracy, operating cost) of the sensory process itself, (2) the use of a common system architecture with shared resources and a common application platform for two or more applications, (3) extending the existing methodologies of multirobot cooperation to multiagent cooperation and sensor sharing/fusion, (4) unified multiobjective optimization and integrated control of multiple networked applications, and (5) self-repair using networked dynamic agents upon automated diagnosis of faults and malfunctions.

Approach: In view of the common characteristics of the implemented engineering applications, innovations in any of the following directions can result in significant technological advances and benefits:

1. Operation of several engineering applications using a common platform and shared resources. The *research innovation* is the use of soft computing, mechatronic design quotient and intelligent decision-making for optimal resource allocation, restructuring, and self-organization of a dynamic networked system.
2. Fusion of information from multiple sensors/nodes to improve accuracy, reliability, and relevance of the decisions based on that information. The *research innovation* is the development of a hybrid approach of Dempster–Shafer (DS) evidence theory and a nonlinear variation of Kalman filter, based on sensor quality criteria, for multisensor data fusion.
3. Navigation and optimal location of dynamic sensor nodes including UAVs and mobile robots. The *research innovation* is the use of feedback control from sensors and optimizing a performance index to determine the best sensor locations and usage. The performance index will be multiobjective and may include such considerations as accuracy, coverage and speed of information, cost, and power usage.
4. Optimal cooperation of agents to carry out tasks. The *research innovation* is the combination of a biology-inspired technique such as artificial immune systems with an optimization technique such as Markov decision process and game theory for optimized cooperative execution of tasks.

12.4 Sensor Fusion

In a particular application, one sensor or one set of sensory data may not provide the required information completely, accurately, and reliably. Then, we will have to rely on data from multiple sensors. Sensor fusion, also called "multisensor data fusion," is the process of combining the data from two or more sensors to improve the sensory decision. The resulting improvements may include accuracy, resolution, reliability, and safety (e.g., to sensor failure, as in applications of aviation), robustness, stability (related to sensor drift, etc.), confidence (reduced uncertainty), usefulness (e.g., broadening of the application range or operating range), level of aggregation (information modeling, compression, combination, etc.), and the level of detail or completeness or comprehensiveness (e.g., combining 2D images to obtain an authentic 3D image, combining complementary frequency responses).

Sensor fusion is a subset of data fusion. In data fusion, the data may not come directly from sensors (e.g., fusion of prior knowledge and experience—also called "indirect" fusion—and model-based data). Also, data fusion is a subset of information fusion. In information fusion, what is fused can be more than just data and can involve qualitative and high-level information (e.g., from "soft" sensors). This association is shown in

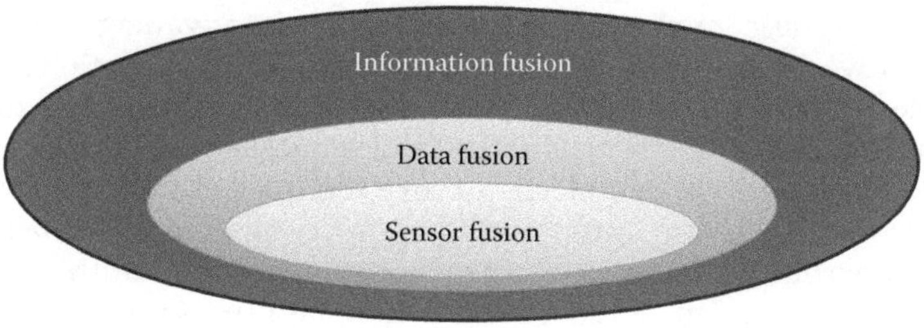

FIGURE 12.8
The placement of sensor fusion.

Figure 12.8. Typically, two or more sensors and large amounts of sensor data are used in sensor fusion.

Human Inspiration of Sensor Fusion

Humans perform sensor fusion in such activities as eating, listening to music, romancing, object recognition, and so on. Data from many biological sensory elements are processed by the brain to make pertinent decisions. The five basic senses of a biological system are sight (visual), hearing (auditory), touch (tactile), smell (olfactory), and taste (flavor). *Note*: Each sense may use many sensory elements. Two or more of the five basic senses may be jointly used (in fusion) in human activities. In particular, information from many sensory elements (e.g., tactile information from many contact locations, images from both eyes, what is heard by both ears) is processed by the brain to make one or more pertinent decisions.

12.4.1 Nature and Types of Fusion

Data into the fusion scheme can be low level (e.g., numerical values of object speed) or high level (e.g., features extracted from sensor data). Furthermore, the fusion outcome may be a low-level numerical value (e.g., angle of orientation of an object) or a high-level inference (e.g., product quality/classification, behavior of an agent). Data from different sensors may be used concurrently—in parallel (more desired) or sequentially one sensor at a time (less desired). It should be noted that sensor networking is not essential for sensor fusion even though networked sensors may be used in sensor fusion.

12.4.1.1 Fusion Architectures

Fusion may be categorized in many ways and into different architectures. They may depend on the types of sensors and sensory information, the nature of the application, and the resources available for the fusion application. Some of these comparative architectures and approaches of fusion are outlined next.

Complementary, competitive, versus cooperative fusion: In complementary fusion, the sensors independently provide complementary (not the same) information, which are then combined. A clear advantage is the reduction of information *incompleteness*, for example, four radars measuring four regions that are not identical (may have some overlap). In competitive fusion, each sensor measures the same property independently, and the separate items of sensory data are comparatively fused.

The better sensor (e.g., more accurate, faster) will be at advantage. The advantages include improvement of accuracy and robustness and the reduction of *uncertainty*, for example, four radars measuring the same region. In cooperative fusion, a sensor measures what another sensor needs (and requested by the sensor itself or by a higher-level monitor) to complete/improve the needed information. *Note*: Two sensors may sense the same property (to improve its accuracy, reliability, etc.) or different properties (to complete the needed information), but this is done cooperatively. For example, in stereo vision, it may be preassigned that one camera acquires images in one plane and another camera acquires images in a different plane. Then the images are combined to form a 3D image. This may be similar to "complementary" fusion. However, in complementary fusion, the sensors are not preassigned to take specific roles of sensing.

Centralized versus distributed (decentralized) fusion: In centralized fusion, data from different sensors are received by a single central processor to carry out fusion. In distributed fusion, a sensor receives information from one or more other sensors and carries out fusion. In this manner, each sensor carries out some level of fusion locally by using some data from other sensors. A hybrid architecture that has both centralized and decentralized clusters of sensors may be implemented as well.

Homogeneous versus heterogeneous fusion: In homogeneous fusion, the involved sensors are identical. In heterogeneous fusion, the involved sensors are disparate (different types, capabilities, etc.).

Waterfall fusion model: According to this model, the fusion procedure is sensing, preprocessing, feature extraction, pattern processing, situation assessment, and then decision-making. The steps are carried out sequentially, from top to bottom. Hence, it is a hierarchical fusion model.

Hierarchical fusion: This architecture is multilayered. Each layer may carry out different levels of fusion, specifically, data fusion, feature fusion, and decision fusion in a bottom-up architecture. The fusion process in a layer may be considered as an independent activity of fusion as indicated next.

Data-level fusion: Sensory data (with minimal preprocessing such as amplification and filtering) are directly fused by the fusion algorithm.

Feature-level fusion: Features (or data attributes) are extracted from different items of sensor data separately. These features are integrated into a feature vector, which is provided to the fusion system for overall decision-making (or estimation of the required quantity).

TABLE 12.1

Possible Inputs and Outputs of a Fusion Process

Fusion Input	Fusion Output	Example
Data	Data	Fusion of multispectral data
Feature	Feature	Fusion of image and nonimage data
Decision	Decision	Incompatible sensors
Data	Feature	Shape extraction
Feature	Decision	Object recognition
Data	Decision	Pattern recognition

Decision-level fusion: Each sensor separately processes its data to arrive at the sensory decision (or estimation of the required quantity). These separate decisions/estimates are evaluated and combined/fused by the fusion system for final decision making (or estimation).

Fusion may be classified as well based on the nature (or level of abstraction) of what goes into the fusion process and what comes out. Clearly, this is relevant to hierarchical fusion. Some possibilities of fusion input and output are given in Table 12.1.

12.4.2 Sensor Fusion Applications

Sensor fusion may be applied in any situation where multiple sensors are used to carry out a specific task. Applications include GPSs; machine learning; business decision-making; intelligent transportation systems; weather prediction; water quality monitoring; medical diagnosis, telemedicine, and telehealth; expert systems; consumer electronics and entertainment; military and defense (target tracking, automated identification of targets, missiles, warning systems, surveillance, navigation, and autonomous vehicle control); automotive (e.g., collision avoidance); aviation and space (e.g., navigation, altitude sensing of an aircraft); robotics (e.g., navigation, object detection, and identification); stereoscopic imaging; process control and automation; diagnostic–prognostic monitoring (machine condition/health monitoring); signal processing; multiresolution image fusion; biomedical applications; and biometric authentication using fingerprints and iris scans.

Example 1. Stereoscopic imaging: In a particular setup, two separate cameras are used at different locations and orientations but focused at the same object. The images from the two cameras can be combined to give a 3D image representation of a view/object. Sensor fusion may be used in this process. *Note*: The ratio of camera-to-camera distance/camera-to-subject distance should be >1/400 to retain the stereoscopic effect; ~1/80 is typical. A relevant application of this procedure is in the estimation of the height of vegetation or structures.

Example 2. Military application: Consider the detection and localization of a submarine. Sensory information from several underwater sonars is transmitted through a satellite system to the onboard computer of a ship. The computer carries out sensor fusion, while incorporating other available information as well in its database, to localize the submarine. This information is then communicated to a military aircraft to take appropriate actions.

Example 3. Multispectral data fusion (complementary sensing) in object recognition: In this application, suppose that Sensor 1 accurately provides image features of low frequency (e.g., large objects of image) and Sensor 2 accurately provides image features of high frequencies (e.g., object edges). By combining (fusing) the information from Sensor 1 and Sensor 2, better object recognition and feature computation (spectral resolution) can be achieved.

Example 4. Filter combination: When the frequency response of a single sensor is not adequate for a given application, two or more sensors of different frequency responses may be used, and their responses are combined to provide a more complete response. This is similar, for example, to the combination of a low-pass filter to a band-pass filter to increase the bandwidth of the filter. This is termed *multispectral data fusion*. Take the following two filters: a low-pass filter $v_o/v_i = (1/(\tau s + 1))$ and a band-pass filter $v_o/v_i = ((\tau_1 s)/((\tau_1 s + 1)(\tau_2 s + 1)))$.

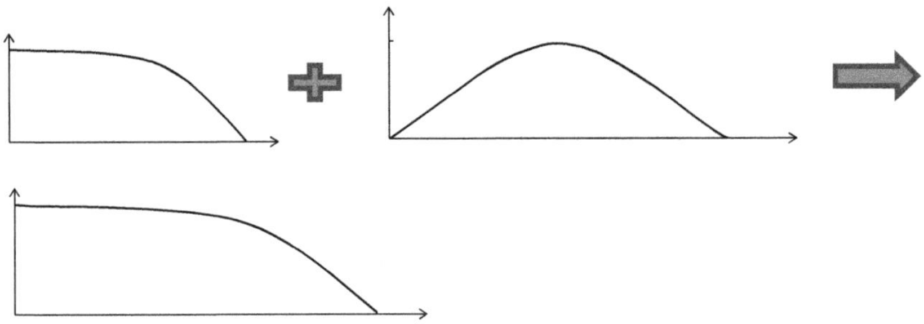

FIGURE 12.9
Multispectral filter fusion.

These two filters may be combined (in parallel) to give

$$\frac{v_o}{v_i} = \frac{\alpha}{(\tau s + 1)} + \frac{\beta \tau_1 s}{(\tau_1 s + 1)(\tau_2 s + 1)}; \quad \alpha + \beta = 1$$

As shown in Figure 12.9, this process generates a low-pass filter that has a larger bandwidth.

Example 5. Sensor combination: Sensors may be integrated to provide additional and enhanced sensory features. An inclinometer and rate gyro combination of this type is commercially available. The inclinometer is a capacitance-based sensor where the tilting of a liquid mass (which is part of the dielectric medium) between the capacitor plates changes the capacitance, which gives the tilt angle. This is a low-bandwidth sensor. Alternatively, the angular speed measurement from the rate gyro is integrated to give the tilt angle. The integrated reading suffers from drift problems. By combining the readings from the two sensors, a more accurate and reliable reading of the tilt angle can be made.

12.4.2.1 Enabling Technologies

Technologies of sensor fusion may incorporate MEMS, digital signal processing, probability and Bayesian methods, statistical estimation including Kalman filter, evidence (belief) theory, soft computing (fuzzy logic, neural networks [NNs], evolutionary computing), artificial intelligence, feature extraction, and pattern recognition and classification. Technologies of sensor networks are applicable as well in the specific situations where fusion involves networked sensors.

12.4.3 Approaches of Sensor Fusion

The procedures of sensor fusion may incorporate one or more of the following techniques: artificial NNs, fuzzy set theory, neurofuzzy systems, kernel methods (support vector machines), and probabilistic methods (Bayesian inference, DS theory [DST], and Kalman filter). The following main methods of sensor fusion are discussed next:

1. Probabilistic (Bayesian) approach
2. Kalman filter
3. DS evidence theory
4. NNs

12.4.3.1 Bayesian Approach to Sensor Fusion

Sensor fusion involves inferring the required information from multisensor data. There exists the possibility of model error and measurement error in the data (see Chapter 7). In this context, "fusion" is an "estimation" problem where the data for estimation come from more than one sensor. Hence, traditional methods of estimation may be used in multisensor estimation problem.

First, consider the case of sensors that generate discrete measurements. This involves the discrete problem of sensor fusion. We will use maximum likelihood estimation (MLE) to carry out sensor fusion. The case of single sensor has been presented in Chapter 7. Now, we will extend the method to multisensor data fusion.

As in Chapter 7, suppose that the estimated quantity m is discrete. Specifically, it takes one of a set of discrete values m_i, $i = 1, 2, \ldots, n$. Hence, it can be represented by the column vector $\boldsymbol{m} = [m_1, m_2, \ldots, m_n]^T$.

For example, each m_i may represent a different state of proximity of an object (near, far, very far, no object, etc.) or several sizes of object measurement (small, medium, large).

Also, suppose that the measurement/observation y is correspondingly discrete. It takes n discrete values y_i, $i = 1, 2, \ldots, n$. It can be represented by the vector $\boldsymbol{y} = [y_1, y_2, \ldots, y_n]$.

In the general case (with an uncertain or noncrisp or "soft" sensor), a measurement or observation may be represented by a probability vector with corresponding probability values for these n discrete quantities (*Note*: The sum of the probabilities in the probability vector = 1). The more common situation is the case of a "crisp" sensor, which would provide a measurement/observation of exactly one of these n discrete quantities, at probability 1.

To make the MLE, we need the *likelihood matrix*

$$L(\boldsymbol{m} \mid \boldsymbol{y}) = P(\boldsymbol{y} \mid \boldsymbol{m}) = \begin{bmatrix} p_{11} & p_{12} & \cdots & p_{1n} \\ p_{21} & p_{22} & \cdots & p_{2n} \\ \cdot & \cdot & \cdot & \cdot \\ p_{n1} & p_{n2} & \cdots & p_{nn} \end{bmatrix} \tag{12.4}$$

Note: The structure of this matrix is

	y_1	y_2	$\cdots$	y_n
m_1	p_{11}	p_{12}	$\cdots$	p_{1n}
m_2	p_{21}	p_{22}	$\cdots$	p_{2n}
	$\cdot$	$\cdot$	$\cdot$	$\cdot$
m_n	p_{n1}	p_{n2}	$\cdots$	p_{nn}

This probability matrix is essentially a "sensor model" that is governed by the characteristics of the sensor. In particular, an element p_{ij} of the matrix indicates the likelihood of existence of the specific discrete state m_i of the measurand m, when the measurement is the discrete value y_j.

Estimation problem: Determine the "probability vector" corresponding to the discrete parameter vector $\boldsymbol{m} = [m_1, m_2, \ldots, m_n]$, given the measured data. The parameter value corresponding to the largest probability value of this vector is the *MLE*.

Suppose that there are $r > 1$ sensors to measure the discrete quantity m. We will have r likelihood matrices $L(\boldsymbol{m} \mid {}^k\boldsymbol{y}) = P({}^k\boldsymbol{y} \mid \boldsymbol{m}),\quad k = 1, 2, \ldots, r$.

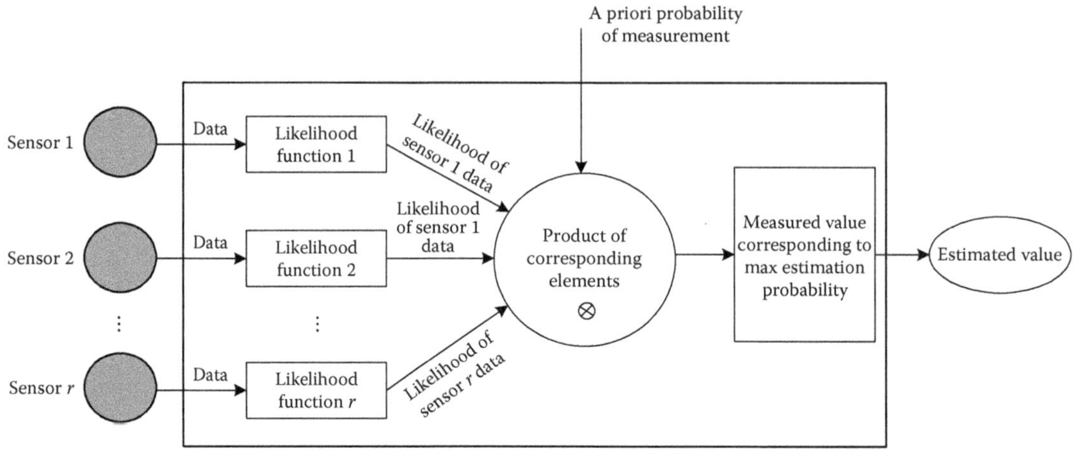

FIGURE 12.10
Bayesian approach to sensor fusion.

Assumption: Given a measurand value m, the measurements of the r sensors are independent.

Note: This is a weaker assumption than that the measurements of the r sensors are independent. The rationale for this "conditional independence" is that, since m is the only common basis for the r sensors, once m is given (i.e., m is no longer random), the randomness of that common basis is removed. Under this condition, the measured data from the r sensors can be assumed independent.

Under this assumption, if the kth sensor gives the measurement y_i and the lth sensor gives the measurement y_j, then the probability vector of the estimate of m is given by (see Chapter 7)

$$P(\boldsymbol{m}\,|\,{}^k y_i, {}^l y_j) = aP({}^k y_i\,|\,\boldsymbol{m}) \otimes P({}^l y_j\,|\,\boldsymbol{m}) \otimes P(\boldsymbol{m}) \tag{12.5}$$

Note: The symbol ⊗ denotes the multiplication of the corresponding elements in the two vectors. The parameter a has to be chosen such that the probability elements of the resulting vector add to 1.

According to the MLE approach (see Chapter 7), we select the element of $\boldsymbol{m}$ that corresponds to the largest probability value in the estimated probability vector $P(\boldsymbol{m}\,|\,{}^k y_i, {}^l y_j)$. This Bayesian approach to sensor fusion is shown in Figure 12.10.

12.4.3.2 Continuous Gaussian Problem

If the sensors are continuous, we can use MLE with the Gaussian assumption as developed in Chapter 7. The recursive formulation for Gaussian MLE with a single sensor, to determine the estimate $\hat{m}$ and the estimation error variance σ_m^2, is as follows (see Chapter 7):

$$\hat{m}_i = \frac{\sigma_w^2}{\left(\sigma_{m_{i-1}}^2 + \sigma_w^2\right)}\hat{m}_{i-1} + \frac{\sigma_{m_{i-1}}^2}{\left(\sigma_{m_{i-1}}^2 + \sigma_w^2\right)} y_i \tag{12.6}$$

$$\frac{1}{\sigma_{m_i}^2} = \frac{1}{\sigma_{m_{i-1}}^2} + \frac{1}{\sigma_w^2} \tag{12.7}$$

For the multisensor data fusion using this formulation, we can simply argue that, in the recursive formulation, it does not matter if the incoming data are from a different sensor as long as the measurements are done in sequence and the corresponding measurement error variance σ_w^2 is used in the computational step. Hence, the same recursive formulation for a single sensor may be used in sensor fusion under the Gaussian probability distribution.

Example 12.2

Two sensors (1 and 2) are used to measure the distance of an object. Since the two sensors have different characteristics, the data from both sensors are used in discrete MLE to determine the distance.

The distance m is treated as a discrete quantity, which can take one of three values: m_1 for near, m_2 for far, and m_3 for very far. They are represented by the vector

$$\boldsymbol{m} = \begin{bmatrix} m_1 \\ m_2 \\ m_3 \end{bmatrix} = \begin{bmatrix} \text{Near} \\ \text{Far} \\ \text{Very far} \end{bmatrix}$$

A sensor will make one of three discrete measurements corresponding to these three distance values, as $\boldsymbol{y} = [y_1 \quad y_2 \quad y_3] = [\text{near} \quad \text{far} \quad \text{very far}]$.

The two sensors have the following likelihood matrices:

Sensor 1: $L(\boldsymbol{m}|{}^1\boldsymbol{y}) = P({}^1\boldsymbol{y}|\,\boldsymbol{m}) =$

	y_1	y_2	y_3
m_1	0.75	0.05	0.20
m_2	0.05	0.55	0.40
m_3	0.20	0.40	0.40

Sensor 2: $L(\boldsymbol{m}|{}^2\boldsymbol{y}) = P({}^2\boldsymbol{y}|\,\boldsymbol{m}) =$

	y_1	y_2	y_3
m_1	0.45	0.35	0.20
m_2	0.35	0.60	0.05
m_3	0.20	0.05	0.75

From these likelihood matrices, it is clear that Sensor 1 is better at sensing distances that are close and Sensor 2 is better at sensing distances that are very far.

Suppose that in the beginning of the sensing process, we do not have a priori information about the distance of an object. Then,

$$P(\boldsymbol{m}) = \begin{bmatrix} 1/3 \\ 1/3 \\ 1/3 \end{bmatrix}$$

Suppose that we use Sensor 1 and then Sensor 2 and finally "fuse" the two measurements using Equation 12.5. Nine possibilities of reading data pairs are possible, and we will consider all these cases now.

Case 1: Sensor 1 measurement is y_1 *and Sensor 2 measurement is* y_1.

A posteriori probability vector of the estimate is

$$P\left(m \mid {}^1y_1, {}^2y_1\right) = aP\left({}^1y_1 \mid m\right) \otimes P\left({}^2y_1 \mid m\right) \otimes P(m) = a\begin{bmatrix}0.75\\0.05\\0.20\end{bmatrix} \otimes \begin{bmatrix}0.45\\0.35\\0.20\end{bmatrix} \otimes \begin{bmatrix}1/3\\1/3\\1/3\end{bmatrix}$$

$$= a\begin{bmatrix}0.3375\\0.0175\\0.04\end{bmatrix} \otimes \begin{bmatrix}1/3\\1/3\\1/3\end{bmatrix} = \begin{bmatrix}0.8544\\0.0443\\0.1013\end{bmatrix}$$

→MLE is m_1.

Case 2: Sensor 1 measurement is y_1 *and Sensor 2 measurement is* y_2.

A posteriori probability vector of the estimate is

$$P\left(m \mid {}^1y_1, {}^2y_2\right) = a\begin{bmatrix}0.75\\0.05\\0.20\end{bmatrix} \otimes \begin{bmatrix}0.35\\0.60\\0.05\end{bmatrix} \otimes \begin{bmatrix}1/3\\1/3\\1/3\end{bmatrix} = a\begin{bmatrix}0.2625\\0.03\\0.01\end{bmatrix} \otimes \begin{bmatrix}1/3\\1/3\\1/3\end{bmatrix} = \begin{bmatrix}0.8678\\0.0992\\0.033\end{bmatrix}$$

→ MLE is m_1 (with stronger probability).

Case 3: Sensor 1 measurement is y_1 *and Sensor 2 measurement is* y_3.

A posteriori probability vector of the estimate is

$$P\left(m \mid {}^1y_1, {}^2y_3\right) = a\begin{bmatrix}0.75\\0.05\\0.20\end{bmatrix} \otimes \begin{bmatrix}0.20\\0.05\\0.75\end{bmatrix} \otimes \begin{bmatrix}1/3\\1/3\\1/3\end{bmatrix} = a\begin{bmatrix}0.15\\0.0025\\0.15\end{bmatrix} \otimes \begin{bmatrix}1/3\\1/3\\1/3\end{bmatrix} = \begin{bmatrix}0.4959\\0.0082\\0.4959\end{bmatrix}$$

→ MLE is not definite (toss-up between m_1 and m_3).

Case 4: Sensor 1 measurement is y_2 *and Sensor 2 measurement is* y_1.

A posteriori probability vector of the estimate is

$$P\left(m \mid {}^1y_2, {}^2y_1\right) = a\begin{bmatrix}0.05\\0.55\\0.40\end{bmatrix} \otimes \begin{bmatrix}0.45\\0.35\\0.20\end{bmatrix} \otimes \begin{bmatrix}1/3\\1/3\\1/3\end{bmatrix} = a\begin{bmatrix}0.0225\\0.1925\\0.08\end{bmatrix} \otimes \begin{bmatrix}1/3\\1/3\\1/3\end{bmatrix} = \begin{bmatrix}0.0763\\0.6525\\0.2712\end{bmatrix}$$

→ MLE is m_2.

Case 5: Sensor 1 measurement is y_2 *and Sensor 2 measurement is* y_2.

A posteriori probability vector of the estimate is

$$P\left(m \mid {}^1y_2, {}^2y_2\right) = a\begin{bmatrix}0.05\\0.55\\0.40\end{bmatrix} \otimes \begin{bmatrix}0.35\\0.60\\0.05\end{bmatrix} \otimes \begin{bmatrix}1/3\\1/3\\1/3\end{bmatrix} = a\begin{bmatrix}0.0175\\0.33\\0.02\end{bmatrix} \otimes \begin{bmatrix}1/3\\1/3\\1/3\end{bmatrix} = \begin{bmatrix}0.0476\\0.8980\\0.0544\end{bmatrix}$$

→ MLE is m_2 (with stronger probability).

Case 6: Sensor 1 measurement is y_2 *and Sensor 2 measurement is* y_3.

A posteriori probability vector of the estimate is

$$P\left(m \mid {}^1y_2, {}^2y_3\right) = a\begin{bmatrix}0.05\\0.55\\0.40\end{bmatrix} \otimes \begin{bmatrix}0.20\\0.05\\0.75\end{bmatrix} \otimes \begin{bmatrix}1/3\\1/3\\1/3\end{bmatrix} = a\begin{bmatrix}0.01\\0.0275\\0.3\end{bmatrix} \otimes \begin{bmatrix}1/3\\1/3\\1/3\end{bmatrix} = \begin{bmatrix}0.0296\\0.0815\\0.8889\end{bmatrix}$$

→ MLE is m_3.

Case 7: Sensor 1 measurement is y_3 and Sensor 2 measurement is y_1.

A posteriori probability vector of the estimate is

$$P\left(\boldsymbol{m} \mid {}^1y_3, {}^2y_1\right) = a\begin{bmatrix} 0.20 \\ 0.40 \\ 0.40 \end{bmatrix} \otimes \begin{bmatrix} 0.45 \\ 0.35 \\ 0.20 \end{bmatrix} \otimes \begin{bmatrix} 1/3 \\ 1/3 \\ 1/3 \end{bmatrix} = a\begin{bmatrix} 0.09 \\ 0.14 \\ 0.08 \end{bmatrix} \otimes \begin{bmatrix} 1/3 \\ 1/3 \\ 1/3 \end{bmatrix} = \begin{bmatrix} 0.2903 \\ 0.4516 \\ 0.2581 \end{bmatrix}$$

→ MLE is m_2.

Case 8: Sensor 1 measurement is y_3 and Sensor 2 measurement is y_2.

A posteriori probability vector of the estimate is

$$P\left(\boldsymbol{m} \mid {}^1y_3, {}^2y_2\right) = a\begin{bmatrix} 0.20 \\ 0.40 \\ 0.40 \end{bmatrix} \otimes \begin{bmatrix} 0.35 \\ 0.60 \\ 0.05 \end{bmatrix} \otimes \begin{bmatrix} 1/3 \\ 1/3 \\ 1/3 \end{bmatrix} = a\begin{bmatrix} 0.07 \\ 0.24 \\ 0.02 \end{bmatrix} \otimes \begin{bmatrix} 1/3 \\ 1/3 \\ 1/3 \end{bmatrix} = \begin{bmatrix} 0.2121 \\ 0.7273 \\ 0.0606 \end{bmatrix}$$

→ MLE is m_2.

Case 9: Sensor 1 measurement is y_3 and Sensor 2 measurement is y_3.

A posteriori probability vector of the estimate is

$$P\left(\boldsymbol{m} \mid {}^1y_3, {}^2y_3\right) = a\begin{bmatrix} 0.20 \\ 0.40 \\ 0.40 \end{bmatrix} \otimes \begin{bmatrix} 0.20 \\ 0.05 \\ 0.75 \end{bmatrix} \otimes \begin{bmatrix} 1/3 \\ 1/3 \\ 1/3 \end{bmatrix} = a\begin{bmatrix} 0.04 \\ 0.02 \\ 0.3 \end{bmatrix} \otimes \begin{bmatrix} 1/3 \\ 1/3 \\ 1/3 \end{bmatrix} = \begin{bmatrix} 0.1111 \\ 0.0556 \\ 0.8333 \end{bmatrix}$$

→ MLE is m_3 (with stronger probability).

12.4.3.3 Sensor Fusion Using Kalman Filter

As studied in Chapter 7, Kalman filter uses "multiple output measurements" to estimate any single state variable. Hence, it inherently uses sensor fusion. Two approaches may be used for sensor fusion with Kalman filter:

1. Use a single measurement model for all r sensors (i.e., r output measurements) where all r measurements are represented as an rth-order measurement vector. Then, no change is needed to the Kalman filter algorithm presented in Chapter 7. We can just apply the Kalman filter to all r measurements simultaneously (i.e., in parallel).
2. Use r different measurement models for the r sensors. Here, the proper output equation has to be used in the Kalman filter, depending on the currently used sensor. Then, the Kalman filter is applied sequentially to the measurements of r sensors.

Note 1: In approach 2, it is better if the system is separately observable for all r measurement models.

Note 2: Any version of Kalman filter (linear, extended, unscented, etc.) may be used in sensor fusion.

A problem based on the monitoring and estimation of a milling machine using Kalman filter has been given in Chapter 7. An extension of this problem, with multisensor monitoring and fusion through Kalman filter, is given as a problem in this chapter. It is seen that the Kalman filter approach to sensor fusion is quite straightforward.

12.4.3.4 Sensor Fusion Using Dempster–Shafer Evidence Theory

As we have seen, a priori probabilities are needed in sensor fusion using the Bayesian approach. This can be a shortcoming of the method, in some situations, because a priori probabilities are generally unknown and it is rather difficult to determine them in a reliable manner. The DS approach overcomes this problem because it does not make assumptions about the past (a priori information) for making decisions and depends entirely on the current evidence (data and perhaps expertise on the problem). The DS approach may be viewed as a generalization of the Bayesian approach. As any other method of data fusion, the DS approach improves the quality of the decision (fused result) when more evidence (data, knowledge) is incorporated.

12.4.3.4.1 DS Approach

The DS approach uses the DST of evidence. This theory involves three components:

1. Frame of discernment Θ
2. Mass function (also called the probability mass function or basic probability assignment function) $m(A)$
3. DS rule of combination (method of reasoning based on belief)

The *frame of discernment* Θ is the set of mutually exclusive (i.e., nonoverlapping) and exhaustive (i.e., complete) propositions concerning the considered problem. It consists of n elements (n propositions). The *power set of* Θ is the set of all subsets of Θ including the empty (null) set Φ. It is denoted by $\Omega(\Theta)$. *Note*: The power set has 2^n sets, including Φ and Θ.

The *mass function* $m(A)$ assigns a value between 0 and 1 to every element (subset) A of the power set Ω. Specifically,

$$m{:}\Omega(\Theta) \rightarrow [0,1] \tag{12.8}$$

The mass function value $m(A)$ represents the level of validity (or confidence or belief) of the proposition A, based on available evidence (data and expertise). In particular, we have

$$m(\Phi) = 0; \quad \sum_{A \subseteq \Omega(\Theta)} m(A) = 1$$

The *DS rule of combination* is the formula of decision-making (or reasoning or "fusion") in the DS method. Specifically, it is the method of combining evidence from independent sources regarding a particular proposition. This is done by combining the mass functions of the independent sources. Specifically, suppose that m_1 and m_2 are the mass functions corresponding to two independent sets of data (evidence), in the same frame of discernment Θ. Then, based on this evidence, the mass function of a proposition A, in the same frame of discernment, is obtained by the DS rule of combination, as follows:

$$m(A) = m_1 \oplus m_2 = \left\{ \begin{array}{ll} 0 & \text{if } A = \Phi \\ \dfrac{\sum_{B \cap C = A} m_1(B) m_2(C)}{1 - K} & \text{if } A \neq \Phi \end{array} \right\} \tag{12.9}$$

where

$$K = \sum_{B \cap C = \Phi} m_1(B) m_2(C) \tag{12.10}$$

Note that, in fusion, we sum the product of the mass function values that correspond to the overlap A of their propositions. Furthermore, we normalize the result using K, which is the sum of the product of the mass functions corresponding to null overlap (i.e., contradictory mass function values). Clearly, K represents the degree of conflict in the two sources of evidence. In particular, when $K = 0$, there is no conflict in the two sets of evidence. If $K = 1$, the two sets of evidence are in complete contradiction, and the combine mass function becomes 0/0, which is not defined.

Note: The combined mass function according to Equation 12.9 is called the "orthogonal sum." This operation is both commutative and associative. The mass functions of three or more sets of evidence can be combined using the same formula (12.9). Specifically, first, we combine two mass functions, and then we combine the result with the third mass function, and so on. The order in which this is done does not matter because the operation given by Equation 12.9 is associative.

Some advantages of the DS approach of data fusion (particularly in comparison to the Bayesian approach) are as follows:

1. It combines the evidence of "propositions." A proposition may consist of multiple hypotheses. The Bayesian approach assigns evidence to a single hypothesis.
2. It handles ignorance (i.e., lack of evidence) and contradiction (i.e., opposing evidence) separately. In the Bayesian approach, any evidence that does not support a hypothesis is assigned to the opposite hypothesis (i.e., negation of the original hypothesis).
3. Unlike the Bayesian approach, it does not rely on a priori probabilities but rather on the current evidence (data and knowledge).

Example 12.3

Two sensors, denoted by 1 and 2, are available for measuring the distance of an object. Each sensor provides its measurement as one of the three states: near (N), far (F), and very far (VF).

Suppose that based on the data from Sensor 1, the following mass function is computed:

$$m_1(N, F, VF) = [0.7 \quad 0.2 \quad 0.1]$$

Similarly, suppose that based on the data from Sensor 2, the following mass function is computed:

$$m_2(N, F, VF) = [0.8 \quad 0.1 \quad 0.1]$$

Using the DS method, combine these two pieces of evidence to obtain a fused mass function. Comment on the result.

TABLE 12.2

Element Products of the Mass Functions

	$m_1(N) = 0.7$	$m_1(F) = 0.2$	$m_1(VF) = 0.1$
$m_2(N) = 0.8$	$m(N) = 0.56$	$m(\Phi) = 0.16$	$m(\Phi) = 0.08$
$m_2(F) = 0.1$	$m(\Phi) = 0.07$	$m(F) = 0.02$	$m(\Phi) = 0.01$
$m_2(VF) = 0.1$	$m(\Phi) = 0.07$	$m(\Phi) = 0.02$	$m(VF) = 0.01$

Solution

The frame of discernment of the problem is

$$\Theta = [N \quad F \quad VF]$$

Note that the elements in Θ are mutually exclusive and exhaustive, as required.

The element products m of the given two mass functions are presented in Table 12.2.

The normalizing parameter (degree of conflict), as given by Equation 12.10, is the sum of the values of $m(\Phi)$. We have

$$K = \sum m(\Phi) = 0.16 + 0.08 + 0.07 + 0.01 + 0.07 + 0.02 = 0.41$$

This is a somewhat low value (<0.5) for the degree of conflict.

Using Equation 12.9, the "fused" mass function values are computed as

$$m_f(N) = \frac{0.56}{1-0.41} = 0.94915; \quad m_f(F) = \frac{0.02}{1-0.41} = 0.03390; \quad m_f(VF) = \frac{0.01}{1-0.41} = 0.01695$$

Hence, the mass function of the combined (fused) evidence is

$$m_f(N,F,VF) = [0.94915 \quad 0.03390 \quad 0.01695]$$

Note: It is seen that the conflict parameter K has normalized the mass function. Specifically, 0.94915 + 0.03390 + 0.01695 = 1.

Furthermore, the combined (fused) mass function has considerably reinforced the belief that the measurement is N.

12.4.3.4.2 Determination of Mass Function

We have addressed the method of data fusion by the DS method when the evidence (data, knowledge) is presented as mass functions. However, the way to convert evidence into mass functions has to be known as well. This topic is addressed now.

It is not possible to present one general method for converting evidence (data, knowledge) into a mass function. The reasons for this include the following:

1. The method of determining a mass function depends on the nature and the requirements of the particular application.
2. Evidence (data, knowledge, etc.) may come from many different forms.
3. Belief level of given evidence may be expressed in different ways.

Now, we present one approach of generating a mass function from sensor data (evidence) and give an illustrative application in the area of water quality determination. The same approach can be applied to a variety of other applications. The present approach may be summarized as follows:

1. A set of specifications (guidelines) is available. A specification indicates what representative data set (measurements) corresponds to a particular class of inference (a proposition in the power set of the frame of discernment). Typically, the guidelines will cover all the possible propositions in the particular problem.
2. A given measurement (data, evidence) is compared with a guideline for a specific class. The closeness of the measurement to each guideline is computed (using a "distance" measure).
3. The mass function value of the given data that supports a particular inference (proposition) is the inverse of the distance value between the data and the guideline corresponding to the particular inference.

Clearly, the method depends on the specific "distance metric" that is used. For example, consider the 2D Cartesian coordinate frame with the axes 1 and 2. This represents a plane (a 2D space), as shown in Figure 12.11. Consider the two points P and Q, which have coordinates (p_1,p_2) and (q_1,q_2), respectively, on the plane. The distance between the two points is given by (using Pythagoras' theorem) $\sqrt{(q_1-p_1)^2+(q_2-p_2)^2}$. This is called the "Euclidean distance" (in the 2D space).

To generalize the concept of distance in an n-D space (i.e., in $\mathbb{R}^n$), consider two vectors $\boldsymbol{p}$ and $\boldsymbol{q}$ with elements $[p_1 \quad \ldots \quad p_n]$ and $[q_1 \quad \ldots \quad q_n]$, respectively. Three possible distance metrics are given here:

1. Euclidean distance:

$$d(\boldsymbol{p},\boldsymbol{q}) = d(\boldsymbol{q},\boldsymbol{p}) = \sqrt{(q_1-p_1)^2+(q_2-p_2)^2+\cdots+(q_n-p_n)^2} = \sqrt{\sum_{i=1}^{n}(q_i-p_i)^2} \quad (12.11)$$

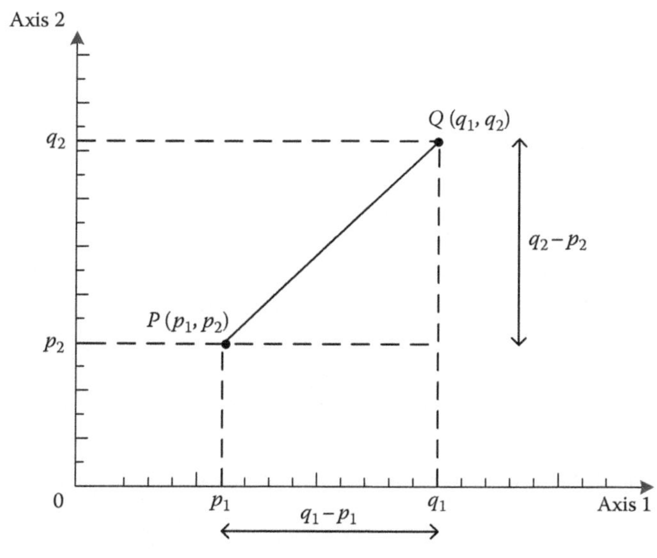

FIGURE 12.11
The distance metric in a 2D Cartesian coordinate system.

2. Manhattan distance: This is also called the taxicab distance and is given by

$$d(p,q) = d(q,p) = |q_1 - p_1| + |q_2 - p_2| + \cdots + |q_n - p_n| = \sum_{i=1}^{n} |q_i - p_i| \tag{12.12}$$

Note that in this case the distance is measured in orthogonal segments and their absolute values are summed. Hence, it does not represent the direct distance (as in the Euclidean distance).

3. Minkowski distance: The Minkowski distance of order r is given by

$$d(p,q) - d(q,p) = \left(\sum_{i=1}^{n} |q_i - p_i|^r \right)^{1/r} \tag{12.13}$$

Clearly, this is generalization of the previous two distance metrics. Specifically, $p = 2$ corresponds to the Euclidean distance and $p = 1$ corresponds to the Manhattan distance.

In what follows, we will use the Euclidean distance. The use of any other distance metric can be done in the same manner.

Suppose that there are m sensors, which generate m data values $[s_1, s_2, \ldots, s_m]$. Let us apply the DS method to classify a particular data set into a specific class. Suppose that Class i is specified by the value set (guideline) $[c_{i1}, c_{i2}, \ldots, c_{im}]$. Then, the distance of the given data set (measurement) to Class i is given by

$$d_i = \left[\sum_{j=1}^{m} \left(\frac{s_j - c_{ij}}{c_{j\max} - c_{j\min}} \right)^2 \right] \tag{12.14}$$

where

$c_{j\max} = \max_i(c_{ij})$
$c_{j\min} = \min_i(c_{ij})$

Note that the terms are divided by $c_{j\max} - c_{j\min}$ for the purpose of normalization. Also, often, $c_{j\min}$ is taken to be 0.

Suppose that there are k inference classes, and the distances of the specific data set $[s_1, s_2, \ldots, s_m]$ (evidence) to all these inference classes are computed as d_i, $i = 1, \ldots, k$. The mass function value m_i of the specific data set, corresponding to the inference Class i, is computed as the normalized inverse of the corresponding distance. Specifically,

$$m_i = \frac{1/d_i}{1/d_1 + 1/d_2 + \cdots + 1/d_k} \tag{12.15}$$

It is seen from this that the following condition is satisfied, as required:

$$\sum_{i=1}^{k} m_i = 1 \tag{12.16}$$

Example 12.4

In a network of water quality monitoring, consider two sensor nodes, each having the same three types of sensors: pH value (pH) sensor, DO sensor, and electrical conductivity (CD) sensor. Suppose that the two sensor nodes generate the data given in Table 12.3.

The water quality is determined as one of the following three categories (quality classes or inferences): good, fair, and poor.

The specifications (guideline sensory values) for these three classes of water quality are given in Table 12.4.

Determine the mass functions corresponding to the two sets of data. Fuse the two mass functions using the DS method. Comment on the result.

Solution

For the data from Node 1,
distance to "good" is

$$d_{1g} = \left[\left(\frac{6.5-7.5}{7.5-0}\right)^2 + \left(\frac{5.0-7.0}{7.0-0}\right)^2 + \left(\frac{600-200}{800-0}\right)^2\right]^{1/2} = (0.01778+0.08163+0.250)^{1/2} = 0.5911$$

distance to "fair" is

$$d_{1f} = \left[\left(\frac{6.5-6.0}{7.5-0}\right)^2 + \left(\frac{5.0-5.0}{7.0-0}\right)^2 + \left(\frac{600-500}{800-0}\right)^2\right]^{1/2} = (0.00444+0+0.01563)^{1/2} = 0.1417$$

and distance to "poor" is

$$d_{1p} = \left[\left(\frac{6.5-4.5}{7.5-0}\right)^2 + \left(\frac{5.0-3.0}{7.0-0}\right)^2 + \left(\frac{600-800}{800-0}\right)^2\right]^{1/2} = (0.07111+0.08163+0.0625)^{1/2} = 0.4639$$

TABLE 12.3

Data from the Two Sensor Nodes

Sensor	pH	DO (mg/L)	CD (μS/cm)
Node 1 data	6.5	5.0	600.0
Node 2 data	6.0	4.5	550.0

TABLE 12.4

Representative (Specified) Sensory Values for the Quality Classes

Quality Class	Guideline pH	Guideline DO	Guideline CD
Good	7.5	7.0	200.0
Fair	6.0	5.0	500.0
Poor	4.5	3.0	800.0

These follow from Equation 12.14.

For the data from Node 2,

distance to "good" is

$$d_{2g}=\left[\left(\frac{6.0-7.5}{7.5-0}\right)^2+\left(\frac{4.5-7.0}{7.0-0}\right)^2+\left(\frac{550-200}{800-0}\right)^2\right]^{1/2}=(0.040+0.12755+0.19141)^{1/2}=0.5991$$

distance to "fair" is

$$d_{2f}=\left[\left(\frac{6.0-6.0}{7.5-0}\right)^2+\left(\frac{4.5-5.0}{7.0-0}\right)^2+\left(\frac{550-500}{800-0}\right)^2\right]^{1/2}=(0+0.00510+0.00391)^{1/2}=0.0949$$

and distance to "poor" is

$$d_{2p}=\left[\left(\frac{6.0-4.5}{7.5-0}\right)^2+\left(\frac{4.5-3.0}{7.0-0}\right)^2+\left(\frac{550-800}{800-0}\right)^2\right]^{1/2}=(0.040+0.04592+0.09766)^{1/2}=0.4285$$

The mass function for the evidence from Node 1 is

$$m_{1g}=\frac{1/d_{1g}}{1/d_{1g}+1/d_{1f}+1/d_{1p}}=\frac{1/0.5911}{1/0.5911+1/0.1417+1/0.4639}=\frac{1.69176}{10.90456}=0.1551$$

$$m_{1f}=\frac{1/d_{1f}}{1/d_{1g}+1/d_{1f}+1/d_{1p}}=\frac{1/0.1417}{1/0.5911+1/0.1417+1/0.4639}=\frac{7.05716}{10.90456}=0.6472$$

$$m_{1p}=\frac{1/d_{1p}}{1/d_{1g}+1/d_{1f}+1/d_{1p}}=\frac{1/0.4639}{1/0.5911+1/0.1417+1/0.4639}=\frac{2.15564}{10.90456}=0.1977$$

The mass function for the evidence from Node 2 is

$$m_{2g}=\frac{1/d_{2g}}{1/d_{2g}+1/d_{2f}+1/d_{2p}}=\frac{1/0.5991}{1/0.5991+1/0.0949+1/0.4285}=\frac{1.66917}{14.5403}=0.1148$$

$$m_{2f}=\frac{1/d_{2f}}{1/d_{2g}+1/d_{2f}+1/d_{2p}}=\frac{1/0.0949}{1/0.5991+1/0.0949+1/0.4285}=\frac{10.53741}{14.5403}=0.7247$$

$$m_{2p}=\frac{1/d_{2p}}{1/d_{2g}+1/d_{2f}+1/d_{2p}}=\frac{1/0.4285}{1/0.5991+1/0.0949+1/0.4285}=\frac{2.33372}{14.5403}=0.1605$$

These follow from Equation 12.15.

Fusion

The mass functions corresponding to the two data sets are fused now, using the DS method.

The element products of the given two mass functions are presented in Table 12.5.

The normalizing parameter (degree of conflict), as given by Equation 12.10, is the sum of the values of $m(\Phi)$. We have

$$K=\sum m(\Phi)=0.07430+0.02270+0.11240+0.14327+0.02489+0.10388=0.48144$$

TABLE 12.5

Element Products of the Mass Functions

	$m_{1g} = 0.1551$	$m_{1f} = 0.6472$	$m_{1p} = 0.1977$
$m_{2g} = 0.1148$	$m(g) = 0.01781$	$m(\Phi) = 0.07430$	$m(\Phi) = 0.02270$
$m_{2f} = 0.7247$	$m(\Phi) = 0.11240$	$m(f) = 0.46903$	$m(\Phi) = 0.14327$
$m_{2p} = 0.1605$	$m(\Phi) = 0.02489$	$m(\Phi) = 0.10388$	$m(p) = 0.03173$

This represents a somewhat low value (<0.5) for the degree of conflict.

Using Equation 12.9, the "fused" membership values are obtained as

$$m_g = \frac{0.01781}{1-0.48144} = 0.0343; \quad m_f = \frac{0.46903}{1-0.48144} = 0.9045; \quad m_p = \frac{0.03173}{1-0.48144} = 0.0612$$

Hence, the mass function of the combined (fused) evidence is

$$m(g,f,p) = \begin{bmatrix} 0.0343 & 0.9045 & 0.0612 \end{bmatrix}$$

Note: The conflict parameter K has normalized the mass function. Specifically, 0.0343 + 0.9045 + 0.0612 = 1.

It is seen that the combined (fused) mass function has considerably reinforced the belief that the measurement is *fair*.

12.4.3.5 Sensor Fusion Using Neural Networks

Inspired by the biological architecture of neurons in the brain, NNs are massively connected networks of computational "neurons." They possess parallel and distributed processing structures. Their key characteristics include the following:

- They can take many inputs (e.g., from many sensors).
- They are able to learn by example. (*Note*: Learning is an attribute of intelligence.)
- They can approximate highly nonlinear functions.
- They have massive computing power.
- They have memory of their processed information.

All these characteristics are useful in sensor fusion.

The *nodes* of an NN, which are connected through weighted pathways called "synapses," are arranged into an *input layer*, one or more *hidden layers*, and an *output layer*. At a node, weighted inputs are summed, thresholded, and passed through an *activation function*, to generate the node output. A variety of functions may be used as the activation function. They include sigmoid function, step function, hyperbolic tangent function, linear function, signum function, and sigmoid derivative function. In sensor fusion, the nodes in the input layer are given data from multiple sensors, and the hidden layers carry out sensor fusion. The fused outputs are provided by the nodes in the output layer. The structure and the operation of a node of an NN are shown in Figure 12.12.

Many types of NNs are available. In a *feedforward network* (static network), the signal flow from a node to another node is in the forward direction only (no feedback paths). In a *feedback network* (dynamic or *recurrent network*), the outputs of one or more nodes are

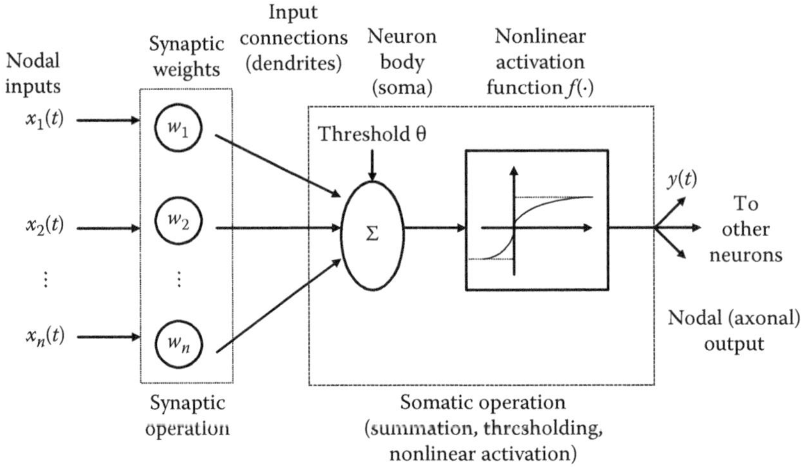

FIGURE 12.12
Operations at a node of a neural network.

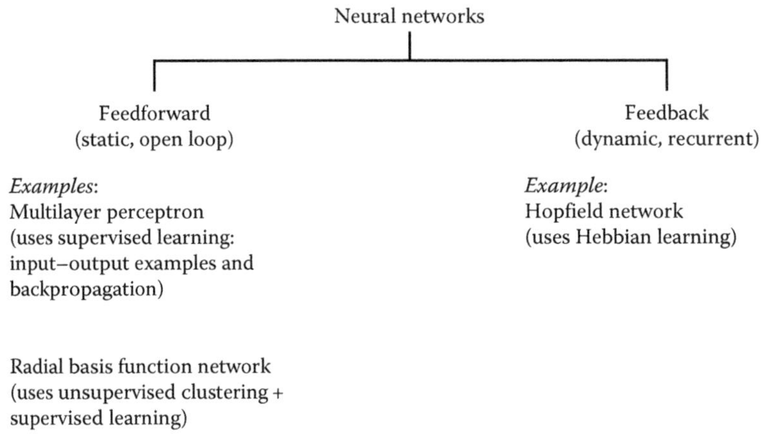

FIGURE 12.13
Classification of neural networks.

fed back into one or more nodes in a previous layer. This classification is summarized in Figure 12.13.

Note: Feedback provides the capability of "memory."

Learning: An NN has to "learn" how to solve the problem that is given to it. This may be done by using examples (and "training" the NN through them) or through experience of carrying out the task where a mechanism is provided to reward correct actions and penalized wrong actions. In supervised learning, an external teacher provides input–output data sets (examples). During training, for a given input, the network output is compared with the desired output. A learning rule (e.g., gradient descent rule) is used to adjust network parameters to minimize the error (e.g., backpropagation algorithm). In unsupervised learning, there is no teacher to provide input–output examples for network training. Instead, prior learned knowledge, guidelines, local information, and internal control are used to update network parameters. *Note*: In this case, for a given input, the correct output is unknown a priori. The associated steps are as follows:

1. Input data are given to the network.
2. The network output is checked based on prior knowledge, guidelines, and internal information.
3. The network parameters are adjusted using Step 2 and an adaptation rule.

Reinforcement learning mimics the adaptive behavior of a human to an environment. It is not supervised in the sense that there is no teacher to give input–output examples. The procedure is as follows:

1. Network connections are modified according to performance and corresponding feedback information from the environment (i.e., correct or wrong response).
2. In a correct response, corresponding connections are strengthened (reward), while in a wrong response, corresponding connections are weakened (penalty).

Multilayer perceptron (MLP) topology: A popular structure of an NN is the MLP, as shown in Figure 12.14. This is a feedforward network (i.e., among the network nodes, information flow is only in the forward direction). It has input layer, one or more hidden layers, and an output layer.

Note: The required number of hidden layers depends on the type of problem (there is no formal theory that governs this choice). An MLP with one hidden layer (with enough nodes and sigmoid activation functions) can approximate any function (i.e., this is a universal approximator).

Backpropagation learning: This is the method of updating weights in an MPL. It is called the "training" of the NN. The method uses input–output training data (i.e., teaching using example) and the gradient descent to minimize squared error (Euclidean norm). The

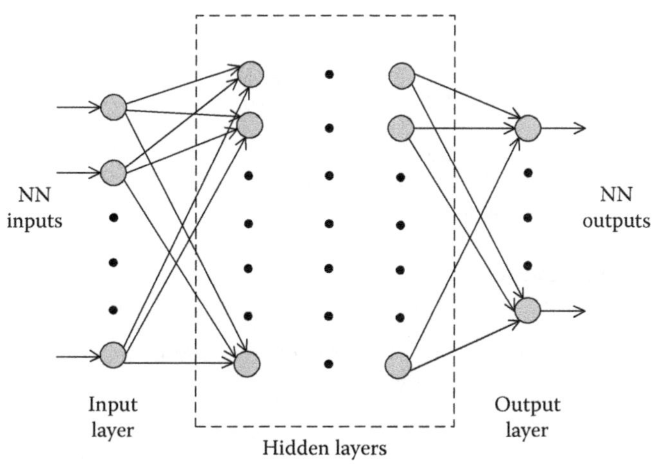

FIGURE 12.14
A multilayer perceptron.

training data are applied at the input of the NN, and the resulting output o of the NN is determined. The correct (required) output t is known (in the training data). One or several sets of training data may be used, online or off-line. For n sets of data, we minimize the cumulative error:

$$e_c = \sum_{k=1}^{n} e(k) = \frac{1}{2}\sum_{k=1}^{n}\sum_{i=1}^{m}[t_i(k) - o_i(k)]^2 \tag{12.17}$$

where
- t is the training (correct) output
- o is the actual output (for the same input) with m elements (output nodes)
- n is the number of data sets

In order to minimize the error, the NN weights are updated in the direction of the "negative" gradient of the error, starting from the output layer and ending at the input layer (i.e., backpropagation). For the kth training data set, the weight update is given by

$$\Delta w^l = -\eta \frac{\partial e(k)}{\partial w^l} \tag{12.18}$$

where
- η is the learning rate parameter (>0)
- w^l are the weights (vector) at the input of the lth layer, for all signal sums coming from the nodes of the l-first layer

Now, drop k, as it is clear that we are considering one data set. Specifically, for the weight w_{ij} for the NN connection (branch) from the jth node of the l-first layer and the ith node of the lth layer, we have

$$\Delta w_{ij}^l = -\eta \frac{\partial e}{\partial w_{ij}^l} = -\eta \frac{\partial e}{\partial o_i^l}\frac{\partial o_i^l}{\partial sum_i^l}\frac{\partial sum_i^l}{\partial w_{ij}^l}$$

Now, since $o_i^l = f(sum_i^l)$ and $sum_i^l = \sum_j w_{ij}^l o_j^{l-1}$, we have

$$\frac{\partial o_i^l}{\partial sum_i^l} = \frac{df\left(sum_i^l\right)}{dsum_i^l} = f'\left(sum_i^l\right) \quad \text{and} \quad \frac{\partial sum_i^l}{\partial w_{ij}^l} = o_j^{l-1}$$

Hence, the weight update is given by

$$\Delta w_{ij}^l = -\eta \frac{\partial e}{\partial o_i^l} f'\left(sum_i^l\right) o_j^{l-1} = \eta \delta_i^l o_j^{l-1} \tag{12.19}$$

where

$$\delta_i^l = -\frac{\partial e}{\partial o_i^l} f'\left(sum_i^l\right) \tag{12.20}$$

Also,

$$\frac{\partial e}{\partial o_i^{l-1}} = \sum_p \frac{\partial e}{\partial o_p^l}\frac{\partial o_p^l}{\partial o_i^{l-1}} \tag{12.21}$$

Since $o_p^l = f\left(\sum_i w_{pi}^l o_i^{l-1}\right)$, we have

$$\frac{\partial o_p^l}{\partial o_i^{l-1}} = f'\left(sum_p^l\right) w_{pi}^l \tag{12.22}$$

Now, substitute this equation and Equation 12.20 into 12.21:

$$\frac{\partial e}{\partial o_i^{l-1}} = \sum_p \frac{\partial e}{\partial o_p^l} f'\left(sum_p^l\right) w_{pi}^l = -\sum_p \delta_p^l w_{pi}^l \tag{12.23}$$

Substituting this equation into Equation 12.20, we get

Backpropagation scheme

$$\delta_i^{l-1} = f'\left(sum_i^{l-1}\right)\sum_p \delta_p^l w_{pi}^l \tag{12.24}$$

Backpropagation Steps:

Step 1: Initialize weights and thresholds to small random values.

Step 2: Choose an input–output training data set.

Step 3: Compute from input to output $o_p^l = f\left(\sum_i w_{pi}^l o_i^{l-1}\right)$.

Step 4: Compute output error e and backpropagation parameter $\delta_i^L = [t_i - o_i^L] f'(sum_i^L)$ at output layer (L).

Step 5: Update weights using $\Delta w_{ij}^l = \eta \delta_i^l o_j^{l-1}$ (Equation 12.19) with backpropagation $\delta_i^{l-1} = f'(sum_i^{l-1})\sum_p \delta_p^l w_{pi}^l$ from $l = L$ to 3 (Equation 12.24).

Step 6: Repeat steps 2 through 5 for another training data set.

Step 7: After using all the training data sets (i.e., one epoch), if final error $e <$ a predetermined tolerance, then network has been trained.

If not, repeat the process for another epoch.

Notes: For the output layer,

$$\delta_i^L = -\frac{\partial e}{\partial o_i^L} f'\left(sum_i^L\right) = \left(t_i - o_i^L\right) f'\left(sum_i^L\right)$$

For sigmoidal function (with $\lambda = 1$), $f' = f(1 - f)$.

Radial basis function (*RBF*) *network*: The topology of an RBF network is shown in Figure 12.15.

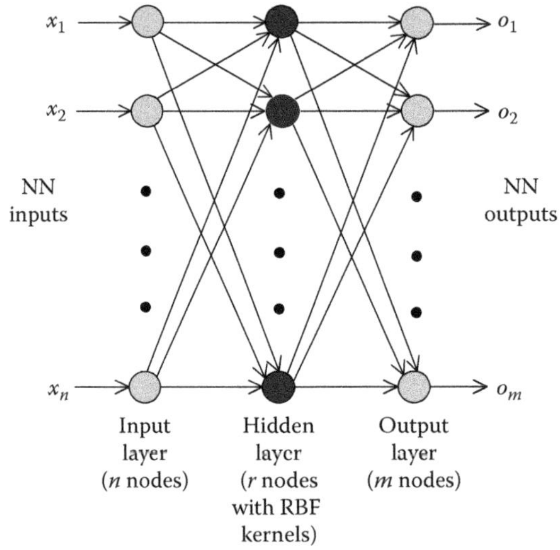

FIGURE 12.15
A radial basis function network.

It has an input layer (n-D), a single hidden layer (r-D), and an output layer (m-D). For the hidden layer, activation function is equal to RBF:

$$g_j(\boldsymbol{x}) = R\left(\frac{\|\boldsymbol{x} - \boldsymbol{v}_j\|}{\sigma_j}\right) \tag{12.25}$$

RBF parameters: $\boldsymbol{v}_j$ is the center vector (n-D); σ_j is the width parameter, $j = 1, \ldots, r$; and $\boldsymbol{x}$ is the network input vector (n-D).

Network output relation:

$$o_i(\boldsymbol{x}) = \sum_{j=1}^{r} w_{ij} g_j(\boldsymbol{x}) \quad \text{for output nodes } i = 1, \ldots, m \tag{12.26}$$

where w_{ij} is the weight from hidden node j to output node i.

Properties of an RBF:

- Is symmetrical
- Reaches maximum at center (at $\boldsymbol{x} = \boldsymbol{v}_i$)
- Goes to zero at the two tail ends (infinite distance from center)
- Has smaller width σ_i and sharper function peak

Typical RBFs:

Gaussian kernel function:

$$g_i(\boldsymbol{x}) = \exp\left(\frac{-\|\boldsymbol{x} - \boldsymbol{v}_i\|^2}{2\sigma_i^2}\right) \tag{12.27}$$

Logistic function:

$$g_i(x) = \frac{1}{1+\exp\left(\|x - v_i\|^2 / \sigma_i^2\right)} \tag{12.28}$$

RBF network learning:

1. Use unsupervised learning to determine activation function parameters (v_i, σ_i).
2. Use supervised learning to determine the weights w_{ij}.

Example

1. Given r pairs of input–output training data vectors: [$\boldsymbol{x}_i$, $\boldsymbol{o}_i$] and $i = 1, 2, \ldots, r$.
2. Use Gaussian activation function with training data $\boldsymbol{x}_i$ as the centers (unsupervised).
3. Select σ_i from past experience, input bandwidth estimate, etc. (unsupervised).
4. Form the two matrices:

$$\boldsymbol{G} = [g_{ij}]_{r\times r}, \quad \text{where } g_{ij} = \exp\left[\frac{-\|\boldsymbol{x}_j - \boldsymbol{x}_i\|^2}{2\sigma_i}\right], i = 1,2,\ldots,r; j = 1,2,\ldots,r \tag{12.29}$$

$$\boldsymbol{O} = [\boldsymbol{o}_1, \boldsymbol{o}_2, \ldots, \boldsymbol{o}_r]_{m\times r} \tag{12.30}$$

   Then, $\boldsymbol{O} = \boldsymbol{WG}$ where $\boldsymbol{W}$ is the matrix of connection weights.

5. Determine the connection weights (supervised learning) using

$$\boldsymbol{W}_{m\times r} = \boldsymbol{O}_{m\times r}\boldsymbol{G}_{r\times r}^{-1} \tag{12.31}$$

Note: With $p \neq r$ pairs of training data vectors $\boldsymbol{x}_i$ and $\boldsymbol{o}_i$, $i = 1, 2, \ldots, p$

$$\boldsymbol{W} = \boldsymbol{OG}^{+} \tag{12.32}$$

where pseudoinverse

$$\boldsymbol{G}^{+} = \boldsymbol{G}^{T}(\boldsymbol{G}^{T}\boldsymbol{G})^{-1} \tag{12.33}$$

Hybrid use with fuzzy logic (neurofuzzy systems): Fuzzy logic, which somewhat mimics the reasoning mechanism of humans, may be integrated with NNs to enhance the overall performance. Such neurofuzzy systems are applicable in sensor fusion. There are three general ways to integrate fuzzy logic and NNs:

1. Incorporate NN as a facilitator/tool in data fusion with fuzzy logic (e.g., learn/train rules and membership functions using NN). Use fuzzy aggregation/inference for sensor fusion.
2. Use fuzzy logic to represent the features of sensor data. Use an NN (possibly with fuzzy neurons, fuzzy weights, etc.) for sensor/feature fusion.

3. Use separate fuzzy subsystems and NN subsystems to carry out different fusion activities (e.g., fuzzy system does "information fusion" for quality assessment, performance assessment, high-level supervisory/tuning actions, etc.; NN fuses low-level sensor data directly, e.g., for feedback control).

Example 12.5 Machine Fault Diagnosis

Sensor fusion is applicable in machine health monitoring, fault detection, and diagnosis. This is quite logical because, typically, several disparate sensors are used in monitoring an industrial process of machine and some of the available information may be noncrisp (qualitative). We have developed an automated industrial machine for fish cutting (see Figure 1.9). In this machine, the potential faults/malfunctions include (1) blocked fish, (2) failure of the hydraulic cylinder system of the hydraulic actuators of the cutter assembly, (3) failure of the conveyor system, (4) failure of the hydraulic pump, (5) failure of the hydraulic servo valves, and (6) failure of the pneumatic controlled cutter.

We have developed a neurofuzzy network for fault diagnosis of the machine, through sensor fusion (see Reference 8). Microphones, cameras, and accelerometers (for vibration sensing at different locations) are the main sensors that are used for the monitoring of the machine (see Figure 12.16). The data from the sensors are preprocessed to extract useful features. These are provided to the input layer of the NN. This information is then transmitted to a fuzzy layer, which uses fuzzy logic to infer the status of the machine based on each sensor data. These sensory inferences are "fused" in the hidden layer of the NN. The output layer then provides the nature of the fault (including the "fault-free" status). The neurofuzzy network that is used for this purpose is shown in Figure 12.17. Initially, the NN has to be trained using input–output data for various fault types.

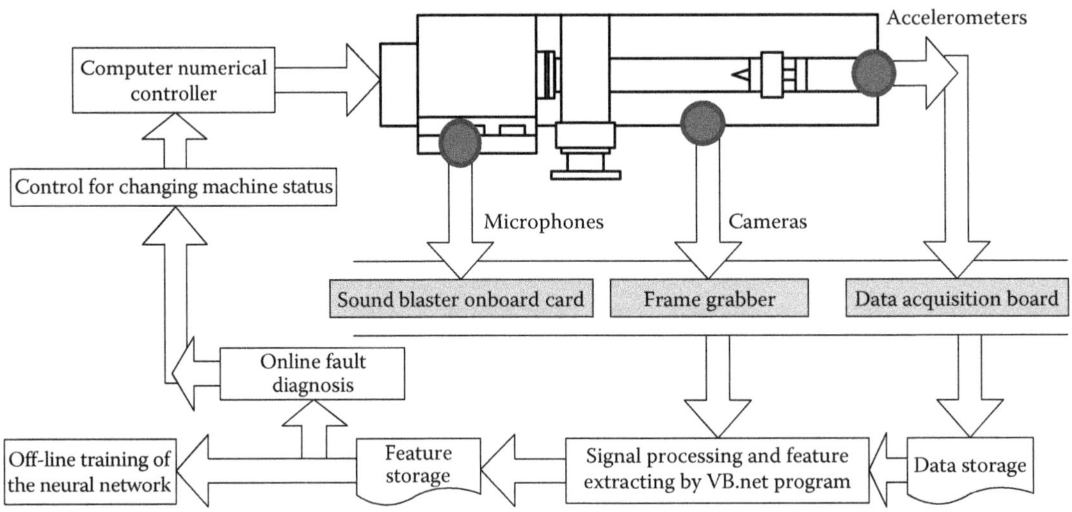

FIGURE 12.16
Architecture of the fault diagnosis system of a machine.

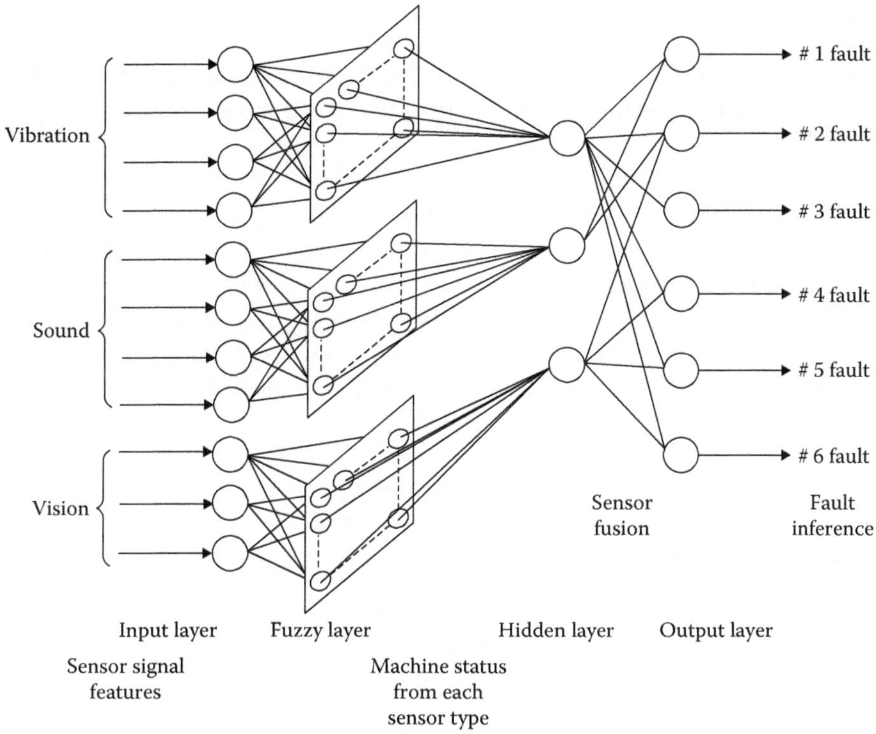

FIGURE 12.17
Neurofuzzy sensor fusion for machine fault diagnosis.

Summary Sheet

Micro-electromechanical systems (MEMS): Microminiature components (sensors, actuators, signal processing, etc.) embedded into a chip. Exploit electrical/electronic and mechanical features. Device size: 0.01–1.0 mm and component size: 0.001–0.1 mm. Uses integrated circuit (IC) technologies in fabrication → many components can be integrated into a single device (e.g., a few to a million).

Advantages of MEMS: Advantages of IC devices. Microminiature size and weight, large surface-area-to-volume ratio, large-scale integration (LSI) of components/circuits, high performance, high speed (20 ns switching speeds), low power consumption, easy mass production, and low cost (in mass production).

MEMS energy conversion mechanism: *Piezoelectric*: mechanical strain → charge separation across piezoelectric material → voltage. Strain energy (mechanical work) is converted into electrostatic energy. Passive device. *Electrostatic*: a voltage → charge separation into capacitor plates → attraction force between plates is supported by external "mechanical" force. Plates move by mechanical work → capacitance is reduced and voltage is increased. Mechanical energy is converted into electrical energy. Passive device. *Electromagnetic*: coil moves in a magnetic foil → current is "induced" in the coil. Mechanical energy is converted into electrical energy. Passive device.

IC fabrication process:

1. *Substrate preparation*: Thin slice of substrate (polished silicon) on which circuit (equivalent of many millions of interconnected transistors and other components) is formed.
2. *Film growth*: On the substrate, a thin film (of silicon, silicon dioxide, silicon nitride, polycrystalline silicon or metal) is deposited.
3. *Doping*: A controlled trace amount of doping material (atomic impurity) is injected into the film (e.g., by thermal diffusion or ion implantation).
4. *Photolithography*: A thin uniform layer of photosensitive material (photoresist) is formed on the substrate through spin coating and prebaking. A pattern corresponding to circuit structure is transferred to the photoresist by applying intense light through a "mask" (a glass plate coated with a circuit pattern of chromium film).
5. *Etching*: A chemical agent (wet or dry) is used to remove regions of film or substrate that are not protected by photoresist pattern.
6. *Photoresist removal*: Ashing.
7. *Dicing*: Wafer containing IC structure is cut into square shape.
8. *Packaging*: Diced wafer is packaged in protective casing, which has electrical contacts that connect IC chip to circuit board.

MEMS fabrication: Same as for IC chip. Deposition (deposition of a film on the substrate; e.g., physical or chemical deposition), patterning (transfer of the pattern or MEMS structure onto the film; typically lithography is used for this purpose), etching (removal of unwanted parts of film or substrate, outside of the MMS structure; wet etching where material is dissolved when immersed in a chemical solution or dry etching where material is sputtered or dissolved using reactive ions or a vapor phase etchant, may be used), die preparation (removal of individual dies that formed MEMS structures on the wafer), and dicing (cutting or grinding the wafer to proper shape; say, a thin square). Complex functional structures (sensing, actuation, signal processing, etc.) are fabricated by the following:

1. *Bulk micromachining*: Structures are etched on substrate in 3D. A wafer may be bonded with other wafers to form special functional structures (e.g., piezoelectric, piezoresistive, and capacitive sensors and bridge circuits).
2. *Surface micromachining*: Structures are formed layer by layer on substrate with multiple deposition and etching (micromachining) processes of film material. Some such layers may form the necessary gaps between structural layers (e.g., plate gap of capacitors).
3. *Micromolding*: Structures are fabricated using molds to deposit the structural layers. Etching is not required (unlike bulk micromachining and surface micromachining). Then mold is dissolved using a chemical that does not affect deposited MEMS structural material.

MEMS accelerometer: Acceleration → inertia force of proof mass → bends a microminiature cantilever. *Arrangement 1*: cantilever with point mass at free end. Associated displacement sensed by capacitive, piezoresistive, or piezoelectric methods. *Arrangement 2*: two "combs," one fixed and the other supported on a cantilever (spring) and carrying a proof mass at other end. Comb teeth: capacitor plates. Fixed plates are in between the movable plates. Measure change in capacitance → acceleration. Range = ±70 g, sensitivity = 16 mV/g, bandwidth (3 dB) = 22 kHz, supply voltage = 3–6 V, and supply current = 5 mA.

MEMS thermal accelerometer: A heated air bubble (proof mass) moves due to acceleration between two thermistor elements. Temperature difference → acceleration.

MEMS gyroscope: Coriolis force or gyroscopic torque as a velocity or angular momentum vector changes orientation. Measure through the capacitance change of a cantilever comb, as in an MEMS accelerometer. For example, three-axis MEMS gyro using Coriolis force. Range = ±250°/s, sensitivity = 7 mV/°/s, bandwidth (3 dB) = 2.5 kHz, supply voltage = 4–6 V, and supply current = 3.5 mA.

MEMS blood cell counter: Two electrodes apply current pulses. Blood sample is injected across the electrode path. Measure electrical resistance changes (resistance pulses) → quantity of blood cells. Pulse height: differentiates between red and white blood cells.

MEMS pressure sensor: *Method 1*: Suspended membrane (Si substrate) between two electrodes. Measure capacitance change. *Method 2*: piezoresistive (strain gauge) cantilever. Deformation due to pressure difference between two sides → change in resistance → pressure. Measurement range of 260–1260 mbar, supply voltage of 1.7–3.6 V, and sensor weight of 10 mg. Used in biomedical applications (diagnosis and treatment of neuromuscular diseases, etc.).

MEMS magnetometer: Uses magnetoresistive property of an MEMS element to measure magnetic field. Applications: electronic compass, GPS navigation, and magnetic field detection.

MEMS temperature sensor: Uses a Zener diode. Breakdown voltage is proportional to the absolute temperature. Sensitivity of 10 mV/K and current range of 450 μA to 5 mA.

MEMS humidity sensor: Capacitance change in polymer dielectric planar capacitor due to humidity.

Wireless sensor network (WSN): Several sensor nodes arranged in a specific *architecture* and in wireless (radio) communication with each other and with a base station (gateway). *Communicated* information is possibly preprocessed (compressed, aggregated, etc.). Base station forwards information (possibly after further processing) to *application server/user.*

Sensor node: One or more sensors, processing (microcontroller with an operating system, CPU, memory, and I/O), data acquisition hardware and software, power source, and radio-frequency (RF) transceiver with antenna (omnidirectional—transmitting uniformly in all direction in 2D). An actuator may be present but is not required. Size from 1 to 10 cm or more.

Advantages of WSNs:

Wireless; Can have many nodes and cover very large areas

Easily scalable to application scale (e.g., thousands of nodes)

Low cost of installation and operation

Reliable connectivity, robust, secure

Flexible node structure (restructuring, mobility, rescaling, etc.)
Autonomous and self-organizing
Can operate in harsh environments
Distributed architecture with distributed processing and decision-making (smart)
Accurate, efficient, and fast

Mote: Low-cost sensor node of mm scale with limited sensing, processing, and transmission capabilities (for defense, environmental monitoring). Deployed over large areas using mobile deployers (e.g., sowed aerially by a helicopter or a drone). Simple power sources (e.g., self-generation, photoelectric).

WSN operating system (OS): Operates node microcontroller. Provides interfaces between applications and hardware, schedules tasks, etc., at the node. Far less complex than a general-purpose OS. An embedded systems OS is suitable (e.g., eCos, uC/OS; application specific; low cost, low power). For example, open-source OS includes TinyOS, LittleOS, Contiki, OpenWSN, RIOT, and ERIKA Enterprise. An OS for a WSN node should be of a computer. Virtual memory and real-time operation may not be needed.

WSN engineering challenges: Component life and network life, response speed, robustness, scalability, autonomous (unattended) operation, unfamiliar and dynamic environments, resource constraints.

Energy management paradigms: Multihop communication, routing control (optimize transmission routes by selecting power-efficient nodes with respect to a performance index), duty cycling, data preprocessing (process/compress data locally before transmission), passive participation (when two nodes have the same information, transmit only one), adaptive sampling, adaptive sensing and transmission (stop sensing and transmission when the sensed quantity does not change), and use of efficient power technologies (e.g., efficient and low-cost batteries, harvesting energy from the environment—solar, vibration, wind, waves, geothermal, etc. Some nodes may use line—AC power).

Communication protocol: Defines format and order of message exchanges, what subsequent actions would be taken among the entities of communication network. Arranged into layers. WSN protocol requirements are constrained by resource limitations (small memory and code size, variable conditions).

MAC protocol layer: Sublayer of WSN protocol model. Functions: coordinate transmission among neighboring nodes (to optimize, avoid packet collision); coordinate actions for a shared channel (steps: test if busy; if not busy, transmit; if busy, wait and try again); communication interactions: request to send (RTS) and clear to send (CTS); sleep mode (for nodes that are not active) → saves energy and simplifies communication for other nodes; low-power "listening mode" to decide sleep or awake; packet backoff (wait if sending is not urgent). Consistent with the IEEE 802.15.4 open standard, widely used, somewhat complex. *Note*: Another version is B-MAC.

IEEE 802.15.4 characteristics (specific for WSN): Transmission frequencies 868 MHz/902–928 MHz/2.48–2.5 GHz; data rates: 20 kbps (868 MHz band), 40 kbps (902 MHz band), and 250 kbps (2.4 GHz band); supports star and peer-to-peer (mesh) network connections; Encryption of transmitted data, for security; determines link quality (useful for multihop mesh networking algorithms); robust data communication.

Routing in WSN: Routing algorithm is used. Does not need complex routing as in the Internet. Typical methods:

1. Discover neighboring nodes (ID and location). *Note*: Node knows its own characteristic (location, capability, remaining power, etc.).
2. Pick the best node.
3. Send message to that destination node.

Note 1: Once node location is known, messages are sent to location coordinates, not node ID (called geographic forwarding or GF).
Note 2: WSN should not send messages to a sleeping node unless it is awakened first.

Issues of routing in WSN: Time delay, reliability, remaining energy, and data aggregation to reduce transmission cost. Multihopping may be used. Unicast semantics: message sent to a specific node; multicast semantics: message sent to several nodes simultaneously; anycast semantics: message sent without specifying any nodes (diffusing or flooding).

Routing protocol: Function: Given a destination address, route data. Should be robust to node failures and unintentional disconnection, power efficient, and not too complex. TCP/IP, a general-purpose protocol, is too complex for WSN and not energy efficient. A routing algorithm is used. Multihopping may be used to optimize power, reliability, etc. IETF (Internet Engineering Task Force) is standardizing ROLL (Routing Over Low power and Lossy networks).

Routing protocols for WSN: LEACH (Low Energy Adaptive Cluster Hierarchy): operation is divided into rounds. Each round uses a different cluster of nodes with cluster heads (CHs). A node selects closest cluster head and joins that cluster to transmit data. PEGASIS (Power-Efficient Gathering in Sensor Information Systems): no cluster formation. Each node communicates only with closest neighbor (by adjusting its power signal to be only heard by the closest neighbor. Signal strength is used to measure the distance of travel). After chain formation, a leader is chosen from chain (which has most residual energy). VGA (virtual grid architecture) utilizes data aggregation and in-network processing to maximize the network lifetime. It is energy efficient.

WSN standards: Needed to achieve component compatibility (interoperability of devices from different manufacturers), proper communication, etc. Examples are:

Wi-Fi: Called WLAN (wireless local area network); uses 2.4 GHz UHF and 5 GHz SHF radio signals; IEEE 802.11 standard; for networking of devices in a local environment. Used for Internet access and for communication between several devices over short distances.

Bluetooth: A wireless technology standard. It uses short-wavelength radio waves in the frequency range 2.4–2.485 GHz. This technology is IEEE standard 802.15 based that uses WPAN. A short-range RF technology for communication among electronic devices and the Internet. User-transparent data synchronization.

ZigBee: *Uses IEEE* 802.15.4 as the physical and MAC layer. More secure. Supports hybrid star-mesh network topology. Cost-effective, low power, and wireless.

6LoWPAN: Addressable as an IPv6 device, for example, by your PC. Standards in progress: RFC4919 (6LoWPAN overview), RFC6775 (neighbor discovery), RFC6282 (compression format for IPv6 datagrams), and RFC6606, 6568 (routing requirements and design space).

WirelessHART/IEC 62591: Alternative to ZigBee for industrial applications, but for a higher cost. Lower power and more robust to interference than ZigBee, but IEEE 802.15.4e will be competitive.

GSM: Global System for Mobile communication. This is a European standard and is applicable for cellular networks (for cell phones). For its incorporation (i.e., for network access), the sensor node needs a *SIM* (subscriber identity module) card.

2G (second generation) GSM networks: Operate in 900 or 1800 MHz bands. Rarely, 400 and 450 MHz frequency bands are used (because *1G* system used those frequencies previously). Most *3G* networks in Europe operate in the 2100 MHz frequency band.

SD (secure digital) card: Is a nonvolatile memory card.

SD card shield: This is the shield that carries the SD card.

Other software of WSN: *Time synchronization*: important because most data are only meaningful with a time reference (time series). *Reprogramming*: needed for updating firmware of all nodes (feature addition, bug/security fix) over air and multiple hops. Security measures are essential to prevent hackers from installing their firmware.

Localization: Determine geographic location of a WSN node. Uses of localization: tracking nodes, monitoring spatial evolution of a WSN (e.g., in spatial data mining and determining spatial statistics), determining quality of node coverage, achieving load balancing of nodes, facilitating routing (e.g., optimal multihop routing), and optimizing communication).

Steps of localization: (1) Establish the location of selected nodes (anchors/beacons/landmarks). (2) Measure the distances to them from the node to be localized.

Issues of localization: Accuracy, speed, communication range, energy requirement whether indoor or outdoor, 2D or 3D, hostile or environment friendly, which nodes to localize, how often to localize, where the computation of localization is performed, and how to localize (i.e., localization method).

Methods of distance: (1) *Time of flight of signal and* (2) *radio signal strength at reception. Note:* Both methods need "beacon nodes (landmarks)" whose locations are known (and a node can send/receive signals to/from them). *Indirect method,* count hops between nodes. Use "average" distance per hop to estimate distance between two nodes.

Multilateration: Estimation of node position of a node using distances from it to three or more landmarks (with known locations). Node position estimate:

$$\beta = \begin{bmatrix} x \\ y \end{bmatrix} = \left[\boldsymbol{X}^T\boldsymbol{X}\right]^{-1}\boldsymbol{X}^T\boldsymbol{y}$$

$$\boldsymbol{X} = \begin{bmatrix} 2(x_1 - x_n) & 2(y_1 - y_n) \\ \vdots & \vdots \\ 2(x_{n-1} - x_n) & 2(y_{n-1} - y_n) \end{bmatrix}, \quad \boldsymbol{y} = \begin{bmatrix} x_1^2 - x_n^2 + y_1^2 - y_n^2 + d_n^2 - d_1^2 \\ \vdots \\ x_{n-1}^2 - x_n^2 + y_{n-1}^2 - y_n^2 + d_n^2 - d_{n-1}^2 \end{bmatrix}$$

(x_i, y_i) = coordinates of the ith landmark, $i = 1, 2, \ldots, n$ for n landmark nodes
d_i = distance of the ith landmark from the node to be localized

In wireless radio-frequency (RF) transmission from node to node, a signal is sent by one node and is received by the second node.

Advantage of RF signals (electromagnetic spectrum): No cabling; can penetrate objects such as walls; can transmit long distances; can accommodate mobile nodes. Use local-area radio channels (for distances 10 m to hundreds of m). Cellular technologies use wide-area radio channels for tens of km.

Signal distortion during transmission: SNR = a measure (dB) of strength of received signal with respect to signal degradation due to transmission; larger SNR → easier (and more faithful) recovery of original signal from received signal. Issues of degradation: (1) Signal strength decreases (signal will disperse) even in free space (path loss). (2) Interference with other signals (particularly transmitted in the same frequency band; environmental electromagnetic noise from other devices, etc.) will decrease SNR. (3) Obstructing objects → reflection, absorption, shadowing, etc. Moving objects → more serious problems. (4) Bit error rate (BER) = probability that transmitted bit is received in error. It decreases with SNR and increases with the transmission rate (Mbps—mega bits per second).

Distance measurement using radio signal strength: Advantage: uses existing communication hardware and resources of the WSN. Method: (1) Power of received signal is determined (from received signal strength indicator [RSSI]) many times, and sample mean $\bar{P}_{i,j}$ is computed. (2) Using a known reference distance d_0 and reference power P_0, use "shadowing model" of signal strength (path loss) to estimate the distance $\hat{d}_{i,j} = d_0(\bar{P}_{i,j}/P_0)^{-(1/\eta)}$, η = path loss exponent ~2.

Use of RSSI (according to IEEE 802.11-1999): RSSI = 0 up to RSSI max. It is provided by a sublayer of radio transmission protocol (8-bit RSSI) as Power (dBm) = RSSI_VAL + RSSI_OFFSET, at typical power ratio accuracy of ±6 dB. *Note*: dBm denotes decibel-milliwatts ← abbreviation for power ratio in decibels (dB) of measured power in milliwatt referenced to one 1 mW. Since signals are "power," we use $10\log_{10}$(power ratio), not $20\log_{10}()$, to convert into dB.

WSN applications: Defense, surveillance, and security; environmental monitoring; transportation (ground, air, water, and underwater); monitoring of machinery and civil engineering structures (e.g., for condition-based maintenance; detecting onset of seismic activity, at low sampling rates; once an activity is detected, sampling is done at much higher rate); industrial automation; robotics; entertainment; intelligent workspaces; medical and assisted living; energy (exploration, production, transmission, management).

Sensor fusion: Multisensor data fusion. Combining data from multiple sensors to improve the sensory decision → accuracy, resolution, reliability, and safety (e.g., sensor failure in aviation), robustness, reduced conflict in information, reduced noise and the influence of unknown factors, stability (sensor drift, etc.), confidence (reduced uncertainty), usefulness (e.g., broadening application range or operating range), level of aggregation (information modeling, compression, combination, etc.), and the level of detail or completeness (combining 2D images to obtain an authentic 3D image; combining complementary frequency responses, etc.). Subset of data fusion (nonsensory data, e.g., prior knowledge, experience, and model-based data, may be used as well). Data fusion is a subset of information fusion where what is fused can involve qualitative and high-level information (e.g., from "soft" sensors).

Complementary fusion: Sensors independently provide complementary (not the same) information, which are combined → information *incompleteness* is one advantage. For example, four radars measuring regions that are not identical (may have some overlap).

Competitive fusion: Each sensor measures the same property independently, and comparatively fused → better sensor (e.g., more accurate, faster) will dominate → improves accuracy and robustness, reduces *uncertainty*, for example, four radars measuring the same region.

Cooperative fusion: Sensor measures what another sensor needs (and requests) → complete/improve needed information. *Note*: Two sensors may sense the same property (to improve its accuracy, reliability, etc.) or different properties (to complete the needed information), but this is done cooperatively. For example, stereo vision: preassign two cameras for two planes. Then combine. *Note*: In complementary fusion, sensors are not preassigned to take specific roles.

Centralized fusion: Data from sensors are fused by a single central processor.

Distributed (decentralized) fusion: A sensor receives information from one or more other sensors and fuses (locally).

Hybrid architecture: Has centralized and decentralized clusters of sensors.

Homogeneous fusion: Sensors are identical.

Heterogeneous fusion: Sensors are disparate (different types, capabilities, etc.).

Hierarchical fusion: Multilayered.

Data-level fusion: Sensory data (with minimal preprocessing, e.g., amplification and filtering) are directly fused.

Feature-level fusion: Features (or data attributes) are extracted from each sensor → a feature vector → fusion system.

Decision-level fusion: Each sensor separately processes and makes sensory decision (or estimation of the required quantity). They are evaluated and combined/fused for final decision-making (or estimation).

Bayesian approach to sensor fusion: (1) *Discrete sensors*: kth sensor gives measurement y_i and lth sensor gives measurement y_j, probability vector of estimate of m, $P(\boldsymbol{m} \mid {}^k y_i, {}^l y_j) = aP({}^k y_i \mid \boldsymbol{m}) \otimes P({}^l y_j \mid \boldsymbol{m}) \otimes P(\boldsymbol{m})$. *Note*: Estimated m takes discrete values $\boldsymbol{m} = [m_1, m_2, \ldots, m_n]^T$. Corresponding discrete sensor readings $\boldsymbol{y} = [y_1, y_2, \ldots, y_n]$. (2) *Continuous sensors (Gaussian)*:

$$\text{Recursive estimate} \quad \hat{m}_i = \frac{\sigma_w^2}{\left(\sigma_{m_{i-1}}^2 + \sigma_w^2\right)} \hat{m}_{i-1} + \frac{\sigma_{m_{i-1}}^2}{\left(\sigma_{m_{i-1}}^2 + \sigma_w^2\right)} y_i$$

$$\text{Estimation error variance} \quad \frac{1}{\sigma_{m_i}^2} = \frac{1}{\sigma_{m_{i-1}}^2} + \frac{1}{\sigma_w^2}$$

Sensor fusion using a Kalman filter: Method 1: Single measurement model for all r sensors. Apply Kalman filter all r measurements simultaneously (i.e., in parallel). Method 2: Use r different measurement models for the r sensors. Proper output equation has to be used in Kalman filter, depending on the currently used sensor → Kalman filter is applied sequentially to r sensors.

Sensor fusion using Dempster–Shafer evidence theory: This combines evidence (current data and other information).

Frame of discernment Θ: Set of mutually exclusive (i.e., nonoverlapping) and exhaustive (i.e., complete) propositions concerning the considered problem. Consists of n elements (propositions).

Power set of Θ: Set of all subsets of Θ including empty (null) set Φ. Denoted by Ω(Θ). Has 2^n sets, including Φ and Θ.

Mass function $m(A)$: Assigns a value between 0 and 1 to every element (subset) A of the power set Ω. $m:\Omega(\Theta)\rightarrow[0,1]$; $m(\Phi)=0$; $\sum_{A\subseteq\Omega(\Theta)} m(A)=1$. $m(A)$ represents the level of validity (confidence, belief) of proposition A, based on available evidence (data and expertise).

DS rule of combination: Formula of decision-making (reasoning, "fusion"). Method of combining evidence from independent sources regarding a particular proposition. Done by combining mass functions of independent sources (orthogonal sum). For mass functions m_1 and m_2 are the mass functions corresponding to two independent sets of data (evidence) in the same frame of discernment Θ, mass function of a proposition A, in the same frame of discernment, is obtained by the DS rule of combination:

$$m(A)=m_1\oplus m_2=\begin{Bmatrix} 0 & \text{if } A=\Phi \\ \dfrac{\sum_{B\cap C=A} m_1(B)m_2(C)}{1-K} & \text{if } A\neq\Phi \end{Bmatrix}$$

Note: $1-K$ normalizes the mass function. K represents the degree of conflict in the two sources of evidence: $K=\sum_{B\cap C=\Phi} m_1(B)m_2(C)$.

Computation of mass function of data: Compute the distance (d) from sensor data (s) to a decision class (c). Mass function = normalized inverse of d.

Distance metrics: *For* two vectors $\boldsymbol{p}$ and $\boldsymbol{q}$ with elements $[p_1 \quad \dots \quad p_n]$ and $[q_1 \quad \dots \quad q_n]$,

Euclidean distance:

$$d(\boldsymbol{p},\boldsymbol{q})=d(\boldsymbol{q},\boldsymbol{p})=\sqrt{(q_1-p_1)^2+(q_2-p_2)^2+\cdots+(q_n-p_n)^2}=\sqrt{\sum_{i=1}^{n}(q_i-p_i)^2}$$

Manhattan (taxicab) distance:

$$d(\boldsymbol{p},\boldsymbol{q})=d(\boldsymbol{q},\boldsymbol{p})=|q_1-p_1|+|q_2-p_2|+\cdots+|q_n-p_n|=\sum_{i=1}^{n}|q_i-p_i|$$

Minkowski distance (of order r):

$$d(\boldsymbol{p},\boldsymbol{q})=d(\boldsymbol{q},\boldsymbol{p})=\left(\sum_{i=1}^{n}|q_i-p_i|^r\right)^{1/r}$$

Note: $r=2\rightarrow$ Euclidean distance; $r=1\rightarrow$ Manhattan distance.

For m data values $[s_1, s_2, \ldots, s_m]$ from m sensors and Class i specified by value set (guideline) $[c_{i1}, c_{i2}, \ldots, c_{im}]$, distance of the data set (measurement) to Class i: $d_i = \left[\sum_{j=1}^{m} ((s_j - c_{ij})/(c_{j\max} - c_{j\min}))^2\right]$; $c_{j\max} = \max_i(c_{ij})$ and $c_{j\min} = \min_i(c_{ij})$. *Note:* $c_{j\max} - c_{j\min}$ normalizes the result. Typically, $c_{j\min} = 0$.

$$\text{Corresponding mass function value:} \quad m_i = \frac{1/d_i}{1/d_1 + 1/d_2 + \cdots + 1/d_k}$$

Sensor fusion using neural networks (NNs): NNs are massively connected networks of computational "neurons." Possess parallel and distributed processing structures. *Nodes* of NN are connected through weighted pathways (*synapses*) and arranged into an *input layer*, one or more *hidden layers*, and an *output layer*. At a node, weighted inputs are summed, thresholded, and passed through an *activation function*, to generate the node output. *Characteristics*: take many inputs (e.g., many sensors). Learn by example (attribute of intelligence); approximate highly nonlinear functions; massive computing power; have memory of processed information. Fusion is done in hidden layers.

Hybrid use with fuzzy logic (neurofuzzy systems): Fuzzy logic mimics the reasoning mechanism of humans. *Three general types*: (1) Incorporate NN as a facilitator/tool in fuzzy logic system (e.g., learn/train rules and membership functions using NN). Use fuzzy inference for sensor fusion. (2) Use fuzzy logic to represent features of sensor data. Use NN (possibly with fuzzy neurons, fuzzy weights, etc.) for sensor/feature fusion. (3) Use separate fuzzy subsystems and NN subsystems to carry out different fusion activities (e.g., fuzzy system does "information fusion" for high-level supervisory/tuning actions; NN fuses low-level sensor data directly for feedback control).

Backpropagation learning: Method of updating weights in a multilayer perceptron (MPL) NN → "training" of NN. Uses input–output training data (i.e., teaching using example) and gradient descent to minimize squared error (Euclidean norm). Method: Apply training input data to NN. Determine output o. For n data sets and m output elements (nodes), minimize cumulative error:

$$e_c = \sum_{k=1}^{n} e(k) = \frac{1}{2}\sum_{k=1}^{n}\sum_{i=1}^{m}[t_i(k) - o_i(k)]^2$$

Weight update (for kth data set):

$$\Delta w^l = -\eta(\partial e(k)/\partial w^l)$$

Backpropagation steps:

Step 1: Initialize weights and thresholds to small random values.

Step 2: Choose an input–output training data set.

Step 3: Compute from input to output $o_p^l = f\left(\sum_i w_{pi}^l o_i^{l-1}\right)$

Step 4: Compute output error e and backpropagation parameter $\delta_i^L = [t_i - o_i^L] f'(sum_i^L)$ at output layer (L).

Step 5: Update weights using $\Delta w_{ij}^{l} = \eta \delta_i^l o_j^{l-1}$ with backpropagation $\delta_i^{l-1} = f'(sum_i^{l-1}) \sum_p \delta_p^l w_{pi}^l$ from $l = L$ to 3.

Step 6: Repeat steps 2 through 5 for another training data set.

Step 7: After using all the training data sets (i.e., one epoch), if final error $e <$ a predetermined tolerance, then network has been trained.

If not, repeat the process for another epoch.

Notes: For the output layer, $\delta_i^L = -(\partial e / \partial o_i^L) f'(sum_i^L) = (t_i - o_i^L) f'(sum_i^L)$.

For sigmoidal function (with $\lambda = 1$), $f' = f(1 - f)$.

Radial basis function network: Three layers: input layer (n-D), hidden layer (r-D), and output layer (m-D). In hidden layer, activation function = RBF:

$$g_j(\boldsymbol{x}) = R\left(\frac{\|\boldsymbol{x} - \boldsymbol{v}_j\|}{\sigma_j}\right)$$

$\boldsymbol{v}_j$ = center vector (n-D)
σ_j = width parameter, $j = 1, \ldots, r$
$\boldsymbol{x}$ = network input vector (n-D)

Network output relation:

$$o_i(\boldsymbol{x}) = \sum_{j=1}^{r} w_{ij} g_j(\boldsymbol{x}) \quad \text{for output nodes } i = 1, \ldots, m$$

w_{ij} = weight from hidden node j to output node i.

Radial basis functions:

Gaussian Kernel function:

$$g_i(\boldsymbol{x}) = \exp\left(\frac{-\|\boldsymbol{x} - \boldsymbol{v}_i\|^2}{2\sigma_i^2}\right)$$

Logistic function:

$$g_i(\boldsymbol{x}) = \frac{1}{1 + \exp\left(\|\boldsymbol{x} - \boldsymbol{v}_i\|^2 / \sigma_i^2\right)}$$

RBF network learning: (1) Use unsupervised learning to determine activation function parameters ($\boldsymbol{v}_i$, σ_i). (2) Use supervised learning to determine the weights w_{ij}.

Example:

1. Given r pairs of input–output training data vectors: $[\boldsymbol{x}_i, \boldsymbol{o}_i]$; $i = 1, 2, \ldots, r$.
2. Use Gaussian activation function with training data $\boldsymbol{x}_i$ as the centers (unsupervised)
3. Select σ_i from past experience, input bandwidth estimate, etc. (unsupervised).

4. Form the two matrices: $G=[g_{ij}]_{r\times r}$; where $g_{ij}=\exp[(-\|x_j-x_i\|^2)/2\sigma_i]$; $i=1,2,\ldots,r; j=1,2,\ldots,r$ $O=[o_1, o_2, \ldots, o_r]_{m\times r} \rightarrow O = WG$, W = matrix of connection weights
5. Determine connection weights (supervised learning) as $W_{m\times r}=O_{m\times r}G_{r\times r}^{-1}$

Note: With $p \neq r$ pairs of training data vectors x_i and o_i, $i = 1, 2, \ldots, p$: $W = OG^+$, where pseudoinverse $G^+ = G^T(G^TG)^{-1}$.

Problems

12.1 Using the usual equation for a DC strain-gauge bridge (see Chapter 4), show that if the resistance elements R_1 and R_2 have the same temperature coefficient of resistance and if R_3 and R_4 have the same temperature coefficient of resistance, the temperature effects are compensated up to first order.

A microminiature (MEMS) strain-gauge accelerometer uses two semiconductor strain gauges, one integral with the cantilever element near the fixed end (root) and the other mounted at an unstrained location of the accelerometer. The entire unit including the cantilever and the strain gauges is constructed as a silicon IC and is smaller than 1 mm in size. Outline the operation of the accelerometer. What is the purpose of the second strain gauge?

12.2 Through a literature search, explore several MEMS sensors in the following categories of applications:

(1) Biomedical
(2) Mechanical
(3) Thermofluid

In each category, describe with sketches an MEMS sensor.

12.3 A method of node localization using landmarks, in WSNs, uses the formula $\beta = [X^TX]^{-1}X^Ty$ where

$$X=\begin{bmatrix} 2(x_1-x_n) & 2(y_1-y_n) \\ \vdots & \vdots \\ 2(x_{n-1}-x_n) & 2(y_{n-1}-y_n) \end{bmatrix}; \quad y=\begin{bmatrix} x_1^2-x_n^2+y_1^2-y_n^2+d_n^2-d_1^2 \\ \vdots \\ x_{n-1}^2-x_n^2+y_{n-1}^2-y_n^2+d_n^2-d_{n-1}^2 \end{bmatrix}; \quad \beta=\begin{bmatrix} x \\ y \end{bmatrix}$$

(a) Define the elements of these three vectors and matrices.
(b) Indicate the principle and the main steps of deriving this equation for node localization.
(c) In a node localization exercise with four landmark nodes, the following four data vectors were obtained:

$$\begin{bmatrix} x_1 \\ y_1 \\ d_1 \end{bmatrix}=\begin{bmatrix} 1 \\ 1 \\ 1 \end{bmatrix}; \quad \begin{bmatrix} x_2 \\ y_2 \\ d_2 \end{bmatrix}=\begin{bmatrix} 2 \\ -1 \\ 1 \end{bmatrix}; \quad \begin{bmatrix} x_3 \\ y_3 \\ d_3 \end{bmatrix}=\begin{bmatrix} -1 \\ 2 \\ 2 \end{bmatrix}; \quad \begin{bmatrix} x_4 \\ y_4 \\ d_4 \end{bmatrix}=\begin{bmatrix} -3 \\ 1 \\ 3 \end{bmatrix}$$

Determine the location (coordinates) of the node that is localized.

(d) Briefly describe (in a few sentences) two methods of ranging (i.e., measurement of the distance of a node) in WSN.

12.4 Search for information on a commercial microcontroller board. Summarize its main features that are relevant for it to be a suitable platform for a sensor node of a WSN.

(a) Select a sensor for a specific WSN application that could be integrated into a sensor node. Describe the sensor, outline its features and specifications, and indicate a practical application where this sensor node could be used.

(b) Indicate what other hardware and software components would be needed to complete the node of the WSN. Search and find information on suitable commercially available products for these components. Provide the performance parameters or attributes of these components that would match your sensor and the microcontroller. These parameters/attributes should meet the requirements of the application as well of your sensor node.

Note: Provide sketches/pictures of the system and the components. Provide numbers and/or descriptions of the relevant performance parameters of all key components of your sensor node.

12.5 Two discrete sensors (1 and 2) are used to measure the size of an object. The size m is treated as a discrete quantity, which can take one of the following three values:

(1) m_1 for small
(2) m_2 for medium
(3) m_3 for large

A sensor will make one of three discrete measurements given in the vector $y = [y_1 \quad y_2 \quad y_3]$ corresponding to these three object size values.

The two sensors have the following likelihood matrices:

Sensor 1:

	y_1	y_2	y_3
m_1	0.75	0.05	0.20
m_2	0.05	0.55	0.40
m_3	0.20	0.40	0.40

Sensor 2:

	y_1	y_2	y_3
m_1	0.45	0.35	0.20
m_2	0.35	0.60	0.05
m_3	0.20	0.05	0.75

(a) Indicate a major difference in the capability of these two sensors.

(b) Suppose that in the beginning of the sensing process, we have no *a priori* information about the size of an object. Consider the following two cases of data:

Case 1: Sensor 1 reads y_1 and Sensor 2 reads y_1.
Case 2: Sensor 1 reads y_1 and Sensor 2 reads y_3.

In each case, what is the fused measurement of the object size?
How would you proceed after obtaining these fused results?

12.6 Explain the meaning of complementary sensing in sensor fusion, with respect to frequency response. Give illustrative examples of complementary fusion of several filters in order to enhance the filtering capability. In particular, indicate the pairing of two low-pass filters and of a band-pass filter and a high-pass filter.

12.7 In this problem, the cutting torque in a CNC milling machine is estimated using multiple sensing and fusion through Kalman filtering (also, see Problem 7.22). The experimental setup is shown in Figure P12.7.

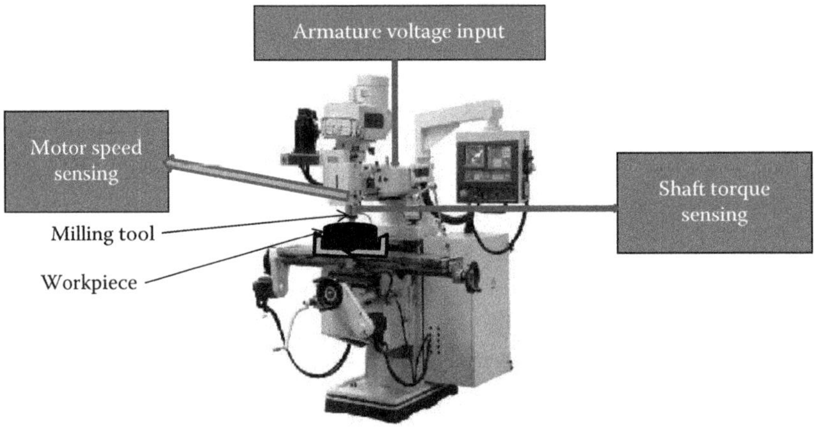

FIGURE P12.7
Experimental setup of a CNC milling machine.

A voltage u is applied to the armature circuit of the drive motor of the milling machine cutter in order to accelerate (ramp-up) the tool to the proper cutting speed. Since it is difficult to directly measure the cutting torque (which is a suitable indicator of cutting quality and cutter performance), two other variables are measured and used with a Kalman filter to estimate the cutting torque. Specifically, the following two variables (which are easier to measure than the cutting torque) are measured:

1. The speed of the drive motor is measured using an encoder.
2. The torque of the drive shaft of the motor is measured using a strain-gauge torque sensor (see Chapter 9).

Both these measurements are used simultaneously in a Kalman filter to estimate the cutting torque. The following information is given:

Discrete-time, nonlinear model of the cutting system of the milling machine is given as follows.

State equations (discrete time):

$$x_1(i) = a_1x_1(i-1) + a_2x_2(i-1) + a_3x_2^2(i-1) - a_4x_3(i-1) + b_1u(i-1) + v_1(\mathrm{i}-1)$$

$$x_2(i) = a_5x_1(i-1) + a_6x_1^2(i-1) + a_7x_2(i-1) + a_8x_3(i-1) + a_9x_3^2(i-1) + b_2u(i-1) + v_2(\mathrm{i}-1)$$

$$x_3(i) = a_{10}x_1(i-1) - a_{11}x_2(i-1) - a_{12}x_2^2(i-1) + a_{13}x_3(i-1) + b_3u(i-1) + v_3(\mathrm{i}-1)$$

where i denotes the time step.
State vector: $\boldsymbol{x} = [x_1, x_2, x_3]^T = [\text{motor speed, cutting torque, shaft torque}]^T$
Note: The shaft connects the motor to the milling cutter.
Input: u is the voltage input to the motor armature.
For ramping up the cutter, use $u = a(1 - \exp(-bt))$ with $a = 2.0$ and $b = 30$.
Measurement vector: $\boldsymbol{y} = [y_1, y_2]^T = [\text{Motor speed, Shaft torque}]^T$
The motor speed measurements and the shaft torque measurements may be simulated using the given nonlinear model with added Gaussian noise. The measurements are made with a sampling period of $T = 2.0 \times 10^{-3}$ s.

Model parameter values: $a_1 = 0.98$; $a_2 = 0.001$; $a_3 = 0.0002$; $a_4 = 0.004$; $a_5 = 0.19$; $a_6 = 0.04$; $a_7 = 0.95$; $a_8 = 0.038$; $a_9 = 0.008$; $a_{10} = 9.9$; $a_{11} = 0.48$; $a_{12} = 0.1$; $a_{13} = 0.97$; $b_1 = 0.004$; $b_2 = 0.0003$; $b_3 = 0.02$.

$$\text{Output matrix:} \quad \boldsymbol{C} = \begin{bmatrix} 1 & 0 & 0 \\ 0 & 0 & 1 \end{bmatrix}$$

Input (disturbance) covariance $\boldsymbol{V}$ and the measurement (noise) covariance $\boldsymbol{W}$ are given as

$$\boldsymbol{V} = \begin{bmatrix} 0.02 & 0 & 0 \\ 0 & 0.05 & 0 \\ 0 & 0 & 0.3 \end{bmatrix}; \quad \boldsymbol{W} = \begin{bmatrix} 0.05 & 0 \\ 0 & 0.1 \end{bmatrix}$$

Note: Both are Gaussian white, with zero mean.

(a) Using the linear continuous-time model of Problem 7.22, check the observability of the system separately for the two measurements.
(b) Using MATLAB®, apply a linear Kalman filter with data from both sensors, to estimate the cutting torque (i.e., state x_2).
(c) Using MATLAB, apply an extended Kalman filter with data from both sensors, to estimate the cutting torque (i.e., state x_2).
(d) Using MATLAB, apply an unscented Kalman filter with data from both sensors, to estimate the cutting torque (i.e., state x_2).
(e) Compare the results from these three approaches. In particular, indicate which approach is appropriate in the present estimation and why.
(f) Repeat a, b, and c, this time using only the data from the speed sensor (i.e., a single sensor, as in Problem 7.22).
(g) Compare the results using both sensors (i.e., items a, b, and c) with the results when only the speed sensor data are used (i.e., item f). Discuss whether sensor fusion is desirable in the present estimation.

Note: Provide plots of the data and the results of Kalman filtering and also the MATLAB script that you used to generate the results.

12.8 An example of multisensor data fusion is the computation of the water quality index (WQI) according to the criteria stipulated by the Canadian Council of Ministers of the Environment. The purpose is to determine a quantitative measure of the quality of water in a water source based on the measurements using several sensors (e.g., pH value, temperature, electrical conductivity, ORP, DO, hardness, turbidity). Several tests are carried out using water samples from the source. Multiple sensors are used, and a test with a particular sensor may be repeated several times. It is clear that, in this method, there is a distinction between the number of failed parameters and the number of failed tests. For each parameter (e.g., pH value), a specification (threshold) is given to represent the acceptable range of values. If a sensor reading falls outside this range, it represents a failed test. If for the same parameter several tests fail, it still is considered as a "single" failed parameter while it represents several failed tests. Using such a series of tests, their numerical attributes of water quality are computed: [Scope, Frequency, Excursions] = $[F_1 \quad F_2 \quad F_3]$.

These three attributes are defined as follows:

$$F_1 \text{ (Scope)} = \left(\frac{\text{Number of failed parameters}}{\text{Total number of parameters}} \right) \times 100$$

$$F_2 \text{ (Frequency)} = \left(\frac{\text{Number of failed tests}}{\text{Total number of tests}} \right) \times 100$$

An excursion is the degree of deviation of a test value from the range of specified values (given as a lower limit and an upper limit). Specifically,

$$\text{Excursion}_i = \left(\frac{\text{Failed test value}_i}{\text{Upper limit}_i} \right) - 1 \quad \text{if failed test value} > \text{upper limit}$$

$$\text{Excursion}_i = \left(\frac{\text{Lower limit}_i}{\text{Failed test value}_i} \right) - 1 \quad \text{if failed test value} < \text{lower limit}$$

Then, the normalized sum of excursions (*nse*) is computed as

$$\text{nse} = \frac{\sum_{i=1}^{n} \text{Excursion}_i}{\text{Total number of tests}}$$

It is then expressed as a percentage:

$$F_3 = \frac{\text{nse}}{\text{nse} + 1} \times 100$$

Finally, the WQI is computed as

$$\text{CCME WQI} = 100 - \left(\frac{F_1^2 + F_2^2 + F_3^2}{3} \right)^{1/2}$$

This gives a "percentage" value in the range 0–100. Comment on this measure of water quality. Particularly, give some of its shortcomings and indicate ways to overcome them.

12.9 The DS approach of data fusion has many advantages over the Bayesian approach. However, it has disadvantages as well. Indicate some disadvantages of the DS approach and give the reasons for these disadvantages.

12.10 Two sensors, denoted as 1 and 2, are available for measuring the distance of an object. Each sensor provides its measurement as one of the three states: near (N), far (F), and very far (VF). Suppose that based on the data from Sensor 1, the following mass function is computed: $m_1(N,F,VF) = [0.4 \quad 0.5 \quad 0.1]$.

Similarly, suppose that based on the data from Sensor 2, the following mass function is computed: $m_2(N,F,VF) = [0.3 \quad 0.1 \quad 0.6]$.

Using the DS method, combine these two pieces of evidence to obtain a fused mass function. Comment on the result. In particular, discuss the degree of conflict in the two pieces of evidence (sensor measurements) and its effect on the fused result.

12.11 Four approaches that may be used in the multisensor data fusion are

(1) Bayesian approach
(2) DS approach
(3) Fuzzy logic (see Reference 4)
(4) NNs

Provide a table that compares the following aspects of these three methods:

(1) The function that is used for knowledge representation
(2) Advantages
(3) Disadvantages

Note: Hybrid methods that integrate two or more methods may be applicable in data fusion, but they need not be included in your table.

12.12 In a network of water quality monitoring, consider three sensor nodes, each having the same three types of sensors: DO sensor, pH value (pH) sensor, and electrical conductivity (CD) sensor. Suppose that the three sensor nodes generate the data given in Table P12.12a.

TABLE P12.12A

Data from the Three Sensor Nodes

Sensor	DO (mg/L)	pH	CD (μS/cm)
Node 1 data	5.0	6.5	600.0
Node 2 data	4.5	6.0	500.0
Node 3 data	5.5	5.5	450.0

The water quality is determined as one of the following five categories (quality classes or inferences): excellent, good, fair, marginal, and poor.

The specifications (guideline sensory values) for these five classes of water quality are given in Table P12.12b.

TABLE P12.12B

Representative (Specified) Sensory Values for the Quality Classes

Quality Class	Guideline DO	Guideline pH	Guideline CD
Excellent	7.5	7.5	200.0
Good	6.0	7.0	300.0
Fair	5.0	6.0	500.0
Marginal	3.0	5.5	700.0
Poor	2.0	4.0	900.0

Determine the mass functions corresponding to the given three sets of data. Fuse the first two mass functions using the DS method. Also determine the corresponding degree of conflict *K*. Then fuse all three mass functions, and determine the corresponding degree of conflict *K*. Comment on the results. Present a MATLAB script (M-file) for obtaining these results.

12.13 Repeat Problem 12.12 using the data given in Table P12.13. The quality classes are kept the same as in Problem 12.12.
Comment on the results.

TABLE P12.13

Data from the Three Sensor Nodes

Sensor	DO (mg/L)	pH	CD (μS/cm)
Node 1 data	5.0	6.1	510.0
Node 2 data	5.1	6.2	520.0
Node 3 data	4.9	6.0	500.0

12.14 List the advantages of multisensor data fusion. What are some applications of data fusion?

12.15 Sensory data fusion is applicable in the automated navigation of aircraft, ships, and ground vehicles. For example, accelerometers may be used to measure rectilinear motions, a ring laser gyro may be used to measure angular (rotatory) motions, and a magnetometer may be used to measure directions (heading). Also, a GPS may be used (where feasible). Discuss how sensor fusion is applicable in this context.

12.16 Hopfield network is a dynamic (or recurrent or feedback) NN. Its topology is shown in Figure P12.16.

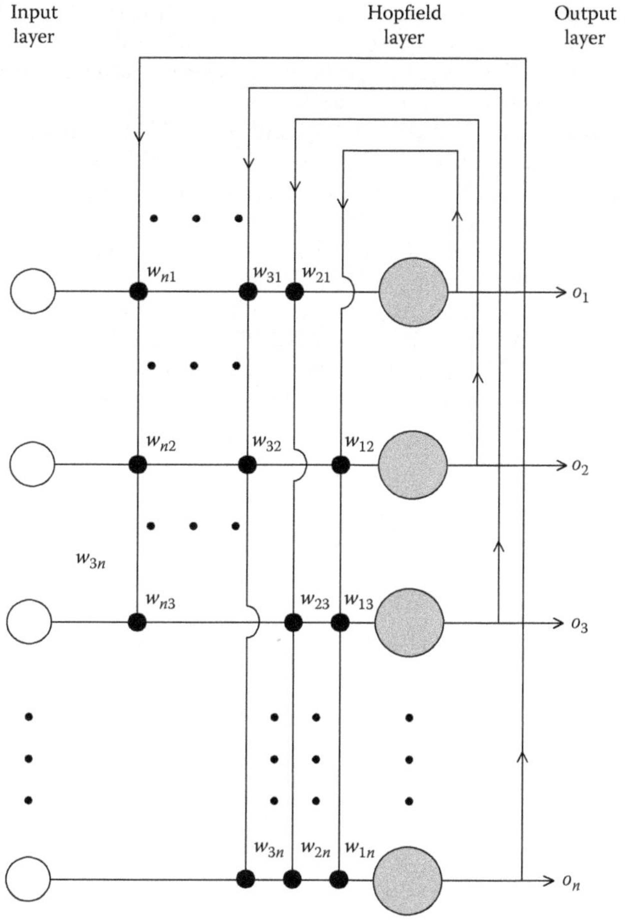

FIGURE P12.16
A Hopfield network.

At each node (i), the feedback signals (the network outputs except the output of the particular node) are weighted and summed, the sum is thresholded (i.e., threshold θ_i is subtracted from the weighted sum), and the sign of the result is generated as the node (network) output.

(a) Analytically express the network operation.
(b) Indicate an application of this network.

Further Reading

1. De Silva, C.W., *Sensors and Actuators—Engineering System Instrumentation*, 2nd edn., Taylor & Francis/CRC Press, Boca Raton, FL, 2015.
2. De Silva, C.W., *Mechatronics—A Foundation Course*, Taylor & Francis/CRC Press, Boca Raton, FL, 2010.

3. Foley, B.G., A Dempster–Shafer method for multi-sensor fusion, Thesis, Air Force Institute of Technology, Wright-Patterson Air Force Base, Dayton, OH, 2012.
4. Karray, F.O. and de Silva, C.W., *Soft Computing and Intelligent Systems Design—Theory, Tools, and Applications*, Addison Wesley, New York, 2004.
5. Khoshnoud, F. and de Silva, C.W., Recent advances in MEMS sensor technology—Thermo-fluid and electro-magnetic devices, *IEEE Instrumentation and Measurement*, 15(3), 16–20, 2012.
6. Khoshnoud, F. and de Silva, C.W., Recent advances in MEMS sensor technology—Mechanical applications, *IEEE Instrumentation and Measurement*, 15(2), 14–24, 2012.
7. Khoshnoud, F. and de Silva, C.W., Recent advances in MEMS sensor technology—Biomedical applications, *IEEE Instrumentation and Measurement*, 15(1), 8–14, 2012.
8. Lang, H. and de Silva, C.W., Fault diagnosis of an industrial machine through sensor fusion, *International Journal of Information Acquisition*, 5(2), 93–110, June 2008.
9. Zhou, J., Liu, L., Guo, J., and Sun, L., Multisensor data fusion for water quality evaluation using Dempster–Shafer evidence theory, *International Journal of Distributed Sensor Networks*, Paper ID 147419, 1–6, 2013.

Appendix A: Laboratory Exercises

A.1 Lab 1: Prepare a Proposal for a Miniproject

You are a mechatronic engineer who has been assigned the task of designing and developing a multisensor architecture and a platform for monitoring a process (a sufficiently complex practical engineering system). In the final project report, you will have to describe the steps of deciding on the type of sensors and transducers, establishing their ratings, selecting appropriate technologies, hardware, and software for signal processing, data communication, and acquisition and the digital platform for data processing and decision making. Keeping this in mind, prepare a project proposal.

In the project proposal, provide the following information:

1. Select a process (plant) that you are interested in, for monitoring. Briefly describe the process, indicating its purpose, how it operates, and what are its important variables (inputs and outputs) and parameters, which should be sensed/measured/monitored. You should consider at least three variables or parameters for monitoring.
2. Give a title for your project.
3. Suggest some types of sensors that are commercially available, which may be used for your monitoring task. Through an electronic search (web browsing), gather the data sheets for the selected sensors, which will provide their characteristics, ratings, and specifications. *Note*: This information need not be complete at this stage. You can add, revise, delete, etc., later, during the course of the semester, particularly during your future lab sessions.
4. Pick the items listed in the data sheets that you have some understanding of. Indicate what the given data mean and why these data are important for your project. *Note*: Some of the ratings may be related to such performance characteristics as accuracy, resolution, dynamic range, speed, linearity, stability, and operating bandwidth.

Note: You should decide on a title for your project.

A.2 Lab 2: Impedance Considerations

1. Select a sensor of interest that is commercially available (this may be a sensor in your miniproject). Establish the rating parameter values for this sensor. In particular, you may determine (from data sheets) the values of such parameters as impedance, output signal range, speed (time constant), operating frequency range

(bandwidth; this is also related to speed), and the nature of possible noise components. *Note*: You may not be able to find all these parameter values. It is okay if you found only some of them.

2. For an application that uses the selected sensor, determine such quantities as signal type (e.g., current or voltage, analog or digital), signal level, operating frequency range (in digital processing, sampling rate may also be determined by this), and noise tolerance (allowable noise level).
3. Determine what interface hardware would be suitable to match the requirements in Item 2 with the sensor parameters in Item 1 (e.g., filters, amplifiers, impedance matching devices, power supplies).
4. As much as you are able, establish the key parameter values of the chosen hardware (e.g., filter pass band, amplifier gain, input impedance, output impedance, slew rate, power supply voltage, and power ratings).
5. Give a sketch of how the components would be interconnected in a possible sensing system.

A.3 Lab 3: Filtering for Signal Conditioning

1. Familiarize yourself with the use of MATLAB® for the filtering of signals/data.
2. Consider a data set (e.g., 501 points sampled at sampling periods of 0.02 s) obtained from a sensor of a setup in your laboratory.
3. Plot the data and identify some characteristics of this signal.
4. Select a suitable filter (say, a Butterworth low-pass filter) to filter out the noise in the signal.
5. Plot the filtered signal.
6. Indicate the reasons for the following: (a) why you used the particular filter, (b) why you chose the particular order for the filter, and (c) why you selected the particular parameters (e.g., cutoff frequency) for the filter.
7. Select a sensor that you may use in your miniproject. Identify the type of noise that may be present in the measured signal from that sensor. Indicate what type of filtering may be used for filtering out that noise.

Note: You may use available MATLAB functions (no need to generate new functions).

A.4 Lab 4: Sensitivity Analysis of Sensors

1. Pick a sensor that will be used in your miniproject.
2. Characterize the following for the chosen sensor:
   a. The *measurand* (primary input x_1) and the *measurement* (required output y).
   b. Many possible secondary inputs that will affect the measurement y. *Note*: These may be the undesirable input variables (e.g., noise signals, disturbance inputs),

parameters (e.g., mass and thermal capacitance that can change due to environmental effects), and needed inputs (e.g., power, carrier, or modulating signal) whose changes/errors will affect the measurement in an undesirable manner.

c. Full-scale range (input/output).
d. Full-scale nonlinearity.
e. Quantization.
f. Signal-to-noise ratio (SNR).

3. Pick at least three from your listed secondary inputs, which you feel are the most important ones with respect to their effect on the measurement. *Note*: This may be done based on your current knowledge or common sense or simply a guess. Call them x_2, x_3, x_4, and so on.
4. Develop an analytical relationship to represent how the errors in x_i will affect y (i.e., a relationship between the error in y and the errors in x_i). *Note*: This may be done based on physical relations, available information, your current knowledge, common sense, guess work, etc. It may be a nonlinear relationship.
5. Determine the nondimensional sensitivities of the input error terms on the output (measurement) error.
6. Arrange the error sources in their order of significance.
7. Discuss whether some of the input error terms among your considered few can be neglected.
8. What other error contributions (of secondary inputs) from your original list may be considered as well for further study (no need to analyze them).

A.5 Lab 5: Two Parts, (a) Sensor Selection and (b) Sensor Modeling

A.5.1 Part (a): Complete the Following Table

Project Title: Give the Title Here

Item	Information
What parameters or variables have to be measured in your application?	
Nature of the information (parameters and variables) needed for the particular application (analog, digital, modulated, demodulated, power level, bandwidth, accuracy, etc.).	
List of sensors needed for the application.	
Signal provided by each sensor (type [analog, digital, modulated, etc.], power level, frequency range, etc.).	
Errors present in sensor output (SNR, etc.).	
Type of signal conditioning or conversion needed with the sensors (filtering, amplification, modulation, demodulation, ADC, DAC, voltage–frequency conversion, frequency–voltage conversion, etc.).	
Any other comments.	

A.5.2 Part (b): Sensor Modeling

For a sensor of your choice (possibly from your project), develop an analytical model in terms of physical parameters. For a set of parameter values, compute some rating parameters (2 or 3) that you may find in the data sheet of that sensor. Comment on how consistent these computed rating values are with what you may find in a commercial data sheet.

A.6 Lab 6: Digital Transducers

The following are the primary considerations in this lab:

a. Selection of commercial hardware and understanding their purpose in the current application.
b. What schemes and logic may be used to perform the processing for obtaining the required results (a pseudocode; not a complete code is adequate)?

Given

You have an incremental encoder to measure the angle of rotation and angular speed (in rad and rad/s) including direction (of a shaft)

The pins of the encoder are as follows:

1. Ground
2. Index
3. A channel
4. +5 V DC power
5. B channel

The A channel and B channel give the quadrature pulse signals (i.e., 90° out of phase). You can detect the direction of rotation using any method described in this book.

In this application, we need to do the following:

1. Acquire both pulse sequences (A and B) from the encoder (into your PC or a microcontroller).
2. Determine the direction of rotation.
3. Compute angle of rotation (in radians, from a reference position).
4. Compute speed (in rad/s) at any time.

In the lab, first you should do the following:

1. Do a search and select other main hardware (in addition to the incremental optical encoder) that you will need.
2. Give a sketch to indicate how these hardware components are connected in the final system.
3. Give a pseudocode (or a C++ code or a MATLAB code if you like) to perform the four operations listed earlier, in the application.

A.7 Lab 7: Two Parts, Part (a) and Part (b)

A.7.1 Part (a): Parameter Estimation through Curve Fitting of Experimental Data

You are asked to estimate the damping parameters of a shock absorber using experimental data. In the experimental setup, one end of the shock absorber is firmly mounted on a load cell. At the other end, a velocity input is applied using a shaker (a linear actuator). The experimental setup is shown in Figure L7.1.

FIGURE L7.1
Experimental setup of a shock absorber.

The velocity v applied by the shaker (m/s) and the resulting force f at the load cell (N) are measured and 41 pairs of data are recorded. Store the obtained data in the two .mat files "velocity" and "force."

1. List possible error sources in estimating the damping parameters.
2. Using MATLAB, curve fit the data (least-squares fit) to the linear viscous damping model $f = b_1v + b_0$ and estimate the damping parameters b_0 and b_1. Give some statistics for estimation error and "goodness of fit."
3. Using MATLAB, curve fit the data (least-squares fit) to the quadratic damping model $f = b_0 + b_1v + b_2v^2$ and estimate the damping parameters b_0, b_1, and b_2. Give some statistics for estimation error and "goodness of fit."
4. Compare the results from the two fits. In particular, is a linear fit adequate or do you recommend quadratic (or still higher order) fit for these data?

Note: Provide plots of the data and the results of curve fitting.

A.7.2 Part (b): Diameter Estimation of a Batch of Bolts

A batch of bolts of nominal diameter 10 mm has been produced by a manufacturer. For the purpose of quality control, a random sample of 50 bolts was taken from the produced batch, and their diameter was accurately measured. You may generate these data by actually measuring a batch of bolts. Alternatively, you may use a random number generator to add error to the nominal diameter.

1. Compute the sample mean and the sample standard deviation of the data using MATLAB (least-squares point estimate).
2. Estimate the mean and the standard deviation of the data using maximum likelihood estimation in MATLAB.
3. Compare the two sets of results. In your opinion, which result is more accurate and why?

A.8 Lab 8: Process Variable Estimation through Kalman Filtering with Experimental Data

In this experiment, the cutting torque in a CNC milling machine is estimated.

Consider a milling machine in your lab. Suppose that a voltage u is applied to the armature circuit of the drive motor of the milling machine cutter in order to accelerate (ramp up) the tool to the proper cutting speed. Since it is difficult to measure the cutting torque (which is a suitable indicator of cutting quality and cutter performance), the motor speed ω_m is measured instead (which much easier to measure). The measurements of motor speed are used in a Kalman filter to estimate the cutting torque. The following information is given:

Discrete-time, nonlinear model of the cutting system of the milling machine is given as follows:

State equations (discrete time):

$$x_1(i) = a_1x_1(i-1) + a_2x_2(i-1) + a_3x_2^2(i-1) - a_4x_3(i-1) + b_1u(i-1) + v_1(\mathrm{i}-1)$$

$$x_2(i) = a_5x_1(i-1) + a_6x_1^2(i-1) + a_7x_2(i-1) + a_8x_3(i-1) + a_9x_3^2(i-1) + b_2u(i-1) + v_2(\mathrm{i}-1)$$

$$x_3(i) = a_{10}x_1(i-1) - a_{11}x_2(i-1) - a_{12}x_2^2(i-1) + a_{13}x_3(i-1) + b_3u(i-1) + v_3(\mathrm{i}-1)$$

where i denotes the time step.

State vector:

$\mathbf{x} = [x_1, x_2, x_3]^T = [\text{Motor speed, Cutter torque, Shaft torque}]^T$

Note: The shaft connects the motor to the milling cutter.

Input: u is the voltage input to the motor armature.

For ramping up the cutter, use $u = a(1 - \exp(-bt))$ with $a = 2.0$ and $b = 30$.

Measurement: Motor speed $y = x_1$

The motor speed y and time t are measured with a sampling period of $T = 2.0 \times 10^{-3}$ s. Generate (through actual measurement in lab or through simulation) 51 points of data and store them in .mat files "speed" and "time."

Note: The "time" file gives the time step (i). Each step is given 0.002 s.

Model parameter values (given):

a1 = 0.98

a2 = 0.001

a3 = 0.0002

a4 = 0.00

a5 = 0.19

a6 = 0.04

a7 = 0.95

a8 = 0.038

a9 = 0.008

a10 = 9.9

a11 = 0.48

a12 = 0.1

a13 = 0.97

b1 = 0.004

b2 = 0.0003

b3 = 0.02

Output matrix $C = [1\ 0\ 0]^T$

Input (disturbance) covariance V and the measurement (noise) covariance W are given as follows:

$$V = \begin{bmatrix} 0.0002 & 0 & 0 \\ 0 & 0.09 & 0 \\ 0 & 0 & 0.1 \end{bmatrix}; W = [0.0004]$$

Note: Both are Gaussian white, with zero mean.

1. Using MATLAB, apply a linear Kalman filter to estimate the cutting torque (i.e., state x_2).
2. Using MATLAB, apply an extended Kalman filter to estimate the cutting torque (i.e., state x_2).
3. Using MATLAB, apply an unscented Kalman filter to estimate the cutting torque (i.e., state x_2).
4. Compare the results from the three approaches. In particular, indicate which approach is appropriate in the present experiment and explain why.

Note: Provide plots of the data and the results of Kalman filtering and also the MATLAB script that you used to generate the results.

A.9 Lab 9: Design of an Intel Galileo–Based Sensor Node for a Wireless Sensor Network

1. Search for information on the Intel Galileo microcontroller board (e.g., at http://arduino.cc/en/ArduinoCertified/IntelGalileo). Summarize its main features that are relevant for it to be a suitable platform for a sensor node of a wireless sensor network (WSN).
2. Select a sensor (possibly from your project, but not necessarily) that would be integrated into the sensor node. Describe the sensor, outline its features and specifications, and indicate a practical application where this sensor node could be used.
3. Indicate what other hardware and software components would be needed to complete the node of the WSN.
4. Search and find information on suitable commercially available products for the components in Part 3. Provide the performance parameters or attributes of these

components that would match your sensor and the Intel Galileo. These parameters/attributes should meet the requirements of the application of your sensor node.

Note: Provide sketches/pictures of the system and the components. Provide numbers and/or descriptions of the relevant performance parameters of all the key components of your sensor node.

A.10 Lab 10: Sensor Fusion through Kalman Filtering

In this experiment, the cutting torque in a CNC milling machine is estimated through sensor fusing using a Kalman filter. The experimental setup is as in Section A.8.

A voltage u is applied to the armature circuit of the drive motor of the milling machine cutter in order to accelerate (ramp up) the tool to the proper cutting speed. Since it is difficult to directly measure the cutting torque (which is a suitable indicator of cutting quality and cutter performance), two other variables are measured and used with a Kalman filter to estimate the cutting torque. Specifically, the following two variables (which are easier to measure than the cutting torque) are measured:

1. The speed of the drive motor is measured using an encoder.
2. The torque of the drive shaft of the motor is measured using a strain gauge.

Both these measurements are used simultaneously in a Kalman filter to estimate the cutting torque. The following information is given:

Discrete-time, nonlinear model of the cutting system of the milling machine is given as follows:

State equations (discrete time):

$$x_1(i) = a_1x_1(i-1) + a_2x_2(i-1) + a_3x_2^2(i-1) - a_4x_3(i-1) + b_1u(i-1) + v_1(i-1)$$

$$x_2(i) = a_5x_1(i-1) + a_6x_1^2(i-1) + a_7x_2(i-1) + a_8x_3(i-1) + a_9x_3^2(i-1) + b_2u(i-1) + v_2(i-1)$$

$$x_3(i) = a_{10}x_1(i-1) - a_{11}x_2(i-1) - a_{12}x_2^2(i-1) + a_{13}x_3(i-1) + b_3u(i-1) + v_3(i-1)$$

where i denotes the time step.

State vector:
$\mathbf{x} = [x_1, x_2, x_3]^T = [\text{Motor speed, Cutting torque, Shaft torque}]^T$

Note: The shaft connects the motor to the milling cutter.

Input: u is the voltage input to the motor armature.

For ramping up the cutter, use $u = a(1 - \exp(-bt))$ with $a = 2.0$ and $b = 30$.

Measurement vector: $\mathbf{y} = [y_1, y_2]^T = [\text{Motor speed, Shaft torque}]^T$

The motor speed measurements may be generated by actual experimentation in laboratory or through simulation and stored in the .mat file "motor-speed." The shaft torque measurements are given in the .mat file "shaft-torque." The corresponding time points of measurement are given the .mat file "time." The measurements are made with a sampling period of $T = 2.0 \times 10^{-3}$ s. Each data file contains 51 values.

Note: The "time" file gives time step (i). Each step is given 0.002 s.

Model parameter values (given):

a1 = 0.98
a2 = 0.001
a3 = 0.0002
a4 = 0.004
a5 = 0.19
a6 = 0.04
a7 = 0.95
a8 = 0.038
a9 = 0.008
a10 = 9.9
a11 = 0.48
a12 = 0.1
a13 = 0.97
b1 = 0.004
b2 = 0.0003
b3 = 0.02

$$\text{Output matrix } \boldsymbol{C} = \begin{bmatrix} 1 & 0 & 0 \\ 0 & 0 & 1 \end{bmatrix}$$

The input (disturbance) covariance $\boldsymbol{V}$ and the measurement (noise) covariance $\boldsymbol{W}$ are given as follows:

$$\boldsymbol{V} = \begin{bmatrix} 0.02 & 0 & 0 \\ 0 & 0.05 & 0 \\ 0 & 0 & 0.3 \end{bmatrix}; \boldsymbol{W} = \begin{bmatrix} 0.05 & 0 \\ 0 & 0.1 \end{bmatrix}$$

Note: Both are Gaussian white, with zero mean.

1. Using MATLAB, apply a linear Kalman filter with data from both sensors, to estimate the cutting torque (i.e., state x_2).
2. Using MATLAB, apply an extended Kalman filter with data from both sensors, to estimate the cutting torque (i.e., state x_2).
3. Using MATLAB, apply an unscented Kalman filter with data from both sensors, to estimate the cutting torque (i.e., state x_2).
4. Compare the results from these three approaches. In particular, indicate which approach is appropriate in the present estimation and explain why.
5. Repeat 1, 2, and 3, this time using only the data from the speed sensor (i.e., a single sensor, as in Section A.8).

6. Compare the results using both sensors (i.e., items 1, 2, and 3) with the results when only the speed sensor data are used (i.e., item 5). Discuss whether sensor fusion is desirable in the present estimation.

Note: Provide plots of the data and the results of Kalman filtering and also the MATLAB script that you used to generate the results.

Appendix B: Projects and Case Studies

B.1 Example Projects

The following are examples of appropriate projects. You may pick one of them or an entirely different project. Even if more than one person picks the same project, the effort and the project report should be individual (one distinct project report per person).

B.1.1 Telemedicine

A wearable jacket with embedded and distributed suite of sensors to monitor the vital signs of a patient is the key device of a telemedicine system. Sensor signals will undergo some level of local processing through external or integrated and embedded hardware and will be transmitted to a host computer at the patient location. The acquired sensory signals will undergo further processing, through the application of signal processing techniques and software in the host computer, for the removal of noise and artifacts and extraction of parameters and vital signs that are important in medical diagnosis. The information will be condensed, formatted, and transmitted in real time to a computer in a doctor's office (hospital) at distance through a public communication link. The transmitted information will be presented to the doctor in a convenient manner using a graphical user interface. Simultaneously, the patient will be connected to the doctor through live video and audio links. The doctor will interact with the patient and obtain additional information and will perform medical assessment, diagnosis, and prescription.

B.1.2 Water Quality Monitoring

The quality of the local drinking water is measured using several sensor nodes that are distributed over a large geographic area. The data acquired from a sensor node may be locally conditioned and then transmitted to a central server, which houses the information and communication technologies (ICT) platform for quality monitoring and assessment of drinking water. The key issues are the following:

1. Local sensing and data acquisition issues of water quality monitoring
2. Signal processing and transmission issues
3. Decision-making on water quality by assessing information from all the sensor nodes
4. Architecture of the ICT platform

B.1.3 Home Care Robotics

One of the ways that can reduce the spending on health care of older people is to exploit the recent technological advancements in sensing and actuation, robotics, and ICT for providing high-quality supportive environments to older people in their own homes. A robotic home care environment may have autonomous robots, which can be augmented with haptic teleoperation capability comprising a haptic-assisted remotely controlled robot to monitor and assist individuals within the home environment. Specifically, the system will have the capability for two modes of operation: (a) more autonomous operation for 24 h routine basic care (mobility, bathing, dressing, toileting, meal preparation, providing medicine, monitoring and seeking external help, etc.) and (b) remote monitoring and haptic teleoperation in emergency situations (until regular help arrives—ambulance, paramedics, police, firefighters, etc.). In addition to the sensors in the robots (including the master and slave units for teleoperation), additional sensors will be necessary for monitoring the environment, which is dynamic, unstructured, and not fully known.

Appendix C: Probability and Statistics

In this appendix, we review some important concepts in probability and statistics.

C.1 Probability Distribution

C.1.1 Cumulative Probability Distribution Function

Consider a real-valued random variable X. The probability that the random variable takes a value equal to or less than a specific value x is a function of x. This function, denoted by $F(x)$, is termed *cumulative probability distribution function*, or simply *distribution function*. Specifically,

$$F(x) = P[X \le x] \tag{C.1}$$

Note that $F(\infty) = 1$ and $F(-\infty) = 0$, because the value of X is always less than infinity and can never be less than negative infinity. Furthermore, $F(x)$ has to be a monotonically increasing function, as shown in Figure C.1a, because a probability is nonnegative.

C.1.2 Probability Density Function

Assuming that random variable X is a continuous variable and, hence, $F(x)$ is a continuous function of x, the probability density function $f(x)$ is given by the slope of $F(x)$, as shown in Figure C.1b. Thus,

$$f(x) = \frac{dF(x)}{dx} \tag{C.2}$$

Hence,

$$F(x) = \int_{-\infty}^{x} f(x)dx \tag{C.3}$$

Note that the area under the density curve is unity. Furthermore, the probability that the random variable falls within two values is given by the area under the density curve within these two limits. This can be easily shown using the definition of $F(x)$ and $f(x)$:

$$P[a < X \le b] = F(b) - F(a) = \int_{-\infty}^{b} f(x)dx - \int_{-\infty}^{a} f(x)dx = \int_{a}^{b} f(x)dx \tag{C.4}$$

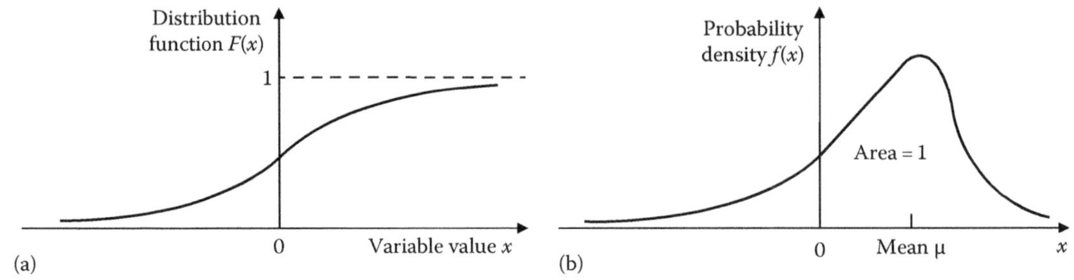

FIGURE C.1
(a) A cumulative probability distribution function. (b) A probability density function.

Note: If X can take only discrete values x_i, then discrete probabilities (point mass probabilities) p_i have to be used in place of the density function, and summation has to be used in place of integration. In particular, $F(x_r) = \sum_{i=-\infty}^{r} p_i$.

C.1.3 Mean Value (Expected Value)

If a random variable X is measured repeatedly for a very large (infinite) number of times, the average of these measurements is the *mean value* μ or *expected value* $E(X)$. It should be easy to see that this may be expressed as the weighted sum of all possible values of the random variable, each value being weighted by the associated probability of its occurrence. Since the probability that X takes the value x is given by $f(x)\delta x$, with δx approaching zero, we have $\mu = E(X) = \lim_{\delta x \to 0} \sum x f(x)\delta x$

Since the right-hand side summation becomes an integral in the limit, we get

$$\mu = E(X) = \int_{-\infty}^{\infty} x f(x) dx \tag{C.5}$$

C.1.4 Root-Mean-Square Value

The mean square value of a random variable X is given by

$$E(X^2) = \int_{-\infty}^{\infty} x^2 f(x) dx \tag{C.6}$$

The root-mean-square value is the square root of the mean square value.

C.1.5 Variance and Standard Deviation

Variance of a random variable is the mean square value of the deviation from mean. This is denoted by Var(X) or σ^2 and is given by

$$\text{Var}(X) = \sigma^2 = \int_{-\infty}^{\infty} (x-\mu)^2 f(x) dx \tag{C.7}$$

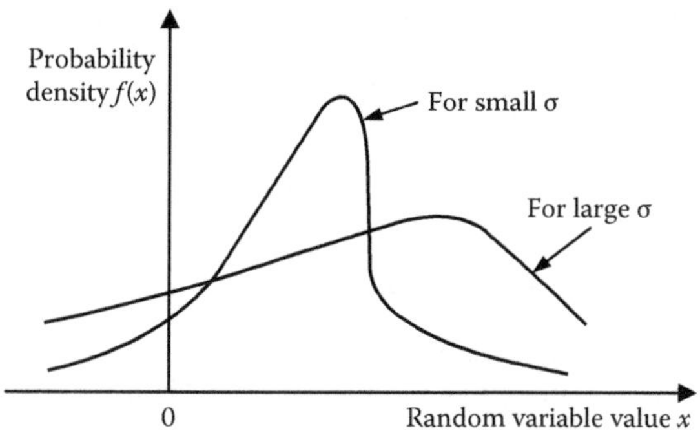

FIGURE C.2
Effect of standard deviation on the shape of a probability density curve

By expanding this, we can show that

$$\sigma^2 = E(X^2) - \mu^2 \tag{C.8}$$

Standard deviation σ is the square root of variance. Note that standard deviation is a measure of statistical "spread" of a random variable. A random variable with smaller σ is less random, and its density curve will exhibit a sharper peak, as shown in Figure C.2.

Some thinking should convince you that if the probability density function of random variable X is $f(x)$, then probability density function of any (well-behaved) function of X is also $f(x)$. In particular, for constants a and b, the probability density function of $(aX+b)$ is also $f(x)$. Note, further, that the mean of $(aX+b)$ is $(a\mu+b)$. Hence, from Equation C.7, it follows that the variance of aX is

$$X^2\ \mathrm{Var}(aX) = \int_{-\infty}^{\infty} (ax - a\mu)^2 f(x)dx = a^2 = \int_{-\infty}^{\infty} (x-\mu)^2 f(x)dx$$

Hence,

$$\mathrm{Var}(aX) = a^2\mathrm{Var}(X) \tag{C.9}$$

C.1.6 Independent Random Variables

Two random variables, X_1 and X_2, are said to be independent if the event "X_1 assumes a certain value" is completely independent of the event "X_2 assumes a certain value." In other words, the processes that generate the responses X_1 and X_2 are completely independent. Furthermore, probability distribution of X_1 and X_2 is completely independent. Hence, it can be shown that for independent random variables X_1 and X_2, the mean value of the product is equal to the product of the mean values. Thus,

$$E\,(X_1X_2) = E\,(X_1)\,E(X_2) \tag{C.10}$$

for independent random variables X_1 and X_2.

Now, using the definition of variance and Equation C.10, it can be shown that

$$\mathrm{Var}(X_1+X_2)=\mathrm{Var}(X_1)+\mathrm{Var}(X_2) \tag{C.11}$$

for independent X_1 and X_2.

C.1.7 Sample Mean and Sample Variance

Consider N measurements $\{X_1, X_2, \ldots, X_N\}$ of random variable X. This set of data is termed a "data sample." It generally is not possible to extract all information about the probability distribution of X from this data sample. We are able, however, to make some useful *estimates*. One would expect that the larger the data sample, the more accurate these statistical estimates would be.

An estimate for the mean value of X would be the *sample mean* $\bar{X}$, which is defined as

$$\bar{X}=\frac{1}{N}\sum_{i=1}^{N}X_i \tag{C.12}$$

An estimate for variance would be the *sample variance* S^2, given by

$$S^2=\frac{1}{(N-1)}\sum_{i=1}^{N}(X_i-\bar{X})^2 \tag{C.13}$$

An estimate for standard deviation would be the *sample standard deviation*, S, which is the square root of the sample variance.

One might be puzzled by the denominator $N-1$ on the right-hand side of Equation C.13. Since we are computing an "average" deviation, the denominator should have been N. But in that case, with just one reading ($N=1$), we get a finite value for S, which is not correct because one cannot talk about a sample standard deviation when only one measurement is available. Since, according to Equation C.13, S is not defined (0/0) when $N=1$, this definition of S^2 is more realistic. Another advantage of Equation C.13 is that this equation gives an *unbiased estimate* of variance. This concept will be discussed next. Note that if we use N instead of $N-1$ in Equation C.13, the computed variance is called "population variance." Its square root is *population standard deviation*. When $N>30$, the difference between sample variance and population variance becomes negligible.

C.1.8 Unbiased Estimates

Note that, prior to measurement, each term X_i in the sample data set $\{X_1, X_2, \ldots, X_N\}$ is itself a random variable just like X, because the measurement process of X_i introduces some randomness. In other words, if N measurements were taken at one time and then the same measurements were repeated, the values in the second set would be different from the first set, since X was random to begin with. It follows that $\bar{X}$ and S in Equations C.12 and C.13 are also random variables. Note that the mean value of $\bar{X}$ is

$$E(\bar{X})=E\left[\frac{1}{N}\sum_{i=1}^{N}X_i\right]=\frac{1}{N}\sum_{i=1}^{N}E(X_i)=\frac{N\mu}{N}$$

Hence,

$$E(\bar{X}) = \mu \tag{C.14}$$

We know that $\bar{X}$ is an estimate for μ. Also, from Equation C.14, we observe that the mean value of $\bar{X}$ is μ. Hence, the sample mean $\bar{X}$ is an *unbiased estimate* of the mean value μ. Similarly, from Equation C.13, we can show that the mean value of S^2 is

$$E(S^2) = \sigma^2 \tag{C.15}$$

assuming that X_i are independent measurements. Thus, the sample variance S^2 is an unbiased estimate of the variance σ^2. In general, if the mean value of an estimate is equal to the exact value of the parameter that is being estimated, the estimate is said to be unbiased. Otherwise, it is a *biased estimate.*

C.1.9 Gaussian Distribution

Gaussian distribution, or *normal distribution,* is probably the most extensively used probability distribution in engineering applications. Apart from its ease of use, another justification for its widespread use is provided by the *central limit theorem.* This theorem states that a random variable that is formed by summing a very large number of independent random variables takes Gaussian distribution in the limit. Since many engineering phenomena are consequences of numerous independent random causes, the assumption of normal distribution is justified in many cases. The validity of Gaussian assumption can be checked by plotting data on *probability graph paper* or by using various tests such as the *chi-square test.*

The Gaussian probability density function is given by

$$f(x) = \frac{1}{\sqrt{2\pi}\sigma} \exp\left[-\frac{(x-\mu)^2}{2\sigma^2}\right] \tag{C.16}$$

Note that only two parameters, mean μ and standard deviation σ, are necessary to determine a Gaussian distribution completely.

A closed algebraic expression cannot be given for the cumulative probability distribution function $F(x)$ of Gaussian distribution. It should be evaluated by numerical integration. Numerical values for the normal distribution curve are available in tabulated form, with the random variable X being normalized with respect to μ and σ according to

$$Z = \frac{X-\mu}{\sigma} \tag{C.17}$$

Note that the mean value of this normalized variable Z is

$$E(Z) = E\left[\frac{(X-\mu)}{\sigma}\right] = \frac{[E(X)-\mu]}{\sigma} = \frac{(\mu-\mu)}{\sigma}$$

or

$$E(Z) = 0 \tag{C.18}$$

and the variance of Z is

$$\mathrm{Var}(Z) = \mathrm{Var}\left[\frac{(X-\mu)}{\sigma}\right] = \mathrm{Var}\frac{(X-\mu)}{\sigma^2} = \mathrm{Var}\frac{(X)}{\sigma^2} = \frac{\sigma^2}{\sigma^2}$$

or

$$\mathrm{Var}(Z) = 1 \tag{C.19}$$

Furthermore, the probability density function of Z is

$$f(z) = \frac{1}{\sqrt{2\pi}}\exp\left(\frac{-z^2}{2}\right) \tag{C.20}$$

What is usually tabulated is the area under the density curve $f(z)$ of the normalized random variable Z for different values of z. A convenient form is presented in Table C.1, where the area under the $f(z)$ curve from 0 to z is tabulated up to four decimal places for different positive values of z up to two decimal places. Since the density curve is symmetric about the mean value (zero for the normalized case), values for negative z do not have to be tabulated. Furthermore, when $z \to \infty$, area A in Table C.1 approaches 0.5. The value for $z=3.09$ is already 0.4990. Hence, for most practical purposes, area A may be taken as 0.5 for z values greater than 3.0. Since Z is normalized with respect to σ, it follows that $z=3$ actually corresponds to three times the standard deviation of the original random variable X. Hence, for a Gaussian random variable, most of the values will fall within $\pm 3\sigma$ about the mean value. It can be stated that approximately

- 68% of the values will fall within $\pm\,\sigma$ about μ
- 95% of the values will fall within $\pm 2\sigma$ about μ
- 99.7% of the values will fall within $\pm 3\sigma$ about μ

These can be easily verified using Table C.1.

C.1.10 Confidence Intervals

The probability that the value of a random variable would fall within a specified interval is called a "confidence level." As an example, consider a Gaussian random variable X that has mean μ and standard deviation σ. This is denoted by

$$X = N(\mu, \sigma) \tag{C.21}$$

Suppose that N measurements $\{X_1, X_2, \ldots, X_N\}$ are made. The sample mean $\bar{X}$ is an unbiased estimate for μ. We also know that the standard deviation of $\bar{X}$ is $\sigma\sqrt{N}$.

Now consider the normalized random variable:

$$Z = \frac{\bar{X}-\mu}{\sigma/\sqrt{N}} \tag{C.22}$$

TABLE C.1

Table of Gaussian Probability Distribution

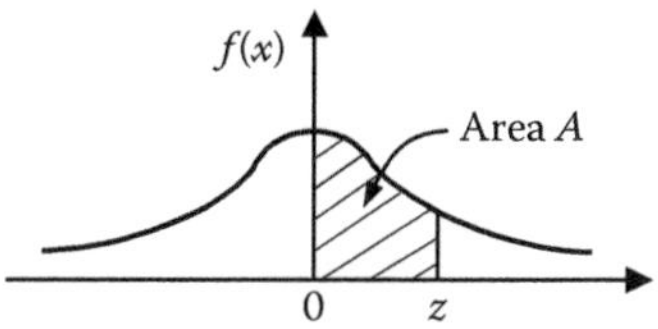

$$f(z) = \frac{1}{\sqrt{2\pi}} \exp\left(\frac{-z^2}{2}\right)$$

$$A = \int_0^z f(z)dz$$

Area *A*

z	0.00	0.01	0.02	0.03	0.04	0.05	0.06	0.07	0.08	0.09
0.0	0.0000	0.0040	0.0080	0.0120	0.0160	0.0199	0.0239	0.0279	0.0319	0.0359
0.1	0.0398	0.0438	0.0478	0.0517	0.0557	0.0596	0.0636	0.0675	0.0714	0.0753
0.2	0.0793	0.0832	0.0871	0.0910	0.0948	0.0987	0.1026	0.1064	0.1103	0.1141
0.3	0.1179	0.1217	0.1255	0.1293	0.1331	0.1368	0.1406	0.1443	0.1480	0.1517
0.4	0.1554	0.1591	0.1628	0.1664	0.1700	0.1736	0.1772	0.1808	0.1844	0.1879
0.5	0.1915	0.1950	0.1985	0.2019	0.2054	0.2088	0.2123	0.2157	0.2190	0.2224
0.6	0.2257	0.2291	0.2324	0.2357	0.2389	0.2422	0.2454	0.2486	0.2517	0.2549
0.7	0.2580	0.2611	0.2642	0.2673	0.2704	0.2734	0.2764	0.2794	0.2823	0.2852
0.8	0.2881	0.2910	0.2939	0.2967	0.2995	0.3023	0.3051	0.3078	0.3106	0.3233
0.9	0.3159	0.3186	0.3212	0.3238	0.3264	0.3289	0.3315	0.3340	0.3365	0.3389
1.0	0.3413	0.3438	0.3461	0.3485	0.3508	0.3531	0.3554	0.3577	0.3599	0.3621
1.1	0.3643	0.3665	0.3686	0.3708	0.3729	0.3749	0.3770	0.3790	0.3810	0.3830
1.2	0.3849	0.3869	0.3888	0.3907	0.3925	0.3944	0.3962	0.3980	0.3997	0.4015
1.3	0.4032	0.4049	0.4066	0.4082	0.4099	0.4115	0.4131	0.4147	0.4162	0.4177
1.4	0.4192	0.4207	0.4222	0.4236	0.4251	0.4265	0.4279	0.4292	0.4306	0.4319
1.5	0.4332	0.4345	0.4357	0.4370	0.4382	0.4394	0.4406	0.4418	0.4429	0.4441
1.6	0.4452	0.4463	0.4474	0.4484	0.4495	0.4505	0.4515	0.4525	0.4535	0.4545
1.7	0.4554	0.4564	0.4573	0.4582	0.4591	0.4599	0.4608	0.4616	0.4625	0.4633
1.8	0.4641	0.4649	0.4656	0.4664	0.4671	0.4678	0.4686	0.4693	0.4699	0.4706
1.9	0.4713	0.4719	0.4726	0.4732	0.4738	0.4744	0.4750	0.4758	0.4761	0.4767
2.0	0.4772	0.4778	0.4783	0.4788	0.4793	0.4799	0.4803	0.4808	0.4812	0.4817
2.1	0.4821	0.4826	0.4830	0.4834	0.4838	0.4842	0.4846	0.4850	0.4854	0.4857
2.2	0.4861	0.4864	0.4868	0.4871	0.4875	0.4878	0.4881	0.4884	0.4887	0.4890
2.3	0.4893	0.4896	0.4898	0.4901	0.4904	0.4906	0.4909	0.4911	0.4913	0.4916
2.4	0.4918	0.4920	0.4922	0.4925	0.4927	0.4929	0.4931	0.4932	0.4934	0.4936
2.5	0.4938	0.4940	0.4941	0.4943	0.4945	0.4946	0.4948	0.4949	0.4951	0.4952
2.6	0.4953	0.4955	0.4956	0.4957	0.4959	0.4960	0.4961	0.4962	0.4963	0.4964
2.7	0.4965	0.4966	0.4967	0.4968	0.4969	0.4970	0.4971	0.4972	0.4973	0.4974
2.8	0.4974	0.4975	0.4976	0.4977	0.4977	0.4978	0.4979	0.4979	0.4980	0.4981
2.9	0.4981	0.4982	0.4982	0.4983	0.4984	0.4984	0.4985	0.4985	0.4986	0.4986
3.0	0.4987	0.4987	0.4987	0.4988	0.4988	0.4988	0.4989	0.4989	0.4989	0.4990

This is a Gaussian random variable with zero mean and unity standard deviation. The probability p that the values of Z fall within $\pm z_o$

$$P(-z_o < Z \le z_o) = p \tag{C.23}$$

can be determined from Table C.1 for a specified value of z_o. Now substituting Equation C.22 in C.23, we get

$$P\left(-z_o < \frac{\bar{X}-\mu}{\sigma/\sqrt{N}} \le z_o\right) = p \rightarrow$$

$$P\left(\bar{X} - \frac{z_o\sigma}{\sqrt{N}} \le \mu < \bar{X} + \frac{z_o\sigma}{\sqrt{N}}\right) = p \tag{C.24}$$

Note that the lower limit has the "≤" sign and the upper limit has the "<" sign within the parentheses. These have been used for mathematical precision, but for practical purposes, either ≤ or < may be used in each limit. Now, from Equation C.24, it follows that the confidence level is p and that the actual mean value μ would fall within $\pm z_o\,\sigma/\sqrt{N}$ of the estimated (sample) mean value $\bar{X}$.

Example C.1

The angular resolution of a resolver (a rotary motion sensor) was tested 16 times, independently, and recorded in degrees as follows:

0.11	0.12	0.09	0.10	0.10	0.14	0.08	0.08
0.13	0.10	0.10	0.12	0.08	0.09	0.11	0.15

If the standard deviation of the angular resolution of this brand of resolvers is known to be 0.01°, what are the odds that the mean resolution would fall within 5% of the sample mean?

Solution

To solve this problem, we assume that resolution is normally distributed. The sample mean is computed as

$$\bar{X} = 1/16(0.11 + 0.12 + \cdots + 0.11 + 0.15) = 0.10625$$

In view of Equation C.26, we must have

$$z_o\sigma/\sqrt{16} = 5\% \text{ of } \bar{X}$$

Hence,

$$(z_o \times 0.01)/\sqrt{16} = (5/100) \times 0.10625 \rightarrow z_o = 2.125$$

Now, from Table C.1,

$$P(-2.125 < Z < 2.125) = 2 \times \frac{(0.4830 + 0.4834)}{2} = 0.9664$$

C.2 Sign Test and Binomial Distribution

Sign test is useful in comparing the accuracies of two similar instruments. First, measurements should be made on the same measurand (i.e., input signal to instrument) using the two devices. Next, the readings of one instrument are subtracted from the corresponding readings of the second instrument, and the results are tabulated. Finally, the probability of getting the number of negative signs (or positive signs) equal to what is present in the tabulated results is computed using *binomial distribution*.

Before discussing binomial distribution, let us introduce some new terminology. First, *factorial r* (denoted by $r!$) of an integer r is defined as the product

$$r! = r \times (r-1) \times (r-2) \times \cdots \times 2 \times 1 \tag{C.25}$$

Now, suppose that there are n distinct articles that are distinguishable from one another. The number of ways in which r articles could be picked from the batch of n, giving proper consideration to the order in which the r articles are picked (or arranged), is called the number of "permutations" of r from n. This is denoted by ${}^nP_r$, which is given by

$${}^nP_r = n \times (n-1) \times (n-2) \times \cdots \times (n-r+2) \times (n-r+1) = \frac{n!}{(n-r)!} \tag{C.26}$$

This can be easily verified, since the first article can be chosen in n ways, the second article can be chosen from the remaining $(n-1)$ articles in $(n-1)$ ways and kept next to the first article, and so on.

If we disregard the order in which the r articles are picked (and arranged), the number of possible choices of r articles is termed the number of "combinations" of r from n. This is denoted by ${}^nC_r$. Now, since each combination can be arranged in $r!$ different ways (if the order of arrangement is considered), we have

$${}^nC_r \times r! = {}^nP_r \tag{C.27}$$

Hence, using Equation C.26, we get

$${}^nC_r = \frac{n \times (n-1) \times (n-2) \times \cdots \times (n-r+2) \times (n-r+1)}{r!} = \frac{n!}{(n-r)!\,r!} \tag{C.28}$$

With the foregoing notation, we can introduce binomial distribution in the context of sign test. Suppose that n pairs of readings are taken from the two instruments. If the probability that a difference in reading would be positive is p, then the probability that the difference would be negative is $1-p$. Note that if the systematic error in the two instruments is the same and if the random error is purely random, then $p = 0.5$.

The probability of getting exactly r positive signs among the n entries in the table is

$$p(r) = {}^nC_r p^r (1-p)^{n-r} \tag{C.29}$$

To verify the previous equation, note that this event is similar to picking exactly r items from n items and constraining each picked item to be positive (having probability p) and also constraining the remaining $(n-r)$ items to be negative (having probability $1-p$).

Note that r is a discrete variable that takes values $r=1, 2, \ldots, n$. Furthermore, it can be easily verified that

$$\sum_{r=1}^{n} p(r) = \sum_{r=1}^{n} {}^{n}C_{r}p^{r}(1-p)^{n-r} = (p+1-p)^{n} = 1 \quad \text{(C.30)}$$

Hence, $p(r)$, $r=1, 2,\ldots, n$, is a discrete function that resembles a continuous probability density function $f(x)$. In fact, $p(r)$ given by Equation C.27 represents *binomial probability distribution*. Using Equation C.27, we can perform the sign test. The details of the test are conveniently explained by means of an example.

Example C.2

To compare the accuracies of two brands of differential transformers (DTs, which are displacement sensors), the same rotation (in degrees) of a robot arm joint was measured using both brands, DT1 and DT2. The following 10 measurement pairs were taken:

DT1	10.3	5.6	20.1	15.2	2.0	7.6	12.1	18.9	22.1	25.2
DT2	9.8	5.8	20.0	16.0	1.9	7.8	12.2	18.7	22.0	25.0

Assuming that both devices are used simultaneously (so that backlash and other types of repeatability errors in manipulators do not enter into our problem), determine whether the two brands are equally accurate at the 70% level of significance.

Solution

First, we form the sign table by taking the difference of corresponding measurements:

DT1 − DT2	0.5	−0.2	0.1	−0.8	0.1	−0.2	−0.1	0.2	0.1	0.2

Note that there are six positive signs and four negative signs. If we had tabulated DT2 − DT1, however, we would get four positive signs and six negative signs. Both these cases should be taken into account in the sign test. Furthermore, more than six positive signs or fewer than four positive signs would make the two devices less similar (in accuracy) than what is indicated by the data. Hence, the probability of getting six or more positive signs or four or fewer positive signs should be computed in this example in order to estimate the possible match (in accuracy) of the two devices.

If the error in both transducers is the same, we should have

$$P\text{ (positive difference)} = p = 0.5$$

This is the hypothesis that we are going to test. Using Equation C.31, the probability of getting six or more positive signs or four or fewer negative signs is calculated as

1 − probability of getting exactly five positive signs

$$= 1 - {}^{10}C_5(0.5)^5 \times (0.5)^{5'} = 1 - \frac{10!}{5!5!} \times (0.5)^{10} = 1 - 0.246 = 0.754$$

Note that the hypothesis of two brands being equally accurate is supported by the test data at a level of significance over 75%, which is better than the specified value of 70%.

Some useful probability distributions (density functions in the continuous case and point mass functions for the discrete case) are listed in Table C.2.

TABLE C.2

Some Useful Probability Density Functions

Name	Function	Parameters
Normal (Gaussian) probability density $f(y)$	$\frac{1}{\sqrt{2\pi\sigma^2}}\exp\left(-\frac{(y-\mu)^2}{2\sigma^2}\right)$	μ, mean; σ, standard deviation
Poisson probability (discrete) p_r	$\frac{\lambda^r \exp(-\lambda)}{r!}$	r, number of successes in a specified duration; λ, mean = standard deviation
Binomial probability (discrete) p_r	${}^nC_r p^r(1-p)^{n-r} = \frac{n!}{(n-r)!r!}p^r(1-p)^{n-r}$	p, probability of success of a trial; n, total number of trials; r, number of successful trials; np, mean; $\sqrt{np(1-p)}$, standard deviation

Appendix D: Reliability Considerations for Multicomponent Devices

In the practice of engineering (e.g., modeling, design, analysis, monitoring, performance evaluation, fault diagnosis, isolation, control, and testing), we depend on the proper component matching, component interconnection, and operation of multicomponent devices and systems. Equipment that has several components that are crucial to its operation can have more than one mode of failure. Each failure mode of the overall system will depend on some combination of failure of the components. Component failure is governed by the laws of probability. In this appendix, we will consider some fundamentals of probability theory that are useful in the reliability and failure analysis of multicomponent systems.

D.1 Failure Analysis

D.1.1 Reliability

The probability that the component will perform satisfactorily over a specified time period t (component age) under given operating conditions is called reliability. It is denoted by R. Hence,

$$R(t) = \wp(\text{survival}) \tag{D.1}$$

in which $\wp\,(\cdot)$ denotes "the probability of."

D.1.2 Unreliability

The probability that the component will malfunction or fail during the time period t is called its unreliability or its probability or failure. It is denoted by F. Hence,

$$F(t) = \wp(\text{failure}) \tag{D.2}$$

Since we know as a certainty that the component will either survive or fail during the specified time period t, we can write

$$R(t) + F(t) = 1 \tag{D.3}$$

The probability of survival of a component usually decreases with age. Consequently, the typical $R(t)$ is a monotonically decreasing function of t, as shown in Figure D.1. If it is known as a certainty that the component is good in the beginning, then $R(0) = 1$. Because of manufacturing defects, damage during shipping, and the like, however, we usually

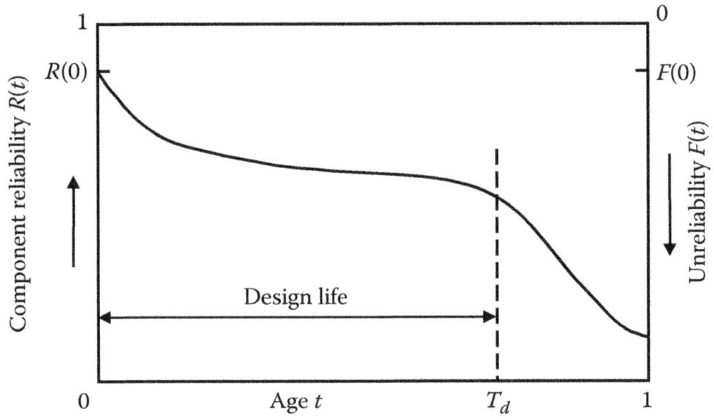

FIGURE D.1
A typical reliability (unreliability) curve.

have $R(0) \leq 1$. For a satisfactory component, $R(t)$ should not drop appreciably during its design life T_d. The drop is faster initially, however, because of infant mortality (again due to manufacturing defects and the like), and later on, as the component exceeds its design life because of old age (wear, fatigue, and so on).

It is clear from Equation D.3 that the unreliability curve is completely defined by the reliability curve. As shown in Figure D.1, transforming one to the other is a simple matter of reversing the axis.

D.1.3 Inclusion–Exclusion Formula

Consider two events A and B that are schematically represented by areas (as in Figure D.2). Each event consists of a set of outcomes. The total area covered by the two sets denoted by A and B is given by adding the area of A to the area of B and subtracting the common area.

This procedure can be expressed as

$$\wp(A \text{ or } B) = \wp(A) + \wp(B) - \wp(A \text{ and } B) \tag{D.4}$$

Example D.1

Consider the rolling of a fair die. The set of total outcomes consists of six elements forming the space: $S = \{1, 2, 3, 4, 5, 6\}$. Each outcome has a probability of 1/6. Now consider the two events: $A = \{\text{outcome is odd}\}$ and $B = \{\text{outcome is divisible by 3}\}$.

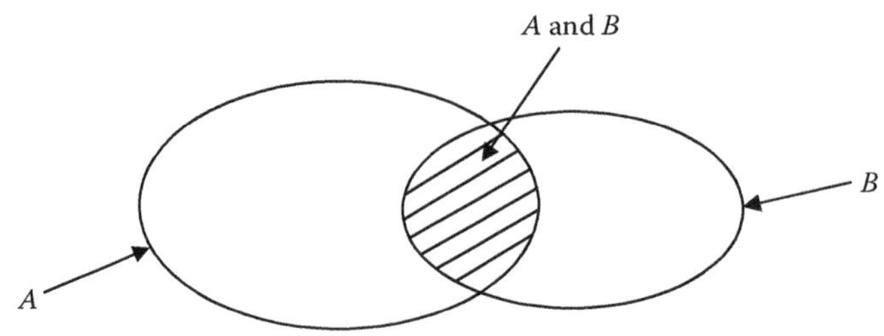

FIGURE D.2
Venn diagram illustrating the inclusion–exclusion formula.

Then, $A = \{1, 3, 5\}$ and $B = \{3, 6\}$.

Consequently, A or $B = \{1, 3, 5, 6\}$ and A and $B = \{3\}$.

It follows that $\wp(A) = 3/6$, $\wp(B) = 2/6$, $\wp(A \text{ or } B) = 4/6$, and $\wp(A \text{ and } B) = 1/6$.

These values satisfy Equation D.4.

If the events A and B do not have common outcomes, they are said to be mutually exclusive. Then, the common area of intersection of sets A and B in Figure D.2 would be zero. Hence, for mutually exclusive events,

$$\wp(A \text{ and } B) = 0 \tag{D.5}$$

D.2 Bayes' Theorem

A simplified version of Bayes' theorem can be expressed as

$$\wp(A \text{ and } B) = \wp(A/B)\wp(B) = \wp(B/A)\wp(A) \tag{D.6}$$

in which $\wp(A/B)$ denotes the conditional probability that event A occurs, given the condition that event B has occurred.

In the previous example of rolling a fair die, if it is known that event B has occurred, the outcome must be either 3 or 6. Then, the probability that event A would occur is simply the probability of picking 3 from the set $\{3, 6\}$. Hence, $\wp(A/B) = 1/2$. Similarly, $\wp(B/A) = 1/3$. It should be noted that Equation D.6 holds for this example.

D.2.1 Product Rule for Independent Events

If the two events A and B are independent of each other, then the occurrence of event B has no effect whatsoever on determining whether event A occurs. Consequently,

$$\wp(A/B) = \wp(A) \tag{D.7}$$

for independent events. Then, it follows from Equation D.6 that

$$\wp(A \text{ and } B) = \wp(A)\ \wp(B) \tag{D.8}$$

for independent events. Equation D.8 is the product rule, which is applicable to independent events.

It should be emphasized that, even though independence implies that the product rule holds, the converse is not necessarily true. In the example on rolling a fair die, $\wp(A/B) = \wp(A) = 1/2$. Suppose, however, that is not a fair die and that the probabilities of the outcomes $\{1, 2, 3, 4, 5, 6\}$ are $\{1/3, 1/6, 1/6, 0, 1/6, 1/6\}$. Then, $\wp(A) = 1/3 + 1/6 + 1/6 = 2/3$ whereas

$$\wp(A/B) = \frac{1/6}{1/6 + 1/6} = 1/2$$

This shows that A and B are not independent events in this sample.

Furthermore, $\wp(B) = 1/6$ and $\wp(A \text{ and } B) = 1/6$. It is seen that Bayes' theorem is satisfied by this example.

D.2.2 Failure Rate

The function $F(t)$ defined by Equation D.2 is the probability–distribution function of the random variable T denoting the time to failure. We will define the following rate functions:

$$r(t) = \frac{dR(t)}{dt} \tag{D.9}$$

$$f(t) = \frac{dF(t)}{dt} \tag{D.10}$$

where $R(t) = \wp(T > t)$ and $F(t) = \wp(T \le t)$.

In Equation D.10, $f(t)$ is the probability–density function corresponding to the time to failure. It follows that

$$\begin{aligned} &\wp(\text{component survived up to } t, \text{ failed within next duration } dt) \\ &= \wp(\text{failed within } t, t + dt) = dF(t) = f(t)dt \end{aligned} \tag{D.11}$$

Also

$$\wp(\text{component survived up to } t) = R(t) \tag{D.12}$$

Let us define the function $\beta(t)$ such that

$$\wp(\text{failed within next duration } dt/\text{survived up to } t) = \beta(t)\, dt \tag{D.13}$$

By substituting Equations D.11 through D.13 into Equation D.6, we obtain

$$f(t)dt = \beta(t)dt\ R(t)$$

or

$$\beta(t) = \frac{f(t)}{R(t)} = \frac{f(t)}{1 - F(t)} \tag{D.14}$$

Let us suppose that there are N components. If they all have survived up to t, then, on the average, $N\beta(t)\delta t$ components will fail during the next δt. Consequently, $N\beta(t)$ corresponds to the rate of failure for the collection of components at time t. For a single component ($N = 1$), the rate of failure is $\beta(t)$. For obvious reasons, $\beta(t)$ is sometimes termed conditional failure. Other names for this function include intensity function and hazard function, but *failure rate* is the most common name.

In view of Equation D.10, we can write Equation D.14 as a first-order linear, ordinary differential equation with variable parameters:

$$\frac{dF(t)}{dt} + \beta(t)F(t) = \beta(t) \tag{D.15}$$

Assuming a good component initially, we have

$$F(0) = 0 \tag{D.16}$$

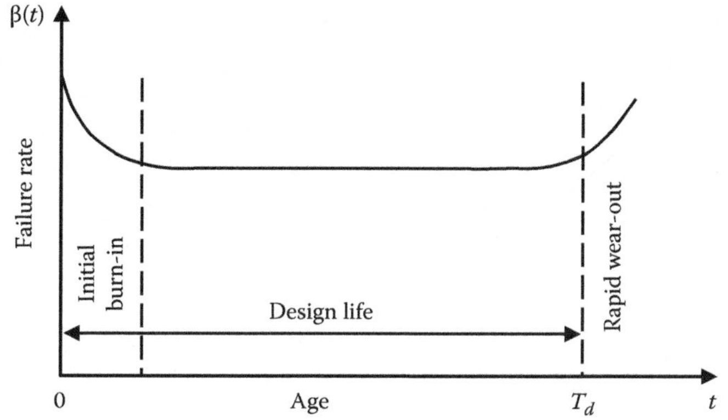

FIGURE D.3
A typical failure rate curve.

The solution of Equation D.15 subject to Equation D.16 is

$$F(t) = 1 - \exp\left(-\int_0^t \beta(\tau)d\tau\right) \tag{D.17}$$

in which τ is a dummy variable. Then, from Equation D.3

$$R(t) = \exp\left(-\int_0^t \beta(\tau)d\tau\right) \tag{D.18}$$

It is observed from Equation D.18 that the reliability curve can be determined from the failure rate curve and the reverse.

A typical failure rate curve for an engineering component is shown in Figure D.3. It has a characteristic "bathtub" shape, which can be divided into three phases, as in the figure. These phases might not be so distinct in a real situation. Initial burn-in period is characterized by a sharp drop in the failure rate. Because of such reasons as poor workmanship, material defects, and poor handling during transportation, a high degree of failure can occur during a short initial period of design life. Following that, the failures typically will be due to random causes. The failure rate is approximately constant in this region. Once the design life is exceeded (third phase), rapid failure can occur because of wear, fatigue, and other types of cumulative damage, and eventual collapse will result.

It is frequently assumed that the failure rate is constant during the design life of a component. In this case, Equation D.18 gives the exponential reliability function:

$$R(t) = \exp(-\beta t) \tag{D.19}$$

This situation is represented in Figure D.4. This curve is not comparable to the general reliability curve shown in Figure D.1. As a result, constant failure rate should not be used for relatively large durations of time (i.e., for a large segment of the design life), unless it has

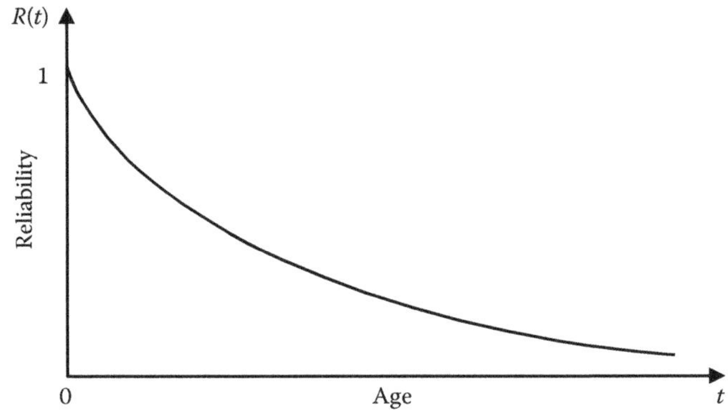

FIGURE D.4
Reliability curve under constant failure rate.

been verified by tests. For short durations, however, this approximation is normally used and it results in considerable analytical simplicity.

D.2.3 Product Rule for Reliability

For multicomponent equipment, if we assume that the failure of one component is independent of the failure of any other, the product rule given by Equation D.8 can be used to determine the overall reliability of the equipment. The reliability of an N-component object with independently failing components is given by

$$R(t) = R_1(t)R_2(t)\ldots R_N(t) \tag{D.20}$$

in which $R_i(t)$ is the reliability of the ith component. If there is no component redundancy, which is assumed in Equation D.20, none of the components should fail (i.e., $R_i(t) \neq 0$ for $i = 1,2,\ldots,N$) for the object to operate properly (i.e., $R(t) \neq 0$). This follows from Equation D.20.

In vibration testing, a primary objective is to maximize the risk of component failure when subjected to the test environment (so that the probability of failure is less in the actual in-service environment). One way of achieving this is by maximizing the test-strength-measure function given by

$$TS = \sum_{i=1}^{r} F_i(T)\Phi_i \tag{D.21}$$

where

$F_i(T)$ is the probability of failure (unreliability) of the ith component for the test duration T
Φ_i is a dynamic response measure at the location of the ith component

The parameters of optimization could be the input direction and the frequency of excitation for a given input intensity.

Regarding component redundancy, consider the simple situation of r_i identical subcomponents connected in parallel (r_ith-order redundancy) to form the ith component. The component failure requires the failure of all r_i subcomponents. The failure of one subcomponent is assumed to be independent of the failure state of other subcomponents. Then, the unreliability of the ith component can be expressed as

$$F_i = (F_{0i})^{ri} \tag{D.22}$$

Here, F_{0i} is the unreliability of each subcomponent in the ith component. This simple model for redundancy may not be valid in some situations.

There are two basic types of redundancy: active redundancy and standby redundancy. In active redundancy, all redundant elements are permanently connected and active during the operation of the equipment. In standby redundancy, only one of the components in a redundant group is active during the equipment operation. If that component fails, an identical second component will be automatically connected.

For standby redundancy, some form of switching mechanism is needed, which means that the reliability of the switching mechanism itself must be accounted for. Component aging is relatively less, however, and the failure of components within the redundant group is mutually independent. In active redundancy, however, there is no need for a switching mechanism. But the failure of one component in the redundant group can overload the rest, thereby increasing their probability of failure (unreliability). Consequently, component failure within the redundant group is not mutually independent in this case. Also, component aging is relatively high because the components are continuously active.

Answers to Numerical Problems

Chapter 1

1.10: 11×10^{-6} Hz

Chapter 2

2.7: (a) 500 MΩ, 2%; (b) 4 Ω, 2.4%

2.10: $b = 6.455 \times 10^3$ N · s/m, at 200 rad/s, transmissibility magnitude = −6 dB

Chapter 3

3.1: 1 MΩ, 50 Ω, 10^6, 10 kHz

3.2(b): Case 1 $v_o = -7.5$ V, not saturated; Case 2 saturated at $v_o = -14$ V

3.5: $f_b =$ 31.8 kHz, $\Delta t = 5$ μs

Chapter 4

4.2(b): 44.5 megabits/s

4.5(a): 3600 rpm, 15 balls

4.7: 0.04 V, 0.02 V

4.13: 1%

4.22: 2.5%

Chapter 5

5.3: (a) 4×10^{-6} m², (b) 2500 N/m², (c) 74.0 dB

5.6: $T_r f_b = 25{\cdot}0$

5.11(b): (iv) <10%; (v) $k_{low} = 6.25 \times 10^4$ N/m, $k_{low} = 6.25 \times 10^4$ N/m, $k_{high} = 25.0 \times 10^4$ N/m

5.13(b): 0.99%

Chapter 6

6.4: $f_s = 512$ Hz, between 200 and 256 Hz

6.5(b): $f_b = 16$ Hz, $f_s > 64$ Hz, say, $f_s = 200$ Hz, $f_c = 50$ Hz

6.11: $e_b = 7.5\%$

6.13(b): $e_m = \pm 0.01 = \pm 1.1\%$, $e_l = \pm 0.11 = \pm 11\%$, $e_r = \pm 0.012 = \pm 1.2\%$, $e_\alpha = \pm 0.01 = \pm 1.0\%$

6.15(b): (ii): $e_V = \pm 13.0\%$

6.17(b): (ii): For $e_v = \pm 1\%$: $e_Q = \pm 1\%$, $e_s = \pm 1.2\%$, $e_f = \pm 3.1\%$

6.19: (a) 110 lbf, (b) $e_{ABS} = (1 - 0.5) \times 2 + 1 + 0.5 \times 1\% = 2.5\%$, (c) $e_w = 1.7\%$, $e_s = 0.8\%$, $e_y = 1.7\%$

6.20: $\sigma_T = 1.32\%$

6.21(b): $e_T = 0.003$

6.22: $P[-z_o < Z \le z_o] = 0.697 < 0.99$

Chapter 7

7.2(b): Estimated $p = 1.988$, estimated $a = 1.606$

7.4: (a)`(With 95% confidence bounds), p1 = 141.9 (138, 145.8), p2 = -0.01645 (-0.0325, -0.000403);` (b)`(with 95% confidence bounds): p1 = 3306 (2825, 3786), p2 = 118.7 (115.2, 122.2), p3 = 0.008617 (0.003353, 0.01388)`

7.5: `samplemean = 100.6646, samplestd = 1.4797, mleest = 100.6646 1.4422`

7.6: (b) `(With 95% confidence bounds): p1 = 2.204 (2.136, 2.271), p2 = 0.01939 (0.01185, 0.02692);` (c) `(with 95% confidence bounds): p1 = 0.7154 (-0.1291, 1.56), p2 = 2.202 (2.135, 2.268), p3 = 0.01055 (-0.00222, 0.02331)`

7.7: `samplemean = 10.0065, samplestd = 0.0776, mleest = 10.0065 0.0768`

7.10: `mley1 = 1.2403 0.5033; mley2 = 1.0326 0.3442; mley12 = 1.1018 0.4159`
LSE results: `mean(y1) = 1.2403, mean(y2) = 1.0326, mean(y) = 1.1018`
`std(y1) = 0.5305, std(y2) = 0.3531, std(y) = 0.4230`

7.12: $P(\boldsymbol{m} \mid y_1) = \begin{bmatrix} 0.75 \\ 0.05 \\ 0.20 \end{bmatrix}$, small

7.13: Laser: `Final estimation m = 4.9795, final estimation = 0.0026`
Ultrasonic ranger: `Final estimation m = 4.9186, final estimation zm = 0.0051`

7.14: (a) `Final estimate of speed, ym(25) = 2.4905 rev/s;` `(b) final estimate of speed and estimation std ×100, m(25), zm(25) = 2.2769 rev/s, 0.1815 rev/s → 0.001815 rev/s`

Chapter 8

8.8: $\alpha = 0.5$

8.11: 0.1%

8.12: $D = 1.25$ cm

8.18: 1000 Hz

8.38: 100 s

8.40: (a) 1.62%; (b) ≃170 Hz

8.43: 9.5 kΩ

8.44: 15.5 s

Chapter 9

9.9: $S_s = 144.0$, $n_p = 5.9\%$

Chapter 11

11.4: ±0.088°

11.9: $1.57 \times 10^{-\cdot}$ rad; 7.9×10^{-4};

11.14: 12 bit buffer, minimum servo < ±1 mm, 12 tracks, $2^{12} = 4096$ sectors

11.15: 0.0072°

11.22: $v = 20.0$ m/s

11.25: (a) 0.088°, (b) 0.0176°

11.27: (a) 10 ms or better (preferably 2 ms), (b) 1 kHz

11.29: MTBF ~ 20,000 h

Chapter 12

12.3(c): $\boldsymbol{\beta} = \begin{bmatrix} 0.0839 \\ -0.4658 \end{bmatrix}$

12.5(b): $\begin{bmatrix} 0.855 \\ 0.044 \\ 0.101 \end{bmatrix}, \begin{bmatrix} 0.496 \\ 0.008 \\ 0.496 \end{bmatrix}$

12.10: $m_f(N, F, VF) = \begin{bmatrix} 0.5217 & 0.2174 & 0.2609 \end{bmatrix}; \quad K = 0.77$

12.12: `m1m2Class1 = 0.0224, m1m2Class2 = 0.0597, m1m2Class3 = 0.8260, m1m2Class4 = 0.0732, m1m2Class5 = 0.0187, K = 0.6660;`

`m1m2m3Class1 = 0.0060, m1m2m3Class2 = 0.0277, m1m2m3Class3 = 0.9420, m1m2m3Class4 = 0.0211, m1m2m3Class5 = 0.0033, K = 0.5716`

12:13: `m1m2Class1 = 0.0024, m1m2Class2 = 0.0074, m1m2Class3 = 0.9836, m1m2Class4 = 0.0051, m1m2Class5 = 0.0015, K = 0.3609;`

`m1m2m3Class1 = 6.3087e-005, m1m2m3Class2 = 3.3552e-004, m1m2m3Class3 = 0.9994, m1m2m3Class4 = 2.0097e-004, m1m2m3Class5 = 3.0881e-005, K= 0.1288`

Units and Conversions (Approximate)

1 cm = 1/2.54 in. = 0.39 in.
1 rad = 57.3°
1 rpm = 0.105 rad/s
1 g = 9.8 m/s^2 = 32.2 ft/s^2 = 386 $in./s^2$
1 kg = 2.205 lb
1 $kg \cdot m^2$ (kilogram-meter-square) = 5.467 $oz \cdot in.^2$ (ounce-inch-square) = 8.85 $lb \cdot in \cdot s^2$
1 N/m = 5.71 × 10^{-3} lbf/in.
1 N/m/s = 5.71 × 10^{-3} lbf/in./s
1 N·m (Newton-meter) = 141.6 oz·in. (ounce-inch)
1 J = 1 N·m = 0.948 × 10^{-3} Btu = 0.278 kWh
1 hp (horse power) = 746 W (watt) = 550 ft·lbf
1 kPa = 1 × 10^3 Pa = 1 × 10^3 N/m^2
= 0.154 psi = 1 × 10^{-2} bar
1 gal/min = 3.8 L/min

Metric Prefixes

Giga	G	10^9
Mega	M	10^6
Kilo	k	10^3
Milli	m	10^{-3}
Micro	μ	10^{-6}
Nano	n	10^{-9}
Pico	p	10^{-12}

Index

E

F

Q

R

W

Z